AF689708

RELIABILITY
OF NUCLEAR POWER PLANTS

PROCEEDINGS SERIES

RELIABILITY OF NUCLEAR POWER PLANTS

PROCEEDINGS OF A SYMPOSIUM
ON THE RELIABILITY OF NUCLEAR POWER PLANTS
ORGANIZED BY THE
INTERNATIONAL ATOMIC ENERGY AGENCY
WITH THE CO-OPERATION OF THE SECRETARIAT
OF THE UNITED NATIONS ECONOMIC COMMISSION FOR EUROPE
AND HELD AT INNSBRUCK, 14 - 18 APRIL 1975

INTERNATIONAL ATOMIC ENERGY AGENCY
VIENNA, 1975

TK1341
S98R38
1975
c.2

RELIABILITY OF NUCLEAR POWER PLANTS
IAEA, VIENNA, 1975
STI/PUB/403
ISBN 92-0-050075-7

Printed by the IAEA in Austria
November 1975

FOREWORD

Nuclear energy is at today's fuel prices fully competitive with conventional energy as a means of electric power generation, even for small plants; it is also an acceptable alternative to fossil fuel for providing district heating and industrial process heat. The increasing use of this energy source must be accompanied by the highest degree of plant safety and of plant availability – a factor which hitherto was rather prone to subjective assessment. An insight into these plant features is now needed primarily in quantitative terms. The knowledge gained will allow us to check whether power plants have been complying with the given safety criteria, and the collection of availability data on a world-wide basis will enable us to establish design requirements and availability targets.

These aims entail the use of the concept of reliability. This may be defined as the probability that a device, system or facility will perform satisfactorily in service for a specified time under stated operating conditions. The probabilistic approach, which is pertinent to the reliability concept, requires the collection of data on the frequency of equipment failure, and their processing into meaningful reliability and failure statistics.

Reliability engineering, based on the reliability concept, is a relatively new discipline, developing and applying appropriate methods and techniques to discover and eliminate hidden causes of equipment and plant failure. In the nuclear power field it is quite novel. Starting as an academic exercise to demonstrate the application of sophisticated methods of reliability analysis to certain power plant systems, it has become indispensable in helping design, operation and maintenance engineers with the practical decisions of their everyday work.

The International Atomic Energy Agency convened the Symposium on Reliability of Nuclear Power Plants in Innsbruck, Austria, from 14 to 18 April 1975 as part of its programme on nuclear power plant technology and reliability. Under this programme information on operating experience with nuclear power plants in the Member States has been collected and analysed, and international working groups on the subjects of pressure components and of power plant control and instrumentation have been set up. The basic purpose has been to promote an exchange of information on research and development aimed at increasing the reliability of nuclear power plant equipment.

The Symposium provided an opportunity for fruitful discussion between reliability analysts and practical engineers in nuclear power plant planning, design, construction, operation and maintenance. It is hoped that the proceedings will encourage the broader use of reliability analysis in obtaining relevant information for practical decisions.

The Agency wishes to thank the Government of Austria, and particularly the Landesregierung of Tyrol, for acting as host and providing the staff and facilities that contributed so much to the success of the Symposium.

EDITORIAL NOTE

The papers and discussions have been edited by the editorial staff of the International Atomic Energy Agency to the extent considered necessary for the reader's assistance. The views expressed and the general style adopted remain, however, the responsibility of the named authors or participants. In addition, the views are not necessarily those of the governments of the nominating Member States or of the nominating organizations.

Where papers have been incorporated into these Proceedings without resetting by the Agency, this has been done with the knowledge of the authors and their government authorities, and their cooperation is gratefully acknowledged. The Proceedings have been printed by composition typing and photo-offset lithography. Within the limitations imposed by this method, every effort has been made to maintain a high editorial standard, in particular to achieve, wherever practicable, consistency of units and symbols and conformity to the standards recommended by competent international bodies.

The use in these Proceedings of particular designations of countries or territories does not imply any judgement by the publisher, the IAEA, as to the legal status of such countries or territories, of their authorities and institutions or of the delimitation of their boundaries.

The mention of specific companies or of their products or brand names does not imply any endorsement or recommendation on the part of the IAEA.

Authors are themselves responsible for obtaining the necessary permission to reproduce copyright material from other sources.

CONTENTS

RELIABILITY ASSESSMENT OF NUCLEAR POWER PLANT SYSTEMS

APPLICATION OF RELIABILITY ANALYSIS IN NUCLEAR POWER PLANT DESIGN

TESTING AND INSPECTION OF NUCLEAR POWER PLANT SYSTEMS AND COMPONENTS

QUALITY ASSURANCE FOR NUCLEAR POWER PLANTS

NUCLEAR TECHNOLOGY STANDARDS AND RELIABILITY

OPERATION AND MAINTENANCE OF NUCLEAR POWER PLANTS

RELIABILITY DATA COLLECTION, STORAGE AND USE

Chairman:

A. NENTWICH, Austria

THE UNITED STATES NUCLEAR PLANT RELIABILITY DATA PROGRAM
Its description and status

M.J. WISE
Southwest Research Institute,
San Antonio, Tex.,
United States of America

Abstract

THE UNITED STATES NUCLEAR PLANT RELIABILITY DATA PROGRAM: ITS DESCRIPTION AND STATUS.

The American National Standards Institute Subcommittee N18-20 has developed and implemented the United States Nuclear Plant Reliability Data System (NPRDS). The NPRDS is designed to accumulate, store, analyse, and report reliability and failure statistics on systems and components of nuclear power plants related to nuclear safety. Input data to the NPRDS consist of engineering, operating, and failure information submitted on a voluntary basis by participating utilities. Prior to entry into the computerized data base, the data are thoroughly checked for accuracy by both the submitting organizations and the NPRDS operating contractor. The data base is the source of various periodic output reports to the nuclear power industry and is utilized to produce special reports upon request. The present data base represents data accumulated from about thirty nuclear units with additional units expected to begin submitting data immediately. The objective is to have essentially all operating nuclear units in the United States of America participating in the program by the end of 1975. The first NPRDS annual reports containing meaningful reliability and failure statistics are expected to be produced following the end of 1975.

1. INTRODUCTION

1.1 Background

Accurate reliability and failure rate data are very critical in the area of reliability engineering. Regarding the U.S. nuclear power industry, there is only a relatively small amount of such data available. Some attempts have been made over the past several years to establish systems for the reporting, storing, analyzing, and distributing of reliability data; but until recently, these attempts have been essentially fruitless.

In the early 1970's, the Equipment Availability Task Force of the Edison Electric Institute (EEI) began the development of the Nuclear Plant Reliability Data System (NPRDS) at the request of the Nuclear Technical Advisory Board (NTAB) of the American National Standards Institute (ANSI). The responsibilities of the Task Force were to develop a standard method for collection and storage of reliability data, to establish data analysis and reporting procedures, and to oversee the operation of the NPRDS.

In the spring of 1973, Subcommittee N18-20, Nuclear Plant Reliability Data, was established under the American National Standards Committee N18, Nuclear Design Criteria. The membership of Subcommittee N18-20 is made up of the EEI Equipment Availability Task Force, which includes representatives from privately and publicly owned electric utilities and nuclear steam system suppliers, and a representative from the U.S. Nuclear Regulatory Commission (NRC). This Subcommittee has assumed and carried out the responsibilities of the EEI Task Force.

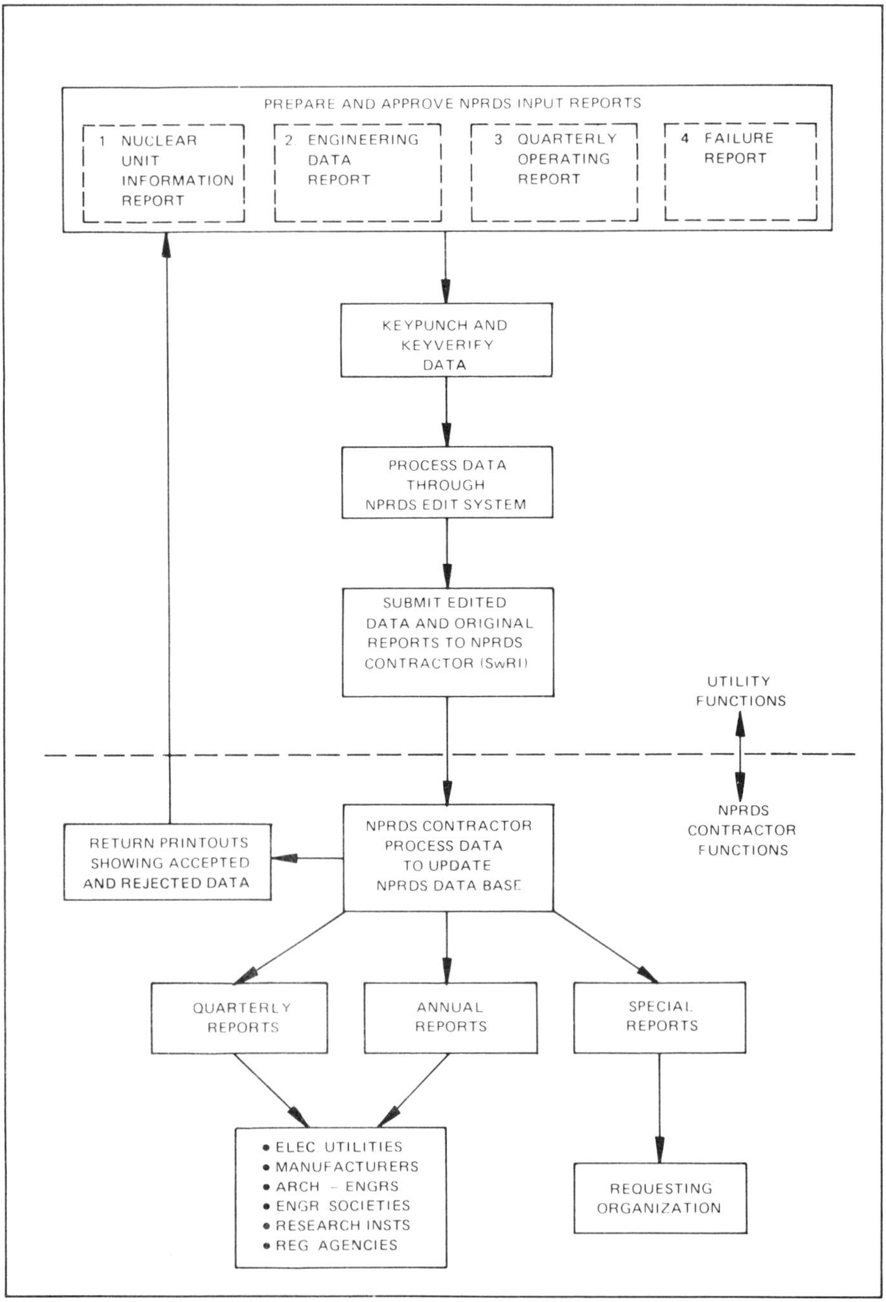

FIG.1. NPRDS functional flow chart.

In July 1973, the NPRDS reporting procedures manual was completed, and a pilot program was initiated with six nuclear units. These units included Maine Yankee, Nine Mile Point Unit One, Oconee Unit One, Point Beach Unit One, Quad Cities Unit One, and San Onofre Unit One. These units represented relatively recent models supplied by each of the four light-water reactor vendors. The objectives of the pilot program were to check out input reporting procedures and forms; to develop the most efficient methods for compiling, submitting and processing the submitted data; and to check out the NPRDS computer programs using real data.

As a result of experience gained during the pilot program, revisions were made to the reporting procedures manual, reporting forms, and the computer programs. Beginning in July 1974, implementation of the NPRDS was begun throughout the nuclear power industry.

1.2 Objectives and Benefits

The primary objective of the NPRDS is to collect, store, and make available reliability and failure statistics for the systems and components of nuclear power plants related to nuclear safety. The data system was designed to be flexible in its ability to extract and analyze data and to provide expansion capabilities to include data on nuclear plant systems and components not presently within the defined scope. Considerable effort was expended with the objective to minimize the paperwork burden on the reporting organizations while still providing the necessary information to the designers and operators of nuclear power plants and to the regulatory agencies.

It is expected that accomplishment of the above objectives will permit the nuclear power industry to:

(1) More accurately assess and improve the reliability of nuclear plant systems and components related to safety. Reliability depends on quality, redundancy and independence of redundant devices, testing, maintenance, and freedom from failures caused by accident. The measurement of system and component reliability by the NPRDS will result in statistical data that can be used to enhance designs and operating performance towards the goals of reliability and safety.

(2) Increase plant availability. It is unfortunate but true that safety and economy are goals which often diverge in system design. To achieve an optimum balance between the two goals, meaningful data must be provided to reduce the level of subjective judgement that is now necessary in establishing these goals.

(3) Optimize the redundancy of safety-related systems. As the NPRDS compiles meaningful information on performance, failures, testing, and reliability of systems and components, less subjective judgement will be required in establishing redundancy to meet the functional objectives required for safety-related systems.

(4) Expedite licensing activities and decrease the overall time and cost required for plant construction. Information provided by the NPRDS will be used to support safety and reliability analyses, thus making such analyses more readily acceptable to licensing and regulatory agencies. Factual operating data will provide for more meaningful standardization of systems and components, alleviating some of the costs and problems in licensing, building, and operating nuclear power plants.

(5) Evaluate and adjust protection system testing schedules based on actual performance. The NPRDS will provide reference data for use in determining optimum test intervals for components and systems, thus increasing system reliability and availability.

Form NPRD-1

NUCLEAR PLANT RELIABILITY DATA SYSTEM;

NUCLEAR UNIT INFORMATION REPORT

This data entry is for:

☒ Original Nuclear Unit Information Form (place a '7' in column two below)

☐ Correction to previous Nuclear Unit Information Form (place a '5' in two below)

INPUT CONTROL		ACTION	UTILITY DESIG.			PLANT			UNIT		REPORT I.D.	
	1	2	3	4	5	6	7	8	9	10	28	29
	A	7	A	C	W	E	Q	A	D	2	A	A

Card	Field (columns 39–80)
01	OPERATING UTILITY NAME: CØMMØNWEALTH EDISØN CØMPANY
02	STATION NAME: QUAD CITIES NUCLEAR PØWER STATIØN
03	REACTOR NAME: UNIT 2
04	MAILING ADDRESS (St. or Box No.): P.Ø. BØX 216
05	MAILING ADDRESS (City, State, Zip Code): CØRDØVA, ILLINØIS 61242
06	UTILITY CONTACT (1) Name, Phone No.: G.J. SPØRER 309/654-2241 EXT. 232
07	UTILITY CONTACT (2) Name, Phone No.: R.A. NEWKIRK 309/654-2241 EXT. 243
08	STATION LOCATION (City, Township, etc), (County), (State), (Zip Code): CØRDØVA, RØCK ISLAND CØUNTY, ILLINØIS 61242
09	NSSS SUPPLIER and TYPE OF REACTOR (BWR, HTGR, PWR, ETC.): GENERAL ELECTRIC - BWR
10	ARCHITECT – ENGINEERING FIRM: SARGENT AND LUNDY
11	REACTOR RATING (MWt): 2511
12	UNIT RATING (MWe): 809
13	AEC DOCKET NO.: 50-265
14	INITIAL CRITICAL DATE: 4/26/72
15	COMMERCIAL SERVICE DATE: 08/72

Form Prepared By: G.J. SPØRER

Date Prepared: JULY 31, 1973

Phone No.: 309/654-2241

(REV. 1, 3/74)

FIG. 2. Nuclear unit information report.

(6) Identify significant failure modes. To obtain meaningful reliability data and provide for proper use of such data, failure rates should be based on actual failure modes. The NPRDS is designed to adequately characterize failures and provide accurate failure mode information.

(7) Identify failure trends and detect wearout patterns. The technical, operating, and failure data provided by the NPRDS will become a valuable source of information over the years for use in improving designs and maintenance programs.

(8) Provide statistics to manufacturers on their product performance. It is important for manufacturers to obtain performance data on their products. The NPRDS will accumulate such data and make them available to manufacturers as well as users and designers.

1.3 Scope

The NPRDS produces failure statistics on selected components and systems related to nuclear safety in nuclear power plants. The present scope of systems and components is, in general, those classified as Safety Class 1 and Safety Class 2 in ANSI Standards N13.2 and N212 and American Nuclear Society (ANS) Standard 23, which specify nuclear safety criteria for nuclear plant design. Some types of systems and components designated as Safety Class 1 or 2 in these standards are excluded from the NPRDS scope of reportable items for water reactor type units. These are:

(1) Spent fuel storage systems and components.

(2) Reactor pressure vessel internals.

(3) Structures and components of structures.

(4) Piping, fittings, and valves in pressurized water reactor systems one inch or less, nominal pipe size.

(5) Piping, fittings, and valves in boiling water reactor systems one and one-quarter inch or less, nominal pipe size.

The selection of these exclusions was primarily based on probability of failure; complexity of data reporting; and, to some extent, the impact on nuclear safety. The actual scope of reportable systems and components can be revised as considered necessary by the nuclear power industry.

Input information for the NPRDS is solicited from all organizations operating nuclear reactors used primarily for generating electric power within the United States. The few operating nuclear power plants licensed to operate prior to January 1967, plus the LaCrosse plant, were excluded from the program because of nontypical size, construction codes and standards, and in some cases, prototypal design.

2. DESCRIPTION OF THE NPRD SYSTEM

2.1 General Description

Figure 1 shows a simplified functional flow chart of the NPRDS. The participating utility is responsible for identifying the systems and components for a particular nuclear unit that are included within the reportable scope of the NPRDS. A "Nuclear Unit Information Report" (Figure 2) is submitted to the NPRDS Contractor for each nuclear unit under the program. Engineering data for each reportable plant system and component are submitted using the "Report of Engineering Data" form (Figure 3). In cases of multiple identical components in a particular system, a supplemental form is used with the Report of Engineering Data form in order to minimize the number of forms to be completed.

Form NPRD-2 (THIS SPACE FOR NPRD OPER. USE)

NUCLEAR PLANT RELIABILITY DATA SYSTEM

REPORT NO. 0023

REPORT OF ENGINEERING DATA

This data entry is for:

☒ Engineering Data Report for original or replacement component or original system (place a '7' in column two below)

☐ Correction to previous Engineering Data Report (place a '5' in column two below)

INPUT CONTROL

Field	Columns	Entry
(Card)	1	C
ACTION	2	7
	3	E
UTILITY DESIG.	4–6	CWE
PLANT	7–9	QAD
UNIT	10	1
SS CODE	11	5
NSSS	12	C
NPRD COMP OR SYSTEM CODE	13–18	PUMPXX
UTILITY COMPONENT IDENTIFICATION NUMBER (* / REPORT NO.)	19–29	1-1302
REPORT I.D.	30	A
DATA START DATE (YR. / MO. / DY.)	31–36	73 07 01
	37	

SYSTEM OR COMPONENT ENGINEERING DATA

CARD I.D.	NPRD SYSTEM CODE (39–44)	UTILITY SYSTEM CODE (45–51)	SAFE. CLS. (52–53)	MODE (54–56)	ENVIRONMENT INT. / EXT. (57–61)	IN-SERVICE DATE YR. / MO. / DY. (62–67)	APPLICABLE MFGR. CODE OR STANDARD (68–80)
A	CEA	1300	2	SBC	ØPXAF	71 10 07	ASME SEC. 3

CARD I.D.	MFGR. REF. NO. (39–42)	MANUFACTURER MODEL NUMBER (43–62)	MANUFACTURER SERIAL NUMBER (63–80)
B	B26D	4X6X9 BMSD	270606

CARD I.D.	SUPP./VEND. REF. NO. (39–42)	SUPPLIER/VENDOR SYSTEM OR COMPONENT IDENTIFICATION NUMBER (43–60)	SOURCE DRAWING OR DOCUMENT NUMBER (61–76)	(77–80)
C	G080	1-1302	RCIC-M-050	

CARD I.D.	A (39–40)	B (41–42)	C (43–44)	D (45–46)	E (47–48)	F (49–50)	(51)	G (52–56)	UNITS (57–60)	(61)	H (62–66)	UNITS (67–70)	(71)	J (72–76)	UNITS (77–80)
D	B	C	G	B	B	A		2800	FTHØ		416	GPM		4500	RPM

OPERATION AND TESTING DATA

CARD I.D.	ESTIMATE OF COMPONENT/SYSTEM OPERATION AS PERCENT OF: 1-REACTOR CRITICAL HOURS (39–43)	2-STANDBY CONDITION HOURS (44–48)	3-REACTOR SHUTDOWN HOURS (49–53)	TESTING – CHECK: FREQ. (54)	INTERVAL (55–56)	OUT OF SERV. (HRS.) (57–59)	FUNCTIONAL: FREQ. (60)	INTERVAL (61–62)	OUT OF SERV. (HRS.) (63–65)	CALIBRATION: FREQ. (66)	INTERVAL (67–68)	OUT OF SERV. (HRS.) (69–71)
E	001.00	001.00	001.00	1	MØ	0.0	1	QT	2.0	0	00	00.0

REMARKS

Date Prepared: 7/15/73

Prepared By: J. B. JØNES

Reviewed/Approved By: R. B. SMITH

Phone No. 512/684-5111

(REV. 1, 3/74)

FIG. 3. Report of engineering data (component).

At the end of each calendar quarter, the participating utility submits a "Quarterly Operating Report" (Figure 4) which includes actual operating data on a particular nuclear unit. Also, the participating utility submits all "Report of Failure" forms (Figure 5) which have been completed and accumulated during the quarter.

Prior to submitting the NPRDS input data forms, the participating utility is responsible for keypunching and keyverifying the data on 80-column cards or magnetic tape. In most cases, the utility processes the punched data through the NPRDS Edit System prior to submittal to the NPRDS Contractor. The NPRDS Edit System is set up on the utility's computer and is operated by utility personnel. Pre-processing through the Edit System performs essentially the same error editing and validating as is performed by the NPRDS Contractor. This function allows the utility to detect and correct many errors which would otherwise result in additional processing transactions.

Upon receipt of the input data, the NPRDS Contractor processes them through the NPRDS computer system, which again checks for errors and updates the NPRDS data base with accepted data. Rejected data are returned to the participating utility for correction and resubmittal.

Various NPRDS output reports are routinely produced and distributed to the nuclear power industry. Descriptions of the output reports are included in Section 2.3.

2.2 NPRDS Input Reports

As mentioned in Section 2.1 above, four basic reports are used to input information to the NPRDS. A brief description of each report is presented below.

2.2.1 Nuclear Unit Information Report (Figure 2).

General descriptive information about the particular nuclear unit and the owner-organization is submitted once for the purpose of identifying subsequent output reports. The information also identifies responsible utility contacts for use by the NPRDS Contractor and others.

2.2.2 Report of Engineering Data (Figure 3).

Specific engineering data for a particular plant system or component are submitted on this form. These data are entered into the NPRDS data base when the equipment is first reported or when a new component is added. The engineering data are changed only if a component is deleted or replaced with a modified or different component having the same function and relative location in a system. Data items included on the form are:

- Utility Designation Code
- Plant Code
- Unit Number
- Nuclear Steam System Supplier Code
- NPRD Component or System Code
- Utility Component Identification Number
- Data Start Date (Commercial operation date or later)
- Utility System Code
- Safety Class
- Normal Mode of Operation
- Internal and External Environment Codes
- In-Service Date (Same as Data Start Date or earlier)

Form NPRD-3

NUCLEAR PLANT RELIABILITY DATA SYSTEM;

QUARTERLY OPERATING REPORT

This data entry is for:

☒ New report (place a '7' in column two below)

☐ Correction to a previously submitted report (place a '5' in column two below)

INPUT CONTROL

Field		ACTION		UTILITY DESIG.			PLANT			UNIT	REPORT I.D.	(columns 12–30)	QUARTER END DATE YR		MO		DY		
Column	1	2	3	4	5	6	7	8	9	10	11	12–30	31	32	33	34	35	36	37
Entry	C	7	E	W	M	P	P	B	H	1	1		7	3	0	9	3	0	

QUARTERLY REPORT UPDATE

	CARD I.D.	QUARTERLY UPDATE DATA: 1-REACTOR CRITICAL (HR)	2-STANDBY CONDITION (HR)	3-REACTOR SHUTDOWN (HR)		QUARTER START DATE YR MO DY	QUARTER END DATE YR MO DY	FAILURE REPORTS THIS QTR. SYST.		COMP		ON-LINE TIME (HR)	
Columns	38	39–43	44–48	49–53	54	55–60	61–66	67–68	69	70–71	72	73–76	77–80
Entry	F	02156	00009	00043	54	730701	730930	00	69	03	72	2093	

REMARKS

Date Prepared 10/20/73

Prepared By: BØB LINK

(REV. 1, 3/74)

Reviewed/Approved By: J. GREENWØØD

Phone No.: 414/755-2321

FIG.4. Quarterly operating report.

- Applicable Manufacture Code or Standard
- Manufacturer Reference Number
- Manufacturer Model Number (Components Only)
- Manufacturer Serial Number (Components Only)
- Supplier/Vendor Reference Number
- Supplier/Vendor Identification Number
- Source Drawing or Document Number
- Classifying Engineering Data - Components Only
- Operation and Testing Data

The classifying engineering data for components are presented by two methods - letter codes for up to six subclassifications and open fields for up to three numerical subclassification values. Figure 6 shows a typical table which includes the classifying data for the various component categories.

Tables are provided for the following major categories of components:

- Accumulators
- Air Dryers
- Annunciator
- Batteries and Chargers
- Blowers
- Circuit Closers/Interrupters
- Control Rods
- Control Rod Drive Mechanisms
- Demineralizers
- Electrical Conductors
- Engines: Internal Combustion
- Filters
- Fuel Elements
- Generators
- Heaters: Electric
- Heat Exchangers
- Instrumentation and Controls
- Mechanical Function Units
- Motors
- Penetrations
- Pipes and Fittings
- Pumps
- Recombiners
- Relays
- Shock Suppressors
- Transformers
- Turbines
- Valves
- Valve Operators
- Vessels: Pressure

In order to compute the component or system service hours, the engineering data report includes estimates of service time as related to reactor service hours at three different conditions: (1) Critical, (2) Standby, and (3) Shutdown. It is assumed that these three service conditions will account for 100 percent of total service time. Actual service times as reported on the Quarterly Operating Reports are used to compute the estimated service hours for each system and component in the NPRDS data base.

Scheduled testing information is reported for each system and component. Specified testing intervals and frequencies are used to accumulate the approximate number of tests conducted for the categories of checks, functional tests, and calibration tests. The approximate out-of-service time is also accumulated for each type of test for each system and component.

2.2.3 Quarterly Operating Report

Figure 4 shows an example of a completed quarterly operating report giving the actual number of hours the reactor was in each of the three operating modes: (1) Critical, (2) Standby, and (3) Shutdown. This report also summarizes the number of system and component failures that were reported for the quarter and the duration that the main generator was connected to the utility's electrical system.

Form NPRD-4

(THIS SPACE FOR NPRD OPER. USE)

NUCLEAR PLANT RELIABILITY DATA SYSTEM;

REPORT OF FAILURE

This data entry is for:

☒ New failure report (place a '7' in column two below)

☐ Correction to previously submitted failure report (place a '5' in column two below)

INPUT CONTROL

	ACTION		UTILITY DESIG.			PLANT			UNIT	S/C CODE	NSSS	NPRD COMP. OR SYSTEM CODE						UTILITY COMPONENT IDENTIFICATION NUMBER											REPORT I.D.	DATE OF FAILURE YR.		MO.		DY.		FAIL NO.
1	2	3	4	5	6	7	8	9	10	11	12	13	14	15	16	17	18	19	20	21	22	23	24	25	26	27	28	29	30	31	32	33	34	35	36	37
C	7	E	C	W	E	Q	A	D	1	5	C	V	A	L	V	E	X	1	-	2	0	3	-	3	E				B	7	3	0	9	2	5	1

EVENT DATA

CARD I.D.	FAILURE EVENT START YR.		MO.		DY.		HR.		MIN.		FAILURE EVENT END YR.		MO.		DY.		HR.		MIN.			NPRD SYSTEM CODE						NPRD COMP. CODE														
38	39	40	41	42	43	44	45	46	47	48	49	50	51	52	53	54	55	56	57	58	59	60	61	62	63	64	65	66	67	68	69	70	71	72	73	74	75	76	77	78	79	80
G	7	3	0	9	2	5	1	3	0	0	7	3	0	9	2	5	1	7	0	0		C	B	A				Y	A	L	V	E	X									

FAILURE DESCRIPTION

H CRACKED SILVER SØLDER WELD ØN VALVE

J SØLENØID CØRE GUIDE PIN.

CAUSE AND CORRECTIVE ACTION

K REWELDED GUIDE PIN WITH SILVER SØLDER

L

NPRD FAILURE DATA

1. CIRCLE COLUMN CODES UNDER EACH APPLICABLE FAILURE CLASSIFICATION
2. CIRCLE AS MANY COLUMN CODE DESIGNATIONS AS NECESSARY TO ADEQUATELY DESCRIBE FAILURE SITUATION.

Card I.D. V (col. 38)

Group	Column	Code	Designation	Circled
MODE OF FAILURE PASSIVE	39	A	LEAK	
	40	B	CRACK	(B)
	41	C	BREACH	
	42	D	PHYS. DISTORTION	
	43	E	PHYS. DISPLACEMENT	
	44	F		
	45	G		
	46	H		
	47	I		
MODE OF FAILURE ACTIVE	48	A	WON'T START	
	49	B	WON'T STOP	
	50	C	WON'T CLOSE	
	51	D	WON'T OPEN	
	52	E	WON'T OPER. PER SPEC.	
	53	F	SPURIOUS OPERATION	
	54	G	FALSE RESPONSE	
	55	H		
	56	I		
	57	J		
TYPES OF FAILURE	58	A	ELECTRICAL	
	59	B	MECHANICAL	(B)
	60	C	METALLURGICAL	(C)
	61	D	CORROSION	
	62	E	OTHER CHEMICAL	
	63	F	INSTRUMENT	
	64	G		
	65	H		
	66	I		
CAUSES OF FAILURE	67	A	ENGR./DESIGN	
	68	B	MANUFACTURING	(B)
	69	C	CONST./INSTALLATION	
	70	D	OPERATING ERROR	
	71	E	MAINT./TESTING	
	72	F	UNUSUAL SERV. COND.	
	73	G	NATURAL END LIFE	
	74	H	ASSOCIATED DEVICES	
	75	I	RANDOM	
	76	J	UNKNOWN	
	77	K	MATERIAL DEFECT	
	78	L		
	79	M		
	80	N		

Card I.D. W (col. 38)

Group	Column	Code	Designation	Circled
FAILURE DETECTION	39	A	TESTING	
	40	B	PREVENTIVE MAINT.	(B)
	41	C	AUDIO/VISUAL ALARM	
	42	D	ROUTINE INSPECTION	
	43	E	ABNORMAL INDICATION	
	44	F		
	45	H		
	46	I		
	47	J		
ACTION TAKEN	48	A	REPLACE PART(S)	
	49	B	REPAIR PART(S)	(B)
	50	C	REPLACE TOTAL COMP.	
	51	D	RECALIB./ADJUST.	
	52	E	REDESIGN/MODIFY	
	53	F	TEMPORARY REPAIR	
	54	G		
	55	H		
	56	I		
EFFECT OF FAILURE	57	A	LOSS OF SYS. FUNC.	
	58	B	LOSS OF SUBSYS./CHAN.	
	59	C	SYS. OPER. DEGRADED	
	60	D	PREVENTS COMP. OPER.	(D)
	61	E	LOAD REDUCTION	
	62	F	DAMAGE TO OTHER COMP.	
	63	G	REACTOR TRIP	
	64	H	OTHER SYS. DEGRADED	
	65	I	GEN. UNIT OFF LINE	
	66	J	NO EFFECT	
	67	K		
	68	L		
	69	M		

Date Prepared: ØCT. 2, 1973 Reviewed/Approved By: G. J. SPØRER

Prepared By: G. M. KRIESER Phone No: 309/654-2241

(REV. 1, 3/74)

FIG. 5. Report of failure (component).

GENERATORS (GENERA)

A. Type

A–Alternator
B–Converter
C–Dynamotor
D–Generator
E–Amplidyne
F–Inverter
X–Other (explain under "Remarks")

B. Voltage Range

A–Less than 100 VDC
B–100-299 VDC
C–300 VDC or higher
D–100-299 VAC single phase
E–100-299 VAC polyphase
F–300-499 VAC polyphase
G–500-699 VAC polyphase
H–2,000-2,599 VAC polyphase
J–3,500-4,999 VAC polyphase
K–6,000-6,999 VAC polyphase
L–13,000-13,999 VAC polyphase
M–Over 14,000 VAC polyphase
X–Other (explain under "Remarks")

C. Capacity

A–Less than 1 KW
B–1-9.9 KW
C–10-49 KW
D–50-249 KW
E–250-999 KW
F–1000-5000 KW
G–Over 5000 KW

D. Output

A–Alternating Current
B–Direct Current
C–Variable Frequency, AC

E. Driver Type

A–Motor
B–Turbine, Steam
C–Internal Combustion Engine
D–Turbine, Hydro
E–Turbine, Combustion
X–Other (explain under "Remarks")

F. Application

A–Emergency, Sole Source
B–Emergency, Redundant Source
C–Emergency, Shared
X–Other (explain under "Remarks")

G. Rated Rotational Speed

(RPM at rated output)

H. Voltage Rating

(VDC or VAC)

J. Power Rating

(KW)

FIG.6. Example of table for classification of components.

2.2.4 Report of Failure

Failure reports are required to be submitted each time system and component failures are experienced. Figure 5 shows a typical report on a component failure. System failure reports are not required with each component failure; they are only completed when an actual system failure occurs.

The information included on the Report of Failure form provides for filing the failure data with the corresponding system or component engineering data. Also included are data on the dates and times the failure occurred, the failure was corrected and the component or system was available for service. These data are used to calculate the out-of-service time due to the failure. At the present, no attempt is made to divide the out-of-service time into the basic mean-time-to-repair (MTTR) elements.

There are provisions on the report form to describe the failure, the cause, and the corrective action taken in narrative form. The failure is further characterized by designating subclassifications of:

- Mode of Failure
- Type of Failure
- Cause of Failure
- Failure Detection
- Action Taken
- Effect of Failure

Failure data can be extracted and reported from the NPRDS data base using any of the various data parameters except the narrative sections.

2.3 NPRDS Output Reports

While recognizing that the data collection system will be of the greatest value to operators, designers, and regulatory agencies who make specific analysis requests, it is also recognized that the nuclear power industry in general requires some summarized information on the subject of safety-related system and component reliability. For this reason, output reports on system and component reliability and failures are made available to the industry. Six different output reports are provided by the system. A brief description of each report is presented below.

2.3.1 Quarterly Report of Engineering Data for Individual Reporting Organization (NPRD Report Q01).

This report is provided each quarter to supply the reporting organization with a list of all engineering data submitted for the current quarter. This report contains the data submitted for components and systems on the reports shown in Figure 3.

2.3.2 Quarterly Component Failure Listing (NPRD Report Q02).

This report, shown in Figure 7, provides engineering and failure data for all components reported as failed during the quarter. The objective is to be able to provide information on all component failures on at least a quarterly basis to interested groups.

2.3.3 Annual Report of System Reliability for Individual Reporting Organizations (NPRD Report A01).

In addition to having the ability to have specific analyses of its own data made and of selected portions of the entire data bank, a participating utility will receive routine analyses and summaries of its data. The NPRD Report A01 is one example of this type of report. The format of this report is similar to that of the Summary

```
NPRD REPORT Q02                          QUARTERLY COMPONENT FAILURE LISTING                                  (SAMPLE)
                                                                                                            (APRIL, 1974)
REPORT DATE: 02/13/74                    NUCLEAR PLANT RELIABILITY DATA SYSTEM

THIS REPORT PRESENTS INFORMATION ON COMPONENT FAILURE REPORTS SUBMITTED FOR QUARTER ENDING 12/31/73
-----------------------------------------------------------------------------------------------------------------------

  NPRD    NPRD                                                              DATE               FAILURE    FAILURE CHARACTERIZATION
FAILURE   COMP                                                               OF       FAIL.   OUTAGE     ----------------------------
RPT.NO.   CODE              COMPONENT DESCRIPTION                          FAILURE     NO.    DURATION   CODES       DESCRIPTION
-------- ------   -----------------------------------------------------   --------   -----  ---------- ----------------------------

73-3-005 VALVEX   VALVES                                                  11/21/73     1    4HR 25MIN  MC MODE    -WONT CLOSE
(REVISED                                                                                                TH TYPE    -MECHANICAL
 REPORT)          01 COMPONENT MANUFACTURER CODE --- A200                                               CB CAUSE   -MANUFACTURING
                  02 MANUFACTURER MODEL NO. -------- 2116-SP                                            DB DETECT  -PREVENTATIVE MAINT
                  03 SAFETY CLASS ------------------ 2                                                  AB ACTION  -REPAIR PART(S)
                  04 MODE OF OPERATION CODE -------- SOC                                                ED EFFECT  -PREVENTS COMP.OPER
                  05 INTERNAL ENVIRONMENT CODE ----- NH
                  06 EXTERNAL ENVIRONMENT CODE ----- FXC
                  07 IN-SERVICE DATE --------------- 12/31/70
                  08 A-TYPE ------------------------ CHECK                        FAILURE DESCRIPTION
                  09 B-OPERATOR -------------------- NONE                         -------------------
                  10 C-FUNC/APPLICATION ------------ ONE-WAY FLOW                 VALVE WOULD NOT CLOSE WHEN RHR SYSTEM OPERATED
                  11 D-MATERIAL(BODY) -------------- TYPE 316 SS                  DURING ANNUAL FUNCTIONAL TEST
                  12 E-MATERIAL(BODY TYPE) --------- CAST
                  13 F-NOMINAL SIZE RANGE ---------- 4-11.99 IN.                  CAUSE AND CORRECTIVE ACTION
                  14 G-SIZE ------------------------ 8 INCH                       ---------------------------
                  15 H-MAX. OPER. PRESSURE --------- 300 PSIG                     VALVE DISMANTLED,CLEANED,INSPECTED, MACHINE
                  16 J-MAX. OPER. TEMPERATURE ------ 1125 F                       CHIPS LODGED BETWEEN DISC AND BODY, RETEST SAT.
                  17 TOTAL CALC. IN-SERV. HOURS ---- 4286
                  18 TOTAL CALC. CHECK TESTS ------- 9
                  19 TOTAL CALC. FUNC. TESTS ------- 3
                  20 TOTAL CALC. CALIB. TESTS ------ 3
                  21 TOTAL CALC. OUT.HOURS FOR TESTS 7.3

73-4-090 VALVOP   VALVE OPERATORS                                         10/06/73     1    6HR 53MIN  ME MODE    -WONT OPER.PER SPEC
                                                                                                        TB TYPE    -MECHANICAL
                  01 COMPONENT MANUFACTURER CODE --- C235                                               CJ CAUSE   -UNKNOWN
                  02 MANUFACTURER MODEL NO.--------- L608                                               DA DETECT  -TESTING
                  03 SAFETY CLASS ------------------ 2                                                  AA ACTION  -REPLACE PART(S)
                  04 MODE OF OPERATION CODE -------- OPC                                                EC EFFECT  -SYS.OPER.DEGRADED
                  05 INTERNAL ENVIRONMENT CODE ----- IM
                  06 EXTERNAL ENVIRONMENT CODE ----- CIT
                  07 IN-SERVICE DATE --------------- 10/07/71
                  08 A-TYPE ------------------------ PNEUMATIC                    FAILURE DESCRIPTION
                  09 B-SUB-TYPE -------------------- PILOT ACTUATED               -------------------
                  10 C-ACTUATION SPEED ------------- LESS THAN 10 SEC.            VALVE WOULD NOT CLOSE WITHIN 10 SEC PER TECH
                  11 D-                                                           SPECS ON ROUTINE CHECK, DIAPHRAGM HAD HOLE IN IT
                  12 E-
                  13 F-NOMINAL SIZE ---------------- 4 TO 11.99 IN.               CAUSE AND CORRECTIVE ACTION
                  14 G-FORCE RATING ---------------- 15000 LB.                    ---------------------------
                  15 H-                                                           REPLACED DIAPHRAGM AND RETESTED SAT. NO INDICA-
                  16 J-                                                           TION AS TO CAUSE OF HOLE, POSSIBLY MATERIAL DEF.
                  17 TOTAL CALC. IN-SERV. HOURS ---- 17363
                  18 TOTAL CALC. CHECK TESTS ------- 7
                  19 TOTAL CALC. FUNC. TESTS ------- 2
                  20 TOTAL CALC. CALIB. TESTS ------ 2
                  21 TOTAL CALC. OUT. HRS.FOR TESTS- 5.4

                              TOTAL COMPONENT FAILURE REPORTS FOR QUARTER -- 90
```

FIG.7. Quarterly component failure listing.

NPRD REPORT A03 — ANNUAL SUMMARY REPORT OF SYSTEM RELIABILITY — (SAMPLE) (APRIL, 1974)

YEAR - 1974 — NUCLEAR PLANT RELIABILITY DATA SYSTEM

NPRD SYSTEM CODE	OPER. MODE	PERIOD	TOTAL SYST. POP.	CALC. SYST. IN-SERV. HOURS: REACTOR CRITICAL (X1000)	CALC. SYST. IN-SERV. HOURS: REACTOR STANDBY (X1000)	CALC. SYST. IN-SERV. HOURS: REACTOR SHUTDOWN (X1000)	TOTAL IN-SERV. HOURS (X1000)	AVG. IN-SERV. HOURS PER SYST.	SYST. OUTAGE HRS. DUE TO FAILURE	NO. OF FAILURES	AVG. SYST. OUT. DUR. PER FAIL. (HOURS)	FAILURES PER MIL. IN-SERV. HRS.	NUMBER OF TESTS
PLANT VENTILATION SYSTEM													
(AAC)	OPER	1/74-12/74	125	788.4	32.8	273.7	1094.9	8759.2	55.00	11	5.00	10.00	13
		1/73-12/74	125	1554.9	87.6	547.5	2190.0	17520.0	164.91	23	7.17	10.00	26
PLANT FIRE PROTECTION													
(ABB)	OPER	1/74-12/74	25	177.4	4.4	37.2	219.0	8760.0	27.00	5	5.40	22.80	2
		1/73-12/74	25	359.1	13.1	65.7	437.9	17516.0	85.04	8	10.63	18.30	5
COMMUNICATIONS													
(ACB)	OPER	1/74-12/74	75	545.3	19.7	41.9	656.9	8758.7	77.01	3	25.67	4.50	40
		1/73-12/74	75	1064.3	39.4	210.2	1313.9	17518.7	115.00	4	26.25	3.00	96
			. . . .	. . . .	. . . .	. . . .	. . . .	. . . .	. . . .	. . . .	. . . .	. . . .	. . . .

RECAP OF REPORTING UNITS

	CURRENT PERIOD	CUMULATIVE PERIOD
TOTAL NO. UNITS REPORTING	25	25
TOTAL REACTOR CRITICAL HOURS (X1000)	157.7	310.9
TOTAL REACTOR STANDBY HOURS (X1000)	6.8	17.5
TOTAL REACTOR SHUTDOWN HOURS (X1000)	54.7	109.5
START DATE	1/1/74	1/1/73
END DATE	12/31/74	12/31/74
HOURS IN PERIOD	8,760	17,520

FIG.8. Annual summary report of system reliability.

Report of System Reliability shown in Figure 8. It will give the recipient utility an analysis of component failures that have occurred in its particular systems. The participating utility then has the necessary information to alter frequencies, calibration, and maintenance schedules and to detect wearout patterns.

2.3.4 Annual System and Component Failure Listing by Individual Reporting Organization. (NPRD Report A02).

A file of detailed information for each system and component that fails during any given year is very valuable to the reporting organization. The format of this report is similar to that of the Quarterly Component Failure Listing shown in Figure 7. Identifying information regarding the failure, such as the component manufacturer, date of failure, outage duration, description of the failure, and the failure characteristics, is all a part of the information on this report.

2.3.5 Annual Summary Report of System Reliability (NPRD Report A03).

Figure 8 contains a Summary Report of System Reliability statistics for generic safety-related systems of all similar nuclear steam supply systems which are reported. The total number of generating units and safety-related systems are indicated. Statistics are provided for the latest or current year and for the total experience recorded in the data base.

2.3.6 Annual Summary Report of Component Reliability (NPRD Report A04).

This report, shown in Figure 9, contains the component code classification by subcategory with the calculated in-service hours for all like components in the data file for the year. The reliability and failure data on the report provide very worthwhile information for use in identifying successful as well as unsuccessful component positions in the nuclear power plant. This should lead to better design information with the ultimate result being better plant availability.

2.3.7 Special Request Reports

In addition to the routine NPRDS output reports described above, participating organizations may request special reports to be produced from the data base. The requesting organization specifies the data base interrogation parameters, calculations required, and format of the desired reports. Following approval of the request by the ANSI Subcommittee N18-20, the NPRDS Contractor provides a cost estimate and schedule for producing the report. The costs for producing all special reports are paid by the requesting organization.

3. PRESENT STATUS OF NPRDS PROGRAM

3.1 Status of Input Data

As of March 1, 1975, input data have been received from about thirty nuclear units. In excess of 73,000 records of the four input types have been entered into the NPRDS data base. Experience to date shows that an average time of about 1.5 man-hours is required to produce one engineering data report. It is estimated that an average of 3,000 to 4,000 systems and components will be reported for each nuclear unit.

PRC REPORT A04 — ANNUAL SUMMARY REPORT OF COMPONENT RELIABILITY — (SAMPLE) (APRIL 1974)

REPORT DATE: 02/15/74 — NUCLEAR PLANT RELIABILITY DATA SYSTEM

THIS REPORT INCLUDES STATISTICS ON ALL COMPONENTS IN THE NPRDS DATA BASE AS OF 12/31/73

NPRD COMP CODE	COMPONENT CLASSIFICATION BY SUBCATEGORY	TOTAL COMP.IN SUBCAT	TOTAL IN-SERV. HOURS (MIL.)	TOTAL NUMBER FAILURES	FAILURES PER MIL. IN-SERV. HOURS	AVG.OUT OF SERV TIME (HOURS)
ACCUMU	A-A TYPE-LIQUID,PRESS.	114	3.316	0	0.000	0.0
ACCUMU	A-B TYPE-LIQUID,UNPRESS	26	1.512	0	0.000	0.0
ACCUMU	A-C TYPE-GAS	11	0.732	0	0.000	0.0
AIRDRY	A-D TYPE-HEATLESS,AUTO.	56	3.710	1	0.269	6.5
AIRDRY	A-G TYPE-REFRIGERATED	32	2.075	0	0.000	0.0
BATTRY	A-A TYPE-CHARGING UNIT C-A APPL-STATION BATTERY	XXX	X.XXX	X	X.XXX	X.X
BATTRY	A-A TYPE-CHARGING UNIT C-B APPL-SWITCHYARD BAT.	XXX	X.XXX	X	X.XXX	X.X
BATTRY	A-C TYPE-WET CELL C-A APPL-STATION BATTERY	XXX	X.XXX	X	X.XXX	X.X
BATTRY	A-C TYPE-WET CELL C-B APPL-SWITCHYARD BAT.	XXX	X.XXX	X	X.XXX	X.X
BLOWER	A-A SUBCAT-COMPRESSOR C-A DRIVER-MOTOR	XXX	X.XXX	X	X.XXX	X.X
BLOWER	A-A SUBCAT-COMPRESSOR C-B DRIVER-TURBINE	XXX	X.XXX	X	X.XXX	X.X
BLOWER	A-A SUBCAT-COMPRESSOR C-C DRIVER-INTER.COMB.ENG.	XXX	X.XXX	X	X.XXX	X.X
BLOWER	A-B SUBCAT-GAS CIRCULATOR C-B DRIVER-TURBINE	XXX	X.XXX	X	X.XXX	X.X
BLOWER	A-C SUBCAT-FAN C-A DRIVER-MOTOR	XXX	X.XXX	X	X.XXX	X.X
BLOWER	A-C SUBCAT-FAN C-B DRIVER-TURBINE	XXX	X.XXX	X	X.XXX	X.X
BLOWER	A-D SUBCAT-VENTILATOR C-A DRIVER-MOTOR	XXX	X.XXX	X	X.XXX	X.X

FIG.9. Annual summary report of component reliability.

3.2 Status of Output Reports

Quarterly reports have been routinely produced since the second quarter of 1974. Annual reports have not yet been produced because of the relatively small amount of data accumulated to date. It is anticipated that the first annual reports will be produced after the end of 1975.

3.3 Plans For NPRDS Modifications

A task force has been established by the ANSI Subcommittee N18-20 to further study the failure reporting forms and procedures. It is realized that the characterization of failures requires additional study and definition in order to more accurately determine the root cause of a particular failure. Also, additional procedures and quality control measures are required for the purpose of making the failure statistics as accurate and meaningful as feasible. It is expected that substantial improvements will be made in the failure reporting portion of the program within the next few months.

4. INFORMATION SOURCES

For additional information on the NPRDS program, contact either of the following sources:

Nuclear Plant Reliability Data System
Edison Electric Institute
90 Park Avenue
New York, New York 10016
Attention: G. A. Olson
Telephone: (212) 573-8759

Southwest Research Institute
Department of Quality Systems Engineering
8500 Culebra Road
San Antonio, Texas 78284
Attention: NPRD System Coordinator
Telephone: (512) 684-5111, Extension 2459

THE COLLECTION, STORAGE AND USE OF EQUIPMENT PERFORMANCE DATA FOR THE SAFETY AND RELIABILITY ASSESSMENT OF NUCLEAR POWER PLANTS

C.D.H. FOTHERGILL
Systems Reliability Service,
United Kingdom Atomic Energy Authority,
Risley, Warrington, Cheshire,
United Kingdom

Abstract

THE COLLECTION, STORAGE AND USE OF EQUIPMENT PERFORMANCE DATA FOR THE SAFETY AND RELIABILITY ASSESSMENT OF NUCLEAR POWER PLANTS.

It has been characteristic of the Nuclear Industry that it should grow up in an atmosphere where reliability and operational safety considerations have been of vital importance. Consequently all aspects of Nuclear Power Reactor design, construction and operation (in the U.K.A.E.A.) are subjected to rigorous reliability assessments, beginning with the automatic protective devices and the safety shut-down systems. This has resulted in the setting up of large and small private data stores to support this upsurgence of Safety and Reliability assessment work. Unfortunately, much of the information being stored and published falls short of the minimum requirements of Safety Assessors and Reliability Analysts who need to make use of it. That there is still an urgent need for more work to be done in the Reliability Data field is universally acknowledged. The characteristics which make up good quality reliability data must be defined and achievable minimum standards must be set for its identification, collection, storage and retrieval. To this end the United Kingdom Atomic Energy Authority have set up the Systems Reliability Service Data Bank. This includes a computerized storage facility comprised of two principal data stores: (i) Reliability Data Store, (ii) Event Data Store. The figures available in the Reliability Data Store range from those relating to the lifetimes of minute components to those obtained from the assessment of whole plants and complete assemblies. These data have been accumulated from many reliable sources both inside and outside the Nuclear Industry, including the transfer of 'live' data generated from the results of reliability surveillance exercises associated with Event Data collection. Computer techniques developed specifically for the Reliability Data Store enable further 'processing' of these data to be carried out. The Event Data Store consists of three discrete computerized data stores, each one providing the necessary storage, retrieval and analysing processes for the data it contains: (i) Installation Operating Record, (ii) Inventory Data Store, (iii) Event Data Store. The contents of these three data stores are processed to produce management information for the benefit of plant operators and maintenance engineers and to generate high quality information for transfer to the Reliability Data Store, from which it is readily available to Design Offices, Reliability Analysts and Safety Assessors. Examples from recent projects will be used as illustrations of the computer processes described.

1. INTRODUCTION

1.1 Nuclear Power Plants, in common with other large scale industrial plants, are constructed from items and devices which themselves should have been designed and built to recognised engineering specifications, in order to meet the functional standards of safety and performance expected of them by the Plant Operators and Insurance Underwriters. If these items and devices do achieve their required operational performance (within the tolerances to which they were made), and they continue to achieve it throughout their useful life, then they may, in a qualitative sense, be classified as "reliable". On the other hand, if any of these items and devices fail completely to perform in the desired manner or frequently fail to perform satisfactorily upon demand then, qualitatively, these may be classified as "unreliable". "Reliability" then is a measure of the degree of successful performance of a functional object within the limits of its' stated conditions of operation. But, before they can be measured, the various degrees of reliability, or, conversely, unreliability, must be described quantitatively.

It has been characteristic of the nuclear industry that it has grown up in an atmosphere where operational safety and reliability considerations have been of paramount importance. To this end all aspects of nuclear power station construction and operation in the United Kingdom have been subjected to special reliability assessments of the design, beginning with the high integrity automatic protective systems and safety shut down systems. However, the nuclear industry is not the only one in which hazards to the public, their employees and the plant can arise due to unreliable operation of that plant or equipment. Dangers are present, to some extent, in all industrial and commercial undertakings. Where such dangers are appreciable it is essential that the degree of reliability, of these equipments and devices which provide the necessary safety protection, is properly evaluated. Consequently the techniques of reliability assessment have found their way into all branches of industry, the Public Utilities, Armed Services, Transport Organisations etc.

2. THE PROBLEM

2.1 The upsurgence of interest in Safety and Reliability Assessment work has resulted in the setting up of large and small private data stores in which equipment performance data, of divers quality, is being amassed. In most of the established Reliability Data Banks good quality information is available on a fairly generous scale on electronic components and a great deal more has been published in open literature. As the benefits from Safety and Reliability assessment work is becoming more universally acknowledged an increasing amount of information is being collected, and made available, on relatively heavy items of plant. Unfortunately, much of this published information is incomplete because many originators have restricted their findings to isolated and often unsupported figures for average failure rates. Any reference to such parameters as "sample size", number of failures and faults, or to such characteristics as types, modes and cause of failure, environmental conditions or failure criteria are disregarded. Furthermore, much of this information is confined to individual components; few data, it appears, are being collected and published relating to complete systems and whole pieces of plant. Some authors quote only the number of fault occurrences without qualitification which renders this information unusable in practice, because it does not yield any failure rate information for the item to which it refers.

During the performance of Safety and Reliability assessment work the need to supplement limited experience failure rate and fault rate data with published data has always, and will always, be necessary in order that the assessment may benefit from the broadest expanse of information possible. The results achieved so far by the Systems Reliability Service have proved this practice to be justified but gaps and inadequacies in the available data have led to difficulties, additional work and long delays for the assessment engineers. This shortage of good quality generic information has led to the development of artificial simulation (or synthesis) techniques which make use of such devices as "Markov chain processes" and "Monte Carlo methods"; necessitating the employment of digital computers. But these mathematical methods, too, cannot be undertaken in complete isolation. They depend upon the existence of some true operational data, however meagre, in order to apply proper "weight" to the state probabilities required for a satisfactory simulation result. This indicates that there is an urgent need for a lot more work to be done in the Reliability Data Bank field so that more good quality information becomes readily available to make the task of Safety and Reliability assessment more simple and much more certain.

3. MINIMUM REQUIREMENTS OF 'GOOD QUALITY' RELIABILITY DATA

3.1 Experience in assessing the reliability of equipment and systems has shown how certain basic items of failure/fault data are necessary before successful analyses can be accomplished. Examples of such items of information are:

(i) Mean failure/fault rate

(ii) Mean repair time

(iii) Number of items involved

(iv) Number of failures/faults recorded

(v) Duration of surveillance period

(vi) Total length of operating time

(vii) Environmental conditions prevailing

(viii) etc.

All this information is required for a wide range of items in the electronics, electrical, mechanical, pneumatic,hydraulic and other fields of engineering. It is often presupposed that these data can be adequately classified under such headings as "components", "equipments" and "systems" but experience has revealed that these terms are not being defined precisely enough. For example, in a particular discipline a certain device is classed as a component whereas others look upon it as an equipment. Similarly, classification under "electronic", "electrical" "mechanical", "pneumatic" and "hydraulic" is just as obscure. The best way of classifying reliability data is by means of its' "item description". This involves drawing up an explicit, yet concise, engineering description for each item of interest. Invariably it is this item description that is recognised first by the assessment engineer when searching for reliability information.

4. THE SYSTEMS RELIABILITY SERVICE DATA BANK

4.1 Quite early in its history the United Kingdom Atomic Energy Authority recognised the need for a central reference of performance information on the component parts and devices of which nuclear reactors are made. This scheme concentrated mainly on the collection and analysis of the failure to danger history of these components and devices. It was soon realised that, although valuable, this data was not in itself enough to constitute information of sufficiently high quality to satisfy the requirements of Safety Assessors and Reliability Engineers engaged in examining new designs and predicting their probable performance. To make up for this deficiency a much improved 'Fault Data Store' was set up. In this the information was assessed and manipulated manually. This was soon transformed into a computer based store of fault data, operated by the then

"Authority Health and Safety Branch". Industry, other Government departments and many other organisations became interested and so the Systems Reliability Service was inaugurated and the Computerised Data Bank built up on a much wider base.[1]

4.2 The Systems Reliability Service Data Bank consists of two discrete data stores, illustrated by Figure 1*:

(i) The Reliability Data Store.

(ii) The Event Data Store.

This Figure shows the close connection between the two data stores and also the general interrelation between the SRS Data Bank and other organisations, with the corresponding information flow.

5. RELIABILITY DATA STORE

5.1 In Reliability Analysis and Safety Assessment work there is a continuous need to store and retrieve data accurately and to process them as required; any processing must not detract from the quality of the original data and it must retain the original raw data intact. These processes must also include facilities for the selection and retrieval of this information based on parameters other than the item description. For example: Type number, manufacturer, identity, size, range of operation, design ratings, design date, environmental conditions etc. These are all of specific interest to different users. In order to store reliability information of this quality, to manipulate it, retrieve it and analyse it effectively, to the Reliability Analyst's and Safety Assessor's requirements, computer techniques must be resorted to. To meet these requirements adequately and to provide an efficient service the partly coded storage system, known as the Reliability Data Store, has been developed by the Systems Reliability Service.

5.2 The information available from the Systems Reliability Service Reliability Data Store ranges from that relating to the failure rates of minute components to those obtained from assessments of presently operating plants, systems and assemblies in the UKAEA, Industry and elsewhere. The most up to date information of this class and quality is that transferred to the Reliability Data Store from the Systems Reliability Service Event Data Store. But much reliability information is being obtained from many other sources; for example;

(i) Published figures in open and classified literature; if the quality (integrity), source and Author are known to be dependable.

(ii) Figures resulting from event data reporting schemes already accumulated in private recording systems and data banks in industry.

(iii) Figures obtained during proving trials and life tests

(iv) The results of laboratory tests, climatic and durability tests and similar exercises.

(v) Direct collection of Reliability information by searching and analysing past history files at processing plants and elsewhere.

(vi) Data transferred from the Systems Reliability Service Event Data Store.

---

* The figures of this paper are on pp. 39-64.

The computer techniques developed for the Reliability Data Store enable further "processing" of the data to take place. Failure rates and other figures can be "pooled" statistically for identical items of, say, different manufacturers, etc. These results are produced, by the computer programs, in the form of printed lists and tables which are dispatched to customers whose requests and enquiries call for searches of the Reliability Data Store to be made.

5.3 All entries for inclusion in the Reliability Data Store list must be reduced to a standard format, without detriment to the original, to facilitate computer operations. This is begun by the donor of the information who is required to complete an information sheet, shown in Figure 2. The complete catalogue of possible entries, which can be accommodated in one item of data, are printed on this sheet. However, it is not necessary to fill in every entry, only those relevant to the particular item of data need be completed. In addition, if the (i) "operating time", (ii)"number of items" in the sample and (iii) the "number of failures" recorded throughout the total operating period are given then the Reliability Data Store computer program will calculate (iv) the "sample size" (i.e. the operating experience in item hours) and (v) the "Mean Failure Rate" together with its associated lower and upper confidence limits based on the Poisson Distribution(1), assuming constant failure rate. In fact, if any of these five parameters is missing and providing that those already given are adequate, then the computer will calculate the missing parameters automatically.

5.4 In order to compress the information further and to reduce it to a suitable form ready for the preparation of punched computer cards the data have to be translated from the Information Sheet to a Punching Form, shown in Figure 3. This process requires special skill and training and it is done by the SRS coding team. The information is then stored on data tape file in the computer ready for retrieval and further processing.

5.5 When a Systems Reliability Service customer asks for Reliability information a search of the Reliability Data Store is made and all entries which appear to fit his requirements are printed out. A typical information print is shown in Figure 4. If several of these data entries are relevant to the enquiry then, in order to enhance the operating experience (sample size), corresponding figures from the several data entries can be "pooled" statistically and a composite set of figures produced. A typical computer print of pooled data is shown in Figure 5.

6. EVENT DATA STORE COMPUTER PROGRAMS

6.1 These consist of a suite of three discrete computer programs, each one enabling the storage, retrieval and analysis of the data it contains:

(i) Inventory Data Store.

(ii) Event Data Store.

(iii) Installation Operating Record.

These three stages of data processing are illustrated in Figure 6. The Installation Operating Record and the Inventory Data Store programs are very important ancillaries to the main Event Data Store program.

6.2 The objectives of the Systems Reliability Service Event Data Store suite of computer programs are threefold:

(i) To provide compact storage of data which has been collected from operating Nuclear Power Stations and other Industrial Plant.

(ii) To facilitate quality assessment and computer processing of this data, to generate Reliability and Availability figures, and some management information, for use by the Plant Operators, Designers and Maintenance Engineers.

(iii) To supply high quality, up to date, Reliability information to the Systems Reliability Service Reliability Data Store.

6.3 The events referred to range from breakdowns of plant and components, through modifications to scheduled and unscheduled maintenance activities. In fact, by the Systems Reliability Service definition, an"event is any occurrence involving items of plant under event data surveillance which interrupts the normal smooth running of that item or installation (plant)". Each one of these events is to be recorded as data. If the information resulting from subsequent analysis is of the right quality then operational failure rates and mean repair times, with their respective statistical distributions, can be obtained and the down time (i.e. fractional dead time) suffered by the operating plant can be assessed. If plant and item outage times are recorded then the mean (and total) outage time following breakdowns and other unscheduled occurrences can be given. By combining this information with the down time attributable to scheduled maintenance and other causes then the total down time and plant (or item) availability can be calculated.

7. THE EVENT REPORTING SCHEME

7.1 In order to obtain satisfactory "live" Reliability Data from operating power plant, equipment and components a well prepared workable scheme of event surveillance has to be set up on the plant. This operation must be tailored to suit local conditions, in order to keep cost and local work load to a minimum without diminishing the standard of data reporting.

7.2 In setting up the scheme the first task is to decide upon what sections of the station are to be made the subject of event reports and determine to what depth these reports are to go. For example on a nuclear power plant, the smallest item considered might be a pump unit (consisting of prime mover, gearbox and pump). Whereas on a motor vehicle items as small as starter motor bushes, wheel bearings and brake linings might be made the basic items subject to event reports. The final outcome is an "Inventory" list of items which are of mutual interest to the plant operator and to the Systems Reliability Service Data Bank. Figures 7 to 9 show a typical Inventory data information sheet, from which a computer card punching form is prepared, an Inventory Data Punching form and a computer print of an inventory data record on data tape file. The items on the Inventory list have to be carefully documented by the plant operators to provide a comprehensive engineering description of each one. These particulars are then coded, to condense them and reduce them to a uniform format, in preparation for storage on the Inventory data tape file as the basic Systems Reliability Service Item Inventory for that project.

7.3 Reportable events result from many occurrences on Plant and Systems and these occurrences are usually documented by operations personnel in the form of Fault Reports, Defect Reports, Unusual Occurrence Reports or

Maintenance Reports; some form of elementary Event Reporting scheme is often already set up by Plant Management. The Systems Reliability Service tries to make use of any existing system, with minimum change, so that no increase in clerical work load is imposed as a result of the introduction of a Systems Reliability Service event data reporting scheme. The documents already being raised on these Plants then take on an extra role in that they become the vehicles by means of which events and other significant occurrences are transmitted to the Event Data Store.

7.4 Every Systems Reliability Service Event Reporting Scheme is set up to satisfy two principal requirements:

(i) They must provide a useful feedback of Management type information to the donors of event data.

(ii) They must form the basis for accumulating a stock of high quality information in the Event Data Store which, after appropriate computer processing, can be transferred with confidence to the Systems Reliability Service Reliability Data Store to supplement the accumulation of up to date information derived directly from operating plant and working components.

However, local fault reporting schemes do not always achieve the necessary standard of information required by the Systems Reliability Service event reporting schemes. In these circumstances the Systems Reliability Service makes recommendations to the local Plant Management concerning improvements to their existing documentation to raise the quality of information contained therein so that it becomes acceptable to the Systems Reliability Service scheme. Alternatively the Systems Reliability Service will design a suitable Event Report Form to either complement or completely supersede those of the existing scheme.

At this stage in setting up the scheme it is advisable to prepare a project specification. This document must describe the scheme in fine detail and it should contain a full Inventory List. The specification should also establish such characteristics as "failure criteria" for each item in the Inventory List. This leads on to the identification of "Types of Event", "Failure Modes", "Causes of Failure", etc. (see Appendix A and reference 3, 4 and 5 for definitions).

8. INVENTORY DATA COLLECTION

8.1 Before an item of plant can be subjected to a period of Systems Reliability Service event data surveillance it must first be fully described in order that it can be identified on its own and, ultimately, together with all similar items stored in the Data Bank. The Systems Reliability Service Inventory Data Store was set up to provide the necessary storage facility for this information. The collection of this inventory data is a "once off" requirement (i.e. once the data has been collected then item documentation is finished). Figure 7 shows the Systems Reliability Service Inventory Data Information Sheet which must be completed as fully as possible by the plant operators for each item that is to be made the subject of event reports. Figure 8 is the Inventory Data Punching form to which the Systems Reliability Service Coding personnel transfer the data in order to concentrate it ready for computer operations.

8.2 The absolute minimum inventory data requirement depends upon the individual scheme being entered into with the Systems Reliability Service.

But for every item in the scheme the most important inventory data entry is the "Item Description". This must explain briefly and concisely in engineering terms exactly what the item is, together with the manufacturer's identity or a drawing number or some form of reference by which the item can be positively identified. The principal entries on the information sheet are as follows.

8.3 The Installation/Location code is allocated by the Systems Reliability Service who keep a confidential register of these codes. It is a unique and confidential identification code raised especially for the particular project. All items included in the inventory list belonging to that project will be given this same code. The inventory number, too, will be supplied by the Systems Reliability Service although very often this is associated with a particular division of the plant or with the item identity or a plant item number, etc.

8.4 Dimensions and other characteristics which distinguish an inventory item from very similar items by bore size, working pressure or safe working load, for example, are entered in the "Size or Range" sections of the form. The order in which these characteristics are entered is important and it can be so arranged that, for similar items, each of the four size or range entries will refer to a different characteristic and the order in which these characteristics are recorded on data tape file can be made the same. When these entries are coded, the quantity and its unit of measurement are reduced to a four character alphabetic entry known as the "SORAY" code (see Figure 27.) If the data entered in size or range 1 and 2 (or 3 and 4) have the same units of measurement then the Inventory Data Store computer program assumes that these two values express a range, the lower value being entered in range 1 (or 3) up to the upper value entered in range 2 (or 4). Figure 9 is a fully documented inventory data record which shows the operating range over which a pressure relief valve may be adjusted to work in lbs per square inch gauge. The Units of Measurement code (SORAY) is stored in the computer program.

8.5 Date of entry into the Systems Reliability Service event reporting scheme is the date on which the surveillance period began for that item. The Item Code is a 14 digit numerical descriptive code, one of which is given to every inventory item under event surveillance. Identical items have the same Item Code. Figure 10 is a restricted selection from the section which shows the way the code describing valves mechanical is built up. This arrangement is typical of all the other sections in this item code list.

8.6 Environment is an eight digit numeric code in which each digit refers to a single word, the words combining to describe the environmental conditions in which the inventory item is working. This code is also used in the Reliability Data Store and Figure 4, the "MAGPIE" print, shows how six of these words have been translated back into literal form from the numeric code stored on data tape file. Application refers to a list of special numerical factors which are used to give quantitatively a broad relation between the failure probability of an item when used in different applications. This code is shown in Figure 11. For example, if an item failure rate has been determined while employed in a normal static industrial application then, if it is proposed to use this same item in a mobile role it is to be expected, by reference to the application code (Figure 11), that the failure probability will worsen by a factor of 4.

8.7 The "Time Function" and "Control Codes" are made use of in a special sub-routine which calculates the operating time since the previous event (which is ultimately used to calculate the item failure rate) using the Installation Operating Record program (see Section 12) together with the time and date of the previous event entry on the event data records (see Section 11). Cycles per year performs a similar duty to the time function in that it enables the calculation of cycles of operation between events which is used, in turn, to compute the mean lifetime of the item in terms of, say, MW hours for a nuclear power station or miles travelled for a motor vehicle.

8.8 The remaining data entries are self explanatory but only those agreed between the plant operators and the Systems Reliability Service need to be entered; although a well documented inventory list adds greatly to the quality of the subsequent reliability data.

8.9 Most of the information stored in the inventory list, for each inventory item, is transferred automatically to the event data tape file to complete the details stored there in the event data records for that item. When event data is being analysed selection and combination of specific inventory items can be accomplished by selecting on any of the data entries stored in the inventory. For example, it is required to analyse the relief valve event data of a particular manufacturer and type. The Event Data Store is interrogated using the code number for the manufacturer together with the corresponding type number. This causes all the events having this combined specification to be selected and presented for event data analysis by the computer, excluding all others. The inventory data itself can be analysed to a limited degree by interrogating the inventory data tape file.

9. COMPILATION OF EVENT REPORTS

9.1 Some typical Event Report forms, of the type used on Systems Reliability Service schemes, are shown in Figures 12, 13, 14 and 15. They were designed jointly by the Systems Reliability Service and the Plant Management and Engineers and they were all tailored to suit local conditions at the plants. The form shown in Figure 12 has a format which involves rather more written material than the Systems Reliability Service likes to impose on the contributors to the scheme; but it does include some "tick box" type entries, which are preferred. The form shown in Figure 13 is composed almost entirely of the "tick box" variety and that of Figure 14 &15 includes some of the Systems Reliability Service codes needed for data compression during preparation for computer operations. In addition to making event reports as simple to fill in as it is possible to do, the Systems Reliability Service try to conform to familiar local colloquialisms so that the process labour and tradesmen, etc., who may be required to complete event report forms, quickly become acquainted with them and so find the additional task less irksome.

9.2 It is an essential requirement of any event reporting scheme that <u>all</u> reportable events, without exception, must be submitted to the data store for recording and accumulation. This insures that a useful and completely dependable return of information can be provided for the exclusive use of the donors of the original data and the value of this feedback is related directly to the excellence of the information on which it was based. Systems Reliability Service standards of quality are met only if a complete information flow is diligently maintained.

10. DATA PREPARATION

10.1 When event reports have been made out, to the satisfaction of the originators, they are sent to the Systems Reliability Service. Upon receipt by the Systems Reliability Service they are given an initial scrutiny and any queries are referred back to the originators. The raw information which these reports contain is then translated, either directly or in code, into a form suitable for computer operation. Each information entry is transferred from the event report to a "Punching Form" (see Figure 16) from which 80 column computer punched cards are prepared. Numeric entries, such as times, dates, manhours, etc., are entered, on the Punching Form as written on the Event Report. Any descriptive information is written, with some abbreviation if it has to be concentrated, just as it is written on the event report. But most of the remaining information must be converted to numeric, or alpha-numeric codes in order to concentrate a large amount of information into the smallest storage area possible, on the data tape file. This process ensures that, during data analysis, the facility for retrieving and decoding the information back into literal form is maintained. To do this the Systems Reliability Service Event Data Store Analysis Programmes make use of two retrieval processes: numerical identification and character recognition.

11. ORGANISATION OF EVENT DATA RECORDS

11.1 The Systems Reliability Service event data records are divided into four hierarchical categories and Figure 17 shows a typical event report submitted to Systems Reliability Service.

(i) GENERAL INFORMATION - consists of data entries which are common to the three other categories. The principal entries are: Type of event code, time and date on which the event took place, the unique INSTALLATION/LOCATION code which identifies the project, a "site reference number" which is a twelve character alpha-numeric entry for use by the originator.

(ii) INSTALLATION CATEGORY - The principal data entries are: an indicator of the operating state of the plant at the outset of the event, an indicator of the effect of the event on the plant (installation), an indicator of the way the plant was made safe (e.g. how it was shutdown), a statement of the outage time of the plant in hours, the total labour and the skilled labour expended in manhours, the materials and parts cost in pounds sterling, the operating hours of the plant since the previous event (Installation lifetime in hours or days), the cycles of operation since the previous event (e.g. MW hours of power generated), the time and date of the previous event to enable calculation of the station operating hours since the previous event when this figure cannot be obtained directly from the plant, the cumulative total of cycles of operation to enable calculation of the cycles since the previous event when this figure cannot be obtained directly from the plant. There is also a generous allocation of space for the storage of a description of the event as it affected the plant as a whole.

(iii) SYSTEM CATEGORY - Nuclear power plants, in common with other industrial plant, are sub-divided into Systems. Each system can be given a unique System Inventory Number and also a System Code in common with all similar systems; for example: the Heat Exchanger System of a steam raising plant has a common purpose within all steam raising plants even though they may be unique in design. The main data entries are: an indication of the effect of the event on system performance, the system outage time and repair time in hours, the operating time (system lifetime in hours or days) since the previous event reported on that system, the cycles of operation (e.g. tons of steam produced, number of starts of a pump unit, etc.,) since the previous event reported on that system, the time and date of the previous event reported on that system to enable calculation of the operating hours since the previous event when this cannot be obtained directly from the plant, the cumulative total of cycles of operation to enable calculation of the cycles since the previous event (on the system) when this figure cannot be obtained directly from the plant. A fairly generous allocation of space for the storage of a description of how the event affected the system is also provided. In addition, up to twenty item codes, representing items (other than the inventory item) of interest which were affected by the event can be accommodated (e.g. component items outside the inventory list of components for that system).

(iv) ITEM CATEGORY - Systems are made up from component items (e.g. the feedwater pumps of a Heat Exchanger system) and these are the items which have been selected to undergo a period of event surveillance. In addition to its unique Inventory Number, which must be entered as data on every event report, each item is also given a Systems Reliability Service Item Code in common with all similar items. The principal data entries under this category are: the Failure Mode and Cause of Failure codes (see Appendix A for definitions), several code entries representing the operating stresses at the outset of the event, the item outage and repair time in hours, the total labour and skilled labour expended in manhours, the materials and parts cost in pounds sterling, a system inventory number or system code corresponding to the parent system, an indicator of the fault category (e.g. fail dangerous fault, unrevealed fault, etc.), the position of the item in the installation (e.g. feedwater pump No. 2), up to eight keywords describing parts which have been renewed together with the quantities of each if more than one was renewed at the time. There is only a limited allocation of space for a description of how the event affected the item (equivalent to twenty 4-character computer words). However, in addition to the foregoing, a large proportion of the information stored in the corresponding inventory data entry, of the Project Item Inventory list, is transferred automatically to each event as it is written on to data tape file. This amount of information ensures that each item category data record is fully documented and can be fully understood in isolation.

12. INSTALLATION OPERATING RECORD

12.1 The Installation Operating Record program routine provides a running account of the Station or Plant operating state throughout the duration of the event surveillance period. It has been developed to perform two functions:

(i) it acts as a continuous operating "log" for information of the Plant Operators, and (ii) it records the time interval between successive changes in the operating state of the station, so avoiding the need for plant operators or Systems Reliability Service Coding personnel to maintain a separate account of the time and date of successive events for each Inventory Item under surveillance.

There is only one report form associated with the program and this is shown in Figure 18. It is a combined information sheet and punching form from which 80 column computer cards are prepared.

Figure 19 shows an extract from the Installation Operating Record of a small research reactor. The Entry Numbers must be maintained in strict numerical sequence to ensure completeness. The time and date entries refer to the time and date on which a change of state took place. "Installation" refers to the unique Installation/Location code given to each Systems Reliability Service Event Surveillance Project. For example H01 refers to the Liverpool and Manchester Universities Research Reactor. Control Codes are introduced to represent up to four different operating states for the Plant under surveillance. For example, in Figure 19:

HP indicates the start of a period of High Power operation.
MP indicates the start of a period of Medium Power operation.
LP indicates the start of a period of Low Power operation.
SH indicates the start of a period during which the Reactor was shut down.

13. EVENT DATA ANALYSIS

13.1 The Event Data Store computer programs include several special routines designed specifically for data analysis. They are, broadly, of two types:

(i) Management information presentation

(ii) Reliability data generating.

They make use of the data stored on the event data tape files to provide listings, statistical and other information forms about the plant items from which the original data was derived. The reliability figures are based on well proven statistical procedures[9] and the data used has been obtained from working plants during their everyday operation. Therefore, Plant Managers and Engineers can rely on these analysis results and use them with confidence when making decisions concerning forward planning and the efficient running of their plant. However, only the Reliability Data Generating routines will be described here (the management type and analyses are all described in reference 6) but, as the latest development, called "LIST", is not included in reference 6 a typical print out is given in Figure 20, showing the general layout.

13.2 There are two main Reliability analysis routines associated with the Event Data Store suite of computer programs. These are:

(i) LIFE analysis

(ii) OUTAGE analysis.

These analyses can be applied to data reported at Installation level (i.e. the Station as a whole), System level or at plant item/component level.

14. LIFE ANALYSIS

14.1 This routine processes "Lifetimes", i.e. the operating times reported between successive events involving the particular item of plant being considered. The routine calculates the "mean interval between events", where the term "events" can range from breakdowns through modifications and calibrations to scheduled and unscheduled maintenance operations. An example of the main data table is shown in Figure 21. As well as the mean interval value, the data table shows the number of events (Failures) taken into consideration in the calculation. It also shows the 95% confidence interval for these values, based on the Poisson distribution[2], assuming constant failure rate. The table includes a statement of the "total Operation" (representing the operating experience or sample size) in item hours and the total population of items contributing "lifetimes" to the sample.

14.2 Events for inclusion in the analysis, between which it is required to calculate the mean interval, are selected by the customer according to "Type of Event" or "Failure Mode" or a combination of both. This is called the "specified criteria". However, the effect of the "lifetimes" recorded on events reported, on the plant item being considered, but outside this specification is not lost to the analysis. The "operating times since the previous event" are all used (if appropriate) to contribute their "weight" to the "Total Operation" figure. But they do not contribute to the value of the divisor of the Total Operation figure, to produce the mean interval. This divisor is made up of that number of events, in the overall sample providing lifetimes, conforming to the specified criteria and it is this criteria which determines the word "Failures", for example, in the case of Figure 21. This divisor is modified when populations are involved: i.e. when more than one identical plant item is described by a single inventory number.

14.3 Failures were specified, in the table of Figure 21, according to the following types of event:

(i) Plant Breakdown (complete failure[3]),

(ii) Unplanned, in situ maintenance due to failure,

(iii) Unplanned, workshop maintenance due to failure,

since these all conform to Failure type events. The word "FAILURES" can be altered (in the three locations shown in Figure 21) at will by the customer to describe the general type of event between which the mean interval is being calculated. (Whatever word is specified this same word will appear in all three locations).

14.4 There is also a statement of the total population of individual plant items contributing to the sample of lifetimes and a brief statement concerning the specified criteria. The largest uncertainty depends upon the most inaccurate lifetime measurement. In Figure 21 all the times used in the analysis had been measured to the nearest hour (i.e. $\pm \frac{1}{2}$ hour). Had there been any "unrevealed" faults reported (i.e. faults which come to light only when the item is maintained, etc.,) then a statement to this effect would have been included at the bottom of the data table.

14.5 The mean lifetime of a plant item can be expressed in terms other than that of time. For example, transport vehicle reliability information is usually expressed in terms of mileage run. The "LIFE" analysis is designed to satisfy this requirement by providing the facility for expressing the mean lifetime in terms of "cycles of operation". The data table produced by this analysis is identical in format to that of Figure 21 with the word "HOURS" replaced with the word "CYCLES".

14.6 If of sufficiently high quality the information given in the LIFE analysis data table could be transferred to the Reliability Data Store through the intermediate of the Punching Form shown in Figure 3. Identification of the item referred to would be obtained from the Inventory Data Store entry for that item and the LIFE data would be entered as follows:

(i) Card A - The words "Plant Failures" would be entered in the "Fault Mode" field and the explanatory comment at the bottom of the data table would be written as an explanation of Plant Failure using a card F (not shown in Figure 3), the Reliability Data Store comment card.

(ii) Card B - The figure against "Total Population" in the data table would become "NO. OF ITEMS". The number of "Failures" would become "NO. OF FAULTS" on the punching form.

(iii) Card D - The item hours figure, shown against "Total Operation" in the data table, would be transferred to the "ITEM TIME/CYCLES" field and in the "UNIT" field the letters "ITTH" would be entered, representing "Item Hours" from the "SORAY" code of Figure 27. The letter "A" would be entered in column 44, T/CODE, to indicate that the time base for the "ITEM TIME/CYCLES" figure is plant operating time. The "Mean Interval Between Failures" together with the confidence limits and the figure 95 would all be put under the "FAILURE RATE, LIFETIME/CYCLES, REPAIR TIME, AVAILABILITY, ETC," section of this card. Since the intervals between failures were measured in hours then, from the "SORAY" code the letters IFTH would be put in the "UNIT" field and the letter "A" in the T/CODE column 78.

14.7 Statistical Analysis of Lifetimes. If events occur in a completely unpredictable manner (i.e. they are randomly distributed with reference to operating time or cycles of operation) then the corresponding distributions of "lifetimes" can be statistically investigated.[9] Figure 22 is a complete list of the lifetimes used in computation of the mean interval between failures of Figure 21. Since these lifetimes occurred at random they can be statistically analysed.

14.8 In Figure 22 an "incomplete" lifetime is one which either does not begin or does not terminate with an event reported from the plant. In the Event Data Store a "dummy event" is used to indicate to the LIFE analysis routine the time and date on which the surveillance period began and ended. The former dummy event is known as an "Initial Entry Event" and the latter is an "End of Analysis Event". The operating hours since the previous event entry, if given, in these two types of event are prefixed with a negative sign to distinguish them from the complete lifetime entries.

14.9 Figure 23 is a selection of values (one out of every five) from the list of cumulative probabilities which is compiled by the statistical analysis sub-routine of the LIFE analysis. There is, above this table, a statement identifying the best fitting unimodal, two parameter, theoretical cumulative distribution function chosen from the Normal, Lognormal, Exponential, Weibull and Special Erlangian models. The left hand column of figures in the table is the reduced list of lifetimes being analysed (the program takes all the lifetimes into consideration). The central column of figures lists the cumulative percentage probability ascribed, by the program routine, to each lifetime and the third column lists the cumulative percentage probability corresponding to the lifetimes listed but obtained from the best fitting mathematical model.

14.10 The "Goodness of Fit" for the theoretical cumulative distribution is assessed using the Kolmogorov-Smirnov test(7)(8) and a statement to this effect is printed below the table together with the "maximum difference" in percentage probability. This test takes into account the number of lifetimes in the sample and the amount by which the most diverse value of lifetime probability differs from the corresponding figure in the list of fitted probabilities. This expression of confidence in the degree of best fit is in the form of the percentage probability that another sample, of equal size to the present sample, selected at random from the parent population of lifetimes, for the item being analysed, would give a fit to the theoretical model as good as (but no better than) that already given by the sample of lifetimes being analysed.

14.11 Two graphs are plotted by this sub-routine, as shown in Figure 24 and 25. The curve of Figure 24 is the graph of the probability density function corresponding to the best fitting theoretical distribution. The ordinate represents the proportion of the maximum frequency (due to the modal lifetime) corresponding to points of interest along the abscissa. The abscissa is graduated in operating hours but, if the analysis is against cycles of operation, then the abscissa is graduated in cycles. A statement relating to the 95% confidence limits, for this graph, is printed beneath it. Figure 25 is a plot of the conditional failure rate (i.e. hazard function) against operating time. If the item concerned has survived to the point selected along the abscissa then, from the ordinate, the probability that the item will fail within the next hour (or before completion of the next cycle of operation) can be read off.

14.12 Furthermore, from the slope of this curve, an indication of the type of maintenance which might be most appropriate for the plant item being analysed, can be obtained. However, this indication must be weighted against other relevant factors known only to the plant operators donating the lifetime information. If the slope has a negative trend then this can infer that regular planned maintenance is not the best form and a system of breakdown maintenance might be more effective and economical. Alternatively, if the graph has a positive trend then this might suggest that planned maintenance is the best form for the item. In addition, the time scale, along the abscissa, can be used as a guide to selection of the best planned maintenance cycle.

15. REPAIR AND OUTAGE TIME ANALYSIS

15.1 These make up the second of the two main Reliability information generating routines associated with the Event Data Store. Both REPAIR times and OUTAGE times are processed by the same computer program subroutine. But different "logical" statements, called up by the individual keywords, direct the program to select either REPAIR time or OUTAGE time data entries for processing. The keywords also choose the appropriate "PRINT" format statements that will effect printing out the correct wording for the main data table (Figure 26a and b).

15.2 Repair time data can be stored,in the Event Data Store, at System level and Inventory Item level only but Outage time can be stored at

Installation (i.e. referring to the Station as a whole, in the case of Nuclear Power Plant), System level and Plant Inventory Item level as well.

15.3 Figure 26a and b show examples of the main data table presentation, provided by the computer program subroutine, for a typical Repair time analysis and the corresponding Outage time analysis for a small sample of identical Inventory items. Both are presented in the form of an "or more" percentage cumulative frequency table of hours to Repair and Outage hours respectively. The main body of the table lists a selection of the proportions of the total frequency with which the Repair time or Outage time exceeds certain values together with a statement of the longest and shortest times reported. Below the main table are a series of short statements concerning the average Repair or Outage time, the total accumulated Repair or Outage time and the total number of Repair or Outage time entries used in calculating the average value. These are followed by a statement concerning the calculation of percentage "AVAILABILITY"(Ref.10) from the Repair time reports or percentage "UTILISATION" calculated from the Outage time reports. The presentation ends with a statement giving the value of the maximum accumulated operating time that is associated with the Repair time or Outage time reports used in the calculation of percentage AVAILABILITY or percentage UTILISATION.

15.4 The value of percentage AVAILABILITY for the System or Inventory item being analysed is calculated as follows:

$$\text{AVAILABILITY \%} = \left\{ 1 - \frac{\Sigma(\text{Repair time})}{\Sigma(\text{Operating time between events}) + \Sigma(\text{Outage time})} \right\} \times 100$$

or, in terms of data stored in the Event Data Store:

$$\text{AVAILABILITY \%} = \left\{ 1 - \frac{\text{Total Repair time}}{\text{Total Operating time between events + Total Outage time}} \right\} \times 100$$

Where Σ(Operating time between events) + Σ(Outage time) represents the total possible operating time that could have been experienced in the absence of any Repair time or Outage time losses. If repair times have been reported but not the corresponding outage times then the Repair time value is taken as the minimum Outage time report for the calculation of percentage AVAILABILITY.

15.5 Outage times are used to calculate the percentage UTILISATION, using a similar relationship to the foregoing, for the Installation (i.e. Station as a whole) the Systems making up the station or the Inventory Items, as follows:

$$\text{UTILISATION \%} = \left\{ 1 - \frac{\text{Total Outage time}}{\text{Total Operating time between events + Total Outage time}} \right\} \times 100$$

(NOTE: these calculations can be performed only if Repair and Outage time data entries are provided from the plant at the appropriate levels of reporting).

15.6 The maximum number of Repair or Outage time reports that can be processed by a single computer analysis run is 5,000. Only those Repair or

Outage time reports which are accompanied by a corresponding report of operating time since the previous event (i.e. the immediate past lifetime) are used in the calculation of AVAILABILITY or UTILISATION. This constraint is imposed to prevent these figures from being unfairly biassed, in small samples, when the occurrence of Repair times and Outage times have not been fully reported. Therefore an occurrence of zero (i.e. 0.0) operating hours since the previous event is significant and must be included as data if the Repair time and Outage time following the zero operating hours is to be included in the corresponding analysis.

15.7 It is recognised that the reverse situation can exist and so the treatment of Repair time and Outage time data must take this into account. In relatively large samples, where the accumulated operating hours greatly exceed the accumulated Repair and Outage hours, the calculations of AVAILABILITY and UTILISATION could be unfairly biassed by the exclusion of any of the associated Repair time or Outage time reports. So there is an alternative method of calculating AVAILABILITY and UTILISATION figures, to accommodate these larger, or longer term, samples. The total operating time, or operating experience, in item hours for the sample undergoing analysis can be obtained from the main data table of the corresponding "LIFE" analysis figure for TOTAL OPERATION (paragraph 14.2). This figure can be put into the analysis as a single additional item of data, thereby overruling accumulation by the computer program of those lifetimes having a corresponding repair or outage time report. The normal relationship for computation of either AVAILABILITY or UTILISATION is then used for the calculation.

15.8 In common with the LIFE analysis, both Outage times and Repair times can be subjected to full statistical treatment. A table of cumulative probability is constructed and the expression for the best fitting theoretical cumulative distribution function is found together with a plot of the corresponding probability density function curve.

15.9 If reports of Repair time are of sufficient quality (i.e. the average Repair time represents the mean fractional dead time of the item being analysed) then this information given in the Repair time data table can be transferred to the Reliability Data Store in a similar manner to the lifetime information. All the Repair time information is put on to a Card D of the Reliability Data Store Punching form (Figure 3). In addition, the figure of AVAILABILITY, derived from the same analysis, can also be transferred by means of a card D entry.

16. CONCLUSION

16.1 This paper sets out to emphasise the benefits to be gained by Plant Operators and Maintenance Engineers of Nuclear Power Plants through the constant examination of the performance characteristics of their plant, systems and items of equipment with a view to generating Management information and Reliability information in particular. Not only can this type of scrutiny highlight unreliability, which can often be concealed by effective (and expensive) planned maintenance and many other activities, but as a bonus a useful stock of plant operating data accumulates. This operating data can be usefully employed when making comparison between the performance of similar areas of the plant and it can be used to monitor changes in operating and maintenance philosophy. It also provides valuable Reliability figures for use by Designers of new plant and also for the Reliability Engineers and Safety Assessors who have the task of appraising these new designs. Another advantage to be gained through the diligent collection of performance data, on newly constructed plant in particular, is the confirmation of predicted failure patterns following simulation exercises, in the absence of real data. Confidence in these techniques can only be strengthened by this type of information feedback.

16.2 The Systems Reliability Service Data Bank has facilities readily available to all who have an interest in furthering the standards of Nuclear Power Station operational safety, efficiency and economy through the collection, storage and use of plant and equipment performance data.

REFERENCES

1. EAMES, A. R., "Data Store Requirements Arising out of Reliability Analyses". UKAEA report AHSB(S)R138 (1967).
2. WOODCOCK, E. R., EAMES, A. R., "Confidence Limits for Numbers from 0 to 1200 Based on the Poisson Distribution". UKAEA report AHSB(S)R179 (1970).
3. HMSO., "Guide on the Reliability of Engineering Equipment and Parts". British Standards Institution, Draft for Development 12 (1971).
4. HMSO., "Guide on the Assessment of Reliability of Engineering Equipment and Parts". British Standards Institution, Draft for Development 13 (1974).
5. HMSO., "Guide on the Production, Flow, Analysis and Interpretation of Reliability Data for Engineering Equipment and Parts". British Standards Institution, Draft for Development 15 (1974).
6. FOTHERGILL, C. D. H., "The Analysis and Presentation of Derived Reliability Data from a Computerised Data Store". UKAEA report SRS/GR/22 (1974).
7. SMIRNOV, N., "On the Estimation of the Discrepancy between Empirical Curves of Distribution for Two Independent Samples. Bulletin Mathematique de l'Universite de Moscow, Vol. 2, pasc 2 (1939).
8. MASSEY, F. J. Jr., "The Kolmogorov-Smirnov Test for Goodness of Fit". Journal of the American Statistical Association, Vol. 46, No. 253, pp 68-78 (1951).
9. THOMASON, R., "Introduction to Reliability and Quality". The Machinery Publishing Co. Ltd. (1969).
10. LEWIS, R., "An Introduction to Reliability Engineering". McGraw-Hill (1970).

APPENDIX A

Failure Terminology - A Definition

The following definitions apply equally to components, items, systems and complete installations.

Failure Mode/Failure Criteria

Every item will have been designed and built such that its operational performance can be measured and assessed by observing one or more specified operational parameters. In order to confirm that the item is functioning satisfactorily it must be seen that every one of the operational parameters is fulfilling the specified design requirements within such tolerances, outside of which its performance is not permitted to deviate. It has failed, from the "reliability" point of view, whenever one (or more) of the specified operational parameters deviates beyond its allowable operational limits to the detriment of its operational performance.

The "Failure Mode" is the description of the degredation of the function monitored by the offending operational parameter. In all cases the following question must be asked:

"What function has been degraded, impaired or lost?"

The description of this loss of function is the "Failure Mode".

All the different ways in which an item can be considered to have failed are called its "Failure Criteria".

Cause of Failure

When an item failure occurs it will always be possible (ideally) to identify one or more component parts, power supplies or operator actions which were responsible for the failure of the parent item. The name of that component part, etc., which was directly responsible for the failure is the "Cause of Failure". When an item itself is a single component then the physical process most directly involved in the component failure is the "Cause of Failure".

Figures 1–27 follow.

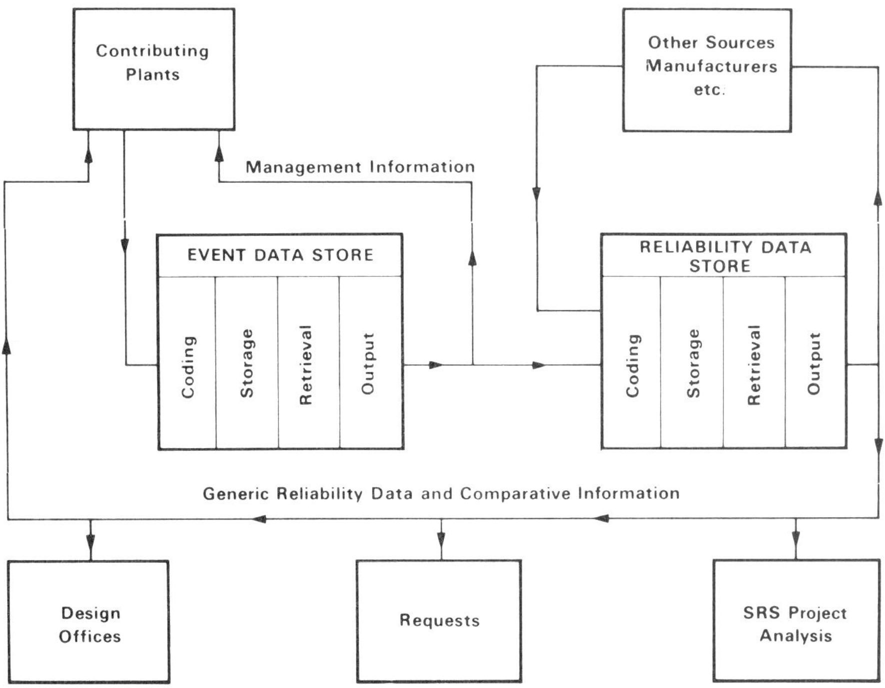

FIG. 1. Systems reliability service data bank information flow.

INFORMATION FOR RELIABILITY DATA STORE

DATA SERIAL NO..........

Please fill-in or tick where applicable where information is known (in RED)
Calculated data is not necessarily required. (See back for notes) Tick box ☐

FAULT MODE (NOTE 1)

MANUFACTURER

DESIGN RATINGS: 1 ____ 2 ____ (NOTE 2) | OPERATING RANGES: 1 ____ 2 ____ (NOTE 2)

INFORMATION SOURCE: (LITERATURE REFERENCE ETC)

ITEM DESCRIPTION (NOTE 3) | DATA YEAR:

IDENTITY (TYPE NO., ETC): | NO. OF ITEMS:

OPERATING TIME/ITEM (MEAN VALUE) | DESIGNER: (NAME OF FIRM)

NO. OF FAULTS: | NO. OF CYCLES OF OPERATION/ITEM (MEAN VALUES)

USAGE: (ELECTRONIC CIRCUITS): TUBES (E) ☐ SOLID STATE (T) ☐ HYBRID (H) ☐ NOT APPLICABLE (Z) ☐

FAULT CATEGORY: DANGEROUS (D) ☐ SAFE (S) ☐ SPURIOUS TRIP (T) ☐ OTHER ....

ENVIRONMENT: CLIMATE ____ TEMPERATURE ____ HUMIDITY ____ ALTITUDE ____
(NOTE 4) VIBRATION RADIATION CORROSION FUNGUS

APPLICATION (NOTE 5) | INFORMATION TYPE (NOTE 6) | INVENTORY NO. (NOTE 7)

MAINTENANCE INTERVAL (NOTE 8) | LOCATION CODE | DESIGN YEAR

QUALITY (NOTE 9) | HISTORY TIME (NOTE 10) | COMPUTER CALC. CODE (NOTE 11)

OPERATING EXPERIENCE: (NOTE 12)

FAILURE RATE: MEAN LOWER UPPER | CONF. BAND

MEDIA: (NOTE 13) | TIME ETC FOR BASIS OF CALC: (T CODE, NOTE 14) | DISTRIBUTION: (NOTE 15)

IF FURTHER DATA IS AVAILABLE PLEASE ENTER HERE:

COMMENT:

Compiled By ....

FIG.2. A reliability data information sheet.

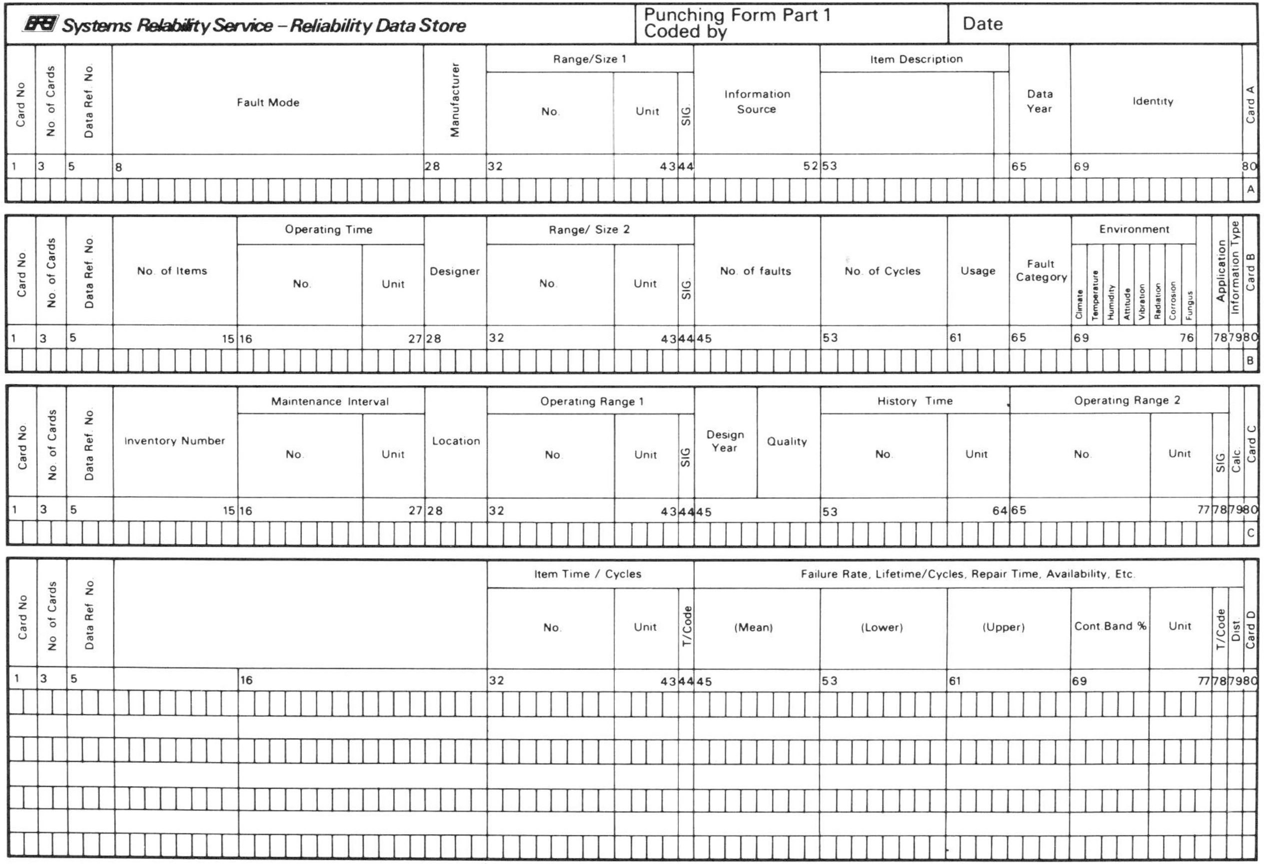

FIG.3. Reliability data store punching form.

```
ITEM DESCRIPTION CODE 1544.20000000          DATA SERIAL NO. 5840

MANUFACTURER                       /5211/
IDENTITY (ITEM TYPE NO.)           WBA 13A
APPLICATION                        AVERAGE (INDUSTRIAL)
LOCATION                           /Z28 /
USAGE CODE                         /Z   /

ENVIRONMENT --
    CLIMATE                        COOL, CONTINENTAL
    TEMPERATURE                    NORMAL
    HUMIDITY                       NORMAL
    ALTITUDE                       SEA LEVEL
    VIBRATION                      GENERAL
    RADIATION                      THERMAL-NUCLEAR

FAULT MODE                         TOTAL
INFORMATION TYPE                   FIELD CORRECTLY REPORTED
DATA YEAR                          1968
MAINTENANCE INTERVAL               6.00E 00  MONTH

NUMBER OF ITEMS                    2.40E  01
NUMBER OF FAULTS                   2.20E  01
OPERATING TIME                     3.52E  04  HOUR
HISTORY TIME                       6.00E  00  YEAR

FAILURE RATE (HISTORY TIME)
    MEAN                           1.74E  01  FAULT/MILLION HR
    LOWER LIMIT                    1.09E  01  FAULT/MILLION HR
    UPPER LIMIT                    2.64E  01  FAULT/MILLION HR
    DISTRIBUTION                   POISSON
    CONFIDENCE BAND                95.00      PER CENT
    SAMPLE SIZE                    1.26E  06  ITEM.HOUR

FAILURE RATE (OPERATING TIME)
    MEAN                           2.60E  01  FAULT/MILLION HR
    LOWER LIMIT                    1.63E  01  FAULT/MILLION HR
    UPPER LIMIT                    3.94E  01  FAULT/MILLION HR
    DISTRIBUTION                   POISSON
    CONFIDENCE BAND                95.00      PER CENT
    SAMPLE SIZE                    8.45E  05  ITEM.HOUR

INT.BETW.FAULTS (HISTORY TIME)
    MEAN                           5.75E  04  HOUR
    LOWER LIMIT                    3.80E  04  HOUR
    UPPER LIMIT                    9.16E  04  HOUR
    DISTRIBUTION                   POISSON
    CONFIDENCE BAND                95.00      PER CENT

INFORMATION SOURCE           210  The full literal source reference is printed.

COMMENT - PWR CONTROL ROD DRIVE ELECTRO/MECH. MAGNETIC JACK CONTROL ROD DRIVE MECHANISM.
```

FIG.4. A typical "MAGPIE" data print.

ITEM	DERIVED FAILURE RATE MEAN	LOWER	UPPER		
ABSORBERS					
RADIO FREQUENCY	6.9E 00	3.3E 00	1.3E 01	FLT/10**6 HR	(O)
ACCELEROMETERS					
LINEAR, DIRECT READING	1.2E 00	6.2E-01	2.1E 00	FLT/10**6 HR	(O)
ACTUATORS					
MANUAL	3.0E 01	2.2E 01	4.1E 01	FLT/10**6 HR	(O)
ROTARY (CAM OR LEVER)	1.4E 01	7.3E 00	2.5E 01	FLT/10**6 HR	(O)
AMPLIFIERS					
GENERAL	1.5E 01	1.2E 01	2.0E 01	FLT/10**6 HR	(O)
AUDIO FREQUENCY	3.4E 01	2.1E 01	5.1E 01	FLT/10**6 HR	(O)
COMPUTER DC	1.7E 02	1.5E 02	1.9E 02	FLT/10**6 HR	(H)
DC	1.6E 00	4.0E-02	8.8E 00	FLT/10**6 HR	(O)
MAGNETIC	8.2E 00	5.0E 00	1.3E 01	FLT/10**6 HR	(O)
SERVO	1.8E 02	5.0E 01	4.7E 02	FLT/10**6 HR	(H)
ANTENNAE					
GENERAL	2.0E 01	9.6E 00	3.7E 01	FLT/10**6 HR	(O)
ATTACHING METHODS AND MATERIALS					
BONDING, CHEMICAL (E.G. EPOXY)	0.0E 00	0.0E 00	2.8E-04	FLT/10**6 HR	(O)
CRIMPING, ELECTRICAL COMPONENTS AND WIRES	4.0E-02	1.9E-02	7.4E-02	FLT/10**6 HR	(O)
SOLDER, ELECTRICAL COMPONENTS AND WIRES (AUTOMATIC)	5.0E-03	2.4E-03	9.2E-03	FLT/10**6 HR	(O)
SOLDER, ELECTRICAL COMPONENTS AND WIRES (MANUAL)	1.7E-02	1.2E-02	2.3E-02	FLT/10**6 HR	(O)
WELDING, ELECTRICAL COMPONENTS AND WIRES	8.0E-02	3.8E-02	1.5E-01	FLT/10**6 HR	(O)
WIRE WRAPPING	1.9E-03	9.2E-04	3.5E-03	FLT/10**6 HR	(O)
AUDIO DEVICES					
BELL, ELECTRIC	1.3E-01	8.1E-02	2.1E-01	FLT/10**6 HR	(O)
BUZZER, ELECTRIC	6.0E 00	2.9E 00	1.1E 01	FLT/10**6 HR	(O)
HORN, ELECTRIC	3.8E 00	2.4E 00	5.9E 00	FLT/10**6 HR	(O)
BATTERIES					
NON-RECHARGEABLE, GENERAL	4.4E-04	2.7E-04	6.7E-04	FLT/10**6 HR	(O)
BATTERIES					
RECHARAGEABLE, GENERAL	4.7E-01	4.0E-01	5.4E-01	FLT/10**6 HR	(O)
BEARINGS					
LINEAR, SLEEVE	2.1E 00	1.0E 00	3.9E 00	FLT/10**6 HR	(O)
RADIAL, BALL	1.2E 01	7.2E 00	1.8E 01	FLT/10**6 HR	(O)
RADIAL, ROLLER	5.0E 00	2.4E 00	9.2E 00	FLT/10**6 HR	(O)
RADIAL, SLEEVE	5.0E 00	2.4E 00	9.2E 00	FLT/10**6 HR	(O)
BLOWERS AND FANS					
GENERAL	1.3E 01	9.9E 00	1.6E 01	FLT/10**6 HR	(O)

FIG. 5. A section from a typical "AVER" print-out.

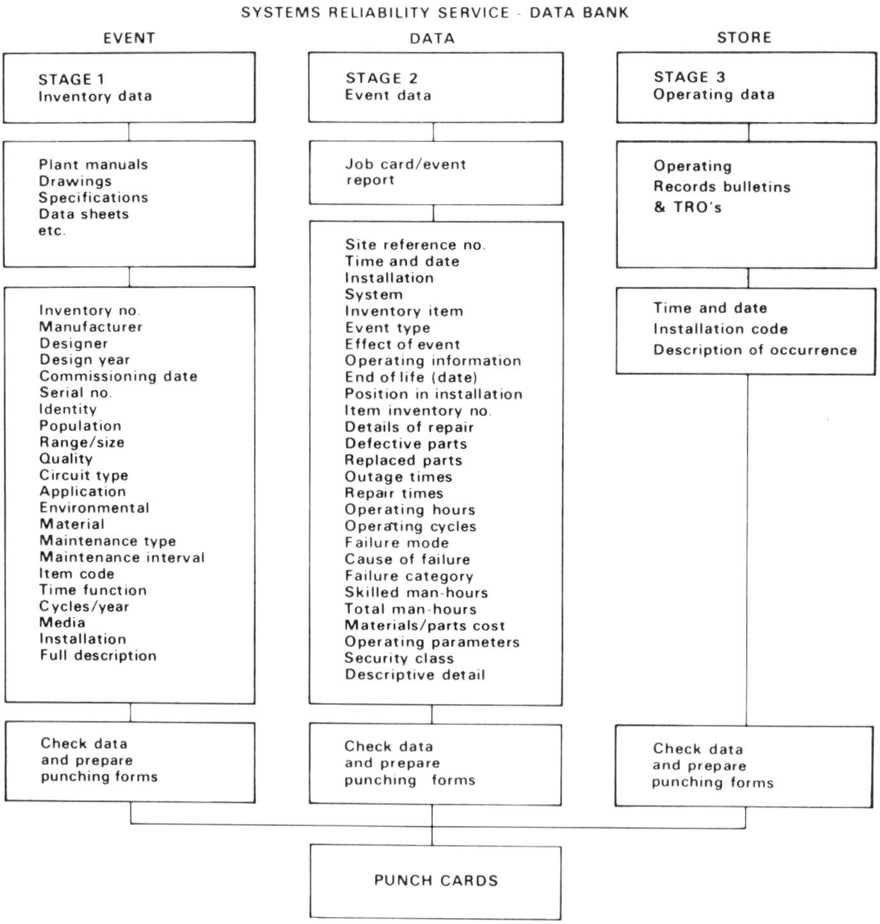

FIG.6. Three stages of data collection and preparation.

SYSTEMS RELIABILITY SERVICE - DATA BANK

INVENTORY ITEM DATA SHEET

Entry No. .... Installation Code ....

Item description ....

Manufacturer ....

Designer .... Design Year ....

Material of composition ....

Inventory No. .... Type No. ....

Identity

Population .... Mark No. ....

Quality (Specification) .... Media ....

Type of maintenance - Planned ☐ at .... intervals

Unplanned ☐

Breakdown ☐

*Size or range 1 .... (Lower Limit) Serial No. ....

2 .... (Upper Limit) Commissioning date ....

3 .... (Lower Limit)

4 .... (Upper Limit) Date of entry into S.R.S. ....

Circuit type - Electronic Valve ☐ Hybrid (valve and transistor) ☐

Solid state ☐ Not applicable ☐

Installation ....

Item Code | | | | | | | | | | | | | | |

Environment | | | | | | | | | Application 1. 2. 3. 4. 5. 6. 7. 8. 9.

Time Function (% of Calendar Time)

A	B	C	D

Control Codes A = ...
B = ...
C = ...
D = ...

Cycles per year ....

GENERAL REMARKS ....

....

....

* Size or range may be used for upper and lower limits of a single parameter as shown.

FIG.7. Typical inventory item data sheet.

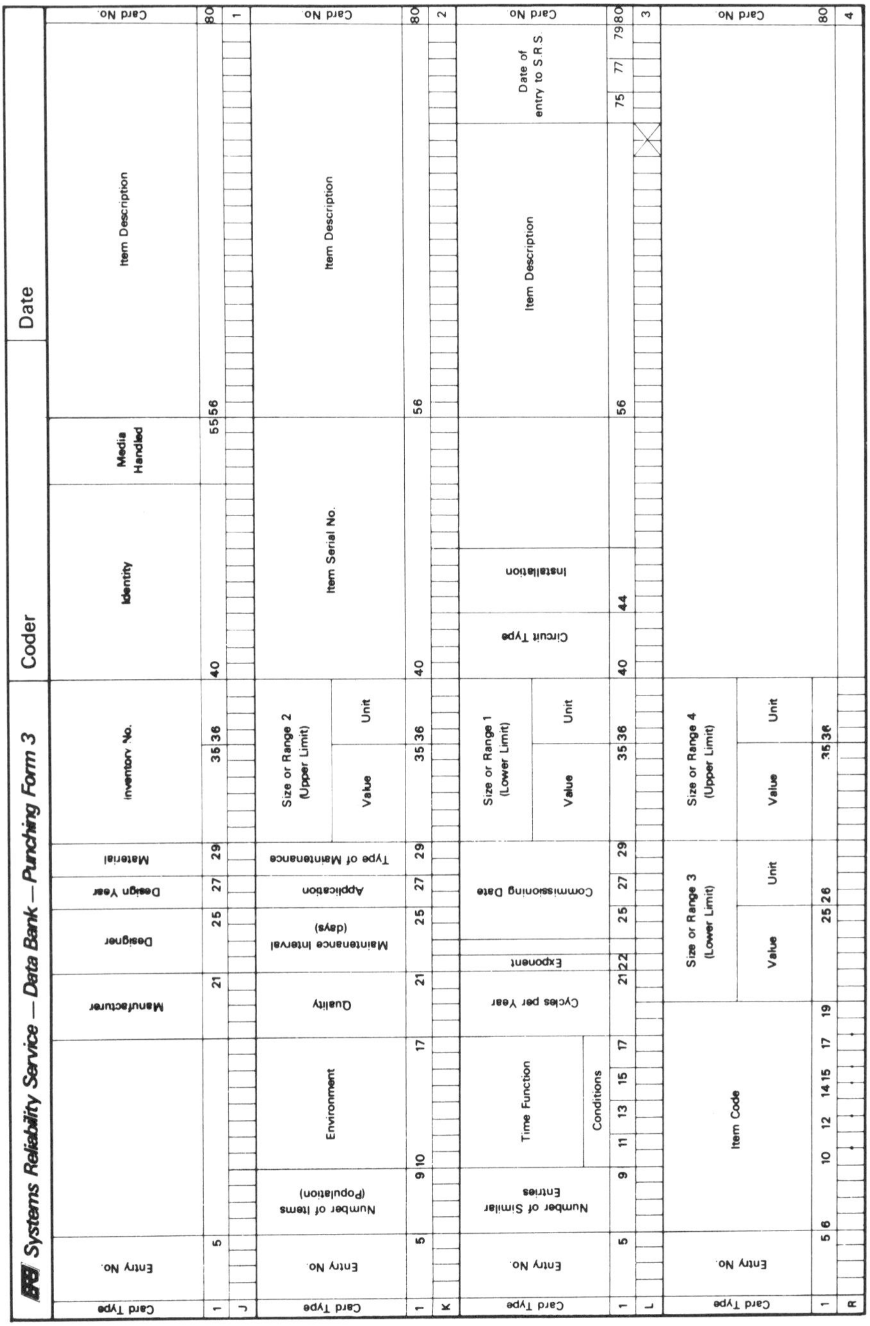

FIG. 8. Inventory data store punching form.

DATA RECORD NUMBER		11412
INSTALLATION		W16
INVENTORY NUMBER		33131W16
IDENTITY		FI/404 6303L
MANUFACTURER	×	2022
DESIGNER	×	1711
YEAR OF DESIGN		1967
SERIAL NUMBER		024102910
COMMISSIONING DATE		0 5 1972
RANGE/SIZE 1	×	28.00 PRSJ TO 34.00 PRSJ
RANGE/SIZE 3	×	0.1020E 06 VFRK
RANGE/SIZE 4	×	5.080 BORC
QUALITY	×	007
CIRCUIT TYPE	×	Z
MEDIA HANDLED	×	0058
MAINTENANCE INTERVAL	#	182
CYCLES PER YEAR	×	****
TIME FUNCTION	×	100.000.000.000
MATERIAL	×	07
ENVIRONMENT	×	31121010
APPLICATION	×	4
CODE FOR ITEM	×	19500.37.26.0.72.40

DESCRIPTION:- 2 INCH FISHER 404 RELIEF REGULATOR VALVE. SERIAL NO. SC 134037 010374

TYPE OF MAINTENANCE IS CODE 2

NUMBER OF ITEMS WITH THIS INVENTORY NUMBER = 1

\# MAINTENANCE INTERVAL IS IN DAYS
× SEE CODE BOOK

FIG. 9. Inventory data print showing design range.

19500 Valves, Mechanical

.00 Construction (Type)

.09 Butterfly
.14 Check, Ball
.21 Cylinder (piston and parts)
.25 Diaphragm

.00.00 Actuation or Control

.12 Cam operated
.32 Float
.54 Manual
.57 Motorised

.00.00.0 Trim Materials (seat and disc)

.1 Cast Iron
.2 Composite (different trim materials)
.5 Nickel based alloys

.00.00.0.00 Function/Application

.08 Bleed
.14 Control/Regulation Flow
.43 Isolation/stop
.72 Relief

.00.00.0.00.0 Pressure

.1 Vacuum
.2 Atmospheric
.3 Up to 30 psig (0 to 206 kN/m^2 g)

.00.00.0.00.0.0 Temperature

.3 $-14^{\circ}C$ to $45^{\circ}C$
.6 $251^{\circ}C$ to $500^{\circ}C$
.9 $1,001^{\circ}C$ and above.

FIG.10. An extract from a typical section of the 14 digit item code.

SYSTEMS RELIABILITY SERVICE - DATA BANK

Code	K =	Application Category
1	0.1	Very good (digital computers, etc)
2	0.2	Good (data handling systems, etc)
3	0.5	Fair (Laboratory, Control room, etc.)
4	1.0	Average (Normal industrial, ground based)
5	2.0	Poor (Shipboard, Chemical, etc)
6	4.0	Mobile (Road, Rail, etc)
7	6.0	Portable (loose bench equipment, carried in the hand equipment)
8	10.0	Severe (Airborne)
9	100.0	Very severe (Missile)
Blank	1.0	Unspecified

FIG.11. The application code.

SYSTEMS RELIABILITY SERVICE - DATA BANK

INSTALLATION:	REPORT NO.
SYSTEM:	DATE OF EVENT: TIME OF EVENT:
ITEM:	EVENT TYPE: ☐ BREAKDOWN ☐ MODIFICATION
ITEM DETAILS: DATA BANK INVENTORY NO: OP. HRS. SINCE LAST EVENT: NO. OF CYCLES SINCE LAST EVENT OUTAGE TIME: REPAIR TIME:	☐ TEST/CALIBRATION ☐ IN SITU MAINTENANCE ☐ WORKSHOP MAINTENANCE ☐ REQUEST ☐ CENSUS
DEFECT/WORK REQUEST: REASON FOR ATTENTION	EFFECT ON SYSTEM: ☐ NO LOSS ☐ PARTIAL LOSS ☐ TOTAL LOSS OF PERFORMANCE STATE OF INSTALLATION WHEN EVENT OCCURRED: ☐ OPERATING ☐ SHUTDOWN
DETAILS OF REPAIR: DEFECTIVE PARTS FAILURE MODE RENEWED ITEMS	EFFECT ON INSTALLATION: ☐ NONE ☐ AUTOMATIC SHUTDOWN ☐ MANUAL SHUTDOWN OUTAGE TIME: TOTAL MANHOURS: SKILLED: UNSKILLED: COST OF MATERIALS AND PARTS:

FIG.12. Typical event report form.

SYSTEMS RELIABILITY SERVICE – DATA BANK

Position No: Machine No. — Report No:

DIECASTING PLANT – EVENT REPORT

Cycles reading: — Time:

SYSTEM	INVENTORY ITEM	Inventory No.	Outage – hrs	Repair time	Date:
☐ Diecasting machine facility	☐ Diecasting machine				EVENT TYPE:
☐ Hydraulic system	☐ Diecasting machine tool				☐ Breakdown
☐ Tool locking system	☐ Fixed platten				☐ Mofidication
☐ Metal injection system	☐ Moving platten				☐ Test
☐ Cold chamber type	☐ Tailstock				☐ Maintenance
☐ Hot chamber type	☐ Toggle mechanism				☐ Request
☐ Hydraulic fluid cooling system	☐ Casting ejector				☐ Census
☐ Pneumatic system	☐ Metal injection plunger				EFFECT ON SYSTEM:
☐ Tool locking system	☐ Metal injection plunger cylinder				☐ No loss
☐ Metal injection system	☐ Gooseneck				☐ Partial loss
☐ Tool heating system	☐ Gooseneck nozzle				☐ Total loss of performance
☐ Tool cooling system	☐ Hydraulic pump (gen.)				OPERATING STATE OF MACHINE:
☐ Inlet section	☐ Hydraulic pump (L.P.)				☐ Operating
☐ Outlet section	☐ Hydraulic pump (H.P.)				☐ Shutdown
☐ Metal heating system	☐ Hydraulic pump motor				☐ Standby
☐ Gas supply	☐ End plate setting motor				EFFECT ON MACHINE:
☐ Air supply	☐ Fan				☐ None
☐ Casting ejection system	☐ Fan motor				☐ Shutdown
☐ Tool section	☐ Clipping press				Outage ....

☐ Machine section
☐ Hydraulic type
☐ Bumper bar type
☐ Operating control system
☐ Sequencing system
☐ Protection system
☐ Guard system
☐ Interlock circuits
☐ Machine lubrication system
☐ Die lubrication system
☐ Electrical supply
☐ Ventilation system
☐ Clipping press system
☐ Conveyor system
☐ Compressed air system

System outage time: ....
Operating hrs. since last event: ....

Operating hrs. since last event ....

Item failure mode:-

Cause of Item failure:-

Repair work carried out:-

Defective parts:	Failure mode:	Renewed
1.		Yes/No
2.		Yes/No
3.		Yes/No
4.		Yes/No
5.		Yes/No
6.		Yes/No
7.		Yes/No

Remarks:-

Signature:
Date:

FIG.13. Typical event report form.

EVENT REPORT

(Characters in parentheses are the Free Format code identifiers)

INSTALLATION/LOCATION CODE(D): ....

EVENT NO: ....

GENERAL DATA (CARD W)

DATE OF EVENT (2): .... TIME OF EVENT (8): ....

TYPE OF EVENT (C):

	Code	✓
Breakdown	10000	
Modification	20000	
Test/Calibration	30000	
Insitu Maintenance	41000	
Workshop Maintenance	42000	

INSTALLATION CATEGORY (Card X)

EFFECT OF THE EVENT (H): (on Ship)

	Code	✓
No effect	33	
Reduced Propulsion	31	
Loss of Propulsion	32	
Loading/Discharging (facilities failure)	82	

OUTAGE TIME OF SHIP (6): .... hours REPAIR TIME (5): .... hours

OPERATING STATE OF SHIP (F): (immediately prior to the event)

	Code	✓
Full ahead	81	
Reduced Speed	83	
Manoeuvring	84	
Loading/Discharging	87	

Description of the way the event affected the ship (:)

FIG.14. Typical event report form (front).

SYSTEM CATEGORY (Card Y)

SYSTEM NAME (K):-

(System code)

EFFECT OF EVENT (J): (on System)

	Code	✓
No loss of performance	1	
Partial loss of performance	3	
Total loss of performance	2	

Outage time of System (6): .... hours: Repair time (5) .... hours

List of items faulty (L):-

Description of the way the event affected the system (:)

ITEM CATEGORY (Card Z)

FAILURE MODE (M):-

CAUSE OF FAILURE (N):-

Outage time of item (6) .... hours: Repair time (5) .... hours

Total Labour (7) .... man hours: Position in Installation (%): ....

Operating hours since last item event (W): ....

Number of starts and stops since last item event (X) ....
(Operating Cycles)

Inventory Item description (=):-
(Code with Inventory Number)

List of Renewed items and parts (including quantity used (;)
(code as keywords, keep register)

Description of action taken (:)

FIG.15. Typical event report form (rear).

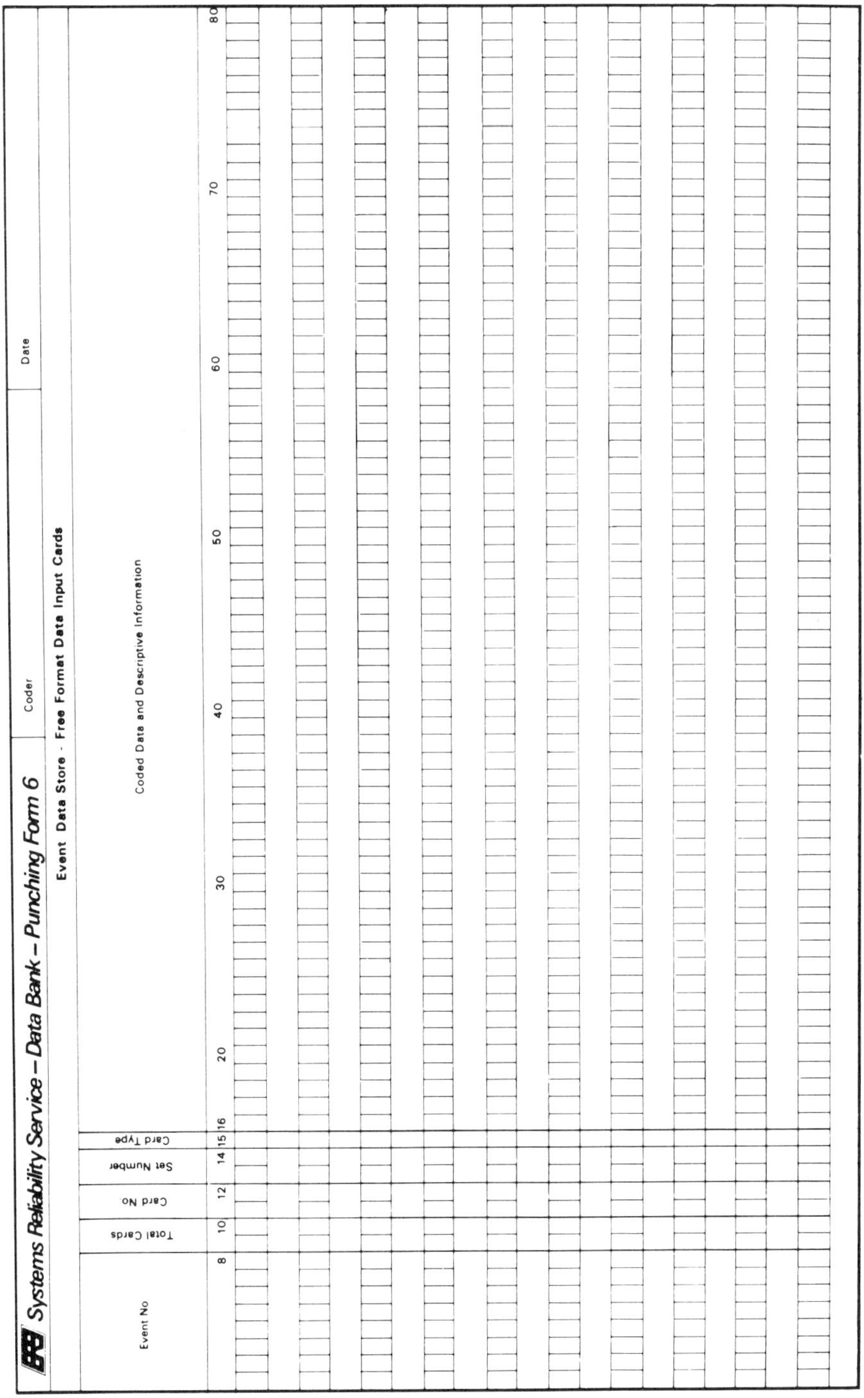

FIG.16. Event data store punching form.

EVENT REPORT

(Characters in parentheses are the Free Format code identifiers)

INSTALLATION/LOCATION CODE(D): CDHF

EVENT NO: 10

GENERAL DATA (CARD W)

DATE OF EVENT (2): 10-7-72 TIME OF EVENT (8): 1345

TYPE OF EVENT (C):

	Code	✓
Breakdown	10000	✓
Modification	20000	
Test/Calibration	30000	
Insitu Maintenance	41000	
Workshop Maintenance	42000	

INSTALLATION CATEGORY (Card X)

EFFECT OF THE EVENT (H): (on Ship)

	Code	✓
No effect	33	
Reduced Propulsion	31	✓
Loss of Propulsion	32	
Loading/Discharging (facilities failure)	82	

OUTAGE TIME OF SHIP (6): – hours REPAIR TIME (5): – hours

OPERATING STATE OF SHIP (F): (immediately prior to the event)

	Code	✓
Full ahead	81	
Reduced Speed	83	✓
Manoeuvring	84	
Loading/Discharging	87	

Description of the way the event affected the ship (:)

REDUCED PROPULSION FOR 24 HOURS.

FIG. 17a. Event report received by SRS (front).

SYSTEM CATEGORY (Card Y)

SYSTEM NAME (K):- BOILER COMBUSTION AIR SUPPLY SYSTEM ON A STEAM POWERED TANKER VESSEL.

(System code)

EFFECT OF EVENT (J): (on System)	Code	✓
No loss of performance	1	
Partial loss of performance	3	✓
Total loss of performance	2	

Outage time of System (6): ...—... hours: Repair time (5) ...20... hours

List of items faulty (L):- COMBUSTION AIR PREHEATER (REGENERATIVE)
VANE TYPE HEATER ELEMENTS
SEALS.

Description of the way the event affected the system (:)

PREHEATER OUT OF SERVICE.

ITEM CATEGORY (Card Z)

FAILURE MODE (M):- DEGRADATION OF PERFORMANCE.

CAUSE OF FAILURE (N):- CHEMICAL CORROSION.

Outage time of item (6) ...20... hours: Repair time (5) ...20... hours

Total Labour (7) ...160... man hours: Position in Installation (%): ...1...

Operating hours since last item event (W): ...948...

Number of starts and stops since last item event (X) ...—...
(Operating Cycles)

Inventory Item description (=):- AIR PREHEATER S. No. 284/1
(Code with Inventory Number)

List of Renewed items and parts (including quantity used (;)
(code as keywords, keep register)

HEATER ELEMENT SEALS - 24

Description of action taken (:) 24 DEFECTIVE HEATER ELEMENT SEALS REPLACED.

FIG.17b. Event report received by SRS (rear).

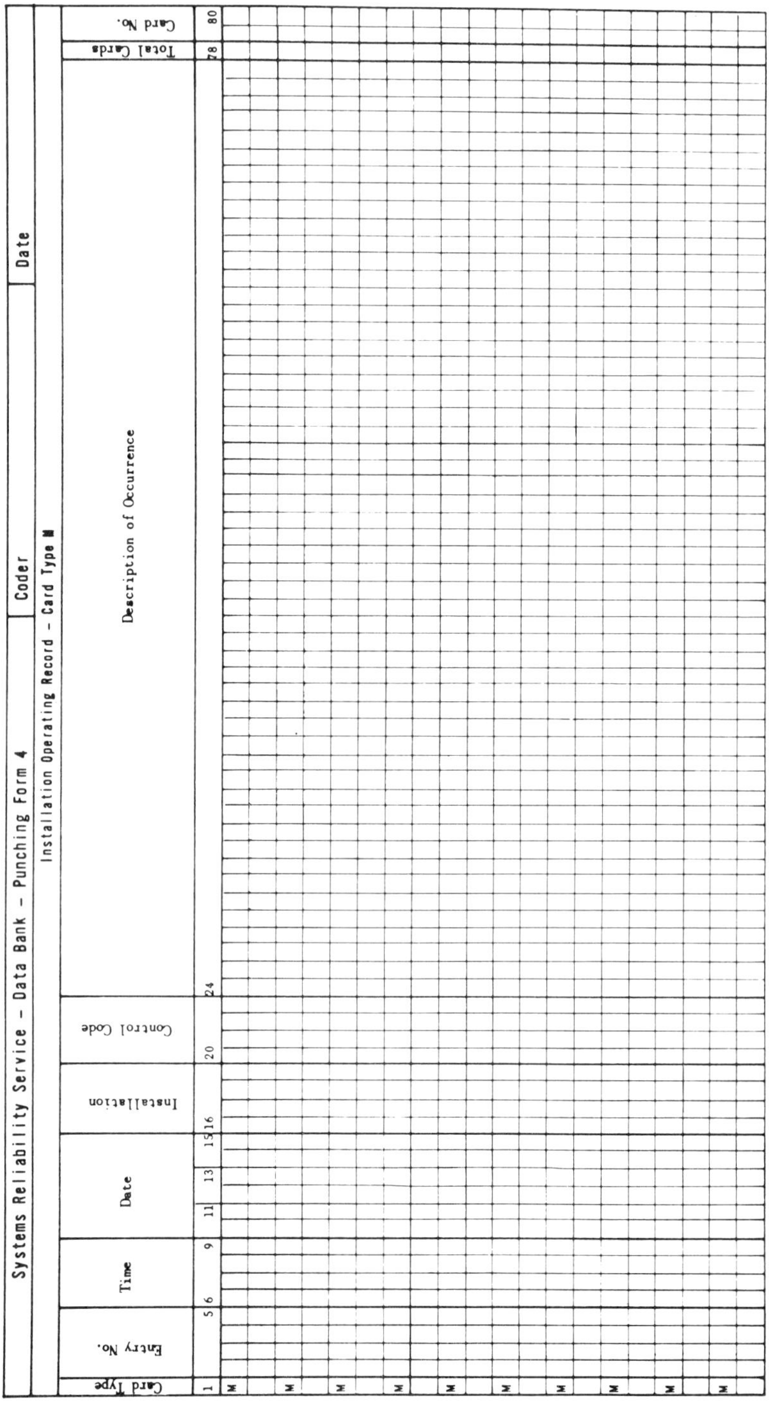

FIG. 18. Installation operating record information sheet/punching form.

INSTALLATION HO1 - ENTRIES AFTER 1256 HOURS ON 13/6/74
INSTALLATION OPERATING RECORD

ENTRY NO.	TIME	DATE	CODE	TEXT
2004	12.56	13.6.74	LP	REACTOR START UP
2005	13.13	13.6.74	HP	REACTOR AT 300KW
2006	16.28	13.6.74	LP	REACTOR SHUTTING DOWN
2007	16.45	13.6.74	SH	REACTOR SHUTDOWN
2008	8.50	17.6.74	LP	REACTOR START UP
2009	9.06	17.6.74	HP	REACTOR AT 300KW
2010	16.36	17.6.74	LP	REACTOR SHUTTING DOWN
2011	16.52	17.6.74	SH	REACTOR SHUTDOWN
2012	13.05	18.6.74	LP	REACTOR START UP
2013	13.19	18.6.74	LP	REACTOR AT LOW POWER FOR 14 MINUTES
2014	16.19	18.6.74	LP	REACTOR SHUTTING DOWN
2015	16.33	18.6.74	SH	REACTOR SHUTDOWN
2016	9.02	19.6.74	LP	REACTOR START UP
2017	9.13	19.6.74	MP	REACTOR AT 100KW
2018	16.43	19.6.74	LP	REACTOR SHUTTING DOWN
2019	16.54	19.6.74	SH	REACTOR SHUTDOWN
2020	10.30	20.6.74	LP	REACTOR START UP
2021	10.40	20.6.74	LP	REACTOR AT LOW POWER FOR 10 MINUTES
2022	12.00	20.6.74	SH	REACTOR SHUTDOWN
2023	13.05	20.6.74	LP	REACTOR START UP, AND RUN AT LOW POWER FOR 1 HOUR
2024	14.05	20.6.74	SH	REACTOR SHUTDOWN
2025	16.05	20.6.74	LP	REACTOR START UP
2026	16.22	20.6.74	HP	REACTOR AT 300KW
2027	16.37	20.6.74	LP	REACTOR SHUTTING DOWN
2028	16.55	20.6.74	SH	REACTOR SHUTDOWN
2029	8.50	21.6.74	LP	REACTOR START UP
2030	10.12	21.6.74	MP	REACTOR AT 100KW
2031	10.42	21.6.74	LP	REACTOR SHUTTING DOWN
2032	12.05	21.6.74	SH	REACTOR SHUTDOWN

FIG.19. An extract from an installation operating record.

ANALYSIS BY MANUFACTURER

	NAME(CODE)	INVENTORY ENTRIES	POPULATION	EVENTS	EVENTS PER ITEM
0316	BLAKEBOROUGH, J & SONS LTD	129	129	78	0.605
0368	BATLEY	4	4	2	0.500
1711	FISHER GOVERNOR CO LTD	6	6	0	0.000
1719	FOXBORO-YOXALL LTD	4	4	0	0.000
1730	FLOWSTREAM I.T.T.	1	1	0	0.000
2318	HOPKINSONS LTD	4	4	31	7.750
2346	HAMMEL DAHL	3	3	0	0.000
2622	INTROL	10	10	1	0.100
2840	MASONEILAN	44	44	30	0.682
3534	PFADLER	1	1	0	0.000
4205	SAUNDERS VALVE CO LTD	8	8	17	2.125
4235	SULZER	4	4	21	5.250
4248	SERCK-GLOCON	104	104	67	0.644
	TOTAL NUMBERS	322	322	247	

FIG.20. Typical computer print of the 'LIST' routine.

RELIABILITY DATA

MEAN INTERVAL BETWEEN FAILURES IS 277.7 HOURS (38 FAILURES)

95% CONFIDENCE LIMITS ARE 202.3 AND 392.4

TOTAL OPERATION IS 10551 ITEM HOURS

TOTAL POPULATION IS 1

FAILURES INCLUDE THOSE EVENTS INVOLVING BREAKDOWN OR UNSCHEDULED MAINTENANCE DUE TO FAILURE.

LARGEST UNCERTAINTY IN TIMES USED IS 0.500 HOURS

FIG.21. Data table from "LIFE ANALYSIS".

THE FOLLOWING VALUES OF LIFETIMES IN HOURS HAVE BEEN FOUND (THE - DENOTES AN INCOMPLETE LIFETIME)
UNREVEALED FAILURES ARE NOT INCLUDED.

-0.50	1.50	2.50	6.50	18.50	22.50	30.50	31.00
42.50	43.50	50.50	-55.50	64.50	102.50	115.00	128.50
132.50	135.00	146.50	149.50	170.50	178.50	190.50	239.50
286.50	306.50	328.50	338.50	384.50	456.50	476.00	504.50
581.50	597.50	600.50	687.50	788.50	936.50	1238.50	

VALUES PRINTED ARE UPPER BOUNDS

FIG.22. Lifetimes used in computation of mean interval between failures and subsequent statistical analysis.

A WEIBULL DISTRIBUTION GIVES THE BEST FIT

DISTRIBUTION IS P = 1 - EXP(-(T/ 2.7802E 02)** 8.5388E-01)

TABLE OF CUMULATIVE PROBABILITY

HOURS	% PROBABILITY	FITTED % PROBABILITY
42.50	18.421	18.220
102.50	29.730	34.724
132.50	37.838	41.203
178.50	51.351	49.590
239.50	59.459	58.539
286.50	59.459	64.155
328.50	64.865	68.434
384.50	70.270	73.259
384.50	72.973	73.259
456.50	75.676	78.285
504.50	81.081	81.048
581.50	83.784	84.707
600.50	86.486	85.486
687.50	89.189	88.541
687.50	91.892	88.541
788.50	91.892	91.244
788.50	94.595	91.244
788.50	94.595	91.244
1238.50	97.297	97.215

CONFIDENCE LEVEL BY KOLMOGOROV/SMIRNOV TEST IS 99.45 PER CENT
(MAXIMUM DIFFERENCE 5.31%)

FIG.23. Cumulative probability table from "LIFE" analysis.

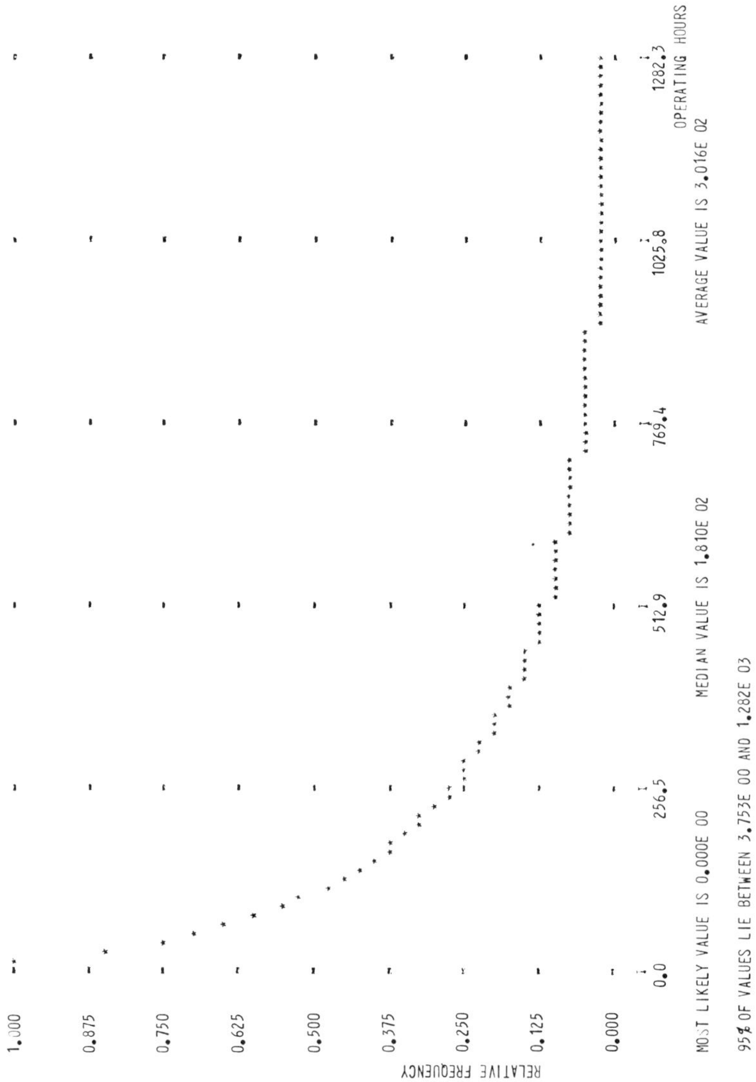

FIG. 24. Graph of probability density function.

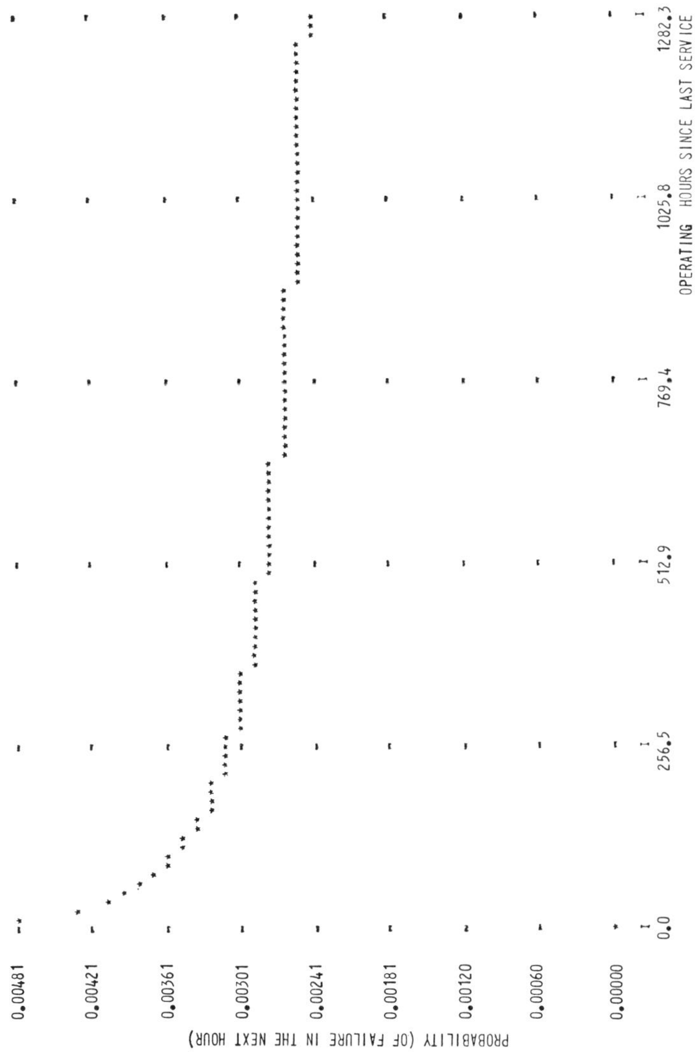

FIG.25. Graph of hazard function (conditional failure rate).

SYSTEMS RELIABILITY SERVICE - DATA BANK

ANALYSIS OF REPAIR TIME

LONGEST REPAIR TIME REPORT IS	10.0 HOURS
ON 5% OF OCCASIONS REPAIR TIME WAS MORE THAN	4.5 HOURS
ON 10% OF OCCASIONS REPAIR TIME WAS MORE THAN	2.8 HOURS
ON 25% OF OCCASIONS REPAIR TIME WAS MORE THAN	1.0 HOURS
ON 50% OF OCCASIONS REPAIR TIME WAS MORE THAN	0.5 HOURS
ON 75% OF OCCASIONS REPAIR TIME WAS MORE THAN	0.5 HOURS
ON 90% OF OCCASIONS REPAIR TIME WAS MORE THAN	0.3 HOURS
ON 95% OF OCCASIONS REPAIR TIME WAS MORE THAN	0.3 HOURS
SHORTEST REPAIR TIME REPORTED IS	0.0 HOURS

AVERAGE REPAIR TIME = 1.1 HOURS

TOTAL REPAIR TIME = 525.7 HOURS

TOTAL NUMBER OF VALUES = 464

AVAILABILITY = 99.73%

BASED ON A MAXIMUM POSSIBLE OPERATING TIME OF 0.195E 06 HOURS

FIG.26a. Proposed layout for the percentage cumulative distribution table from the "REPAIR" time analysis.

SYSTEMS RELIABILITY SERVICE - DATA BANK

ANALYSIS OF OUTAGE TIME

LONGEST OUTAGE TIME REPORTED IS	70.0 HOURS
ON 5% OCCASIONS OUTAGE TIME WAS MORE THAN	33.0 HOURS
ON 10% OCCASIONS OUTAGE TIME WAS MORE THAN	24.0 HOURS
ON 25% OCCASIONS OUTAGE TIME WAS MORE THAN	13.0 HOURS
ON 50% OCCASIONS OUTAGE TIME WAS MORE THAN	5.0 HOURS
ON 75% OCCASIONS OUTAGE TIME WAS MORE THAN	2.0 HOURS
ON 90% OCCASIONS OUTAGE TIME WAS MORE THAN	0.5 HOURS
ON 95% OCCASIONS OUTAGE TIME WAS MORE THAN	0.2 HOURS
SHORTEST OUTAGE TIME REPORTED IS	0.1 HOURS

AVERAGE OUTAGE TIME = 8.7 HOURS

TOTAL OUTAGE TIME = 4036.8 HOURS

TOTAL NUMBER OF VALUES = 464

UTILISATION FACTOR = 97.93%

BASED ON A MAXIMUM POSSIBLE OPERATING TIME OF 0.195E 06 HOURS

FIG.26b. Proposed layout for the percentage cumulative distribution table from the "OUTAGE" time analysis.

QUANTITY	QUANTITY CODE	UNIT	UNIT CODE	UNIT NUMBER	CONVERSION FACTOR
INT.BETW.FAULTS	/IFT /				
		MINUTE	/IFTJ/	482	6.000E 01
		HOUR	/IFTH/	483	3.600E 03
		DAY	/IFTV/	484	8.640E 04
		WEEK	/IFTW/	485	6.048E 05
		MONTH	/IFTQ/	486	2.630E 06
		YEAR	/IFTA/	487	3.156E 07
		SEC	/IFTE/	650	1.000E 00
INT.BETW.FAULTS	/IFL /				
		KM	/IFLK/	634	1.000E 03
		METRE	/IFLE/	635	1.000E 00
		MILE(STATUTE)	/IFLS/	636	1.609E 03
INT.BETW.FAULTS	/IFZ /				
		CYCLE	/IFZE/	633	1.000E 00
ITEM.LENGTH.TIME	/ILT /				
		10**6KM.HR	/ILTK/	672	3.600E 12
		10**6STAT.ML.HR	/ILTM/	673	5.792E 12
		10**9YARD.HR	/ILTY/	674	3.292E 12
		METRE.SEC	/ILTE/	675	1.000E 00
		KM.YEAR	/ILTA/	693	3.156E 10
ITEM.LENGTH.CYC	/ILZ /				
		KM.CYCLE	/ILZK/	676	1.000E 03
		METRE.CYCLE	/ILZE/	677	1.000E 00
ITEM.TIME	/ITT /				
		ITEM.SEC	/ITTE/	614	1.000E 00
		ITEM.MIN	/ITTJ/	615	6.000E 01
		ITEM.HOUR	/ITTH/	616	3.600E 03
		ITEM.DAY	/ITTV/	617	8.640E 04
		ITEM.WEEK	/ITTW/	618	6.048E 05
		ITEM.MONTH	/ITTQ/	619	2.630E 06
		ITEM.YEAR	/ITTA/	620	3.156E 07

FIG.27. An extract from the "SORAY" code.

DISCUSSION

(on papers IAEA-SM-195/44 and IAEA-SM-195/37)

S. HARTWIG: I should like to ask Mr. Wise whether it is intended to make a systematic comparison of the predictions resulting from fault-tree calculations (which are presumably relied on during the construction of a plant) with actual failure data subsequently collected.

M.J. WISE: At present, the information provided by the Nuclear Plant Reliability Data System (NPRDS) in the United States of America is considered, for the most part, to consist of raw reliability data. These are data that reliability analysts can apply to their assessments, due consideration being given to the quality and quantity of the data, environmental factors, operating modes and so on. There are other factors that will have to be specified and taken into account, too, before really sophisticated reliability calculations can be made. I must emphasize that the NRPDS is in its infancy; the full scope of the future programme is not yet known.

N.R. McDONALD: I am concerned about the manpower requirements quoted by Mr. Wise for input data collection. In Section 3.1 of his paper we read that the average time required to produce one engineering data report is 1.5 man-hours, and that on average 3000 to 4000 systems and components have to be reported on for each nuclear unit. Does this really mean that up to 6000 man-hours will be reqired per unit? How is this manpower supplied?

M.J. WISE: The calculation of 1.5 man-hours per engineering data report emerged from the pilot phase of the NPRDS programme, when single units of nuclear plants were being studied, and this was the first time all concerned had been involved in the programme. More recent experience suggests that not so much time will be required, in view of duplicate nuclear units and more efficient procedures. Even so, it is expected that between one and two man-years of effort will be required to prepare and submit the technical data for each nuclear unit. In the United States this manpower is supplied by the utilities, and most appear to consider it worth wile.

A.V. KARPOV: I should like to ask Mr. Wise whether the reporting programme includes data on fast reactors.

M.J. WISE: No, not at present; the reason for this is that the data programme is designed for power reactors which are operational at the present time, and the United States do not have any fast reactors in that category. However, the programme will be capable of handling fast reactor data when they become available.

A.V. KARPOV: Do you have ways of familiarizing yourself with the data programmes of other countries?

M.J. WISE: Yes, the Committee for the United States data programme is taking an active part in international programmes. It is familiar with the reliability programmes of other countries and will welcome any useful exchange of data.

G. BREILING: Do you plan to treat your collected data by the more sophisticated statistical methods, such as calculation of failure rates per 10^6 hours, calculation of the most likely failure rate distribution and of the probability parameters for this distribution together with its confidence intervals?

M.J. WISE: As I said in my reply to Mr. Hartwig, the data provided under the NPRDS programme are considered for the most part to be raw reliability data which analysts can apply to their assessments with due consideration for the quality and quantity of the data, environmental factors, operating modes and so on. Other factors will have to be specified and taken into account before sophisticated reliability calculations can be made with the requisite accuracy. One example of how NPRDS reliability data can be effectively used is to be seen in the work which the Institute of Electrical and Electronic Engineers is doing on the Failure Rate Handbook. Other examples will become evident as more data are accumulated.

J.M. BUTTERFIELD: I see that Mr. Fothergill uses the Poisson distribution to estimate confidence limits on observed failure rate data, whereas Mr. Andersson (paper IAEA-SM-195/52) has used the χ^2 distribution for a similar purpose. Perhaps Mr. Fothergill would like to comment on this difference.

C.D.H. FOTHERGILL: The Reliability Data Store of the Systems Reliability Service assumes a constant failure rate (that is, that the components of interest are functioning in the most useful phase of their life). Thus, a derivation of confidence limits based on the Poisson distribution is considered reasonable.

D. NORMAN: Poisson confidence limits and χ^2 are in fact equivalent here.

L. CAVE: Some interesting and rather disconcerting results have been given which seem to show the bad effect of maintenance on reliability in a particular instance (instrumentation and control in chemical plants). Could Mr. Fothergill say whether these results are at all typical of the effects of maintenance on a wider range of plant?

C.D.H. FOTHERGILL: The slides in question were made from a very small amount of data, only to demonstrate the form of presentation used in the LIFE analyses of the Event Data Store of the Systems Reliability Service. The figures shown have no bearing on the effects of maintenance on reliability.

J.-C. ARLAUD: What is the precise definition of 'reactor outage time', according to Mr. Wise and Mr. Fothergill?

M.J. WISE: The generally accepted definition of 'reactor outage time', as used in the United States data system and as applied to the whole nuclear plant, is time during which the reactor is completely shut down for some reason. A reduction in thermal output is not usually considered an 'outage'; it is called 'reduced capability' or 'derated capability'.

C.D.H. FOTHERGILL: For the purpose of the Data Bank of the Systems Reliability Service the definition of 'outage time', for nuclear reactors or any other type of installation, is agreed on between the plant operator and the Systems Reliability Service when they plan the collection of operating data for that plant. In general the definition depends on what is meant by 'total loss of reactor performance'.

J.-C. ARLAUD: I should also like to ask Mr. Wise and Mr.Fothergill whether it is not awkward to mix all the data for a given country, without taking account of the fact that maintenance policy, in particular, is likely to be reflected in the failure rate at different installations.

M.J. WISE: The United States data system relies heavily on the fact that numerous nuclear units of the same type are included in the data base. It is considered that individual utility maintenance practices will not be significant when the aggregate amount of data is assessed. Ideally, in such a reliability data programme, the analyst, in making his assessment, should not have to question the validity of each and every source of data but only that of the composite figures.

C.D.H. FOTHERGILL: I quite agree that it would be awkward to mix all the data for a given country! Data received by the Data Bank of the Systems Reliability Service from various contributors are not mixed but stored under a separate unique source reference or location code which identifies the contributor.

D. NORMAN: Although as head of a reliability engineering group I have a keen interest in collecting reliability data, I agree with Mr. Andersson (paper IAEA-SM-195/52) in his warning against the uncritical use of data obtained in various operating conditions. It would be interesting to hear from Mr. Wise and Mr. Fothergill whether they feel it might be better to offer an equivalent to 'computer dating' services, putting the suppliers of data in touch with users so that important engineering details and reservations could also be exchanged. In this way the mixing of data which are incompatible could be avoided.

C.D.H. FOTHERGILL: In our Data Bank, as I said, reliability information is stored under individual source reference numbers. Failure rate statistics are combined by special computer sub-routine only after a critical engineering assessment has been made of the individual sources selected.

M.J. WISE: At present, the source of data — the particular nuclear plant operating company — is not identified to all recipients of the routine quarterly and annual reports issued under the United States data system. The nuclear system supplier and architect-engineer receive detailed information on the nuclear units which they supply or for which they provide design services. It is usually possible also, by special request, to put other suppliers and manufacturers in contact with specific users.

J. KRASNODEBSKI: I would like to comment on data collection from the standpoint of the plant designer. Designers are interested not in sophisticated statistical analyses but in identifying systems and equipment with high failure rates and long repair times — in other words, in information that shows the 'soft spots' where improvements are required. Equipment suppliers must be informed about failures of their equipment so that they can improve it when and where necessary. Hence I think it is important to be sure that the data collection systems described here provide for regular feedback of information to equipment suppliers.

C.D.H. FOTHERGILL: Yes, there is provision for feedback to manufacturers who supply data to the Systems Reliability Service Data Bank. These manufacturers would be given limited information about their own products.

M.J. WISE: The United States data system provides for identifying the manufacturer, model number, serial number and so on; the supplier of the nuclear system (NSSS) of the plant is also designated. Essentially all failure data reported by the utilities are available to manufacturers each calendar quarter (see Fig. 7 in paper IAEA-SM-195/44). Manufacturers and designers can obtain information on specific models of components by requesting interrogation of the total data base.

D. OKRENT: I have a number of questions which bear on safety. First of all, are modes of failure as well as failure rates recorded in your data stores?

C.D.H. FOTHERGILL: Both modes of failure and failure rates are included in the Systems Reliability Service Data Bank. These can be used, as in the case of the Event Data Store, to facilitate the retrieval of information relating to specific failure modes and/or types of event, for analyses which yield reliability figures based on these parameters. I have in fact included a definition of 'failure mode' in my paper (IAEA-SM-195/37).

M.J. WISE: Our data system records the failure mode for each event but does not include a calculation of the actual failure rate. The number of failures per 10^6 calculated in-service hours for a particular type of system or component is, however, calculated and reported.

D. OKRENT: Are the effects of system interactions (such as fires) on failure rates included?

C.D.H. FOTHERGILL: Consequential events are included as data in the Event Data Store. These do not contribute to the reliability statistics of the item affected by such circumstances, but they do add to the feedback of management information.

M.J. WISE: The United States data system does record incidents where interactions occur — in both coded and narrative form. At present, however, this information is not applied to failure rate calculations. Persons performing reliability assessments can be provided with raw data and information for their use, as necessary.

D. OKRENT: How is the question of non-prototypic components handled?

C.D.H. FOTHERGILL: The reliability figures obtained for 'older' equipment often contribute valuable information during the re-design or modification of such devices, particularly when no better data are available. Prototype information is collected from manufacturers, and these quality data are clearly identified when supplied as information from the Reliability Data Store.

M.J. WISE: In the United States data system no attempt is made to determine whether a system or component is prototypic or not. This question will be brought to the attention of the NPRDS Committee.

D. OKRENT: How is an evaluation of failures incorporated?

C.D.H. FOTHERGILL: The evaluation, i.e. definition of failure criteria, is made at the specification stage of an event reporting project between the field engineers representing the Systems Reliability Service and the plant operators and maintenance engineers contributing the data.

M.J. WISE: Our data system, as constituted at present, is designed to collect and distribute raw data and information rather than evaluated data. Evaluations are for the most part left to the user of the data.

FAILURE DATA COLLECTION FROM A SWEDISH NUCLEAR POWER PLANT

B. ANDERSSON, A. BHATTACHARYYA
AB ASEA-ATOM, Västerås

S. HILDING
Oskarshamnsverkets Kraftgrupp AB,
Stockholm, Sweden

Abstract

FAILURE DATA COLLECTION FROM A SWEDISH NUCLEAR POWER PLANT.

The Swedish nuclear utilities have formed a joint working group in the field of reliability data of thermal power plants, nuclear and fossil fuelled. The primary task of the working group is to create a standard procedure of collecting failure data from the Swedish nuclear power plants in operation. The failure data will be stored in a joint data bank. A first test collection of such data has been implemented on Oskarshamn I, and the experience with this work is discussed in this report. Reliability analysis of an engineering system is based on the availability of pertinent information on the system components. Right from the beginning within the Swedish nuclear industry the consensus has been that such data can be suitably obtained by monitoring the operating power stations. This has led to a co-operative arrangement between the vendor, ASEA-ATOM and a utility, Oskarshamnsverkets Kraftgrupp AB (OKG) to utilize information from component malfunctions in the reliability analysis. The utility prepares component failure reports which are sent to the vendor for further treatment. Experience gathered to date indicates that this arrangement is effective although many persons are involved in this process of information transmittal. The present set-up is flexible enough to accommodate necessary changes in view of problems which arise now and then in monitoring a complex system like a nuclear power station. This report briefly describes the structure of the failure data collection system. The way in which the raw data collection is done in the station by the owner and the subsequent data processing by the vendor is discussed. A brief status report of the information collected since 1971 is given. It can be concluded that valuable reliability data can be obtained by monitoring component failure reports from an operating power plant. Two requirements are, however, that all the parties involved in the arrangement follow given instructions carefully and that the assumed analytical models are sufficient to produce the desired reliability information.

1. INTRODUCTION

The present report describes the procedures for collecting and processing component failure data from a nuclear power plant. Also, the purpose of such a data collection system is discussed. One approach is to discuss the various areas where need for such operational data exists and how the raw data is converted to a usable form.

Both the supplier and the user have the common interest that a particular product should have high reliability to assure personnel safety and good economy. Availability of operating data enhances the possibility of achieving this goal. With the help of the operating data the owner can plan his maintenance work better, optimize the repair requirements and the size of the maintenance crew. Also, it helps in effective use of the station and follow-up of the costs in the station.

The availability of failure data helps the vendor to meet more easily the requirements imposed by the utility or the official agencies. Reliability analysis is used for this purpose. This is coupled to such items like failure rate data, duration of repairs and types of failures. With the help of these data it is possible to recommend suitable interval for replacing components and inspection intervals.

It is feasible to collect some data by laboratory tests but it is difficult to simulate all the operating situations which occur in a power station. Such tests with a large number of identical components are prohibitively expensive.

To avoid this problem, data is collected from an operating station. However, a certain care must be exercised in using the data. An uncritical use of the data is not advisable since there might be sometimes significant differences in the operating conditions of the component on which a certain failure rate data is based and the component to be analyzed. It should also be remembered that for certain components, e.g. mechanical equipments, failure rate is affected by such typically time dependent phenomena as fatigue, creep and wear.

2. REPORTING PROCEDURE FOR MALFUNCTIONS

It has always been apparent that an effective data collection system must be based on cooperation between the vendor and the user. To this date the main part of the data comes from the Oskarshamn units 1 & 2 owned by OKG [1]. Only the data collection system based on a cooperative arrangement between OKG, the owner, and AB Asea-Atom, the supplier, of the BWR units is described.

Activities of the operational crew in the station is almost totally guided by written operation manuals and instructions. A library of instructions is available in the control room. These instructions cover all aspects of operation and changes in various operational modes. All operational incidents as well as all actions taken by the repair crew are recorded centrally.

2.1. Log book, operational and failure reports

Operational activities in the power station are registered in various ways. Central registration of operational history is done in the log book in the control room. This contains brief chronological reports of instructions and operating orders issued. A more detailed documentation is done in the form of a large number of operating reports. If an operational incident takes place and a component malfunction is discovered then a failure report is written. This is processed by the maintenance crew.

The main purpose of the failure report is to initiate action for restoring the operating capability of the component. Also, on various occasions it is necessary to document the circumstances in which the malfunction occurred. This enables one to check whether there are any systematic malfunction tendencies.

---

1) Oskarshamnsverkets Kraftgrupp AB

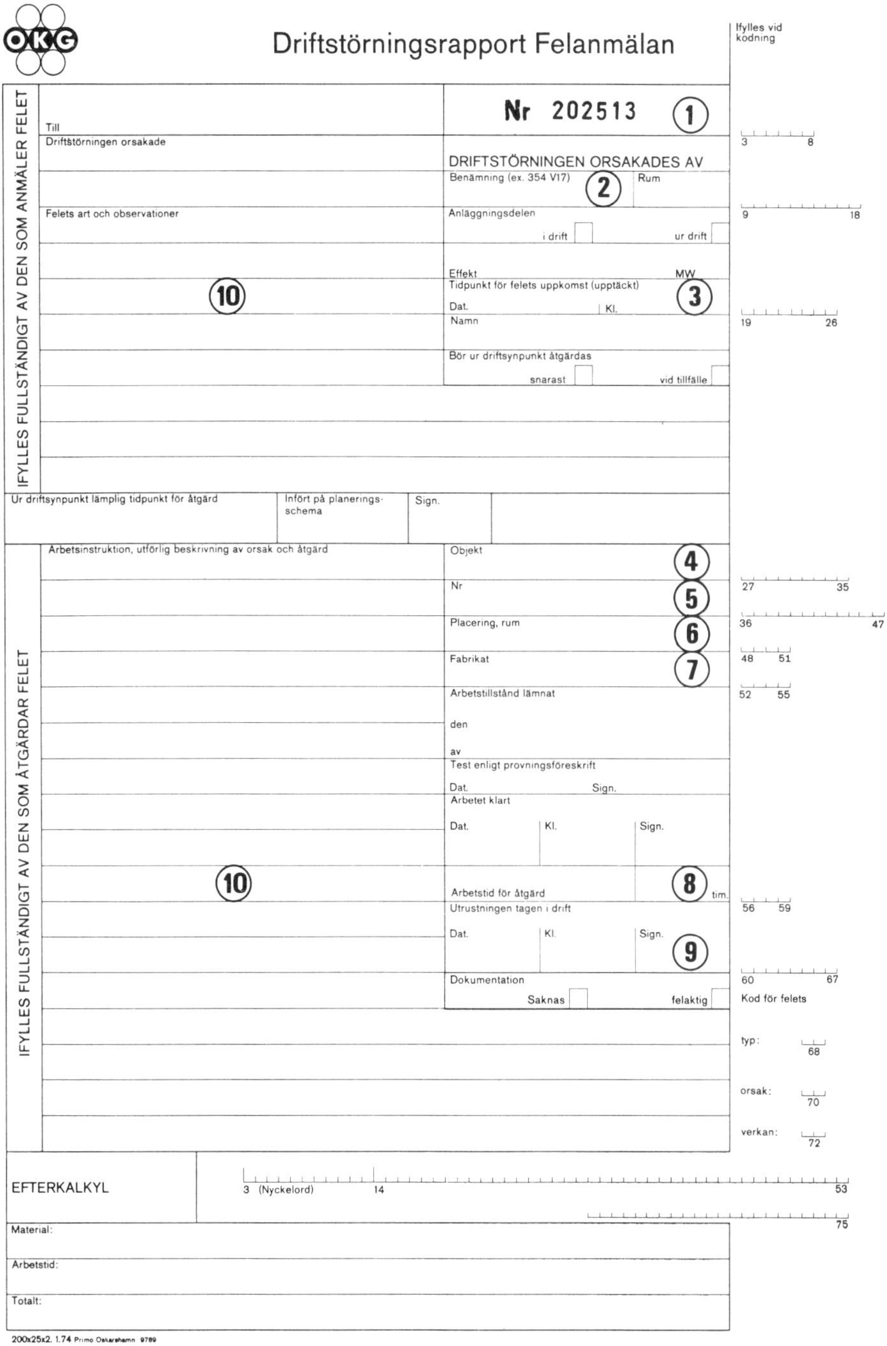

OKG

Driftstörningsrapport Felanmälan

Ifylles vid kodning

IFYLLES FULLSTÄNDIGT AV DEN SOM ANMÄLER FELET

Till

Nr 202513 (1)

3 8

Driftstörningen orsakade

DRIFTSTÖRNINGEN ORSAKADES AV

Benämning (ex. 354 V17) (2) Rum

9 18

Felets art och observationer

Anläggningsdelen

i drift ☐ ur drift ☐

Effekt MW

Tidpunkt för felets uppkomst (upptäckt) (3)

Dat. Kl.

19 26

Namn

(10)

Bör ur driftsynpunkt åtgärdas

snarast ☐ vid tillfälle ☐

Ur driftsynpunkt lämplig tidpunkt för åtgärd | Infört på planeringsschema | Sign.

IFYLLES FULLSTÄNDIGT AV DEN SOM ÅTGÄRDAR FELET

Arbetsinstruktion, utförlig beskrivning av orsak och åtgärd

Objekt (4)

27 35

Nr (5)

36 47

Placering, rum (6)

48 51

Fabrikat (7)

52 55

Arbetstillstånd lämnat

den

av

Test enligt provningsföreskrift

Dat. Sign.

Arbetet klart

Dat. Kl. Sign.

(10)

Arbetstid för åtgärd (8) tim.

56 59

Utrustningen tagen i drift

Dat. Kl. Sign. (9)

60 67

Dokumentation

Saknas ☐ felaktig ☐

Kod för felets

typ: 68

orsak: 70

verkan: 72

EFTERKALKYL

3 (Nyckelord) 14 53

75

Material:

Arbetstid:

Totalt:

200x25x2. 1.74 Primo Oskarshamn 9789

FIG. 1. Failure report form.

2.2. Failure Report Form

The form filled up for reporting malfunctions is called "Operational Disturbance and Failure Report". It is shown in fig.1. It has four main sections. (Numbers 1 - 10 under A and B below refer to the numbering scheme in fig 1.)

A. Reporting of the malfunction with such details as nature of the malfunction, time of discovery, operational mode etc.

1. Identification number of the failure report

2. Identification of the malfunctioning object, in general, with the system number and component number

3. Date and time when the malfunction is detected

B. Report from the person or persons involved in the repair work including such details as: what actions were taken, location of the object in the station, subsequent testing, duration of repair etc.

4. Object identification which allows one to prepare general tables for analyzing the operating disturbances

5. Serial number identification of the object

6. Information regarding physical location of the object allows tabulation of malfunctions with reference to radiological classification of various rooms

7. Manufacturer's name

8. Time required to perform the repair work

9. Date and time when the object is again in working order

10. Comments on any special observations made, cause of the malfunction and steps taken during the disruption. This is written both by the person detecting the malfunction and person putting back the object in working order. This information is needed for proper coding of the mode, cause and effect of the failure. This coding is used in subsequent sorting of the data.

C. Putting together data which allows cost calculation after the repair work is completed.

D. Columns for data processing.

2.3. Failure Data Collection in the Power Station

The first step in the documentation procedure is taken when a malfunction of a component or a system occurs and it is reported in the way described above. Fig.2 shows schematically a flow diagram.

Failures which lead to operational disturbances are detected and reported directly by the control room personnel. Failures are

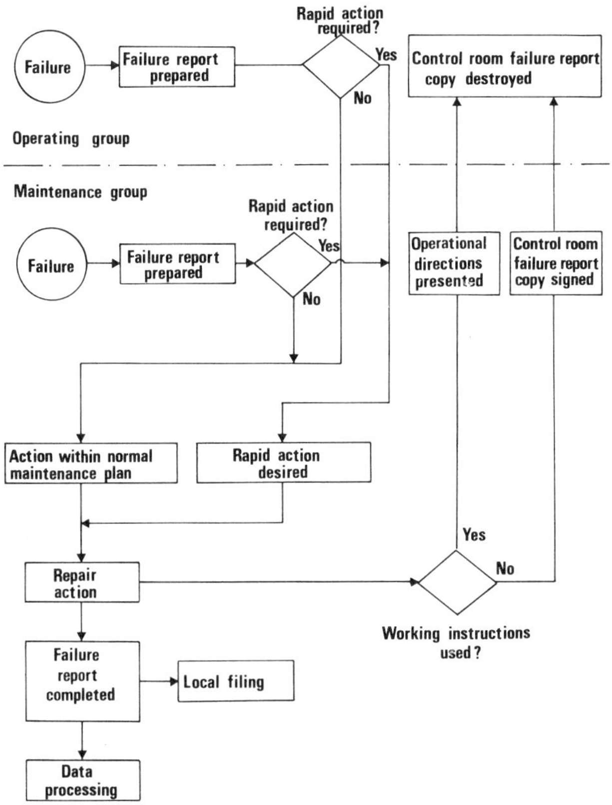

FIG.2. Failure report routine in the power station.

also detected during the inspection rounds undertaken both by the operational crew working in shift and the maintenance crew. After reporting the failure an assessment is made about how immediate is the need for repair. If a delay can be tolerated then it is included in the routine maintenance work. If the problem requires immediate attention then a speedy repair work is undertaken by the specialists concerned in the maintenance department.

Failure reports are written in duplicate. One copy is temporarily placed in the control room. After the fault is corrected the control room copy is destroyed. The original copy of the failure report is completed and checked by the maintenance section and stored in their archives. At the end of each month photostat copies of the reports which have been dealt with are sent for coding and data processing.

3. CODING AND PROCESSING OF FAILURE DATA

With the above mentioned procedure a preliminary local data bank is achieved. This provides the basis for a more comprehensive data bank which is managed by the vendor in cooperation with the owner.

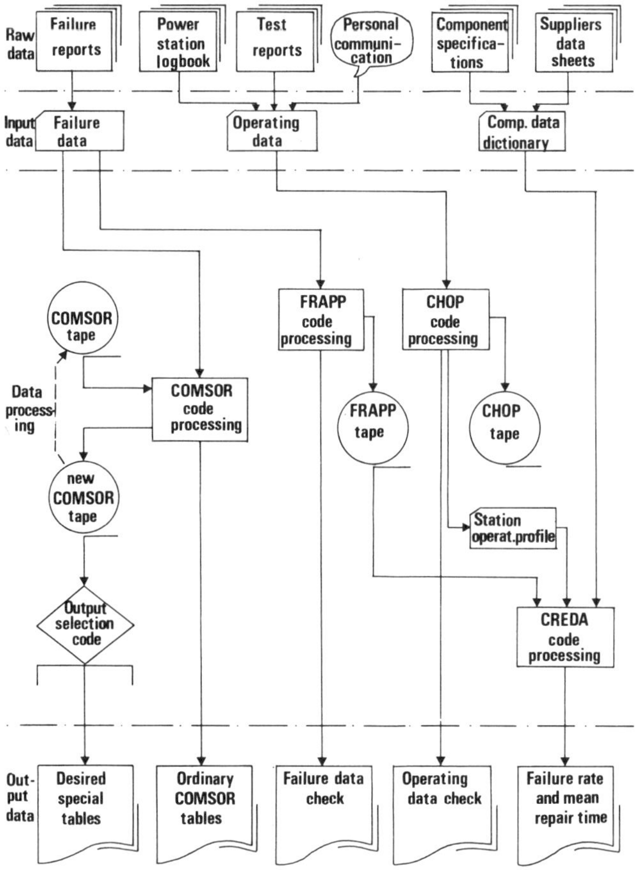

FIG.3. The vendor's failure data processing routine.

At the end of 1972 the then existing form for failure report was modified to take into account the needs of the more comprehensive data bank. Personnel from the power station and the vendor cooperated to design the reporting form which is in use since 1973.

Fig.3 shows how the failure data is processed by the vendor. It consists of two main parts, namely a qualitative part (COMSOR code) and a quantitative part (FAILDARE-S code). The latter is a further development of the principles outlined in the FAILDARE code (ref. 1, 2) written by B. J. Garrick. The COMSOR code has been developed by AB Asea-Atom.

3.1. COMSOR Code

The COMSOR code was written to facilitate handling and interpretation of the information about operational disturbances collected in the power station. This is achieved by sorting all new information in specified categories. One byproduct of this code is simple statistics regarding down time for the object as a result of the malfunction, actual repair time, etc. These are summed up for every component type, system and room.

The above mentioned three parameters are chosen to simplify the complex problem of collecting data from a nuclear power station. The reasoning behind this is as follows: mapping of the behaviour of a particular component type helps one with the future selection of components, information about the system behaviour allows one to judge the suitability of a particular system configuration, and spatial location of the faults gives valuable information for maintenance planning e.g. with reference to radiation doses.

Before the raw data from the failure reports are fed into the COMSOR code it is necessary to preprocess the data for cross-checking and coding. Some reports are found to be insufficient in this preprocessing. An attempt is made to complete these reports by direct contact with the maintenance section in the power station. As shown in fig.1 the completed forms are prepared for punched card input. All the information is available in the failure reports possibly with the exception of the manufacturer code. Coding tables for names of manufacturers, mode, cause and effect of the failure are available.

3.2. FAILDARE-S Code

FAILDARE-S code is used for a rational statistical processing of the information from the malfunctions which occur in the power station. Aim is to obtain failure rate data and mean repair times for the objects studied on the basis of the data obtained from the station. The FAILDARE-S code consists of three sub-codes called FRAPP, CHOP and CREDA which are described below.

3.2.1. FRAPP Code

FRAPP code is used for reading in the raw data into a magnetic tape. All new failure reports are fed into the same magnetic tape. This allows to have an up to date tape always available for further statistical treatment with the CREDA code. Also, the data from the tape can be sorted in other ways than discussed under the description of the COMSOR code.

The data is stored in a magnetic tape not only for convenience but also to cope with the large volume of data.

The FRAPP code provides an output of the fed failure data after the punched cards are read into the magnetic tape. This is helpful in cross-checking of the input data.

3.2.2. CHOP Code

The CHOP code is used for storing information regarding the operational profile of the station. Primary sources of such information are control room log book and test reports.

Main purpose is to get enough information about the operational profile of the station. This is later used in calculating failure rate data for specific objects. By definition the time in operation is a function of both the operational profile of the station and other factors such as test and repairs. At this time, the last two factors can routinely be monitored but do not significantly influence the operational time input as discussed above.

All pertinent information from the control room log book is fed into a magnetic tape via punched card inputs. Simultaneously, a listing of the input information is printed out.

Time intervals between various changes in the operational modes of the stations are categorized by nine different codes for defining operational modes.

3.2.3. CREDA Code

The CREDA code is used for calculating failure rates and mean repair times for specific components in the station on the basis of the data from the FRAPP and CHOP codes.

6800 VALVES

680x	Type
1	Ball
2	Butterfly
3	Check
4	Gate
5	Globe
6	Needle
7	Plug
8	Diaphragm
9	Relief

680x.x — Service, Media Handled (Common)

680x.xx	Function
A	One Way Flow
B	Pressure Relief
C	Vacuum Relief
D	Shutoff
E	Three-Way Selector
F	Four-Way Selector
G	Control
H	Pressure Reducing
I	Isolation
J	Multifunction or NOC

680x.xx.x	Operation
A	Manual
B	Electric Motor/Servo
C	Hydraulic
D	Pneumatic, Diaphragm/Cylinder
E	Solenoid
F	Thermal
G	Self-Actuated

680x.xx.xx — Size, In. (Common)

680x.xx.xx.x — Pressure Rating, psig, USA Standards (Common)

680x.xx.xx.xx — Material/Composition (Common)

FIG.4. Examples of the generic component classification code used.

The presently available routines are suitable for processing data in the apparatus level. It is feasible to go into level of parts or blocks of components with minor modifications. One basic requirement is that the failure reporting is done down to the same level.

Methods used for obtaining the desired results are based on studying a predefined object. An object is defined as a functional unit in the apparatus level. The object is identified with a coding number from the component data dictionary. In this dictionary all similar objects are identified by a classification code. A sample structure of the classification code which is obtained from ref. 1 is shown in fig. 4. Every object is also allotted a functional code which is an important factor in estimating operational time of the object. This is coupled to the operational profile data from the CHOP code as mentioned earlier.

Information about malfunctions in the objects concerned is needed for failure rate calculations. This is obtained by selective searching of the tape produced by the FRAPP code. When all the pertinent information about operational disturbances are collected the CREDA code calculates the failure rate with the help of a statistical model based on χ^2-distribution built into the code. This is done partly for individual objects and for classes of objects. In parallel mean repair times for individual objects and classes of objects are obtained by calculating an arithmetic mean of all repair times attributed to the object or class of objects in question.

3.2.4. Statistical Model for Failure Rate Estimation

In the presently available version of the CREDA code failure rate estimation uses statistical model based on χ^2-distribution. It is a very simple model with the limitation that the failure rates are assumed to be constant.

In the description which follows the following definitions of parameters are used:

r	= number of faults
t	= time in operation (for the object) (h)
n	= number of units in the sample
$(T_{r,n})$	= n.t = cumulative operational time (h)
$100\ (1 - \alpha)$	= confidence level (%)
λ	= failure rate (fault/h)

If it is assumed that the estimation is based on a predetermined sampling time then it can be shown that the following is valid for λ with an upper confidence limit

$$\lambda < \frac{\chi^2_\alpha\,(2r+2)}{2\cdot(T_{r,n})} \qquad (1)$$

This analytical model allows one also to arrive at a "zero-error" estimation by putting r = 0. This gives

$$\lambda < \frac{\chi^2_\alpha(2)}{2 \cdot (T_{r,n})} \qquad (2)$$

"Zero-error" estimation is used primarily for evaluating objects with low failure frequencies or limited sampling size.

4. EXPERIENCE FROM THE EXISTING SYSTEM

Collection of malfunction data has been a natural part of the fixed administrative routines in the operation of the Oskarshamn units. Need for this lies in the effective utilization of the station with due consideration to safety and availability requirements. Use of the existing reporting forms and other administrative routines are continuously reviewed for an effective utilization of personnel and equipment.

4.1. Present Volume of Failure Reports

The systematic processing of the failure reports was started in 1971. Over the last four years approximately 3 400 failure reports have been registered. In the first two years the average number of reports per month was 100. Subsequently, this rate has reduced to 75 reports per month on average. Too much significance should not be attached to these numbers because Oskarshamn unit 1 is the first commercial LWR in Sweden. By virtue of the complexities involved in a kind of a first nuclear station it is natural that the volume of failure reports is somewhat large.

4.2. Data Quality

To obtain meaningful failure rates it is necessary that the raw data is of a certain minimum quality. It is necessary that the malfunction is reported correctly and that all malfunctions are reported. It is also important to separate actual repair work from preventive maintenance. Many persons are involved in the reporting procedure.

Quality of the raw data can be improved by having a selected group of personnel who checks and completes the failure reports. Number of people in such a group is determined by the volume of reports and the desire to obtain a reasonably even data processing. This reporting system was not applied strictly in the early phase of the start-up. Vendor or the subcontractors to the vendor detected various start-up problems and these were corrected without any formal report to the maintenance section of the owner. At this time, however, the centralized reporting system controlled by the owner's maintenance section is in operation. This has resulted in an orderly documentation. Further efforts are being made to improve the quality of the raw data by obtaining information about deficiencies in the filling out of the forms. This is then fed back to the maintenance crew.

A positive attitude of the personnel towards improving the quality of the data is definitely an advantage. The operating crew is conscious of the fact that if by simple means one can locate any eventual deficiencies on time in the station it can only ease their burden in the long run. An impersonal, external and very inquisitive data bank can give rise to irritations because of the extra work involved in writing down good quality failure reports. Here, the need for good communication between the operating crew and the vendor's personnel who deal with the data bank cannot be over-emphasized. Value of the extra work done by the maintenance crew lies in the systematic data processing which generates useful information for the maintenance crew in tabular forms.

4.3. Work Load on Station Personnel

Reporting malfunction and collection of malfunction data is thus an integrated part of the operational administrative procedure in the Oskarshamn units. A quantitative assessment of the relative values of the extra load on the maintenance crew in reporting the malfunction in a suitable form for a data bank and the feedback from the data bank is not possible.

It is difficult to assess accurately how much extra work is involved in preparing the failure reports in a suitable form for the data bank. It is also a function of the total volume of the failure reports. It is estimated as a few man-days of a qualified engineer's work in each of the mechanical, electrical and the instruments groups in the maintenance section. This applies to the early operational phase of the Oskarshamn 1 unit which is a first-of-a-kind unit. In later units it is expected that this work load can be reduced to less than a day's work per month for one qualified engineer in each group.

5. RESULT OF DATA PROCESSING

Computerized data processing of the operational disturbances has led to more rational bookkeeping of the maintenance actions. The manually prepared service log has been replaced by output from the COMSOR code. The information is sorted in the following three categories:

system and component
component type
radiological room classification

The data can be analyzed in other ways, if desired, after the basic categorization as shown above is done. This working procedure gives more information more quickly than the manual procedure used earlier.

The statistical treatment of the data is at this limited only to the mechanical components. Special attention has been paid to the valves which because of their large population in the station are of great importance for the total availability of the station. Data from pumps and other rotating components are being processed. It should be realized that the sampling time is still too small for any definite failure rate data. Table I shows some examples of failure rates obtained. The figures shown include all failure modes.

TABLE I
Examples of failure rates obtained from the reliability data collection system.

Component	Failure rate (f/h)	Mean repair time (h)
Gate valve	$20 \cdot 10^{-6}$	17.0
Globe valve	$6 \cdot 10^{-6}$	5.1
Diaphragm valve	$3 \cdot 10^{-6}$	6.0
Ball valve	$7 \cdot 10^{-6}$	5.6
Check valve	$10 \cdot 10^{-6}$	2.8
Solenoid valve	$8 \cdot 10^{-6}$	1.8

TABLE II
Observed failure mode distribution of pumps

Failure mode	Centrifugal pumps (%)	Reciprocating pumps (%)
Failure to continue operation	3	-
Erratic operation	13	25
Failure to start	9	2
External leakage	67	59
Internal leakage	-	12
Clogging	8	2

TABLE III
Observed failure mode distribution of valves

Failure mode	Shut-off valve (%)	Check valve (%)	Solenoid valve (%)
Spurious operation	1	-	1
Erratic operation	27	1	11
Failure to operate	11	4	3
External leakage	42	57	5
Internal leakage	18	29	12
Clogging	2	9	-
Solenoid failure	-	-	58

Special studies have been carried out to look into various failure modes in pumps and valves. Some examples are shown in table II and III.

6. FUTURE DEVELOPMENT

In the nuclear power stations the tendency is towards greater automation in operational surveillance and documentation of operation. At present the aim is to reduce the number of instrument readings which require manual surveillance and at the same time retain the features which allow the operating personnel to get a "personal" feel for the operating condition of the station.

In a power station equipped with central data processing units a more accurate operational history is available. At this time it is difficult to judge to what extent such improved accuracy in recording operational history will affect the nature of the reliability data.

Automation does not exclude the need for qualified reexamination of the failure reports before these are fed into the computer. It is feasible that by using display units one can improve the quality of the failure reports by looking at automatically registered information available in the station.

In order to answer these and other questions in the field of reliability data the Swedish nuclear utilities have formed a joint working group. The primary task of this group is to create a standard procedure of collecting failure data from the Swedish nuclear power plants in operation. Because of the complexities involved in the reliability data collection, the work of the group till now has been concentrated on the question of a standardized failure report form and general failure classification code. In the next stage the group will formulate the basic structure of a joint data bank. In that period the group will also outline models for the statistical treatment of the collected data. It is hoped that the experiences gathered from the data collection work in Oskarshamn 1 will be useful in the new system.

Finally, one should reemphasize the intimate relationship between reliability and safety of equipments in a nuclear power station and over-all availability of the power generating unit.

LIST OF REFERENCES

1 Garrick, B.J., et al.,
Classification and Processing of Reliability Data from Nuclear Power Plants,
HN-193 (1968)

2 Garrick, B.J., et al.,
Collection of Reliability Data at Nuclear Power Plants
HN-199 (1968)

COLLECTION OF RELIABILITY DATA FOR SODIUM COMPONENTS IN KNK

H. ZEIBIG, W. SCHMIDT
INTERATOM,
Bensberg

W. MARTH
Gesellschaft für Kernforschung mbH,
Karlsruhe

G. BRUDERMÜLLER
Kernkraftwerk Betriebsgesellschaft mbH,
Leopoldshafen,
Federal Republic of Germany

Abstract

COLLECTION OF RELIABILITY DATA FOR SODIUM COMPONENTS IN KNK.

With the start of the test reactor KNK I the collection and evaluation of reliability data was performed up to the final shut-down on 2 September 1974, the beginning of the reconstruction phase for the insertion of a fast core (KNK II). Reliability data useful for applications must contain the components that have not failed. To estimate, for example, the failure rates it is necessary to know the total operating time of the population as well as the number of failures. Data collection that included information on all components in the plant was not feasible. Thus an important step was to select a limited number of sodium component categories. By agreement with the operating company — the Kernkraftwerk Betriebsgesellschaft mbH, Leopoldshafen — the order forms for repairs were changed slightly to include the most important reliability information. For the evaluation a computer program was used. Results were obtained especially for valves, pumps, heat exchangers, cold traps and other sodium components. It is shown that useful reliability data can be achieved with limited effort if one concentrates on selected components.

1. INTRODUCTION

December 1970 saw the start of operation of the test reactor KNK I (Kompakte Natriumgekühlte Kernreaktoranlage), the first sodium-cooled reactor in the Federal Republic of Germany. The operating time consists of a start-up phase (December 1970 - July 1972) and an experimentation phase (1 August 1972 - 2 September 1974). The purpose of the project was to gain experience with sodium systems and components necessary for the sodium-cooled fast-breeder reactor SNR 300. Whereas for LWRs statistical data are available from field experience and from conventional plants, there is a lack of statistical data for components of sodium-cooled reactors. Thus there was a clear case for gaining as much experience as possible on KNK I with a view to the statistical evaluation of reliability data.

2. RECORDING STRATEGY

A data recording system was developed which would meet the conditions of:

having no effect on the operation and repair processes within the plant

placing no additional strain on the operating crew

requiring no separate reporting forms.

By agreement with the operating company, the order forms for repairs were slightly changed to cover the most important reliability information, such as:

Component inventory number
Item description
Cause and mode of failure
Method of failure detection
Report of repair
Instant of failure
Time to repair

The recording system was restricted to sodium components. The components that are provided for recording belong to the following parts of the plant:

Reactor equipment, e.g. reactor vessel, core components and auxiliary cooling and heating systems

Fuel handling system

Reactor coolant system

Reactor auxiliary systems.

The data of the components of these parts were stored at the operating company and only about 200 components were selected for the determination of statistical reliability data. However, it was possible to change the selected items if it were desirable.

3. EVALUATION PROGRAM

For evaluating the failure and repair data, a computer program was used at INTERATOM. In this program the basic and the event data of the selected components are stored. It is possible to get an output of the lifetime history of each component from the entry date into the system up to the moment of interest.

To determine reliability data, all components of the same type and under the same operating conditions were put together by the program in one group. For these samples from the population the total operating time is computed, and mean time to failure and mean time to repair are determined. The following data are stored or computed by the program:

Type of component
Number of items
Total operating time for all the items
Number of failures

Failure modes
Mean time to failure
Mean time to repair

With regard to different failure modes the population of components can be subdivided and the reliability data can be computed for each failure mode separately.

4. FAILURE DETERMINATION

To describe the failure of a component a constant failure rate is used. This seems to be useful for several reasons:

The exponential distribution — which leads to a constant failure rate — can not be rejected in most cases if there are less than about 20 failures in a sample.

There is no reason to use another distribution with more than one parameter as long as the exponential distribution cannot be rejected by statistical data.

An estimate of the parameter of the exponential distribution is possible even if a few failures only are observed.

For a sample in which k failures occurred the estimate of the failure rate is

$$\lambda = \frac{k}{t_{tot}}$$

where t_{tot} is the total operating time of the sample and the 'mean time to failure' is given by MTTF = $1/\lambda$.

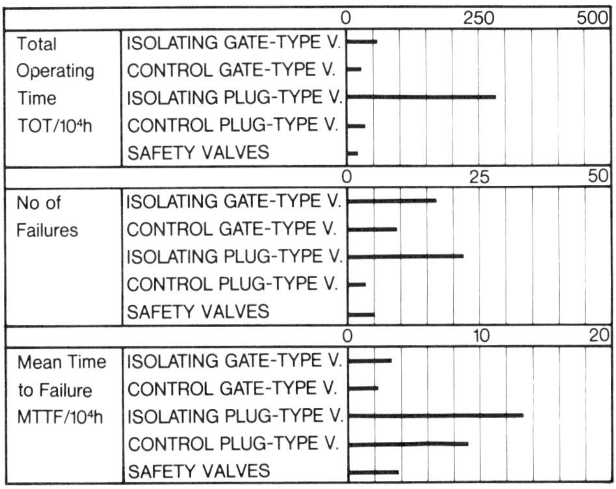

FIG.1. Reliability data for Na valves, divided up according to valve function.

5. RESULTS

Some results are shown in Figs 1-6, where the total operating time of the samples, the number of failures and the 'mean time to failure' are listed. The data should be discussed with respect to the following aspects:

To determine reliability data the whole operating time of KNK I — from 1970 up to 1974 — was used. This time includes the start-up phase.

The plant was operated as a test reactor. A typical temperature versus time curve of the reactor tank is given in Fig. 7. It can be seen, that the components are operated in the temperature range from 200 to 520°C.

Within some groups of components no failure occurred during the operating time. These groups are

Electromechanical pumps
Sodium-sodium heat exchangers
Coolers
Cold traps
Heaters
Oxygen detectors

To compute the 'mean time to failure' for these cases the pessimistic assumption was made that three failures might have occurred during the total operating time.

Where only one failure occurred during the operating time — as was in the case with the steam generators — the 'mean time to failure' was computed under the assumption that four failures might have occurred. For two and three failures the same procedure was used.

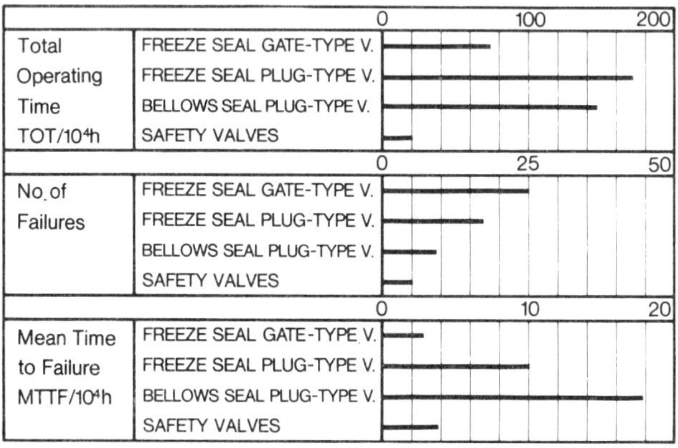

FIG.2. Reliability data for Na valves, divided up according to valve type.

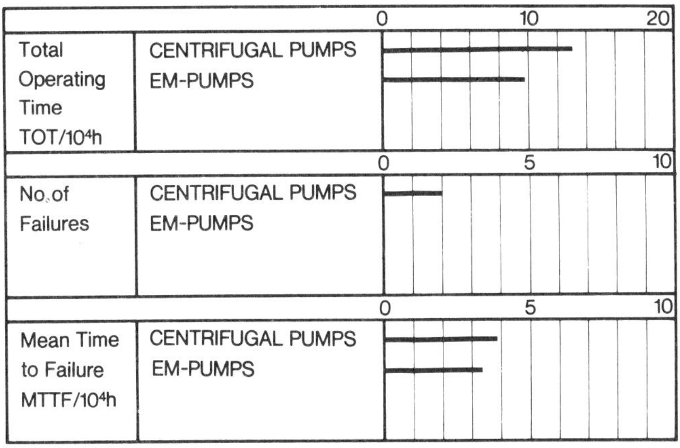

FIG.3. Reliability data for Na pumps.

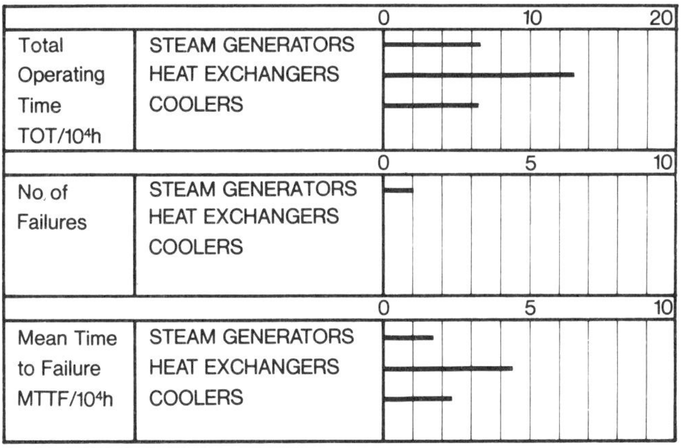

FIG.4. Reliability data for Na heat exchangers.

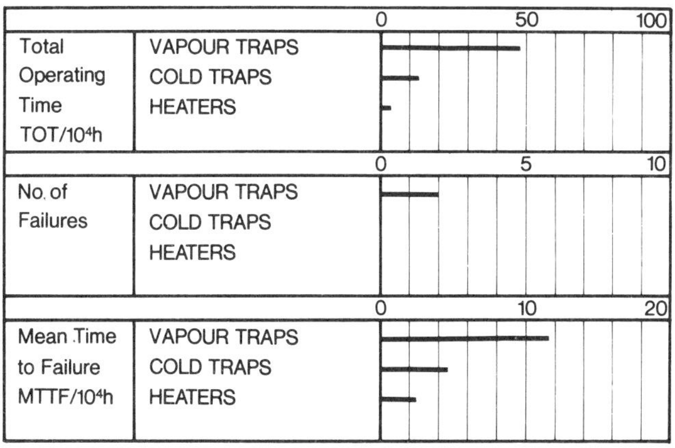

FIG.5. Reliability data for Na traps and heaters.

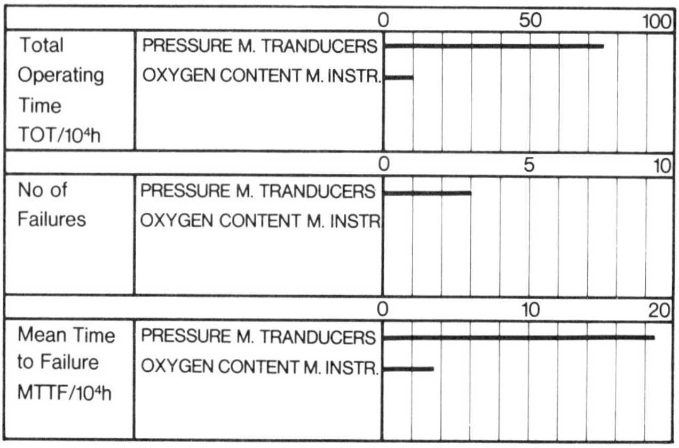

FIG.6. Reliability data for Na instrumentation.

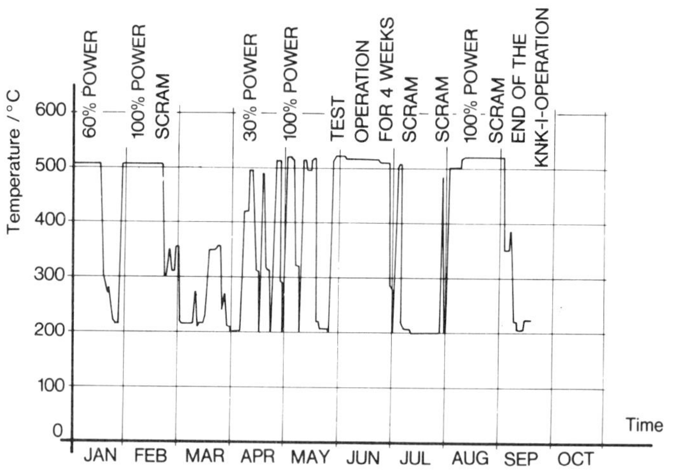

FIG.7. Temperature versus time curve of KNK reactor tank for 1974.

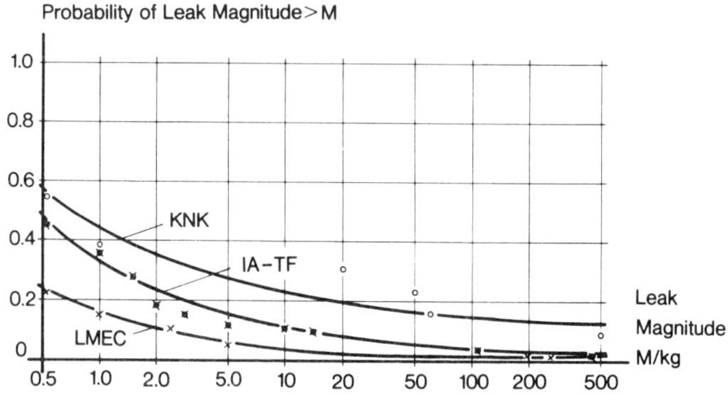

FIG.8. Cumulative probability distribution of sodium leakages.

Because of their importance, experience about sodium leakages was recorded separately. In the time during which the plant was filled with sodium, 12 sodium leakages were reported. The smallest leaks observed were approximately 0.1 kg in magnitude. The largest leak was about 500 kg and occurred at an auxiliary vessel which was only needed during the non-nuclear heat-up phase. Six out of the 12 leakages led to a sodium fire.

Figure 8 shows the cumulative probability distribution of sodium leakages. The upper curve gives the leakages which occurred at KNK. The second curve is the result of 55 leakages in the test facilities at INTERATOM. These data are compared with 80 leakages at six plants reported by the Liquid Metal Engineering Center [1].

6. CONCLUSIONS

The data collection system used at KNK I, which was integrated in the operating process, demonstrates that useful reliability data can be achieved with only a small effort. Because of the short operating period the reliability data are not very exact. To apply the results one has to take into account that the reliability data include the start-up phase and the experimentation phase and that some of the components are prototypes. It is planned to continue the work at KNK II.

REFERENCE

[1] LIQUID METAL ENGINEERING CENTER, Failure Data Handbook for Nuclear Power Facilities, LMEC - Memo 69-7 (Aug.1969), Canooga Park, Calif.

PERFORMANCE ANALYSIS ON NUCLEAR POWER PLANTS FROM OPERATING EXPERIENCE DATA

A. FATTAH, R. SKJOELDEBRAND
International Atomic Energy Agency,
Vienna

Abstract

PERFORMANCE ANALYSIS ON NUCLEAR POWER PLANTS FROM OPERATING EXPERIENCE DATA.

The authors discuss the role of the annual reports on "Operating Experience with Nuclear Power Stations in Member States", produced by the IAEA from 1970 onwards. An analysis of the performance of nuclear power plants has been made from operating experience data. An attempt made to find out the reasons of unavailabilities is also described.

INTRODUCTION

Since 1970 the IAEA has issued annual reports on "Operating Experience with Nuclear Power Stations in Member States" [1]. These reports have developed over the years and the format has now been brought in close conformity with the most recent reporting practice for power stations used, e.g. by UNIPEDE, to permit comparison with the performance of conventional thermal plant.

During 1973 the first analytical reports were prepared on the basis of the data given in the annual reports [2, 3] and it is intended that these studies shall be made each year to supplement the report itself.

The IAEA report uses strictly defined performance factors, firstly the load and operation factors:

> The load factor (LF) is the ratio between the energy that a power plant produced during the period and the energy that it could have produced at maximum capacity under continuous operation during the whole of that period.
>
> The operation factor (OF) is the ratio between the number of hours the unit was on line and the total number of hours in the reference period.

In addition, unavailability factors (UF) are used, based on an energy definition of unavailability as follows:

> Unavailability means that the available capacity is lower than the maximum capacity. Unavailability is classified as planned if it is foreseen long in advance, generally at the time when the annual overhaul programme is established, and if the beginning of the unavailability period can largely be controlled and deferred by management. All other unavailability is classified as unplanned.

The definition of the unavailability factor then becomes:

$$UF = E_u / E_m$$

The unavailability factor over a specified period is the ratio of the energy E_u that could have been produced during this period by a capacity equal to the unavailable capacity C, and the energy E_m that could have been produced during the same period by the maximum capacity.

The unavailability factor UF over a specified period can be separated into:

PUF = unavailability factor due to planned outages such as refuelling and maintenance work, and
UUF = unavailability factor due to all other reasons.

LOAD FACTOR

The report issued in 1974 [1(e)], covering operating experience with nuclear power plants in 1973, contains the annual data for 107 reactors in 15 Member States of the IAEA with an average overall load factor of 62.1% and operation factor of 72.9%.

These relatively low values would appear to confirm the pessimistic data on the performance of nuclear power plants which have been quoted in the public debate. It may, however, be interesting to study more in detail what is behind these two averaged basic performance factors. If one plant, KKN, which ceased operation in 1973 for economic reasons,

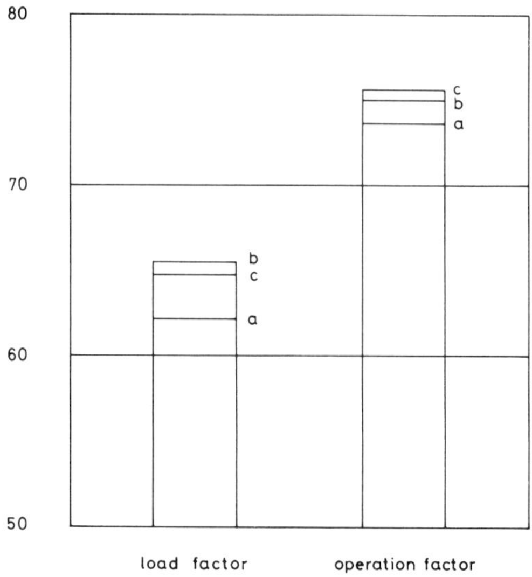

FIG. 1. Average load and operation factors from Ref. 1(e). (a) All 107 reactors in the report. (b) Excluding KKN and reactors which started operation after 1 July 1973. (c) Excluding KKN and reactors which started operation after 1 July 1973, and reactor plants with at least 100 MW(e) output.

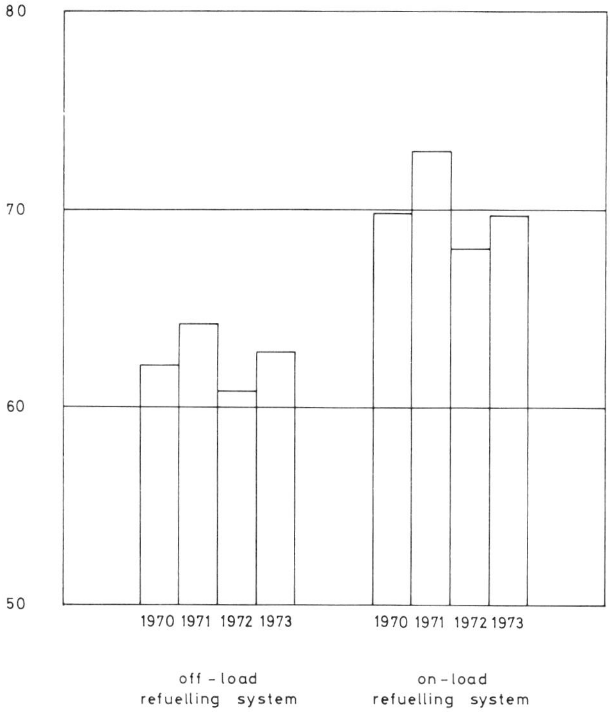

FIG. 2. Comparison of load factor between off-load and on-load refuelling systems 1970-1973.

and those plants that started commercial operation in the second half of 1973 are excluded, the average load factor for the remaining 99 plants becomes 65.4% and the operation factor 74.9%. If, in addition, all plants with an output of less than 100 MW(e) are excluded, so as not to include pilot or test plants, this leaves 74 reactors with an average load factor of 64.7% and operation factor of 75.6% (Fig.1). There is thus an indication that performance is better for the commercial plants after an initial 'shake down' period but the data are still far from the often quoted design objective of a load factor in the range 75-80%.

For a more detailed study of the data a distinction should first be made between those commercial reactor systems which are designed for off-load refuelling (essentially the light water reactors, LWRs) and those which are refuelled on load (gas-cooled and heavy water reactors, GCRs, PHWRs, HWGCRs). Figure 2 shows that over the last four years there has, as expected, been an apparent significant difference in performance between these two major groups of plants.

Several authors have also reported a clear dependence of performance as measured by the load factor and age of the plant [2, 4]. Such data from the Agency's annual reports are shown in Figs 3 - 5. The selection of the plants, for which the data have been averaged, has been made as follows:

Off-load refuelling plants (LWRs) have been separated from on-load refuelling plants (GCRs, HWRs).

A somewhat arbitrary distinction has been made between first and second generation commercial LWRs, the second generation starting with the Haddam Neck and Oyster Creek plants.

Only plants with an output of minimum 100 MW(e) have been included.

The number of plants in each of the groups is too small to permit drawing significant conclusions from the statistics but the trends shown in the figures are clear, namely, that there is a definite maturing process, in which the power plants improve their performance over an initial period of 3-4 years after which the plants have achieved load factors in the range 70-80%. The total averages are thus misleading; this would appear to be due to the prevalence of new reactor plants in the rapidly increasing population.

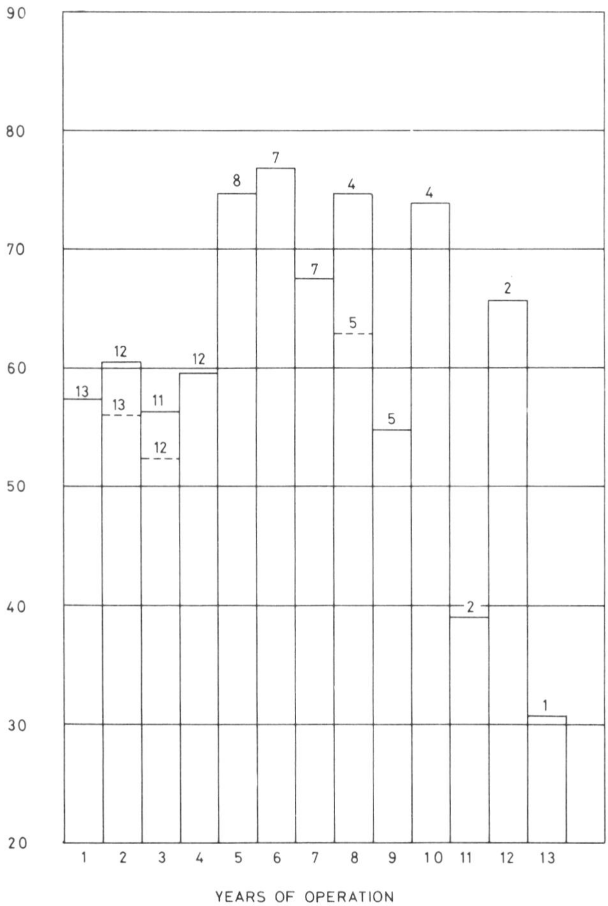

FIG. 3. Load factors for first generation off-load refuelling systems type reactors (LWRs), at least 100 MW(e).

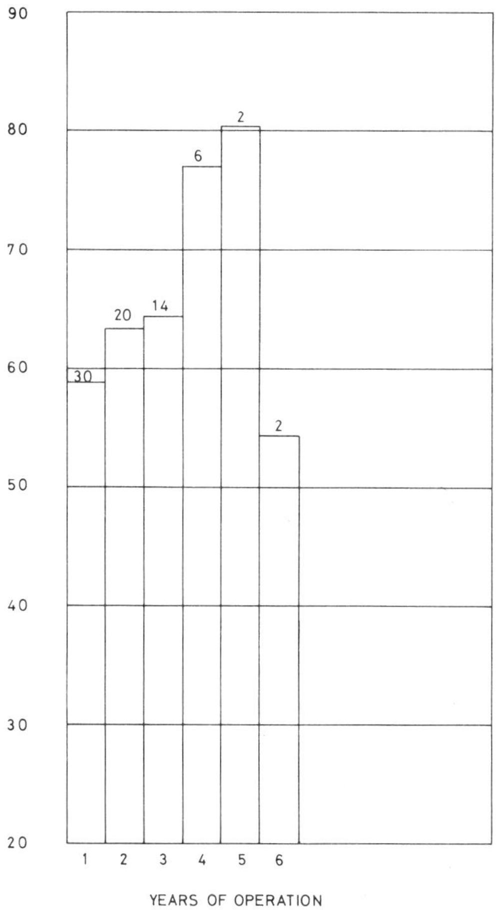

FIG. 4. Average load factors for second generation off-load refuelling systems type reactors (LWRs).

It is worth noting that fossil-fuelled plants also have shown marked decreases in availability as their sizes have increased. The data published by Edison Electric Institute for U.S. plants [5], as shown in Fig.6, give an indication of the actual average operating availability for a 10-year period from 1964 to 1973. Actually there are 59 plants reported in the USA in the range of 600 MW(e) and above, while the data only show 14 nuclear plants in that range. These 14 plants, according to IAEA data, had a load factor of 66.5% with an operation factor of 71.8% in 1973, as compared to the overall LF = 64.7 and OF = 75.6% reported earlier in this paper. For the 59 fossil-fuelled plants quoted in the size range of 600 MW(e) and above, the availability factor is 72.9, which is not directly comparable with the IAEA operation factor which refers to the fraction of hours on line rather than of hours available. It appears, however, that the large-size fossil plants have shown performance which is of the same order and not significantly better than the nuclear plants.

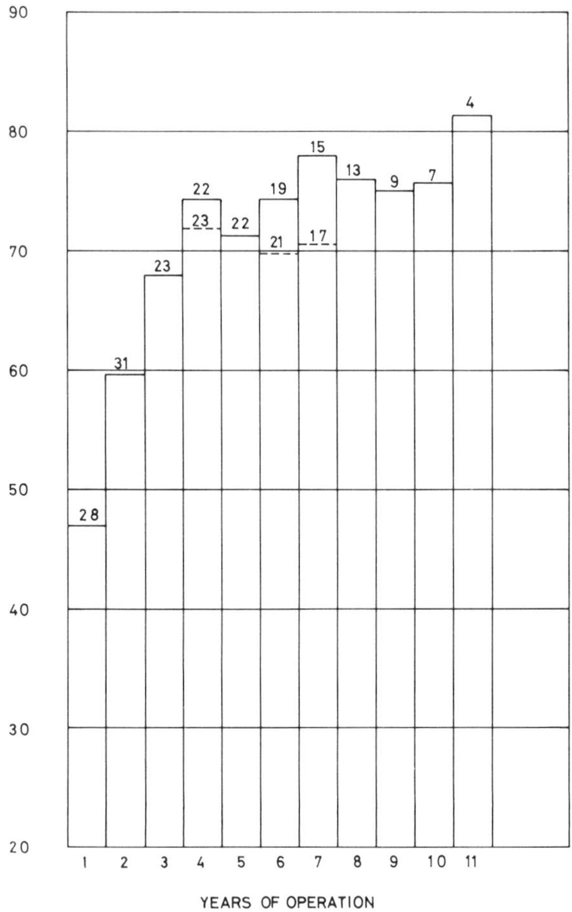

FIG. 5. Load factors for on-load refuelling system at least 100 MW(e).

UNAVAILABILITY

In the report on operating experience for 1973, for the first time an effort was made to include data on unavailability of plants classified not only in relation to planned and unplanned unavailability but also with a classification as to the reason of the unavailability. The classification, which was recommended by a consultants' meeting in January 1974, is reproduced in Table I. The response to our questionnaires generally showed a great willingness to contribute to the report but there have been many misunderstandings of the definitions of the unavailability factors. This is partly no doubt due to the prevailing practice in many countries of defining unavailability in time rather than energy and the widely varying

FIG. 6. Fossil unit operating availability versus unit size; 10-year period 1964-1973.

TABLE I. CLASSIFICATION OF OUTAGES AND UNAVAILABILITY

REASONS FOR OUTAGES:

A. Equipment failure
B. Operation error
C. Refuelling
D. Maintenance and repair
E. Testing of plant systems/Components
F. Training and licensing
G. Regulatory limitation
H. Other

Equipment failures can be categorized as those occurring in:

1. Reactor and accessories: (a) Pressure vessel
   (b) Reactor internals
   (c) Shielding

2. Fuel: (a) Fuel rod and cladding
   (b) Fuel bundle structure

3. Reactor control system and instrumentation:
   (a) Control rods and mechanism
   (b) Control rod drives
   (c) Chemical control system
   (d) Radiation monitoring
   (e) Fluid flow control instrumentation
   (f) Temperature measuring devices
   (g) Pressure control instrumentation
   (h) In-core instrumentation
   (i) Other instrumentation

TABLE I. (cont.)

4. Nuclear auxiliary and emergency systems:
   (a) Primary and secondary containment
   (b) Gaseous and liquid radwaste and radwaste handling systems
   (c) Fuel handling facilities and systems
   (d) Emergency shutdown cooling system
   (e) Decay heat removal system
   (f) Core spray cooling system
   (g) Post-incident cooling system

5. Main heat removal system:
   (a) Main coolant circulating pumps
   (b) Main coolant pump drives
   (c) Prime circuit, piping and fittings
   (d) Auxiliary cooling systems
   (e) Main steam valves
   (f) Steam piping

6. Steam generators

7. Feedwater, condenser and circulating water systems:
   (a) Raw water supply system
   (b) Make-up water supply system
   (c) Feedwater heaters
   (d) Feedwater pumps
   (e) Feedwater pump drives
   (f) Feedwater piping and fittings
   (g) Condensers
   (h) Condensate piping and fittings

8. Turbine generator system:
   (a) Turbine system
   (b) Generator

9. Electrical power and supply system

10. Miscellaneous:
   (a) Ventilation system
   (b) Compressed air and vacuum system
   (c) Other systems and components

TABLE II. SUMMARY OF UNAVAILABILITY FACTORS

	No. of reactor plants	PUF (%)	UUF (%)	UF (%)
Off-load refuelling plants of first generation	10	12.1	18.0	30.1
Off-load refuelling plants of second generation	15	21	12.9	33.9
On-load refuelling plants	26	13.7	13.6	27.3

TABLE III. REASONS OF UNAVAILABILITY OF FIRST GENERATION LWRs (min. 100 MW(e), 1 year old)

Category	Reasons	Total hours shutdown	Percentage of total outage time
PLANNED			
C	Refuelling	4 114.0	16.4
D	Maintenance & repair	4 811.5	19.2
CD	Refuelling and maintenance & repair	3 536.5	14.1
E	Testing of plant system/comp.	107.0	0.4
EF	Testing of plant system/comp. and training and licensing	36.5	0.2
	TOTAL PLANNED	12 605.5	50.3
UNPLANNED			
A	Equipment failure, total	10 240.7	40.8
A2	Fuel	852.0	3.4
A3	Reactor control and instrumentation	576.0	2.3
A5	Main heat removal system	363.0	1.4
	TOTAL NUCLEAR PART	1 791.0	7.1
A6	Steam generators	3 653.0	14.6
A7	Feedwater, condenser and circulating water system	879.0	3.5
A8	Turbine generator system		
	(a) Turbine	1 065.0	4.2
	(b) Generator	2 081.0	8.3
A9	Electrical power and supply system	391.7	1.6
A10	Miscellaneous	380.0	1.5
	TOTAL CONVENTIONAL PART	8 449.7	33.7
E	Testing of plant system/Components	20.0	0.1
H	Others	2 194.0	8.8
	TOTAL UNPLANNED	12 454.7	49.7
	GRAND TOTAL	25 060.2	100.0

practices in defining planned and unplanned unavailability. As a result, the statistical data quoted in the following tables are imprecise and must be used with caution, except with regard to the trends, which are clearly apparent.

It should also be noted that, as a result of the recommendations of the consultants, the reporting of unavailability has been limited to "significant unavailability or outages", defined as corresponding to at least 10 hours' operation at maximum electrical capacity.

TABLE IV. REASONS OF UNAVAILABILITY OF SECOND GENERATION LWRs (min. 1 year old)

Category	Reasons	Total hours shutdown	Percentage of total outage time
PLANNED			
C	Refuelling	2798.4	7.0
D	Maintenance and repair	12428.8	31.2
CD	Refuelling and Maintenance & repair	8985.9	22.6
E	Testing of plant system/Components	2005.9	5.0
F	Training and licensing	75.7	0.2
H	Others	179.0	0.5
GD	Regulatory limitation and Maintenance and repair	262.0	0.7
	TOTAL PLANNED	26735.7	67.2
UNPLANNED			
A	Equipment failure, total	10842.0	27.2
A1	Reactor and accessories	3201.0	8.0
A3	Reactor control system and instrumentation	523.0	1.3
A5	Main heat removal system	1103.8	2.8
	TOTAL NUCLEAR PART	4827.8	12.1
A6	Steam generators	267.8	0.7
A7	Feedwater, condenser and circulating water systems	631.6	1.6
A8	Turbine generator system		
	(a) Turbine	5054.8	12.7
	(b) Generator	19.0	0.1
A9	Electrical power & supply system	26.0	0.1
A10	Miscellaneous	15.0	0.0
	TOTAL CONVENTIONAL PART	6014.2	15.1
B	Operation error	35.0	0.1
D	Maintenance and repair	2028.0	5.0
E	Testing of plant system/Components	73.0	0.2
F	Training and licensing	14.0	0.0
G	Regulatory limitation	73.0	0.2
	TOTAL UNPLANNED	13065.0	32.8
	GRAND TOTAL	39800.7	100.0

TABLE V. REASONS OF UNAVAILABILITY OF ON-LOAD REFUELLING TYPE REACTORS (min. 100 MW(e), 1 year old)

Category	Reasons	Total hours shutdown	Percentage of total outage time
PLANNED			
D	Maintenance and repair	29 808.2	41.3
E	Testing of plant systems/Components	4721.0	6.5
H	Others	4 986.0	6.9
	TOTAL PLANNED	39 515.2	54.7
UNPLANNED			
A	Equipment failure, total	23 983.2	33.2
A1	Reactor and accessories	968.0	1.3
A2	Fuel	630.0	0.9
A3	Reactor control system & instrumentation	1 120.0	1.6
A4	Nuclear auxiliary & emergency systems	300.0	0.4
A5	Main heat removal system	8 015.7	11.1
	TOTAL NUCLEAR PART	11 033.7	15.3
A6	Steam generators	5 940.0	8.2
A7	Feedwater, condenser and circulating water systems	191.0	0.3
A8	Turbine generator system		
	(a) Turbine	5 380.0	7.4
	(b) Generator	410.0	0.6
A9	Electrical power and supply system	473.5	0.7
A10	Miscellaneous	555.0	0.8
	TOTAL CONVENTIONAL PART	12 949.5	17.9
B	Operation error	161.0	0.2
D	Maintenance and repair	2 955.0	4.1
H	Other	5 606.0	7.8
	TOTAL UNPLANNED	32 705.2	45.3
	GRAND TOTAL	72 220.4	100.00

In discussing the unavailability factors the same criteria as for the basic performance factors (LF and OF) have been used in selection of the plants and also the same plant classification, i.e. off-load refuelling commercial plants of the first and second generations and on-load refuelling plants. Table II gives a summary of planned and unplanned unavailability factors and Tables III to V give the detailed breakdown as to the occurrence for the unavailabilities.

The data in Tables II to V are self-explanatory but a few comments may be made:

The planned and unplanned unavailabilities have for all three plant categories been of the same general magnitude with a possibly significant difference between the first and second generations of off-load refuelling plants, indicating some design maturity but also increasing plannable maintenance problems between the first and second generation LWRs. This trend would still have to be confirmed in coming years.

The often quoted statement that the major reasons for unplanned outages are in the conventional, non-nuclear part of the plant is only confirmed by the data, and in that case mainly for the first generation of LWRs. It is still generally true that steam-generator and turbo-generator systems have been the two systems causing the unit unplanned unavailability, in all cases having together given a higher total unavailability than the total "nuclear" parts of the plants.

It is evident that maintenance and repair work, although often amenable to planning, have caused a major part of the unavailability of all plant categories; this stresses the need for efforts by the Agency to improve plant reliability and technology, and indeed the need for the present symposium.

CONCLUSION

The Agency's annual Report on Operating Experience with Nuclear Power Stations in Member States has now been brought into a format which permits some general conclusions to be drawn on the performance and reliability of nuclear power plant systems. It is obvious that for economic reasons it will pay to spend more effort on achieving a higher degree of reliability. It is also very desirable for the plant operators to go into more detailed studies of reliability data from the different systems and components than can be done by the Agency, whose contribution of necessity is limited to comparison of generally classifiable plants and systems. The report has met with considerable interest in recent years and it is now intended to supplement it with a short analytical summary each year.

REFERENCES

[1] IAEA, Operating Experience with Nuclear Power Stations in Member States, (a) IAEA-127 (1970); (b) IAEA-137 (1971); (c) IAEA-150 (1972); (d) IAEA-155 (1973); (e) IAEA-168 (1974). Unpriced documents.

[2] CHASE, D.B.A., ILJAS, J., "Measuring performance of nuclear power station units", Experience from Operating and Fuelling Nuclear Power Plants (Proc. Symp. Vienna, 1973), IAEA, Vienna (1974) 65.

[3] ILJAS, J., CHASE, D.B.A., ROBERTS, J.T., SKJOELDEBRAND, R., "Experience of nuclear power station availability and reliability", Effects of Introducing Nuclear Power Stations on the Structure and Reliability of Supply of Electricity Generating Systems in the ECE region (Proc. Symp. Madrid, 1973).

[4] SKOVHOLT, D.J., "Performance and reliability of operating United States nuclear power plants", Experience from Operating and Fuelling Nuclear Power Plants (Proc. Symp. Vienna, 1973), IAEA, Vienna (1974) 3.

[5] EDISON ELECTRIC INSTITUTE, Report on Equipment Availability for the ten year period, 1964-1973, No. 74-57, December 1974.

DISCUSSION

(on papers IAEA-SM-195/52, IAEA-SM-195/31 and IAEA-SM-195/36)

J.R. TAYLOR: I should like to ask Mr. Fattah whether he has any further data on availability factors for coal- and oil-fired power plants which would allow a more meaningful comparison with the data for nuclear plants.

A. FATTAH: The International Atomic Energy Agency does not collect data for coal- and oil-fired plants. Edison Electric Institute (EEI) and the International Union of Producers and Distributors of Electricity (UNIPEDE) do collect such data, and in Fig.6 we have plotted a curve for fossil-fired units (operating availability versus size), based on data released by EEI, which to some extent enables us to compare their performance with that of nuclear power plants.

Y.-F. ONG: It would be interesting and useful to have the information contained in your Figs 3, 4 and 5 broken down into separate bar charts for BWRs and PWRs (Figs 3 and 4) and for GCRs and HWRs (Fig.5), showing the trend of the load factor against years of operation.

A. FATTAH: As an international organization the IAEA cannot make this kind of distinction because it might affect the dealings of contractors and suppliers. The kind of identification you want can be derived, however, from the reports on operating experience which we have published annually since 1970, and anyone can get copies of these reports simply by dropping us a letter. Let me emphasize once again, however, that the data available are not sufficient for statistical analysis or for the drawing of definite conclusions.

W.A. WOLFE: Have you, in your analyses of unplanned unavailability, considered a category such as 'sudden forced outages'? Such outages are often the most serious in their effects on the system, and hence, frequently, the most costly.

A. FATTAH: No, I am sorry; we do not single out data on 'sudden forced outages' and we are therefore not in a position to specify their economic consequences.

J.-C. ARLAUD: I should just like to ask Mr. Zeibig whether he does not think it a bit dangerous to do statistical analyses when the amount of information available is not really sufficient for the purpose. Would it not be better to wait a while before drawing conclusions, rather than do draw conclusions and make the reservation that they are not conclusive?

H. ZEIBIG: For our purposes the failure data were sufficient, to the extent that they provided useful information for the KNK II and SNR 300 fast reactor projects. Another purpose of the work was to devise a data collection strategy for the fast reactor plants as early as possible. We consciously and deliberately accept, and take account of, the handicap that our failure data are so far rather sparse.

L. CAVE: I might just comment that for the designer even poor data are better than no data at all when it comes to designing the next plant.

H. ZEIBIG: Being on the manufacturing side, I completely agree with you.

A.V. KARPOV: Could you quote us the load factor and the power factor for the KNK reactor?

H. ZEIBIG: During the power campaign of 1973/74 the load factor was 55%, the power factor 45%. Interruptions due to experimenters' needs were properly taken into account.

S. HARTWIG: As I mentioned earlier, I am interested in being able to compare theoretical calculations of failure rates, based on fault trees and the like, with actual failure rate data. I should therefore like to ask Mr. Andersson whether the suppliers' data sheets which he showed us give theoretical failure rates that have been compared with the rates quoted in Table I.

B. ANDERSSON: The suppliers' data sheets are intended to provide technical information as a support to component classification. The purpose of this classification is to state statistical populations for the sub-code CREDA. Thus we have not obtained any comparable reliability data from component suppliers. We have, however, compared some of our calculated data with data derived from the literature and have found good agreement when our data provided enough material for statistical analysis.

J.K. VAURIO: I should like to emphasize how important it is that the collection and publishing of failure data should be done in an open and generous spirit. Much information will have to be fed back to the utilities and industry before we can hope to take any important steps forward, for example towards the establishment of safety criteria. I hope that this Symposium may mark the beginning of a new, more liberal attitude towards the publishing of such data. Could Mr. Andersson tell us, for example, how freely the Swedish data will be published in the open literature?

B. ANDERSSON: Under our present system of data collection, the processed data are fed back to the utility member of the team. They can thus be used by the utilities for their special purposes. But open publication of the data cannot come about, I think, until we have acquired more experience with the behaviour of the statistically treated data.

W. HAWICKHORST: I am very interested in failure data on the instrumentation installed in nuclear power plants. Are the instruments considered in failure data collection, and, if so, are any data of this kind available yet?

B. ANDERSSON: All kinds of nuclear power plant equipment are considered in our data collection system. We have, for the time being, been concentrating our statistical treatment on mechanical components because of the very obvious need for data in this area, but in future we plan to give equal treatment to the raw failure data on instrumentation.

C.D.H. FOTHERGILL: I might just add that the Reliability Data Store of our Systems Reliability Service also holds some failure rate data on instrumentation.

A.J. SCHEPENS: It is a pleasure for me to extend to all of you the greetings of the International Union of Producers and Distributors of Electrical Energy (UNIPEDE), and I am grateful that UNIPEDE has been given the opportunity of being represented at this Syposium.

I wish not to ask a question but to make a brief statement. No meaningful comparison and utilization of statistical data is possible unless we establish our criteria clearly, agree to accept the same criteria, and apply the same definitions correctly. For conventional power plants we have, in Europe, the data of UNIPEDE, and, in the United States, the data of the Edison Electric Institute. Thanks to the efforts of UNIPEDE's Committee on Statistics and to the personal action of Mr. Vibert of Electricité de France, there now exists a well established relationship between the two which makes a compilation of European and American statistics possible. Let me recall once again that the data on European power plants are published by UNIPEDE.

ВОПРОСЫ НАДЕЖНОСТИ РАБОТЫ ОБОРУДОВАНИЯ И СИСТЕМ РЕАКТОРОВ НА БЫСТРЫХ НЕЙТРОНАХ

А.В.КАРПОВ, А.А.РИНЕЙСКИЙ, И.А.ЕФИМОВ,
А.И.ГОРЛОВ
Физико-энергетический институт,
Обнинск

В.В.ЕГОРОВ, В.О.ОРЛОВА
Научно-исследовательский институт
атомных реакторов,
Димитровград,
Союз Советских Социалистических Республик

Abstract–Аннотация

OPERATIONAL RELIABILITY OF FAST-NEUTRON REACTOR EQUIPMENT AND SYSTEMS.

The paper quotes statistical data on the operation of BR-5 and BOR-60 reactor equipment and systems. The data are analysed, and reference is made to the items of equipment requiring special operational attention and to the sophisticated design features. The requirements to be met by equipment quality, assembly and testing during start-up and commissioning are likewise discussed. Designs are proposed for the thermal and hydraulic systems of a fast power reactor.

ВОПРОСЫ НАДЕЖНОСТИ РАБОТЫ ОБОРУДОВАНИЯ И СИСТЕМ РЕАКТОРОВ НА БЫСТРЫХ НЕЙТРОНАХ.

В докладе приводятся статистические данные по работе основного оборудования и систем реакторов БР-5 и БОР-60. Приводится анализ данных, указывается оборудование, требующее повышенного внимания при эксплуатации и усовершенствования конструкции. Обсуждаются требования, предъявляемые к качеству изготовления оборудования, монтажу и испытаниям в процессе пуско-наладочных работ. Даются предложения по выбору тепловой и гидравлической схемы быстрого энергетического реактора.

1. РЕАКТОР БР-5

Экспериментальный реактор на быстрых нейтронах БР-5 мощностью 5000 кВт с натриевым теплоносителем был сооружен в г.Обнинске в 1958 году [1]. Физический пуск реактора с топливом из двуокиси плутония был произведен в январе 1959 г. 21 июля 1959 г. реактор БР-5 был выведен на номинальную мощность. К 1964 году максимальное выгорание в топливе из двуокиси плутония достигло ~ 6,7%. В 1965 г. реактор БР-5 был загружен топливом из монокарбида урана. К 1971 г. максимальное выгорание монокарбида урана достигло 6,2%. В мае 1971 г. реактор БР-5 был остановлен для проведения реконструкции с целью повышения его мощности до 10 МВт.

Опыт работы реактора БР-5 описан в работах [2-4]. В табл.I приведены основные показатели работы реактора БР-5 за 12 лет, с 1 июля 1959 г. по 30 апреля 1971 г. При вычислении коэффициента работы реактора на мощности из календарного времени были исключены остановки реактора для полной перегрузки активной зоны в 1961-62 гг. и в 1964-65 гг. Следует иметь в виду многоплановость задач по эксплуатации реактора БР-5. Это – испытание и отработка натриевого оборудования, освоение работ с радиоактивным натрием, испытание опытных

ТАБЛИЦА I. ОСНОВНЫЕ ПОКАЗАТЕЛИ РАБОТЫ РЕАКТОРА БР-5

Год	Время работы реактора на мощности (сут)	Коэффициент работы на мощности (%)	Отработано эффективных суток	Коэффициент нагрузки (%)	Эффективный коэффициент работы на мощности (%)
1959	51	27	15,9	8,7	31
1960	282	77	168,8	46,2	78
1961	182	75	146,6	60,4	77
1962	185	67	24,3	8,4	73
1963	220	60	46,0	12,6	65
1964	164	54	75,6	24,8	54
1965	102	47	77,3	36,0	49
1966	207	57	161,8	44,4	64
1967	270	74	209,7	57,4	75
1968	232	63	182,5	50,0	66
1969	198	54	173,4	47,5	61
1970	228	63	153,0	42,0	68
1971	57	48	45,7	38,1	76
Всего за 12 лет	2377	62	1485,7	39,0	67

тепловыделяющих элементов, обеспечение проведения физических исследований на нейтронных пучках реактора и др. Это безусловно наложило отпечаток на основные показатели работы реактора, в особенности на коэффициент нагрузки. Правильнее было бы характеризовать работу опытного, исследовательского реактора не коэффициентом нагрузки, а коэффициентом работы на мощности.

Как видно из табл.I, в среднем за 1959-1971 годы реактор БР-5 работал на различных уровнях мощности 62% времени. Этот коэффициент не может достаточно полно характеризовать эффективность работы реактора БР-5, так как часть исследований на реакторе проводились без мощности. В связи с этим в табл.I введен условный коэффициент работы реактора на мощности, при вычислении которого время остановок, связанных с экспериментальными работами, вычиталось из общего календарного времени.

В табл.II приведена статистика остановок реактора БР-5 в 1959-1971 гг. с анализом причин остановок. Как видно из таблицы, 16% времени остановок вызваны плановыми догрузками активной зоны. Большинство этих остановок было вызвано не необходимостью догрузки топлива для компенсации выгорания, а необходимостью загрузки в активную зону или выгрузки пакетов с экспериментальными образцами твэл и конструкционных материалов.

25% времени остановок были связаны с проведением экспериментальных работ на реакторе. Половина времени всех остановок реактора БР-5 была вызвана неисправностью оборудования основных контуров, в основном циркуляционных насосов I и II контуров.

Значительная величина времени остановок при ремонтах оборудования I контура связана с необходимостью выжидания распада изотопа ^{24}Na, определяющего активность теплоносителя (как правило ~ 8 дней). При реконструкции реактора в 1971-73 гг. имевшие низкую надежность механические насосы I и II контуров были заменены на электромагнитные.

В табл.III представлена статистика срабатываний аварийной защиты реактора БР-5 с анализом причин срабатывания. За 12 лет работы реактора БР-5 было 341 срабатывание аварийной защиты, в том числе 197 срабатываний БАЗ (быстрой аварийной защиты) и 144 срабатываний МАЗ (медленной аварийной защиты). Более половины всех случаев срабатывания аварийной защиты приходится на мощность до 1000 кВт. 30% всех срабатываний аварийной защиты происходило в процессе вывода реактора на мощность на уровнях до 2% от номинала. Это объясняется большим числом операций переключения и блокировок аппаратуры СУЗ, при выходе на мощность. Как видно из таблицы, из-за неисправности системы управления и защиты реактора произошла треть срабатываний аварийной защиты. Четверть срабатываний была вызвана неисправностями технологического оборудования, в основном натриевых циркуляционных насосов.

2. РЕАКТОР БОР-60

БОР-60 – экспериментальная атомная электростанция с реактором на быстрых нейтронах мощностью до 60 МВт с натриевым теплоносителем и топливом в виде двуокиси урана [5]. Сооружен БОР-60 в г.Димитровграде в 1968 г. Физический пуск реактора с теплоносителем и выход на энергетический уровень мощности со съемом тепла воздушным теплообменником были произведены в декабре 1969 г. В 1970 году был произведен пуск парогенератора. Первое подключение турбины к электросети было осуществлено в декабре 1970 г. К 1 июля 1974 г. максимальное выгорание двуокиси урана в реакторе достигло 11%.

Опыт работы реактора БОР-60 изложен в работах [6,7]. В табл.IV приведены основные показатели работы реактора БОР-60 с 1 января 1970 г. по 30 июня 1974 г. При вычислении коэффициента нагрузки за номинальную мощность реактора была принята мощность 50 МВт. Реактор БОР-60, также как реактор БР-5, является экспериментальным и поэтому эффективность работы реактора лучше характеризовать коэффициентом работы на мощности. В связи с существенным временем остановок реактора, связанных с экспериментальными работами, в табл.IV, аналогично реактору БР-5, введен условный коэффициент работы реактора на мощности. Высокие значения этого коэффициента говорят о надежной работе оборудования реактора БОР-60.

В табл.V приведена статистика остановок реактора БОР-60 с анализом их причин. 66% времени остановок реактора были вызваны плановым профилактическим ремонтом оборудования и догрузками активной зоны. 26% времени связаны с экспериментальными работами. Только около 8% времени остановок реактора были вызваны неисправностями оборудования.

ТАБЛИЦА II. СТАТИСТИКА ОСТАНОВОК РЕАКТОРА БР-5

№ п/п	Причины остановок реактора	Продолжительность остановок (час)														
		1959	1960	1961	1962	1963	1964	1965	1966	1967	1968	1969	1970	1971	Всего	% от всех остановок
1.	Плановая догрузка активной зоны (а.з.)	127	241	649	756	67	34	260	13	1350	253	670	769	252	5440	15,8
2.	Экспериментальные работы	444	173	250	601	851	34	1594	130	151	449	1027	738	1080	8522	24,7
3.	Неисправность оборудования I контура	837	137	410	77	2463	1548	1	2061	703	1403	1300	550	-	11726	34,0
4.	Неисправность оборудования II контура	500	120	166	-	-	1077	265	47	185	615	812	100	-	3887	11,3
5.	Неисправность оборудования III контура	53	470	-	-	150	-	-	8	-	-	2	52	-	735	2,1
6.	Неисправность оборудования вспомогательного	36	425	24	204	23	129	9	18	665	65	196	198	1	1993	5,8
7.	Неисправность системы СУЗ	80	80	30	45	15	4	11	37	28	72	-	57	1	460	1,3
8.	Неисправность системы КИП	134	-	1	3	-	562	140	298	12	10	-	-	2	1162	3,4
9.	Ошибочные действия персонала	5	53	9	5	92	8	56	4	7	4	3	3	2	251	0,7
10.	Неисправность системы электроснабжения	1	5	4	121	17	-	12	8	41	34	2	53	-	299	0,9
11.	Всего	2217	1740	1543	1811	3678	3396	2348	3624	3141	2905	4012	2520	1338	34474	100

ТАБЛИЦА III. СТАТИСТИКА СРАБАТЫВАНИЙ АВАРИЙНОЙ ЗАЩИТЫ РЕАКТОРА БР-5

№ п/п	Причина срабатывания аварийной защиты	Количество срабатываний а.з.	В том числе		% по всем а.з.
			БАЗ	МАЗ	
1.	Неисправность системы СУЗ	116	66	50	34,0
2.	Неисправность технологического оборудования	84	43	41	24,6
3.	Ошибочные действия персонала	60	34	26	17,6
4.	Неисправность электроснабжения	58	37	21	17,0
5.	Неисправность системы КИП	23	17	6	6,8
6.	Всего	341	197	144	100

ТАБЛИЦА IV. ПОКАЗАТЕЛИ РАБОТЫ РЕАКТОРА БОР-60 в 1970-74 гг.

Год	Время работы реактора на мощности (сут)	Коэффициент работы на мощности (%)	Отработано эффективных суток	Коэффициент нагрузки (%)	Эффективный коэффициент работы на мощности (%)
1970	273	75	50	13,6	81
1971	236	65	142	39,0	75
1972	242	66	176	48,3	70
1973	246	68	216	59,2	74
1974 до 30 июня	133	74	117	64,6	78
Всего за 4,5 года	1130	69	701	42,7	76

Нужно отметить, что за все время работы реактора не было ни одного случая остановки из-за неисправности оборудования I контура. Это говорит о высокой надежности основного оборудования реактора. Следует особо отметить высокую надежность работы основных натриевых насосов с гидростатическими подшипниками. За время эксплуатации не было ни одного случая поломки насосов I и II контуров. В обычном серийном оборудовании III контура (питательная вода и пар) возникало больше неисправностей, чем в оборудовании, работающем в среде натрия.

В табл.VI дана статистика срабатываний аварийной защиты с анализом причин срабатываний. За 4,5 года работы реактора произошло 39 срабатываний аварийной защиты, что значительно меньше, чем на

ТАБЛИЦА V. СТАТИСТИКА ОСТАНОВОК РЕАКТОРА БОР-60

№ п/п Причины остановок реактора	1970		1971		1972		1973		1974 по 30.06.74		Всего за 1970-74 гг.		
	число оста-новок	время (час)	число оста-новок	время (час)	число оста-новок	время (час)	число оста-новок	время (час)	число оста-новок	время (час)	число оста-новок	время (час)	% от суммарного времени
1. Плановая догрузка а.з. и ППР	2	1512	2	1372	2	2390	2	1780	1	988	9	8042	65,5
2. Остановки, связанные с экспериментальными работами	21	683	3	1086	6	501	8	793	2	133	40	3196	26,0
3. Неисправность оборудования II контура	-	-	2	242	-	-	1	48	-	-	3	290	2,4
4. Неисправность оборудования III контура	-	-	4	283	2	8	8	163	2	19	16	473	3,8
5. Неисправность системы СУЗ	7	35	3	59	2	2	2	20	-	-	14	116	0,9
6. Неисправность системы КИП	1	2	1	20	3	26	1	5	-	-	6	53	0,5
7. Неисправность внешнего электроснабжения	-	-	1	12	1	3	2	25	2	16	6	56	0,5
8. Ошибочные действия персонала	4	4	1	16	2	16	1	8	-	-	8	44	0,4
9. Всего	35	2236	17	3090	18	2946	25	2842	7	1156	102	12270	100%

ТАБЛИЦА VI. СТАТИСТИКА СРАБАТЫВАНИЙ АВАРИЙНОЙ ЗАЩИТЫ РЕАКТОРА БОР-60

№ п/п	Причины срабатывания а.з.	1970		1971		1972		1973		1974 по 30.06.74		Всего за 1970-74 гг.	
		МАЗ	БАЗ	МАЗ	БАЗ	МАЗ	БАЗ	МАЗ	БАЗ	МАЗ	БАЗ	колич.	%
1.	Неисправность системы СУЗ	3	3	2	1	1	1	2	-	-	-	13	33,4
2.	Неисправность системы КИП	1	-	1	-	2	1	1	-	-	-	6	15,3
3.	Неисправность оборудования II контура	-	-	1	-	-	-	-	-	-	-	1	2,6
4.	Неисправность оборудования III контура	-	-	1	-	-	-	2	-	1	-	4	10,0
5.	Неисправность внешнего электро-снабжения	1	-	1	-	1	-	2	-	2	-	7	18,0
6.	Ошибочные действия персонала	1	3	1	-	2	-	1	-	-	-	8	20,7
7.	Всего за год	12		8		8		8		3		39	100

реакторе БР-5. Особенно это заметно по срабатываниям аварийной защиты из-за неисправности системы СУЗ. Таких срабатываний произошло всего 13 раз, да и то половина в первый год эксплуатации реактора. Это говорит о значительном повышении надежности системы СУЗ реактора БОР-60 по сравнению с реактором БР-5, благодаря применению более совершенной аппаратуры и уделению большего внимания отстройки аппаратуры от наводок.

3. ТРЕБОВАНИЯ К НАДЕЖНОСТИ ОБОРУДОВАНИЯ, КАЧЕСТВУ МОНТАЖА И ПУСКО-НАЛАДОЧНЫХ РАБОТ

Эффективность работы реактора определяется прежде всего надежной работой оборудования, особенно труднодоступного для ремонта оборудования I контура. Опыт эксплуатации реакторов БР-5 и БОР-60 говорит о том, что при достаточном внимании к качеству изготовления и стендовой отработке натриевого оборудования можно добиться надежной работы быстрого реактора. Это хорошо видно на примере натриевых насосов. Низкая надежность работы насосов реактора БР-5 объяснялась неудовлетворительной работой подшипников качения в среде газа, содержащего пары натрия, что заставило перейти на электромагнитные насосы. Применение на реакторе БОР-60 насосов с нижним гидростатическим подшипником и верхним подшипником, изолированным от натриевой полости, существенно повысило рабочий ресурс насосов и эффективность работы реактора в целом.

Использование смазок, содержащих водород и углерод в уплотнениях и подшипниках натриевых насосов, ставит весьма остро проблему исключения попадания смазки в натрий. Так на реакторе БОР-60 имел место случай заброса масла из масляной системы циркуляционного насоса сначала в газовые, а затем и в натриевые коммуникации II контура [7]. В результате существенно увеличилось содержание углерода в натрии II контура, а также ухудшился коэффициент теплопередачи в промежуточном теплообменнике натрий-натрий.

Анализ возможных и гипотетических аварийных ситуаций, связанных с выбросом масла, указывает на то, что для перспективных установок целесообразно рассмотреть вопрос о разработке газовых уплотнений: газостатических или газодинамических.

Большое значение для надежной работы реактора имеет качество монтажа и пуско-наладочных работ. В связи с применением натриевого теплоносителя, в быстрых реакторах повышенное внимание должно уделяться обеспечению герметичности и чистоты контуров. Эти требования были выполнены при пуске реакторов БР-5 и БОР-60 благодаря осуществлению следующих мероприятий:

а) тщательная очистка оборудования, арматуры и трубопроводов перед монтажем;

б) отказ от гидравлических (с водой) испытаний контуров с заменой их пневматическими испытаниями;

в) контроль герметичности контуров путем проведения гелиевых и вакуумных испытаний;

г) сушка и обезгаживание контуров после монтажа путем разогрева штатным электрообогревом с одновременным вакуумированием;

д) тщательная очистка натрия перед его подачей в контур;

е) применение инертного газа с низким содержанием примесей. Эти требования должны выполняться при пуске и более мощных быстрых реакторов.

4. ТЕПЛОВАЯ И ГИДРАВЛИЧЕСКАЯ СХЕМА ЭНЕРГЕТИЧЕСКОГО РЕАКТОРА

Существенными особенностями тепловой схемы АЭС с быстрым реактором является сравнительно большое количество последовательно работающего оборудования, входящего в состав теплоотводящих петель I, II и III контуров, а также наличие разветвленной системы крупногабаритных трубопроводов с высокотемпературным активным теплоносителем. В теплоотводящей нитке, по которой тепловая энергия реактора передается к турбоустановке, имеется большое количество последовательно включенных звеньев: промтеплообменник – насос и трубопроводы I контура – насос и трубопроводы II контура – парогенератор – турбоустановка со своими системами. Отказ любого из элементов этой последовательной цепи приводит к выключению соответствующей теплоотводящей нитки или реактора в целом. Это обстоятельство заставляет с особой тщательностью подходить к вопросам выбора количества теплопередающих петель и однотипных элементов и взаимных связей между ними, а также предъявлять высокие требования к надежной работе оборудования. Наличие разветвленной системы натриевых трубопроводов выдвигает серьезные проблемы при проектировании. Сюда в первую очередь относятся вопросы герметизации жидкометаллических контуров, компенсации температурных расширений трубопроводов, а также обеспечение приемлемых условий работы узлов примыкания трубопроводов к корпусам реактора и оборудования.

Отмеченные выше особенности эксплуатационного и инженерно-конструкторского характера, а также экономика АЭС заставляют особое внимание уделять вопросам упрощения и повышения надежности в работе системы теплоотвода быстрых реакторов.

В ряде работ предлагался переход на двухконтурную схему: натрий в I контуре, пароводяной или газожидкостный цикл во II контуре. Для двухконтурных тепловых схем изыскивались специальные конструкции парогенераторов: с двухстенными концентрическими трубками; с передачей тепла через натриевый объем в корпусе, который пронизывают водопаровые трубки и трубки с натрием I контура. Обеспечение высокой степени герметичности парогенераторов в условиях больших давлений со стороны пара, подбор теплоносителя, который не вызывал бы опасных эффектов в случае взаимодействия с теплоносителем I контура и не оказывал опасного влияния на физику реактора, представляет собой пока сложную проблему. Поэтому все работающие и сооружаемые в настоящее время, а также проектируемые перспективные установки с реакторами на быстрых нейтронах используют трехконтурную схему теплосъема.

Избыточное давление внутри корпусов реакторов с натриевым охлаждением не превышает 0,5÷1,0 ати, что не препятствует собрать на площадке бак реактора большого диаметра с небольшой толщиной стенки и расположить в этом баке все оборудование I контура. Преимуществами интегральных конструкций реактора являются компактное расположение

I контура, исключение внешних трубопроводов с активным теплоносителем, снижение затрат на строительную часть. Интегральная компоновка оказывает определенное влияние на организацию движения потоков теплоносителя внутри бака, на количество и взаимные связи между компонентами тепловой схемы. Опыт проектирования показывает, что размеры бака существенно зависят от количества теплоотводящих петель и тепломеханического оборудования. Уменьшение их количества ведет к уменьшению размеров бака.

Были оценены сравнительные показатели надежности установки в зависимости от числа теплоотводящих ниток. В результате рассмотрения большого числа взаимозависимых факторов было получено, что для мощного энергетического реактора оптимальным числом теплоотводящих ниток является 3-4. Корректировку в сторону увеличения числа теплоотводящих петель может внести парогенератор в зависимости от продолжительности ликвидации последствий аварии (восстановление ПГ, очистка петли II контура, промежуточного теплообменника). Рассмотрения показали, что для технологической схемы с небольшим числом петель весьма важно, чтобы труднодоступные, склонные к выходу из работы элементы имели соответствующее резервирование в схеме – "горячий" резерв, т.к. иметь автоматический "холодный резерв" нецелесообразно из-за невозможности доступа и ремонта при работающем реакторе.

Анализ работы оборудования I контура показал, что таким элементом в интегральном реакторе может быть насос I контура – труднодоступный, работающий в напряженных условиях, имеющий динамические системы: подшипники, уплотнение, являющиеся причиной остановки насосов. В связи с этим гидравлическую схему реактора целесообразно строить таким образом, чтобы выход из работы одного из насосов I контура не затрагивал работу теплообменников, насосов II контура, турбоустановки и др.оборудования; остановка любого из агрегатов II или III контуров не должна приводить к автоматической остановке насосов I контура, поскольку это сопряжено с быстротечными процессами в активной зоне, с возникновением нежелательных эффектов реактивности, что может привести к остановке реактора и АЭС в целом. С этой точки зрения желательно исключить имеющуюся "жесткую" связь между технологическим оборудованием I контура. Сделать это можно за счет включения промтеплообменников и насосов на общий коллектор. В петлевом варианте компоновки I контура это связано с усложнением схемы, трудностями с разводкой трубопроводов и установкой дополнительной арматуры, что снижает надежность АЭС. Значительно проще решается этот вопрос в интегральных конструкциях реакторов.

Было проведено сравнение схем интегральных конструкций реакторов: с автономными петлями в баке и с включением насосов и теплообменников на общий коллектор. Эти схемы были проанализированы с точки зрения режимов работы реактора в системе АЭС и требуемых для этого дополнительных мер конструктивного и схемного характера, а также была предпринята попытка определить показатели надежности и коэффициент нагрузки АЭС для каждой из упомянутых схем. Для того, чтобы получить количественные показатели надежности анализируемых гидравлических схем, необходимо располагать данными по интенсивности отказов и времени восстановления, входящих в состав АЭС компонентов. Характеристики надежности энергооборудования были получены путем обработки данных по установкам других типов, а также однотипным или подобным

ТАБЛИЦА VII. ХАРАКТЕРИСТИКИ НАДЕЖНОСТИ ОБОРУДОВАНИЯ АЭС

№ п/п	Тип оборудования	Интенсивность отказов (1/час · 10^4)	Интенсивность восстановления (1/час · 10^2)
1.	Насос I контура	0,5	0,05
2.	Промтеплообменник	0,09	0,05
3.	Насос II контура	0,33	1,0
4.	Испаритель-экономайзер	0,5	0,5
5.	Пароперегреватель	0,28	1,0
6.	Натриевые задвижки	0,86	3,3
7.	Натриевые трубопроводы (петля)	0,28	0,5
8.	Турбина	1,0	0,5
9.	Электрогенератор	0,5	0,5
10.	Питательный насос	0,66	1,0
11.	Конденсатный насос	0,66	0,4
12.	Регенеративная система	1,43	0,4
13.	Вспомогательные системы	0,83	0,62

ТАБЛИЦА VIII. ПОКАЗАТЕЛИ НАДЕЖНОСТИ ИНТЕГРАЛЬНОГО РЕАКТОРА

Характеристика режима	Автономные петли в баке	Все теплообменники включены на общий коллектор
Продолжительность работы на 3-х насосах (полная мощность)(%)	70	70
Продолжительность работы на 2-х насосах (%)	27	27
Мощность установки при работе на 2-х насосах (%)	67	90
Продолжительность работы с одним насосом (%)	3	3
Мощность установки при работе с 1 насосом (%)	33	50
Расчетный коэффициент готовности АЭС (%)	89	95

элементам, используемым на теплоэнергетических установках, а по некоторым типам оборудования приняты произвольно. Используемые в расчетах характеристики надежности тепломеханического оборудования приведены в табл.VII.

Обсуждаемые типы схем достаточно проанализировать с точки зрения выхода из работы насосов I контура, поскольку на неисправности других компонентов обе схемы реагируют примерно одинаково. Показатели надежности интегрального реактора с тремя теплоотводящими нитками и одной турбиной в зависимости от типов обсуждаемых здесь тепловых схем представлены в табл.VIII. Из таблицы видно, что для установки с тремя теплоотводящими нитками и одной турбиной на реактор применение гидравлической схемы с включением всех теплообменников на общий коллектор позволяет увеличить коэффициент готовности АЭС примерно на 6%, что в условиях низкой топливной составляющей расчетных затрат позволяет уменьшить суммарные расчетные затраты на производство электроэнергии примерно на 4%.

Аналогичные расчеты были проведены для установки, в каждой из 3-х теплоотводящих петель которой есть своя турбина. Эти расчеты показали, что организация общей смесительной камеры на выходе из теплообменников – на входе в насосы для такой схемы является менее эффективной. Объясняется это тем, что большая доля простоев имеет место из-за выхода из строя турбин. Получаемое увеличение коэффициента готовности установки при такой схеме (1,5÷2%) лежит в пределах точности расчетов и не позволяет сделать однозначный вывод об использовании схемы с разомкнутыми теплоотводящими петлями в баке, если на каждую теплоотводящую нитку имеется своя турбина, что, собственно говоря, и определило использование такой схемы в АЭС БН-600.

ЛИТЕРАТУРА

[1] ЛЕЙПУНСКИЙ, А.И., КАЗАЧКОВСКИЙ, О.Д., ПИНХАСИК, М.С., Атомная Энергия 11 4(1961) 370.

[2] ПИНХАСИК, М.С., АРИСТАРХОВ, Н.Н., КАРПОВ, А.В., "Некоторые вопросы опыта эксплуатации реактора на быстрых нейтронах БР-5", Power Reactor Experiments (Proc. Symp. Vienna, 1961) 1, IAEA, Vienna (1962) 375.

[3] ЛЕЙПУНСКИЙ, А.И. и др., "Итоги восьмилетней эксплуатации реактора БР-5", доклад на симпозиуме стран СЭВ по быстрым реакторам, Обнинск, 1967 г.

[4] ЛЕЙПУНСКИЙ, А.И. и др., Атомная Энергия 31 4 (1971) 344.

[5] ЛЕЙПУНСКИЙ, А.И. и др., Атомная Энергия 21 6 (1966) 450.

[6] ЛЕЙПУНСКИЙ, А.И. и др., Атомная Энергия 30 2 (1971) 165.

[7] КАЗАЧКОВСКИЙ, О.Д. и др., "Основные результаты эксплуатации реактора БОР-60", доклад на симпозиуме стран СЭВ по быстрым реакторам, Обнинск, 1973 г.

DISCUSSION

R. Di SAPIA: If I understood you properly, there have been no sodium pump failures in BOR-60 in about 4.5 years of operation. If this is so, I wonder whether you would care to comment on this result and compare it with the one failure per year which, I believe, Mr. Procaccia of Electricité de France mentions in his paper (IAEA-SM-195/13), since this is the figure that has usually been adopted, up to now, for sodium pump failures.

Secondly, have you detected any sodium leaks in the BR-5 and BOR-60 plants? If so, could you tell us the order of magnitude of the sodium flow rate and comment on the consequences of the failures for plant operation and safety?

A.V. KARPOV: It is a fact that in the BOR-60 installation there has not been a single unplanned sodium pump stoppage in 4.5 years of operation. However, in our plant reliability calculations we have normally assumed one failure in two years for the primary circuit pump and one failure in three years for the secondary circuit pump.

Neither BR-5 nor BOR-60 has been affected by any large sodium leakages. There have been instances of small leaks involving a few litres of sodium as a result of which some damage, though not serious, was done to the electrical heating equipment and thermal insulation.

J.-C. ARLAUD: It is important to specify — and this again, I think, is a point that Mr. Procaccia makes in his paper IAEA-SM-195/13 — exactly what one means by an incident. On Rapsodie we have in fact had 26 incidents in 8 years on the 4 primary and secondary pumps taken together, which means an average of one failure per pump per year. But these were incidents involving not just the mechanical parts of the pumps but also the speed selectors, drive motors and associated electronics. In Rapsodie the mechanical part has been responsible, I believe, for only one incident. This shows how important it is to agree on precise definitions if we wish to make meaningful comparisons of our plants.

A.V. KARPOV: Yes. When I said that there had been no failures of the BOR-60 pumps I meant that there had been no failures of the electrical equipment associated with the pumps either.

IAEA-SM-195/16

A STUDY OF FAILURE CAUSES BASED ON U.S. POWER REACTOR ABNORMAL OCCURRENCE REPORTS

J.R. TAYLOR
Danish Atomic Energy Commission,
Research Establishment Risø, Roskilde,
Denmark

Abstract

A STUDY OF FAILURE CAUSES BASED ON U.S. POWER REACTOR ABNORMAL OCCURRENCE REPORTS.

Abnormal occurrence reports submitted to USAEC provide a detailed account of the types of failure treated in reliability and safety analyses. A study of the causes and consequences of failure was made, to support reliability analyses and to indicate the areas of most significance for the future development of techniques. The reports indicate that a large proportion of failures involve design, installation, and operation errors, and an unexpectedly large proportion of incidents involve multiple failures. It has also been possible to gain some insight into the nature and relative importance of different kinds of common mode failure.

INTRODUCTION

Abnormal occurrence reports from U.S. commercial power reactors provide a detailed and consistent source of failure incident data. For reliability analysts these records present a good opportunity for checking the assumptions built into their models, and to indicate those areas where further development is most useful.

In three separate studies, an attempt has been made to extract as much detailed information from abnormal occurrence reports as is possible. Information concerning cause of failure, the number of failures detected, and the relationship between several different failures in a particular occurrence was extracted from the records and classified. In many instances classification required a cross check between occurrence reports and safety analyses reports.

The first study concerned just 'design errors', and included data for all BWR abnormal occurrences during one year. The second study involved all abnormal occurrences for just two reactors during one year. The third study concerned five reactors, all available abnormal occurrence reports being classified, so that some indication of the evolution of error causes and/or reporting practices could be obtained.

'Abnormal occurrences' generally involve incidents at a fairly low level of risk. To provide a cross check on results, incidents reported in Reactor Operating Experiences for one year were also classified, on the assumption that only the 'more serious' incidents would be reported here. In practice ROE reports include incidents involving relatively high risk potential, and also cases of exceptionally frequent failure, where often the risk involved was not so great.

In all 513 separate incidents were analysed and classified involving approximately 700 separate failure reports and 1012 separate component failures.

Before starting the study considerable effort was placed on developing objective criteria for classification of failures. Nevertheless, it was necessary to broaden and refine the classifications in the later studies, particularly to include more information about the causal structure of incidents, and about common mode failure.

CLASSIFICATION OF FAILURE CAUSES

A failure of a pump may be due to an installation error; the error may be due to lack of information on the part of the installer; the lack of information may be due to organizational defects. It is generally possible to trace long chains or 'trees' of failure causes in this way.

In this study a 'failure' is a failure of a component to function correctly after it has been installed and has begun to be used. The number of 'failures' observed depends on how finely the plant is divided into components. The level of classification used here was the highest level of detail, based on the available information. This point is important especially in discussing common mode failure, because one large component may fail as a result of failure in several of its sub-components.

The causes of failure in this study have been traced at one or two levels, giving the classification shown in Tables I and II. A combination of two or more causes could be traced for 10% of the individual component failures, and in these cases a 'half failure' was attributed to each cause class (contrast this with the approach used in Cranston Research Inc.'s report of 1972) [1].

In classifying failure causes, three groups give difficulty and require special definitions, that is, design, procedural and operator errors.

TABLE I. DISTRIBUTION OF FAILURE CAUSES, FOR 490 FAILURES TAKEN FROM ABNORMAL OCCURRENCE REPORTS
(See also Ref. [3])

Design		36%
of which component selection	5%	
other	31%	
Component failure		19%
Maintenance or installation error		12%
Operator error		11%
Procedural or administrative error		10%
Fabrication error		0.8%
Unknown or unrecorded		11%

TABLE II. DESIGN ERROR SECONDARY CAUSES, FOR DESIGN ERROR FAILURES TAKEN FROM ABNORMAL OCCURRENCE REPORTS
(See also Ref. [3])

S	component selection	16%
O	oversight	15%
U	error due to effect unknown at design time	23%
Z	sizing, dimensioning error	12%
C	error due to lack of recognition of complex system interactions	11%
K	error due to communication problems	1%
?	error with cause unknown or unrecorded	22%

A failure is classified as arising from a design error if

(a) The abnormal occurrence report ascribes failure to a design error.
(b) The frequency of failure is very high, several times per year for the same component.
(c) The plant design is changed as a result of the failure.

This definition allows an objective classification and avoids problems in classifying failures involving completely novel phenomena.

A failure is attributed to procedural error if standard procedures are changed after a failure. An operator error is recorded if an operation error is described in the abnormal occurrence report, and no change in operating procedure is described in the report.

For the purpose of deriving a picture of cause distributions, common mode failures involving several components are regarded as single failures.

As can be seen from Table I, the largest proportion of failures arise from design errors. However, a large number of such errors involve misapplication or selection problems with instruments, seals, bearing materials etc. It is worth while to separate out such failures. They are of a different character from other errors in their effects. And there is a direct solution to such problems, in terms of component and materials testing prior to use.

Design error failures were further classified as shown in Table II with the objective of discovering the areas in which improved design procedures could be effective. A large number of design errors involve problems which were unrecognized or unknown at design time.

'Component failures' of Table I are the traditional kinds of failure treated by reliability theory, that is random failures, due to wear, corrosion, material defects etc.

Operator errors were classified further according to their effect on the plant, because the type of error has a significant bearing on reliability analysis procedures. Of forty operator errors which could be classified 55% involved omission of a single operation or procedure affecting one component. 30% involved omission of operations or misoperations involving several components all of the same type. And 15% involved execution of a wrong procedure, or misdiagnosis of a situation leading to wrong actions, which involved several different types of component.

An attempt was made to discover whether there was any trend in the kinds of failure, either as individual plants became older, or from plant to plant. The data for five reactors were plotted for each six month period from reactor start up. Although records were available only since 1969, the difference in start up dates for the five reactors gives a coverage of twelve years of plant life. No significant variation in the proportion of various failure causes could be detected, although the older plants had fewer failures.

No significant differences in distributions of causes could be detected for the ('more serious') Reactor Operating Experiences failure data, except perhaps a higher proportion of design errors involving complex system interaction, and a very low proportion of design errors involving component selection problems.

A small proportion of failures could not be attributed to a single component, but involved a whole system of components. As mentioned earlier, some 15% of operator errors were of this type. In addition, out of a total of 1012 failures, there were six cases of wiring errors and sequence control system design problems; four cases of water hammer: and five cases of control system oscillation, which were of the kind which are difficult to discover in any systematic way in failure mode analysis.

MULTIPLE FAILURES

Abnormal occurrences involving three or more failures are much more frequent than one would expect from a naive application of fault tree analysis methods with constant failure rate data. As Fig. 1 shows, six-fold, and seven-fold failures are almost as frequent as four-fold failures. In arriving at the histogram of Fig. 1, failures which were a direct consequence of earlier failures, such as motor failure due to power supply failure, were ignored. Also, common mode failures involving components of identical type were regarded as single failures. And only 'actual' failures were recorded, failures found during testing being ignored. ('Actual' failures are failures found during operation, not by testing.)

Individual failures were classified as spontaneous, gradual, latent, consequent, or 'misoperation'. Spontaneous and gradual failures are those which initiate a failure sequence, the 'gradual' classification being introduced to cover problems such as slow leaks, or corrosion, where effects build up gradually.

Latent failures are those which develop in inactive components, and are 'revealed' when the components come into use. The earlier term, 'unrevealed faults' was discarded, because there are often several different ways in which latent failure may be discovered. By far the largest proportion of failures in abnormal occurrence reports involve latent failures discovered during testing. However, even for 'actual' incidents over half the failures were 'latent'.

Consequent failures are those which arise as a direct consequence of some earlier failure, where there is no latent defect in the affected component. Consequent failures can further be classified as 'command

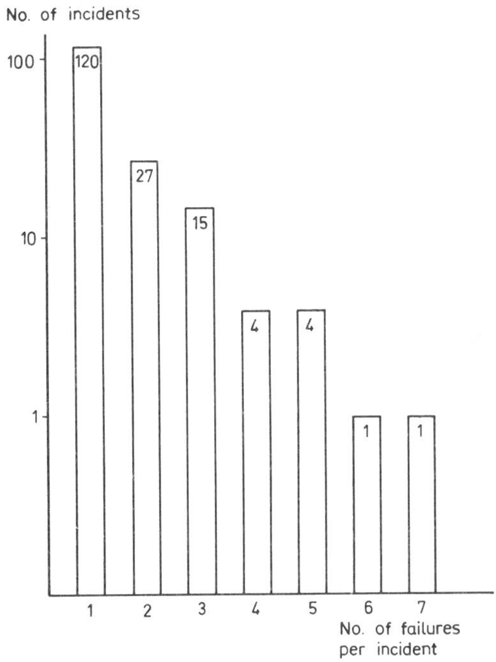

FIG.1. Multiple failure incidents for 172 'actual' abnormal occurrences.

failures' (resulting from failure of inputs such as power or control signals); secondary failures (caused by flood, fire, broken rotor missiles etc.); overload failures (failure of one component places an excessive load on others in parallel); and failures in redundant channels due to failures in switching to standby components.

All but secondary consequent failures were ignored in collecting data, in order to concentrate on the probabilistic rather than the deterministic aspects of failure analysis. Of the secondary failures recorded, two involved steam impingement, two involved water hammer, a sequence of several problems at one plant involved corrosion, which may have been due to incorrect chemical dosing procedures. Two secondary failures arose as a result of lightning strikes. Twenty-four secondary failures involved sticking or jamming of control rods and valves due to loose parts in oil, air and water flow streams.

Misoperation failures are those which arise when a component fails during an operation in spite of being in good condition at the start of the operation; or an error on the part of an operator in carrying out a single operation. In principle, misoperation probabilities should be relatively insensitive to the time between operations, when compared to latent failures. In 47 multi-failure incidents (The 'five-plant' study), with 131 individual failures 13% were due to misoperations, all operator errors.

It is, perhaps, self-evident that multi-failure incidents will tend to be more serious than single-failure incidents. In Fig. 2, the numbers

FIG.2. Multiple failure incidents — 'actual' failures from ROE bulletins 1972.

of failures in 26 'actual' occurrences drawn from Reactor Operating Experience bulletins are shown. They show even more markedly the high proportion of multiple failures (the 'flatness' of the failures per incident histogram). The main explanation for this effect seems to be that some initial failures create unusual plant circumstances and subject a wide variety of components to situations and conditions not experienced in testing.

COMMON MODE FAILURE

'Common mode failure' can be defined in many ways. Rasmussen et al. [2] write

> 'Common mode failures can be defined as multiple failures which are dependent. The multiple failures are common mode, or dependent because they result from a single initiating cause. The initiating cause can be any one of a number of possibilities: a common property, a common process, a common environment, or a common external event. The multiple failures which are dependent can likewise encompass a spectrum of possibilities: multiple system failures caused by a common component failure, system failures caused by a common external event, multiple component failures caused by a common defective manufacturing process, a sequence of failures caused by a human operator etc.'

Because the phrase 'common mode failure' can mean many things to different people, the term 'coupled failure' is used here.

Coupled failures are defined as simultaneous failures of several components as a result of a common cause. By 'simultaneous' is meant that the components are in a failed state at the same time.

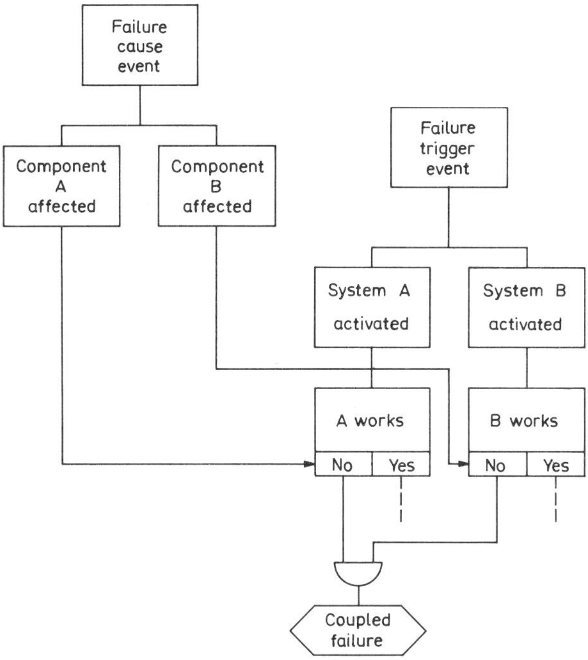

FIG.3. A typical 'similar component' coupled failure.

One should note that failure frequency is dependent on the degree to which a system is divided into components.

This definition of coupled failure is very broad. It includes what were called 'consequent failures' in the previous section.

It is desirable to obtain data which can be used in estimating coupled failure probabilities. The need for data arises especially in establishing levels of redundancy and diversity. In this study interest was focussed on coupled failure in similar components. (See Fig. 3.)

Detailed common mode failure data were collected for five reactors, using all of the available abnormal occurrence reports. Of 379 component failures or groups of failures arising from independent causes, 78 involved common mode failure of two or more components. Most such failures were detected during testing.

In addition there were five sequences of 'repeated' failures in which several components failed repeatedly as a result of a common cause. In one case there were fifteen groups of 'simultaneous' failures, giving a total of thirty-eight individual failures for one group of four components.

There were twenty-four cases in which just one component failed but in which a common cause was involved which affected several components to the extent that repair was required.

Depending on how these results are interpreted, the percentage of coupled failures was between 18.5 and 34%.

Coupled failure statistics are only of use, however, if probabilities for coupled failure can be given for particular component types. Data were collected for particular component types in those cases where sufficient data were available to give a meaningful result.

The coupled failures recorded could be classified into four distinct types, on the basis of the time at which the original failure 'cause' occurred, and the time at which the failure was revealed.

(1) Failures due to design, installation errors etc. which cannot be detected by testing, but which reveal themselves under unusual conditions.

(2) Failures resulting from effects which greatly increase failure rates, so that there is a high probability that several components will fail between activations or tests.

(3) Failures arising from misadjustment during maintenance, or mis-operation, of a group of components, so that all fail shortly afterwards. For this type of failure, 'cause' and 'trigger' are the same event.

(4) Failures arising from effects which give an unusually high failure rate in operation, so that a second failure occurs while a first failure is being repaired (only one instance of this type was present in the occurrences studied).

Many types of coupled failure involve an original failure cause (e.g. design error) and a trigger event (e.g. unusual mode of use), so that a coupled failure can occur immediately. In such cases, the frequency of coupled failure, for situations in which two components are involved can be expressed approximately as

$$F_{CF} = F_A P_S P_F P_F$$

P_S is the a priori probability that the particular components are susceptible to a common failure cause. F_A is the average frequency with which the common failure mode is activated or triggered. P_F is the probability that a single component will fail, given that it is susceptible and that a triggering event occurs.

The form of expression is also appropriate for failures in which the trigger (e.g. vibration, corrosion) works continuously while the component is in standby, where the failure can be detected by testing (Type 2 failures). Here F_A becomes test frequency, and P_F becomes the probability that a single component fails within the test period.

For type 4 failures, a different model is required.

For continuously operating failure 'triggers', the frequency of coupled failure will vary depending on test frequency. In practice, the distinction between continuously operating and intermittent instantaneously operating triggers, is not so clear cut. For example, the problem of environmental sensitivity of components may be important over a short period of days during the winter (see Fig. 4).

Typical values for F_A, P_S and P_F are given in Table III for different types of failure. Proportions of different failure types are shown in Table IV,

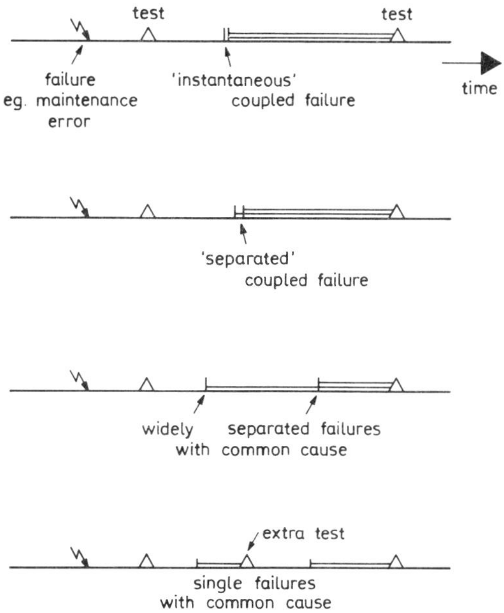

FIG.4. Test interval dependence of coupled failure frequency.

TABLE III. TYPICAL VALUES FOR PROBABILITIES DETERMINING COUPLED FAILURE FREQUENCY

	P_S	F_A	P_F
Type 1	< 1	$\ll 1$ per year	≈ 1
Type 2	$\ll 1$	12-14 per year	< 1
Type 3	≈ 1	1	< 1

TABLE IV. PERCENTAGES OF DIFFERENT COUPLED FAILURE TYPES, FOR 83 COUPLED FAILURES IN FIVE REACTORS

Type 1	Type 2	Type 3
16%	77%	7%

TABLE V. NUMBER OF 'COUPLED' FAILURES IN RELATION TO NUMBER OF SINGLE FAILURES (FROM ABNORMAL OCCURRENCE REPORTS FOR FIVE REACTORS)

Component type	Failure types			
	Type 1	Type 2	Type 3	Single failure
Flow switch	1	4		2
Control rod (jam on withdrawal)	2	6		5
Motorized valve	1	2		21
Timer relay	2	2		
Main steam isolation valve		4		15
Pressure switch (drift)		11	1	12
Solenoid valve		4		7
Relief valve		3		3
Relay		3		16
Generator		4		14
Sequence or logic controller	5			

for 83 coupled failures (the 'repeated' failures were included as single instances of coupled failure). It is generally difficult to distinguish between intermittently acting instantaneous, short period continuous acting, and constant continuous acting failure triggers. However, over 50% of the coupled failures were clearly instantaneous in their effect. Type 3 failures are necessarily instantaneously triggered and type 4 failures are necessarily continuously triggered.

Type 1 failures involve a wide spectrum of effects in design, installation, and environmental influences. It seems doubtful whether any statistical basis for type 1 coupled failure probability estimates can be established. The best that can be achieved seems to be to provide some bound on a priori probabilities, based on experience. For type 2 failures, it appears feasible to establish coupled failure rate values, with reasonably narrow confidence intervals, for some types of components.

Table V gives the number of coupled failures, for those components for which a reasonably high number of coupled failures were found.

COMMENT

The objective of the various studies of abnormal occurrences was to provide a basis for development in reliability and failure analysis. Particular problems were investigated from this point of view. Definitions and classifications were made in order to study failure causes, rather than to compare particular power plant components.

Although the techniques used were quantitative, the conclusions must be qualitative. The quantities of data are insufficient to allow confident estimates of failure statistics.

The main features to be noted from the study of failure causes is the large number of varied failure mechanisms involved, and the importance of a priori factors such as design error probability. One would not expect there to be detailed agreement between practical component failure rates for a particular plant and average failure rates for many plants.

The large proportion of failures due to design errors emphasizes the value of component qualification programmes and standardization. There are some types of design errors involving systems rather than component design, for which failure mode analysis is extremely difficult. (For example, errors in control systems design.) Such errors play a minor, but not insignificant role in abnormal occurrences, and indicate the value of special techniques for their removal.

With respect to human errors, techniques designed to treat 'single operation' errors are clearly valuable, but need to be supported by studies of the effects of procedural and 'decision' errors affecting many components.

The data for multi-failure incidents indicate the value of the 'event tree' technique in dividing failure analysis problems into separate stages. Event tree techniques allow situations with relatively large numbers of independent failures to be analysed, and it appears that such situations occur frequently in practice.

For the author one of the most useful aspects of the study was to shed some light on the meaning of 'common mode failure'. The definition 'simultaneous failure of several components due to a common cause' raises the problem of how close together failures should be to be 'simultaneous', and how the failures may be causally related. The objective of the definition is to isolate those cases where the probability of double failure is significantly greater than the product of single-failure probabilities.

In the question of how close together failures should be before being considered 'simultaneous', a coupled failure was recorded if two components were in a failed state at the same time. It is perhaps worrying to some that by defining common mode failure in this way, virtually all double failures are classified as common mode failures. It would perhaps be preferable to collect statistics concerning all double failures (not just those arising from a common cause) and in this way arrive at empirical values for the probability of double failures, irrespective of classification, definition, etc.

For the purposes of this study, common mode or 'coupled' failures were grouped first separating out those situations where one failure was a direct consequence of another failure, and then focussing attention on coupled failures in similar components. Similar component coupled failures were further classified into those which are detectable during normal operation or testing, those revealed only during special operating circumstances, and those arising directly from misoperation, misadjustment etc.

It seems possible to collect statistical data for some kinds of coupled failure in some kinds of component. For other classes of coupled failure involving problems not generally detected during testing, and with a low a priori probability of occurrence, the prospects of acquiring adequate statistical data seem slim. Fortunately such failures seem less frequent than other types.

There seem to be sufficient series of 'repeated' failures in practice, to warrant the use of reliability models incorporating 'imperfect repair'.

REFERENCES

[1] Cranston Research Inc., Manned Space Program Accident/Incident Summaries 1963-69, NASA-CR-120998 (1970).

[2] RASMUSSEN, N.C., et al., An Assessment of Accident Risks in Commercial Nuclear Power Plants (Draft) WASH-1400 (1974).

[3] TAYLOR, J.R., Design Errors in Nuclear Power Plant, RISØ-M-1742 (1974).

DISCUSSION

R. SARTORI: One of the design bases of safety systems is the single-failure criterion, but you appear to have been concerned with large numbers of coupled failures. Do you think that we ought to change from the single-failure to two- or three-failure criteria?

J.R. TAYLOR: Both the information concerning coupled failures and the information concerning multiple failures is relevant to this question. The true choice is not between a single-failure criterion and n-fold failure criteria, but between these criteria and probabilistic studies. The advantage of the single-failure criterion is that it is very simple to apply; but a designer must be prepared to use all techniques and criteria if he wishes to produce high-reliability systems.

K. EFFAT: What is the source of the figures used in your analysis? Are the percentages calculated figures or are they based on practical experience?

J.R. TAYLOR: The figures which I showed are derived from the abnormal occurrence reports submitted to the American NRC by plant managers and operating engineers. I would like to say that we should be grateful for this information which the American engineers make available, as it is among the most useful sources of reliability data available. The responsibility for classification and interpretation of this information is our own.

METHODS AND TECHNIQUES
OF RELIABILITY ANALYSIS

Chairman:

D. OKRENT, United States of America

REDIS, A COMPUTER PROGRAM FOR SYSTEM RELIABILITY ANALYSIS BY DIRECT SIMULATION

H.E. KONGSØ
Danish Atomic Energy Commission,
Research Establishment Risø,
Roskilde,
Denmark

Abstract

REDIS, A COMPUTER PROGRAM FOR SYSTEM RELIABILITY ANALYSIS BY DIRECT SIMULATION.

The REDIS program is based upon simulation of the direct type (no variance reduction). The program can be used for any system in principle, and it is particularly designed for standby systems taking into consideration the various kinds of errors that can cause failure during both operating and standby conditions. The program is described with emphasis on the particular aspects of the simulation models, illustrating the ability of the program to consider details of systems and its flexibility. The different kinds of faults that occur in practice are described and categorized. The main categories of faults are STOPPAGE faults, which can lead to stoppage of operation, and BLOCKING faults, which can cause failure to start on demand. Various types of faults can occur within each of the main categories, the two main types being the time dependent and the activation dependent types. The method used to incorporate the various kinds of faults, dependencies between faults and uncertainties in the data, are described. In standby systems, groups of components have common operation and standby periods. A description is presented of the model used for simulation of STOPPAGE and BLOCKING of such groups. The program includes a model for the simulation of periodic testing of specified groups of faults, and for renewal of worn-out components. An example calculation on a simple system is presented, including input and output. The experience with the program is summarized.

1. INTRODUCTION

In order to perform meaningful reliability analyses of systems it is mandatory not to overlook any detail of component and system behaviour of importance.

A computer program, REDIS, based upon simulation was developed for the purpose of the detailed reliability analysis of systems.

It is the primary purpose of this paper to present a description of the special features and applications of the REDIS program. A detailed description of how the program works is given in another report (RISÖ-M-1781).

The program makes possible realistic simulation with models, taking into account such details as standby, repair, failure during start, failure during operation, failure during switching to a standby unit and periodic testing.

Unfortunately, any program providing detailed simulation is necessarily slow. However for large, complicated systems the analyses can often be carried through without losing any detail, by analysing complicated sub-systems separately. Afterwards the entire system can be simulated by a faster model, in which the sub-systems are blocks, having failure characteristics, found by the preceding sub-system analyses.

Although the program is designed to analyse several detailed aspects, it cannot be equipped in advance with models to analyse every aspect of any system. Emphasis has therefore been placed on flexibility throughout the program in order to facilitate the modelling to the large variety of details in the design and operation of a general system.

The programme calculates a series of failure characteristics for the system, as seen from the example of program output, shown in Fig. 5. The program is written in FORTRAN IV, and it has been used on both the IBM 370/165 and Burroughs B6700/7700 computers.

2. PARTICULAR ASPECTS OF THE SIMULATION MODELS

The following particular aspects of the simulation models are presented in order to give an illustration of the two principal features of the program: its ability to consider details of systems and its flexibility.

2.1. Simulation type

The simulation type used is direct simulation. With this method, models are based upon the original failure data. With other types of simulation a technique is used to reduce the variance of the results. This is done by distorting the failure data into areas of more frequent system failures, followed by a correction of the results for the data distortions.

The direct simulation is much slower than simulation with variance reduction, but it offers a much greater flexibility due to a much simpler modelling procedure.

2.2. Incorporation of fault categories

Studies of failures occurring in practice reveal several different types of faults that can occur in complex systems. Too often, however, one finds that a well established terminology for a given fault type does not exist.

In the following an attempt has been made to categorize the faults and to establish new terms, where such did not already exist.

The concept of fault must in this paper be conceived as covering also the concept of 'inactive' condition, for example a component or a sub-system being tested.

The specification of fault data is done by means of two matrices of input data, one for fault indicators and one for fault data. One row of indicators and one row of data are specified for each fault. The indicators are used for qualitative specifications like type of fault and relations between faults, while the data provide information on the frequency with which the faults occur.

2.2.1. Consequence categories

All faults are divided into the following two main categories, according to the possible functional consequences of their occurrence: stoppage faults and blocking faults.

The specification of the particular functional consequences for the system or the sub-systems affected by these two categories of faults takes place in a special sub-routine of the program (FAILSY) as described in Section 3.

2.2.1.1. Stoppage faults. This kind of fault causes failures of equipment in the condition of operation that can lead to stoppage of operation. An example of this kind of fault is a failure of a bearing in a running pump, which can cause an overheating of the bearing and a manual or automatic stoppage of the pump.

Depending upon the nature of the mechanisms that cause this kind of fault, stoppage faults may occur during periods of operation or during periods of standby.

It is also possible to establish a model for faults having different probability density functions during periods of operation and standby.

2.2.1.2. Blocking faults. This is a kind of fault which can cause failure of equipment to start on demand, i.e. a blocking. The blocking faults can occur during periods of standby or during periods of operation.

2.2.2. Fault types

Within each of the two main 'consequence' categories, described in Section 2.2.1., several different types of faults can occur. In the following sub-sections these different faults are categorized according to failure probability dependence and the way in which the fault is revealed.

A review of the various fault types is shown in Table I. As can be seen from this Table all combinations of failure probability dependence and failure revealing categories are possible and can be taken into account in this program.

TABLE I. FAULT TYPES

Fault probability / Fault revealed	Depends on operating time	Depends on standby time	Depends on total time	Depends on no. of activations	Depends on no. of activ. and time
Immediately	+ *)	+	+	+	+
By test or start attempt	+	+	+	+	+
Only by start attempt	+	+	+	+	+

*) + = possible combination

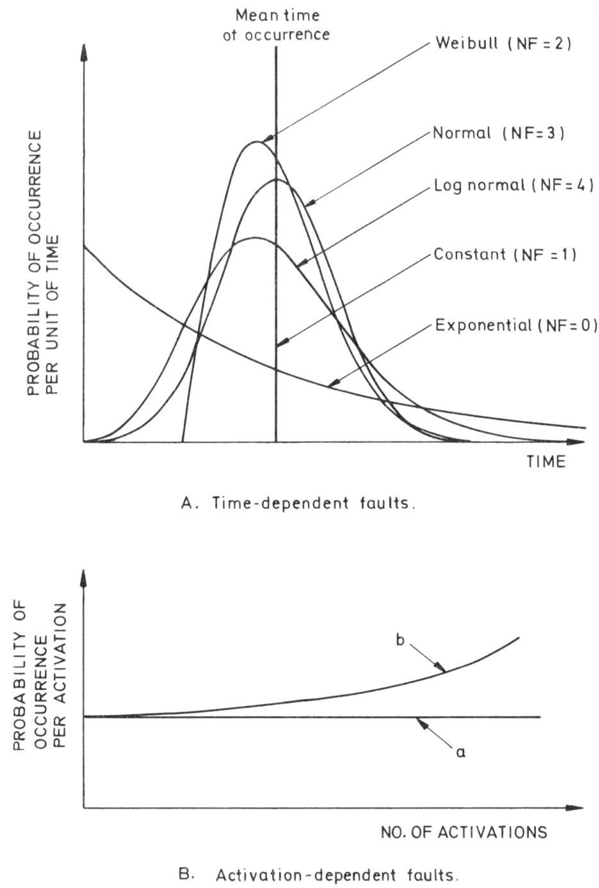

FIG.1. Time-dependent and activation-dependent faults.

In the following each of these different fault types are described in a separate sub-section.

2.2.2.1. Time-dependent faults. Time-dependent faults are faults caused by continuously working failure mechanisms. The time-dependent faults can thus be characterized by probability density functions for the time to occurrence, as illustrated in Fig. 1A.

The program can accommodate any time-dependent fault for which the times to failure can be expressed as a probability density function, either by giving parameters for standard functions or by defining the functions pointwise.

Fig. 1A shows the probability density functions for the times to failure, presently incorporated in the program. Also shown on this figure are the indicator values applied for each of the probability density functions.

Repair times are simulated using similar types of probability density functions, as shown in Fig. 1A.

2.2.2.2. Activation-dependent faults. Activation-dependent faults are caused by failure mechanisms that work when starting or stopping of equipment takes place. Activation-dependent faults can thus be characterized by means of a certain probability of occurrence per activation, independent of time, as shown in Fig. 1B.

In many cases a model is applicable where the probability of occurrence of the fault is independent of the number of activations (curve a, Fig. 1B). In other cases the most realistic model is based upon a probability of occurrence per activation, which is an (increasing) function of the total number of activations since installation or overhaul (curve b, Fig. 1B). Both these models can be used in this program since a continuous record is kept, as the program runs, of the total number of activations for each component since overhaul. Where appropriate, models where the probability of occurrence is a function of both the number of activations and of time can be applied. This is done by incorporating the mathematical expression for the probability of occurrence as a function of the number of activations and time, into the sub-routine of the program where the simulation of activation dependent faults takes place.

In some cases activation dependent faults must be simulated by models where repair is relevant. This is for instance the case if it is a failure of a switch as illustrated in Fig. 2A, where the fault condition is considered to prevail during a certain period (dead time) after appearance. In other cases repair is not relevant, for instance in the case of an operator error.

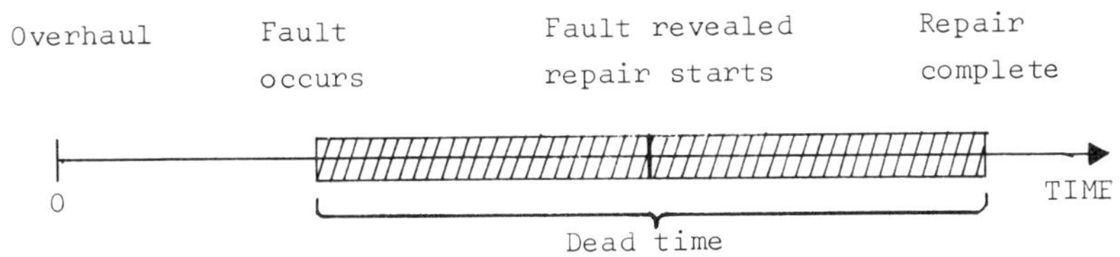

A. Switch - type of fault. The faulty condition exists within a certain period (dead time).

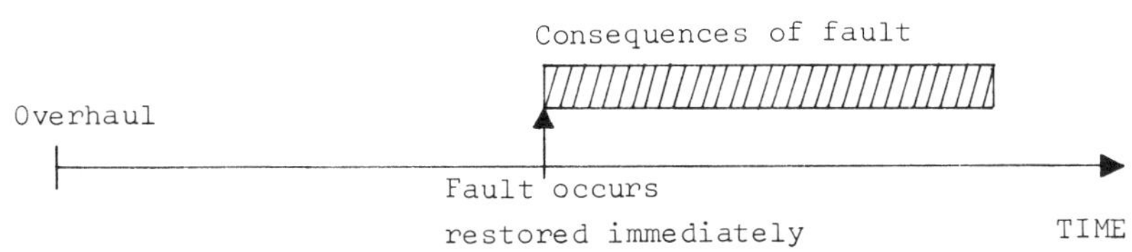

B. Operator-error type of fault. No dead time.

FIG.2. Two kinds of activation-dependent faults.

In this case the fault consists of an erroneous action at a certain time, and the faulty condition disappears immediately (Fig. 2B), but of course the consequences of such a fault could be failure of a component or a subsystem, lasting for a certain period, as indicated in the figure.

Where appropriate, simulation of the activation-dependent faults takes place at those moments of the program run when the switching-on or -off of the corresponding component group takes place, as described in Section 2.3.

2.2.2.3. Faults revealed immediately. This is the kind of fault that appears to the operating personnel immediately after the occurrence, because the consequences are apparent either directly or via a control system.

This kind of fault is indicated in the program by means of a positive sign of a particular fault indicator, and this will cause a repair simulation to be initiated at the time of occurrence of the fault.

2.2.2.4. Unrevealed faults in components subject to tests. This is a kind of fault that does not appear to the operating personnel in any way at the time of occurrence, but the faulty condition remains 'latent' until discovered by a test or on demand of the operating unit concerned.

In the program this kind of fault is indicated by means of a negative sign of a particular fault indicator, and repair simulation is not initiated until the appropriate time for routine testing or until component activation.

2.2.2.5. Unrevealed faults in components not subject to tests. This is a kind of fault that does not appear to the operating personnel at the time of occurrence. The faulty condition remains 'latent', until discovered by a demand of the operating unit concerned. This kind of fault is also characterized by a negative sign of the above fault indicator, and repair is only initiated when the fault has been revealed by an attempt to activate the corresponding operating unit.

2.2.3. Interactions between faults

The following two different kinds of interactions between faults can be accounted for:

2.2.3.1. Interdependent faults. A model is included to account for interdependencies or couplings between faults. The model simulates the situation, where the occurrence of one fault (fault 1) influences the probability of occurrence of other faults (the dependent faults). The model works this way:

Two different probability density functions are provided for each of the dependent faults. If a fault occurs on which their probability of occurrence depends (fault 1), simulation of the dependent faults is switched from one probability density function to the other. Similarly, for activation-dependent faults, the probability of occurrence of dependent faults is switched from the normal value to another value.

The present model implies that as soon as fault 1 has been repaired, the simulation of the dependent faults will again be based upon the 'normal' data. However, the model can be modified so that the return to the 'normal' data does not take place in cases where that seems unrealistic. Although the mechanism for representing coupling used in the program is specific to one kind of failure dependence, it can be used to approximate a wide range of dependencies.

2.2.3.2. 'Cascade' faults. A model is included to take into consideration faults that do not occur spontaneously but, with a specified probability, as the result of the occurrence of another, 'originating' fault. This facility is very convenient, because it removes the need for 'unrealistic' modifications to the fault tree models.

2.2.4. Uncertainties in data

The most widely used probability density function for time-dependent faults is the exponential distribution. In most cases however, due to uncertainties in the origin of the data or specifications for the systems to be analysed,the most realistic way to describe the failure rate is by giving a distribution for failure rate rather than a single figure. That is, one provides a probability density function for a probability density function parameter. A model is included for exponentially distributed times to failure, where probability density functions of the normal, log normal or Weibull types are superimposed on the failure rate to account for uncertainties in this parameter.

For similar reasons a model is included for the activation-dependent faults, to account for uncertainties in the data. The model can consider activation dependent faults where the probability of occurrence per activation is defined by probability density functions of the normal, log normal or Weibull types.

2.2.5. Misoperation

A special kind of fault could adequately be designated as misoperation.

This kind of fault often attracts attention in safety analyses. It can be characterized as a wrong action, in other words an occurrence of an alternative action to the desired one.

Examples of this kind of fault are: an operator presses button A instead of button B, and a switch closes contact 4 instead of contact 3 on activation.

Faults of this kind can also be included into calculation models in this program. Depending on the circumstances they must be considered as stoppage faults or blocking faults of the activation dependent type.

2.3. Simulation of system performance

2.3.1. Grouping of faults

In systems comprising standby units, series of components have common operating and standby periods. An example is a system consisting of two feed pumps, one of which is operating and one is standby, each feed pump unit comprising pump, motor, valves, power supply etc.

To make a realistic simulation of such systems, a grouping of the corresponding faults is established by means of a common group number indicator (see sample calculation, Section 3).

On the basis of a failure-mode-and-effect analysis of the system to be analysed, the consequences of faults and combinations of faults for the operation of each group and for the entire system are specified. These specifications of fault consequences are written as boolean expressions in one of the sub-routines of the program as shown in Section 3. Included in the specifications are boolean expressions for the faults or combinations of faults that will cause stoppage and blocking of each individual group by means of logic parameters SBD and SBB, respectively.

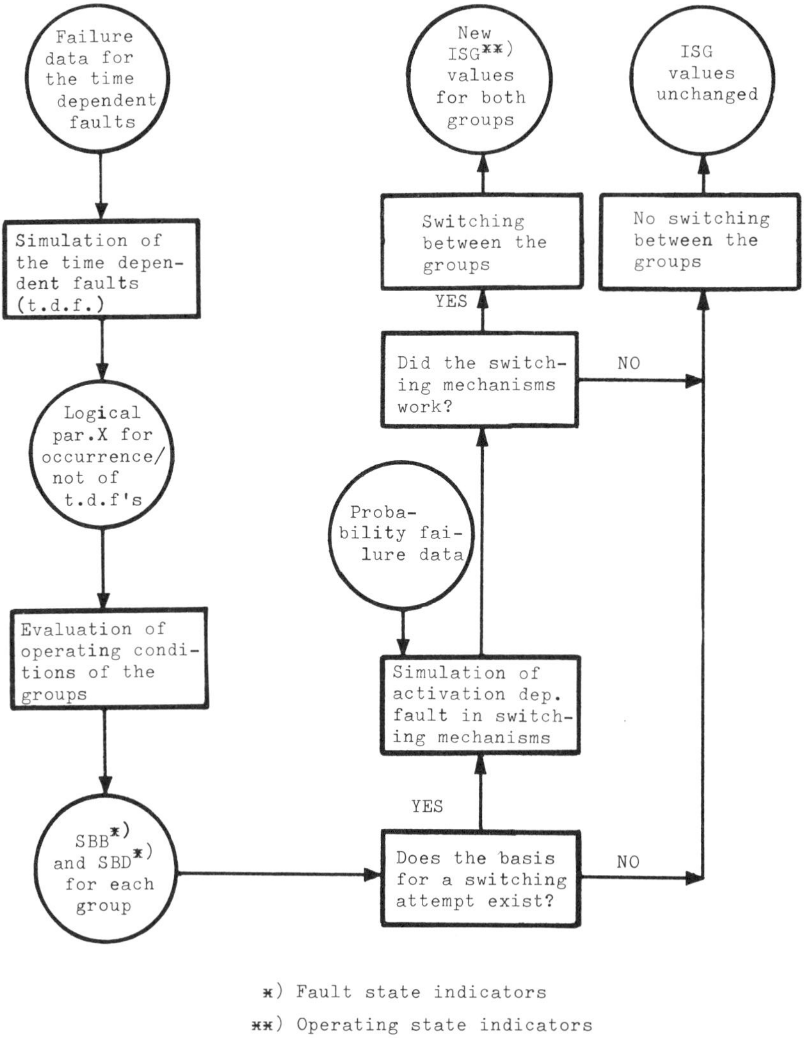

FIG.3. Data flow in the model applied for switching between standby groups.

A continuous record is being kept by means of the indicators, ISG, of which of the groups are operating and which are not. Switching between groups is simulated, when relevant, based upon evaluations of the conditions for stoppage and blocking of each individual group.

The grouping of faults represents a great simplification in the specification of models for complex systems with standby groups. The data flow in the model applied in the switching operations is illustrated in the diagram shown in Fig. 3.

2.3.2. Testing

Periodic testing is a well-known way of improving the reliability of standby equipment. Two different kinds of periodic testing can be simulated by this program. Both kinds of testing involve initiation of repair of the unrevealed faults that might exist in the components being tested.

The first of the models involve a periodic switching between operating groups and the corresponding standby group. The other models for testing involve testing at specified intervals of time of specified groups of faults.

2.3.3. Renewal of worn-out components

In connection with preventive maintenance of practical systems a renewal is carried out of such components that are revealed as being close to wear out.

A model is included to enable a simulation of this kind of maintenance.

3. SAMPLE CALCULATION

In the following section a sample calculation is presented.

The system analysed is a very simple one for the purpose of clarity. The system consists of two pumps in parallel, each with a capacity of 100% of rated flow. One of the pumps is in operation, the other is a standby. The testing of the standby unit is carried out by periodic switching between the operating and the standby unit every week (168 h).

The input for the program consists of two parts: the fault data and the fault consequence specifications.

Table II shows part of the input data for two of the faults in pump unit 1 (group 1). The table shows the principles in the specification of fault indicators and fault data.

Table III presents a definition of the parameters used in the fault consequence specifications. These specifications provide the part of the input for the program that contains the specifications for the system structure. The specifications are written in a special sub-routine called FAILSY by means of the statements shown in Table IV. The output from this sub-routine are logical parameters which show when STOPPAGE or BLOCKING occurs in each group and when the system is failed, depending on the occurrence or not of the individual faults.

By means of the grouping of faults (see Section 2.3.1) the condition for failure of the example system becomes the very simple one that both groups are not in operation, as shown in Table IV.

TABLE II. EXAMPLE INPUT DATA FOR TWO OF THE FAULTS. THE FIGURES ARE INTERPRETED BELOW THE TABLES.

Fault indicators NF
for faults in pump unit 1 (group 1) [a)]

Fault No., n	NF(n,1)	NF(n,2)	NF(n,3)	NF(n,4)	-	-	-	-	-	-
1	1[b)]	0[c)]	3[d)]	1[a)]	-	-	-	-	-	-
2	-4[e)]	0	3	1	-	-	-	-	-	-
- [*)]	-	-	-	-	-	-	-	-	-	-
-	-	-	-	-	-	-	-	-	-	-

Fault data F

	Lifetime p.d.f.			Repair time p.d.f.		
Fault No., n	F(n,1)	F(n,2)	F(n,3)	F(n,4)	F(n,5)	F(n,6)
1	2.00×10^{-4} [f)]	0	0	29[g)]	5[h)]	
2	3.34×10^{-4} [f)]	0	0	32[g)]	7[h)]	
- [*)]						

a) Group numbers are indicated in NF (n,4)

b) Fault no. 1 occurs only during periods of operation

c) Exponential lifetime p.d.f.

d) Normal repair time p.d.f.

e) Fault no. 2 occurs only during periods of standby; unrevealed (subject to tests)

f) The failure rate (n^{-1})

g) The mean repair time (h)

n) The standard deviation on the repair time (h).

*) Similar data for faults in pump unit 2

TABLE III. DEFINITIONS OF PARAMETERS USED IN THE FAULT CONSEQUENCE SPECIFICATIONS

Parameters in the subroutine for fault consequence specifications	
Name	Definition
X(N)	Logical parameter having the value TRUE when fault no. N occurs and the value FALSE when it does not.
1SGN=(=ISG(N))	Another subroutine lets ISG(N) = 1 if group no. N is operating and ISG(N) = 0 if it is not.
SBD(N)	Logical parameter, having the value TRUE if STOPPAGE occurs in group no. N and the value FALSE when STOPPAGE does not occur.
SBB(N)	Logical parameters having the value TRUE if BLOCKING occurs in group no. N, and the value FALSE when blocking does not occur.
SYFAIL	Logical parameter, having the value TRUE if the system has failed and the value FALSE if it has not failed.

Figure 4 shows the fault tree corresponding to the fault consequence specifications in Table IV.

Figure 5 shows the output from the calculation, where the listing of input data has been deleted.

4. EXPERIENCE

Experience with the REDIS program has been limited, but sufficient to show that the detailed models incorporated can be used on smaller systems with complicated operation without unacceptable increase in computer time.

The computer time depends on the size and complexity of the system and also on the failure probability data. An example to indicate the speed of the program, on a system having 25 possible faults with constant failure rates of 5×10^{-7} h^{-1} and no periodic testing during a period of 10 000 h between overhauls, the IBM 370/165 computer can run 17 000 trials per minute, corresponding to an accuracy on the result of $\pm$ 2%. For systems

TABLE IV. FORTRAN STATEMENTS USED FOR THE SPECIFICATIONS OF FAULT CONSEQUENCES, WHICH ARE PART OF THE INPUT IN THE SUB-ROUTINE FAILSY

Fortran statements used for fault consequence specifications	
Statement	Interpretation
SBD(1) = X(1)*)	Stoppage of group 1 occurs when fault no. 1 occurs, and does not occur when fault no. 1 does not occur.
SBB(1) = X(2)*)	Blocking of group 1 occurs when fault no. 2 occurs, and does not occur when fault no.2 does not occur.
SYFAIL = .FALSE. IF (ISG1.EQ.0.AND. ISG2.EQ.0) SYFAIL = .TRUE.	The system has failed if and only if both groups are not operating.

*) Similar statements are written for group no. 2.

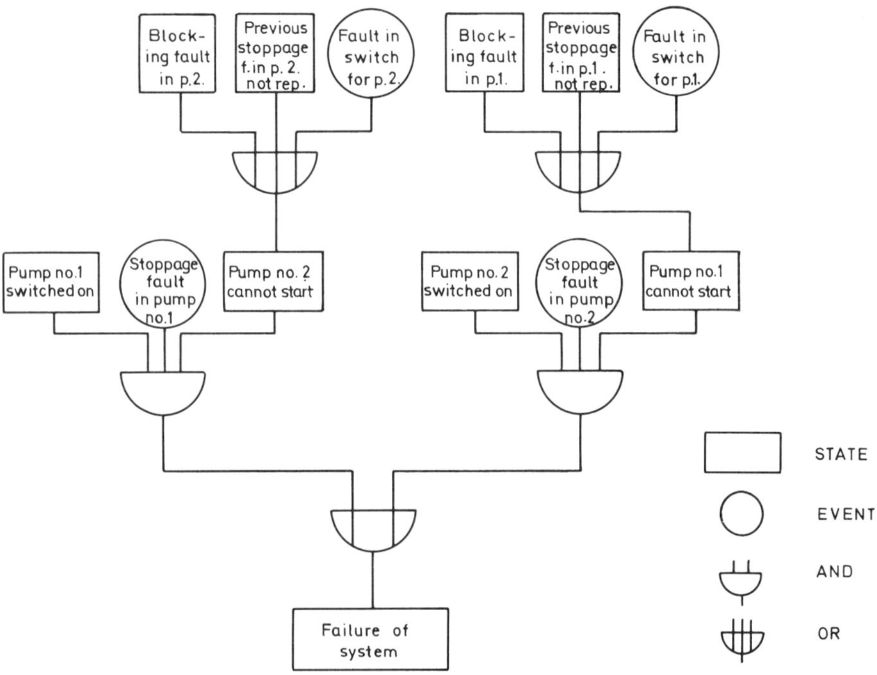

FIG.4. Fault tree corresponding to fault consequence specifications in Table IV.

Computer time, min.	10
Total Number of Trials, actually run	8719
Total number of cut sets registered	769
The failure probability over the period of t_{max} (≏10000 h)	8.5×10^{-2}
with the standard deviation	3.0×10^{-3}
The unavailability of the system at the time of t_{max}	1.1×10^{-4}
With the standard deviation	1.1×10^{-4}
The reliability of the system over the period of t_{max}	9.15×10^{-1}
With the standard deviation	3×10^{-3}
Average number of system failures per period of duration t_{max}	8.8×10^{-2}
Average outage time for each failure of the system	2.3×10^{1}
The average unavailability of the system	2.1×10^{-4}
Average time to first failure of the system (h)	4.9×10^{3}

Cut Sets:

1	2	3
2	1	1
3	4	3
0	0	0
0	0	0
Total no.		
355	346	68
Resp. downtimes (total)		
8.5×10^{3}	8.5×10^{3}	9.8×10^{2}

FIG.5. Output from sample calculation.

with more complex operation the number of trials per minute decreases considerably, as seen from the example calculation.

Testing of the program is continuing. So far it has included, among others, two boiler feed pump systems. The results have agreed closely with calculations by other methods.

A computer program of the direct simulation type like this one will always be limited by computer time to smaller sub-systems with failure probability data above a certain minimum level, as will be seen from the examples given.

Experience has shown that the program to a reasonable extent is able to account for the many details which are encountered in reliability analyses of practical systems.

ACKNOWLEDGEMENTS

The author would like to thank Mr. J.R. Taylor for proof-reading this paper and Mr. R.K. Larsen for his assistance in the testing and debugging of the program.

INTERPRETING FAILURE DATA IN A NUCLEAR PLANT BY THE HAZARD FUNCTION

W.A. WOLFE
Atomic Energy of Canada Ltd,
Chalk River Nuclear Laboratories,
Chalk River, Ont.,
Canada

Abstract

INTERPRETING FAILURE DATA IN A NUCLEAR PLANT BY THE HAZARD FUNCTION.

Field failure data are often dirty and incomplete. Many items survive the reporting period. The failure rate may be changing because of better operating conditions, better maintenance, or improvements. The hazard function as developed by Mr. Wayne Nelson of the General Electric Company provides a useful tool for the interpretation of incomplete data and failure from different causes. Failure data on pump seals for slightly more than three years' experience are examined and interpreted by the hazard function.

1. INTRODUCTION

Located about 20 miles east of Toronto on Lake Ontario, the Pickering Nuclear Generating Station (Fig. 1) comprises four 540 MW(e) units. Each unit has 16 pumps in the main heat transport circuits. The pumps are arranged in two banks of eight (Fig. 2), one bank on each side of the reactor. Each bank is further sub-divided into two groups of four, each group in a circuit independent of the adjacent group. Normal operation is three pumps of a group running with one as standby; hence 12 pumps are running per reactor during normal operation, or a total of 48 for the 4 reactors.

Early during the evolution of CANDU reactors, designers foresaw the need for large main heat transport circulating pumps, larger, in fact, than any in service. Heavy water is expensive and, if lost, contributes to radiation fields and must be replaced. As reactor power increases, costs of lost production due to replacing or repairing faulty components become more onerous. Although designers considered hermetically sealed (canned rotor) pumps, they selected shaft seal pumps because development to larger sizes seemed easier and their inherent efficiency higher. Pickering pumps (Fig. 3) are a stage in the evolutionary process.

They have 1900 h.p. motors and deliver 10 100 gal (imp) per minute against 480 ft of heavy water. Our next generation (Bruce Nuclear Generating Station at Douglas Point on Lake Huron) is ten times larger.

Obviously, reliability data are generated at a substantial rate in an operating nuclear generating station. These data are somewhat dirty, reflecting as they do, actual operating conditions and human error. Moreover, the test is never finished. To treat data on pump seals, the hazard function procedure of Nelson [1] was selected. This method is easily understood and simple to apply. It was first developed as a graphical process [2], but is readily adapted to numerical analysis and computer programs are simple and easy to prepare.

FIG.1. Pickering Nuclear Power Station.

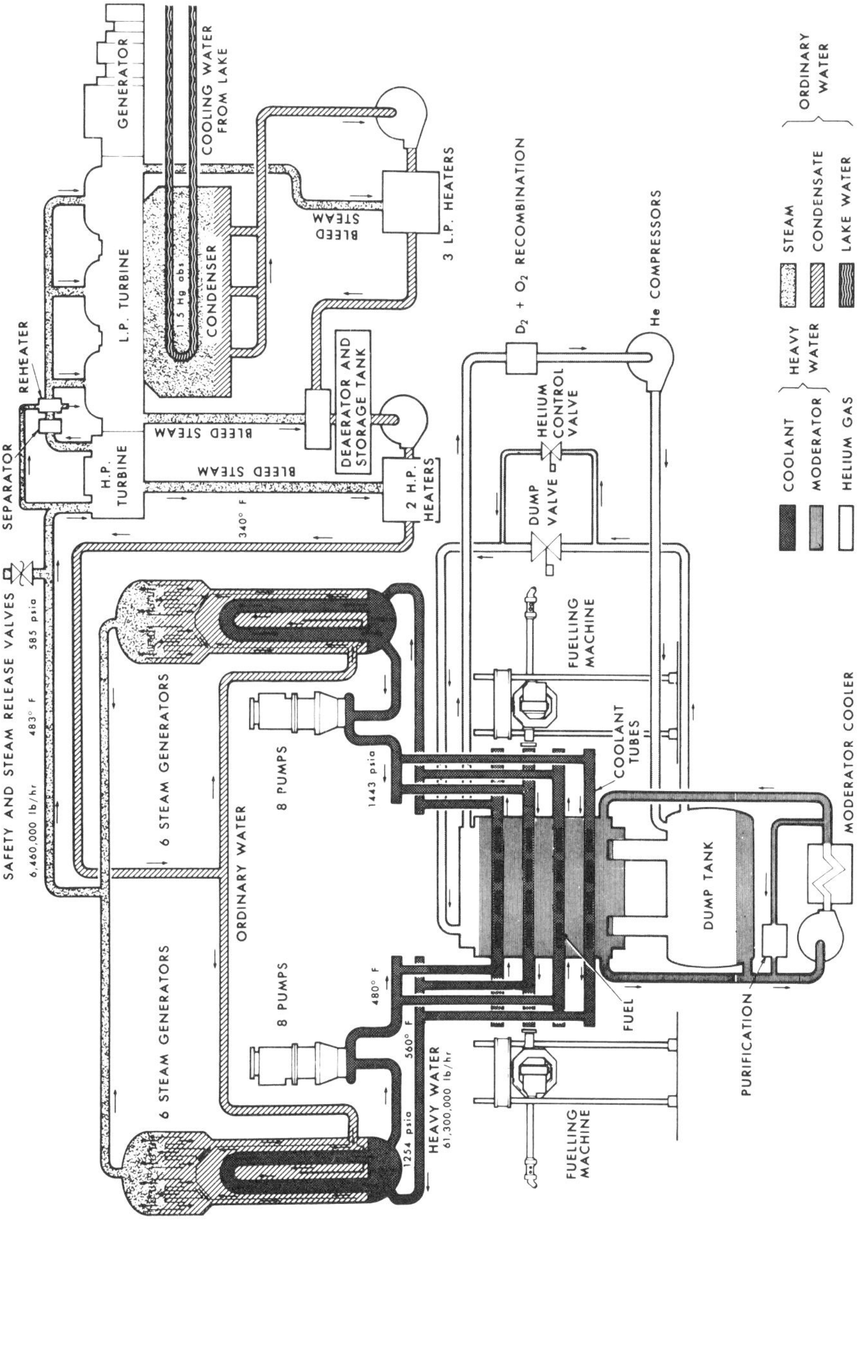

FIG.2. Simplified station flow diagram.

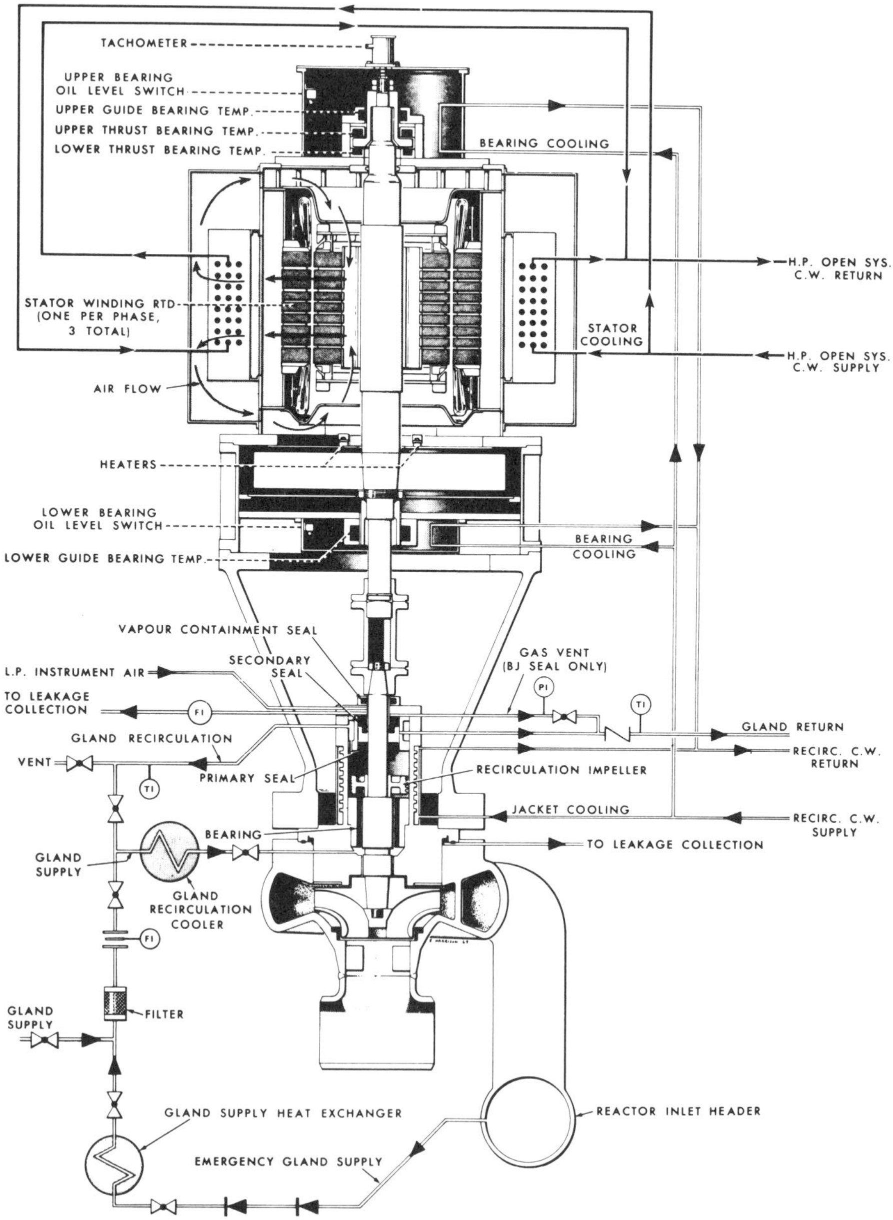

FIG.3. Heat transport pump.

2. HAZARD FUNCTION AND RELIABILITY

Before proceeding to typical field data, a review of the relation between the hazard function and reliability of a component will provide the background for details of the method.

Consider the probability of a device surviving to time t + dt. We can write

$$R(t+dt) = R(t)\prod_i (1 - h_i(t)\,dt) \tag{1}$$

where $R(t)$ = probability of surviving to time t,
$h_i(t)$ = failure rate at time t by an identifiable mechanism or component of the device.

If one assumes simultaneous failures by two or more mechanisms or components impossible, Eq. (1) may be written as

$$R\,(t+dt) = R(t)\left(1 - dt\sum_i h_i(t)\right) \tag{2}$$

or

$$R(t) = \exp\left(-\int dt \sum h_i(t)\right) \tag{3}$$

$$= \exp\,(-H_i(t))$$

where $H_i(t) = \int \sum_i h_i(t)\,dt$

In our terminology we call $\sum_i h_i(t)$ the instantaneous failure rate or hazard function and $H_i(t)$ the cumulative hazard function. Obviously a knowledge of the hazard function is equivalent to that of the reliability as one is readily derived from the other. In the practical sense, however, the hazard function and cumulative hazard function are readily obtained from field data, particularly when the data are censored, as they almost always are.

One way of relating the hazard function to censored data is:

Assume K units survive to time t and one unit fails at time t + Δt. As all K units are equally likely to fail, the conditional probability that a specific unit fails in t + Δt is 1/K. We may then use the relation

$$P(A/B) \times P(B) = P(A) \cdot P(B) \tag{4}$$

where $P(A/B)$ = probability of a unit failing at t + Δt given it survived to t,
$P(B)$ = probability of surviving to time (t), R(t),
$P(A)$ = probability of failing in interval Δt, (h $\times$ Δt).

Hence, Eq. (4) may be rewritten as

$$\frac{1}{K} R(t) = (h \times \Delta t) \times R(t) \tag{5}$$

or

$$\frac{1}{K} = h \times \Delta t \tag{6}$$

If more than one failure mode is present, the same relation holds,

$\frac{1}{K} = h_i \times \Delta t$, where 'i' represents the failure mode;

the cumulative hazard function for the ith mode is $\sum h_i \times \Delta t$.

3. SUSPENDED ITEMS

With suspensions (a failure in the ith mode represents a suspended item with respect to other modes) the value of K must be adjusted.

Consider the situation where N items are on test and the first item is removed before failure. Our results take the form with failures arranged in ascending time order:

Test	Reverse order number	Adjusted reverse order number
S	N	
F_1	N-1	
F_2	N-2	
F_3	N-3	
F_P	N-P	$N + 1 - \frac{P(N+1)}{N}$
F_{N-1}	1	

What number K should we assign to the pth failure? Were the suspended item left on test, it could fail in any of the N positions; if it failed before the pth failure then K = N-P, if it failed after the pth failure then K = N - P + 1. By assuming the suspended item equally likely to fail in any position, we obtain the expected value for K.

$$\exp(K_P) = \frac{(N-P) \times P + (N-P+1)(N-P)}{N}$$

$$= (N-P)\frac{(N+1)}{N}$$

$$= N + 1 - \frac{P}{N}(N+1) \tag{7}$$

If two suspended items precede a series of failures, i.e. $(S_1 S_2 F_1 F_1 F_2 \ldots F_p \ldots F_{N-2})$ one can show by the same argument that

$$\exp(K_p) = N + 1 - P\frac{(N+1)}{(N-1)} \tag{8}$$

and generally

$$\exp(K_p) = N + 1 - P\frac{(N+1)}{(N+1-q)} \tag{9}$$

where q are the suspended items preceding failures. In Appendix A this formulation is developed further for ease of computation.

To determine the fraction of population surviving to time 't' we should know the true rank of the failure associated with order number K_p. However, we do not know it; the best we can do is estimate. Johnson [4] has shown the 'median rank', one equally likely to be high or low, should be used.

He developed exact expressions for median ranks, but these are complicated and tedious to compute. Simpson [5] suggests median ranks can be estimated with sufficient accuracy by

$$\text{Median rank} = \frac{K_p - 0.3}{N + 0.4}$$

where K_p = order number,
N = sample size.

Perhaps a simple numerical example will clarify matters. Consider the test sequence:

Test order	Order number	Expected order number	Median rank
F_1	5	5	
S_1	4		
F_2	3		
S_2	2		
F_3	1		

The calculation for $\exp(K_2)$ is

$$\exp(K_2) = \frac{3(1) + 4(3)}{4} = 3.75$$

and for exp (K_3)

$$\exp(K_3) = \frac{1(3) + 2(1+2) + 3(2)}{8} = 1.875$$

Hence, our table in final form is:

Test order	Order number	Expected order number	Median rank	Hazard	Cum. hazard
F_1	5	5	0.87	0.20	0.20
S_1	4				
F_2	3	3.75	0.64	0.267	0.467
S_2	2				
F_3	1	1.875	0.29	0.538	1.00

The median rank of the first failure estimates the probability of survival to time of first failure, that of the second failure, survival to time of second failure, and so on. Thus, the median rank is a non-parametric estimate of the reliability or survival function (3).

4. THE WEIBULL DISTRIBUTION

We should like to fit our data to a parametric distribution curve. For our applications we find the Weibull relation [3]

$$R(t) = e^{-(t/\alpha)^{\beta}} \tag{10}$$

and its cumulative hazard function

$$H(t) = (t/\alpha)^{\beta} \tag{11}$$

adequate. The significance of β is readily apparent from the hazard function

$$h(t) = \frac{\beta}{\alpha}\left(\frac{t}{\alpha}\right)^{\beta-1} \tag{12}$$

Simple inspection shows:

$\beta < 1$ - decreasing hazard function,
$\beta = 1$ - constant hazard function,
$\beta > 1$ - increasing hazard function.

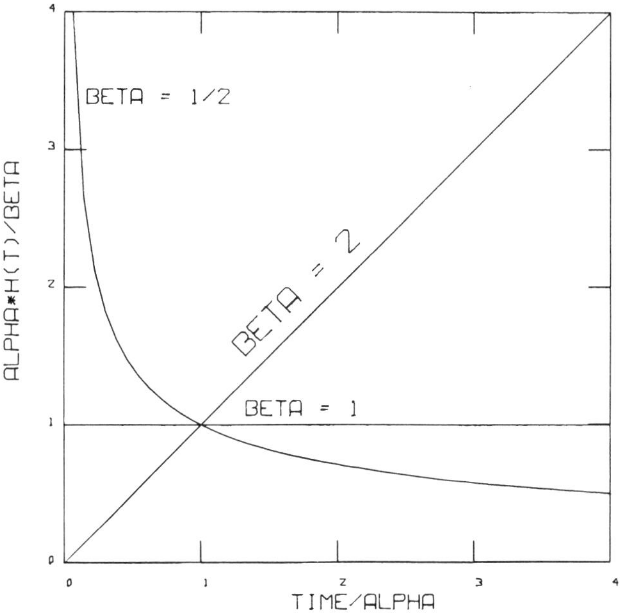

FIG.4. Hazard function versus time.

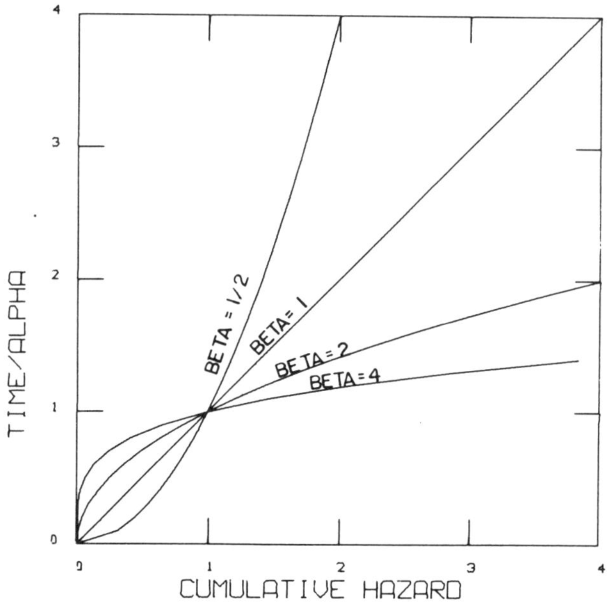

FIG.5. Time versus cumulative hazard function.

A plot of $\alpha h(t)/\beta$ versus $(t/\alpha)^{\beta-1}$ (Fig. 4) illustrates these relations. Normally, however, we plot experimental values of time versus cumulative hazard function. Figure 5 shows how β affects the shape of the curve. For $\beta = 1$ we have a constant slope, for $\beta < 1$ an increasing slope, and for $\beta > 1$ a decreasing slope.

5. APPLICATION TO PUMP SEALS

Initially Pickering heat transport pumps had two kinds of seals; type 1, a primary hydrostatic seal (Fig. 6) with a secondary elliptical carbon face seal, and type 2, two face seals in series (Fig. 7).

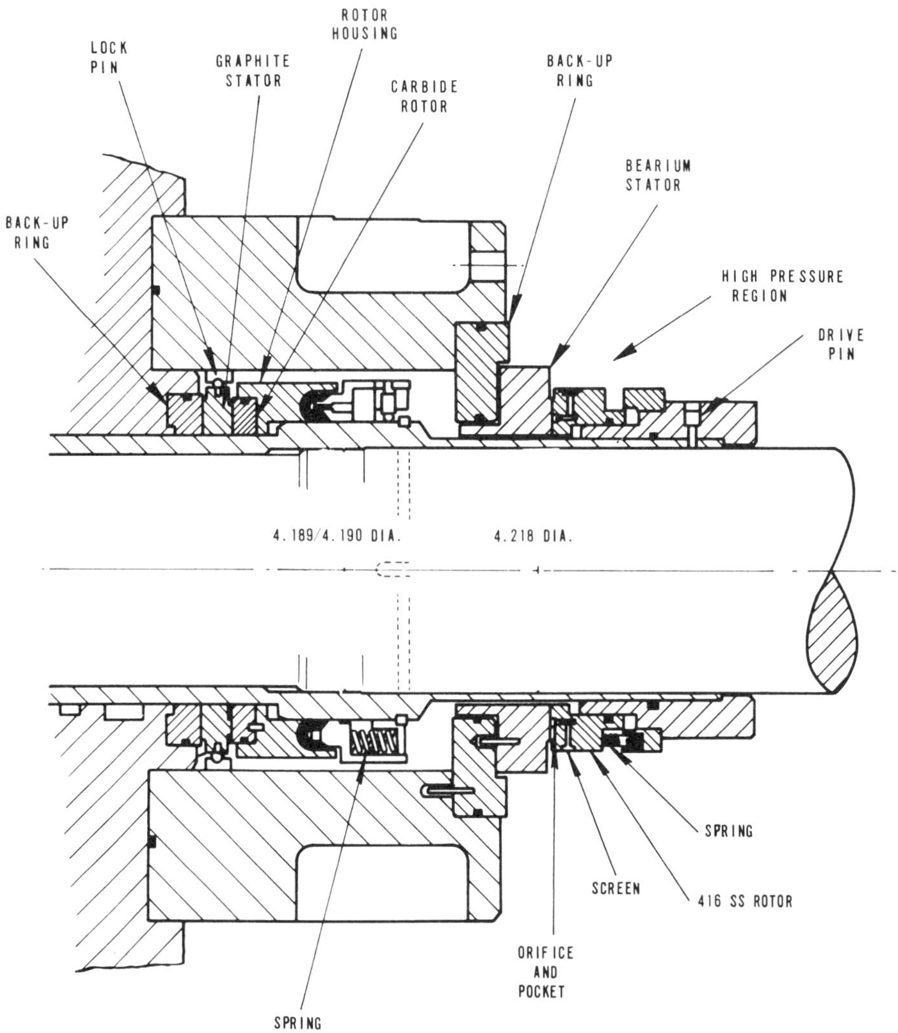

FIG.6. Hydrostatic seal.

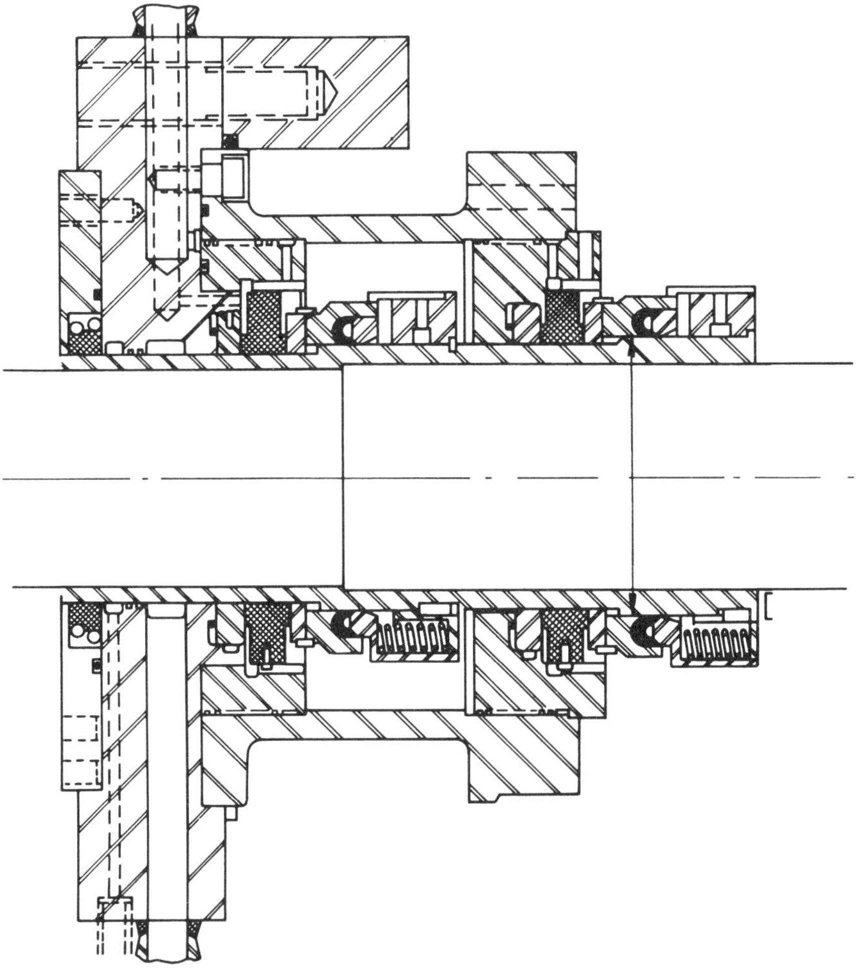

FIG.7. Face seal.

Type 1 seals exhibited two failure modes, (a) failure of the primary seal, and (b) failure of the secondary seal. Table I shows results for about 70 000 h of running. Reverse order numbers are adjusted for suspended tests. A primary seal failure is considered suspended with respect to secondary seal failures and vice versa. Table II gives calculated non-parametric reliability of primary and secondary seals.

Figure 8 is a plot of time to failure against cumulative hazard for type 1 primary and secondary seals. Both seals exhibit anomalies with the last points; these depart markedly from previous trends. In fact, they suggest a significant change in failure rate. Figure 9 is a log-log plot of cumulative hazard against time.

TABLE I. HYDROSTATIC SEAL HAZARD CALCULATIONS

PUMP	HOURS	FAILURE PRI	FAILURE SEC	REVERSE ORDER	CUM.HAZARD PRI	CUM.HAZARD SEC
2P1B	0.		1	35.00		2.86
2P10B	0.		1	34.00		5.80
2P16D	0.	1		34.94	2.86	
2P16C	3.	1		33.88	5.81	
2P1A	117.		1	32.94		8.83
2P4A	150.		1	31.88		11.97
2P16A	171.	1		32.75	8.87	
2P2	187.	SUSPENDED				
2P1C	275.		1	30.74		15.23
2P10A	327.		1	29.60		18.60
2P6B	385.		1	28.46		22.12
2P4B	432.	NEITHER				
3P14B	530.		1	27.27		25.78
2P8A	581.	1		31.33	12.06	
2P12A	623.		1	26.03		29.62
2P16B	737.	1		29.84	15.41	
2P12B	747.		1	24.73		33.67
2P13B	754.		1	23.43		37.94
2P14B	768.		1	22.13		42.46
2P14A	804.		1	20.83		47.26
2P9	807.	SUSPENDED				
2P13A	814.	1		27.85	19.00	
2P15A	843.	1		25.86	22.87	
2P2A	889.		1	19.23		52.46
2P6A	915.	1		23.70	27.09	
3P16A	1258.	1		21.55	31.73	
2P9A	3383.	1		19.39	36.88	
2P11A	5131.		1	17.09		58.31
2P4	6638.					
2P7	6847.					
2P8	6856.					
2P10	6893.					
2P1	6973.					
2P5	7403.					
2P3	7538.					

PRIMARY CORRELATION COEFFICIENT = .8729

SECONDARY CORRELATION COEFFICIENT = .9093

PRIMARY ALPHA = 338129.53
PRIMARY BETA = .26
SECONDARY ALPHA = 5423.56
SECONDARY BETA = .57

To give us good Weibull representations these points should fall on straight lines, which is obviously not the case. However, the Weibull parameters were estimated by a least square fit of cumulative hazard on time.

	Alpha (h)	Beta	Correlation coefficient
Primary seal	338 129.53	0.26	0.8729
Secondary seal	5 923.56	0.57	0.9093

TABLE II. NON-PARAMETRIC RELIABILITY

HOURS	PRIMARY	SECONDARY
0.		.9802
0.		.9520
0.	.9786	
3.	.9487	
117.		.9220
150.		.8919
171.	.9167	
275.		.8598
327.		.8276
385.		.7955
530.		.7620
581.	.8765	
623.		.7270
737.	.8344	
747.		.6902
754.		.6534
768.		.6166
804.		.5799
814.	.7782	
843.	.7220	
889.		.5346
915.	.6611	
1258.	.6003	
3383.	.5394	
5131.		.4743

Figure 10 shows plots of reliability calculated with these parameters; non-parametric values are superimposed. Calculated values differ considerably from non-parametric results. Inspection of Figs 8 to 10 will reveal notable departures of the last points from preceding trends. Table I indicates no failures after 5200 h to the end of the record.

All these seals were original installations (the letter suffix at the end of the seal identifier indicates replacement: A = 1st, B = 2nd and so forth). The record indicates, for example, removal of seal 2P1 in the unfailed state at 6973 h, its replacement, 2P1A, failed at 117 h, seal 2P1B failed at 0 h. We will return to this point after examination of type 2 seals. Type 1 seals were replaced with type 2 (face seals) either on failure or as part of scheduled maintenance. Type 2 (face seals) were retained and we have information on 73 seals. Table III shows the hazard calculations. The computed results were:

Alpha = 150116 h
Beta = 0.238

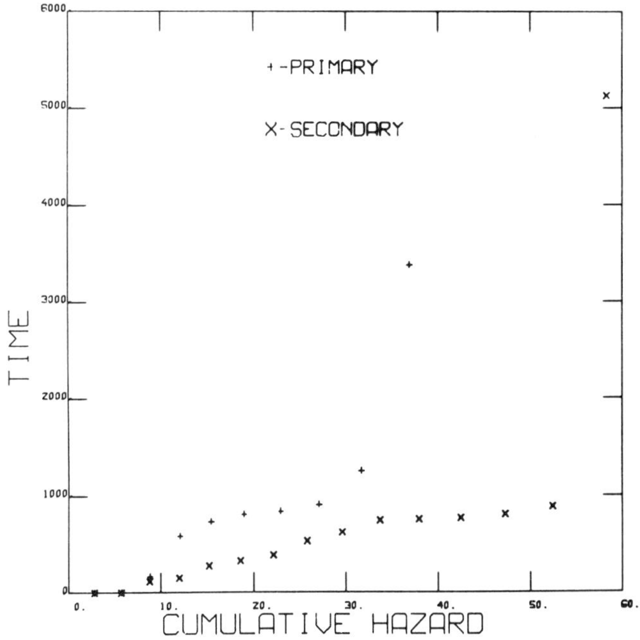

FIG.8. Time versus cumulative hazard hydrostatic seal.

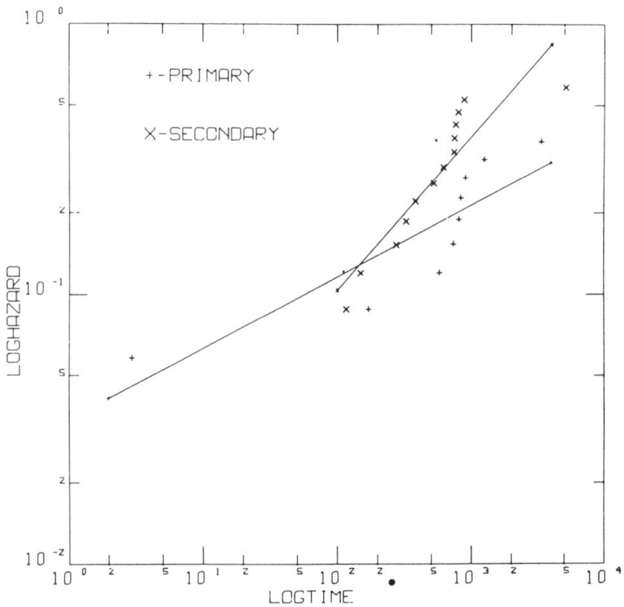

FIG.9. Plot of log — cumulative hazard versus log time.

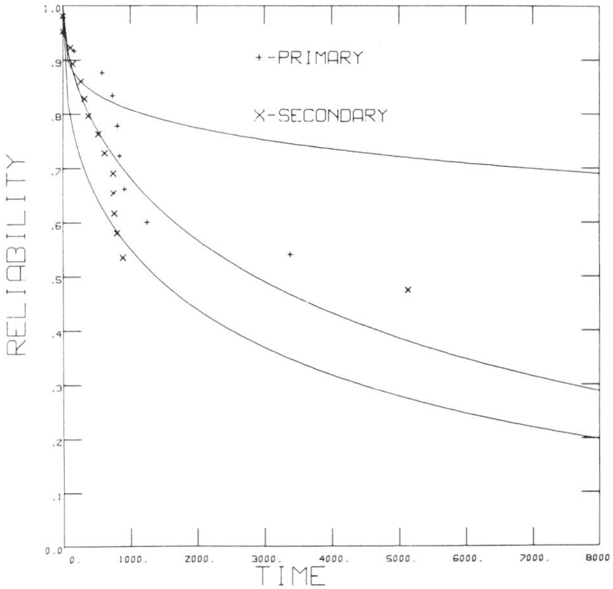

FIG.10. Calculated reliability of hydrostatic seal.

6. INTERPRETATION OF DATA

It is interesting to note there are no failures after 6254 h. Some 31 seals survived to the end of the reporting period. The low value of beta indicates a decreasing failure rate in agreement with the hours versus cumulative hazard plot (Fig. 11). Comparison of the face seal reliability plot (Fig. 12) with that of the hydrostatic seal (Fig. 10) will reveal the significantly better performance of the face seal. Also worthy of note is the agreement between calculated and non-parametric reliability in contrast to the hydrostatic seal. The predominance of early failures (before, say, 500 h of operation) suggests some change in seal behaviour. Assuming such was the situation, we recalculated the parameters, including only those seals with lives greater than 500 h. Table IV shows the results. The parameters are:

Alpha = 105077 h
Beta = 0.7168

If our assumption can be accepted, the failure rate is still decreasing (Beta $<$ 1) but is closer to a constant failure rate (Beta = 1). Figure 13 shows the reliability calculated from these parameters with non-parametric reliability superimposed. Figure 14 shows the time versus cumulative hazard.

Unquestionably, rejection of data in this manner is controversial and requires justification. We do have reason to suspect the early failures were caused more by human errors in both assembly and operation of the seals than by random causes. Unfortunately we cannot quantify this supposition. An area of investigation would be a method of treating the data to indicate a changed distribution of failures and to reject the appropriate early failures. It is interesting to examine the time versus cumulative hazard plots (Figs 8 to 11). In both cases the curves exhibit sharp changes of slope after about 500 h of operation. Seal 3P3 (Table III) is especially interesting. The suffixes A, B and C after the identifying symbol indicate replacements in the 3P3 position. The seal 3P3 was removed after 4464 h (unfailed) and replaced with seal 3P3A which failed after 25 h, then 3P3B which failed after 17 h, and finally 3P3C which failed after 2 h. Again 1P6 was removed at 17877 h and its replacement failed at 0 h. Perhaps this

TABLE III. SEAL FAILURE DATA AND HAZARD CALCULATIONS

PUMP	HOURS	FAILURE	HAZARD	CUM. HAZARD	REVERSE ORDER	
1P6A	0.	1	1.37	1.37	1	73.00
1P7A	0.	1	1.39	2.76	2	72.00
2P11B	0.	1	1.41	4.17	3	71.00
3P3D	0.	1	1.43	5.60	4	70.00
1P12	1.				5	
3P3C	2.	1	1.45	7.05	6	68.99
1P5A	3.	1	1.47	8.52	7	67.97
4P14	3.				8	
3P3B	17.	1	1.49	10.01	9	66.94
3P3A	25.	1	1.52	11.53	10	65.91
3P11A	38.	1	1.54	13.07	11	64.88
2P11C	48.	1	1.57	14.64	12	63.85
4P15A	187.				13	
1P7B	197.	1	1.59	16.23	14	62.80
1P7C	200.	1	1.62	17.85	15	61.76
4P14A	213.	1	1.65	19.49	16	60.71
4P16A	251.	1	1.68	21.17	17	59.66
4P13A	267.	1	1.71	22.88	18	58.62
3P5A	285.	1	1.74	24.61	19	57.57
3P8A	288.	1	1.77	26.38	20	56.52
3P9A	290.	1	1.80	28.18	21	55.48
2P9	361.				22	
3P9B	485.	1	1.84	30.02	23	54.41
4P13B	628.	1	1.87	31.90	24	53.34
3P6A	1114.	1	1.91	33.81	25	52.28
1P12A	1382.	1	1.95	35.76	26	51.21
3P10	2056.				27	
3P5	2359.				28	
1P10A	3126.	1	2.00	37.76	29	50.10
2P6	3955.				30	
1P16	4165.				31	
3P3	4464.				32	
3P16	4507.				33	
2P10	4751.				34	
2P11	4860.				35	
2P14	4886.				36	
4P12	5003.				37	
2P6C	5200.	1	2.05	39.81	38	48.74
4P5	5639.				39	
4P13	5734.				40	
3P14	5741.				41	
2P14C	6254.	1	2.12	41.93	42	47.27
4P15	6369.				43	
4P16	6375.				44	

TABLE III (cont.)

PUMP	HOURS	FAILURE	HAZARD	CUM. HAZARD	REVERSE ORDER
3P1	7745.				45
2P16	8501.				46
3P9	9668.				47
3P6	10189.				48
3P15	10206.				49
3P13	10225.				50
3P7	10426.				51
2P13	10535.				52
3P8	11854.				53
3P11	11874.				54
3P4	12075.				55
3P12	12125.				56
3P2	12194.				57
1P3	12749.				58
1P7	13864.				59
2P12	13883.				60
1P1	15038.				61
2P15	15082.				62
1P8	16763.				63
1P6	17877.				64
1P5	17909.				65
1P10	18779.				66
1P4	19332.				67
1P2	19517.				68
1P14	19650.				69
1P15	20328.				70
1P9	21527.				71
1P11	22067.				72
1P13	22209.				73

NOTE-1-DENOTES FAILURE

TOTAL OPERATING HOURS = 515920.

TOTAL FAILURES = 25

MEAN TIME BETWEEN FAILURES =20636.8 HRS

SUFFIX A = SECOND FAILURE, B= THIRD FAILURE IN SAME PUMP

CORRELATION COEFFICIENT = .9720

says something about replacement procedures, as one notes the preponderance of early failures with suffixes after the identifier. Of the first 25 seals, 22 have suffixes; of the remaining 48, 4 have suffixes. One would tend to believe some statistical significance exists in this division, but it has not been investigated. However, if one assumes original and replaced seals are equally likely to fail, then such a result should occur about once in 10 000 trials.

One further attempt was made to obtain information by classifying failures of type 2 seals into three categories: Maintenance failures; operator failures; and random failures.

By maintenance failures we mean such faults as pinched U-cups, O-rings leaking at stamp marks on shafts, O-rings twisted on shafts; by operator failures, vent valves not closed, and high seal temperatures

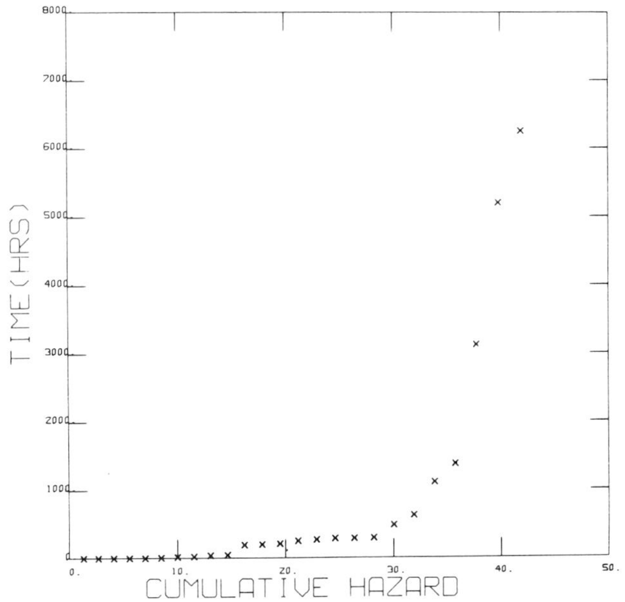

FIG.11. Time versus cumulative hazard face seal.

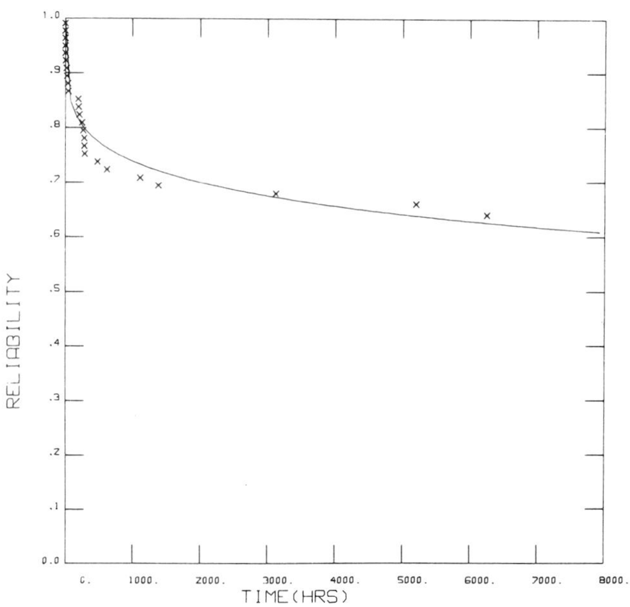

FIG.12. Calculated reliability of face seal.

due to insufficient cooling; by random failures, those failures for which no obvious cause could be determined. The results are:

	Number of failures	Per cent of total failures	Cumulative per cent
Maintenance	14	56	56
Operator	6	24	80
Random	5	20	100
Total	25		

TABLE IV. SEAL FAILURE DATA AND HAZARD CALCULATIONS

PUMP	HOURS	FAILURE	HAZARD	CUM. HAZARD	REVERSE ORDER	
4P13B	628.	1	2.00	2.00	1	50.00
3P6A	1114.	1	2.04	4.04	2	49.00
1P12A	1382.	1	2.08	6.12	3	48.00
3P10	2056.				4	
3P5	2359.				5	
1P10A	3126.	1	2.13	8.25	6	46.96
2P6	3955.				7	
1P16	4165.				8	
3P3	4464.				9	
3P16	4507.				10	
2P10	4751.				11	
2P11	4860.				12	
2P14	4886.				13	
4P12	5003.				14	
2P6C	5200.	1	2.19	10.44	15	45.69
4P5	5639.				16	
4P13	5734.				17	
3P14	5741.				18	
2P14C	6254.	1	2.26	12.70	19	44.30
4P15	6369.				20	
4P16	6375.				21	
3P1	7745.				22	
2P16	8501.				23	
3P9	9668.				24	
3P6	10189.				25	
3P15	10206.				26	
3P13	10225.				27	
3P7	10426.				28	
2P13	10535.				29	
3P8	11854.				30	
3P11	11874.				31	
3P4	12075.				32	
3P12	12125.				33	
3P2	12194.				34	
1P3	12749.				35	
1P7	13864.				36	
2P12	13883.				37	
1P1	15038.				38	
2P15	15082.				39	
1P8	16763.				40	
1P6	17877.				41	
1P5	17909.				42	
1P10	18779.				43	
1P4	19332.				44	

TABLE IV (cont.)

PUMP	HOURS	FAILURE	HAZARD	CUM. HAZARD	REVERSE ORDER
1P2	19517.				45
1P14	19650.				46
1P15	20328.				47
1P9	21527.				48
1P11	22067.				49
1P13	22209.				50

NOTE-1-DENOTES FAILURE

TOTAL OPERATING HOURS = 512759.

TOTAL FAILURES = 6

MEAN TIME BETWEEN FAILURES =85459.8 HRS

SUFFIX A = SECOND FAILURE, B= THIRD FAILURE IN SAME PUMP

CORRELATION COEFFICIENT = .9625

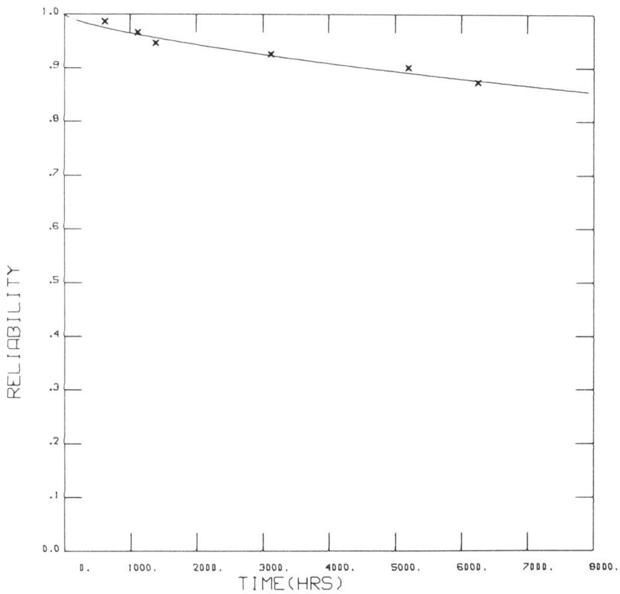

FIG.13. Calculated reliability face seal. (failure times > 500 h).

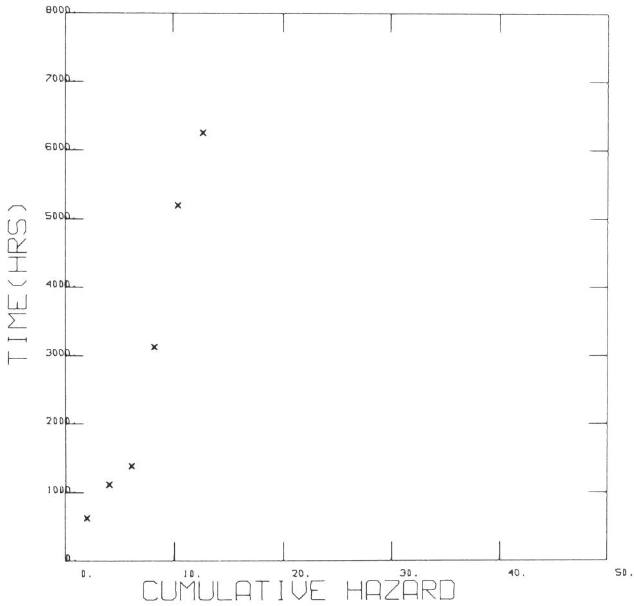

FIG.14. Time versus cumulative hazard face seal (failure times > 500 h).

The last column suggests 80% of failures resulted from human error. This analysis together with the previous observation on the early failures after maintenance suggests the importance of good maintenance and operating procedures as well as design of seals for easy and correct replacement. The frequently noted discrepancy of behaviour of devices in laboratory and field tests often results from differences between laboratory and field procedures.

7. CONCLUSIONS

Some tentative conclusions are possible:

(1) Hazard function analysis of incomplete and censored field data is readily applied and provides some insight into the behaviour of devices. One is as much interested in failure rates as in reliability.

(2) The Weibull relation seems adequate to represent the distribution and gives a recognizable indication of the rate of change of the failure rate (hazard function). For example, with the type 2 seals the decreasing hazard function indicated they had not yet reached a wearout mode of failure. There is no incentive, as yet, to schedule replacement on the basis of running time. In fact, replacement was often deleterious to performance.

(3) Classification of failures by causes revealed about 80% were related to human error.

(4) The time span to accumulate significant data is somewhat surprising when one realizes some seals exceed 20 000 h running time and more than 500 000 h in total were accumulated.

(5) Both seals indicated possible changes in the character of their failure distribution functions. Some criteria are necessary to determine whether or not apparent changes are real. Of course, more complete recording of failures and their causes would enable independent judgements.

(6) By identifying two or more modes of failure from physical arguments one can apply the hazard function technique by treating failures in one mode as suspended items in other modes. In the type 1 seal the face seal was by far the most unreliable; whereas in the type 2 seal it behaved well.

Appendix A

CALCULATION OF REVERSE ORDER NUMBER WITH SUSPENDED ITEMS IN THE LIST

The first step in hazard analysis is to arrange failures and suspensions in a sequence from earliest to latest event and then assign numbers in reverse order to the list, the largest number to the earliest event. However, when suspensions precede failures, reverse order numbers must be modified because suspended items may or may not fail before the first failure. For example, if three suspensions occur before a failure, the reverse order number can be any one of four different numbers depending on the times when suspended items failed, were their tests continued. The best we can do is to calculate the expected value of the reverse order number by assuming suspended items are equally likely to fail in any order number subsequent to the last preceding failure.

Consider the following example:

Event	Reverse order number	
	Unmodified	Modified
S_1	N	
S_2	N-1	
F_1	N-2	
F_2	N-3	
-		
-		
F_P	(N-P-1)	$(N-P-1)\frac{(N+1)}{N-1}$
F_{N-2}	1	

The expected value of reverse order number for failure F_P may be calculated by summing permutations (multiplied by appropriate reverse order number) for two suspensions, one suspension, and no suspensions preceding $F_{(P)}$ and dividing by possible number of permutations of suspensions.

$$\text{exp (rev. order)} = \frac{(N-P-1)(p+1)\,P + 2(N-P)(P)(N-P-1) + (N-P+1)(N-P)(N-P-1)}{N(N-1)}$$

$$= (N-P-1)\left\{\frac{N+1}{N-1}\right\}$$

The decrement between two consecutive failures following suspended items is

$$\left\{(N-P) - (N-P-1)\right\}\frac{N+1}{N-1} = \frac{N+1}{N-1}$$

Hence the decrement in reverse order number following a suspension (until we meet another suspension) is constant. Note suspensions following a failure cannot affect its reverse order number because the suspension can only fail at rank subsequent to its time of removal or at ranks of suspensions immediately preceding without intervening failures.

Consider an example with a failure preceding a suspension:

Event	Reverse order number	
	Unmodified	Modified
F_1	N	N
S_1	N-1	
F_2	N-2	
F_3	N-3	
-		
-		
F_P	N-P	$\frac{(N-P)N}{N-1}$
-		
-		
F_{N-1}	1	

$$\text{exp (rev. order)} = \frac{(N-P)(P-1) + (N-P+1)\,(N-P)}{N-1}$$

$$= (N-P)\,\frac{(N)}{N-1}$$

The decrement between two successive failures is $N/(N-1)$. Here we note the decrement equals the reverse order number of the preceding failure divided by the number of items remaining plus one.

Another way of looking at the problem is to realize that originally we wish to order N items between positions N + 1 and O. Now if 'q' suspensions precede the first failure, there are left N-q items. To place these uniformly between the end positions we will require N - q + 1 equal spaces. Hence the decrement will be

$$\frac{N+1}{N-q+1}$$

and the reverse order number of the first failure

$$N+1-\frac{N+1}{N-q+1}$$

Programming is very simple. We make the initial value of the reverse order number equal N + 1, calculate the decrement at each failure by the relation: Decrement minus reverse order number of preceding failure, divided by number of items remaining plus one. The next reverse order number is the preceding minus the decrement.

The expression

$$K_p = N+1-P(N+1)/(N-q+1)$$

where

K_p = reverse order number of the pth failure
N = number of items starting the test,
q = number of suspensions preceding first failure,

is developed as follows:

Consider the sequence

$$S_1, S_2, -, S_q, F_1, F_2, F_3 -, -, F_p -- F_{N-q}$$

An expression for the expected value of K_p, is

$$\exp(K_p) = \sum_{s=0}^{q} (N-P-q+1+S)\binom{q}{S} P_{p+q-1-s,\, q+s} \cdot P_{N-p-q+s,}\, S/P_{N,q}$$

where $P_{n,K} = n(n-1)-(n-K+1) = \binom{n}{K} K!$

(number of ways of selecting K objects out of n)

$$\binom{n}{k} = \frac{n!}{K!\,(n-K)!}$$

(combination of K objects selected from n)

Rewriting, one finds

$$\exp(K_p) = \sum_{s=0}^{q} (N-p-q+1+S)\binom{q}{S}\binom{p+q-1-S}{q+S}\binom{N-p-q+S}{S} S!\,(q+s)!\Big/\binom{N}{q} q!$$

$$= \sum_{s=0}^{q} (N-p-q+1+S)\binom{q}{S}\binom{p+q-1-S}{p-1}\binom{N-p+S-1}{S}\Big/\binom{N}{q}\binom{q}{S}$$

$$= \sum_{s=0}^{q} (N-q-p+1)\binom{p+q-1-S}{p-1}\binom{N-p-q+S}{N-p-q+1}\Big/\binom{N}{p}$$

Summing this sequence by the relation [6],

$$\sum \binom{r-S}{m} \binom{\ell+S}{n} = \binom{r+\ell+1}{m+n+1}$$

one gets

$$= (N-q-p+1)\binom{N+1}{N-q+1} \Big/ \binom{N}{q}$$

and finally upon expressing the binomial coefficients in terms of factorials

$$= (N+q+1-p)\frac{(N+1)}{N-q+1}$$

$$= N+1-\frac{P(N+1)}{N-q+1}$$

REFERENCES

[1] NELSON, W., "Hazard plot analysis of incomplete failure data", Proc. IEEE Annu. Symp. on Reliability, Chicago, Ill., Institute of Electrical and Electronics Engineers, New York (1969).

[2] KING, J.R., Probability Charts for Decision Making, Industrial Press Inc., New York (1971).

[3] ARNIC Research Corporation, Reliability Engineering, Prentice-Hall, Inc., Englewood Cliffs, N.J. (1964).

[4] JOHNSON, L.G., "The median ranks of sample values in their population with an application to certain fatigue studies", Industrial Mathematics 2 1, Industrial Mathematics Soc., Roseville, Mich. (1951).

[5] SIMPSON, B.H., "Reliability methods and applications at Ford Motor Company", Product Design and Value Engineering, Wallace Press, Montreal, Canada (Oct. 1966).

[6] KNUTH, D.E., The Art of Computer Programming 1, 2nd Edn, Addison Welsey Inc., Reading, Mass.

DISCUSSION

(on papers IAEA-SM-195/17 and IAEA-SM-195/51)

R. GONNOT: I want to ask Mr. Kongsø whether his programme is designed to take 'operator error' into account — errors which might occur during normal operation or in the course of a test procedure.

H.E. KONGSØ: Yes, it is designed to take both kinds of operator error into account.

Nadia KONTOLEON: I wonder whether Mr. Kongsø could describe the switching mechanisms to which he referred.

H.E. KONGSØ: The switching mechanisms consist of all the mechanisms that are activated during a process of switching off an operating unit and subsequently switching on the corresponding standby unit. Such switching operations are carried out in cases of failure of operating units or as a means of testing standby units.

C.D.H. FOTHERGILL: I should like to ask Mr. Wolfe whether the theoretical predictions resulting from the mathematical techniques he has described have been proved right subsequently by event surveillance, once the equipment was put into operation.

W.A. WOLFE: Quantitatively, no: I have no more numerical data. But in a practical sense they have, because the prediction of decreasing failure rate has been verified. Seal failures in the main heat transfer system pumps no longer concern the operator. A mean time to failure exceeding 20 000 hours has been achieved.

The expected mean time to failure from the Weibull relation is MTTF = $\alpha\Gamma(1 + \beta^{-1})$; this is quite sensitive to β, which in turn depends on how one fits the curve.

My feeling is that such analyses are more valuable for identifying areas where improvements are possible than in providing data for general use. Extrapolation is risky.

D. OKRENT (Chairman): I should like to ask — not only the authors of the papers we have heard so far this morning but anyone who cares to reply — whether you feel that probabilistic safety assessments should include the burn-in and burn-out phases or only the frequently-used constant failure rate.

C.D.H. FOTHERGILL: It is very difficult to obtain dependable information on equipment malfunction during the commissioning stage of plant construction; one problem is that the plant may still be in the hands of contractors who are reluctant or unable to provide such data. However, since most of the 'teething troubles' occur during the early life of a plant, and these are quickly dealt with, reliability engineers and safety assessors are particularly interested in performance data relating to the useful life phase. Burn-out or wear-out data on components are difficult to obtain, and when this stage is reached it is often masked by planned maintenance.

H.E. KONGSØ: Our REDIS programme can take both phases into account, provided it is possible to define the corresponding density functions for the times to failure.

D. NORMAN: I believe that burn-in and wear-out regimes do matter. In the first place, burn-in phenomena often occur at intervals throughout the life of the plant (following maintenance) and therefore must be considered when maintenance and inspection rates are being chosen. Secondly, these regimes indicate phenomena which have practical engineering relevance and which demand practical action. Lastly, we should bear in mind that, in practice, no one ever knows when burn-in stops or wear-out starts.

J.O. TATTERSALL: What is the ratio between the burn-in and random constant failure rates, and also the ratio between the latter and the wear-out failure rate? Are these ratios sigificant in terms of the accuracy required for safety assessment?

E.R. SNAITH: Random failure rate data collected from field experience must inevitably include the failures associated with the burn-in phase of equipment operation — due to replaced components for example — as well as the constant random failure rate depicted by the plateau of the bathtub curve. Thus in a practical analysis the data do take account of both random constant failures and failures during the burn-in phase.

J.R. TAYLOR: I have some information, derived from a study of the causes of failure, which may help to elucidate the question of burn-in failure. In my paper (IAEA-SM-195/16) some of this information is presented in greater detail. Certain causes, such as 'design error', are of a kind which one might associate with the burn-in phase. I studied the proportions of different failure causes for five reactors, with the data sampling period chosen so that twelve years of reactor operating experience were covered. No change in the proportion of different failure causes could be detected, although the total number of failures appeared to fall. The amount of data used in the study was probably not sufficient to justify any definite conclusions, but a study of failure causes can certainly help in studying changes in failure rates.

FIABILITE ET DISPONIBILITE DE SYSTEMES REDONDANTS: PROGRAMME DE CALCUL ET UTILISATION D'ABAQUES

J.-P. SIGNORET
CEA, Centre d'études nucléaires de Saclay,
Gif-sur-Yvette,
France

Abstract–Résumé

RELIABILITY AND AVAILABILITY OF REDUNDANT SYSTEMS: COMPUTATIONAL PROGRAM AND THE USE OF NOMOGRAMS.

A rigorous mathematical approach to determining the reliability and availability of repairable actively redundant systems — (r/m) systems — is considered for the case where the m units comprising the system are identical and the failure and repair rates, λ and μ respectively, are constant. The method used involves the Markov processes, operator calculus and matrix calculus. All the results of the study are handled by the FIDIAS program, which is a practical tool for calculating with a high degree of precision the reliability or availability of such (r/m) systems whatever the values of m and r. In the FIDIAS-TC version of FIDIAS it is possible to plot curves with a Benson plotter, so that nomograms are produced for rapid and simple determination of the probabilities of failure or non-availability of the (r/m) systems considered. The practical application of nomograms is of interest because (2/3) and (2/4) actively redundant systems are very often used in the control circuits of power reactors. It is shown how easily one can compare these two systems using nomograms and how one can determine λ or μ as a function of the anticipated result.

FIABILITE ET DISPONIBILITE DE SYSTEMES REDONDANTS: PROGRAMME DE CALCUL ET UTILISATION D'ABAQUES.

Le traitement mathématique rigoureux de la détermination de la fiabilité et de la disponibilité de systèmes en redondance active (r/m) réparable est effectué dans le cas où les m dispositifs sont identiques avec des taux de défaillances λ et de réparations μ constants. La méthode utilisée fait appel aux processus de Markov, au calcul symbolique et au calcul matriciel. Le programme FIDIAS reprend tous les résultats de l'étude et constitue un outil pratique pour calculer avec beaucoup de précision la fiabilité ou la disponibilité de tels systèmes (r/m) quels que soient m et r. La version FIDIAS-TC du programme FIDIAS permet le traçage de courbes au traceur Benson. Des abaques ont ainsi été réalisés permettant une détermination graphique rapide et simple des probabilités de défaillances ou de l'indisponibilité des systèmes (r/m) considérés. L'utilisation de systèmes en redondance active (2/3) et (2/4) étant très fréquente dans les circuits de contrôle-commande des réacteurs nucléaires, l'application pratique des abaques les concerne. On montre comment à l'aide de ces abaques la comparaison de ces deux systèmes est facile et comment on peut déterminer les paramètres λ ou μ en fonction du résultat escompté.

INTRODUCTION :

Un problème fréquemment rencontré par l'ingénieur fiabiliste travaillant pour le Nucléaire est celui du traitement de systèmes à redondance active (r/m). Si on trouve dans la littérature [1,2,3,4,6] des traitements rigoureux pour les systèmes non réparables, il n'en est pas de même lorsqu'il est nécessaire de prendre en compte une politique de maintenance. Néanmoins, de nombreuses approches sont proposées [1,2,3,4,6] et notre but est d'en développer une particulièrement bien adaptée. Cette méthode est analytique et fait appel aux processus de Markov, au calcul matriciel et au calcul symbolique. Un programme - FIDIAS - réalisé en Fortran à partir de cette méthode permet le calcul de la fiabilité, de la disponibilité et des probabilités de défaillances associées pour un système à redondance active (r/m) dont les évènements suivent des lois de défaillances et de réparation exponentielles. Des abaques générales ont été tracées par FIDIAS et leur utilisation pratique au traitement d'un cas concret montre la facilité avec laquelle ils permettent la comparaison de la fiabilité ou de la disponibilité de divers systèmes en (r/m).

Le développement mathématique intégral, ainsi que le listage et les spécifications du programme FIDIAS se trouvent dans les rapports cités en référence [7,8] .

1 - SYSTEMES EN REDONDANCE ACTIVE (r/m) :

Un tel système se compose de m dispositifs identiques actifs et il est nécessaire qu'au moins r d'entre eux soient en état de bon fonctionnement pour que le système puisse assurer sa fonction.

Lorsque k dispositifs sont inhibés et si $k>m-r$, le système est inhibé et on dit qu'il est en état de défaillance non sûre [5].

Lorsque le système (r/m) est un dispositif de déclenchement d'une action de sécurité un autre type de défaillance est à envisager, celui des déclenchements intempestifs. Par opposition au cas précédent, on parle ici de défaillance sûre. Lorsque k dispositifs sont en défaillance sûre et si $k \geq r$, alors le système est en état de défaillance sûre [5].
Seules les pannes non sûres seront traitées ici.

2 - ETATS D'UN SYSTEME REDONDANT D'ORDRE m - TRANSITION ENTRE ETATS - PROBABILITES ASSOCIEES :

Tout ensemble redondant comportant m dispositifs identiques peut être caractérisé à un instant quelconque par le nombre i des dispositifs fonctionnant correctement à cet instant.

On note "état i" : exactement i dispositifs en bon fonctionnement,
$i \to j$: transition de l'état i à l'état j,
$P(i,t)$: probabilité d'être à l'état i à l'instant t,
$d(i,t)dt$: probabilité de la transition i à i-1 entre t et t+dt,
$S(i,t)dt$: probabilité de la transition i à i+1 entre t et t+dt.

On suppose pour la suite que les transitions entre les différents états du système suivent un processus Markov , c'est-à-dire que la probabilité $P(i,t+dt)$ ne dépend que de l'état du système à l'instant t.

On ne conserve que les transitions $i \to i \pm 1$ donnant des termes en dt car les autres donnent des termes en dt^n négligeables devant les premiers.

3 - PROBABILITE DE SE TROUVER A L'ETAT i A L'INSTANT t+dt :

Trois possibilités de transition se présentent entre t et t+dt :

1° - $(i+1) \to i$: un dispositif tombe en panne entre t et t+dt et aucun n'est remis en service.

2° - $i \to i$: aucun dispositif ne tombe en panne entre t et t+dt et aucun n'est remis en service.

3° - $i-1 \to i$: aucun dispositif ne tombe en panne entre t et t+dt et un est remis en service.

Les probabilités de chacune de ces transitions s'expriment facilement à partir des probabilités définies au paragraphe précédent et on obtient la relation fondamentale suivante :

$$P(i,t+dt)=P(i+1,t)d(i+1,t)dt+P(i,t)\left[1-d(i,t)dt-S(i,t)dt\right]+P(i-1,t)S(i-1,t)dt \quad (1)$$

Soit en faisant apparaître la dérivée :

$$\frac{d}{dt}P(i,t)=P(i+1,t)d(i+1,t)-P(i,t)\left[d(i,t)+S(i,t)\right]+P(i-1,t)S(i-1,t) \quad (2)$$

A ces relations générales, il nous faut ajouter $\sum_{i=o}^{m} P(i,t)=1$ qui traduit l'incompatibilité des états i qui forment un système complet d'états.

4 - FIABILITE ET DISPONIBILITE D'UN SYSTEME (r/m) A REDONDANCE ACTIVE :

La relation fondamentale ci-dessus permet théoriquement de déterminer les probabilitésde tous les états i du système. Elle décrit aussi l'évolution du système au cours du temps.

Fiabilité :

C'est la probabilité pour le système de n'être jamais défaillant pendant une période donnée. Il ne faut donc conserver dans le processus que les états de bon fonctionnement provenant eux-mêmes d'états de bon fonctionnement. Pour éliminer les états de bon fonctionnement provenant d'états de panne, il suffit d'écrire que la probabilité de transition d'un état de panne vers un état de bon fonctionnement est nulle. On traduit, de ce fait, en disant que les états de pannes sont des états absorbants pour le processus de Markov, car il doit être interrompu dans le sens de la réparation dès qu'il atteint un état de panne.

On a donc : $S(i,t)=o \quad i<r$

et la fiabilité s'exprime par : $R(t)=\sum_{i=r}^{m} P(i,t)=1-\sum_{i=o}^{r-1} P(i,t)$

On notera $P_{ns}(t)$ la probabilité complémentaire qui est la probabilité de défaillance du système vis-à-vis des pannes non sûres.

Disponibilité :

C'est la probabilité pour le système d'être en bon état de fonctionnement à un instant donné. Il n'y a pas lieu d'introduire ici la notion d'état absorbant.

La disponibilité s'exprime par : $A(t)=\sum_{i=r}^{m} P(i,t)=1-\sum_{i=o}^{r-1} P(i,t)$

Seule la fiabilité sera étudiée par la suite, mais la disponibilité se traite de manière analogue.

5 - APPLICATION DE LA RELATION FONDAMENTALE AUX LOIS EXPONENTIELLES - CAS DE LA FIABILITE :

Nous supposons par la suite que les lois de défaillances et de réparations sont de la forme exponentielle et qu'il n'y a qu'une seule équipe de maintenance. Ces hypothèses conduisent à poser :

$d(i,t)=i.\lambda \quad o<i\leqslant m$

$S(i,t)=\begin{cases}\mu & \text{si i n'est pas absorbant}\\ o & \text{ailleurs.}\end{cases}$

λ est le taux de défaillances, μ le taux de réparations.

Introduisons les notations suivantes pour simplifier l'écriture :

$P_i=P(i,t)$, $\star(k,r)$ t.q $\star(k,r)=\begin{cases}1 & r\leqslant k<m\\ o & \text{ailleurs.}\end{cases}$

$d_i=d(i,t)$,

$S_i=S(i,t)$, Q_k t.q $Q_k=\begin{cases}1 & o<k\leqslant m\\ o & \text{ailleurs.}\end{cases}$

$\delta^i_m=\begin{cases}1 & i=m\\ o & i\neq m\end{cases}$

Par transformation de Laplace de la formule n° 2 et en utilisant p comme variable conjuguée de t on obtient :

$$\boxed{-\delta^i_m=\star(i-1,r)\mu\bar{P}_{i-1}-\left[Q_i.i.\lambda+\star(i,r)\mu+p\right]\bar{P}_i+Q_{i+1}(i+1)\lambda\;\bar{P}_{i+1}} \qquad (3)$$

$(\bar{P}_i=\mathcal{L}(P_i))$

La formule (3) décrit un système de Kramer de (m+1) équations à (m+1) inconnues (les $\bar{P}_i, i\in[o,m]$). La matrice (a^i_j) de ce système est tridiagonale et définie par :

$a^i_j=o \qquad i\neq j+1,j,j-1\ ;\ (i,j)\notin[o,m]^2$

$a^i_{i-1}=\begin{cases}\mu & r\leqslant i-1<m\\ o & \text{ailleurs}\end{cases}$

$a^i_i=\begin{cases}-p & i=o\\ -(i\lambda+p) & o<i<r\ ;\ i=m\\ -(i\lambda+\mu+\rho) & r\leqslant i<m\end{cases}$

$a^i_{i+1}=\begin{cases}(i+1)\lambda & o<i+1\leqslant m\\ o & \text{ailleurs}\end{cases}$

Si on définit **la matrice** $(b^i_j)_k$ par :

$b^i_j=a^i_j\ \forall\, j\neq k$ et $b^i_k=-\delta^i_m$

la solution du système est alors : $\bar{P}_k=\dfrac{\left|(b^i_j)_k\right|}{\left|(a^i_j)\right|}$ (4)

Le retour à l'original donne les P_k et la fiabilité s'exprime par :

$$R(t)=\sum_{k=r}^{m}P(k,t)=1-\sum_{k=o}^{r-1}P(k,t)$$

6 - ETUDE MATHEMATIQUE GENERALE DE LA FIABILITE - 1ère METHODE :

6.1 Déterminant de la matrice (a_j^i) :

Posons $\lambda_i = i.\lambda \ \forall i \in [o,m]$

$$\theta_i = \begin{cases} -(p+i\lambda) & \forall i \in [o,r-1] \ ; \ i=m \\ -(p+i\lambda+\mu) & \forall i \in [r,m[\end{cases}$$

$$\delta_i = \begin{cases} o & \forall i \in [o,r-1] \\ \mu & \forall i \in [r,m[\end{cases}$$

(a_j^i) s'écrit alors :

$$\begin{pmatrix} \theta_o & \lambda_1 & & 0 & \\ \delta_o & \theta_1 & \ddots & \lambda_{m-1} & \\ o & \delta_1 & \ddots & \theta_{m-1} & \lambda_m \\ & 0 & \ddots & \delta_{m-1} & \theta_m \end{pmatrix}$$

Les δ_i étant nuls jusqu'au rang (r-1) le développement du déterminant donne :

$$|(a_j^i)| = \theta_o.\theta_1 \ldots . \theta_{r-1} \begin{vmatrix} \theta_r & \ddots & \\ \lambda_r & \ddots & \lambda_m \\ & \ddots & \theta_m \end{vmatrix}$$ soit en posant Δ_r le déterminant

partiel mis en évidence :

$$\boxed{|(a_j^i)| = \theta_o.\theta_1 \ldots . \theta_{r-1}\Delta_r} \qquad (5)$$

6.2 Déterminants de la forme $|(b_j^i)_k|$:

La matrice $(b_j^i)_k$ est identique à la matrice (a_j^i) sauf en ce qui concerne la colonne n° k. On peut développer les $|(b_j^i)|_k$ et on trouve facilement que l'on a :

$$\boxed{|(b_j^i)_k| = (-1)^{m+k+1}\theta_o.\theta_1 \ldots . \theta_{k-1}.\lambda_{k+1} \ldots . \lambda_m} \quad \text{pour } k \in [o,r-1] \qquad (6)$$

Les $|(b_j^i)_k|$ sont les numérateurs des $\bar{P}_k$ qui par ailleurs ont tous le même dénominateur $|a_j^i|$. $\bar{P}_{ns}(t) = 1 - R(t)$ est égale à $\sum_{k=o}^{r-1} \bar{P}_k$. Il faut donc calculer $\sum_{k=o}^{r-1} |(b_j^i)_k|$.

A titre d'exemple : $|(b_j^i)_o| + |(b_j^i)_1| = \lambda_2.\lambda_3 \ldots . \lambda_m.(-1)^{m+1}[\lambda_1 - \theta_o]$

Mais $\lambda_1 - \theta_o = \lambda + p = -\theta_1$

d'où $|(b_j^i)_o| + |(b_j^i)_1| = (-1)^{m+2}\lambda_2.\lambda_3 \ldots . \lambda_m.\theta_1$

En procédant de proche en proche on arrive à :

$$\sum_{k=o}^{r-1} |(b_j^i)_k| = \lambda_r \ldots \lambda_m (-1)^{m+r} \theta_1 . \theta_2 \ldots \theta_{r-1} \tag{7}$$

6.3 Transformée de Laplace $\bar{P}_{ns}$ de la probabilité de défaillance sûre :

C'est le quotient de la formule (7) par la formule (5). Si on remplace θ_o par sa valeur on obtient :

$$\bar{P}_{ns} = (-1)^{m-r+1} \frac{r(r+1)\ldots m\lambda^{m-r+1}}{p\,\Delta_r} \tag{8}$$

Il reste à déterminer Δ_r pour avoir une formule exploitable.

6.4 Etude du déterminant Δ_r :

$$\Delta_r = \begin{vmatrix} \theta_r & & 0 \\ \delta_r^r & \ddots & \lambda_m \\ 0 & & \theta_m \end{vmatrix} \quad \text{soit en posant } r=m-i \quad \Delta_{m-i} = \begin{vmatrix} \theta_{m-i} & & 0 \\ \delta_{m-i} & \ddots & \lambda_m \\ 0 & & \theta_m \end{vmatrix}$$

Si on développe partiellement Δ_{m-i}, on trouve :

$$\Delta_{m-i} = \theta_{m-i} \Delta_{m-i+1} - \delta_{m-i} \lambda_{m-i+1} \Delta_{m-i+2} \tag{9}$$

La formule (9) est une formule de récurrence permettant de déterminer les Δ_{m-i} dès lors que Δ_o et Δ_1 sont connus.

Des considérations sur l'homogénéité de Δ_{m-i} par rapport à λ, μ et p montrent que c'est un polynôme de degré i+1, soit (m-r+1), et qu'on peut le mettre sous la forme :

$$\Delta_{m-i} \equiv \sum_{t=o}^{i+1} p^t \sum_{l=o}^{i+1-t} \alpha_{i,t}^{\,l} \lambda^l \mu^{i+1-t-l} \tag{10}$$

De plus cette forme est générale $\forall i=o,1,\ldots m-r$.

En exprimant Δ_{m-i+1} et Δ_{m-i+2} sous la même forme et en procédant à une identification terme à terme la relation de récurrence suivante sur les $\alpha_{i,k}^{\,u}$ est trouvée :

$$-\alpha_{i,k}^{\,u} = \alpha_{i-1,k-1}^{\,u} + (m-i)\alpha_{i-1,k}^{\,u-1} + \alpha_{i-1,k}^{\,u} + (m-i+1)\alpha_{i-2,k}^{\,u-1} \tag{11}$$

Il suffit dès lors de déterminer les $\alpha_{o,k}^{u}$ et les $\alpha_{1,k}^{u}$

$$\Delta_{m-o} = |\theta_m| = \theta_m = -(p+m\lambda) = \sum_{t=o}^{1} p^t \sum_{l=o}^{1-t} \alpha_{o,t}^{1} \lambda^1 \mu^{1-1-t} = (\alpha_{o,o}^{o}\mu + \alpha_{o,o}^{1}\lambda) + p\alpha_{o,1}^{o}$$

d'où $\boxed{\alpha_{o,o}^{o}=o \quad \alpha_{o,o}^{1}=-m \quad \alpha_{o,1}^{o}=-1}$ (12)

en procédant de même, on trouve :

$$\begin{array}{lll} \alpha_{1,o}^{o}=o & \alpha_{1,o}^{1}=o & \alpha_{1,o}^{2}=m(m-1) \\ \alpha_{1,1}^{o}=1 & \alpha_{1,1}^{1}=2m-1 & \\ \alpha_{1,2}^{o}=1 & & \end{array} \qquad (13)$$

Remarque : Le problème est alors résolu et peut être programmé facilement.
- Les $\alpha_{o,i}^{j}$ et $\alpha_{1,i}^{j}$ ne dépendent que de m et le degré de Δ_{m-i} ne dépend que de m-r.

6.5 Inversion de la formule n° 8 - Probabilité de pannes non sûres :

$\bar{P}_{ns}$ est le quotient d'un terme constant par un polynôme $p\Delta_r$ de degré m-r+2 avec son terme constant nul (<=> une racine est nulle).

$$\bar{P}_{ns} = \frac{N}{a_1 p^{m-r+2} + a_2 p^{m-r+1} + \ldots a_{m-r+3}}$$

Sous l'hypothèse que $p\Delta_r$ possède m-r+2 racines $(p_o,\ldots,p_{m-r+1})$ réelles distinctes, on peut écrire :

$$\bar{P}_{ns} = \frac{C_o}{(p-p_o)} + \frac{C_1}{(p-p_1)} + \ldots + \frac{C_{m-r+1}}{(p-p_{m-r+1})}$$

$$\text{avec } C_i = \frac{N}{a_1(p_i-p_o)(p_i-p_1)\ldots(p_i-p_{i-1})(p_i-p_{i+1})\ldots(p_i-p_{m-r+1})}$$

La transformation inverse de Laplace permet d'exprimer $P_{ns}(t)$ par :

$$\boxed{P_{ns}(t) = \sum_{i=o}^{m-r+1} C_i e^{p_i t}} \qquad (14)$$

Remarque : Le retour à l'original repose sur l'existence de racines réelles distinctes.

6.6 Nouvelle formulation du numérateur N :

Le retour à l'original repose sur le calcul des racines de Δ_r, seule une solution numérique par ordinateur peut nous donner ces racines, aussi il est intéressant de rechercher s'il est possible d'exprimer N en fonction de ces racines.

On démontre que $N = \prod_{i=o}^{m-r} p_i$ si les $p_i, i \in [o, m-r]$, sont les m-r+1 racines de Δ_r.

Cette formule permet le calcul des C_i avec plus de précision.:

$$\boxed{N = \prod_{i=o}^{m-r} p_i}$$

6.7 Forme asymptotique de $P_{ns}(t)$ au voisinage de zéro :

On démontre que $P_{ns}(t) \simeq \frac{t^i}{i!} \lim_{p \to \infty} p^{i+1} \bar{P}_{ns}$ au voisinage de t=o.

L'utilisation de cette formule conduit à :

$$\boxed{P_{ns}(r/m)(t)_{as} = \frac{r(r+1)\ldots m\lambda^{m-r+1}}{(m-r+1)!} t^{m-r+1}} \qquad (15)$$

6.8 Moyenne des temps de bon fonctionnement (M.T.B.F.) :

On démontre que C_{m-r+1} est égal à 1 et correspond à la racine nulle :

$$P_{ns}(t) = 1 + \sum_{i=o}^{m-r} C_i \, 1^{p_i t}$$

$$\text{donc} \quad R(t) = - \sum_{i=o}^{m-r} C_i \, 1^{p_i t}$$

Comme M.T.B.F. $= - \int_o^{\infty} R(t)dt$, par intégration :

$$\boxed{M.T.B.F.(r/m) = \sum_{i=o}^{m-r} \frac{C_i}{p_i}} \qquad (16)$$

7 - ETUDE MATHEMATIQUE GENERALE DE LA FIABILITE - 2ème METHODE :

7.1 Eléments de Mathématiques :

Si l'on considère la relation n° 2 donnant la dérivée de P(i,t) en fonction des probabilités P(i-1,t),P(i,t),P(i+1,t) on voit que l'on peut exprimer cette relation sous forme matricielle :

$$\frac{d}{dt} P_i = \beta_i^o P_o + \beta_i^1 P_1 + \dots\dots + \beta_i^m P_m$$

soit $D(P)=\beta(P)$ où (P) est le vecteur des probabilités, β la matrice des β_j^i et D l'opérateur de dérivation.

Il suffit d'intégrer ce système pour avoir la solution du problème. Il faut bien entendu une condition initiale pour t=o ; nous prenons :

$P_o = \begin{pmatrix} o \\ o \\ \vdots \\ o \\ 1 \end{pmatrix}$ ce qui signifie qu'à t=o on est dans l'état m c'est-à-dire que tous les dispositifs sont en état de bon fonctionnement.

$$P(t)=P_o + \int_o^t D(P) = P_o + \int_o^t \beta P(\tau) d\tau$$

$$\text{Soit } Q = \int_o^t .d\tau \text{ alors } Q\beta P = \int_o^t \beta P(\tau) d\tau$$

$$\text{d'où } P=P_o+Q\beta P=P_o+Q\beta(P_o+Q\beta P)= \dots$$

$$P=P_o\left[1+Q\beta+Q\beta Q\beta+Q\beta Q\beta Q\beta+\dots\dots\right] \tag{17}$$

Etudions $\underbrace{Q\beta . Q\beta \dots Q\beta}_{n \text{ fois}}$

$$Q\beta Q\beta \dots Q\beta = \int_o^t \beta d\tau \int_o^t \beta d\tau \dots \int_o^t \beta d\tau$$

Dans le cas présent, il est facile de voir que l'hypothèse des lois exponentielles de défaillance et de réparation fait que la matrice β est une matrice de termes constants :

$$\text{donc } Q\beta Q\beta \dots Q\beta = \beta^n \int_o^t d\tau \int_o^t d\tau \dots \int_o^t d\tau = \beta^n \frac{t^n}{n!}$$

La formule (17) s'écrit alors :

$$P(t)=P_o(1+\beta t+\beta^2 \frac{t^2}{2} + \dots + \beta^n \frac{t^n}{n!} + \dots\dots) \tag{18}$$

Soit, par définition de l'exponentielle de matrice $\boxed{P(t)=P_o e^{\beta t}}$ (19)

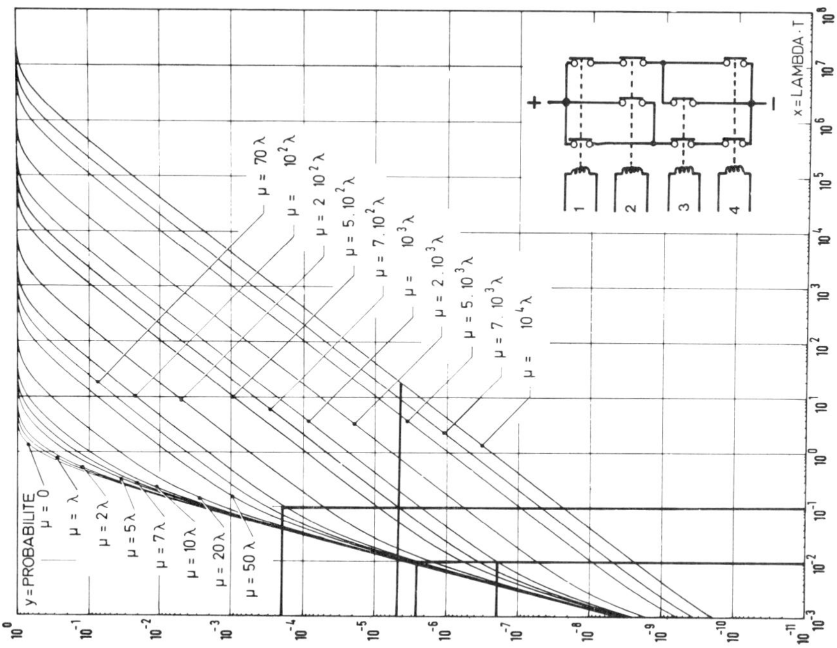

FIG.2. Probabilité de défaillances non sûres du système redondant 2/4.

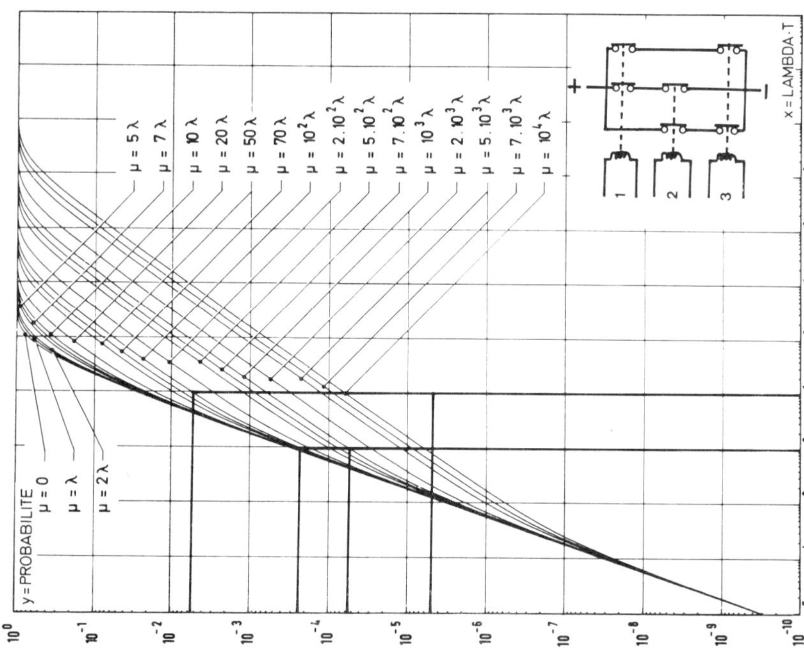

FIG.1. Probabilité de défaillances non sûres du système redondant 2/3.

7.2 Application à notre cas :

β est identique à (a^i_j) dans laquelle la variable p a disparu et le calcul doit se faire par la formule (18). Il est évident que cette méthode ne présente un intérêt que lorsque la série $\sum_{k=o}^{\infty} \beta^k \frac{t^k}{k!}$ converge rapidement. Pour cela il est nécessaire que les produits $\beta^i_j .t$ soient assez petits.

8 - PROGRAMME FIDIAS :

FIDIAS (pour Fiabilité et DIsponibilité par une méthode Analytique de Systèmes redondants (r/m) est un programme Fortran en double précision. Il reprend tous les résultats de l'étude ci-dessus.

Une étude analogue à celle de la Fiabilité a été faite pour la Disponibilité et FIDIAS en utilise aussi les résultats.

FIDIAS calcule : la fiabilité et la probabilité de défaillance associée
la moyenne des temps de bon fonctionnement.(MTBF)
la disponibilité et la probabilité de défaillances associées (indisponibilité).
la disponibilité moyenne.

FIDIAS donne aussi les formes asymptotiques au voisinage de zéro.

Les données sont entrées sans format grâce à l'appel de sous-programmes de lecture. Le calcul des racines du polynôme est effectué par un sous-programme bibliothèque.

La version FIDIAS-TC possède en plus un sous-programme de traçage de courbe grâce auquel des abaques ont été réalisés.

9 - ABAQUES :

Les diverses probabilités calculées par FIDIAS dépendent des paramètres r,m,λ,μ,t. Pour un système (r/m) donné, si $\lambda_1 t_1=\lambda_2 t_2$ et $\mu_1/\lambda_1=\mu_2/\lambda_2$, on a $P(r,m,\lambda_1,\mu_1,t_1)=P(r,m,\lambda_2,\mu_2,t_2)$. Des réseaux de courbes ont été tracés à r et m fixés en fonction de λt ou de $1-e^{-\lambda t}$,μ/λ servant de paramètre. Ces réseaux sont des abaques généraux permettant de déterminer facilement et graphiquement à partir des paramètres r,m,λ,μ,t la probabilité recherchée avec une précision satisfaisante. A l'heure actuelle des abaques sont disponibles pour les systèmes (1/1),(1/2),(1/3),(2/2),(2,3),(2/4),(3/3), en ce qui concerne les défaillances non sûres et l'indisponibilité [8]. La fiabilité et la disponibilité s'obtiennent par soustraction à 1 des valeurs précédentes. La transformation (r/m) $\rightarrow$(m-r+1/m) permet de traiter les défaillances sûres du système (r/m) en utilisant les abaques concernant les défaillances non sûres du système (m-r+1/m).

10 - UTILISATION DES ABAQUES DANS UN CAS PRATIQUE :

On rencontre fréquemment dans le circuit de contrôle commande des réacteurs nucléaires des chaînes de protection comportant des circuits logiques chargés d'exécuter des coïncidences en 2/3 ou 2/4 sur des signaux provenant de chaînes de mesures. Ces circuits logiques fonctionnent à manque de tension et sont réalisés à partir de relais à deux contacts (fig. n° 1 et 2). Il suffit qu'il y ait manque de tension sur les bobines de 2 relais pour que l'action de sécurité soit déclenchée. Ces circuits correspondent très exactement aux systèmes en redondance active (r/m) pour lesquels les abaques ont été tracés, aussi, nous allons montrer comment on peut déterminer les probabilités de défaillances de ces systèmes logiques (2/3) et (2/4).

10.1 Comparaison des probabilités de défaillances non sûres des (2/3) et (2/4) :

Pour de tels circuits à manque de tension les défaillances non sûres correspondent aux contacts immobilisés en position fermée (contacts collés, bobine maintenue exitée etc...).

- Un taux de défaillance non sûre (λ) de $10^{-6}h^{-1}$ par relais est une valeur courante. Un taux de réparation de $10^{-3}h^{-1}$ et une durée T de 10^4h étant choisis quelle est la probabilité de défaillances non sûres de chacun de ces circuits ?

  $\lambda T=10^{-2}$, $\mu/\lambda=10^{-3}$les abaques 1 et 2 donnent directement après report de ces valeurs : $P_{ns}(2/3)=5,6.10^{-5}$
  $P_{ns}(2/4)= 2.10^{-7}$

- Imaginons maintenant que nous ayons des relais avec un λ de $10^{-5}h^{-1}$ seulement, en conservant le même μ de $10^{-3}h^{-1}$ et la même durée T de 10^4h on trouve :

  $$\lambda T=10^{-1},\ \mu/\lambda=10^2 \Rightarrow \begin{array}{l} P_{ns}(2/3)=5,6.10^{-3} \\ P_{ns}(2/4)=2\ .10^{-4} \end{array}$$

  Ceci montre que le passage de $10^{-5}h^{-1}$ à $10^{-6}h^{-1}$ (facteur 10 sur le λ) fait gagner un facteur 100 sur la fiabilité du (2/3) et un facteur 1000 sur celle du (2/4).

- Conservons maintenant $\lambda=10^{-6}h^{-1}$ mais prenons $\mu=10^{-4}h^{-1}$, toujours avec $T=10^4$h on trouve :

  $$\lambda T=10^{-2}, \mu/\lambda=10^2 \Rightarrow \begin{array}{l} P_{ns}(2/3)=2,5.10^{-4} \\ P_{ns}(2/4)=2,7.10^{-6} \end{array}$$

  On constate que pour ces valeurs de λ et de T les courbes $\mu/\lambda=10^2$ sont pratiquement confondues avec celle à μ/λ=o qui caractérise un système sans maintenance. $1/\mu$ représente une moyenne de la somme, durée de l'intervalle entre test + temps de réparation. Comme on a $1/\mu=10^4$h , $T=10^4$h, il est évident que, dans ce cas, le paramètre μ n'a pratiquement aucune influence. En fait dès que $\frac{1}{\mu}>T$ on doit se retrouver sur la courbe à μ=o. Ici l'influence de μ ne commence à se faire sentir qu'à partir de $10^{-4}h^{-1}$. Et on voit qu'un facteur 10 sur μ (passage de 10^{-4} à $10^{-3}h^{-1}$) fait gagner un facteur 4,4 pour le (2/3) et un facteur 14 pour le (2/4).

- On veut maintenant obtenir $P_{ns}(2/4)=5.10^{-6}$ avec $T=10^4$h et $\lambda=10^{-5}h^{-1}$

  $\lambda T=10^{-1}$; on trouve $\mu/\lambda= 7,1\cdot10^2$ $\quad \mu = 7,1\cdot10^{-3}h^{-1}$

  De même pour $P_{ns}(2/3)=5.10^{-6}$ avec $T=10^4$h et $\lambda=10^{-5}h^{-1}$.
  On voit qu'il est nécessaire que μ/λ soit supérieur à 10^4 car on est en dehors de l'abaque.

- Comment choisir λ et μ en fonction du résultat escompté?
  On trace une droite parallèle à l'axe des λT à la valeur de la probabilité désirée. Elle coupe les courbes à μ/λ=cte et chacun de ces points d'intersection donne un couple de valeur (λ,μ). On trace ensuite la courbe λ=f(μ) à l'aide de ces points. Cette courbe permet facilement de choisir le meilleur compromis (λ,μ).

Sur la figure n° 3 on a représenté λ=f(μ) dans le cas des (2/3) et (2/4) pour une probabilité de défaillance de 5.10^{-6} à $T=10^4$h. En coordonnées log-log ces courbes incurvées au départ deviennent ensuite des droites ce qui permet de les extrapoler facilement.

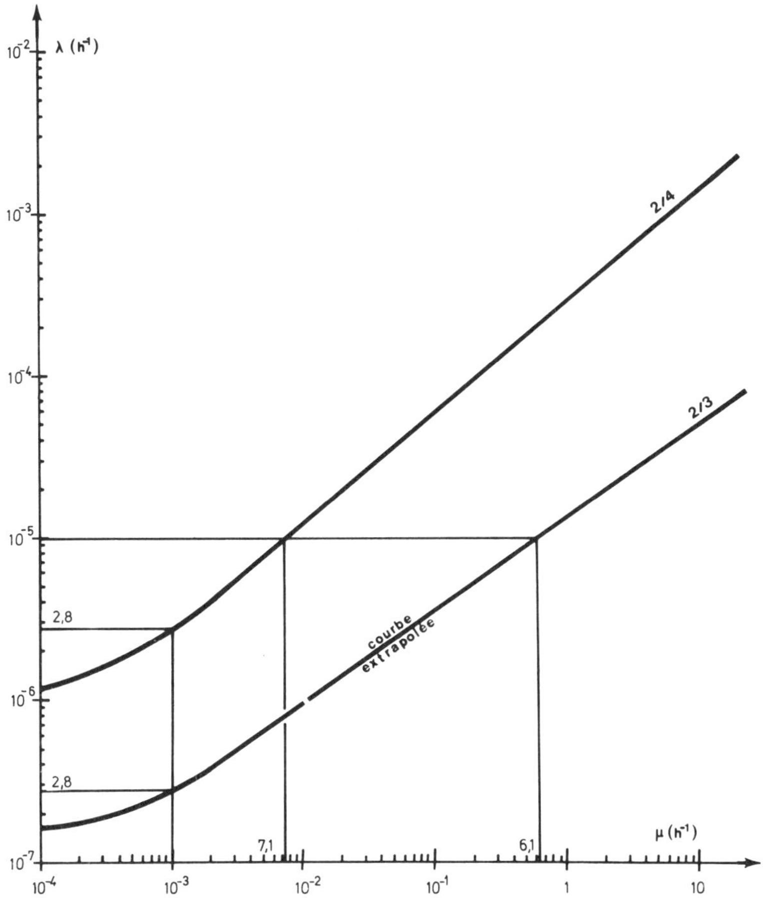

FIG.3. Courbes $\lambda = f(\mu)$ pour $P_{ns} = 5 \cdot 10^{-6}$ et $T = 10^4$ h.

Soit le problème suivant à résoudre : comment obtenir $P_{ns}(2/4)=P_{ns}(2/3)=5.10^{-6}$ pour $T=10^4h$ et $\mu=10^{-3}h^{-1}$?

On trouve immédiatement $\lambda=2,8.10^{-7}h^{-1}$ pour le (2/3)
$\lambda=2,8.10^{-6}h^{-1}$ pour le (2/4)

D'où un gain d'un facteur 10 sur les λ en faveur du (2/4).

- Voyons maintenant un autre problème : comment obtenir :
$P_{ns}(2/3)=P_{ns}(2/4)=5.10^{-6}$ avec $\lambda=10^{-5}$ et $T=10^4h$?
Comme ci-dessus on retrouve $\mu=7,1.10^{-3}h^{-1}$ pour le (2/4), mais alors qu' avec l'abaque n° 1, on ne pouvait pas trouver le μ correspondant pour le (2/3) l'extrapolation de la courbe $\lambda=f(\mu)$ relative au (2/3) donne $\mu=6,1.10^{-1}h^{-1}$ pour le (2/3).
Donc alors qu'un facteur 10 suffisait sur les λ (à μ égal) il faut un facteur 87 sur les μ (à λ égal) pour obtenir le même résultat pour le (2/3) que pour le (2/4).

Il ressort de tout cela, que, vis-à-vis des pannes non sûres, le système en (2/4) est meilleur que le (2/3). Il est possible de faire les mêmes études vis-à-vis de la disponibilité et vis-à-vis des pannes sûres, mais nous ne les développerons pas ici.

11 - CONCLUSION :

L'utilisation des abaques tracés à l'aide du programme FIDIAS-TC est facile et permet de déterminer rapidement les probabilités de défaillances vis-à-vis des pannes non sûres ou sûres, ainsi que la disponibilité des systèmes en redondance active (r/m). Ils permettent aussi facilement de trouver la valeur des paramètres à adopter en fonction du résultat escompté. La précision de ces résultats est acceptable, mais dès lors que l'on désire plus de précision, il suffit de passer au calcul par le programme FIDIAS.

Pour l'ingénieur fiabiliste du domaine Nucléaire, qui doit traiter couramment de tels systèmes, FIDIAS et les abaques qui en découlent, forment un outil de travail très intéressant et donnent très rapidement la réponse recherchée.

REFERENCES

[1] SCHWOB, M., PEYRACHE, G., Traité de fiabilité, Masson, Paris.

[2] GARNIER, J., Fiabilité, Cours de l'Ecole nationale supérieure des techniques avancées (1971).

[3] GREEN, A.E., BOURNE, A.J., Reliability Technology, Wiley Intersciences.

[4] BAZOVSKY, I., Fiabilité, Théorie et pratique de la sûreté de fonctionnement, Dunod, Paris.

[5] CARNINO, A., Terminologie et méthodologie utilisées en analyse de fiabilité, CEA, Rapport DSN/SETS n° 11 (février 1973).

[6] QUENEE, R., Sûreté et disponibilité d'une centrale nucléaire, Logiques à vote majoritaire, CEA, Note technique DSN/SETS n° 17 (juin 1973).

[7] SIGNORET, J.P., Fiabilité et disponibilité d'un système à redondance active (r/m) réparable, CEA, Rapport DSN/SETS n° 20 (août 1973).

[8] SIGNORET, J.P., BLIN, A., Fiabilité et disponibilité de systèmes en redondance active (r/m) réparable, Solution générale et programme, CEA, suite du rapport DSN/SETS n° 20 (à paraître).

MODELLING MAXIMUM LIKELIHOOD ESTIMATION OF AVAILABILITY*

R. A. WALLER, G. L. TIETJEN
Los Alamos Scientific Laboratory,
Los Alamos, N. Mex.

G. W. ROCK
Lafayette College,
Easton, Pa.,
United States of America

Abstract

MODELLING MAXIMUM LIKELIHOOD ESTIMATION OF AVAILABILITY.

Suppose the performance of a nuclear powered electrical generating power plant is continuously monitored to record the sequence of failure and repairs during sustained operation. The purpose of this study is to assess one method of estimating the performance of the power plant when the measure of performance is availability. That is, we determine the probability that the plant is operational at time t. To study the availability of a power plant, we first assume statistical models for the variables, X and Y, which denote the time-to-failure and the time-to-repair variables, respectively. Once those statistical models are specified, the availability, A(t), can be expressed as a function of some or all of their parameters. Usually those parameters are unknown in practice and so A(t) is unknown. This paper discusses the maximum likelihood estimator of A(t), say $\widehat{A(t)}$, when the time-to-failure model for X is an exponential density with parameter, λ, and the time-to-repair model for Y is an exponential density with parameter, θ. Under the assumption of exponential models for X and Y, it follows that the instantaneous availability at time t is $A(t) = \lambda/(\lambda+\theta) + \theta/(\lambda+\theta)\exp\{-[(1/\lambda)+(1/\theta)]t\}$ with $t > 0$. Also, the steady-state availability is $A(\infty) = \lambda/(\lambda+\theta)$. We use the observations from n failure-repair cycles of the power plant, say $X_1, X_2, \ldots, X_n, Y_1, Y_2, \ldots, Y_n$ to present the maximum likelihood estimators of A(t) and $A(\infty)$. The exact sampling distributions for those estimators and some statistical properties are discussed before a simulation model is used to determine 95% simulation intervals for A(t). The methodology is applied to two examples which approximate the operating history of two nuclear power plants.

1. INTRODUCTION AND DEFINITIONS

We suppose a nuclear powered electrical generating plant (hereafter called a power plant) is placed into operation and that its performance is continually monitored to decide whether or not it is operating properly. Any power plant is a complex system composed of many widely varying components. However for the purposes of this study, we consider a power plant as a single entity under investigation rather than a composite of several components. Although our presentation is directed toward the operation of a power plant, the techniques contained herein are applicable to the operation of any system or component of interest.

A variety of measures such as reliability, hazard rate, mean-time-to-failure, and availability may be used to investigate the performance of a power plant, however, we restrict the discussion to instantaneous availability and steady-state (long-run) availa-

* Under the auspices of the Reactor Research and Development Division of the U.S. Energy Research and Development Administration.

bility. By instantaneous availability, $A(t)$, we mean the probability that the power plant is operating successfully at time t. By steady state or long-run availability, $A(\infty)$, we mean the probability of successful operation for large values of t. That is,

$$A(t) = P(\text{successful operation at time } t),$$

and

$$A(\infty) = \lim_{t\to\infty} A(t).$$

Specification of probability models for two variables associated with the power plant operation is the first step in an availability study. The first variable, X, denotes the time-to-failure for the power plant, while the second variable, Y, represents the time-to-repair for the same power plant. Independent exponential models are assumed for X and Y. The second step in an availability analysis is the derivation of $A(t)$ and $A(\infty)$ which in general provides functions of the parameters of the probability models specified for X and Y.

In practice, the parameters of the models assumed for X and Y are usually unknown. Thus, $A(t)$ and $A(\infty)$ are unknown quantities for most practical problems. To estimate either $A(t)$ or $A(\infty)$, we observe the power plant through n failure-repair cycles and let $(X_1, X_2, \ldots, X_n)$ and $(Y_1, Y_2, \ldots, Y_n)$ denote the time-to-failure and time-to-repair data, respectively. Those data are then used to calculate estimates for $A(t)$ and $A(\infty)$. Although there are a variety of possible methods for estimating availability, this presentation is limited to an investigation of the sampling distributions for the maximum likelihood estimators, $\hat{A}(t)$ and $\hat{A}(\infty)$. The development consists of three parts. First, explicit descriptions of $A(t)$ and $A(\infty)$ for the exponential time-to-failure and time-to-repair models are provided in Section 2. Second, Section 3 is devoted to a study of the statistical properties of the estimators, $\hat{A}(t)$ and $\hat{A}(\infty)$. Third, examples are used to illustrate the application of the proposed availability analyses to two power plants in Section 4.

2. MODEL DESCRIPTION AND MAXIMUM LIKELIHOOD ESTIMATOR

We assume that the time-to-failure, X, is an exponential variable with parameter, λ. Therefore, the density function of X is given by

$$f(x) = \frac{1}{\lambda}e^{-x/\lambda}, \quad \text{for } x>0.$$

The parameter λ is the mean time-to-failure for the power plant being studied and may assume any positive value. The hazard rate for the object is $1/\lambda$. Similarly, we assume an exponential model with the parameter θ for the time-to-repair variable, Y, so that the density function of Y is

$$g(y) = \frac{1}{\theta}e^{-y/\theta}, \quad \text{for } y>0.$$

Here θ denotes the mean time-to-repair for Y while $1/\theta$ is the hazard rate of Y. Thus, θ may be any positive number.

Shooman [1] and Gnedenko, Belyayev, and Solovyev [2] show that for the exponential models assumed above, the instantaneous availability of the power plant is

$$A(t) = \frac{\lambda}{\lambda+\theta} + \frac{\theta}{\lambda+\theta} \exp\left\{ -\left[\frac{t}{\lambda} + \frac{t}{\theta}\right]\right\}. \quad (1)$$

Further the steady-state availability is obtained by taking the limit of (1) as t goes to infinity. Thus

$$A(\infty) = \frac{\lambda}{\lambda+\theta} . \quad (2)$$

Note that both A(t) and $A(\infty)$ are functions of λ and θ, the parameters of the exponential models for X and Y. Thus, we can use the result of Zehna [3] to write the maximum likelihood estimators of A(t) and $A(\infty)$ by substituting the maximum likelihood estimators of λ and θ into Equations (1) and (2).

To get the maximum likelihood estimators of λ and θ, we suppose that the power plant has been on test for n failure-repair cycles. Then the available data are the n times-to-failure, $(X_1, X_2, \ldots, X_n)$, and the n times-to-repair, $(Y_1, Y_2, \ldots, Y_n)$. Actually, the data are two independent exponential samples.

The maximum likelihood procedure as developed in Kendall and Stuart [4] yields the following estimators:

$\hat{\lambda} = \frac{\Sigma X_i}{n}$ is the maximum likelihood estimator of λ.

$\hat{\theta} = \frac{\Sigma Y_i}{n}$ is the maximum likelihood estimator of θ.

Therefore the maximum likelihood estimators of A(t) and $A(\infty)$ are

$$\widehat{A(t)} = \frac{\Sigma X_i}{\Sigma X_i + \Sigma Y_i} + \frac{\Sigma Y_i}{\Sigma X_i + \Sigma Y_i} \exp\left\{-nt\left(\frac{1}{\Sigma X_i} + \frac{1}{\Sigma Y_i}\right)\right\} \quad (3)$$

and

$$\widehat{A(\infty)} = \frac{\Sigma Y_i}{\Sigma X_i + \Sigma Y_i} . \quad (4)$$

Note that the second term of $\widehat{A(t)}$ goes to zero exponentially as t increases. Thus, $\widehat{A(t)}$ approaches $\widehat{A(\infty)}$ as t increases. Also, as n increases, ΣX_i and ΣY_i approach $n\lambda$ and $n\theta$, respectively. Therefore, as n increases, $\widehat{A(t)}$ and $\widehat{A(\infty)}$ approach the true values of A(t) and $A(\infty)$, respectively.

Some statistical properties of the maximum likelihood estimators, $\widehat{A(t)}$ and $\widehat{A(\infty)}$, are developed in the next section.

3. STATISTICAL PROPERTIES OF THE MAXIMUM LIKELIHOOD ESTIMATORS OF AVAILABILITY

The development in this section is divided into two parts. We first discuss the distributional properties of $\hat{A}(\infty)$, the steady-state availability. Second, we investigate the distributional properties of $\hat{A}(t)$, the instantaneous availability.

3.1. Distributional properties of $\hat{A}(\infty)$.

Since the data, $(X_1, X_2, \ldots, X_n)$ and $(Y_1, Y_2, \ldots, Y_n)$, represent two independent samples of size n from exponential models having parameters λ and θ, respectively, it follows

$$W = \frac{2\Sigma X_i}{\lambda} \text{ and } Z = \frac{2\Sigma Y_i}{\theta}$$

are independent chi-square variables with 2n degrees of freedom. Thus, by starting with the product of two independent chi-square densities as the joint density of W and Z, the usual change of variable techniques (see Kendall and Stuart [5]) show that the density of the steady-state availability estimator,

$$\hat{A}(\infty) = \frac{\lambda W}{\lambda W + \theta Z}$$

is given by

$$h(a) = \frac{\Gamma(2n)}{[\Gamma(n)]^2}\left(\frac{\theta}{\lambda}\right)^n (1-a)^{n-1}\, a^{n-1}\left(1-\left[\frac{\lambda-\theta}{\lambda}\right]a\right)^{-2n}, \quad 0<a<1. \tag{5}$$

For most practical examples λ is much greater than θ. For example, we expect the mean time-to-failure, λ, to exceed the mean time-to-repair, θ, for a power plant. With that assumption and with the aid of Euler's integral representation of the Gauss hypergeometric series (see Abramowitz and Stegun [6]) it follows that the r-th ordinary moment of $\hat{A}(\infty)$ is given by

$$E\left[(\hat{A}(\infty))^r\right] = \int_0^1 a^r h(a)\, da = \frac{\Gamma(2n)\Gamma(n+r)}{\Gamma(2n+r)\Gamma(n)}\left(\frac{\theta}{\lambda}\right)^n \sum_{i=0}^{\infty} \frac{(2n)_i (n+r)_i}{(2n+r)_i} \frac{\left[\frac{(\lambda-\theta)}{\lambda}\right]}{i!} \tag{6}$$

where $(a)_i = a(a+1)\ldots(a+i-1)$. Using (6) for r = 1 and 2 provides the mean and variance to be

$$E(\hat{A}(\infty)) = n\left(\frac{\theta}{\lambda}\right)^n \sum_{i=0}^{\infty} \frac{(n+1)_i}{(2n+i)} \frac{\left[\frac{\lambda-\theta}{\lambda}\right]^i}{i!} \tag{7}$$

and

$$V(\hat{A}(\infty)) = (n+1)n\left(\frac{\theta}{\lambda}\right)^n \sum_{i=0}^{\infty} \frac{(n+2)_i}{(2n+i+1)(2n+i)} \frac{\left[\frac{\lambda-\theta}{\lambda}\right]^i}{i!} - [E(\hat{A}(\infty))]^2 \tag{8}$$

It is not easy to evaluate the moments expressed in (6) and (7). In fact the series in (7) converges very slowly for moderate values (say 20 or more) of n unless θ and λ are nearly equal. That characteristic is true for the two examples given in Section 4.

3.2. Distributional properties of $\hat{A}(t)$

We now turn to the estimation of A(t) and investigate the properties of $\hat{A}(t)$. Note that

$$A(t) = A(\infty) + \frac{\Sigma Y_i}{\Sigma X_i + \Sigma Y_i} \exp\left\{-nt\left[\frac{1}{\Sigma X_i} + \frac{1}{\Sigma Y_i}\right]\right\}.$$

Therefore, we need only study the properties of the second term. However, distributional theory does not appear to yield a useful form for the density of $\hat{A}(t)$. Likewise, the moments of A(t) are more cumbersome than those for $\hat{A}(\infty)$ and require bivariate numerical integration for solution. In particular, one expression for the moments of $\hat{A}(t)$ includes powers of the right-hand-side of (6) and powers of a double integral which must be evaluated numerically. Numerical integration of the first and second moments of A(t) for a selected set of λ and θ values gave results which agreed very closely (at least two and usually three decimal places) to the empirical moments of 5000 simulations. Hence, simulation is a tractable and practical method for studying the distributional properties of $\hat{A}(t)$. We believe that the empirical distribution produced by 5000 simulations is adequate for all practical purposes and can be used to obtain bounds on $\hat{A}(t)$. Some applications of the simulation approach for availability analysis of specific examples follow in the next section.

4. EXAMPLES

The United States Atomic Energy Commission document WASH-1203-73 [7] provides graphical summaries for the performance of nuclear powered electrical generating plants operating in the United States of America. The data for the two examples in this section closely approximate the operating history of two of those power plants. In the examples which follow, n denotes the number of failure-repair cycles observed; x_i denotes the ith time-to-failure; y_i denoted the ith time-to-repair; $\tilde{A}(t)$ is the estimate of the availability provided by the ratio of "successful operating time" to "total time" at the end of each failure-repair cycle; and $\hat{A}(t)$ is the maximum likelihood estimate of A(t).

4.1. Example 1

The (x_i, y_i) data in Table I approximate the times-to-failure and time-to-repair observed for 18 failure-repair cycles of the Yankee Nuclear Power Plant in Rowe, Massachusetts, U.S.A., from 1960 to 1973. The observations are measured in years.

For these data n=18, Σx_i = 9.279 and Σy_i = 2.287, so the maximum likelihood estimates of A(∞) and A(t) are

$$\hat{A}(\infty) = \frac{9.279}{9.279+2.287} = .802$$

TABLE I

DATA AND AVAILABILITY ESTIMATES FOR 18 FAILURE-REPAIR CYCLES OF A POWER PLANT

i	x_i	y_i	$\widetilde{A_i(t)}$	$\widehat{A_i(t)}$
1	.156	.033	.825	.833
2	.033	.056	.680	.815
3	.189	.044	.740	.803
4	.989	.322	.750	.802
5	.856	.267	.755	.802
6	.678	.111	.777	.802
7	.844	.256	.774	.802
8	.789	.100	.792	.802
9	.644	.044	.808	.802
10	.611	.122	.810	.802
11	.456	.022	.819	.802
12	.589	.100	.822	.802
13	.467	.033	.829	.802
14	.367	.033	.832	.802
15	.133	.089	.827	.802
16	.878	.022	.840	.802
17	.211	.233	.825	.802
18	.389	.400	.802	.802

and

$$\widehat{A(t)} = .802 + .198\ e^{-9.81t},$$

respectively.

The next point of interest for study is the distributional behavior of the estimator, $\widehat{A(t)}$. To do that, we use the estimates $\hat{\lambda} = \frac{\Sigma x_i}{18} = .515$ and $\hat{\theta} = \frac{\Sigma y_i}{18} = .127$ as λ and θ in a simulation model to generate 5000 simulations of 18 failure-repair cycles of the power plant. From those simulations we obtained the 2.5 and 97.5 percentiles of the simulated distribution of $\widehat{A(t)}$ for the 18 time points representing the end of the 18 failure-repair cycles. Also, since $\widehat{A(t)}$ converges to $\widehat{A(\infty)}$ very rapidly for this example we investigate the behavior of $\widehat{A(t)}$ for very small t by evaluating the simulation results for t = .001, .005, .010, .050, and .100. Figure 1 graphs $\widetilde{A(t)}$, $\widehat{A(t)}$ and 95% simulation intervals for the data in Table I. Note that the scale is broken at t = .5 to indicate an expanded scale for small values of t. Also, the values at successive values of t are joined to produce continuous functions.

4.2. Example 2

The data in Table II approximate the 16 failure-repair cycles observed for the Humboldt Bay Power Plant, Unit No. 3, Eureka,

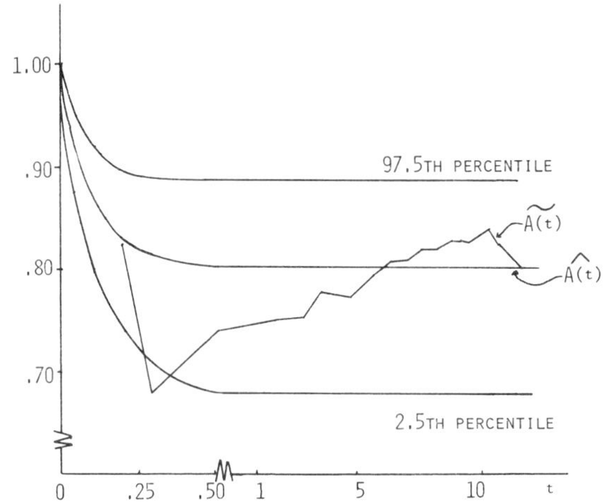

FIG. 1. Graphic representation of $\tilde{A}(t)$, $\hat{A}(t)$ and 95% simulation intervals for the data in Table I.

TABLE II

DATA AND AVAILABILITY ESTIMATES FOR 16 FAILURE-REPAIR CYCLES OF A POWER PLANT

i	x_i	y_i	$\tilde{A}_i(t)$	$\hat{A}_i(t)$
1	.467	.067	.875	.865
2	.178	.033	.866	.865
3	.356	.100	.834	.865
4	1.022	.155	.851	.865
5	.133	.044	.844	.865
6	.756	.067	.862	.865
7	.256	.033	.864	.865
8	.456	.033	.872	.865
9	.867	.044	.886	.865
10	.533	.100	.881	.865
11	.678	.067	.885	.865
12	.144	.044	.881	.865
13	.267	.022	.883	.865
14	.322	.189	.866	.865
15	.811	.156	.863	.865
16	.767	.100	.865	.865

California, U.S.A., from 1963 to 1973. The data are measured in years.

For the data in Table II, $\Sigma x_i = 8.013$ and $\Sigma y_i = 1.254$, so the maximum likelihood estimates of $A(\infty)$ and $A(t)$ as provided by (4) and (3) are

$$\widehat{A(\infty)} = \frac{8.013}{8.013+1.254} = .865$$

and

$$\widehat{A(t)} = .865 + .135\ e^{-14.76t},$$

respectively.

Next we simulated 5000 observations for 16 failure-repair cycles of a power plant. The sample values, $\hat{\lambda} = \frac{\Sigma x_i}{16} = .501$ and $\hat{\theta} = \frac{\Sigma y_i}{16} = .078$, were used for λ and θ in the simulation model. The 2.5 and 97.5 percentiles produced by those simulations are graphed in Figure 2 along with $\widetilde{A(t)}$ and $\widehat{A(t)}$. As for Figure 1, the scale near zero is expanded to better illustrate the behavior of $\widehat{A(t)}$ for small values of t because $\widehat{A(t)}$ converges to $\widehat{A(\infty)}$ very rapidly for this example.

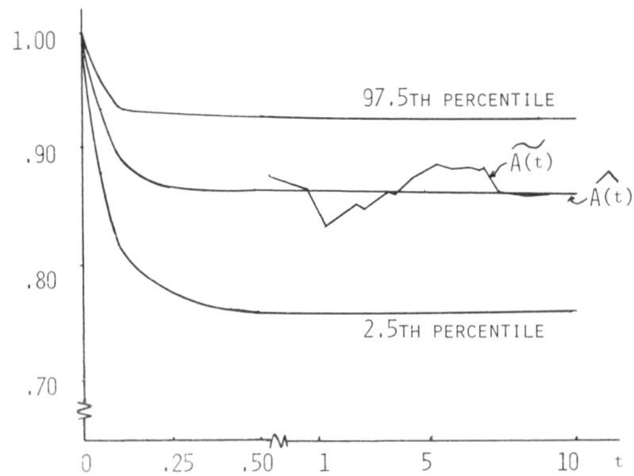

FIG. 2. Graphic representation of $\widetilde{A(t)}$, $\widehat{A(t)}$ and 95% simulation intervals for the data in Table II.

A brief discussion of the preceding examples and the methods of availability analyses proposed in this paper follow in the next section.

5. DISCUSSION

The major proposal in this paper is that we use simulation models to construct lower bounds or interval estimates for the availability of power plants. Direct analytic methods are not tractable in that numerical integration is required for most if not all, practical problems. The application and performance of the simulation approach is illustrated by two examples in Section 4.

One alternative to the simulation modeling approach proposed would be to use the first four moments to select a member of the

Pearson family of density functions to model the sampling distribution of $\hat{A}(t)$. After some consideration, that approach was rejected for two reasons: First, a considerable amount of numerical integration is required to obtain the first four moments of $\hat{A}(t)$. Second, for most choices of λ and θ the type of curve changes as t varies. Those properties make it very difficult to use Pearson type curves for modeling $\hat{A}(t)$.

In constrast to the difficulties encountered when selecting and fitting Pearson models for $\hat{A}(t)$, the simulation program runs very efficiently. The complete simulation results for Examples 1 and 2 required 65 seconds of computer time on a CDC 7600 machine. That efficiency, along with the simplicity of calculating the estimates of λ and θ for use in the simulation model make it possible to simulate the distribution of $\hat{A}(t)$ and print out percentiles, the median, the mean, the variance and any other statistics of interest to the user. We believe that the simulation model is more useful to an experimenter than any model obtained by a method which requires numerical integration and analytic approximations such as a Pearson type model.

In closing, we observe that the data in the two examples in Section 4 agreed quite well with the simulation model. Only one of the eighteen observed values of $\tilde{A}(t)$ fell outside of the 95% simulation intervals in Example 1, while all sixteen of the $\tilde{A}(t)$ values in Example 2 fell inside the 95% simulation bounds. That type of performance is enough to encourage the exploration of simulation models for studying availability estimators when assuming other than exponential models for the time-to-failure and/or time-to-repair variables.

REFERENCES

[1] Shooman, M. L. (1968). Probabilistic Reliability, An Engineering Approach. McGraw-Hill Book Company, New York.

[2] Gnedenko, B. V., Belyayev, Y. K., Solovyev, A. D. (1969). Mathematical Method of Reliability Theory. Academic Press, New York.

[3] Zehna, P. (1966). "Invariance of Maximum Likelihood Estimation." Annals of Mathematical Statistics, 37, p. 744.

[4] Kendall, M. G. and Stuart, A., (1967). The Advanced Theory of Statistics, Volume II, 2nd Edition. Hafner, New York. Chapter 18.

[5] Kendall, M. G. and Stuart, A., (1963). The Advanced Theory of Statistics, Volume I, 2nd Edition. Hafner, New York. pp. 23-24, 247-248.

[6] Abramowitz, M. and Stegun, I. A., (1970). Handbook of Mathematical Functions with Formulas, Graphs and Mathematical Tables. U. S. Government Printing Office, Washington, D.C.

[7] "Operating History: U. S. Nuclear Power Reactors" WASH 1203-73, Division of Reactor Development and Technology, U. S. Atomic Energy Commission, 1973.

RECHERCHE D'UNE FONCTION D'ACCELERATION POUR LE CALCUL DE LA FIABILITE DE SYSTEMES REDONDANTS
Application aux défaillances de mode commun

R. GONNOT
FRAMATOME, Courbevoie,
France

Abstract–Résumé

FINDING AN ACCELERATION FUNCTION FOR CALCULATING THE RELIABILITY OF REDUNDANT SYSTEMS – APPLICATION TO COMMON MODE FAILURES.

While it may be reasonable to assume that the reliability of a system – the design of which is perfectly known – can be evaluated, it seems less easy to be sure that overall reliability is correctly estimated in the case of multiple redundancies arranged in sequence. Framatome is trying to develop a method of evaluating overall reliability correctly for its installations. For example, the protection systems in its power stations considered as a whole are such that several scram signals may be relayed in sequence when an incident occurs. These signals all involve the same components for a given type of action, but the components themselves are in fact subject to different stresses and constraints, which tend to reduce their reliability. Whatever the sequence in which these signals are transmitted (in a fast-developing accident, for example), it is possible to evaluate the actual reliability of a given system (or component) for different constraints, as the latter are generally obtained via the transient codes. By applying the so-called "equal probability" hypothesis one can estimate a reliability acceleration function taking into account the constraints imposed. This function is linear for the principal failure probability distribution laws. By generalizing such a method one can: (1) Perform failure calculations for redundant systems (or components) in a more general way than is possible with event trees, since one of the main parameters is the constraint exercised on that system (or component); (2) Determine failure rates of components on the basis of accelerated tests (up to complete failure of the component) which are quicker than the normal long-term tests (statistical results of operation); (3) Evaluate the multiplication factor for the reliability of a system or component in the case of common mode failures. The author presents the mathematical tools required for such a method and describes their application in the cases mentioned above.

RECHERCHE D'UNE FONCTION D'ACCELERATION POUR LE CALCUL DE LA FIABILITE DE SYSTEMES REDONDANTS – APPLICATION AUX DEFAILLANCES DE MODE COMMUN.

S'il est raisonnable de penser que la fiabilité d'un système (la conception en étant connue parfaitement) peut être évaluée, il apparaît moins aisé de s'assurer que la fiabilité globale, en cas de redondances multiples intervenant en séquence, est correcte. Pour ses installations, Framatome essaie de développer une méthode d'approche de l'évaluation correcte de cette fiabilité. Par exemple, pour l'ensemble des systèmes de protection de ses centrales, plusieurs signaux d'arrêt d'urgence peuvent intervenir en séquence lors d'un accident. Ces signaux sollicitent les mêmes composants pour le même type d'action, alors que ceux-ci sont en fait soumis à des contraintes différentes, ce qui a pour effet de diminuer leur fiabilité. Quel que soit l'ordre séquentiel d'apparition de ces signaux (lors d'un accident d'évolution rapide, par exemple), il est possible d'évaluer pour différentes contraintes la fiabilité réelle d'un système donné (ou d'un composant), ces contraintes étant généralement obtenues par les codes de transitoires. L'application de l'hypothèse dite ≪des probabilités égales≫ permet d'estimer une fonction d'accélération de la fiabilité tenant compte des contraintes exercées. Cette fonction est linéaire pour les principales lois de distribution de probabilités de défaillance. La généralisation d'une telle méthode permet: 1°) d'effectuer les calculs de la défaillance de systèmes (ou de composants) redondants, dans un cas plus général que ceux que l'on peut étudier à l'aide d'arbres d'événements, puisque l'un des paramètres principaux est la contrainte exercée sur ce système (ou ce composant); 2°) de rechercher des taux de défaillance de composants à partir d'essais accélérés (jusqu'à la mort du composant) plus rapides que les essais de longue durée conventionnels (résultats statistiques d'exploitation); 3°) d'évaluer le facteur multiplicatif de la fiabilité d'un système ou d'un composant, en cas de défaillance de mode commun. Le mémoire présente «l'arsenal mathématique» nécessaire à une telle méthode, ainsi que son application aux divers points énumérés ci-dessus.

POSITION DU PROBLEME

La croissance rapide de l'industrie nucléaire, stimulée récemment par la crise de l'énergie, provoque une recrudescence des critiques à son sujet. De plus, une population grandissante, se sentant directement concernée, exige une évaluation du risque associé.

Des travaux ont été effectués récemment dans ce domaine tant en Grande-Bretagne qu'aux Etats-Unis d'Amérique, principalement par l'équipe du professeur Rasmussen sous l'égide de l'USAEC. Les statistiques manquant, de nouvelles techniques ont été mises au point, développant les analyses d'arbres d'événements et d'arbres de fautes, compte tenu de l'évaluation de la fiabilité des systèmes de protection et de sauvegarde, afin d'établir la probabilité d'occurrence de séquences accidentelles. Ceci permet d'estimer les risques associés à une installation nucléaire donnée par rapport à ceux de la vie quotidienne.

Lors d'une telle évaluation, de nombreuses questions restent en suspens:

- Avec quelle précision peut-on connaître l'émission de produits radioactifs consécutive à une séquence accidentelle?
- Comment évaluer les effets à long terme de la radioactivité ainsi relâchée sur l'environnement et la population?
- Peut-on être sûr de n'avoir oublié aucune séquence accidentelle possible?
- A-t-on évalué correctement la fiabilité des systèmes et de leurs composants intervenant lors des séquences accidentelles?

S'il est raisonnable de penser que la fiabilité d'un système (la conception en étant connue parfaitement) peut être évaluée, il apparaît moins aisé de s'assurer que la fiabilité globale, en cas de redondances multiples intervenant en séquences, est correcte.

Framatome étudie pour ses installations une méthode d'approche de l'évaluation correcte de cette fiabilité.

Par exemple, pour l'ensemble des systèmes de protection de ses centrales, plusieurs signaux d'arrêt d'urgence peuvent intervenir en séquence lors d'un accident:

- flux nucléaire élevé, niveau source
- flux nucléaire élevé, niveau intermédiaire
- flux nucléaire élevé, niveau puissance, seuil bas
- flux nucléaire élevé, niveau puissance, seuil haut
- ΔT température élevée
- ΔT surpuissance.

Tous ces signaux sollicitent les mêmes composants, pour le même type d'action, alors que ceux-ci sont en fait soumis à des contraintes différentes, ce qui a pour effet de diminuer leur fiabilité. De plus, dans quel ordre interviennent ces signaux pour un accident d'évolution rapide?

Par application de l'hypothèse des «probabilités égales», quel que soit l'ordre séquentiel d'apparition de ces signaux, il est possible d'évaluer pour différentes contraintes la fiabilité réelle d'un système donné (ou d'un composant), ces contraintes étant généralement obtenues à l'aide des codes de transitoires.

La généralisation de cette méthode permet:

1°) d'effectuer les calculs de la défaillance de systèmes (ou de composants) redondants dans un cas plus général que ceux que l'on peut étudier à l'aide d'arbres d'événements ou de la théorie des processus de Markov appliquée aux «stress-strengths», puisque l'un des paramètres principaux est la contrainte exercée sur ce système (ou ce composant);

2°) de rechercher des taux de défaillance de composants à partir d'essais accélérés (jusqu'à la mort du composant) plus rapides que les essais de longue durée conventionnels (résultats statistiques d'exploitation);

3°) d'évaluer le facteur multiplicatif de la fiabilité d'un système ou d'un composant, en cas de défaillance de mode commun.

Dans ce mémoire «l'arsenal mathématique» nécessaire à une telle méthode sera présenté, ainsi que son application aux divers points énumérés ci-dessus.

1. HYPOTHESE DES PROBABILITES EGALES

Soit

X	une contrainte donnée
t	un temps
$F_x(t)$	la probabilité d'occurrence d'une défaillance dans l'intervalle de temps (0,t) sous la contrainte X
$F_x(\Delta t/t, X^*)$	la probabilité d'occurrence d'une défaillance dans l'intervalle de temps $(t, t+\Delta t)$ sous la contrainte X, lorsque durant l'intervalle de temps (0, t) le composant a été soumis à la contrainte X*.

D'après l'abaque théorique des probabilités de défaillance d'un composant sous différentes contraintes (voir fig.1), on peut formuler l'hypothèse dite «des probabilités égales» suivante:

$$F_x(t_2) = F_y(t_1) = P \tag{1}$$

On peut alors écrire:

$$F_x(\Delta t/t_2, X) = F_y(\Delta t/t_1, Y) \tag{2}$$

2. RECHERCHE D'UNE «FONCTION D'ACCELERATION»

2.1. Définition de la «fonction d'accélération»

Soit $P_1 = F_x(t)$ et $P_2 = F_y(t)$ deux fonctions de probabilités de défaillance d'un dispositif (composant ou système) sous les contraintes X et Y.

Supposons, d'après la figure 1:

- que quel que soit le temps t considéré

$$F_x(t) < F_y(t) \tag{3}$$

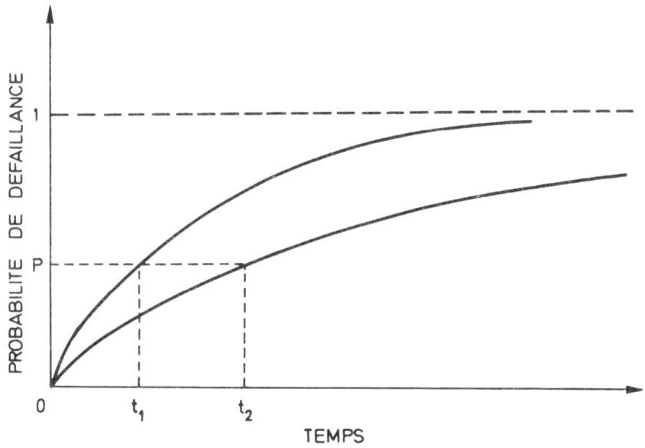

FIG. 1. Probabilité de défaillance sous les contraintes X et Y.

- que les contraintes X et Y vérifient l'inégalité

$$Y > X \tag{4}$$

Soit t_1 le temps qui permet d'obtenir une certaine probabilité de défaillance sous la contrainte Y et soit t_2 le temps qui correspond à la même probabilité de défaillance sous la contrainte X.

Il est évident qu'à chaque valeur du temps t_1 on peut associer une valeur définie par

$$t_2 = g(t_1) \tag{5}$$

telle que

$$F_y(t_1) = F_x[g(t_1)] \tag{6}$$

ceci d'après l'équation (1).

La fonction $g(t_1)$ ainsi définie sera appelée «fonction d'accélération».

D'après les équations (3) et (4), la fonction d'accélération $g(t_1)$ est une fonction monotone croissante de t_1, ce qui permet d'écrire

$$g(0) = 0$$

$$\lim_{t\to\infty} g(t) = \infty$$

On peut ramener le problème posé à celui de la détermination d'une probabilité de défaillance $F_x(t)$ d'après la probabilité de défaillance supposée connue $F_y(t)$. Un tel problème sera résolu à condition de définir la fonction d'accélération $g(t)$.

2.2. Exemples de fonction d'accélération

Dans ce paragraphe, nous allons étudier les cas où les probabilités de défaillance $F_x(t)$ et $F_y(t)$ suivent les mêmes lois de distribution, mais avec des paramètres différents, pour les distributions les plus couramment rencontrées: Weibull, lognormale, exponentielle.

2.2.1. Distribution de Weibull

Dans ce cas

$$\left.\begin{aligned} F_x(t) &= 1 - \exp\left(-\frac{t^{\alpha_x}}{\beta_x}\right) \\ F_y(t) &= 1 - \exp\left(-\frac{t^{\alpha_y}}{\beta_y}\right) \end{aligned}\right\} \quad (7)$$

où α_x, α_y et β_x, β_y sont les paramètres des fonctions (paramètre de forme et paramètre d'échelle).

D'après les équations (5) et (6):

$$F_x(t_2) = F_y(t_1)$$

$$t_2 = g(t_1) = \left(\frac{\beta_x}{\beta_y}\right)^{1/\alpha_x} t_1^{\alpha_y/\alpha_x} \quad (8)$$

Dans le cas où $\alpha_x = \alpha_y = 1$ (si la loi de distribution est une loi exponentielle), l'équation (8) devient

$$t_2 = g(t_1) = \frac{\beta_x}{\beta_y}\, t_1 \quad (9)$$

Dans le cas où $\alpha_x = \alpha_y = 2$ (si la loi de distribution est une loi de Raleigh), l'équation (8) devient

$$t_2 = g(t_1) = \sqrt{\frac{\beta_x}{\beta_y}}\, t_1 \quad (10)$$

Pour ces deux cas particuliers, la loi d'accélération est une loi linéaire. On peut donc écrire $\underline{g(t) = Ct}$, C étant une constante dépendant des paramètres des lois de distribution sous les contraintes X et Y.

2.2.2. Distribution lognormale

Dans ce cas:

$$\left.\begin{aligned} F_x(t) &= \frac{1}{\sqrt{2\pi}} \int_{-\infty}^{U_x} \exp\left(-\frac{Z^2}{2}\right) dZ \\ F_y(t) &= \frac{1}{\sqrt{2\pi}} \int_{-\infty}^{U_y} \exp\left(-\frac{Z^2}{2}\right) dZ \end{aligned}\right\} \quad (11)$$

où

$$\left.\begin{aligned} U_x &= \frac{\log t - \log a_x}{\sigma_x} \\ U_y &= \frac{\log t - \log a_y}{\sigma_y} \end{aligned}\right\} \quad (12)$$

a_x, a_y, σ_x, σ_y étant les paramètres des distributions lognormales considérées; $\log a_x$ et $\log a_y$ l'espérance mathématique de la variable aléatoire, durée de vie du composant ou du système; σ_x et σ_y les variances de cette variable dans les mêmes conditions.

D'après l'équation (6) on peut écrire

$$U_x(t_2) = U_y(t_1)$$

$$t_2 = g(t_1) = \frac{a_x}{a_y^{\sigma_x/\sigma_y}} \; t_1^{\sigma_x/\sigma_y} \tag{13}$$

Lorsque les variances σ_x et σ_y sont égales, l'équation (13) devient

$$t_2 = g(t_1) = \frac{a_x}{a_y} \, t_1 \tag{14}$$

La loi d'accélération est encore une loi d'accélération linéaire: $g(t) = Ct$, C dépendant des paramètres des lois de distribution sous les contraintes X et Y.

3. PROPRIETES DES FONCTIONS LINEAIRES D'ACCELERATION

3.1. Vis-à-vis du type de la loi de distribution

D'après ce qui précède, on peut voir que la conservation du type de distribution sous les contraintes X et Y n'est pas une condition suffisante pour que la fonction d'accélération soit linéaire.

Cependant, on peut affirmer qu'une fonction d'accélération linéaire ne modifie aucunement le type de la loi de distribution. En effet, lorsque la fonction d'accélération suit une loi linéaire:

$$g(t) = Ct \tag{15}$$

D'après l'équation (6) on obtient

$$F_y(t) = F_x(Ct) \tag{16}$$

Donc les fonctions $F_y(t) = F_x(Ct)$ et $F_x(t)$ décrivent les mêmes types de loi de distribution.

3.2. Vis-à-vis des moments centrés et non centrés des distributions

De même lorsque la fonction d'accélération est linéaire, il existe certaines relations entre les moments des distributions de $F_x(t)$ et $F_y(t)$. En effet:

- pour les moments non centrés $\alpha_x^{(k)}$, $\alpha_y^{(k)}$ d'ordre k:

$$\left.\begin{aligned} \alpha_x^{(k)} &= \int_0^\infty t^k \, d\,\{F_x(t)\} \\ \alpha_y^{(k)} &= \int_0^\infty t^k \, d\,\{F_y(t)\} = \int_0^\infty t^k \, d\,\{F_x[g(t)]\} \end{aligned}\right\} \tag{17}$$

– pour les moments centrés $\beta_x^{(k)}$, $\beta_y^{(k)}$ d'ordre k:

$$\left.\begin{aligned} \beta_x^{(k)} &= \int_0^\infty (t - \alpha_1^{(1)})^k \, d\{F_x(t)\} \\ \beta_y^{(k)} &= \int_0^\infty (t - \alpha_y^{(1)})^k \, d\{F_y(t)\} \\ &= \int_0^\infty (t - \alpha_y^{(1)})^k \, d\{F_x[g(t)]\} \end{aligned}\right\} \qquad (18)$$

D'où, en utilisant la relation d'accélération linéaire (15), on obtient

$$\frac{\alpha_x^{(k)}}{\alpha_y^{(k)}} = \frac{\beta_x^{(k)}}{\beta_y^{(k)}} = C^k = \left(\frac{t_2}{t_1}\right)^k \qquad (19)$$

3.3. Vis-à-vis de la durée de vie

Si on a une fonction d'accélération linéaire, l'équation (6) devient, d'après (15)

$$F_y(t) = F_x(Ct)$$

Par conséquent, la durée de vie du dispositif sous la contrainte Y suit la même loi $F_y(t)$ que la durée de vie sous la contrainte X (d'après le paragraphe 3.1), à un changement de variable près. On peut donc déterminer $F_x(t_2)$ connaissant C et $F_y(t_1)$.

Or la constante C peut être déterminée expérimentalement. En effet, si on limite l'essai sous une contrainte X à un temps (qui peut être très court) t_Q, qui correspond à une probabilité de défaillance Q, la connaissance de la répartition de la durée de vie sous la contrainte Y permet de déterminer le temps t'_Q qui correspond à une probabilité de défaillance Q sous la contrainte Y, raisonnement qui s'explique par la figure 2.

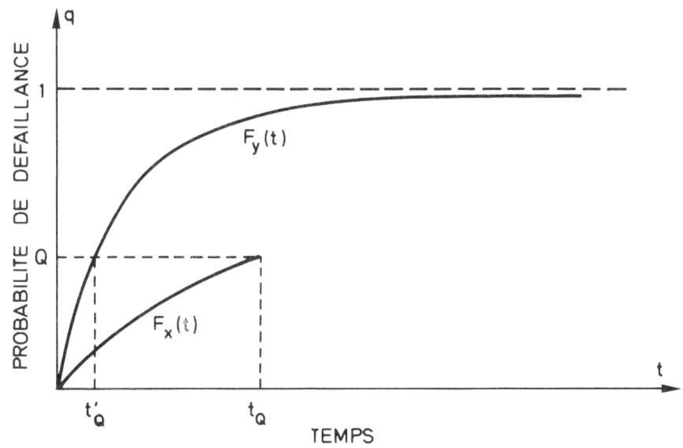

FIG. 2. Détermination de la durée de vie sous les contraintes X et Y.

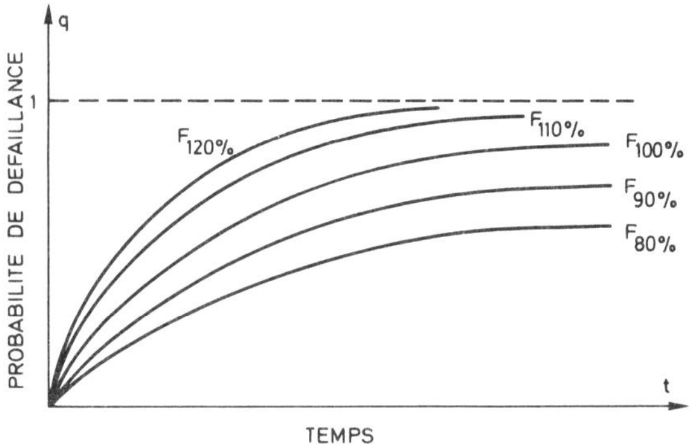

FIG. 3. Abaque des probabilités de défaillance suivant les contraintes exercées. (Les contraintes sont chiffrées en pourcentage du fonctionnement normal du dispositif considéré, pris égal à 100.)

Graphiquement la valeur de la constante C d'accélération est alors déterminée par $C = t_Q/t'_Q$, d'où l'équation (5) devient

$$t_2 = \frac{t_Q}{t'_Q}\, t_1 \tag{20}$$

équation qui permet de connaître, quel que soit le temps t_1, la loi de répartition de la durée de vie sous la contrainte X, et d'établir l'abaque (fig.3) des probabilités de défaillance d'un dispositif.

3.4. Relation de sommation linéaire de défaillances

Supposons qu'une expérience se fasse de la façon suivante: l'ensemble des systèmes est essayé

- d'abord sous une contrainte X durant un temps t_1
- puis sous une contrainte Y (Y > X) durant un temps t_2.

Ainsi:

- à l'instant t_1 où la première étape est terminée, on a la probabilité de défaillance Q_1
- à l'instant $t_1 + t_2$ où l'essai est achevé, on a la probabilité de défaillance Q_2.

Si l'on désigne par t'_1 la durée au cours de laquelle on aurait la probabilité de défaillance Q_1 sous la contrainte Y d'après la définition même de la fonction d'accélération, on peut écrire:

$$t_1 = g(t'_1) \tag{21}$$

Désormais appelons:

$\Theta_x(Q_2)$ la durée au cours de laquelle on obtiendrait la probabilité de défaillance Q_2 sous la contrainte X

$\Theta_y(Q_2)$ la durée au cours de laquelle on aurait la même probabilité de défaillance sous la contrainte Y.

On a alors:

$$\Theta_x(Q_2) = g[\Theta_y(Q_2)] \tag{22}$$

Lorsque l'hypothèse des probabilités égales est vérifiée, on peut écrire

$$\Theta_y(Q_2) = t'_1 + t_2 \tag{23}$$

soit, d'après l'équation (22):

$$\Theta_x(Q_2) = g[\Theta_y(Q_2)] = g(t'_1 + t_2) \tag{24}$$

De plus, si on suppose que la fonction d'accélération g(t) est linéaire, g(t) = Ct, les équations (21) et (24) deviennent

$$t_1 = Ct'_1 \tag{25}$$

$$\Theta_x(Q_2) = C\,\Theta_y(Q_2) = Ct'_1 + Ct_2 \tag{26}$$

d'où la relation de sommation linéaire de défaillance

$$\frac{t_1}{\Theta_x(Q_2)} + \frac{t_2}{\Theta_y(Q_2)} = 1 \tag{27}$$

Lorsque les valeurs t_1, t_2 et $\Theta_y(Q_2)$ sont connues, ceci permet d'effectuer un essai accéléré pour déterminer $\Theta_x(Q_2)$.

Il est possible de changer l'expérience et d'essayer l'ensemble des systèmes sous la contrainte Y jusqu'à l'apparition de défaillances. Dans ce cas la relation de sommation linéaire de défaillance (27) devient

$$\frac{t_1}{t_x} + \frac{\overline{t_2}}{t_y} = 1 \tag{28}$$

où

$\overline{t_2}$ est la moyenne de la durée de vie sous la contrainte Y après la contrainte X
t_x est la moyenne de la durée de vie sous la contrainte X constante
t_y est la moyenne de la durée de vie sous la contrainte Y constante.

4. APPLICATION AU CALCUL DE LA PROBABILITE DE DEFAILLANCE DE SYSTEMES REDONDANTS

4.1. Hypothèses, expression générale de la probabilité de défaillance

Soit $f_1(x)$ la probabilité de défaillance du système sous la contrainte normale de fonctionnement et $f_2(x)$ la probabilité de défaillance du système «redondant» sous la même contrainte.

On suppose que les deux systèmes sont identiques et que par conséquent $f_1(x) = f_2(x)$.

La probabilité de défaillance de l'ensemble des deux systèmes à un instant t donné est

$$Q_2(t) = \int_0^t f_1(\tau)\ q_2(t/\tau)\ d\tau \tag{29}$$

où $q_2(t/\tau)$ est la probabilité conditionnelle que la défaillance du système redondant se réalise à l'instant t, la défaillance du système initial étant réalisée à l'instant τ.

Cette probabilité $q_2(t/\tau)$ est, d'après le théorème de Bayles,

$$q_2(t/\tau) = \int_0^t f_2(\tau,x)\,dx \tag{30}$$

où $f_2(\tau,x)$ est la densité conditionnelle de probabilité de défaillance du système redondant.

Cette densité conditionnelle de probabilité de défaillance du système redondant est déterminée séparément pour les deux intervalles de temps $(0,\tau)$ et (τ,t) par:

$$f_2(\tau,x) = \begin{cases} f_2^{(1)}(x) & \text{pour} \quad x < \tau \\ f_2^{(2)}(x) & \text{pour} \quad x \geqq \tau \end{cases} \tag{31}$$

En utilisant l'équation (30) dans (29) et en utilisant l'équation (31) on obtient:

$$Q_2(t) = \int_0^t f_1(\tau) \left[\int_0^t f_2(\tau,x) \right] d\tau$$

soit

$$Q_2(t) = \int_0^t f_1(\tau) \left[\int_0^\tau f_2^{(1)}(x)\,dx + \int_\tau^t f_2^{(2)}(x)\,dx \right] d\tau \tag{32}$$

Or la probabilité de défaillance du système «normal» sous la contrainte normale de fonctionnement est donnée par

$$q_1(\tau) = \int_0^\tau f_1(y)\,dy$$

d'où

$$f_1(\tau) = \frac{d}{d\tau}\, q_1(\tau) \tag{33}$$

Par conséquent, l'équation (32) devient

$$Q_2(t) = \int_0^t \left[\int_0^\tau f_2^{(1)}(x)\,dx + \int_\tau^t f_2^{(2)}(x)\,dx \right] \frac{d}{d\tau} \{q_1(\tau)\}\, d\tau \qquad (34)$$

Cette dernière relation nous montre que la fiabilité d'un ensemble utilisant la redondance dépend de la manière dont cette redondance fonctionne. On peut de manière générale distinguer trois catégories de fonctionnement du système à redondance.

4.2. Redondance de première catégorie

Supposons que le système redondant fonctionne en permanence sous la contrainte normale de fonctionnement (redondance dite de première catégorie). La fiabilité est alors indépendante de l'instant τ lorsqu'il devient le système de service.

Dans ce cas, la densité conditionnelle de probabilité de défaillance s'exprime par

$$f_2(\tau,x) = \begin{cases} f_2^{(1)}(x) = f_2(x) & \text{pour} \quad x < \tau \\ f_2^{(2)}(x) = f_2(x) & \text{pour} \quad x \geqq \tau \end{cases} \qquad (35)$$

L'équation (32) est alors le produit de deux intégrales indépendantes. Une de ces intégrales est la probabilité de défaillance à un instant t du système normal, la seconde correspond à la probabilité de défaillance du système redondant au même instant, ce qui peut être représenté par la figure 4.

4.3. Redondance de deuxième catégorie

Si le système redondant fonctionne jusqu'au moment de son entrée en service sous une contrainte si faible que sa «ressource de vie» n'est

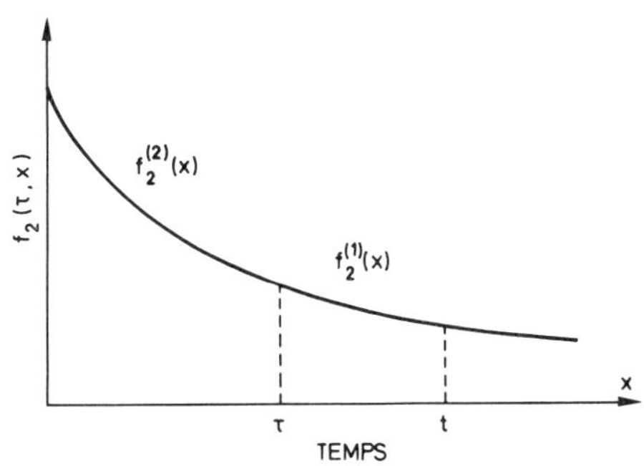

FIG. 4. Densité de probabilité de défaillance du système redondant dans le cas d'une redondance de première catégorie.

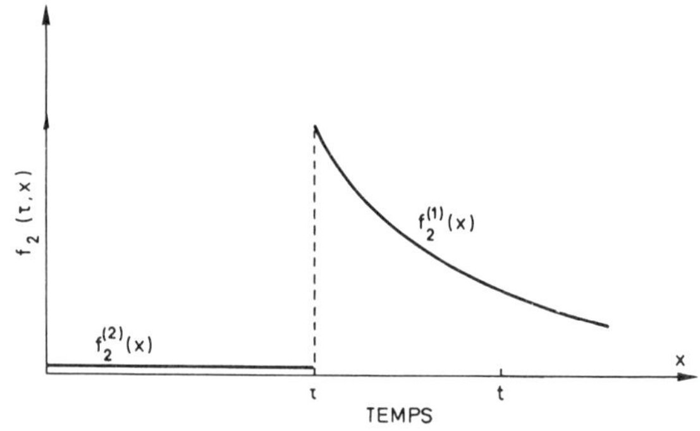

FIG. 5. Densité de probabilité de défaillance du système redondant dans le cas d'une redondance de deuxième catégorie.

pas dépensée (redondance dite de deuxième catégorie), l'équation (31) devient

$$f_2(\tau,x) = \begin{cases} f_2^{(1)}(x) = 0 & \text{pour } x < \tau \\ f_2^{(2)}(x) = f_2(x-\tau) & \text{pour } x \geqq \tau \end{cases} \tag{36}$$

La courbe de densité de probabilité de défaillance du système redondant a alors l'allure de la figure 5.

4.4. Redondance de troisième catégorie

Elle regroupe les redondances n'appartenant à aucune des deux autres catégories, ce qui est en fait le cas le plus général: le système redondant fonctionne jusqu'à sa mise en service lors de la défaillance du système normal sous une contrainte d'utilisation plus faible que la contrainte normale.

Dans ce cas, l'emploi des équations (32) et (34) est difficile, puisque $f_2(\tau,x)$ n'est pas connu habituellement. A vrai dire, la fonction $f_2^{(1)}(x)$ est toujours connue, mais la fonction $f_2^{(2)}(x)$ ne l'est pas. Il semble difficile de déterminer $f_2(\tau,x)$ à partir des données statistiques puisque cette fonction dépend des deux paramètres τ et x.

Il apparaît donc honnête de formuler l'hypothèse que le séjour du système redondant sous une contrainte d'utilisation moins rude que la contrainte normale ne modifie pas sa loi de fiabilité, c'est-à-dire:

$$f_2(\tau,x) = \begin{cases} f_2^{(1)}(x) & \text{pour } x < \tau \\ f_2^{(2)}(x-\tau) & \text{pour } x \geqq \tau \end{cases} \tag{37}$$

ce qui peut être représenté par la figure 6.

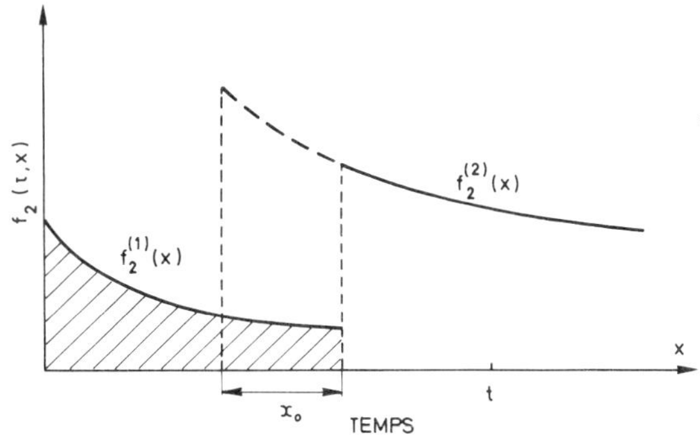

FIG. 6. Densité de probabilité de défaillance du système redondant, dans le cas d'une redondance de troisième catégorie.

L'équation (37) peut se simplifier par l'emploi de l'hypothèse des probabilités égales du chapitre 1 sous la forme suivante:

$$f_2(\tau,x) = \begin{cases} f_2^{(1)}(x) & \text{pour } x < \tau \\ f_2(x+x_0-\tau) & \text{pour } x \geqq \tau \end{cases} \quad (38)$$

où x_0 est déterminé par l'équation

$$\int_0^{\tau} \lambda_2^{(1)}(x)\,dx = \int_0^{x_0} \lambda_2(x)\,dx \quad (39)$$

où $\lambda_2^{(1)}(x)$ et $\lambda_2(x)$ sont les taux de défaillance correspondant respectivement aux densités $f_2^{(1)}(x)$ et $f_2(x)$.

Cette dernière équation est équivalente à la suivante:

$$\int_0^{\tau} f_2^{(1)}(x)\,dx = \int_0^{\tau} f_2(x)\,dx \quad (40)$$

4.5. Redondance de troisième catégorie pour une probabilité de défaillance à loi exponentielle

4.5.1. Recherche de la fiabilité des systèmes

Dans le cas où l'ensemble des deux systèmes normal et redondant suivent des lois exponentielles

$$\left.\begin{aligned} f_1(x) &= f_2(x) = \lambda_1 \exp(-\lambda_1 x) \\ f_2^{(1)}(x) &= \lambda_2 \exp(-\lambda_2 x) \quad \text{pour } x < \tau \end{aligned}\right\} \quad (41)$$

Supposons que $\lambda_1 = \lambda$ (ce qui n'est rien d'autre que l'hypothèse du paragraphe 4.4), et posons $\mu = \lambda_2/\lambda_1$ (il est aisé de vérifier que $0 \leq \mu \leq 1$).

L'équation (41) devient sous ces notations

$$\left.\begin{aligned} f_1(x) &= f_2(x) = \lambda \exp(-\lambda x) \\ f_2^{(1)}(x) &= \mu\lambda \exp(-\mu\lambda x) \quad \text{pour } x < \tau \end{aligned}\right\} \tag{42}$$

Pour déterminer la fiabilité de l'ensemble des deux systèmes (équation (32)), il suffit de connaître la fonction $f_2^{(2)}(x)$. Pour cela, on utilise l'équation (38) avec $x_0 = \mu t$ (39). D'après les équations (40) et (42):

$$\int_0^{\tau} \mu\lambda \exp(-\mu\lambda x)\, dx = \int_0^{x_0} \lambda \exp(-\lambda x)\, dx \tag{43}$$

On peut donc écrire, d'après (38)

$$f_1(x) = \lambda \exp(-\lambda x)$$

$$f_2(\tau,x) = \begin{cases} \mu\lambda \exp(-\mu\lambda x) & \text{pour } x < \tau \\ \lambda \exp[-\lambda(x + \mu\tau - \tau)] & \text{pour } x \geqq \tau \end{cases} \tag{44}$$

L'équation (37) s'écrit pous sa part:

$$f_1(x) = \lambda \exp(-\lambda x)$$

$$f_2(\tau,x) = \begin{cases} \mu\lambda \exp(-\mu\lambda x) & \text{pour } x < \tau \\ \lambda \exp[-\lambda(x - \tau)] & \text{pour } x \geqq \tau \end{cases} \tag{45}$$

Après intégration de l'équation (32) en tenant compte de (44) on trouve la fiabilité:

$$R_2(t) = 1 - Q_2(t) = \frac{1}{\mu}\,[1 + \mu - \exp(-\mu\lambda t)]\, \exp(-\mu t) \tag{46}$$

tandis que la même intégration à partir de (45) donne un résultat différent:

$$R_2(t) = (2 + \lambda t) \exp(-\lambda t) - \frac{1}{\mu + 1}\, \{\mu + \exp[-(1+\mu)\lambda t]\} \tag{47}$$

En utilisant la théorie des processus stochastiques homogènes de Markov avec un nombre d'états finis et un temps continu, il est possible de trouver la solution exacte.

4.5.2. Recherche de la fiabilité exacte – Emploi des théories de Markov

Désignons par:

- H_j l'état de l'ensemble à l'instant considéré, s'il y a exactement j systèmes fonctionnant incorrectement à cet instant

- $P_j(t)$ la probabilité d'être dans l'état H_j à un certain instant t, c'est-à-dire pour qu'il y ait, à cet instant, j systèmes défaillants.

Lorsque j = 0, 1 ou 2, le graphe des états de l'ensemble des systèmes est le suivant:

$$H_0 \xrightarrow[(\mu+1)\lambda]{} H_1 \xrightarrow[\lambda]{} H_2$$

Ce graphe est décrit par la matrice de passage:

$$P = \begin{matrix} \\ H_0 \\ H_1 \\ H_2 \end{matrix} \begin{Vmatrix} H_0 & H_1 & H_2 \\ 1-(\mu+1)\lambda & (\mu+1)\lambda & 0 \\ 0 & 1-\lambda & \lambda \\ 0 & 0 & 1 \end{Vmatrix} \qquad (48)$$

Ainsi, la probabilité d'être dans l'état H_0 à l'instant t + dt est la probabilité d'y être à l'instant t et d'y rester pendant l'intervalle dt.

La probabilité d'être dans l'état H_1 à l'instant t + dt est égale à la probabilité d'être à l'état H_0 à l'instant t et de passer de l'état H_0 à l'état H_1 pendant l'intervalle dt plus la probabilité d'être à l'état H_1 à l'instant t et d'y rester pendant dt.

La probabilité d'être dans l'état H_2 à l'instant t + dt est égale à la probabilité d'y être déjà à l'instant t plus la probabilité d'être dans l'état H_1 à l'instant t et de passer à l'état H_2 pendant l'intervalle de temps dt.

Ceci s'écrit, en utilisant la matrice de passage (48):

$$P_0(t+dt) = [1-(\mu+1)\lambda\, dt]\, P_0(t)$$

$$P_1(t+dt) = (\mu+1)\lambda\, dt\, P_0(t) + [1-\lambda\, dt]\, P_1(t)$$

$$P_2(t+dt) = \lambda\, dt\, P_1(t) + P_2(t)$$

soit

$$P_0(t+dt) - P_0(t) = -(\mu+1)\lambda\, dt\, P_0(t)$$

$$P_1(t+dt) - P_1(t) = (\mu+1)\lambda\, dt\, P_0(t) - \lambda\, dt\, P_1(t)$$

$$P_2(t+dt) - P_2(t) = \lambda\, dt\, P_1(t)$$

En divisant par dt, et en passant à la limite lorsque dt → 0, on obtient le système différentiel

$$\left.\begin{aligned} \frac{d}{dt}\{P_0(t)\} &= -(\mu+1)\lambda\, P_0(t) \\ \frac{d}{dt}\{P_1(t)\} &= (\mu+1)\lambda\, P_0(t) - \lambda\, P_1(t) \\ \frac{d}{dt}\{P_2(t)\} &= \lambda\, P_1(t) \end{aligned}\right\} \qquad (49)$$

où les conditions aux limites sont données par

$$\left.\begin{aligned} P_0(0) &= 1 \\ P_1(0) &= P_2(0) = 0 \end{aligned}\right\} \qquad (50)$$

Il suffit alors d'utiliser les transformées de Laplace pour obtenir le système d'équations linéaires suivant:

$$\left.\begin{aligned} \overline{P_0(S)} &= \frac{1}{S+(\mu+1)\lambda} \\ \overline{P_1(S)} &= (\mu+1)\lambda \frac{1}{(S+\lambda)[S+(\mu+1)\lambda]} \\ \overline{P_2(S)} &= (\mu+1)\lambda^2 \frac{1}{S(S+\lambda)[S+(\mu+1)\lambda]} \end{aligned}\right\} \qquad (51)$$

où $\overline{P_j(S)}$ est la transformée de $P_j(t)$ et S est l'opérateur laplacien

$$\overline{P_j(S)} = \int_0^\infty \exp(-St)\, P_j(t)\, dt$$

Par transformation inverse, on obtient

$$\left.\begin{aligned} P_0(t) &= \exp[-(\mu+1)\lambda t] \\ P_1(t) &= \frac{\mu+1}{\mu}\{\exp[-\lambda t] - \exp[-(\mu+1)\lambda t]\} \\ P_2(t) &= 1 - \frac{1}{\mu}\{1+\mu - \exp[-\mu\lambda t]\}\exp[-\lambda t] \end{aligned}\right\} \qquad (52)$$

La fiabilité de l'ensemble est donc donnée par

$$R_2(t) = 1 - P_2(t) \qquad (53)$$

soit

$$R_2(t) = \frac{1}{\mu}\{1+\mu - \exp[-\mu\lambda t]\}\exp[-\lambda t] \qquad (54)$$

Ainsi le résultat exact donné par (54) est identique à celui donné par (46), ce qui justifie a priori l'emploi de l'hypothèse des probabilités égales.

On peut remarquer que le résultat de (47) ne concorde pas avec le résultat exact de la fiabilité sauf dans le cas où $\mu = 0$ qui correspond à la redondance de deuxième catégorie, ce qui est normal puisque

$$\lim_{\mu \to 0} R_2(t) = (1 + \lambda t)\exp(-\lambda t) \qquad (55)$$

4.6. Généralisation

Le problème exposé ci-dessus peut se généraliser à un ensemble de m systèmes redondants.

La probabilité de défaillance $Q_m(t)$ de cet ensemble à un instant t donné est

$$Q_m(t) = \int_0^t \left[\int_0^\tau f_m^{(1)}(x)\,dx + \int_\tau^t f_m^{(2)}(x)\,dx \right] \frac{d}{d\tau}\,[Q_{m-1}(\tau)]\,d\tau$$

où $Q_1(t) = q_1(t)$.

Cette relation de récurrence s'obtient:

- en considérant un système normal et les m - 2 premiers systèmes redondants comme un seul dispositif qui est secouru par le dernier système
- en utilisant l'équation (34) et en appliquant chaque fois le postulat des probabilités égales.

5. APPLICATION AU CAS DES ESSAIS ACCELERES

5.1. Formulation de l'hypothèse des probabilités égales dans le cas des essais accélérés

Supposons que les lois $F_x(t)$ et $F_y(t)$ de distribution des probabilités de défaillance sous les contraintes X et Y (Y > X) soient trouvées au cours de deux expériences (voir figure 1).

Soit une troisième expérience conduite de la manière suivante: le système est essayé sous la contrainte X durant le temps t_1, puis est soumis à la contrainte Y. Désignons alors par $F_{xy}(t/t_1)$ la fonction de distribution de la probabilité de défaillance dans ces conditions. Dans ce cas:

$$F_{xy}(t/t_1) = \begin{cases} F_x(t) & \text{lorsque } t \leq t_1 \\ F_{t_1}(t) & \text{lorsque } t > t_1 \end{cases} \tag{56}$$

La fonction $F_{t_1}(t)$ n'est pas connue en général, mais l'emploi de l'hypothèse des probabilités égales permet de l'expliciter. Cette hypothèse peut s'énoncer ainsi:

1°) Le séjour du système sous la contrainte X ne modifie pas sa loi de fiabilité sous la contrainte Y (voir paragraphe 3.1), ce qui signifie que $F_{t_1}(t)$ est une certaine partie de la courbe $F_y(t)$.

2°) De plus, lorsque $t > t_1$

$$F_{t_1}(t) = F_y(t + t'_1 - t_1) \tag{57}$$

où t'_1 satisfait l'équation

$$F_y(t'_1) = F_x(t_1) = Q_1 \tag{58}$$

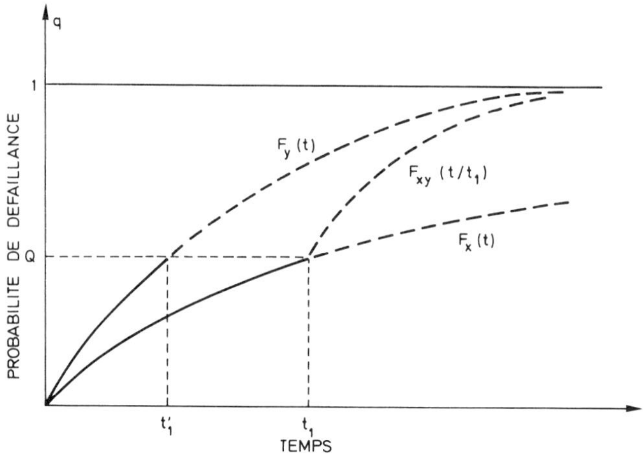

FIG. 7. Probabilité de défaillance de la fonction $F_{xy}(t/t_1)$.

D'après la définition même de la fonction d'accélération, cette équation est équivalente à

$$t_1 = g(t'_1) \tag{59}$$

Dans ces conditions, l'équation (56) devient

$$F_{xy}(t/t_1) = \begin{cases} F_x(t) & \text{lorsque } t \leq t_1 \\ F_y(t + t'_1 - t_1) & \text{lorsque } t > t_1 \end{cases} \tag{60}$$

L'illustration graphique de cette équation est donnée par la figure 7. Ainsi la solution des problèmes des essais accélérés est simplifiée par la linéarité de la fonction d'accélération.

L'objet du paragraphe suivant est de déterminer un critère de linéarité de la fonction d'accélération en utilisant encore l'hypothèse des probabilités égales sous la forme de l'équation (56).

5.2. Critère de linéarité de la fonction d'accélération

Admettons que les équations (59) et (60) soient vérifiées. Dans ce cas, on peut énoncer le critère suivant.

5.2.1. Critère de linéarité

L'hypothèse des probabilités égales étant vérifiée, d'après les essais n°1 et n°2 (voir leur énoncé ci-dessous) on peut affirmer que la fonction d'accélération g(t) est linéaire si $Q_2 = Q'_2$ et réciproquement si la fonction d'accélération est linéaire les probabilités Q_2 et Q'_2 sont égales.

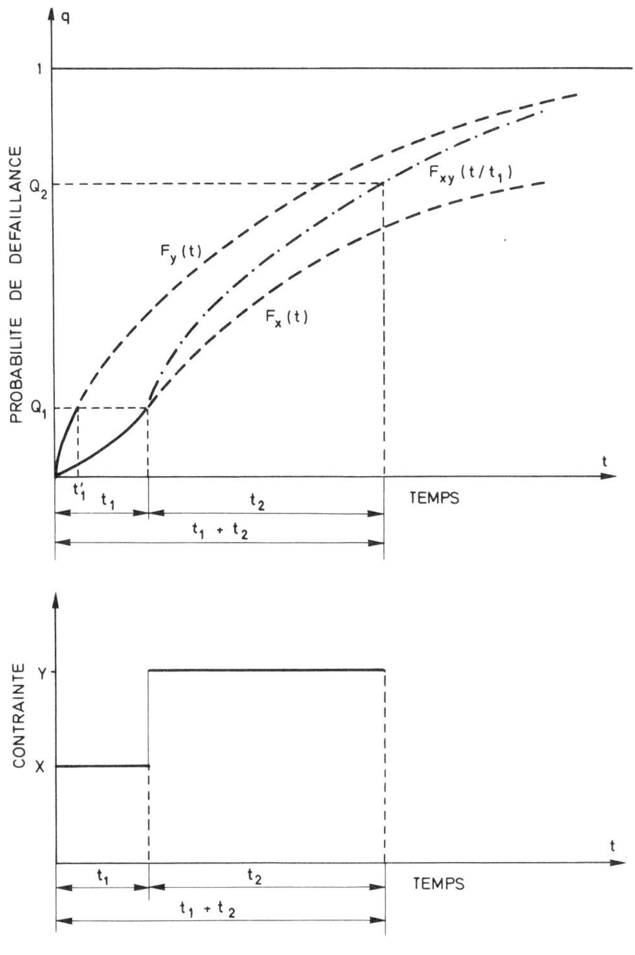

FIG. 8. Essai n°1.

5.2.2. Essai n° 1

L'un des deux échantillons (système ou composants) est essayé sous la contrainte X durant un temps t_1, puis sous la contrainte Y ($Y > X$) durant un temps t_2. Au moment $t_1 + t_2$ où l'essai est achevé, on a la probabilité de défaillance Q_2 (voir figure 8).

5.2.3. Essai n° 2

L'autre échantillon est essayé d'abord sous la contrainte Y pendant le temps t_2 puis sous la contrainte X durant le temps t_1. Au moment $t_1 + t_2$ où l'essai est achevé, on a la probabilité de défaillance Q'_2 (voir figure 9).

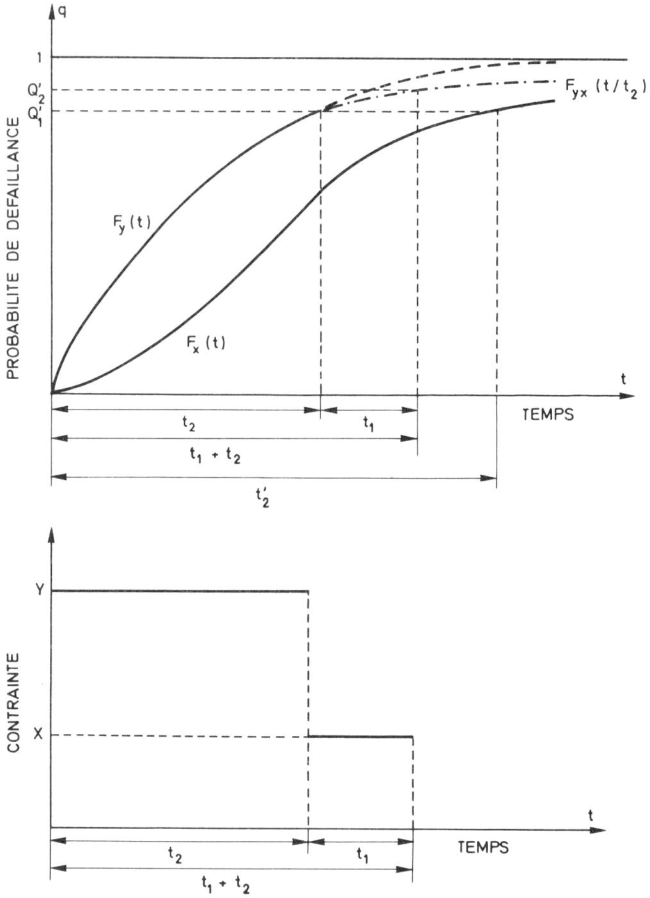

FIG. 9. Essai n°2.

5.2.4. Démonstration du critère de linéarité

La fonction de distribution de probabilité de défaillance $F_{xy}(t/t_1)$ dans l'essai n°1 est donnée par l'équation

$$F_{xy}(t/t_1) = \begin{cases} F_x(t) & \text{pour } t \leq t_1 \\ F_y(t + t'_1 - t_1) & \text{pour } t > t_1 \end{cases} \tag{61}$$

où

$$t_1 = g(t'_1) \tag{62}$$

avec

$$Q_2 = F_{xy}(t = t_1 + t_2/t_1) = F_y(t'_1 + t_2) \tag{63}$$

La fonction de distribution de probabilités de défaillance $F_{yx}(t/t_2)$ dans l'essai n°2 est donnée par l'équation

$$F_{yx}(t/t_2) = \begin{cases} F_y(t) & \text{pour } t \leq t_2 \\ F_x(t + t_2' - t_2) & \text{pour } t > t_2 \end{cases} \qquad (64)$$

où

$$t_2' = g(t_2) \qquad (65)$$

avec

$$Q_2' = F_{yx}(t = t_2 + t_1/t_2) = F_y(t_1 + t_2') \qquad (66)$$

d'après les données du critère $Q_2 = Q_2'$, d'où

$$F_y(t_1' + t_2) = F_x(t_1 + t_2') \qquad (67)$$

Par définition même de la fonction linéaire d'accélération (voir équation (6)), on peut écrire

$$t_1 + t_2' = g(t_1' + t_2) \qquad (68)$$

soit, d'après les équations (62) et (65):

$$g(t_1') + g(t_2) = g(t_1' + t_2) \qquad (69)$$

ce qui démontre que la fonction d'accélération g(t) est linéaire.

Réciproquement, démontrons que si cette fonction d'accélération est linéaire, les probabilités Q_2 et Q_2' sont égales. On sait que

$$\begin{aligned} Q_2 &= F_y(t_1' + t_2) \\ Q_2' &= F_x(t_1 + t_2') \end{aligned} \qquad (70)$$

Si la fonction d'accélération g(t) est linéaire, on a

$$t_1 + t_2' = g(t_1') + g(t_2) = g(t_1' + t_2) \qquad (71)$$

par conséquent

$$Q_2' = F_x[g(t_1' + t_2)] \qquad (72)$$

D'après la définition d'une fonction d'accélération, on a

$$F_y(t_1' + t_2) = F_x[g(t_1' + t_2)]$$

d'où:

$$Q_2 = Q_2'$$

Ainsi l'hypothèse des probabilités égales permet de définir une condition nécessaire et suffisante pour que la fonction d'accélération soit linéaire.

5.3. Application pratique

D'un point de vue pratique, ce résultat permet de connaître la loi de distribution de la probabilité de défaillance $F_x(t)$ (sous la contrainte X) à partir des trois expériences suivantes:

Première expérience: L'essai sous contrainte constante Y permet d'obtenir la loi de distribution $F_y(t)$ (la contrainte Y étant supérieure à la contrainte X, le temps de l'essai étant court).

Deuxième et troisième expériences: Deux essais sous contraintes variables (X durant t_1 suivi de Y durant t_2 et Y durant t_2 suivi de X durant t_1) permettent de montrer la linéarité de la fonction d'accélération g(t) et de déterminer la constante d'accélération C. Ces deux essais, utilisant une partie du temps sous la contrainte Y, peuvent être aussi de courte durée.

Ces trois expériences permettent de déduire $F_x(Ct)$ d'après l'équation $F_x(Ct) = F_y(t)$, en moins de temps qu'il n'en faudrait pour cette détermination sous la contrainte X.

DISCUSSION

(on papers IAEA-SM-195/33 and IAEA-SM-195/14)

J.-C. ARLAUD: I have a question for Mr. Gonnot. You say that if, with stress levels X and Y, the patterns of incident distribution remain essentially of the same kind, the acceleration functions are linear. Are there in practice, however, many cases where a change of stress level does not entail a modification of the distribution patterns?

R. GONNOT: In most cases a change of stress level does not appreciably modify the distribution patterns. In circumstances where there is in fact a substantial change in the patterns of distribution, the study I have presented here is no longer sufficient; improvements are required, and we propose to work on them.

L. CAVE: I wonder whether Mr. Gonnot has considered the application of his method to more complicated components, such as pumps and valves. In such cases the definition of 'stress' to provide the acceleration of failure would seem to present some difficulty.

R. GONNOT: So far the method has been developed only theoretically, for the most general cases. It is of course true that the definition of an acceleration function for more complicated components would be difficult.

W.A. WOLFE: A question for Mr. Waller. I assume that if you have $A(\infty)$ a function defined by $\lambda/(\lambda+\theta) = 1/(1+\lambda/\theta)$, λ determines the frequency of repair and should also be considered in assessing the relative worths of plants.

R.A. WALLER: It is true that $A(\infty)$ is a function of λ/θ only, but it is not possible to reduce $A(t)$ to a function of λ/θ alone. Thus the instantaneous availability, $A(t)$, varies with changes in the values of λ and/or θ, while the long-run availability varies only with changes in the ratio λ/θ.

D. OKRENT (Chairman): Again I would like to ask a more general question, to which a number of you might wish to venture an answer. Do you consider the method employed in the draft WASH 1400 (the Rasmussen Report) valid for its purpose, or is it in your opinion out of date?

R.A. WALLER: It seems clear that the whole area of risk assessment is in its infancy; the discussion of the Rasmussen Report has highlighted that fact.

E.R. SNAITH: At the moment there is no single technique which can be shown to be the best or the only correct method. If one wishes to consider reactor faults and their causes, and then to calculate the overall probability of occurrence, a fault-tree and/or cut-set analysis may be used. For a normally dormant system, a logic flow diagram synthesizing system operation may be the approach to use. However, both methods could be used in each of the foregoing cases. A Monte Carlo approach may be the most appropriate for determining the mean time to failure for an operating system. Thus we cannot say, for the time being at least, that one specific method is the best. As far as the Rasmussen Report is concerned, the event or fault tree approach achieved the aim of the study, so it can be considered adequate. However, it must be accepted that other techniques are available for such studies and may be equally valid.

L. CAVE: Obviously one can criticize the methods used for the Rasmussen Report, but in my view the end justifies the means. The most important result of the study was to show that the probability of core melt-down in the LWRs examined is substantially greater than had been supposed. Nevertheless, because of the protection afforded by the containment, the overall risk is low enough to be acceptable.

Refinement of the reliability analysis of the reactor or the containment might alter the probability of core melt-down, or the effective 'decontamination factor' for the containment, by factors of, say, three in either direction. However, much larger changes would be necessary to invalidate the main result. Moreover, it is doubtful whether the accuracy of the failure rate data which are available would justify any further refinement of the methods.

L.A. HOPKINS: I think the greatest weakness of the Rasmussen Report lies in the limitations of the probability data that are used for the lowest-level events. This is fairly obvious. With regard to the appropriateness of the fault-tree or event-tree technique, I should say it is no more limited than any other technique that could be applied to define the failure event logic.

D. OKRENT (Chairman): Occasionally an event actually occurs in a reactor whose probability of occurrence, if calculated by the usual techniques, would emerge as something like 10^{-9} or 10^{-12} per year. What is the significance of such an occurrence? Does it raise serious questions as to the basic validity of probabilistic safety assessments?

J.R. TAYLOR: The most usual cause of such problems is that one overlooks some source of coupling between individual failures in an event, so that the calculation is simply incorrect. Techniques of common mode failure analysis have been devised to solve just this problem.

Another source of this kind of problem, which could in principle be important, lies in the possible existence of a large number of complex failure combinations. The low probability of individual event combinations could then be counterbalanced by the large number of possible combinations to give an appreciable total probability. Whether such complex failure possibilities really exist, and are significant compared with simpler failure combinations, can be determined only from experience with a particular kind of plant.

The importance of this problem depends on the adequacy of the failure analysis techniques used. The recently developed combination of event-tree and fault-tree techniques, and common mode failure analysis techniques, allow many low-probability event combinations to be treated. In the studies of abnormal occurrence reports which I have described, there were very few failures which were so complex as to present problems for common-mode, fault-tree and event-tree techniques. Other techniques must then be used to deal with these more complex event combinations.

One of the major strengths of the fault-tree and event-tree techniques is their ability to take account of a whole class of different failure combinations, as soon as a single example of that class is known. For example, the number of causes of loss of external power to a power plant is potentially very large; but from experience we have a good estimate of the frequency of the whole class of these occurrences, and if a new and rare cause of this kind of occurrence is found, its significance will be swamped by the existing experience. By combining analyses of individual component failures and experience of more diffuse classes of event, one can make the best use of all information available. But reliability analysis techniques must be supplemented by other techniques such as quality control, and must be checked against experience of the particular kind of plant involved, to ensure that the basis for analysis holds.

L. CAVE: In my view the relatively high frequency of potentially dangerous events which has been observed in power reactors, as compared with the frequencies predicted by reliability analysis, reflects a lack of imagination on the part of the reliability engineer when seeking ways in which a particular event might occur; often these events represent a form of common mode failure which has not been recognized. For example, in the recent incident at Brown's Ferry it is obvious, with hindsight, that there was not enough segregation between the cabling of the various emergency systems to justify the assumption that the diverse systems were in fact completely independent of one another. A similar example of lack of imagination (or foresight), which was not detected until the plants had been in operation, is provided by certain other LWR stations in which the steam mains ran beneath rooms containing emergency plant.

The problem presented by the limits of human imagination is a serious one; experience suggests that at a frequency of about 10^{-3} to 10^{-4} per year it is difficult to be sure that all possible interactions have been foreseen. Consequently, it is necessary to view with caution claims that any one system has a reliability better than one failure per 1000 demands, even though it may have a super-abundance of redundancy giving a theoretical failure rate of, say, one per 10^6 demands. A study of the literature in the parallel field of industrial safety might be helpful as a means of improving our imaginative powers; we will find there reports of accidents of a bizarre nature which occur with what appears to be a surprisingly high frequency.

J.K. VAURIO: I should like to try to relate the subjects of the first two sessions to the presentation of Mr. Kongsø (IAEA-SM-195/17), who defined the terms 'stoppage faults' and 'blocking faults'. There are two kinds of blocking faults, those caused by activation itself and those caused by random failures between activations. I wonder whether anyone here can say which mode is more important. If the first mode were dominant we would gain nothing by testing equipment like pumps, but if the second mode is dominant we should test them as often as possible. This is a problem for the operators of nuclear power plants.

H.E. KONGSØ: The answer depends on the circumstances, I think, and so does the optimum test interval. WASH 1400 (Draft) has emphasized that human error is an important factor to be considered in this connection.

DIAKOPTICAL RELIABILITY ANALYSIS OF TRANSISTORIZED SYSTEMS

J.M. KONTOLEON*
Reactor Department,
Nuclear Research Center Democritos,
Athens, Greece

J.W. LYNN
Department of Electrical Engineering and Electronics,
The University,
Liverpool

A.E. GREEN
Systems Reliability Service,
United Kingdom Atomic Energy Authority,
Culcheth, Warrington, Cheshire,
United Kingdom

Presented by Nadia Kontoleon

Abstract

DIAKOPTICAL RELIABILITY ANALYSIS OF TRANSISTORIZED SYSTEMS.

Limitations both on high-speed core availability and computation time required for assessing the reliability of large-sized and complex electronic systems, such as used for the protection of nuclear reactors, are very serious restrictions which continuously confront the reliability analyst. Diakoptic methods simplify the solution of the electrical-network problem by subdividing a given network into a number of independent subnetworks and then interconnecting the solutions of these smaller parts by a systematic process involving transformations based on connection-matrix elements associated with the interconnecting links. However, the interconnection process is very complicated and it may be used only if the original system has been cut in such a manner that a relation can be established between the constraints appearing at both sides of the cut. Also, in dealing with transistorized systems, one of the difficulties encountered is that of modelling adequately their performance under various operating conditions, since their parameters are strongly affected by the imposed voltage and current levels. In this paper a new interconnection approach is presented which may be of use in the reliability analysis of large-sized transistorized systems. This is based on the partial optimization of the subdivisions of the torn network as well as on the optimization of the torn paths. The solution of the subdivisions is based on the principles of algebraic topology, with an algebraic structure relating the physical variables in a topological structure which defines the interconnection of the discrete elements. Transistors, and other nonlinear devices, are modelled using their actual characteristics, under normal and abnormal operating conditions. Use of so-called k factors is made to facilitate accounting for use of electrical stresses. The approach is demonstrated by way of an example.

1. INTRODUCTION

Overall computer-aided reliability analysis of large-sized and complex electronic systems, such as used for the protection of nuclear reactors, usually becomes prohibited due to the requirements in high-speed memory capacity and computation time.

---

* Present address: Department of Electrical Engineering, Wollongong University, Wollongong, N.S.W., Australia 2500.

Decomposition or "diakoptic" methods [1, 2] greatly facilitate the analysis of large systems and results have already been published on computer prediction of the behaviour of highly nonlinear amplifier units and on the investigation of electro-mechanical instability conditions in a nuclear power station. All these studies have included extensive optimization in regard to reduction of computer time and storage.

In general, diakoptic methods simplify the solution of the electrical-network problem by subdividing a given network into a number of independent subnetworks and then interconnecting the solution equations of these smaller parts by a systematic process involving transformations based on connection-matrix elements associated with the interconnecting links. The economics of computer time involved in using tearing methods depends on the topological structure of the problem itself. However, the maximum number of subdivisions to gain advantage in time and storage is limited by the size of the connection-matrix.

Physical systems, more generally, can be handled in a similar manner to obtain their overall solution from the solution of the smaller component systems. These may include a wide variety of phenomena, such as electrical, mechanical, chemical or thermal. The difficulty associated with the general case of a physical system lies in the fact that the formulation of an appropriate model is not an easy task. A model should be either an algebraic or a topological structure. Usually, a circuit may link the gap between the purely mathematical formulation and the abstract topological structure of a system, and therefore it can be successfully used in a wide variety of problems.

In this work, an approach is presented which has been successfully used in the reliability analysis of large-size transistorized protective systems for nuclear reactors; here, the original system is modelled by standard circuit elements and then is torn apart into two smaller parts which are solved separately in a number of successive steps. The method involves partial optimization of the subdivisions as well as optimization of the torn paths [3, 4].

The solution of the subnetworks is based on the principles of algebraic topology, with an algebraic structure relating the physical variables in a topological structure which defines the interconnection of the discrete elements. The topological characteristics are based on Kirchhoff's voltage and current laws, while Ohm's law serves as a transformation between the dual sets of physical variables associated with the impedance elements. At each step only one subdivision is solved and optimized with respect to the previous solution.

Both the interconnection process and the modelling technique for representing the behaviour of transistors under normal and abnormal conditions are described in the following, and a simple example is given for demonstration purposes.

2. NOTATION

A, A^t : Branch-node matrix and its transpose
B^t : Node-datum path matrix
C, C^t : Circuit matrix and its transpose

C_T : Tree-branch matrix
E, I : Arbitrary source vectors
I_b : Base current
I_c : Collector current
I_e : Emitter current
Z : Impedance matrix
Z_L, Z_T: Tree impedance matrices
V, J : Voltage and current-coil variables
Y : Admittance matrix
Y_L, Y_T: Tree admittance matrices
E_i^* : Voltage at the i-th node of the original network
$E_i^{(x)}$: Voltage at the i-th intersection of subnetwork (x)
X_i : Parameter vector at the i-th step of transistor optimization
$\hat{X}_i$: Decided parameter vector at the i-th step
e, i : Branch voltage and current vectors
e' : Node-datum voltage vector
e_i : Objective function at the i-th step
$z_i^{(x)}$: Connected impedance at the i-th intersection of subnetwork (x)
β : h_{FE}
ϵ : Accepted error vector of e_i

3. DESCRIPTION OF THE INTERCONNECTION PROCESS

From linear analysis the following expressions are well known

$$V = E + e \tag{1}$$

$$J = I + i \tag{2}$$

$$J = YV \tag{3}$$

$$V = ZJ \tag{4}$$

$$e' = Ae \tag{5}$$

$$e = B^t e_T \tag{6}$$

$$C^t e = 0 \tag{7}$$

$$A^t i = 0 \tag{8}$$

From Eq. (7) it follows that

$$[C_T^t \; U_L] \begin{bmatrix} e_T \\ e_L \end{bmatrix} = 0 \tag{9}$$

which gives

$$e_L = - C_T^t e_T \tag{10}$$

Equations (1)-(8) give the following expression

$$e = (A^t YA)^{-1} A^t (I - YE) \quad (11)$$

Assuming that Y and Z are mutually inverse and that there are no mutual admittances between the tree and link portions of Y, i.e.

$$Y = \begin{bmatrix} Y_T & 0 \\ 0 & Y_L \end{bmatrix}$$

then one may tear the network into two parts consisting of the tree and link part, respectively. Let all the E and I sources be considered as equivalent nodal current sources applied to the nodes of the tree. Then

$$e = (B^t Z_T B) I = Z_1 I \quad (12)$$

where

$$I = A^t I - YE \quad (13)$$

The solution for the link portion is

$$e_L = Z_L i_L \quad (14)$$

and the interconnection formula is given as

$$(A^t YA)^{-1} = Z_1 - Z_1 A_L^t (A_L Z_1 A_L^t + Z_L)^{-1} A_L Z_1 \quad (15)$$

From Eqs (11) and (13) we obtain

$$e = (A^t YA)^{-1} I \quad (16)$$

which means that $(A^t YA)^{-1}$ is a transformation matrix that relates node voltage response to nodal current excitation. The interconnection formula provides an explicit formula for $(A^t YA)^{-1}$. Thus, in Eq. (15) it is possible to start with the network tree and connect one link at a time. Therefore, instead of inverting a large matrix, the inverses of several smaller matrices of subnetworks are required.

In general, when tearing a network there may exist nonlinear interrelationships between the network elements, and therefore only with the use of iterative techniques does it become possible to establish the actual dependence of the network parameters on the flowing currents and imposed voltages. Also, the conditions at the intersections as well as in the resulting subnetworks are interrelated in a nonlinear way. Therefore, optimization at the interconnection level must also be carried out.

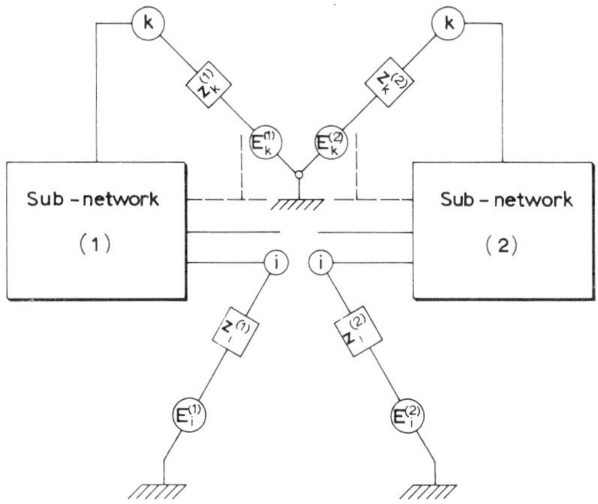

FIG.1. Network tearing.

Let us consider a network which is subdivided into two parts as shown in Fig. 1, and assume branches with impedance (or admittance) elements and voltage sources associated with the torn paths. Let $E_i^{(1)}$ and $E_i^{(2)}$ be the node-voltages at the i-th intersection point of the resulting subnetworks, (1) and (2), respectively, and E_i^* the corresponding voltage of the original network (before tearing). Denoting by $z_i^{(1)}$ and $z_i^{(2)}$ the connected equivalent impedances at the i-th intersection point of the two subnetworks, the admittance matrix of subnetwork (1) will be:

$$Y_1 = \left[\begin{array}{cccc|cccc} 1/z_1 & \cdots & 0 & \cdot & \cdot & \cdots & & 0 \\ \cdot & & & & & & & \cdot \\ \cdot & & & & & & & \cdot \\ 0 & & & 1/z_n & & & & \cdot \\ \hline \cdot & & & & 1/z_1^{(1)} & & & \cdot \\ \cdot & & & & & & & \cdot \\ 0 & \cdots & & 0 & \cdot & \cdots & & 1/z_k^{(1)} \end{array}\right] \begin{array}{l} \text{subnetwork (1)} \\ \\ \text{connected equivalent} \\ \text{impedances at the} \\ \text{intersections of (1)} \end{array}$$

where Y_1 is the admittance matrix of subnetwork (1), containing n internal nodes and k intersection points. Similarly, the admittance matrix of subnetwork (2) will be of the form:

$$Y_2 = \left[\begin{array}{c|c} Y_r & 0 \\ \hline 0 & Y^{(2)} \end{array}\right]$$

Then, the partial solutions for the two subnetworks are obtained from Eq. (11) as:

$$e_1 = (A_1^t Y_1 A_1)^{-1} A_1^t (I_1 - Y_1 E_1) \tag{17}$$

$$e_2 = (A_2^t Y_2 A_2)^{-1} A_2^t (I_2 - Y_2 E_2) \tag{18}$$

where the subscripts 1 and 2 indicate the subnetworks (1) and (2), respectively. Thus, in order to obtain the partial solutions of the subnetworks, the parameters must be adjusted in a number of steps. At each step one subnetwork is analysed and its parameters are optimized using the previous solution. This is an optimization at the subnetwork level. When the partial solution of one subnetwork has been obtained, then the equivalent admittances in the torn paths are calculated as well as the corresponding voltage sources. These values are then introduced into the other subnetwork which is analysed in a similar manner. This is an optimization at the intersection level. This process is repeated until the equilibrium state is reached, where

$$E_i^{(1)} = E_i^{(2)} = E_i^* \qquad i = 1, 2, \ldots k$$

and

$$z_i^{(1)} = z_i^{(2)} \qquad i = 1, 2, \ldots k$$

where k is the number of the torn paths.

Conditions for convergence vary with the particular problem. In some cases it can be demonstrated mathematically that convergence is assured, but in others, usually the multivariable cases, it is difficult to prove convergence and various constraint techniques are used to give optimal or sub-optimal solutions within the required ranges.

It is clear that at each step only one subnetwork is needed, i.e. half of the computer memory required to analyse the original network. However, the question of modelling appropriately nonlinear devices remains to be examined; under tearing conditions, the behaviour of such devices may change completely. In the following a transistor model is described which has been found of use in reliability studies.

4. TRANSISTOR MODELLING FOR RELIABILITY STUDIES

Various transistor models exist in the literature [5-8] describing the device behaviour in linear and nonlinear operation. From the point of view of a computer circuit analysis, a model should contain parameters which can be fairly easily calculated or are standard parameters given in the data sheets. However, only a few models meet these requirements.

There are two approaches that can be used in the modelling of non-linear devices: (1) an equivalent network or (2) a mathematical model consisting of a set of nonlinear equations. The equations derived from the physics take no account of numerical difficulties, and the resultant system of equations may be quite difficult to solve. Whatever form of model is chosen, it must be possible to obtain the model parameters fairly easily from measurements.

In dealing with transistor circuits one of the difficulties encountered is that of modelling adequately their performance under various d. c. conditions, since transistor parameters are strongly affected by the imposed voltage and current levels. Also, in various applications the model parameters may have a different contribution to the overall circuit performance. Therefore, when one choses approximate values for the parameters of one's model, as for example the values given in the data sheets, it is possible to obtain erroneous results.

The determination of the performance of a transistor model under various d. c. conditions is of importance in reliability studies in order to assess the effect of component failures on circuit behaviour. Also, in such studies it is desired to know the stresses acting on the involved components under various fault conditions. Although a variety of modelling techniques exist, as far as the authors are aware, no attention has so far been paid to this problem.

Here, a transistor model is presented which can accurately simulate in a digital computer the operation under a wide variety of operating conditions. The basis of the model is the hybrid-π model and an optimization algorithm, which optimizes the model parameters according to the imposed voltages and currents, using common emitter (CE) transistor output characteristics. The topological structure of the model is shown in Fig. 2.

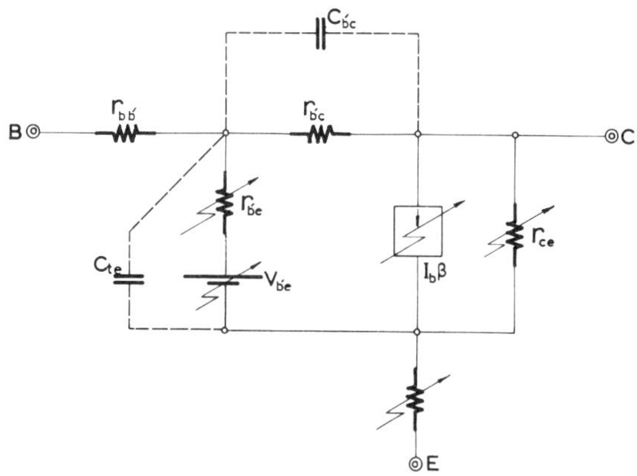

FIG.2. Topological structure of the optimized model.

Two methods of describing the CE transistor output characteristics have been tried successfully. In the first, a piecewise linear representation is used which provides a high computation speed and a high accuracy in the active region of operation. In the second, curve-fitting is employed to obtain higher accuracy near the saturation region. However, the price paid for this is an increase in computation time. In Table I the conditions for the model states are given.

TABLE I. CONDITIONS FOR THE MODEL STATES FOR PNP AND NPN TRANSISTORS

Type	Active	Cutoff	Saturation
PNP	I_b 0	I_b 0	I_b 0
	V_{ce} V_{sat}	V_{ce} V_{sat}	V_{sat} V_{ce} 0
NPN	I_b 0	I_b 0	I_b 0
	V_{ce} V_{sat}	V_{ce} V_{sat}	0 V_{ce} V_{sat}

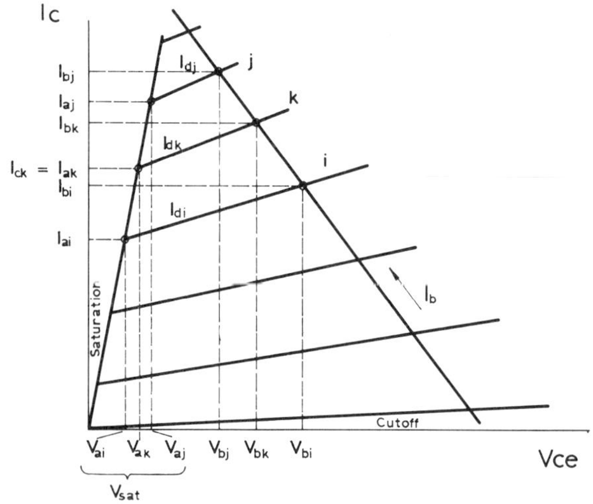

FIG.3. Representation of the model characteristics.

Using this approach it becomes possible to analyse accurately circuits by considering the exact characteristics of the transistors involved. Thus, the model used here does not suffer from the spread of the device parameters due to the manufacturing process.

5. THE OPTIMIZATION PROCESS

The optimization problem here is that of finding the model parameters which best simulate measurements made on the actual device. Such measurements are the CE transistor output characteristics, normally given in the specification sheets.

Two methods of describing these characteristics have been tried here. The first is based on the piecewise linear representation which is illustrated in Fig. 3. Using this method, only the coordinates of the constant base-current characteristics are required in order to describe the whole family.

Thus, from Fig. 3, the characteristic for the i-th line is

$$I_c = I_{ai} + \frac{I_{bi} - I_{ai}}{V_{bi} - V_{ai}} V_{ce}, \quad \text{for} \quad V_{ai} \leqq V_{ce} \leqq V_{bi} \tag{19}$$

and

$$I_c = \frac{I_{ai}}{V_{ai}} V_{ce}, \quad \text{for} \quad 0 \leqq V_{ce} \leqq V_{ai} \tag{20}$$

Considering a characteristic k between two given characteristics i and j, lying quite close to one another with base current $I_{dj} > I_{di}$, a very close approximation to the coordinates of these characteristics is:

$$I_{ak} = I_{ai} + \frac{I_{dk} - I_{di}}{I_{dj} - I_{di}} (I_{aj} - I_{ai}) \tag{21}$$

$$I_{bk} = I_{bi} + \frac{I_{dk} - I_{di}}{I_{dj} - I_{di}} (I_{bj} - I_{bi}) \tag{22}$$

$$V_{ak} = \frac{V_{aj}}{I_{aj}} I_{ak} \tag{23}$$

$$V_{bk} = \frac{V_{bj}}{I_{bj}} I_{bk} \tag{24}$$

These expressions are used at each iteration to generate a new characteristic corresponding to the base-current obtained. The new characteristic defines a new set for the model parameters. Let this set be represented at the i-th iteration, $i \geqq 2$, by a vector X_i, and denote by $\hat{X}_{i-1}$, $\hat{X}_{i-2}$ the computed vectors from the decision portion of the program at the i-1 and i-2 iterative steps, respectively. The decided vector at the i-th iteration is obtained by considering the experience gained from the previous explorations, as described below.

Denote by e_i the value of the assumed objective function at the i-th step and by ϵ the maximum accepted error. The objective function adopted here is based on the sum of squares criterion, but other expressions can also be used. Then, the following steps are considered:

(a) If $e_i \leqq \epsilon$, then stop iterations.

(b) If $e_i > \epsilon$ and the state has not been changed, then calculate:

$A_i = X_i - \hat{X}_{i-1}$ and $B_i = \hat{X}_{i-1} - \hat{X}_{i-2}$. Let a_i^k and b_i^k be the k-th elements of A_i and B_i; then, if

$$a_i^k b_i^k \begin{cases} > 0 & \text{let } \hat{x}_i^k = x_i^k \\ > 0 & \text{let } \hat{x}_i^k = x_i^k/2 \end{cases}$$

where $\hat{x}_i^k$ and x_i^k are the k-th elements of $\hat{X}_i$ and X_i, respectively.

(c) If $e_i > \epsilon$ and the state has been changed, then set $X_i = \hat{X}_{i-1}$ and continue iterations.

At the beginning, the procedure starts at a guess point X_0 and for the next two iterations the vectors $\hat{X}_1$ and $\hat{X}_2$ are obtained neglecting step (b) in the procedure described above, since there is no past history about the nature of the best pattern move.

The algorithm has been found to be very effective and in most cases the final solution is reached in a few iterations. Problems have not arisen in extreme cases such as the catastrophic failures of components.

The second method which has been tried employs curve-fitting in order to gain extra accuracy near the saturation region of operation. In this case a number of points (usually 10 to 15) lying in each constant base-current characteristic are taken and their coordinates are used. Then, the CE output characteristics are reproduced with the aid of interpolation formulas. When the characteristic corresponding to the computed base-current has been reproduced, the procedure described before is followed.

6. AN ILLUSTRATIVE EXAMPLE

An example of a two-stage simple transistor feedback amplifier is given to illustrate the application of the diakoptical analysis with optimization at the subdivision and intersection levels. The circuit, torn into two parts, is shown in Fig. 4, while in Fig. 5 the node-voltages of the torn paths are plotted during the various steps of the interconnection process. Results obtained with this technique are compared with those obtained from the overall circuit analysis in Table II.

With the use of the method presented above, the computer-aided reliability analysis of large systems can be very much facilitated. Instead of subdividing a system into parts which are assumed to be statistically independent, it is necessary only to tear the original system into two parts; the dependence of the subnetworks is taken into account through the iterative interconnection process and the optimization of the individual subnetworks. Fault conditions, which are due to single or multiple element failures, can be also investigated in this way, without the statistical independence assumption between the elements involved in the various parts of the system.

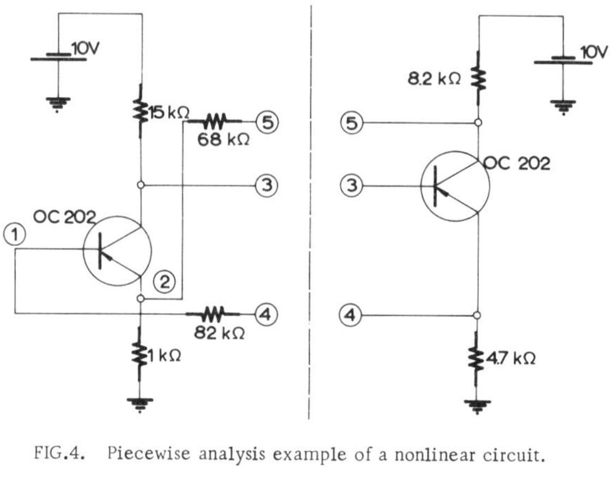

FIG.4. Piecewise analysis example of a nonlinear circuit.

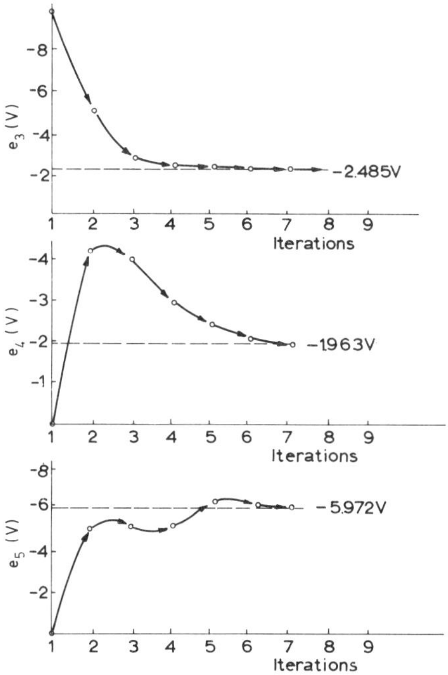

FIG.5. Node-voltage of the torn paths during the interconnection process.

TABLE II. NODE-VOLTAGES [a] OF THE CIRCUIT OF FIG. 3 OBTAINED FROM THE OVERALL AND PIECEWISE ANALYSES

Node	Analysis of the overall circuit	Piecewise analysis
1	- 1.107	- 1.106
2	- 0.590	- 0.586
3	- 2.437	- 2.485
4	- 1.930	- 1.963
5	- 5.994	- 5.972

[a] The values are given in volts.

Following this approach, the reliability analysis, or prediction, of large-scale systems can be seen in a more realistic manner; here use of the k factors is assumed in order to proceed from the electrical stress or temperature stress domain to the failure rate domain and then to reliability through combinatorial probabilistic calculations. This leads to the conclusion that diakoptical analysis for the reliability prediction has much to offer in the investigation of the interrelations between various systems, which is of great importance in reliability studies.

7. CONCLUSIONS

An approach has been presented for analysing large-size nonlinear circuits in a diakoptical manner, which applies optimization in both the subdivision and intersection levels. Transistors have been modelled using their actual characteristics and an iterative process is used to adjust the circuit parameters according to the imposed voltage and current levels. The method can be applied in the reliability analysis of large-sized systems without necessarily imposing the statistical independence assumption between the various parts of the system.

REFERENCES

[1] KRON, G., Diakoptics – Piecewise Solutions of Large Scale Systems, McDonald, London (1963).
[2] HAPP, H.H., Diakoptics and Networks, Academic Press, London (1971).
[3] DASANI, G.J., KONTOLEON, J.M., LYNN, J.W., "Digital analysis of an electronic amplifier for reliability studies", Systems Reliability Assessment, NATO Advanced Study Institute, Liverpool,Jul.1973 (HENLEY, E.J., LYNN, J.W., Eds), Noordhoff International Publishing, Leiden (1974).
[4] KONTOLEON, J.M., Digital Analysis of Electronic Networks for Protective and Control Systems, Ph. D. Thesis, Liverpool Univ. (1971).
[5] EBERS, J.J., MOLL, J.L., Large-signal behavior of junction transistors, Proc. IRE 42 (1954) 1761.
[6] BEAUFOY, R., SPARKES, J.J., The junction transistor as a charge-controlled device, Automatic Telephone Electric Co. J. 13 (1957) 310.
[7] LINVILL, J.G., Lumped models of transistors and diodes, Proc. IRE 46 (1958) 949.
[8] NANAVATI, R.P., An Introduction to Semiconductor Electronics, McGraw-Hill, New York (1963).

RELIABILITY ASSESSMENT OF NUCLEAR POWER PLANT SYSTEMS

Chairmen:

S. E. BARDEN, United Kingdom
J. A. VAUJOUR, France

PROJETS DE REACTEURS A NEUTRONS RAPIDES – AIDE APPORTEE PAR LA FIABILITE A LA CONCEPTION DU SYSTEME D'ARRET D'URGENCE

R. QUENEE, Annick CARNINO
CEA, Centre d'études nucléaires de Saclay,
Gif-sur-Yvette

P. VERRIERE
CEA, Centre d'études nucléaires de Cadarache,
St-Paul-lez-Durance

A. JEANNOT, J. ALLANIC
CIRNA, CEA, Centre d'études nucléaires de Saclay,
Gif-sur-Yvette,
France

Abstract–Résumé

FAST REACTOR PROJECTS: THE CONTRIBUTION OF RELIABILITY ANALYSIS TO THE DESIGN OF EMERGENCY SHUT-DOWN SYSTEMS.

The authors describe a theoretical study carried out with a view to providing the designers of fast reactor emergency shut-down systems with data on the relative importance of the reliability required of different materials used in such systems. Three types of system are considered: one with two independent safety circuits (preliminary design); one with three independent safety circuits and unequal distribution of the control rods; one with three independent safety circuits and equal distribution of the control rods. The probability of failure affecting the three circuits is calculated with a view to meeting a specified reliability objective, which implies a certain reliability of the safety system materials. These system analyses took into account common mode failures due to internal and external occurrences. Parallel to the theoretical study, an investigation of materials of proven stability is to be carried out, involving thorough inquiries with builders, so as to ensure that the specified objectives can be achieved.

PROJETS DE REACTEURS A NEUTRONS RAPIDES – AIDE APPORTEE PAR LA FIABILITE A LA CONCEPTION DU SYSTEME D'ARRET D'URGENCE.

Dans le cadre des projets de réacteurs à neutrons rapides, une étude théorique a été entreprise afin de fournir aux équipes chargées de la conception des systèmes d'arrêt d'urgence, des données sur l'importance relative du niveau de fiabilité requis entre les différents matériels composant les systèmes d'arrêt d'urgence. Dans cet esprit, trois types d'organisation possibles ont été examinés: deux lignes de sécurité indépendantes, schéma d'avant-projet; trois lignes de sécurité indépendantes, répartition des barres de commande non équitable; trois lignes de sécurité indépendantes, mais répartition équitable des barres sur les trois lignes. La probabilité de défaillance relative aux trois schémas a été calculée de manière à répondre à un objectif de fiabilité fixé, ce qui implique une fiabilité déterminée au niveau des matériels. Ces analyses de schémas ont été assorties de la prise en compte de défaillances de mode commun dues à des phénomènes internes et externes. Parallèlement à cette étude théorique, une recherche des matériels éprouvés et stables, avec enquête approfondie auprès des constructeurs, sera effectuée afin de s'assurer que la technologie actuelle permet d'atteindre les objectifs fixés.

INTRODUCTION

Le but de l'étude était de mettre à la disposition des différents services concernés par les projets un «outil» permettant une comparaison rationnelle de différentes organisations possibles du système d'arrêt d'urgence du réacteur, ceci afin de permettre un choix objectif et en toute connaissance de cause d'une solution tenant compte des impératifs techniques, géographiques ou économiques.

Nous nous sommes fixé (hypothèse de calcul) comme objectif de fiabilité, un taux de défaillance global annuel du système d'arrêt d'urgence de 10^{-8}, la défaillance du système ne permettant pas de chuter le nombre minimal de barres de commande nécessaire à l'arrêt du réacteur:

$$\lambda \leqq 10^{-8}/\text{an}$$

1. CONCEPTION DU SYSTEME D'ARRET D'URGENCE (AU)

Pour cette étude, nous avons retenu comme élément de base le fait que l'arrêt d'urgence du réacteur (AU) est provoqué par la chute de 24 barres neutrophages qui comprennent:
- 3 barres dites barres SAC (système arrêt complémentaire)
- 21 barres dites barres SAP (système arrêt principal).

Les barres SAC et SAP sont d'une constitution mécanique différente.

2. HYPOTHESES

Nous disposons des données suivantes:
- 3 barres SAC suffisent à arrêter le réacteur
- 4 barres SAP suffisent à arrêter le réacteur
- 1 barre SAC représente une antiréactivité moyenne de 500 pcm
- 1 barre SAP représente une antiréactivité moyenne de 400 pcm.

Sachant que la chute des barres SAC valant chacune 500 pcm (soit 1500 pcm au total) suffit à provoquer l'arrêt du réacteur, on peut retenir les hypothèses suivantes:

– Toute combinaison C de barres ayant chuté et totalisant une antiréactivité A supérieure ou égale à 1500 pcm est un arrêt d'urgence réussi (événement AU).

– Toute combinaison C de barres ayant chuté et totalisant une antiréactivité A inférieure à 1500 pcm est un échec de l'arrêt d'urgence (événement $\overline{\text{AU}}$).

$$A \geqq 1500 \text{ pcm} \rightarrow \text{AU} \qquad (1)$$

$$A < 1500 \text{ pcm} \rightarrow \overline{\text{AU}} \qquad (2)$$

Les expressions (1) et (2) peuvent être présentées sous forme de tableaux donnant respectivement les combinaisons de barres réalisant les événements AU et $\overline{\text{AU}}$.

TABLEAU I. COMBINAISONS DE CHUTES DE BARRES REALISANT L'EVENEMENT AU

Nombre de barres SAC ayant chuté (1 barre = 500 pcm)	Nombre de barres SAP ayant chuté (1 barre = 400 pcm)	Antiréactivité introduite (pcm)
3	0	1500
2	2	1800
1	3	1700
0	4	1600

TABLEAU II. COMBINAISONS DE CHUTES DE BARRES REALISANT L'EVENEMENT $\overline{AU}$

Nombre de barres SAC ayant chuté (1 barre = 500 pcm)	Nombre de barres SAP ayant chuté (1 barre = 400 pcm)	Antiréactivité introduite (pcm)
2	1	1400
1	2	1300
0	3	1200

Tableau I

Ce tableau donne l'ensemble des éléments (combinaisons de chutes simultanées de barres SAC et SAP) ayant la propriété commune d'être nécessaires et suffisants à la réalisation de l'événement AU.

Toute combinaison d'un plus grand nombre de barres est incluse.

Toute combinaison d'un plus petit nombre de barres n'appartient pas à l'ensemble (elle réalise l'événement complémentaire $\overline{AU}$).

Tableau II

Ce tableau (complémentaire du précédent) donne l'ensemble des éléments réalisant l'événement $\overline{AU}$.

Toute combinaison d'un plus petit nombre de barres est incluse.

Toute combinaison d'un plus grand nombre de barres n'appartient pas à l'ensemble.

Tableau III

Dans les tableaux I et II, on considère les barres qui ont chuté. Ainsi qu'on le verra par la suite, il est plus commode (possibilité de négliger certains termes) de considérer le complément, c'est-à-dire le nombre

TABLEAU III. COMBINAISONS DE BARRES DEFAILLANTES REALISANT L'EVENEMENT $\overline{AU}$

Nombre de barres SAC ne chutant pas (sur un total de 3)	ET	Nombre de barres SAP ne chutant pas (sur un total de 21)	⇒	Antiréactivité introduite (pcm)	⇒	Résultat
1		20		1400		$\overline{AU}$
2		19		1300		$\overline{AU}$
3		18		1200		$\overline{AU}$

de barres qui n'ont pas chuté sachant que l'on a en tout 3 barres SAC et 21 barres SAP. On obtient ainsi le tableau III qui donne l'ensemble des combinaisons de barres SAC et SAP défaillantes, nécessaires et suffisantes pour réaliser l'événement $\overline{AU}$.

Toute combinaison d'un plus grand nombre de barres défaillantes est incluse.

Toute combinaison d'un plus petit nombre de barres défaillantes n'appartient pas à l'ensemble.

C'est ce tableau qui sera utilisé par la suite.

3. PROBABILITE DE DEFAILLANCE DU SYSTEME D'ARRET D'URGENCE

3.1. Généralités

Dans ce qui suit, on pose que:

- les éléments composant le système sont absolument indépendants;
- les seules défaillances prises en compte sont des défaillances catalectiques, c'est-à-dire soudaines et imprévues (les défaillances par dérive sont exclues);
- les défaillances de mode commun ne sont pas envisagées dans ce chapitre;
- dans la plupart des cas, lorsqu'une probabilité est de la forme $P(a + b) = P(a) + P(b) - P(a) \cdot P(b)$, on néglige le terme produit, ce qui est légitime car on travaille sur des probabilités de défaillance ($P(a)$, $P(b)$) faibles, ce qui implique que le produit $P(a) \cdot P(b)$ est négligeable par rapport à $P(a) + P(b)$.

3.2. Schémas (fig.1a)

Un dispositif L élabore et envoie le signal d'AU:

- d'une part sur 11 blocs (c_1 à c_{11}) commandant les 21 barres SAP (b_1 à b_{21}); chaque bloc commande deux barres sauf c_{11} qui n'en commande qu'une (b_{21});
- d'autre part sur le bloc (d) commandant les trois barres SAC (f_1 à f_3).

Le diagramme logique des défaillances (fig.1b) est établi très facilement à partir du tableau III.

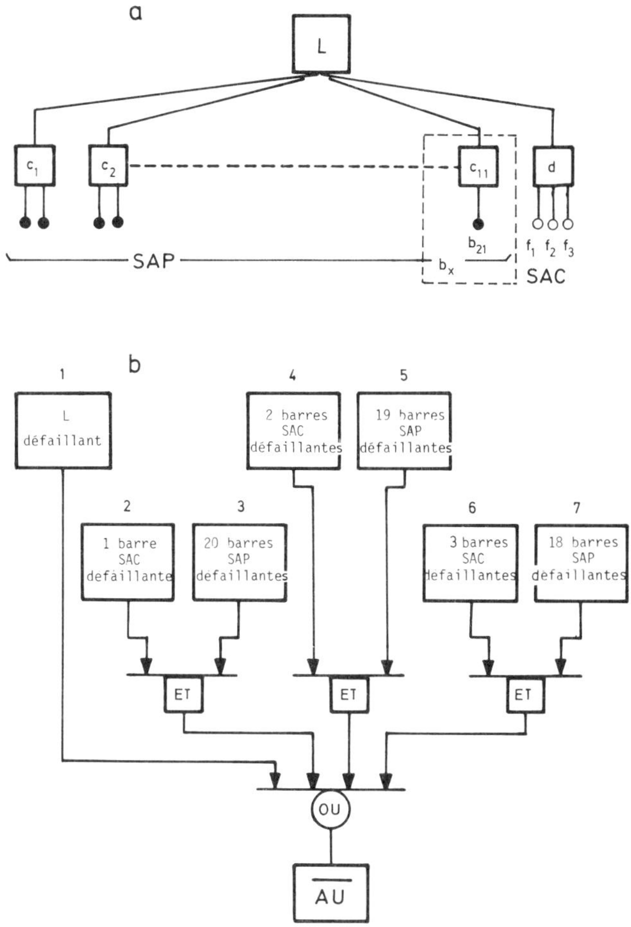

FIG.1. a — Schéma; b — arbre de défaillance.

Les autres modèles de schémas étudiés par la suite ne diffèrent du schéma type que par la redondance des circuits L qui y est introduite et par la répartition des barres en groupes.

3.3. Probabilité de défaillance de l'AU

Posons:

a = probabilité de défaillance d'un dispositif L
b = probabilité de défaillance d'une barre SAP
f = probabilité de défaillance d'une barre SAC
c = d = probabilité de défaillance d'un bloc de commande.

La probabilité de défaillance de chacun des éléments du système étant supposée connue, on peut, à partir des schémas et des diagrammes précédents, calculer les probabilités partielles d'apparition des événements numérotés de 2 à 7 et de là, calculer la probabilité de défaillance de l'AU.

Le détail des calculs n'est pas donné ici. On peut le trouver dans le rapport DSN n° 57, «Etude probabiliste du système d'arrêt d'urgence du réacteur Super Phénix» (janvier 1975).

4. APPLICATIONS NUMERIQUES

La probabilité de défaillance du système d'AU étant obtenue sous forme littérale ainsi qu'on vient de le voir, un calcul programmé sur ordinateur permet d'obtenir la valeur finale pour les probabilités de défaillance de chaque composant variant de 10^{-2} à 10^{-5}, par pas de 10.

5. ETUDE DES SCHEMAS PROPOSES

Pour chaque schéma (3 au total) on trouve successivement:
- le schéma
- un tableau récapitulant les défaillances sous une forme explicite
- un tableau de probabilité de défaillance sous forme littérale (pour le détail des calculs, voir rapport DSN n° 57)
- un tableau de valeurs numériques.

5.1. Etude du schéma n° 1

5.1.1. Schéma (voir fig.2a)

Un circuit L_1 envoie l'ordre de chute à 5 blocs commandant 10 barres SAP1 (b_1 à b_{10}) et à un bloc d commandant 3 barres SAC (f_1 à f_3).

Un circuit L_2 identique à L_1 envoie l'ordre de chute à 6 blocs commandant 11 barres SAP2 (b_{11} à b_{21}) et au bloc d commandant les 3 barres SAC.

Etant donné la redondance des barres SAP, nous ne ferons, en ne tenant pas compte de la onzième barre SAP2 (b_{21}), qu'une erreur négligeable (et pessimiste) sur le résultat final. La symétrisation apportée de cette manière nous permet de traiter le problème de manière beaucoup plus simple et plus claire sans influer sur la valeur de la probabilité finale de manière significative.

5.1.2. Tableau des défaillances réalisant $\overline{AU}$ (tableau A)[1]

Ce tableau montre, en fonction de chacun des états possibles des circuits L, les différentes combinaisons minimales de non-chute de barres SAC et SAP réalisant $\overline{AU}$.

Rappelons que l'on considère un circuit symétrique, c'est-à-dire deux groupes de 10 barres SAP.

Les lignes horizontales représentent des ET logiques.

Les numéros d'ordre indiquent des OU logiques.

---

[1] Les tableaux A à J figurent dans l'annexe.

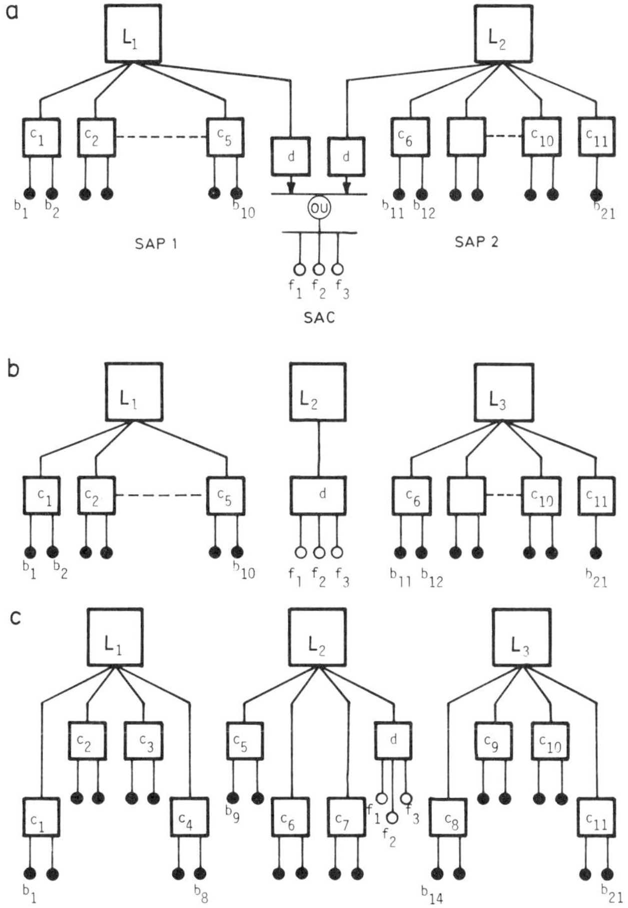

FIG.2. Schémas n° 1 (a), n° 2 (b) et n° 3 (c).

5.1.3. Tableau des probabilités de défaillance (tableau B)

Ce tableau donne les probabilités de réalisation des événements (défaillances) portés dans le tableau A calculées conformément au rapport DSN n° 57. A partir de là, on peut calculer la probabilité de défaillance du système:

$$P = P_1 + P_2 + P_3 + P_4$$

(en négligeant les termes «produit»).

5.1.4. Tableau des valeurs numériques (tableau C)

Ce tableau, obtenu à partir du précédent, donne les probabilités de $\overline{AU}$ pour des valeurs successives de 10^{-2}, 10^{-3}, 10^{-4}, 10^{-5} des paramètres a, b, c, f.

La dernière colonne indique en pour cent l'importance relative du paramètre a dans le résultat final P.

5.2. Etude du schéma n° 2 (fig.2b)

Trois circuits L_1, L_2, L_3 commandent respectivement:

- 1 groupe de 10 barres SAP (SAP1)
- 1 groupe de 3 barres SAC
- 1 groupe de 11 barres SAP (SAP2).

De même que pour le schéma n° 1, on ne tient pas compte de la onzième barre SAP2 (b_{21}).

De même que pour le schéma n° 1, on établit successivement les tableaux D, E, F.

5.3. Etude du schéma n° 3 (fig.2c)

Les 24 barres sont réparties en trois groupes de 8 barres.

Les circuits L_1, L_2, L_3 commandent respectivement:

- 1 groupe de 8 barres SAP
- 1 groupe comprenant 5 barres SAP et 3 barres SAC
- 1 groupe de 8 barres SAP.

De même que précédemment, on trouve les tableaux G, H, J.

6. EXPLOITATION DES RESULTATS NUMERIQUES

Les résultats numériques des tableaux C, F, J relatifs respectivement aux schémas 1, 2, 3 sont consignés sous forme de courbes (fig. 3, 4, 5).

Sur la figure 3, on a représenté la probabilité de l'événement $\overline{AU}$ (échec de l'arrêt d'urgence) en fonction de la probabilité de défaillance a d'un circuit L (circuit générant le signal d'arrêt d'urgence) pour les trois schémas étudiés. Dans les trois cas, la probabilité de défaillance d'une barre SAP, d'une barre SAC, d'un bloc de commande est prise égale à 10^{-2} (paramètres b, c, f).

Si on associe la notion de temps aux probabilités, on obtient des taux de défaillance et la ligne horizontale représente le critère sur les transitoires sans chute de barre (anticipated transients without scram) de 10^{-8}/an.

On voit que le respect de ce critère impose les taux de défaillance maximaux des circuits L, soit:

Schéma 1: $\lambda_L \leq 10^{-4}$/an
Schéma 2: $\lambda_L \leq 5 \cdot 10^{-4}$/an
Schéma 3: $\lambda_L \leq 2 \cdot 10^{-3}$/an

Dans le cas du schéma 1, les paramètres b, c, f n'influent pas sur la forme de la courbe représentative (dans les limites étudiées).

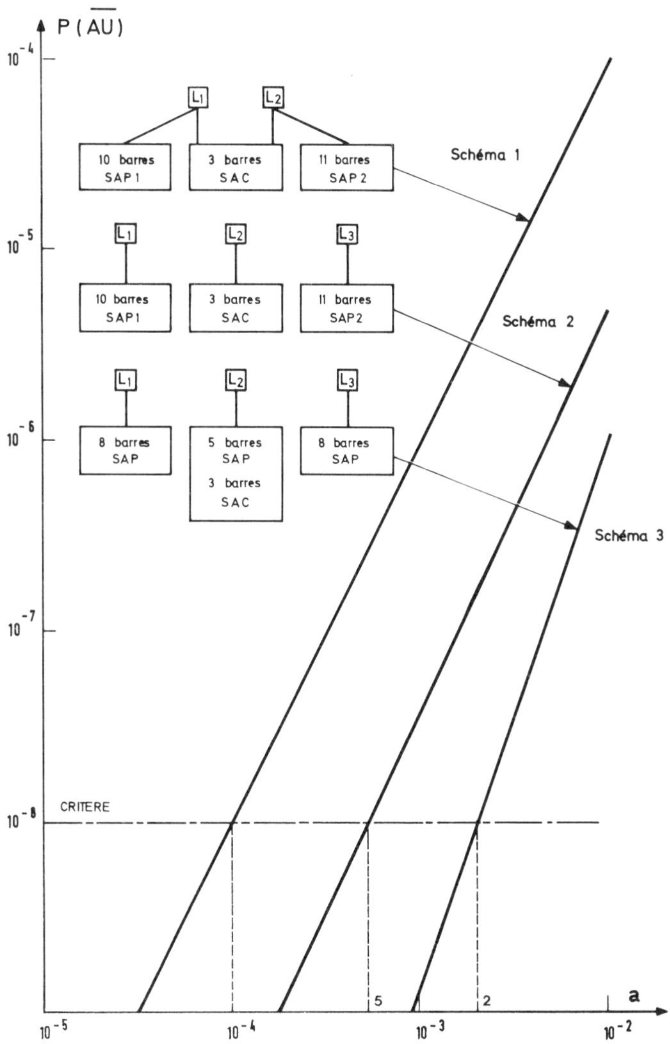

FIG.3. Probabilité de l'événement $\overline{AU}$ en fonction de la probabilité de défaillance a d'un circuit L (paramètres b, c, f = 10^{-2}).

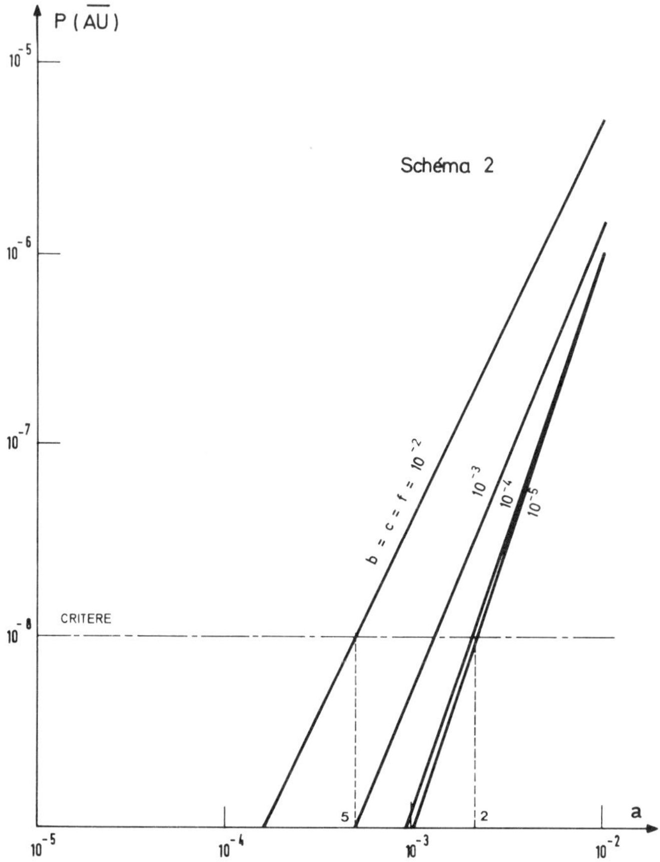

FIG.4. Schéma n° 2: influence des paramètres b, c, f.

Dans le cas du schéma 2, on peut voir sur la figure 4 l'influence de ces paramètres. On constate en particulier que pour $b=c=f=10^{-5}$, un taux de défaillance de $2 \cdot 10^{-3}$/an du circuit L est acceptable.

Dans le cas du schéma 3 (fig.5), les paramètres b, c, f ont peu d'influence et le circuit L peut avoir un taux de défaillance de $2 \cdot 10^{-3}$/an.

7. DEFAILLANCE DE MODE COMMUN

Dans ce qui précède, nous avons établi les relations existant entre les taux de défaillance catalectique des éléments constitutifs du système d'arrêt d'urgence et le taux de défaillance du système dans son ensemble.

Nous avons considéré comme respectivement identiques:

- les circuits L
- les blocs de commande
- les barres SAC
- les barres SAP.

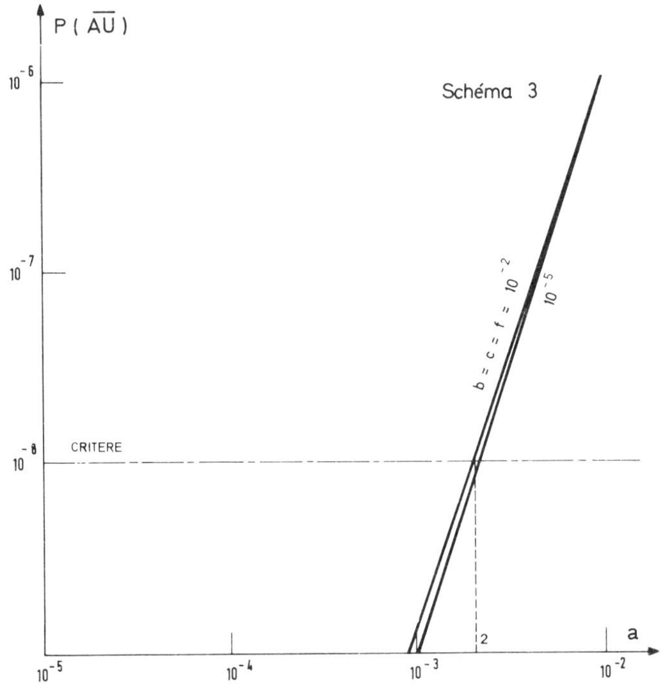

FIG.5. Schéma n° 3: influence des paramètres b, c, f.

Du fait même de leur identité, ces éléments sont sensibles à des événements (erreurs de conception, de réalisation, événements extérieurs, etc.) capables de générer leur défaillance simultanée. Un arbre de défaillance plus réaliste peut ainsi être dessiné (fig.6).

Les taux de défaillance de mode commun des circuits L et des blocs de commande viennent simplement s'ajouter aux taux de défaillance catalectique du système.

Il est intéressant d'examiner le cas d'une défaillance de mode commun des barres SAP.

Ainsi que le montre l'arbre, il y a échec de l'arrêt d'urgence si une défaillance de mode commun affecte les barres SAP et si, simultanément, une des 3 barres SAC ne chute pas.

Compte tenu des remarques des chapitres précédents (et du rapport DSN n° 57) les événements conduisant à la non-chute d'une barre SAC parmi 3 sont respectivement dans le cas des trois schémas:

	Schéma 1	Schéma 2	Schéma 3
Circuit L (a)	*	×	×
Bloc de commande (c)	×	×	×
1 barre SAC intrinsèquement défaillante	×	×	×

[* 1 seul circuit L défaillant n'est pas suffisant; le cas de 2 circuits L défaillants déjà traité conduit à lui seul à l'événement $\overline{AU}$.]

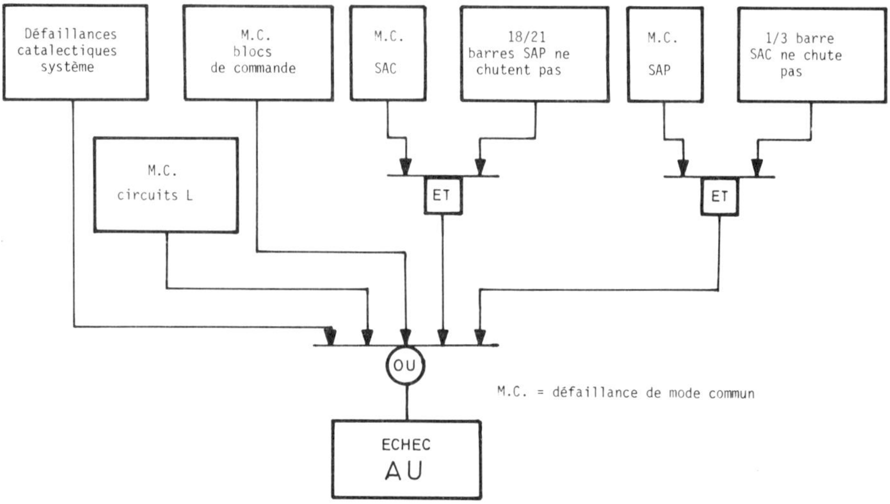

FIG.6. Arbre de défaillance avec modes communs.

Les probabilités de ces événements sont:

Schéma 1: c + 3 f

Schéma 2: a + c + 3 f

Schéma 3: a + c + 3 f

Les probabilités de ces causes supplémentaires d'échec de l'AU sont alors (en posant MC_{SAP} = probabilité de défaillance de mode commun des barres SAP):

Schéma 1: (c + 3 f) MC_{SAP}

Schéma 2: (a + c + 3 f) MC_{SAP}

Schéma 3: (a + c + 3 f) MC_{SAP}

Les courbes des figures 7, 8, 9 donnent (pour les trois schémas) la probabilité d'échec de l'arrêt d'urgence en fonction des paramètres a, b, c, f, pour trois valeurs de la probabilité d'une défaillance de mode commun affectant les barres SAP.

Si l'on considère les taux de défaillance par an, on voit que pour les trois schémas, le respect du critère 10^{-8}/an impose un taux de défaillance de mode commun très inférieur à 10^{-3} par an.

Les schémas 2 et 3 sont pratiquement équivalents. Ils permettent de respecter le critère 10^{-8}/an pour des valeurs plus élevées des paramètres a, b, c, f que dans le cas du schéma 1 (à taux de défaillance de mode commun égaux).

8. CONCLUSIONS DE L'ETUDE THEORIQUE

Les résultats obtenus permettent de comparer quantitativement plusieurs schémas du système d'arrêt d'urgence et mettent en lumière l'importance relative des éléments constitutifs et des types de défaillances.

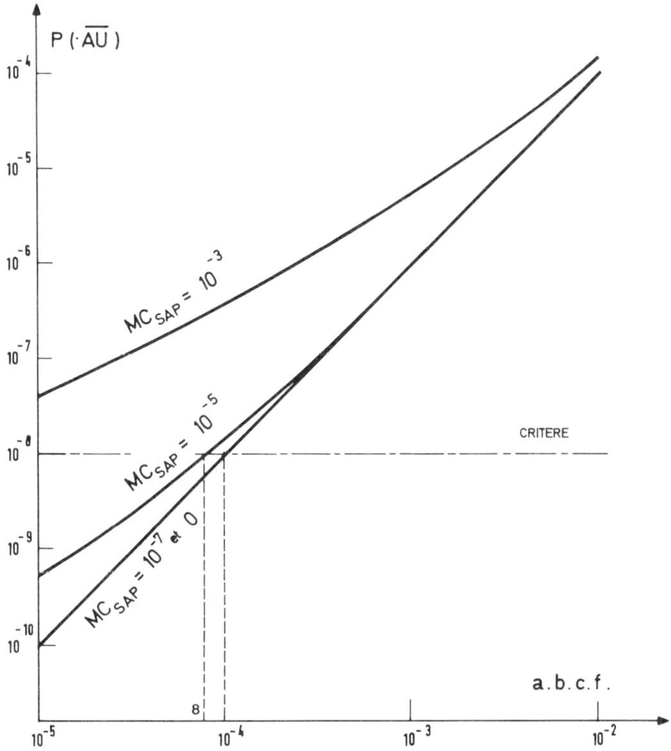

FIG.7. Schéma n° 1: influence d'une défaillance de mode commun affectant les barres SAP.

Les considérations techniques, géographiques ou économiques conduisant à préférer telle ou telle solution peuvent ainsi être examinées avec plus d'objectivité.

Les résultats obtenus, qui ne sont en fait qu'un exemple d'utilisation des formules générales établies, pourront être complétés si la suite de l'étude l'exige. Ils sont rappelés dans deux tableaux qui donnent les conditions nécessaires pour réaliser un système dont le taux de défaillance annuel $\lambda_{\overline{AU}}$ est inférieur à 10^{-8}. Le tableau IV résume la première partie de l'étude où seules les défaillances catalectiques sont prises en compte. Les taux de défaillance résultent bien entendu de la fiabilité intrinsèque des éléments, mais aussi d'un politique de test bien conçue. Le tableau V, relatif à la deuxième partie de l'étude, prend en compte les défaillances de mode commun.

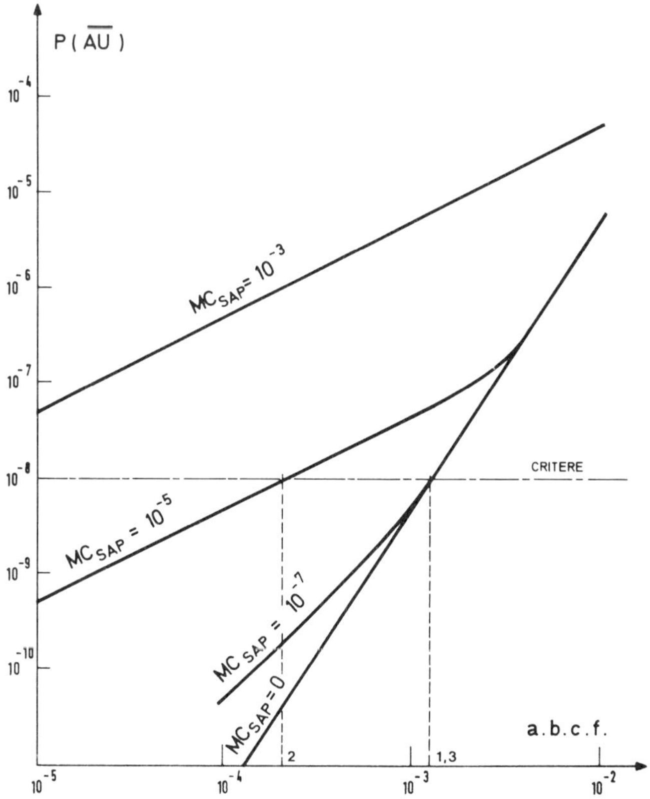

FIG.8. Schéma n° 2: influence d'une défaillance de mode commun affectant les barres SAP.

Le tableau V, plus réaliste que le tableau IV, montre bien l'attention que l'on doit apporter à l'étude des défaillances de mode commun. On voit par ailleurs que les schémas n° 2 et n° 3, sensiblement équivalents, sont meilleurs que le schéma n° 1. En effet, toutes choses égales par ailleurs, les fiabilités requises des circuits L sont dans un rapport 2,5 pouvant même atteindre 25 si le taux de défaillance de mode commun des barres SAP est de l'ordre de 10^{-7}.

La rareté de données opérationnelles de fiabilité rend délicate l'utilisation dans l'absolu des valeurs précédentes, mais elles gardent toute leur importance en relatif.

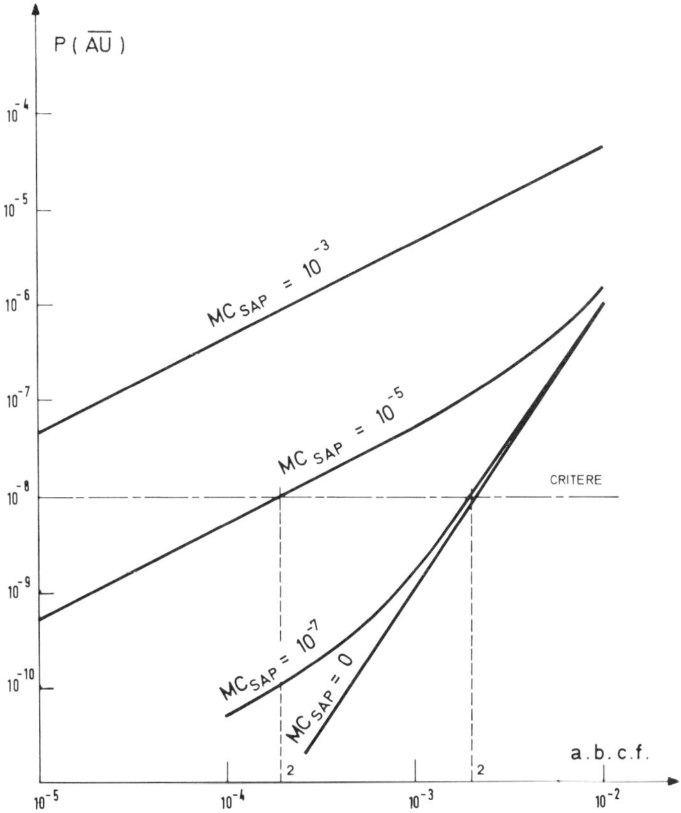

FIG.9. Schéma n° 3: influence d'une défaillance de mode commun affectant les barres SAP.

TABLEAU IV. REPARTITION DES TAUX ANNUELS DE DEFAILLANCE CATALECTIQUE SELON LES SCHEMAS CONSIDERES POUR OBTENIR $\lambda_{\overline{AU}} < 10^{-8}$ PAR AN

N° du schéma	λ/an, circuit L	λ/an, bloc de commande	λ/an, barre de sécurité
1	$\leq 10^{-4}$	$\leq 10^{-2}$	$\leq 10^{-2}$
2	$\leq 5 \cdot 10^{-4}$ $\leq 2 \cdot 10^{-3}$	$\leq 10^{-2}$ $\leq 10^{-5}$	$\leq 10^{-2}$ $\leq 10^{-5}$
3	$\leq 2 \cdot 10^{-3}$	$\leq 10^{-2}$	$\leq 10^{-2}$

TABLEAU V. REPARTITION DES TAUX ANNUELS DE DEFAILLANCE CATALECTIQUE ET DE MODE COMMUN SELON LES SCHEMAS CONSIDERES POUR OBTENIR $\lambda_{\overline{AU}} < 10^{-8}$ PAR AN

N° du schéma	Taux de défaillance de mode commun			Taux de défaillance catalectique d'un circuit L (λ/an)
	Circuits	Bloc de commande	Barres SAP	
1	$< 10^{-8}$	$< 10^{-8}$	$\leq 10^{-5}$	$\leq 8\cdot10^{-5}$
2	$< 10^{-8}$	$< 10^{-8}$	$\leq 10^{-5}$	$\leq 2\cdot10^{-4}$
	$< 10^{-8}$	$< 10^{-8}$	$\leq 10^{-7}$	$\leq 1{,}3\cdot10^{-3}$
3	$< 10^{-8}$	$< 10^{-8}$	$\leq 10^{-5}$	$\leq 2\cdot10^{-4}$
	$< 10^{-8}$	$< 10^{-8}$	$\leq 10^{-7}$	$\leq 2\cdot10^{-3}$

9. UTILISATION PRATIQUE DES RESULTATS

A partir des résultats de l'étude probabiliste, les équipes chargées de la conception des systèmes d'arrêt d'urgence disposent d'un «outil» leur permettant de choisir l'organisation et les technologies les mieux adaptées aux projets en études.

ANNEXE

SCHEMA N° 1

Tableau A. Défaillances réalisant $\overline{AU}$

N° d'ordre de la configuration	Etat des circuits L 0 = L défaillant 1 = L fonctionnant ⇒		Nombre de barres indisponibles du fait des circuits L ⇒	Nombre de barres restantes à prendre en compte	Nombre nécessaire et suffisant de barres intrinsèquement défaillantes pour réaliser $\overline{AU}$	
	L_1	L_2			SAC ET	SAP
(1)	0	0	3 SAC + 20 SAP	0		
(2)	1	0	10 SAP 2	3 SAC+10 SAP1	1/3	9/10
(3)	1	0	"	"	2/3	8/10
(4)	1	0	"	"	3/3	7/10
(5)	0	1	10 SAP 1	3 SAC+10 SAP2	1/3	9/10
(6)	0	1	"	"	2/3	8/10
(7)	0	1	"	"	3/3	7/10
(8)	1	1	0	3 SAC+20 SAP	1/3	19/20
(9)	1	1	0	"	2/3	18/20
(10)	1	1	0	"	3/3	17/20

SCHEMA N° 1

Tableau B. Probabilités de défaillance

N° d'ordre	Indices des probabilités		Probabilité état circuit L ×	Probabilité défaillance barres SAC ×	Probabilité défaillances barres SAP (Evénement Z)
(1)	P_1	p_1	a^2	1	1
(2)		p_2	(1-a) a	3 f	$10\,c^4 b + 40\,c^3 b^3 + 60\,c^2 b^5 + 40\,cb^7 + 10\,b^9$
(3)	P_2	p_3	(1-a) a	$3\,f^2$	$5\,c^4 + 60\,c^3 b^2 + 150\,c^2 b^4 + 140\,cb^6 + 45\,b^8$
(4)		p_4	(1-a) a	$c + f^3$	$40\,c^3 b + 200\,c^2 b^3 + 280\,cb^5 + 120\,b^7$
(5)		p_5	a (1-a)	3 f	
(6)	P_3	p_6	a (1-a)	$3\,f^2$	identique à (2), (3), (4)
(7)		p_7	a (1-a)	$c + f^3$	
(8)		p_8	$(1-a)^2$	3 f	$20\,c^9 b + 180\,c^8 b^3 + 720\,c^7 b^5 + 1680\,c^6 b^7 + 2520\,c^5 b^9 + 2520\,c^4 b^{11} + 1680\,c^3 b^{13} + 720\,c^2 b^{15} + 180\,c\,b^{17} + 20\,b^{19}$
(9)	P_4	p_9	$(1-a)^2$	$3\,f^2$	$10\,c^9 + 270\,c^8 b^2 + 1800\,c^7 b^4 + 5880\,c^6 b^6 + 11340\,c^5 b^8 + 13860\,c^4 b^{10} + 10920\,c^3 b^{12} + 5400\,c^2 b^{14} + 1530\,c\,b^{16} + 20\,b^{18}$
(10)		p_{10}	$(1-a)^2$	$c + f^3$	$180\,c^8 b + 2400\,c^7 b^3 + 11760\,c^6 b^5 + 30240\,c^5 b^7 + 46200\,c^4 b^9 + 43680\,c^3 b^{11} + 25200\,c^2 b^{13} + 8160\,c\,b^{15} + 1140\,b^{17}$

a = probabilité de défaillance d'un circuit L
b = probabilité de défaillance d'une barre SAP
c = probabilité de défaillance d'un bloc de commande (SAC ou SAP)
f = probabilité de défaillance d'une barre SAC.

Chaque ligne correspond au produit de 3 termes.

SCHEMA N° 1

Tableau C. Valeurs numériques

Probabilité défaillance circuit L a	Probabilité défaillance bloc c	Probabilité défaillance barre SAP b	Probabilité défaillance barre SAC f	Probabilité événement $\overline{AU}$ P	a / P %
10^{-2}	10^{-2}	10^{-2}	10^{-2}	$\simeq 10^{-4}$	$\simeq$100
	10^{-3}	10^{-3}	10^{-3}	$\simeq 10^{-4}$	100
	10^{-4}	10^{-4}	10^{-4}	$\simeq 10^{-4}$	100
	10^{-5}	10^{-5}	10^{-5}	$\simeq 10^{-4}$	100
10^{-3}	10^{-2}	10^{-2}	10^{-2}	$\simeq 10^{-6}$	100
	10^{-3}	10^{-3}	10^{-3}	$\simeq 10^{-6}$	100
	10^{-4}	10^{-4}	10^{-4}	$\simeq 10^{-6}$	100
	10^{-5}	10^{-5}	10^{-5}	10^{-6}	100
10^{-4}	10^{-2}	10^{-2}	10^{-2}	$\simeq 10^{-8}$	100
	10^{-3}	10^{-3}	10^{-3}	$\simeq 10^{-8}$	100
	10^{-4}	10^{-4}	10^{-4}	$\simeq 10^{-8}$	100
	10^{-5}	10^{-5}	10^{-5}	$\simeq 10^{-8}$	100
10^{-5}	10^{-2}	10^{-2}	10^{-2}	$\simeq 10^{-10}$	100
	10^{-3}	10^{-3}	10^{-3}	$\simeq 10^{-10}$	100
	10^{-4}	10^{-4}	10^{-4}	$\simeq 10^{-10}$	100
	10^{-5}	10^{-5}	10^{-5}	$\simeq 10^{-10}$	100

SCHEMA N° 2

Tableau D. Défaillances réalisant $\overline{AU}$

N° d'ordre de la configuration	Etat des circuits L ⇒			Nombre de barres indisponibles du fait des circuits L ⇒	Nombre de barres restantes à prendre en compte	Nombre nécessaire et suffisant de barres restantes défaillantes pour réaliser $\overline{AU}$	
	L_1	L_2	L_3			SAC	SAP
(1)	0	0	0				
(2)	1	0	0	3 SAC+10 SAP2	10 SAP 1		7/10
(3)	0	1	0	10 SAP1+10 SAP2	2 SAC	1/3	
(4)	0	0	1	10 SAP1+ 3 SAC	10 SAP 2		7/10
(5)	1	1	0	10 SAP 2	10 SAP1+ 3 SAC	1/3	9/10
(6)						2/3	8/10
(7)						3/3	7/10
(8)	1	0	1	3 SAC	10 SAP1 + 10 SAP2		17/20
(9)	0	1	1	10 SAP1	3 SAC + 10 SAP2	1/3	9/10
(10)						2/3	8/10
(11)						3/3	7/10
(12)	1	1	1	0	10 SAP1 + 3 SAC + 10 SAP2	1/3	19/20
(13)						2/3	18/20
(14)						3/3	17/20

SCHEMA N° 2

Tableau E. Probabilités de défaillance

N° d'ordre	Indices des probabilités		Probabilité état circuit L	Probabilité défaillance barres SAC	Probabilité défaillance barres SAP (Evénement Z)
(1)	P_1	p_1	a^3	1	1
(2)		p_2	$a^2(1-a)$	1	$40c^3b+200c^2b^3+280cb^5+120b^7+5c^4$
(3)	P_2	p_3	$a^2(1-a)$	$c+3f$	1
(4)		p_4	$a^2(1-a)$	1	$40c^3B+200c^2b^3+280cb^5+120b^7+5c^4$
(5)		p_5	$a(1-a)^2$	$3f$	$c^5+10c^4b+40c^3b^3+60c^2b^5+40cb^7+10b^9$
(6)		p_6	"	$3f^2$	$5c^4+60c^3b^2+150c^2b^4+140cb^6+45b^8$
(7)		p_7	"	$(c+f^3)$	$40c^3b+200c^2b^3+280cb^5+120b^7$
(8)	P_3	p_8	"	identique à (5),(6),(7)	identique à (5), (6), (7)
(9)		p_9	"		
(10)		p_{10}	"		
(11)		p_{11}	"		$10c^9+180c^8b+2400c^7b^3+11760c^6b^5+$ $30240c^5b^7+46200c^4b^9+43680c^3b^{11}+$ $25200c^2b^{13}+8160cb^{15}+1140b^{17}$
(12)		p_{12}	$(1-a)^3$	$3f$	$c^{10}+20c^9b+180c^8b^3+720c^7b^5+1680c^5b^7+$ $2520c^5b^9+2520c^4b^{11}+1680c^3b^{13}+720c^2b^{15}+$ $180cb^{17}+20b^{19}$
(13)	P_4	p_{13}	"	$3f^2$	$10c^9+270c^8b^2+1800c^7b^4+5880c^6b^6+$ $11340c^5b^8+13860c^4b^{10}+10920c^3b^{12}+$ $5400c^2b^{14}+1530cb^{16}+20b^{18}$
(14)		p_{14}	"	$c+f^3$	$180c^8b+2400c^7b^3+11760c^6b^5+30240c^5b^7+$ $46200c^4b^9+43680c^3b^{11}+25200c^2b^{13}+$ $8160cb^{15}+1140b^{17}$

SCHEMA N° 2

Tableau F. Valeurs numériques

Probabilité défaillance circuit L a	Probabilité défaillance bloc c	Probabilité défaillance barre SAP b	Probabilité défaillance barre SAC f	Probabilité échec $\overline{AU}$ P	a / P %
10^{-2}	10^{-2}	10^{-2}	10^{-2}	$4,9\ .10^{-6}$	20,16
	10^{-3}	10^{-3}	10^{-3}	$1,39.10^{-6}$	71,63
	10^{-4}	10^{-4}	10^{-4}	$1,03.10^{-6}$	96,19
	10^{-5}	10^{-5}	10^{-5}	10^{-6}	99,6
10^{-3}	10^{-2}	10^{-2}	10^{-2}	$4,09.10^{-8}$	2,44
	10^{-3}	10^{-3}	10^{-3}	$4,99.10^{-9}$	20,01
	10^{-4}	10^{-4}	10^{-4}	$1,3\ .10^{-9}$	71,44
	10^{-5}	10^{-5}	10^{-5}	$1,03.10^{-9}$	96,15
10^{-4}	10^{-2}	10^{-2}	10^{-2}	$4,02.10^{-10}$	0,24
	10^{-3}	10^{-3}	10^{-3}	$4,09.10^{-11}$	2,43
	10^{-4}	10^{-4}	10^{-4}	$4,99.10^{-12}$	20
	10^{-5}	10^{-5}	10^{-5}	$1,39.10^{-12}$	71,43
10^{-5}	10^{-2}	10^{-2}	10^{-2}	$4,11.10^{-12}$	0,024
	10^{-3}	10^{-3}	10^{-3}	$4\ \ .10^{-13}$	0,25
	10^{-4}	10^{-4}	10^{-4}	$4\ \ .10^{-14}$	2,44
	10^{-5}	10^{-5}	10^{-5}	$4,9\ .10^{-15}$	20

SCHEMA N° 3

Tableau G. Défaillances réalisant $\overline{AU}$

N° d'ordre de la configuration	Etat des circuits L ⇒			Nombre de barres indisponibles du fait des circuits L ⇒	Nombre de barres restantes à prendre en compte	Nombre nécessaire et suffisant de barres restantes défaillantes pour réaliser $\overline{AU}$	
	L_1	L_2	L_3			SAC	SAP
(1)	0	0	0	21 SAP+3 SAC	0		
(2)	1	0	0	13 SAP+3 SAC	8 SAP		5/8
(3)	0	1	0	16 SAC	5 SAP+3 SAC	1/3	4/5
(4)						2/3	3/5
(5)						3/3	2/5
(6)	0	0	1	13 SAP+3 SAC	8 SAP		5/8
(7)	1	1	0	8 SAP	13 SAP+3 SAC	1/3	12/13
(8)						2/3	11/13
(9)						3/3	10/13
(10)	1	0	1	5 SAP+3 SAC	16 SAP		13/16
(11)	0	1	1	8 SAP	13 SAP+3 SAC	1/3	12/13
(12)						2/3	11/13
(13)						3/3	10/13
(14)	1	1	1	0	21 SAP+3 SAC	1/3	20/21
(15)						2/3	19/21
(16)						3/3	18/21

SCHEMA N° 3

Tableau H. Probabilités de défaillance

N° d'ordre	Indices des probabilités		Probabilité état circuit L	Probabilité défaillance barres SAC	Probabilité défaillance barres SAP (Evénement Z)
(1)	P_1	P_1	a^3		
(2)	P_2	P_2	$a^2 (1-a)$		$4 c^3 + 24 c^2 b + 80 cb^3 + 56 b^5$
(3)		P_3	"	$3 f$	$c^2 + 5 b^4 + 4 b^3 c + 6 cb^2 + 4 bc^2$
(4)		P_4	"	$3 f^2$	$2 c^2 + 6 cb + 10 b^3 + 6 b^2 c$
(5)		P_5	"	$c + f^3$	$2 c + 10 b^2 + 4 bc$
(6)		P_6	"		identique à (2)
(7)	P_3	P_7	$a (1-a)^2$	$3 f$	$c^6 + 18 c^5 b^2 + 12 c^6 b + 75 c^4 b^4 + 60 c^5 b^3 + 140 c^3 b^6 + 120 c^4 b^5 + 135 c^2 b^8 + 120 c^3 b^7 + 66 cb^{10} + 60 c^2 b^9 + 13 b^{12} + 12 cb^{11}$
(8)		P_8	"	$3 f^2$	$18 c^5 b + 6 c^6 + 150 c^4 b^3 + 90 c^5 b^2 + 420 c^3 b^5 + 300 c^4 b^4 + 540 c^2 b^7 + 420 c^3 b^6 + 324 cb^9 + 270 c^2 b^8 + 77 b^{11} + 66 cb^{10}$
(9)		P_9	"	$c + f^3$	$6 c^5 + 150 c^4 b^2 + 60 c^5 b + 700 c^3 b^4 + 400 c^4 b^3 + 1260 c^2 b^6 + 840 c^3 b^5 + 990 cb^8 + 720 c^2 b^7 + 286 b^{10} + 220 cb^9$
(10)		P_{10}	"		$8 c^7 + 112 c^6 b + 1120 c^5 b^3 + 3920 c^4 b^5 + 6720 c^3 b^7 + 6160 c^2 b^9 + 2912 cb^{11} + 560 b^{13}$
(11)		P_{11}	"		identique à (7)
(12)		P_{12}	"		identique à (8)
(13)		P_{13}	"		identique à (9)
(14)	P_4	P_{14}	$(1-a)^3$	$3 f$	$c^{10} + 20 c^9 b + 180 c^8 b^3 + 720 c^3 b^5 + 1680 c^6 b^7 + 2520 c^5 b^9 + 2520 c^4 b^{11} + 1680 c^3 b^{13} + 720 c^2 b^{15} + 180 cb^{17} + 20 b^{19}$
(15)		P_{15}	"	$3 f^2$	$10 c^9 + 270 c^8 b^2 + 1800 c^7 b^4 + 5880 c^6 b^6 + 11340 c^5 b^8 + 13260 c^4 b^{10} + 10920 c^3 b^{12} + 5400 c^2 b^{14} + 1530 cb^{16} + 20 b^{18}$
(16)		P_{16}	"	$c + f^3$	$180 c^8 b + 2400 c^7 b^3 + 11760 c^6 b^5 + 30240 c^5 b^7 + 46200 c^4 b^9 + 43680 c^8 b^{11} + 25200 c^2 b^{13} + 8160 cb^{15} + 1140 b^{17}$

Nota: On a pris ici les mêmes valeurs de probabilité que dans les cas précédents, c'est-à-dire qu'on néglige la 21e barre.

SCHEMA N° 3

Tableau J. Valeurs numériques

Probabilité défaillance circuit L a	Probabilité défaillance bloc c	Probabilité défaillance barre SAP b	Probabilité défaillance barre SAC f	Probabilité échec AU P	a / P %
10^{-2}	10^{-2} 10^{-3} 10^{-4} 10^{-5}	10^{-2} 10^{-3} 10^{-4} 10^{-5}	10^{-2} 10^{-3} 10^{-4} 10^{-5}	10^{-6} 10^{-6} 10^{-6} 10^{-6}	97,25 99,97 99,99 100
10^{-3}	10^{-2} 10^{-3} 10^{-4} 10^{-5}	10^{-2} 10^{-3} 10^{-4} 10^{-5}	10^{-2} 10^{-3} 10^{-4} 10^{-5}	$1,88.10^{-9}$ 10^{-9} 10^{-9} 10^{-9}	77,80 99,79 99,99 100
10^{-4}	10^{-2} 10^{-3} 10^{-4} 10^{-5}	10^{-2} 10^{-3} 10^{-4} 10^{-5}	10^{-2} 10^{-3} 10^{-4} 10^{-5}	$3,8\ .10^{-12}$ 10^{-12} 10^{-12} 10^{-12}	25,93 97,95 99,97 99,99
10^{-5}	10^{-2} 10^{-3} 10^{-4} 10^{-5}	10^{-2} 10^{-3} 10^{-4} 10^{-5}	10^{-2} 10^{-3} 10^{-4} 10^{-5}	$3\ .10^{-14}$ $1,2\ .10^{-15}$ 10^{-15} 10^{-15}	3,36 82,75 99,79 99,99

IAEA-SM-195/32

OPTIMIZING THE DESIGN AND OPERATION OF REACTOR EMERGENCY SYSTEMS USING RELIABILITY ANALYSIS TECHNIQUES

E.R. SNAITH
Safety and Reliability Directorate,
United Kingdom Atomic Energy Authority,
Culcheth, Warrington, Cheshire,
United Kingdom

Abstract

OPTIMIZING THE DESIGN AND OPERATION OF REACTOR EMERGENCY SYSTEMS USING RELIABILITY ANALYSIS TECHNIQUES.

Following a reactor trip various reactor emergency systems, e.g. essential power supplies, emergency core cooling and boiler feed water arrangements are required to operate with a high degree of reliability. These systems must therefore be critically assessed to confirm their capability of operation and determine their reliability of performance. The use of probability analysis techniques enables the potential operating reliability of the systems to be calculated and this can then be compared with the overall reliability requirements. However, a system reliability analysis does much more than calculate an overall reliability value for the system. It establishes the reliability of all parts of the system and thus identifies the most sensitive areas of unreliability. This indicates the areas where any required improvements should be made and enables the overall systems' designs and modes of operation to be optimized, to meet the system and hence the overall reactor safety criteria. This paper gives specific examples of sensitive areas of unreliability that were identified as a result of a reliability analysis that was carried out on a reactor emergency core cooling system. Details are given of modifications to design and operation that were implemented with a resulting improvement in reliability of various reactor sub-systems. The report concludes that an initial calculation of system reliability should represent only the beginning of continuing process of system assessment. Data on equipment and system performance, particularly in those areas shown to be sensitive in their effect on the overall nuclear power plant reliability, should be collected and processed to give reliability data. These data should then be applied in further probabilistic analyses and the results correlated with the original analysis. This will demonstrate whether the required and the originally predicted system reliability is likely to be achieved, in the light of the actual history to date of equipment performance.

1. Introduction

That a Nuclear Power Plant must be reliable in all aspects i.e. by design, construction and operation, is an undisputed requirement. The reliability requirement furthermore may be justified by both safety and economic considerations. The safety consideration is that of ensuring that the probability of a hazard to the public as a result of any mal-function on the reactor is reduced to an acceptable minimum. Thus it must be demonstrated that the plant can be operated in a safe controlled manner, and that for all foreseeable accidents it can be safely shutdown and maintained in a safe shutdown state for an indefinite period.

The economic considerations arise from the fact that a nuclear power plant is a generating station, a manufacturer of electricity, and thus to be economic it must have a high load or availability factor. A shutdown reactor may be safe, but it is not fulfilling its prime function, and it is thus uneconomic.

It is therefore necessary that a nuclear power plant be assessed to determine and demonstrate, as far as possible, its capability and reliability, particularly under fault conditions. This requires a detailed examination of all the reactor sub-systems which go to make up the whole reactor unit, particularly those systems associated with the ultimate safety of the reactor e.g. protective systems, emergency core cooling, emergency power supplies etc. The reliability of these dictate the overall reliability of the nuclear power plant.

2. Assessment Method

Let us then examine the methods of assessing the reliability of these plants. The assessment will confirm basically that the various sub-systems are capable of performing in the desired manner thereby confirming the overall design capability of the whole nuclear power plant. It will also aim at demonstrating that the degree of reliability (or availability) actually required can potentially be achieved.

An assessment may therefore be arbitrarily divided into two distinct analyses

(a) a qualitative analysis
(b) a quantitative analysis

(a) Qualitative Analysis

A major part of the qualitative analysis is ensuring that the appropriate design and operating requirements have been satisfied, and for those systems forming part of the reactor safety services e.g. protective system, emergency power and emergency core cooling, that they follow good emergency system practice. This part of the reliability assessment is deterministic and can be based on emergency codes of practice and safety principles which form the safety criteria for the reactor. Examples of such safety principles are:

(a) Emergency systems should be automatically initiated and capable of operation within the prescribed timescale.

(b) Emergency systems, such as emergency core cooling, should be independent of normal core cooling systems and thus not subject to common mode faults.

(c) Emergency power supplies should be provided for all essential plant. They should be independent of normal supplies and capable of providing a continuous independent supply for such time as is necessary to keep the reactor in a safe shut-down state.

(d) No single fault should invalidate a system's operation and any such faults should be alarmed and if necessary automatically isolated to prevent cascade failure effects.

(e) Duplicate equipment should be independent in all respects, thus avoiding the chance of total failure due to common mode faults.

(f) Facilities should be provided for regular testing of emergency and/or safety systems, to demonstrate their continuing capability and reliability. Such tests should demonstrate the individual performance capability of each independent channel forming part of a redundant system.

The foregoing examples of safety principles enable a qualitative assessment to be carried out.

(b) Quantitative Analysis

A Quantitative analysis represents a method whereby the potential reliability of the various reactor sub-systems is calculated in numerical terms.

Where an overall criterion for the plant or system is available, the work of the assessor may be eased since, if the derived probability of failure can be compared with that required, then clearly the system can be judged adequate or otherwise.

In the UKAEA nuclear industry the Farmer criterion [1] defined objectives which set the scene for considerable activity in the formulation and adoption of reliability analysis techniques and these techniques are now finding increasing application in nuclear reactor assessments [2,3,4]

Various methods such as fault trees, truth tables, Monte Carlo analysis and network synthesis using logic diagrams have all been tried and much has been written discussing these different methods. [5,6,7,8] In terms of their application to reactor systems, past papers have, in general, merely quoted the overall results obtained from an analysis and reconciled these with the achievable reliability for a well designed and well maintained system. The end result quoted has been a specific value for the system e.g. it has shown that for a protective system a probability of failure of say 10^{-3} to 10^{-4} for a single system is achievable or that for an emergency core cooling system a failure probability of 10^{-3} may be demonstrated. The conclusion has then been that by applying logical probability techniques the reliability number for the system can be determined and used in the overall reactor safety equations to be measured against some specific criteria.

However, what must be recognised is that a reliability analysis does much more than determine the numerical reliability of a system. Indeed such an analysis carried out at the design stage may not always show that the system or systems can achieve the required reliability. The analysis may show that a system falls short of the reliability target. However, a real virtue of the analysis is that of identifying the vulnerable areas of unreliability, i.e. those areas where either by design modification or by different test arrangements or even by alternative operating modes, the overall target reliability figure is more likely to be achieved. To demonstrate this, therefore, brief details are given of a part of an analysis that was carried out on the emergency core cooling arrangements for an Advanced Gas Cooled Reactor (AGR) in the UK. The original analysis identified specific sensitive areas of unreliability and details are given of some modifications that were carried out both to the design and the mode of operation and with regard to test arrangements, which resulted in an acceptable target reliability being achieved. This was demonstrated by a subsequent analysis which took into account the various modifications.

3. Reliability Analysis - Case Study

The case considered is that of the decay heat rejection arrangements for a current design of a 1000MW(e) advanced gas-cooled reactor following a reactor trip.

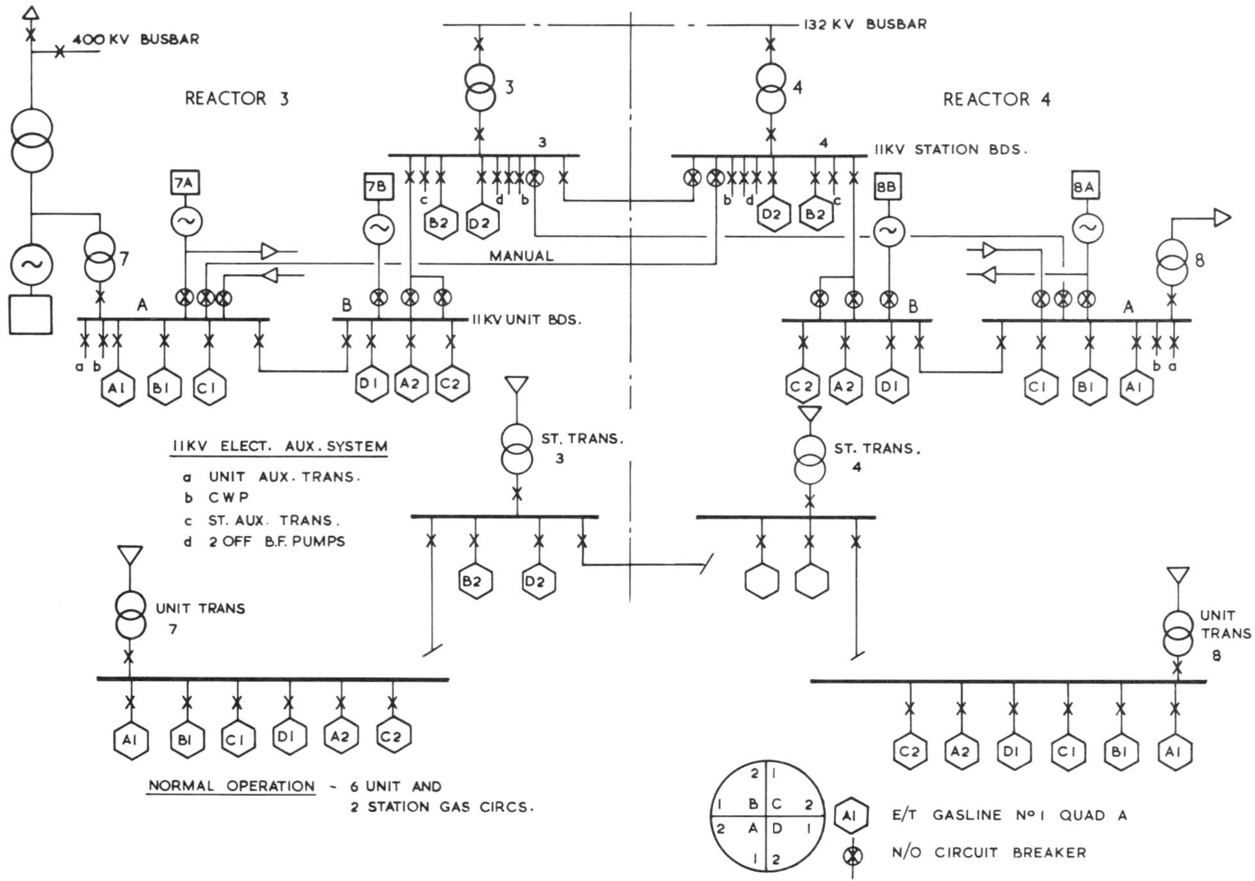

FIG. 1. Schematic diagram of electrical supply arrangements for a nuclear power station.

While some degree of natural circulation of the primary coolant gas from the reactor is available in the AGR design, it is necessary to restore forced gas circulation and sufficient boiler feed to maintain the heat removal capacity of the boilers for all reactor shut-down conditions.

The requirements for removal of fission product heating are met by several independent mechanical systems of primary coolant fans, boiler feedwater pumps and valves integrated with electrical power sources, switch-gear and automatic control equipment.

The timescale of requirement and amount of forced cooling is dependent on the temperature transient during and following a reactor trip condition. This in turn varies with the circulator run-down times, fuel rating and condition, and the rate of any reactor depressurisation.

The analysis represents the calculation of the probability of failure to provide the minimum required cooling, following a reactor trip for both the pressurised and the depressurised reactor situation and also assuming that all normal power supplies may be lost.

Fig. 1 is a schematic diagram showing the alternative arrangements for providing electrical supplies to reactor essential loads following a reactor trip. The control arrangements are such that each section of switchboard may be supplied either from the 132 kV supply, if available, or from any one of three gas turbines.

Fig. 2 shows a schematic arrangement of the various systems and items of plant that are required to automatically sequentially operate to remove primary and secondary heat e.g. primary flow guide vanes, boiler feed pumps and associated boiler vent systems and power supplies.

Fig. 3 represents a typical logic flow diagram, for a specific reactor requirement. Each of the blocks in this diagram represent a sub-system and the diagram combines these in the correct logic pattern to meet the overall requirement. The probability of failure of each of these blocks is separately calculated, either manually or by computer analysis, depending on the complexity of the sub-system and the overall system reliability determined. With the calculated value of each block inserted on the diagram the contribution of each to the overall power plant reliability can be seen. In the initial calculations certain blocks were found to be dominant. Design modifications and revised test frequencies were proposed and the calculated improvement in reliability was shown to meet the required value.

For example box p_4 in Fig. 3 represents the sequence permission timer logic which gives the final initiating signal to start the relevant gas circulators, providing all pre-requisites i.e. boiler feed, power supplies etc. are satisfied. This circuit originally consisted of three control channels of electrical timers and relays, operating on a two from three principle and it represented a dominant area of unreliability. To an extent it nullified the virtues of redundancy employed in other parts of the system and set a limiting value on the overall system reliability. By changing the design to a two out of four configuration and increasing the frequency of testing to once a month the revised probability of failure, due to random faults, was reduced to 10^{-6} from an original calculated value of 10^{-3}. Thus this common area of dependency no longer represented the sensitive area of unreliability.

A similar situation existed with the boiler venting arrangements. A vent system is provided for each boiler but these are all interconnected.

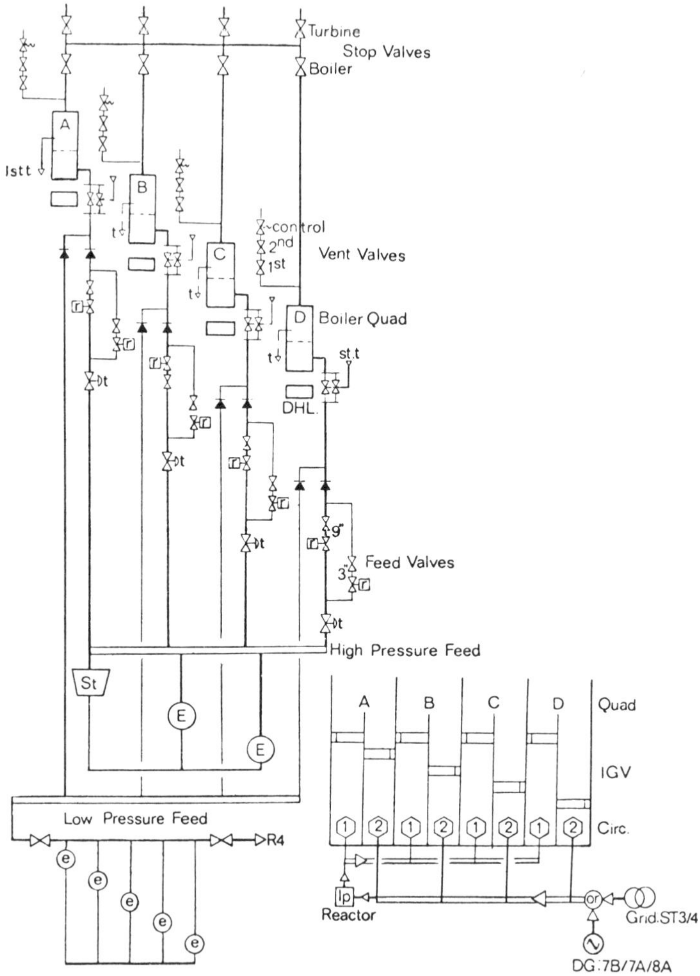

FIG. 2. Arrangement of reactor sub-systems for emergency core cooling.

It was thus deemed that all boilers could be vented, to permit either H.P. or L.P. feed operation, provided any one boiler vent system operated, thus giving a redundant arrangement of any one from four systems. However the overall probability of failure to vent due to failure of all four vent systems, was limited by the reliability of the common initiating logic, again comprising a two out of three combination of timers and relays. This was also modified to a two out of four arrangement with a resulting improvement and an optimising of overall system reliability i.e. the elimination of a single sensitive area of unreliability.

A further area where considerable improvement in the overall reliability of the plant operation has been made is that associated with restoration of the alternative source of electrical supply. For the situation where all normal electrical supplies are lost following a reactor trip any

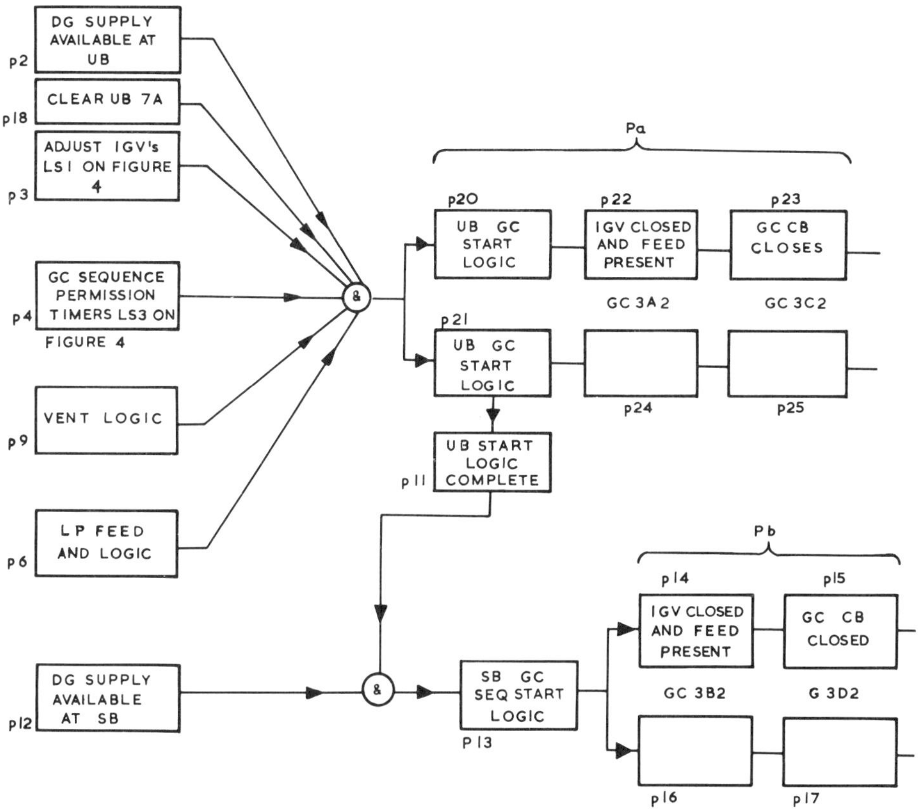

FIG. 3. Logic diagram of sub-systems required for emergency core cooling.

one of three available diesel generators is capable of supplying the reactor essential loads. From figure 1 which shows the schematic arrangement of power supplies for two reactors, the unit board B loads of Reactor 3 and the Station Board 3 loads may be supplied from the 132 kV supply, if it is available, or from either DG7B OR DG7A OR DG8A, in this priority order of selection. Using a simple approach, if the probability of failure of a single DG to start and auto-close in the bus-bar is $\bar{p}_1$ then the probability of failure of all three (considering random independent faults) is $\bar{p}_1^3$.

With a value for $\bar{p}_1$ of say 2×10^{-2}, this would give a probability of failure to obtain a supply, due to all three failing of 8×10^{-6}.

However, if one now draws the logic flow diagram (Fig.4) the interdependence between the different sections of the system can be seen. Since UB7A and 7B are operated with the interconnector closed then before any supply can be connected to either section of the board, the normal supply circuit bracket e.g. UT7 would require to be tripped together with any other connected loads. Thus, despite the redundant supply sources, the overall system may be no better than a value attributed to the probability of failure of this circuit breaker to trip.

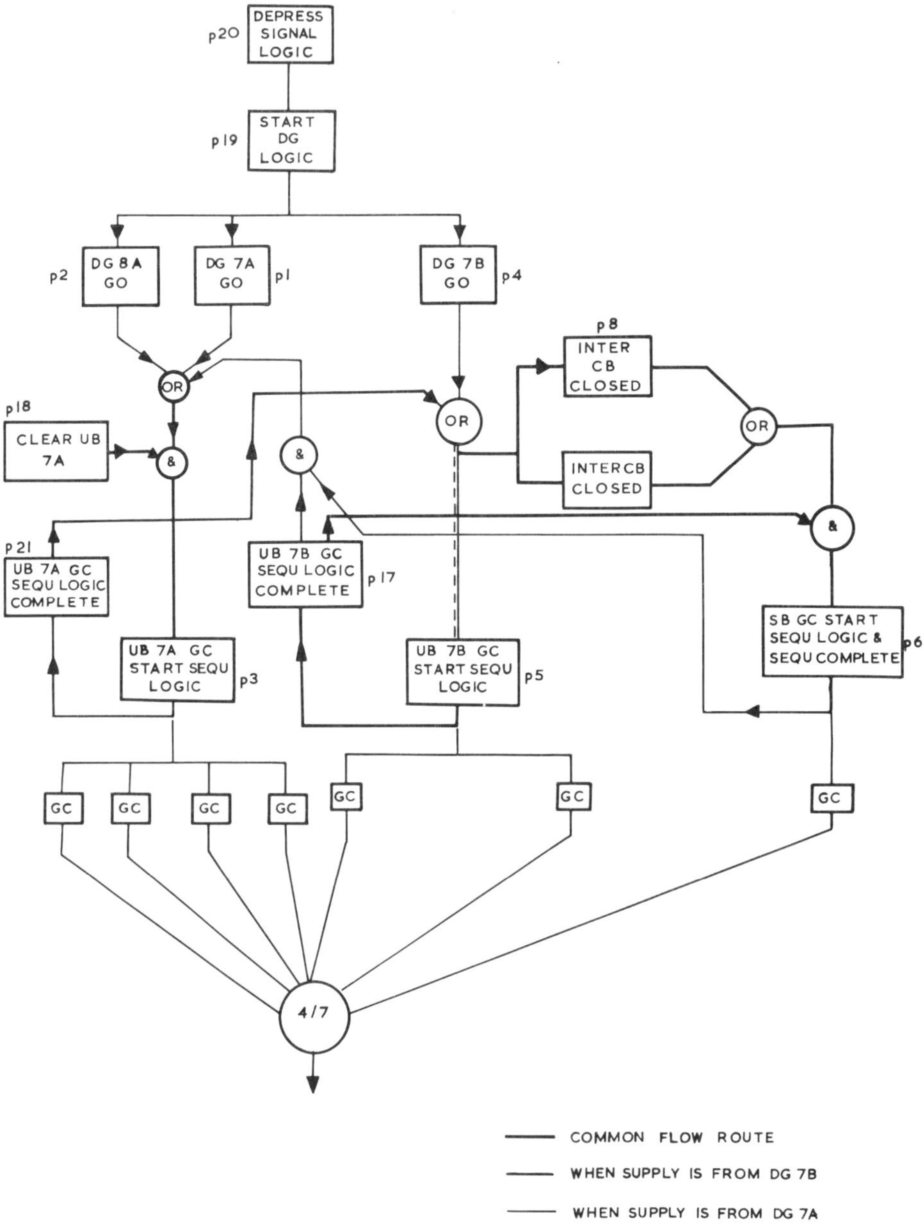

FIG. 4. Logic flow diagrams of arrangements for restoring electrical power supplies.

Similarly if the station board supply is available, then this becomes the priority source of supply and in order to supply the Unit Boards the SB/UB interconnector CB at the UB would require to close. These areas were identified as being sensitive in the overall system analysis and calculations were carried out for various system configurations and operating modes to determine the optimum arrangement. Improvements have now been made which remove these over sensitive areas.

The virtue of a reliability analysis which makes use of a logical flow diagram depicting the system operation is the identification of these sensitive areas which may be common to the overall system, thereby having a limiting influence on the effect of other redundant sections of plant. Several such areas were identified in the analysis resulting in an overall improvement in plant reliability and optimisation of the design, of operating methods and of testing arrangements.

This type of analysis thus enables the overall reliability of a system to be derived and quoted with full justification that an optimised system design has been achieved. Many projects have been examined in this way and in all cases known to the author, some areas of potential weakness have been identified - areas in which improvement has been possible. In some cases recommendations for additional testing have been made as a means of establishing an initial reliability target. Continuing testing is then advised in order to build up a picture of the overall reliability of areas which have been shown up to be particularly sensitive in the overall reliability analysis of the complete plant.

4. Conclusions

Having completed a reliability analysis and hopefully achieved the sub-systems target reliability which in turn demonstrates the reliability requirement for the complete nuclear power plant to be satisfactory, let us summarise the information obtained from the analysis.

(a) the reliability of the complete plant has been calculated.

(b) the reliability of the individual sub-systems which combine to form the overall power plant has been calculated.

(c) the exact method of operation of the plant is known.

(d) The sensitive areas of unreliability are identified.

(e) the facilities for testing and the capability of being able to test the various parts of the system have been examined.

(f) the effects of frequency of testing and/or planned maintenance on overall reliability have been demonstrated.

(g) the effects of possible operator involvement may have been highlighted.

(h) common mode fault areas are identified.

Although the combination of the above establishes the potential achievable reliable of the power plant - only actual operation of the plant will determine whether this predicted reliability can be achieved. Thus, the analysis rather than representing the completion of the assessment of a system must surely be only the beginning of a continuing commitment. A test now remains to be carried out during the lifetime operation of the plant, the test being to determine whether the systems do perform with the reliability as predicted; to determine whether those potential vulnerable areas of failure are in fact as sensitive as the analysis showed.

Data on plant and system performance must therefore be recorded. Periodic test results must be analysed. It is suggested that since it is generally not possible, nor justified, to collect and process all failure rate data, the areas be limited to those shown up in the analysis to be particularly sensitive. The values obtained from the operating plant should then be compared with the values used in the initial analysis. A further analysis should then be carried out and compared with the original analysis to determine whether the systems are as reliable as predicted. The results of this will determine whether any design or operating changes are necessary.

Thus a reliability analysis must be a continuing process. This seems to be an area where little work has been done i.e. a later analysis using actual component and plant performance data from the plant, to be compared with the original predicted analysis. The initial analysis is very necessary, indeed it may be a very essential step, before operating approval is given. However, continuing work to demonstrate the operating reliability of the plant is a necessary step to make the original analysis a viable one.

The techniques of reliability analysis as applied to a nuclear power plant may be well known but the virtues of such an analysis are still not yet fully appreciated. However the feeding back of operational data, albeit in limited areas, to enable subsequent analyses to be carried out can only give more enlightenment of system performance and enable optimum reliability to be achieved and maintained.

REFERENCES

1. FARMER, F. R., "Siting criteria - A new approach", Containment and Siting of Nuclear Power Plants (Proc. Symp. Vienna, 1967), IAEA, Vienna (1967) 303.
2. GREEN, A. E. and BOURNE, A. J., Safety Assessment With Reference to Automatic Protective Systems for Nuclear Reactors. UKAEA Report No. AHSB(S)R117, 1966.
3. GREEN, A. E., "Safety assessment of systems", Principles and Standards of Reactor Safety (Proc. Symp. Jülich, 1973), IAEA, Vienna (1973) 105.
4. KIRK, J., TAYLOR, R. S., "Design of safety of gas cooled reactors", 4th Int. Conf. Peaceful Uses At. Energy (Proc. Conf. Geneva, 1971) 3, IAEA, Vienna (1972) 151.
5. BOURNE, A. J., "Considerations in the Use of Systems Reliability Models" Proceedings of CREST Specialist Meeting on Applicability of Quantitative Reliability Analysis of Complex Systems and Nuclear Plants in its Relation to Safety. Munich May 1971.

6. COLOMBO, A. G., RICCHENA, R., and VOLTA, G., "Survey and Critical Analysis of Programs for System Reliability Computation", Proceedings of the CREST Specialist Meeting, Munich, 26-28 May, 1971.

7. FUSSEIL, J. B., and VESELY, W. E., "A New Methodology for Obtaining Cut Sets for Fault Trees", Transactions of the American Nuclear Society, 15(1), 1972.

8. DANIELSON, A. E., SNAITH, E. R., "The Use of Reliability Analysis Techniques Applied to Nuclear Power Station Emergency Core Cooling Systems" CSNI Meeting on the Development and Application of Reliability Techniques to Nuclear Plant. Liverpool, April 1974.

FIABILITE OPERATIONNELLE DU POSTE DE SECURITE «ECHAUFFEMENT ANORMAL D'ASSEMBLAGE» SUR LE REACTEUR PHENIX

R. QUENEE*, J. GRISOLLET**
CEA, Centre d'études nucléaires de Saclay,
Gif-sur-Yvette,
France

Abstract–Résumé

OPERATIONAL RELIABILITY OF THE SAFETY SYSTEM TO DETECT ABNORMAL HEATING OF FUEL ASSEMBLIES IN THE PHENIX REACTOR.

In fast reactors, the surveillance of safety barriers requires a large number of temperature measurements, so that any anomalous heating of fuel assemblies can be detected very rapidly. The data can be handled within the time available only by means of computers. For dealing with overheating, the Phénix reactor has a safety system to which information is transmitted from the thermocouples on the fuel assemblies; the information passes through industrial mini-computers to a relay system which triggers the control rod drop if necessary. During normal operation the computers serving each safety circuit operate in pairs in a coincidence regime – in other words, the information from the two thermocouples on a fuel assembly must coincide, as must the information passed on by the two computers, if an emergency shut-down is to be triggered. In view of the importance of this safety system and the use of small computers for surveillance and initiation of safety measures, knowledge of the operational history of the system components is essential. Since a record has been kept of breakdowns affecting Phénix safety components, operational reliability estimates covering a period of over 18 months have been made for the components of the safety system for detection of anomalous heating of fuel assemblies. On the basis of these estimates it has been possible to verify the reliability and availability of the system in real operating conditions. The results of this study will be used in designing and constructing the corresponding safety system for the next French fast reactor, Super Phénix.

FIABILITE OPERATIONNELLE DU POSTE DE SECURITE «ECHAUFFEMENT ANORMAL D'ASSEMBLAGE» SUR LE REACTEUR PHENIX.

La fonction surveillance des barrières sur les réacteurs rapides impose un nombre important de mesures dans le domaine des températures. La connaissance de ces températures permet de détecter dans un laps de temps très court l'échauffement anormal d'un assemblage combustible. Seuls des ordinateurs permettent le traitement de ces données dans le temps imparti. Pour assurer la fonction «détection d'échauffement anormal», le réacteur Phénix comporte un poste de sécurité partant des informations délivrées par les thermocouples des assemblages puis transitant par des mini-ordinateurs industriels et enfin aboutissant au relayage ordonnant la chute des barres si nécessaire. Cette fonction de détection est assurée par deux lignes. En fonctionnement normal, les ordinateurs de chacune des lignes sont utilisés en deux sur deux, c'est-à-dire qu'il faut qu'il y ait coïncidence des informations sur les deux thermocouples d'un même assemblage et sur les deux calculateurs pour provoquer l'arrêt d'urgence. Devant l'importance de ce poste de sécurité et l'utilisation de petits ordinateurs en surveillance et action de sécurité, le suivi opérationnel de ce matériel est fondamental. Comme il existe sur Phénix une collecte des pannes survenant sur les matériels liés à la sûreté de l'installation, des données opérationnelles de fiabilité ont été estimées sur plus de 18 mois de fonctionnement pour les composants du poste de sécurité de détection d'échauffement anormal d'assemblage. Grâce à ces données, on a pu vérifier la fiabilité et la disponibilité de cette fonction de sécurité dans son fonctionnement réel. Les enseignements retirés de cette étude vont servir à la conception et à la réalisation du même poste de sécurité du réacteur suivant de la filière rapide Super Phénix.

---

* Département de sûreté nucléaire.
** Technicatome.

INTRODUCTION

Les réacteurs rapides refroidis au sodium sont dotés d'un nombre d'assemblages combustibles important. Chacun de ces assemblages peut présenter un risque de bouchage partiel en cas de formation de dépôt d'oxyde au niveau de l'admission du fluide de refroidissement. Il y a alors risque d'échauffement rapide de l'élément combustible, gonflement de la gaine, voire fusion du combustible dans le cœur.

De plus, un échauffement au niveau d'un assemblage entraîne rapidement l'échauffement des assemblages voisins (voir les conséquences de l'accident survenu à Enrico Fermi le 5 octobre 1966).

Il fut donc décidé de surveiller la température de sortie des assemblages combustibles du réacteur Phénix à l'aide de deux thermocouples disposés sur un doigt de gant au-dessus de chaque assemblage, l'ensemble étant supporté par le couvercle cœur.

Cette mesure est aussi représentative de l'état de l'assemblage, pour un niveau de flux déterminé, que la mesure du flux neutronique ou du rapport débit/puissance. On doit donc répondre aux mêmes exigences de sûreté.

En collaboration étroite avec le constructeur et l'exploitant, les responsables de la sûreté concentrèrent toute leur attention sur les calculateurs affectés à cette tâche, au niveau des projets, de la construction, de la mise en service et de l'exploitation. Grâce à ce travail d'équipe, les points importants pour la sûreté furent progressivement mis en lumière et on dispose maintenant d'un document précisant les critères auxquels doivent répondre les calculateurs utilisés dans les systèmes de protection des réacteurs.

Le présent exposé donne des précisions sur les méthodes de traitement et les dispositions retenues pour Phénix et envisagées pour l'avenir.

Une étude de la fiabilité opérationnelle de la fonction «échauffement anormal d'assemblage» montre comment cette technique permet d'estimer avec le maximum d'objectivité le comportement dans le temps d'un dispositif lorsqu'on dispose d'une banque de données.

1. PRINCIPE DU TRAITEMENT, SURVEILLANCE DES CRITERES PHYSIQUES

1.1. Détection de bouchage

Le calculateur évalue cycliquement pour chaque mesure le gradient thermique $\Delta T_i = T_{si} - T_e$.

Ce gradient est comparé, à chaque cycle, à la valeur de seuil:

$$a_i \, \Delta T_m \left(\frac{100 + \alpha}{100}\right) + \beta$$

Cette valeur concerne chaque assemblage puisque a_i est un coefficient qui personnalise l'élément combustible, compte tenu de sa position géographique dans le cœur.

T_{si} = température de sortie de chaque assemblage

T_e = température d'entrée du sodium dans le cœur

ΔT_m = valeur moyenne du gradient de température des assemblages du groupe considéré
α = coefficient de seuil relatif en %
β = coefficient de seuil absolu en °C.

Valeur moyenne. Pour un nombre de groupes d'assemblages variable de 1 à 10, suivant la dispersion de flux dans le cœur considéré, le calculateur réactualise en fin de cycle le ΔT_m, soit:

$$\Delta T_m = \frac{1}{n} \sum_{1}^{n} \Delta T_i$$

avec n = nombre d'éléments combustibles dans le groupe considéré.

Coefficient de seuil. Ce coefficient est obtenu par tarage en début de chargement ou à chaque rechargement du cœur.

En régime stable de puissance, à demi-puissance nominale par exemple, on évalue sur une dizaine de cycles les valeurs de chaque a_i.

Ces valeurs moyennes sont ensuite stockées en mémoire et comparées aux anciennes.

Si l'opérateur est d'accord sur les chiffres obtenus, il valide le tarage et ce sont les nouveaux coefficients qui sont automatiquement pris en compte. Sinon il fait manuellement rentrer en mémoire les valeurs calculées.

1.2. Surveillance de la température de gaine

Une augmentation de 5% de la puissance nominale du réacteur entraîne un dépassement de 15% de la température de la fibre moyenne de la gaine de l'élément le plus chaud.

Le calculateur évalue donc à chaque cycle la température théorique de gaine qui est actuellement une donnée nécessaire à la conduite du réacteur:

$$T_G = T_e + A\Delta T_m + B$$

A Phénix, un seuil d'alarme sur cette valeur est fixé à 690°C et un seuil d'arrêt d'urgence à 700°C (aux conditions nominales de fonctionnement).
A = coefficient relatif
B = coefficient en degrés celsius.

2. ORGANISATION MATERIELLE DES SYSTEMES

Phénix. Ce réacteur rapide de 250 MW(e) comporte 121 assemblages instrumentés.

La figure 1 donne le schéma synoptique du système. Chaque calculateur surveille deux mesures par assemblage, ce qui permet de traiter ces mesures en discordance par rapport à la valeur moyenne ΔT_m calculée.

Lorsque le seuil de sécurité est atteint, les deux calculateurs commandent simultanément l'arrêt d'urgence du réacteur au moyen d'une combinaison logique en 2/2.

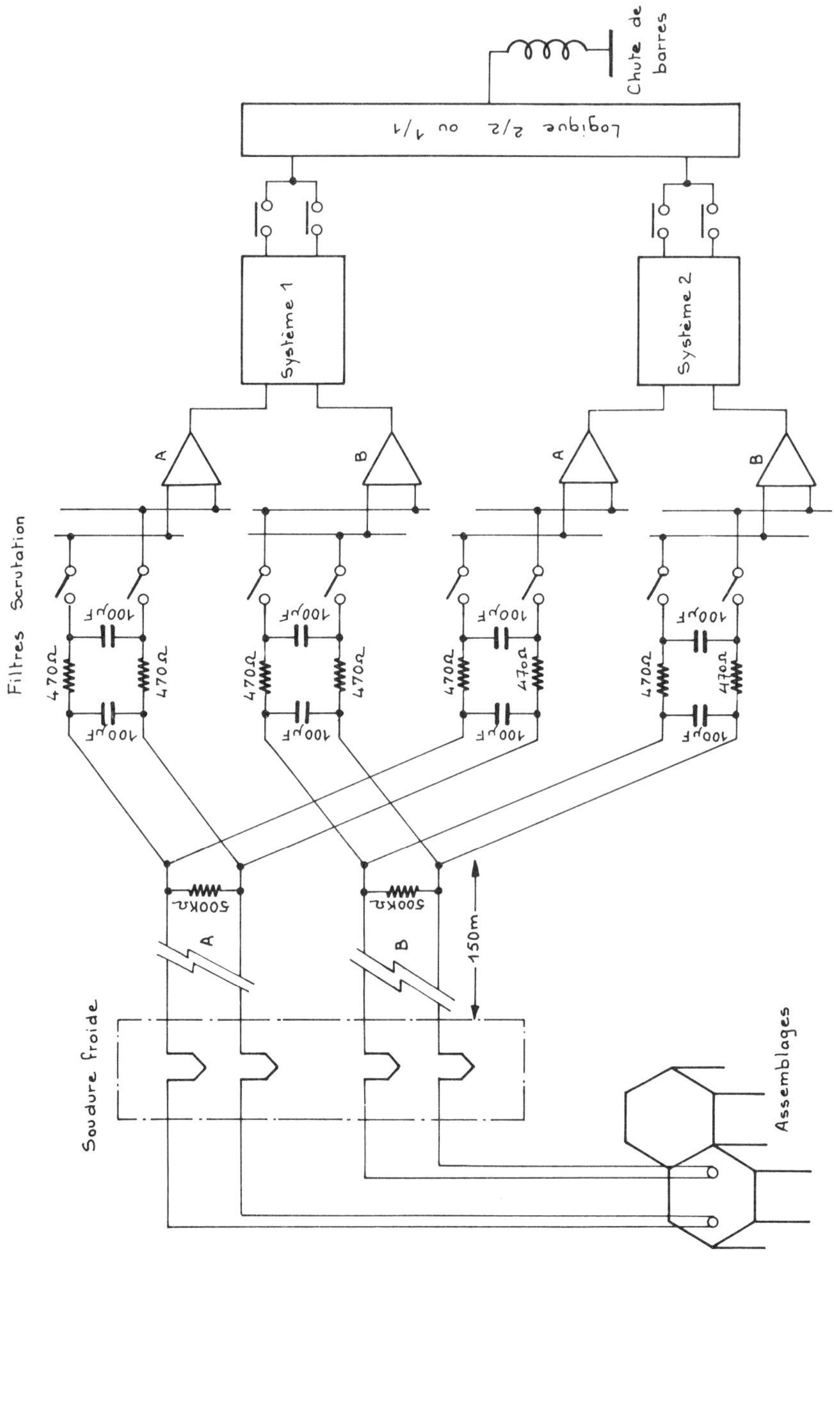

FIG.1. Organisation générale du système de surveillance du réacteur Phénix.

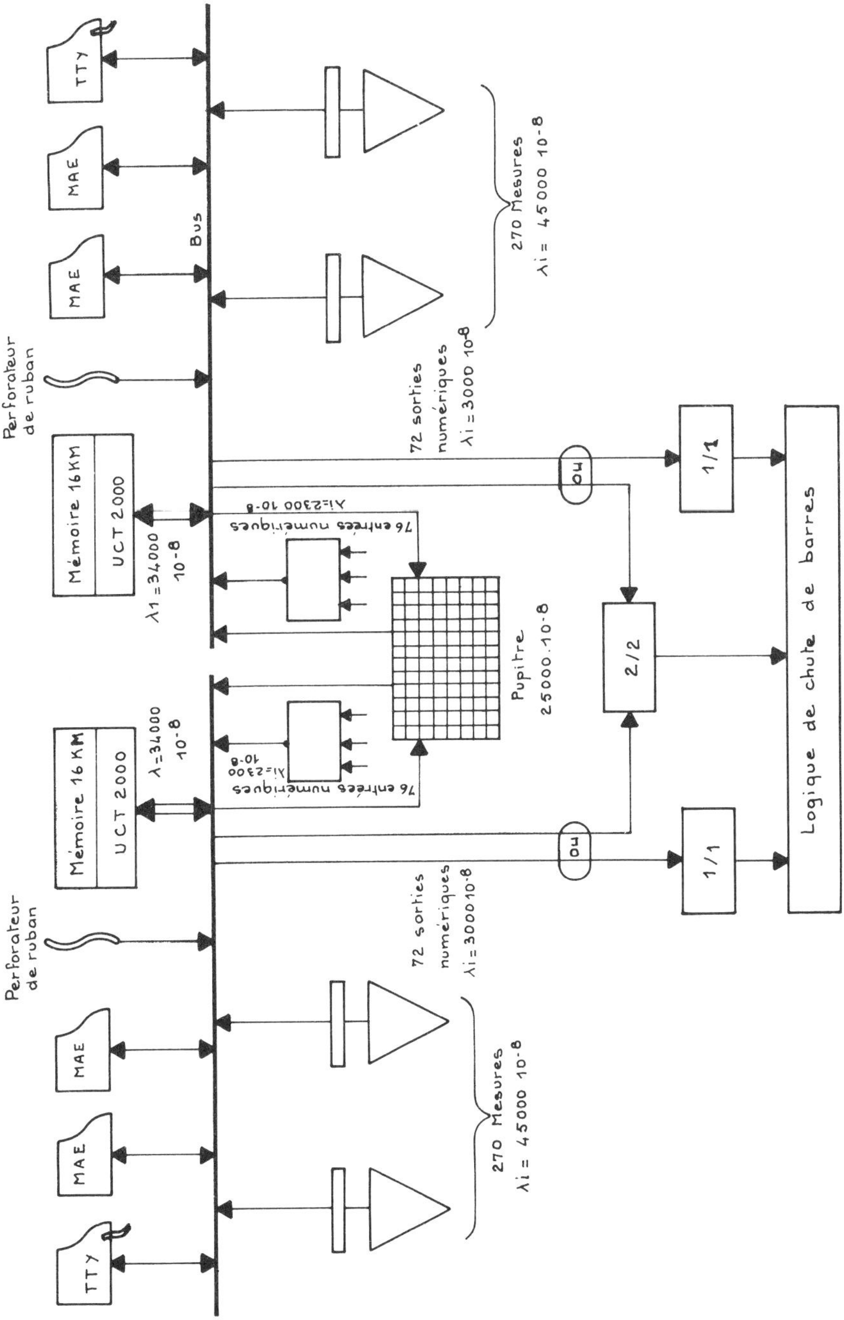

FIG.2. Diagramme du traitement des températures cœur (TRTC) de Phénix.

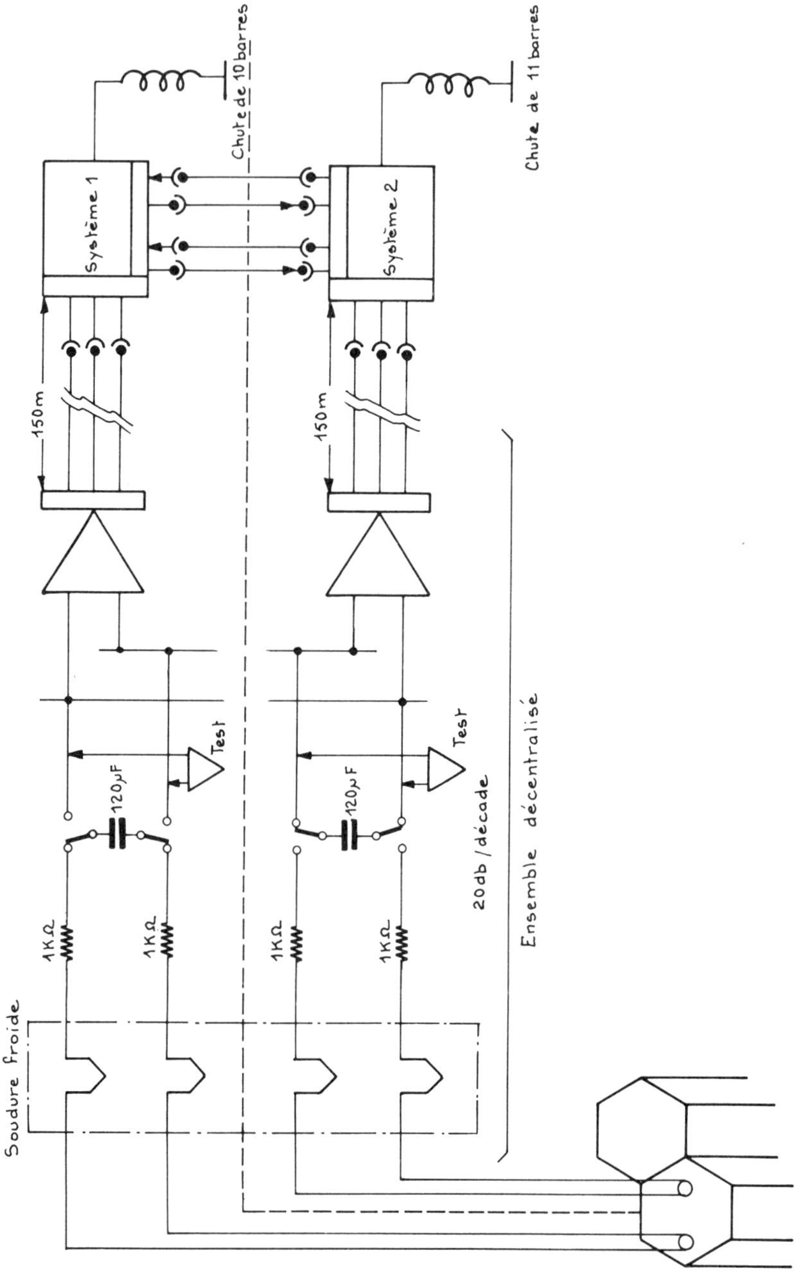

FIG.3. Nouvelle structure envisagée.

En cas de défaut d'un thermocouple, celui-ci est automatiquement inhibé et est considéré «en dépassement» vis-à-vis du traitement de sécurité. Si le deuxième thermocouple homologue de l'assemblage tombe en défaut, l'action de sécurité n'est pas déclenchée. Aucune modification de puissance du réacteur n'est autorisée.

En cas de défaut d'un calculateur, la sortie logique est positionnée en «panne sûre» et le fonctionnement en coïncidence 2/2 du système est transféré en fonctionnement 1/1. Si le deuxième calculateur est hors service ou tombe en défaut, il y a inhibition de l'arrêt d'urgence, mais dans cette situation, l'opérateur a pour consigne l'arrêt manuel du réacteur (voir fig.1 et 2).

3. AMELIORATIONS POSSIBLES DE LA FABRICATION SUR LES REALISATIONS FUTURES

Les normes en vigueur — normes IEEE 279 et recommandations DSN — conduisent à revoir complètement les structures de manière à séparer entièrement les fonctions.

Il y a lieu de profiter des deux prises de mesure indépendantes par assemblage pour commander la chute par deux jeux de barres, chaque jeu de barres permettant de commander l'arrêt d'urgence.

La puissance élevée des futurs réacteurs rapides de 1200 MW(e) nécessite de surveiller de l'ordre de 400 à 500 assemblages (voir fig. 3 et 4). A Super Phénix, sur l'ensemble des 21 barres de commande, 4 ou 5 suffisent à arrêter le réacteur.

Toutefois, pour garantir une bonne fiabilité du système et éviter les ordres de chute de barres intempestifs qui, par les contraintes qu'ils imposent aux structures, sont également nuisibles à la sûreté, il paraît intéressant de disposer d'un dialogue mémoire à mémoire permettant de renseigner le calculateur homologue de la validité des ordres élaborés par le dispositif adjacent.

Les informations échangées seront par assemblage:
- thermocouple adjacent en bon état ou hors service
- chaîne analogique d'acquisition en défaut ou en service
- dialogue mémoire opérationnel ou non
- autocontrôle du calculateur adjacent bon ou non
- bouchage sur l'assemblage i oui ou non.

Ce dialogue bilatéral ne concerne que l'état des assemblages, il est entièrement découplé par des dispositifs opto-électroniques.

De plus, un dispositif de test global entrée-sortie de chaque calculateur est à l'étude.

4. FIABILITE OPERATIONNELLE DU POSTE «ECHAUFFEMENT ANORMAL D'ASSEMBLAGE»

Ainsi que nous venons de le voir, le poste ou fonction de sécurité «échauffement anormal d'assemblage» comprend:
- les calculateurs et les relais d'acquisition
- les thermocouples
- les relais terminaux.

Un dispositif de collecte de données mis en place à la Centrale permet entre autres de suivre le comportement du matériel concerné.

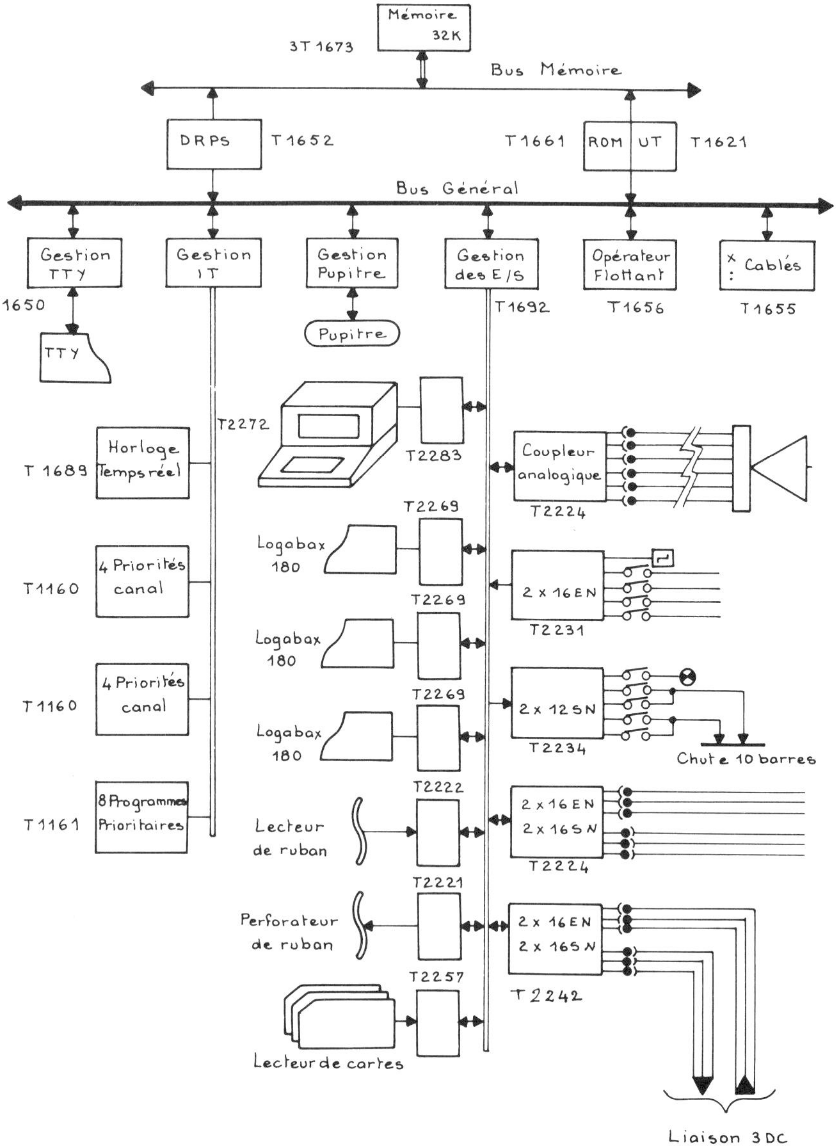
Mémoire 32K
3T 1673
Bus Mémoire
DRPS
T1652
T1661
ROM
UT
T1621
Bus Général
Gestion TTY
Gestion IT
Gestion Pupitre
Gestion des E/S
Opérateur Flottant
Câblés
T1650
T1692
T1656
T1655
TTY
Pupitre
T2272
T2283
Horloge Temps réel
T 1689
Coupleur analogique
T2224
T2269
Logabax 180
4 Priorités canal
T1160
2 x 16 EN
T2231
T2269
Logabax 180
4 Priorités canal
T1160
T2269
Logabax 180
2 x 12 SN
T2234
Chute 10 barres
T2222
8 Programmes Prioritaires
T1161
Lecteur de ruban
2 x 16 EN
2 x 16 SN
T2224
T2221
Perforateur de ruban
2 x 16 EN
2 x 16 SN
T2242
T2257
Lecteur de cartes
Liaison 3DC

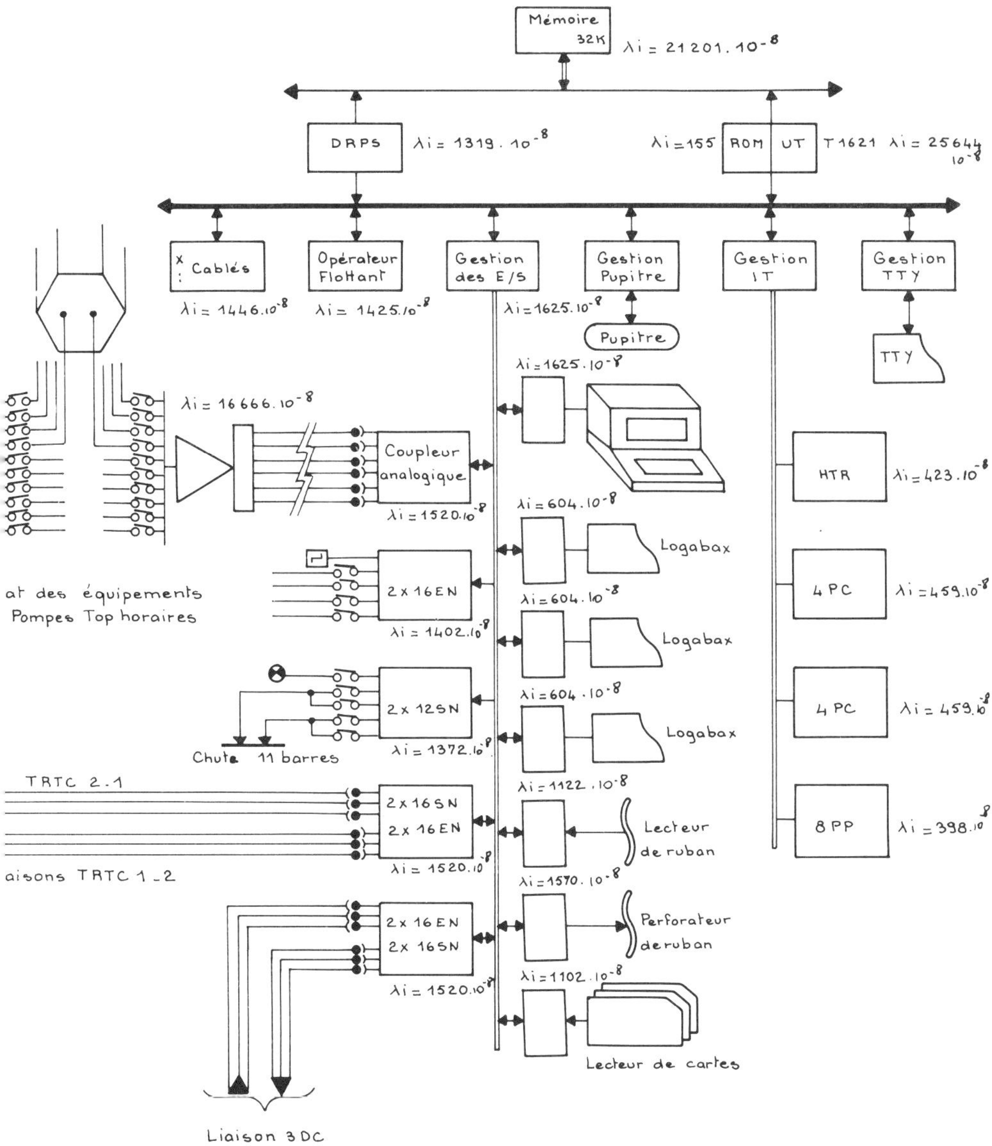

FIG.4. Solution future.

4.1. Calculateurs

Le tableau I illustre le comportement des deux calculateurs sur un intervalle de 18 mois de marche continue (≃ 13 000 heures). Les défaillances des périphériques d'édition ne sont pas prises en compte du fait qu'elles ne constituent en réalité qu'une gêne pour l'exploitation et que ces matériels n'ont aucune incidence sur l'action de sûreté.

On constate sur ce tableau que la majorité des défaillances proviennent des entrées analogiques (particulièrement les relais d'acquisition). Les défaillances ont toutes été révélées grâce aux dispositifs de test interne.

Ces chiffres permettent de calculer les valeurs ponctuelles suivantes (en supposant un taux de défaillance constant, c'est-à-dire une loi de survie exponentielle):

nombre d'heures · appareil: $13\,000 \cdot 2 = 26\,000$
nombre de défaillances: 53

soit:

MTBF[1] = 26 000/53 ≃ 490 heures
λ/heure = 53/26 000 ≃ $2 \cdot 10^{-3}$

Par ailleurs, les renseignements collectés font apparaître un temps moyen de réparation que l'on peut estimer à 30 minutes.

Relais d'acquisition. La fiabilité des calculateurs est fortement pénalisée par les relais d'acquisition des données de température des assemblages. La cadence de fonctionnement de ces relais est élevée (une manœuvre toutes les trois secondes). Sans compter les relais de groupe, on a sur chaque calculateur 246 relais[2] qui exécutent chacun 10^7 manœuvres par an. L'examen de la répartition dans le temps des défaillances de ces relais n'apporte rien de significatif (tableau II). Par contre, les derniers relevés en notre possession semblent faire apparaître un phénomène d'usure et les relais vont être prochainement remplacés par un modèle plus élaboré.

4.2. Thermocouples

Sur les 242 thermocouples d'assemblage, on a constaté une défaillance au cours de la période, soit un taux de défaillance ponctuel λ de $3{,}2 \cdot 10^{-7}\ h^{-1}$.

4.3. Relais terminaux

Aucune défaillance n'a été constatée au cours de la période, ce qui permet d'estimer la borne inférieure de la MTBF à 13 000 heures pour un niveau de confiance de 60%, c'est-à-dire un taux de défaillance pessimiste:

$$\lambda \cong 7 \cdot 10^{-5}\ h^{-1}$$

4.4. Fiabilité et disponibilité de la fonction

Si on se réfère au schéma d'organisation, l'arbre de défaillance de la fonction est celui de la figure 5 (vis-à-vis de la sûreté).

---

[1] MTBF = moyenne des temps de bon fonctionnement.

[2] 242 relais servent à l'acquisition des températures de sortie des assemblages, 4 relais servent à l'acquisition des températures d'entrée cœur.

TABLEAU I. COMPORTEMENT DES CALCULATEURS SUR UN INTERVALLE DE 18 MOIS

Calculateur	Nombre de défaillances et localisation					Révélation de défaillances	
	Entrées analogiques			Autres	Total général	Révélées	Non révélées
	Relais Aquisition	Amplificateur de mesure	Total				
C1	30	2	32	3	35	35	0
C2	12	3	15	3	18	18	0
C1 + C2	42	5	47	7	53	53	0

TABLEAU II. REPARTITION MENSUELLE DES DEFAILLANCES DES RELAIS D'ACQUISITION

Calculateur \ Mois	1	2	3	4	5	6	7	8	9	10	11	12	13	14	15	16	17	18
C1	1	0	1	0	0	2	4	4	1	2	3	1	0	0	1	1	6	3
C2	1	0	0	0	0	1	1	1	0	1	2	1	1	0	0	0	2	1
C1 + C2	2	0	1	0	0	3	5	5	1	3	5	2	1	0	1	1	8	4

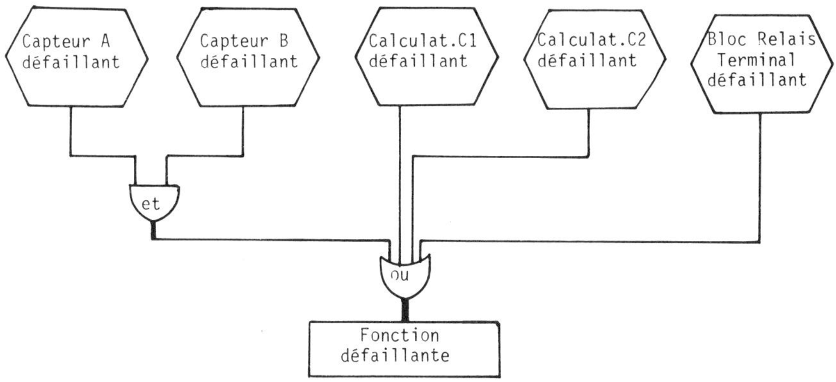

FIG.5. Arbre de défaillance de la fonction.

Mais en réalité, le système est exploité de la manière suivante: les informations élaborées par les deux calculateurs sont traitées par une logique en 2/2, ce qui limite les arrêts d'urgence intempestifs, mais est défavorable du point de vue de la sûreté (courbe 3, fig.6); mais des dispositions (dispositifs automatiques ou consignes) font que, dès qu'une défaillance de quelque nature que ce soit est révélée, on transforme immédiatement la logique 2/2 en une logique 1/1. Ceci autorise à dire que

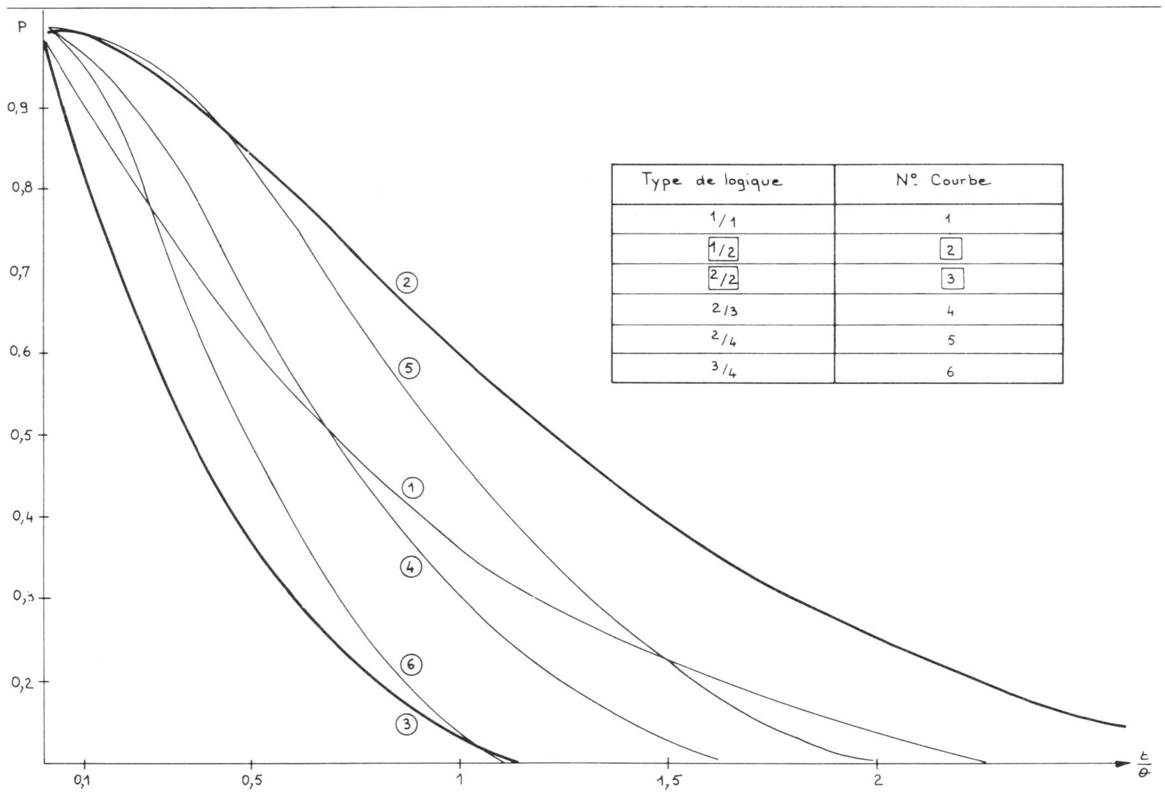

FIG.6. Courbes de probabilité de survie d'un système en fonction du temps. (Le temps est exprimé en valeurs relatives par rapport à la MTBF, θ). Le système comprend des ensembles redondants identiques et une logique en 2/m.

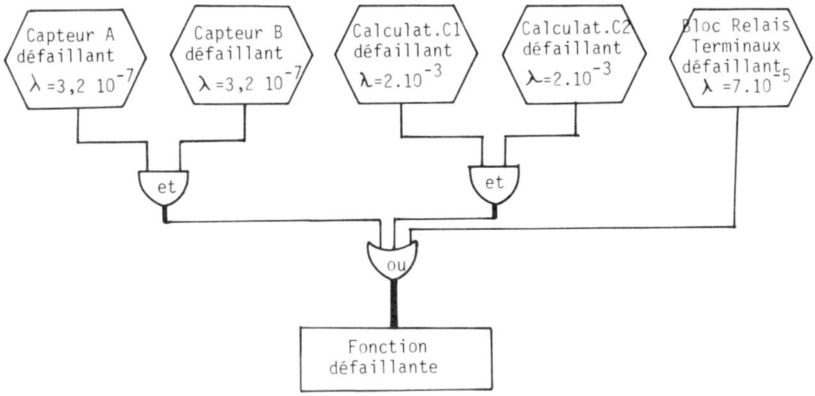

FIG.7. Arbre de défaillance plus complet.

dans le cas idéal où toutes les défaillances sont détectées et suivies immédiatement de la commutation de la logique, il suffit qu'un des calculateurs soit disponible pour que la fonction soit disponible (à condition bien sûr que les capteurs et les relais terminaux ne soient pas défaillants). De cette manière, on a deux calculateurs regroupés en 1/2 et (en supposant la fiabilité des capteurs et des relais terminaux égale à 1) la courbe de survie du système est représentée par la courbe 2 de la figure 6. La courbe de survie réelle doit donc se situer entre les courbes 2 et 3 de la figure 6.

Le tableau I montre que sur un intervalle de temps de 18 mois, toutes les défaillances ont été révélées et qu'il n'y a pas eu de pannes de commutation de logique de 2/2 en 1/2. On notera à ce propos que les défaillances n'ont pas été ventilées en défaillances dangereuses et non dangereuses, ce qui signifie que les valeurs que nous avons retenues sont des valeurs pessimistes.

Compte tenu de ces considérations, l'arbre de défaillance proche de la réalité est celui de la figure 7.

Un programme de calcul[3] utilisant la méthode des processus de Markov permet d'estimer la disponibilité moyenne du système, c'est-à-dire la probabilité de trouver ce système en bon état de fonctionnement à un instant t. Il est possible de négliger la contribution des capteurs étant donné leur faible taux de défaillance.

Dans le graphe des états (fig.8), on utilise les notations suivantes:

2 c: les deux calculateurs sont en bon état de fonctionnement
$\overline{c}$: un calculateur fonctionne, l'autre est en panne
$\overline{2\ c}$: les deux calculateurs sont défaillants
L: le bloc relais terminaux fonctionne
$\overline{L}$: le bloc relais terminaux est défaillant
λ c: taux de défaillance d'un calculateur
μ c: taux de réparation d'un calculateur
λ L: taux de défaillance d'un bloc relais
μ L: taux de réparation d'un bloc relais.

---

[3] BLIN, A., et al., ces comptes rendus, IAEA-SM-195/11.

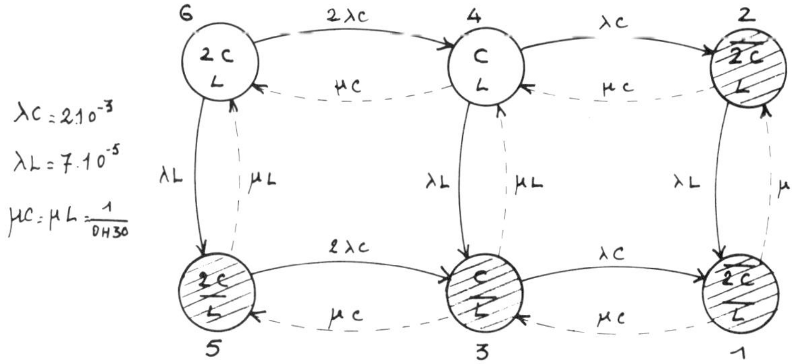

FIG.8. Graphe des états.

Les états sont numérotés de 1 à 6, et les états correspondant à une indisponibilité du système sont hachurés.

L'état 6 est l'état normal du système, l'état 4 est un état dégradé, mais correspond à une disponibilité. Les états 1, 2, 3 et 5 correspondent à une indisponibilité.

On passe de l'état 6 à l'état 4 dans un intervalle de temps dt avec une probabilité $2\lambda c\,dt$ et on repasse de 4 en 6 avec une probabilité $\mu c\,dt$, etc.

Le calcul donne la probabilité de chacun des états, à savoir:

P de l'état 1: $0{,}6985769 \cdot 10^{-10}$
P de l'état 2: $0{,}1995934 \cdot 10^{-5}$
P de l'état 3: $0{,}6985769 \cdot 10^{-7}$
P de l'état 4: $0{,}1995934 \cdot 10^{-2}$
P de l'état 5: $0{,}3492885 \cdot 10^{-4}$
P de l'état 6: $0{,}9979670$

La disponibilité moyenne du système est la somme des probabilités d'être dans l'état 6 ou dans l'état 4 (ou exclusif), soit:

$$0{,}1995934 \cdot 10^{-2} + 0{,}9979670 = \underline{0{,}999963}$$

Cette valeur, déjà excellente en elle-même, est très pessimiste puisque les relais d'acquisition des calculateurs vont être remplacés par des modèles plus fiables et qu'en fait, aucune défaillance du bloc relais terminal n'a été constatée au cours de la période (13 000 heures).

5. SOLUTIONS FUTURES

Dans le cadre des centrales futures, l'exemple d'une réalisation possible est donné aux figures 3 et 4. Il résulte d'une première estimation prévisionnelle que le temps moyen de bon fonctionnement (pour un calculateur) serait de l'ordre de 1200 heures.

On peut constater que le schéma du système correspond à une logique 1/2 réelle, à condition toutefois de faire la preuve que le dialogue au niveau des unités centrales des calculateurs ne constitue pas un point commun qui doit être pris en compte. De toute façon, la configuration proposée constitue une nette amélioration. Si l'on ajoute qu'un dispositif de test global (simulation à l'entrée du calculateur et vérification de l'ordre en sortie) est prévu, on est déjà bien plus proche des exigences actuelles.

CONCLUSION

L'étude probabiliste du système étaye la solution technique choisie. Au cours des dix-huit premiers mois de fonctionnement de la Centrale, la disponibilité obtenue a été satisfaisante. Les solutions futures envisagées permettront d'améliorer la fiabilité du système de surveillance des centrales de la filière rapide française.

BIBLIOGRAPHIE

BAZOVSKY, I., Fiabilité, Théorie et pratique de la sûreté de fonctionnement, Dunod, Paris.

GREEN, A.E., BOURNE, A.J., Reliability Technology, Wiley Interscience.

GRISOLLET, J., Modalités de mise en œuvre des critères de fiabilité des équipements de traitement des températures cœur, Technicatome RR.67308 PA 567 RD 005 (déc. 1972).

QUENEE, R., Sûreté et disponibilité d'une centrale nucléaire, CEA, Note DSN/SETS n° 17 (juin 1973).

SIGNORET, J.P., Fiabilité et disponibilité d'un système à redondance active r/m, réparable, CEA, Rapport DSN/SETS n° 20 (août 1973).

SIGNORET, J.P., Calcul sur ordinateur de la fiabilité et de la disponibilité d'un système à redondance active r/m, réparable, CEA, suite du rapport DSN/SETS n° 20 (à paraître).

ANNEXE

RESUME DU PRINCIPE DE TRAITEMENT

1. Evaluation des différents $\Delta T_i = T_{si} - T_e$

2. Calcul de la moyenne $\Delta T_m = \frac{1}{n} \sum_1^n \Delta T_i$ à chaque cycle

3. Tarage en début de campagne de fonctionnement

   Stockage des différents coefficients $a_i = \frac{\Delta T_{i0}}{\Delta T_m}$

4. Calcul des valeurs de seuil:

$$S_1 = a_i \, \Delta T_m \left(\frac{100 + \alpha_1}{100}\right) + \beta$$

$$S_2 = a_i \, \Delta T_m \left(\frac{100 + \alpha_2}{100}\right) + \beta$$

5. Détection de bouchage

$\Delta T_i \geqq S_1$	Alarme à + 5°C
$\Delta T_i \geqq S_2$	Bouchage à + 10°C

6. Surveillance de ΔT_m par rapport à 2 seuils

7. Evaluation de $T_G = T_e + A \Delta T_m + B$

8. Surveillance de T_G par rapport à 2 seuils.

DISCUSSION

(on papers IAEA-SM-195/8, IAEA-SM-195/32 and IAEA-SM-195/7)

J.M. BUTTERFIELD: I should like to ask Mr. Quénée how the design of the three SAC (complementary shut-down system) rods differs from that of the 21 SAP (main shut-down system) rods.

R. QUENEE: The SAC rods are of a somewhat simpler and more robust design than the SAP rods. They serve only safety purposes.

R. GONNOT: I believe you specified a failure rate of 10^{-8} per year for the emergency shut-down system. When you take common mode failures into account you consider them separately for the SAC and SAP rods, and unless I am much mistaken you find common mode failure rates of 10^{-5} to 10^{-3} per year. What is the relative weight of these common mode failures, as compared with the overall probability of failure of the emergency shut-down system? Could you say what the principal causes of common mode failure are, and what their probability of occurrence is?

R. QUENEE: The equipment has not yet been specified, and the common mode failures considered in our paper are purely hypothetical.

R. SARTORI: Have you divided the input signals to the different logic circuits as well, so that you have two completely independent protection systems?

R. QUENEE: We have not yet completed our study of this problem.

G.F. LUCENET: I get the impression that in one of the solutions considered by Mr. Quénée a failure rate of the order of 10^{-2} per year is sufficient to attain the overall goal of 10^{-8} that has been established. Is it not true that this failure rate is relatively easy to attain, a fact which suggests that you can look to the future with a certain amount of optimism?

R. QUENEE: This is above all a parametric study, but other studies conducted parallel to our own do in fact indicate that a certain degree of optimism is justified.

G.F. LUCENET: Is the figure of 10^{-8} per year not rather conservative compared, say, with the American standards, which, if I remember rightly, are closer to 10^{-7} per year?

R. QUENEE: The failure rate of 10^{-8} per year was selected arbitrarily. As to the rate of 10^{-7} per year which you mention — well, he who can do more can also do less!

G.F. LUCENET: Is the method applicable, and has it in fact been applied, to other types of reactors?

R. QUENEE: Yes, the method is very general and can be applied to other reactor types.

G.F. LUCENET: What is the advantage of considering the failure rate per year and not per trip? Is the annual failure rate independent of the number of trips?

R. QUENEE: The purpose of routine testing and maintenance is to maintain a constant failure rate for the equipment considered. Wear thus is left out of account, and it is perfectly legitimate to consider an annual failure rate.

Let me remind everyone, once again, that this study is supposed to be an aid to design; the equipment is not specified and the common mode failures are purely hypothetical.

R. ZAOUI: I have a question for Mr. Grisollet. Is the cladding temperature internal or external?

J. GRISOLLET: The cladding temperature is a theoretical value calculated by each computer and corresponds to the temperature of the mean fibre of the hottest can. This value is calculated during each cycle of 3 s; the expression for it is $T_G = T_e + A \Delta T_m + B$, where A is a relative coefficient and B the absolute coefficient in °C. At present this temperature is one of the basic operating parameters of the Phénix reactor.

R. ZAOUI: How are the thermocouples arranged? Are they placed all along the cladding to measure the temperature at the hottest point?

J. GRISOLLET: The thermocouples are arranged in sleeves, each of which is fixed at the core cover plate, 10 cm above the fuel assemblies. It is the sodium flow emerging from the assemblies that bathes the sleeves, so the thermocouples measure the outlet temperature from each assembly. Their position is fixed — there is no shifting of thermocouples.

R. ZAOUI: Could you quote a failure rate for the thermocouples?

J. GRISOLLET: In 18 months there has been one failure out of 242 thermocouples; this means a failure rate per point of 3.2×10^{-7} per hour.

L. CAVE: My question is for Mr. Snaith. A safety feature of the AGR is that the reactor can be deprived of forced emergency core cooling for half an hour to an hour before reaching an irretrievable condition which would terminate in damage to the core. Thus in the event of a failure of the automatic system it should be possible for the operator to take remedial action. Have you considered how the reliability of the operator in such circumstances might be assessed, and have you carried out such an assessment?

E.R. SNAITH: The period for which an AGR can be deprived of emergency core cooling following a trip under fault conditions depends on the fault that initiates the trip and hence on the reactor transients characteristic of the situation. The possibility of the operator taking remedial action in the event of a partial failure of the automatic system is considered in our analysis, but it is assumed that such action would be taken only after a lapse of, say, half an hour. It is difficult to quantify the behaviour of an operator, but the problem has been analysed by means of a boundary approach — that is, with the assignment of upper and lower limits.

The extent of the action which an operator might be expected to take would depend, however, on a number of things, including (a) his awareness of the fault, (b) his knowledge of the sequence of functions performed by the automatic emergency systems and (c) his ability to identify the fault, and to assess and interpret correctly the information presented to him in the form of alarms which could assist his decision-making process. It will thus be appreciated that, if one wishes to consider the improvement in reliability that might be expected to accrue through operator action following a fault, each situation must be individually considered.

L. CAVE: It is only in relatively rare situations that the operator would have less than half an hour available to take remedial action, whereas the frequency of other situations requiring the operation of the emergency feed system is quite high — say 2 or 3 per reactor year. As the automatic system, by its very complexity, cannot be expected to have a reliability much better than, say, 10^{-4} failures per demand, it does seem to me most important that the operator should be in the best possible position to supplement the automatic system.

D. OKRENT: What is the potential role of the guide vanes in increasing the probability of a failure of shut-down cooling?

E.R. SNAITH: The guide vanes are normally fully open at 100% power. In the worst fault condition they would thus not have to operate, because they would be required to be 100% open. In the course of a normal reactor trip specific vanes would be required to close partially, so their failure mode could be stiction; and this has a low failure rate. It should also be observed that because of the redundancy of circulated loops required under trip conditions, the failure rate value for the guide vanes did not represent a significant contribution to the overall calculated failure probability for the complete system.

R. SARTORI: As fas as I can see, you do not mention the probability of one system being down for repair when an accident occurs. Is this taken into account in your analysis at all?

E.R. SNAITH: We would expect to be given specific information about the extent to which reactor operation is permitted at full power, or at reduced power, when less than 100% of the emergency plant is available. When this information is provided, we would certainly consider the effect of one or more redundancy loops, or of parts of the system being down and under repair, on system reliability.

R. SARTORI: Do you, in your analysis, consider an aeroplane crash as an accident to be taken into account in the design of the systems?

E.R. SNAITH: This type of accident, and similar external influences such as lightning strikes, would be considered in the overall hazards analysis for a reactor, which would take into account the consequences and the probability of all types of extraneous influence.

R. SARTORI: Do you consider failures of passive components such as feedwater ducts in your analysis?

E.R. SNAITH: Incidents such as the fracture of feedwater pipes would be revealed faults, and the reactor operating conditions would be modified accordingly. The occurrence of such failures simultaneously with a further reactor fault implies a situation marked by two independent, simultaneous faults, and this has low probability. The redundancy provided also enables a low probability value to be demonstrated.

P.G. SMITH: You refer in your paper to increased frequency of testing as a means of improving the reliability of a sub-system. Do you assume that after a routine test the sub-system has been shown to be fully serviceable, or do you calculate a probability that the test may have failed to detect an unrevealed unsafe failure?

E.R. SNAITH: The assumption is made, in carrying out a reliability analysis, that, following the test of a sub-system, the system in question is in a fully operational state. However, if the test involves equipment operating which then must return to some standby "healthy" state, the probability of unrevealed faults existing as a result of the test is considered.

P.G. SMITH : Your reference to increasing the frequency of testing to reduce the probability of failure in normally dormant sub-systems illustrates one use of your method. Could it also be used to show that less testing would be satisfactory? One often feels that a reduction in testing would be acceptable, provided it could be shown to be justified numerically in relation to observed failure rates.

E.R. SNAITH: The consideration of increased frequency of testing to improve the reliability of operation of normally dormant systems is realistic in the example given in my paper, since monthly testing was being considered in place of testing at three-monthly or longer intervals. However, it is agreed that test frequencies could be increased so much that reliability of operation would actually be reduced. This problem has to be closely examined for each situation, which means that test frequencies must be considered specifically and separately for each individual system and component.

D. NORMAN: I would like to change the emphasis of the discussions slightly. Yesterday's proceedings gave us, first, data collection schemes but few detailed reasons for their implementation and, secondly, analyses — some of them of great practical importance but, again, with no description of their practical applications.

This morning we have in fact heard about some of the practical applications of these techniques, but we are still concentrating on the details of each illustration. I believe we should discuss the practical value of these techniques more generally, because practical engineers are in danger of underestimating their value and because analysts are in danger of becoming too enchanted by the elegance of their techniques to question their utility.

My view is that the importance of analysis (and data collection) is not just that it quantifies hitherto qualitative judgements but that the actual performance of the analysis (and interpretation of the data) helps design, maintenance and operational engineers to take practical decisions. Some important techniques in this connection — cost/benefit analysis and decision theory — have so far not been dealt with. They would serve to emphasize that we should compute reliability and availability only if we think we may need to change them and if we have to decide how to change them. I hope we will find time to consider these management questions as the work progresses. In the meantime I would like to elicit from this morning's speakers a few comments on how they influence designers and operators, how their efforts are received and understood, and how much consideration safety analysts give to the effect of decisions on plant operating availability.

E.R. SNAITH: In general I agree with Mr. Norman's remarks. Probabilistic techniques are certainly used in the Safety and Reliability Directorate of the United Kingdom Atomic Energy Authority to demonstrate where design weaknesses may exist which could reduce system reliability; indeed, my paper gives details of some design modifications that were carried out, and of how the mode of operation was modified to give optimum reliability under emergency conditions. Thus, the results of the analysis were accepted as a positive aid to both the designer and the operator.

By and large the experience in our organization is that design groups have come to recognize and appreciate the value of a reliability analysis. We find that operating engineers are somewhat less able to appreciate the direct advantages. However, these analyses are being accepted by plant operators and are welcomed particularly when an unsophisticated approach can be used to demonstrate how they can lead to optimization of operation and/or maintenance.

Our organization has done some work aimed at applying reliability analysis techniques to the demonstration of cost/benefit analyses for alternative system arrangements. I quite agree that they should in future be applied more intensively in this type of work.

J. KRASNODEBSKI: The papers we have heard this morning give good examples of how reliability analyses can be applied to reactor safety systems. However, I am a little bit concerned that there should have been no examples of how R & M analysis can be used in the design of power plants to improve their availability. The design of our plants must be based on considerations of safety and availability; this is rapidly being recognized both in Canada and in the United States. And to improve the availability of our plants, we must improve our process systems and equipment where necessary. The data collection systems should show us where the problem areas lie. Thus R & M activities must be an integral part of the design process; plant and equipment designers, reliability engineers and operators must work in very close co-operation if we are to be assured of high availability in our future plants.

L. CAVE: Getting the engineers responsible for the actual design process to employ the techniques of reliability engineering and to make use of the reliability data now being collected at many centres is a real problem. The extent of it was demonstrated recently in the United States 'Nerva' rocket programme, in which the aerospace and nuclear industries collaborated and in which it was decided, through the influence of the aerospace element, that design should proceed on the basis of reliability engineering. From discussions with senior reliability engineers involved in the project I have gathered that there was considerable difficulty in getting the design engineers, who came mainly from the nuclear side, to understand and apply the techniques of reliability engineering, which to them were quite novel. The reliability engineers found that in practice much of their time was spent in an educational rôle, though eventually the communication problems were largely solved.

Accordingly, I rather fear that the continued development of more and more complex methods of reliability analysis might prove to be self-defeating unless a determined effort is also made to present them in a readily understandable form to the ordinary 'nuts-and-bolts' mechanical and electrical engineers who have the task of preparing the detailed design. Perhaps there is a need for a new type of 'generalist' who can interpret the work of the reliability engineers and couch it in terms which the design engineer can understand and apply.

D. NORMAN: In the Central Electricity Generating Board we train graduate apprentices, and also older engineers, in the rudiments of reliability analysis. Few of them, even new graduates, have ever had any training in it previously, though the new graduates have sometimes been taught statistics (including reliability). In most United Kingdom universities the study of reliability is confined to post-graduate courses. However, some efforts are being made to include reliability in first degree courses and in the Higher National Diploma course.

J. KRASNODEBSKI: We have also found that, by and large, engineers in the electric power industry are not familiar with R & M techniques, and we have run a number of courses aimed at introducing design and operating engineers to reliability as a distinct discipline. The introductory course lasts two days and has been attended by some 300 engineers. Another course, consisting of 10 two-hour lectures, was organized to give design engineers the opportunity of performing basic reliability analyses, and about 150 engineers attended this course.

As to the application of R & M techniques developed in other areas of engineering, it is important to ensure that practitioners understand the conditions and customs prevailing in the electric power industry and adapt their techniques accordingly. In general a disciplined approach to design is required, consisting in the preparation of system specifications, the application of basic reliability analysis and of operational feedback, and design reviews. Sophisticated R & M analyses are required only for a few major complex systems.

USE OF STANDARD RELIABILITY LEVELS IN DESIGN AND SAFETY ASSESSMENT OF IN-PILE LOOPS

G. BOGANI, A. VERRE
Centro Informazioni Studi Esperienze,
Segrate, Milan

S. BALESTRERI, A.G. COLOMBO, T. LUISI
Centro Comune Ricerche,
Ispra, Varese

B. BOTTELLI
Comitato Nazionale per l'Energia Nucleare,
Ispra, Varese,
Italy

Abstract

USE OF STANDARD RELIABILITY LEVELS IN DESIGN AND SAFETY ASSESSMENT OF IN-PILE LOOPS.

This paper describes a logic-probabilistic analysis technique for a critical design review and safety assessment of in-pile loops. The examples in this paper refer to the analysis performed for the experimental loops already constructed or under construction in the ESSOR reactor of the Joint Research Centre of Ispra, as irradiation facilities for fuel element research and development tests. The proposed technique is based on the classification into categories of components and protective device malfunctions. Such subdivision into categories was agreed upon by the Italian Safety Authority and Euratom JRC, and adopted for the safety assessment of the ESSOR reactor in-pile loops. For each category, the method makes a link with a corresponding malfunction probability range (probability level). This probability level is defined taking into account design, construction, inspection and maintenance criteria as well as periodic controls; therefore the quality level and consequently the reliability level are thus also defined. The analysis is developed in the following stages: (1) definition of the analysis object (top event) and drawing of the relative fault-tree; (2) loop design analysis and preliminary optimization based on logic criteria; (3) classification into categories of the fault-tree primary events; (4) final loop design analysis and optimization based on defined component quality requirements. Stages 2 and 4 are quite different since stage 2 mainly consists of a redundance optimization, while stage 4 acts on the component quality level in such a way that each minimum cut-set leading to the top has an acceptable probability level. During analysis development, use is made of computer codes which, among other things enable the verification of fault-tree logic makeup, the listing of the minimum cut-sets with and without event categorization, and the evaluation of each cut-set order.

1. INTRODUCTION

In the framework of the development of the ESSOR reactor experimental facilities, CISE is constructing two in-pile facilities for LWR and HWR fuel experiments.

This paper reports the work developed in strict cooperation with Euratom (ESSOR, CETIS, and Technology Division) with the aim of supplying each plant with an adequate degree of safety, considering that the experimental facilities even if aimed at experiments, are an essential part of the reactor assembly involving a large number of circuits and primary and auxiliary systems, including containment in the case of fission product release.

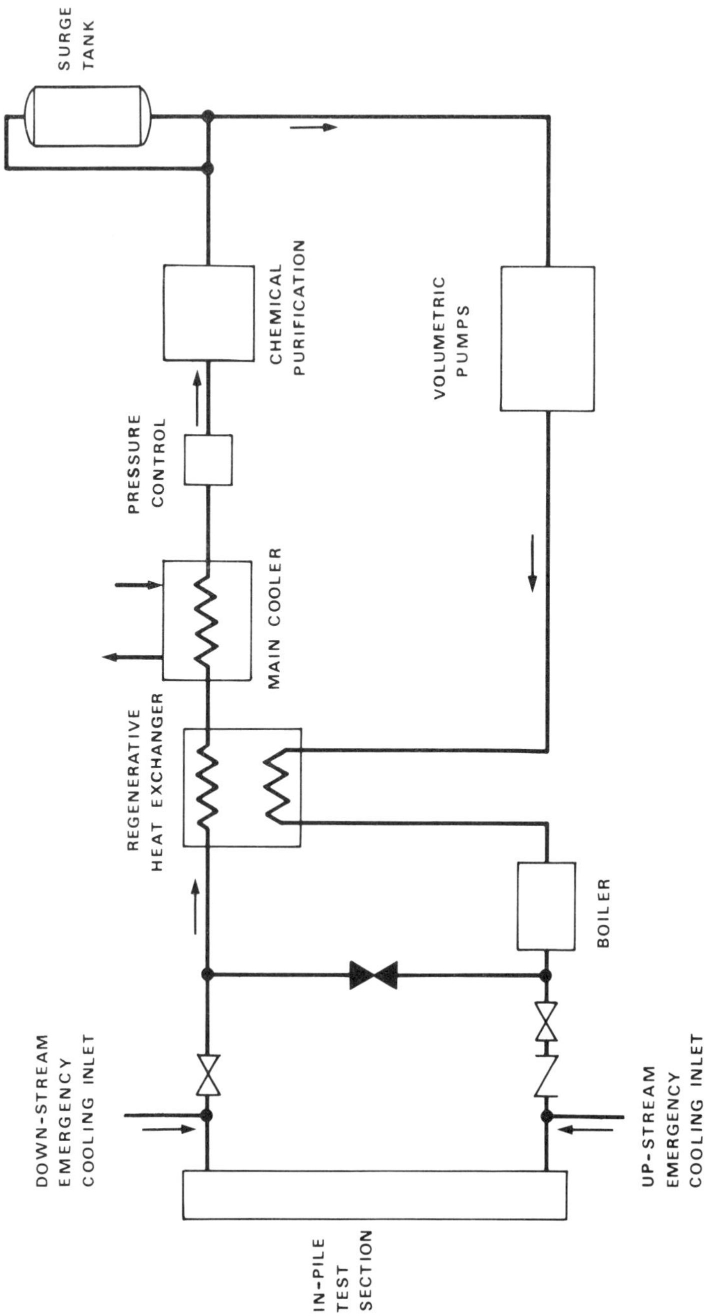

FIG.1. Scheme of the process applied to cart, cart TC (twin-channel) and IRA loops.

It is worth noting that this method is usable not only as an a posteriori validity assessment but also as an actual system design tool. The method can be generalized for all types of more or less complex plants — nuclear and non-nuclear — allowing in every case the singling out of the system's critical components, and consequently of adopting suitable construction criteria.

As will be discussed in the following, the method proves to be particularly valid as far as safety, functionality and economy are concerned.

2. BRIEF DESCRIPTION OF THE LOOPS TO WHICH THE METHOD WAS APPLIED

Figure 1 schematically shows the cooling primary loop flowsheet typical of the facilities considered in this paper.

This process, adopted for CART (1.5 MW), is now adopted also for CART TC (Twin Channel 3 MW) and IRA (0.3 MW) for tests on fuel elements in assemblies and rods, in the water reactor experimental range. As an example we mention that IRA is a multi-purpose loop reaching a 180 kg/cm^2 operating pressure.

This type of circuit is basically a generator of thermal and flow dynamic conditions required by the experiments; it can be connected with one or more in-pile once-through or plunging (U-tubes) power channels. The injection of the emergency cooling circuit can occur upstream of and/or downstream of channel according to the type of loop. Emergency cooling generally operates in two stages: one of short duration and close to the nominal operating pressure, and the other of longer duration but at low pressure.

3. METHOD DESCRIPTION

As shown in Fig. 2, first of all the loop is examined and one or more final or top accidents (e.g. fission product release outside the airtight container or loop-room contamination) are defined, as well as the possible elementary accidents or primary events (e.g. spring rupture, remote control switch failure etc.). The possible evolution of elementary accidents gives rise to a logic diagram of the AND-OR type (fault-tree).

The fault-tree must be continuously updated to design modifications and developments.

The fault-tree permits one to single out the system cut-sets, each cut-set being the whole of the primary events which, occurring simultaneously, involve the assumed top accident. The cut-sets may be of different "n" order (1st, 2nd, 3rd,... nth) according to the number, n (1, 2, 3, ..., n) primary events which are concerned.

Probability concepts are not introduced into the definition of cut-set. The maximum cut-set order considered as useful for the analysis is fixed at the beginning, and is higher the lower the quality level of the components and systems appearing as primary events.

This work stage considers the number of cut-sets and their subdivision versus order. If necessary, the design is revised, introducing or modifying functions and organs.

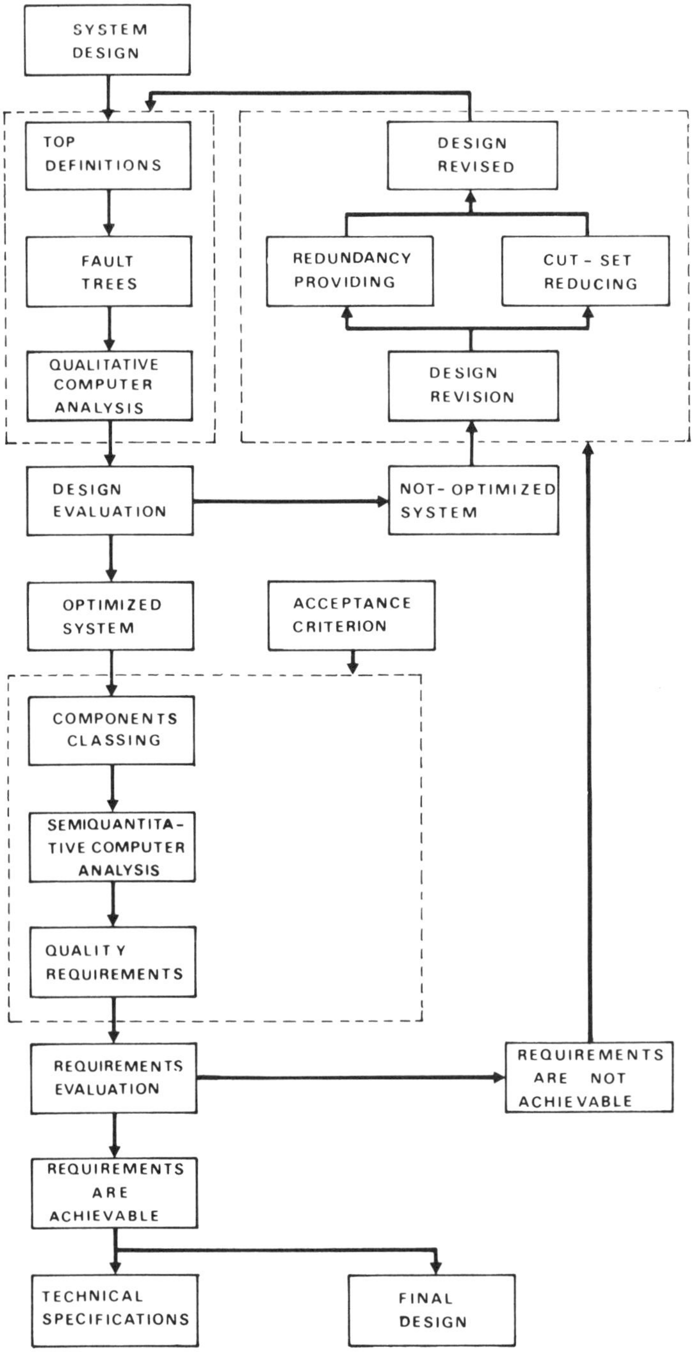

FIG.2. Logic flow diagram for the design.

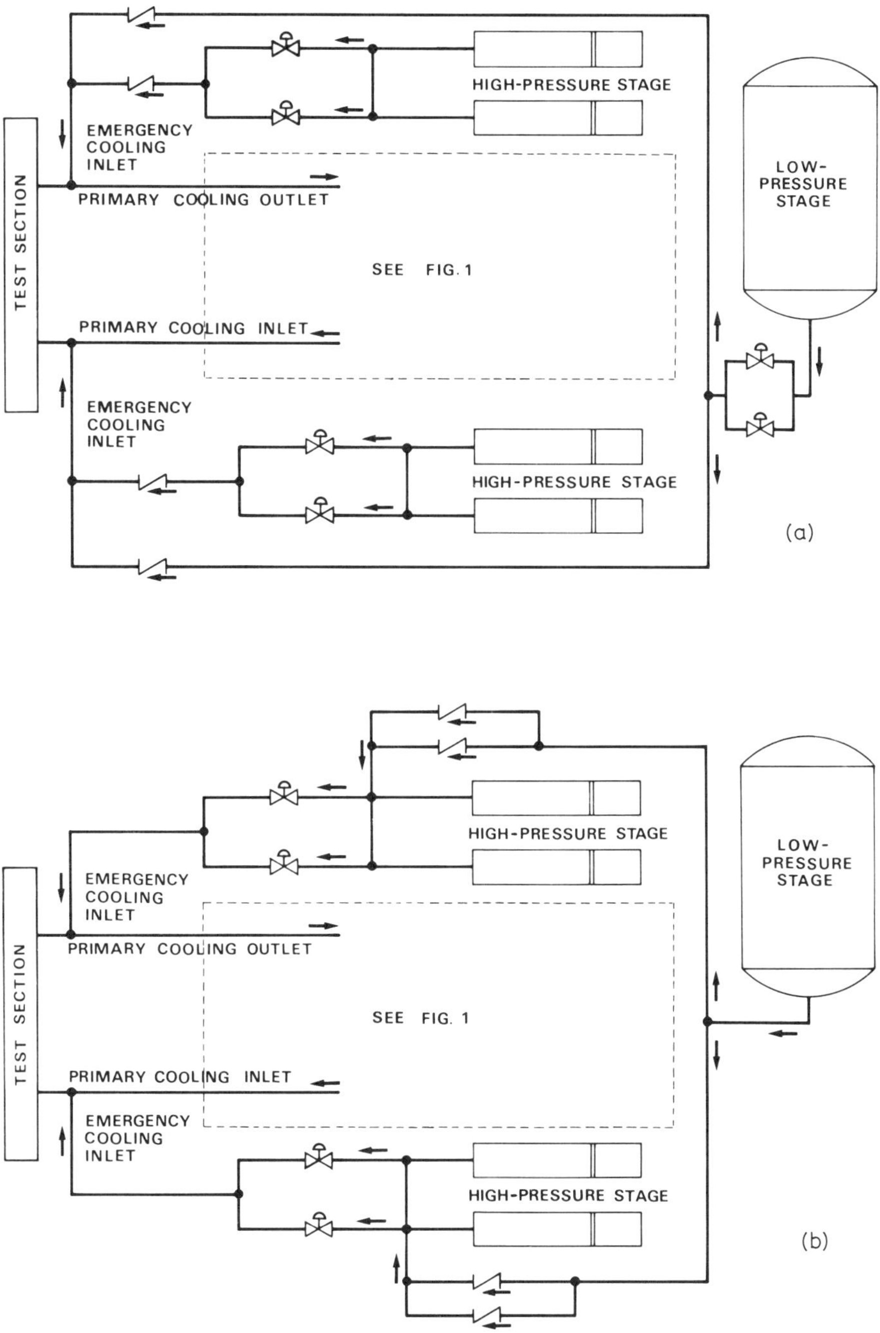

FIG.3. (a) Emergency cooling circuit first solution; (b) emergency cooling circuit revised solution.

This procedure is iterative up to the achievement of a system-optimized version: thus cut-sets and primary events are known.

The case shown in Fig. 3 is a typical example of the effectiveness of just one iteration (see Fig. 3a as compared with Fig. 3b); in fact the configuration of the two emergency cooling circuits connected upstream and downstream of the power channel was modified during the revision stage (Fig. 3b). Thus the system is better protected against piping rupture, and against water leakages on the check valve seat between the high-pressure and low-pressure stages. Figure 3b also shows the doubling in parallel of the check valves between the high and low pressure stages as a protection against the dangerous event "valve jammed in closed position".

These modifications cause the following in the systems:

(a) Elimination of
- 7 primary events
- 14 cut-sets of 2nd order
- 16 cut-sets of 3rd order
- 32 cut-sets of 4th order

(b) Shifting of 129 cut-sets from 2nd to 3rd order.

At this point an acceptability criterion is necessary to account for the risk concept usually associated with nuclear plant operation. Practically, the acceptance criterion adopted was the one proposed by the Safety and Control Division of the Comitato Nazionale per l'Energia Nucleare (CNEN) [1].

It can be summarized as follows:

Subdivision of primary events into two categories:

(1) Initial causes of accidents
(2) Unavailability of protective systems

and their subdivision into reliability levels and unavailability classes, respectively, according to the definitions listed below.

Initial causes

L 0: Failures of normal-type industrial components or the like.

L 1: Failures of components which meet the requirements of high industrial quality (i.e. Class 3 of ASME Sec. III) for design, construction, testing, inspection, maintenance.

L 2: Failure of components meeting the requirements of very high industrial quality (i.e. Class 2 of ASME, Sec. III) for design, construction, testing, inspection, maintenance.

L 3: Failures of components subjected to more severe design, construction, testing requirements than foreseen by the most advanced industrial standards and subjected to programmes of preventive maintenance (i.e. quality level obtainable by Class 1 of ASME Sec. III).

L S: Special failures of a practical negligible frequency; this level includes failures from multiple single components of level 3, if independent and separately checked.

Protective systems

C 0: Systems (1) for protective systems for whose availability particular efforts were not made, and (2) whose standards correspond to those of a normal industrial construction.

C I: Systems for which the probability of a non-detected failure is non-negligible, and whose industrial standard is high and subjected to particularly careful maintenance.

C 2: Systems for which the unavailability is very low, thanks also to preventive maintenance and to the high degree of redundancy of fail-safe self-control systems.

C S: Systems, whose protection actions require nuclear-grade characteristics, or which are obtained by redundancy of differentiated and independent Class C2 systems.

Table I summarizes the data concerning the above classification criteria.

For each cut-set of any order n, an overall index is computed, being the sum of the indexes m_i pertaining to each primary event forming the cut-set.

If $\sum m_i \geq 7$, the accident likelihood is considered as practically negligible and it is unnecessary to evaluate its consequences.

If $\sum m_i = 6$, the accident is considered to be at the credibility limit, and the consequences acceptable outside the container are those of Maximum Acceptable Release.

If $\sum m_i < 6$, the accident is even more probable, and therefore the radiological consequences are evaluated in a more restrictive way.

For the loops to which this method has so far been applied in ESSOR, the event "significant disequilibrium between generated power and power extracted from in-pile channel" was assumed as "top", the acceptance criterion $\sum m_i = 6$ being chosen. Hence the loop acceptance criterion is verified by the $\sum m_i = 5$ relationship for each cut-set, since a merit index m = 1 was conservatively assumed for the four following barriers opposing in series the release of fission products:

(a) Fuel claddings
(b) Primary containment
(c) Loop room (or upper lower reactor room)
(d) Reactor containment

TABLE I. CLASSIFICATION CRITERIA

Reliability	Yearly failure rate	Failure rate for plant lifetime	Index number (m_i)	Protective systems class	Unavailability
L 0	10^{-1}	3	0	C 0	10^{-1}
L I	$10^{-2} - 10^{-1}$	$3 - 3 \times 10^{-1}$	1	C 1	$10^{-2} - 10^{-1}$
L 2	$10^{-3} - 10^{-2}$	$3 \times 10^{-1} - 3 \times 10^{-2}$	2	C 2	$10^{-3} - 10^{-2}$
L 3	$10^{-4} - 10^{-3}$	$3 \times 10^{-2} - 3 \times 10^{-3}$	3		
L S	$\ll 10^{-4}$	$3 \times 10^{-3} - 3 \times 10^{-7}$	7	C S	$\sim 10^{-7}$

The above reliability analysis is automatically performed by the CADI [2, 3] code normally supplying the following outputs:

(a) The listing of all cut-sets for the cut-set orders of interest.
(b) The listing of all cut-sets whose $\sum m_i$ is equal to or lower than the chosen $\sum m_i$.
(c) The number of times each primary event, independently of its index number m_i, takes part in the combinations of various order leading to the top.

The result of the activities described above is a list of component functions, subsystems, each being identified and classified from the quality standard viewpoint.

The requirement of adequate quality results in design, construction, operation and maintenance specifications.

Should the required specifications not be achievable because of delivery times, cost, supplying or other reasons, a new iteration is then necessary which again brings the work back to the beginning.

4. ENGINEERING CONSIDERATIONS AND CONCLUSIONS

The applied method is quite simple since the failure probability is quantized in few standard levels — therefore it can be rapidly applied by designers themselves.

The analysis results give:

(a) Indications on risk acceptability
(b) Objective factors which can be used in the design also for alternative solutions
(c) Mandatory data to establish the quality level of single components and subsystems

These results have a direct impact on the design, in terms of safety, economy and functionality and on construction aspects.

With a view to safety analysis, a general conclusion can be given which accounts for a larger or smaller number of degenerative processes leading to accidents and requires, if necessary, their reduction. Because of the strict connection between safety analysis and design ever since the preliminary stage, the most apparent result is a sound, simple and systematic procedure for presentation and evaluation of the report to be discussed with the Safety Authorities.

Also, the components which are significant for safety assessment are identified, their requirements having been previously fixed.

It is worth noting that the component requirements are not fixed a priori according to whether they belong to some subsystems (e.g. primary circuit, emergency cooling circuit etc.) but on account of the actual importance as to the residual risk associated with the "top". Hence, for example, parts of the primary cooling circuit may have different standards while the maximum standard might be required for components of a secondary circuit.

From the economic viewpoint, it is evident that, owing to the close relationship between quality and cost, the method allows cost optimization:

in fact, it is evident that sometimes only a few component functions are critical. Therefore, investments with a view to overall quality may be a waste in terms of safety, on the other hand it is doubtful whether availability will increase. As an example, a primary pump, typical of the systems presented, can be considered: body integrity is more essential for containment than is piston jamming or electric supply loss for flowrate continuity. Therefore, it is better to increase the quality and cost of the pump body alone more than anything else. Another feature of the method which might be of interest from an economic point of view is the possibility of proving the importance of the system single components, thus justifying investments in some, instead of in other, components as far as safety or availability are concerned.

From the point of view of construction, one of the advantages of the method is the practical and immediate language (specifications for calculation, checking, testing, installation etc.) expressing the class ascribed to components. In fact, contacts with suppliers at this stage are made easier and more productive using these means instead of the specialized language of failure rates.

In conclusion, the presented method can be used as a simple and flexible working tool for different kinds of evaluations; it is interesting, at any rate, since it strictly relates plant reliability assessment to construction and operation quality standards.

The natural outcome of the use of this method is a careful wellweighted application of the Quality Assurance engineering practice.

ACKNOWLEDGEMENTS

The authors wish to thank Mr. S. Ragusa (ESSOR Division) and Mr. G. Volta (Technology Division) for their helpful contributions.

REFERENCES

[1] BANDIERA, P.G., CESARI, F., GALVAGNI, R., MISLEY, P., RIGHINI, R., SANTANDREA, E., "Effects due to thermal transients originated by plant incidents and some operating procedures", Paper presented to the Specialist's Meeting on Handling of the Design for an Integration of Thermal Transients in LMFBR Plant, Canoga Park, June 1974 (in press).

[2] COLOMBO, A.G., CADI, A Computer Code for System Availability and Reliability Evaluation, Rep. EUR 4940e (1973).

[3] COLOMBO, A.G., VOLTA, G., "Logic probabilistic approaches for nuclear safety assessment", Paper presented to the Conference on Reliability and Fault Tree Analysis, University of California, Berkeley, September 1974 (in press).

IAEA-SM-195/10

DETERMINATION DES SEQUENCES ACCIDENTELLES IMPORTANTES POUR LA SURETE D'UNE CENTRALE NUCLEAIRE DU TYPE A EAU ORDINAIRE

B. GACHOT
Electricité de France,
Paris

J.-P. SIGNORET, Annick CARNINO
CEA, Centre d'études nucléaires de Saclay,
Gif-sur-Yvette,
France

Abstract–Résumé

DETERMINATION OF INCIDENT SEQUENCES WITH IMPORTANT CONSEQUENCES FOR SAFETY IN A LIGHT-WATER REACTOR.

To assess the safety of a nuclear power station by probability methods one must first draw up as comprehensive a list as possible of the incidents and chains of incidents which can give rise to a radiological hazard for the public or the operator. To this end the authors propose a method yielding results which can be used by reliability engineers, based on two complementary approaches: (1) determination of fault trees associated with the different barriers for preventing the spread of radionuclides; and (2) determination of event trees consequent on failure of each of the main safety systems in a nuclear power station and culminating in strains on the preceding safety barriers.

DETERMINATION DES SEQUENCES ACCIDENTELLES IMPORTANTES POUR LA SURETE D'UNE CENTRALE NUCLEAIRE DU TYPE A EAU ORDINAIRE.

Pour évaluer la sûreté d'une centrale nucléaire selon des méthodes probabilistes, il est nécessaire, au préalable, de dresser une liste aussi complète que possible des accidents et enchaînements d'accidents pouvant présenter un risque radiologique pour le public ou l'exploitant. Dans ce but, les auteurs proposent une méthode de travail aboutissant à des résultats utilisables par l'ingénieur fiabiliste et qui est basée sur deux démarches complémentaires, à savoir: 1) la détermination des arbres de défaillance des différentes barrières qui s'opposent à la diffusion des radioéléments; 2) la détermination des arbres d'événements consécutifs à la défaillance de chacun des systèmes importants pour la sûreté dans une centrale nucléaire et dont l'aboutissement est une agression des barrières précédentes.

INTRODUCTION

Dans l'évaluation de la sûreté d'une tranche nucléaire selon des méthodes probabilistes, un problème important est de s'assurer que l'on a bien envisagé toutes les séquences d'accidents plausibles.

Il est donc nécessaire, avant toute évaluation probabiliste, de dresser une liste aussi complète que possible de tous les accidents et de toutes les séquences pouvant conduire à un accident et d'effectuer un premier classement qualitatif de ces séquences par ordre de gravité. Bien que l'expérience vienne souvent déjouer les prévisions les plus sérieuses, nous proposons ici une méthode de recherche de toutes les séquences d'événements importantes pour la sécurité du public et de l'exploitant, méthode que nous souhaitons rendre aussi exhaustive qu'il est possible de l'imaginer. C'est pourquoi nous espérons recueillir le plus grand nombre de critiques et de conseils nous permettant de perfectionner cet outil de travail.

1. PRINCIPES DE LA METHODE PROPOSEE

1.1. Le premier principe utilisé est que tout incident pouvant avoir des conséquences radiologiques graves dans une centrale nucléaire a pour origine la défaillance de l'une au moins des trois barrières successives interposées entre les radioéléments et les individus, à savoir:

- le gainage du combustible (1ère barrière)
- l'enveloppe du système primaire (2e barrière)
- l'enceinte de confinement (3e barrière).

Ajoutons que quelques systèmes ont également un rôle de barrière car ils véhiculent des fluides qui peuvent être fortement contaminés:

- le système de contrôle volumétrique
- les systèmes de traitement d'effluents.

1.2. Le deuxième principe utilisé est qu'un accident affectant une barrière peut se produire de deux manières différentes:

- perte d'étanchéité de la barrière se produisant de façon imprévisible alors que la tranche est en régime stable, cet incident étant évidemment le résultat de toute l'histoire antérieure de la barrière en question
- perte d'étanchéité de la barrière à la suite d'un accident affectant un autre matériel ou à la suite d'un transitoire de fonctionnement de la tranche.

Compte tenu des deux principes précédents, la recherche des séquences accidentelles plausibles peut s'effectuer en deux étapes:

a) recherche de toutes les causes physiques possibles de défaillance de chacune des trois barrières définies plus haut
b) recherche de tous les accidents et de toutes les séquences de fonctionnement anormal pouvant conduire à la défaillance d'une ou plusieurs barrières, compte tenu des causes mises en évidence sous a).

Une fois dressée la liste présumée exhaustive des accidents et séquences d'accidents, il ne reste plus qu'à les classer en première approximation par ordre de gravité, classement que l'analyse quantitative (évaluation des probabilités) permettra ensuite de rectifier et de préciser.

2. RECHERCHE DES CAUSES POSSIBLES DE DEFAILLANCE DES BARRIERES

Nous avons effectué une recherche systématique de toutes les causes possibles de défaillance de chacune des trois barrières.

Dans l'analyse concernant la 1ère barrière (gainage du combustible) nous avons distingué deux états de la tranche car ils conduisent à des séquences d'accidents assez différentes:

– Tranche à l'arrêt, cuve ouverte pour chargement de combustible. Dans cet état de la tranche, on peut considérer que les barrières 2 et 3 sont ouvertes.

— Tous les autres états de la tranche sont regroupés sous la désignation de «tranche en fonctionnement».

Nous avons représenté sous forme des arbres de défaillance 1A et 2A (fig. 1 et 2)[1] les résultats de notre analyse. Les événements de base de ces arbres représentent autant de causes physiques de défaillance dont il faut ensuite examiner de quelle manière elles pourront se manifester.

En ce qui concerne la 2e barrière (enveloppe du système primaire) nous avons répertorié dans l'arbre de défaillance 3A (fig.3) tous les phénomènes physiques pouvant conduire à une perte de l'étanchéité ou au by-pass de cette barrière.

De la même façon les phénomènes physiques pouvant conduire à la perte d'étanchéité ou au by-pass de la 3e barrière (enceinte de confinement) ont été représentés dans l'arbre 4A (fig.4).

3. RECHERCHE DE TOUS LES ACCIDENTS ET DE TOUTES LES SEQUENCES DE FONCTIONNEMENT ANORMAL POUVANT CONDUIRE A LA DEFAILLANCE D'AU MOINS UNE BARRIERE OU A SON BY-PASS

A partir des arbres de défaillance 1A à 4A on peut imaginer assez facilement tous les accidents en «enchaînement» d'accidents actuellement étudiés dans les rapports de sûreté ou sur le point de l'être.

Cela conduit aux arbres de défaillance 1B à 4B (fig. 5 à 8), qui sont un développement plus poussé des arbres 1A à 4A.

Cependant, il est difficile d'imaginer par ce seul moyen tous les cas de fonctionnements anormaux pouvant conduire à une défaillance des barrières. Un recoupement est nécessaire par une méthode qui applique la démarche inverse de celle exposée dans le paragraphe 2 précédent en partant de la défaillance d'un système ou d'un composant et en recherchant les séquences qui, à partir de cette défaillance, peuvent conduire à la perte d'une ou plusieurs barrières. D'une manière plus détaillée, une analyse qui nous paraît exhaustive pourrait être conduite de la manière suivante:

3.1. Définition d'une liste d'états de fonctionnement caractéristiques d'un réacteur à eau sous pression. Nous proposons la liste suivante:
- tranche à l'arrêt, cuve ouverte
- tranche couplée au réseau, quelle que soit sa puissance
- tranche à l'arrêt à chaud: réacteur sous-critique ou critique à puissance faible (quelques pour cent de la puissance nominale), fluide réfrigérant à température > 180°C et à pression > 30 bar (circuit de refroidissement du réacteur à l'arrêt non en service)
- tranche à l'arrêt à froid: réacteur sous-critique, fluide réfrigérant à température ≤ 180°C et à pression ≤ 30 bar (conditions de passage sur le circuit de refroidissement du réacteur à l'arrêt réalisées).

3.2. Définition d'une liste des systèmes ayant une importance non négligeable pour la sûreté. Cette liste peut fort bien être la liste des systèmes rangés en classes de sûreté 1, 2 et 3 selon les critères de l'ANS. La répartition en classes de sûreté représente déjà l'idée que l'on se fait a priori de l'importance de chaque système.

[1] Voir annexe.

3.3. Pour chacun des systèmes de la liste précédente et dans des états définis ci-dessus analyser toutes les conséquences des événements suivants:

- perte d'étanchéité du système s'il a un rôle de barrière ou si cet événement peut avoir une influence directe sur une barrière
- perte de la fonction par la suite de l'indisponibilité du système
- fonctionnement intempestif du système.

Dans chaque cas on ne retiendra que les conséquences ayant une influence sur l'intégrité des barrières en se référant si nécessaire aux arbres de défaillance de ces barrières et on tiendra compte de tous les moyens dont dispose l'exploitant pour:

- être informé de la défaillance du système
- limiter les conséquences de cette défaillance lorsqu'elles peuvent avoir une répercussion sur l'intégrité des barrières.

La définition des séquences dangereuses pour les barrières, séquences qui débutent par une des défaillances décrites ci-dessus, peut se faire ensuite en utilisant la méthode des arbres d'événements. Le chemin le plus favorable sera celui où la défaillance initiale du système étudié ne sera suivie par aucune défaillance et le chemin le plus défavorable celui qui correspond à l'enchaînement de la défaillance du système et de celle de tous les systèmes limitatifs associés qui auraient protégé les barrières.

Nota: a) Chacune des études précédentes devra tenir compte de la cause possible de défaillance du système étudié en distinguant les causes internes ou limitées au système et les causes externes qui peuvent être communes à plusieurs systèmes et donner une probabilité non négligeable aux séquences les plus favorables.

b) La connaissance du fonctionnement des systèmes permet d'éliminer un grand nombre de cas de l'étude précédente. C'est ainsi, par exemple, qu'il est inutile d'étudier les conséquences de l'indisponibilité des systèmes d'injection de sécurité quand la tranche est à l'arrêt, cuve ouverte.

c) Selon les systèmes étudiés, on peut être conduit à subdiviser davantage les états de la tranche dans lesquels la défaillance du système est à considérer. En effet, certains systèmes sont inhibés en dessous ou en dessus de certaines valeurs de la puissance de la tranche ou de la pression ou de la température du primaire. Les conséquences de l'inhibition d'un système sont d'ailleurs à examiner de près.

4. CLASSEMENT QUALITATIF DES SEQUENCES D'ACCIDENT PAR ORDRE D'IMPORTANCE

La méthode exposée précédemment permet de connaître toutes les séquences d'accident où intervient un même système.

Pour mieux juger de la gravité relative de telle ou telle séquence ou de l'importance de tel ou tel système pour la sûreté, il est intéressant de faire un premier classement des séquences par ordre de gravité.

Lorsqu'une seule barrière est concernée, il est relativement simple de classer les séquences accidentelles qui lui correspondent par ordre de gravité.

Lorsque plusieurs barrières risquent d'être touchées, il est intéressant de classer les combinaisons de pertes de barrière par ordre de gravité ce qui permet ensuite de classer les séquences accidentelles correspondantes.

Nous avons représenté sous forme d'un arbre d'événements (fig.9) les $2^3=8$ combinaisons possibles de perte des barrières.

Parmi ces huit cas, le cas 1 correspond à l'état normal d'intégrité des trois barrières. Les sept autres cas correspondent à des situations accidentelles que l'on peut ranger par ordre d'importance en tenant compte des remarques suivantes:

– La perte de la 3e barrière seule (cas 2) n'a pas de conséquences pour la sûreté dans la mesure où les barrières 1 et 2 sont étanches.

– La perte de la 1ère barrière seule (cas 5) nous paraît plus grave que le cas précédent du fait qu'en fonctionnement normal, le système primaire a un débit de fuite non nul.

– La 2e barrière est la plus importante de toute évidence et nous estimons que le cas où cette 2e barrière seule est perdue (cas 3) est plus grave que le cas où les barrières 1 et 3 sont perdues (cas 6).

– Si la 2e barrière est rompue, il est très probable que la 1ère sera touchée. On peut donc considérer que le cas 3 (2e barrière seule perdue) aboutira bien souvent au cas 7 (barrières 1 et 2 perdues) et que le cas 4 (barrières 2 et 3 perdues) aboutira au cas le plus défavorable qui est le cas 8 (toutes barrières perdues).

Compte tenu des remarques précédentes, le classement des sept situations accidentelles ci-dessus peut s'écrire dans l'ordre de gravité croissante de la façon suivante:

2 - 5 - 6 - 3 - 7 - 4 - 8.

Ce classement peut faciliter ensuite le classement des séquences accidentelles par ordre de gravité croissante.

CONCLUSION

L'application de la méthode de travail proposée plus haut devrait permettre une détermination plus exhaustive que jusqu'à présent de tous les accidents et enchaînements d'accidents pouvant conduire à une perte d'étanchéité des barrières de confinement des radioéléments et un premier classement qualitatif de ces séquences d'accidents par ordre de gravité.

Ce premier classement permet ensuite de fixer des ordres de priorité pour l'évaluation de la probabilité de chaque accident ou séquence d'accident.

Cette évaluation quantitative, qui peut être conduite en utilisant par exemple la méthode des arbres de défaillance, permet ensuite de connaître l'importance réelle d'un composant ou d'un système pour la sûreté de la tranche et d'en déduire des règles de décision en matière de conception des centrales nucléaires.

L'évaluation quantitative aura d'autant plus de valeur que la méthode de recherche des causes possibles d'accident aura été plus efficace et exhaustive, ce qui justifie l'étude précédente.

ANNEXE

FIGURES 1 A 9

LEGENDE

CONDITION LOGIQUE "OU"

CONDITION LOGIQUE "ET"

L'EVENEMENT DECRIT AU-DESSUS DE CE SYMBOLE DOIT ETRE ANALYSE EN DETAIL

RENVOI

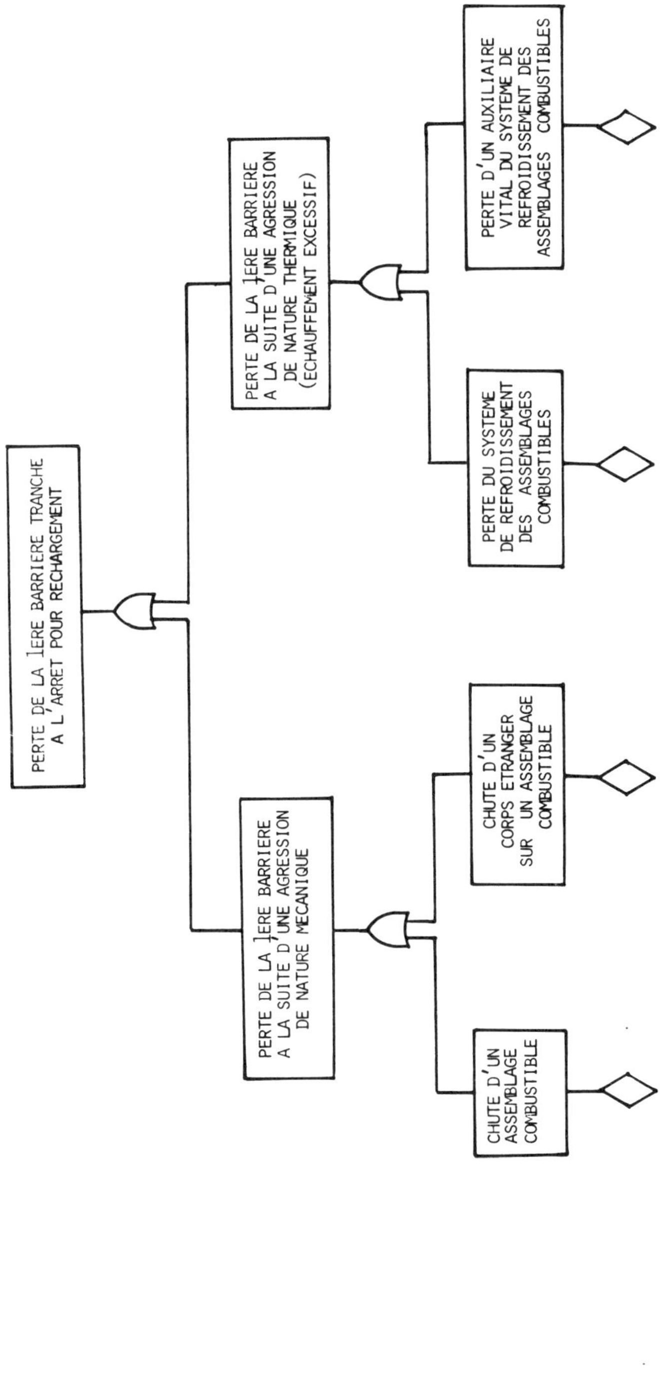

FIG.1. Arbre de défaillance 1A.

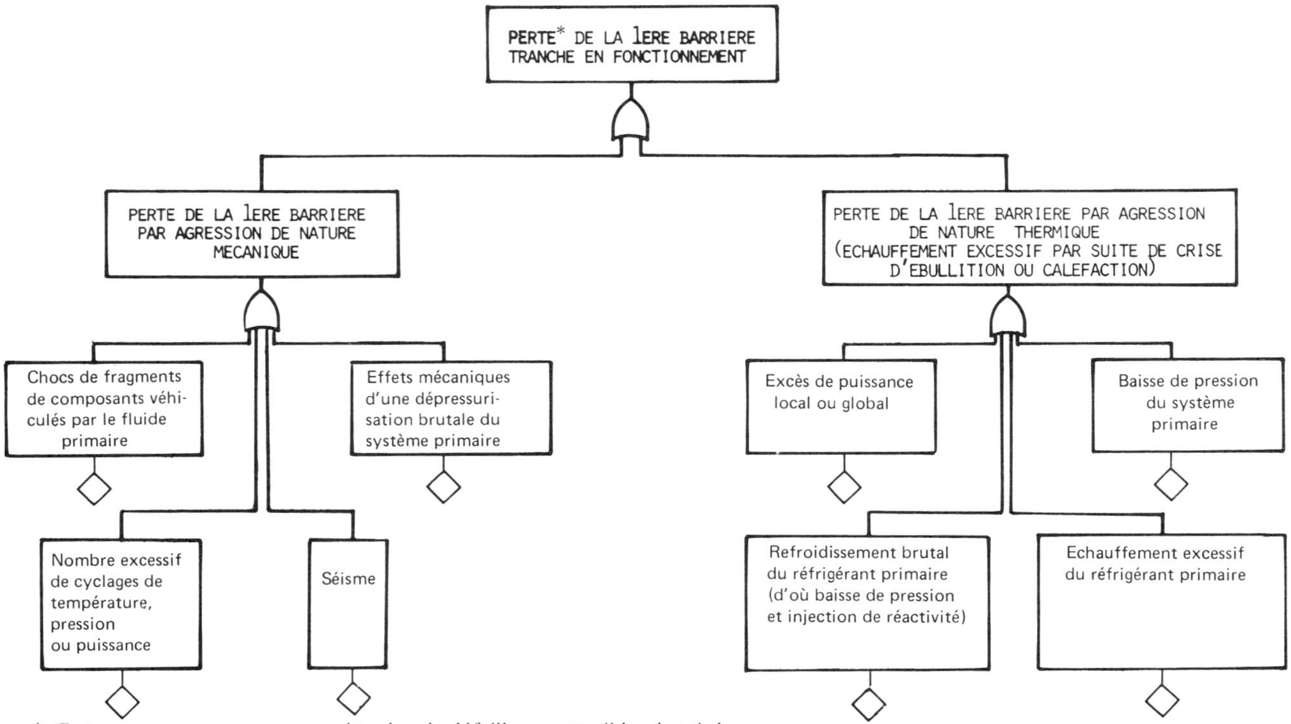

* Ce terme recouvre une gamme étendue de défaillances possibles depuis la fissuration locale d'un tube d'assemblage combustible jusqu'à la rupture ou la fusion d'un grand nombre de gainages.

FIG.2. Arbre de défaillance 2A.

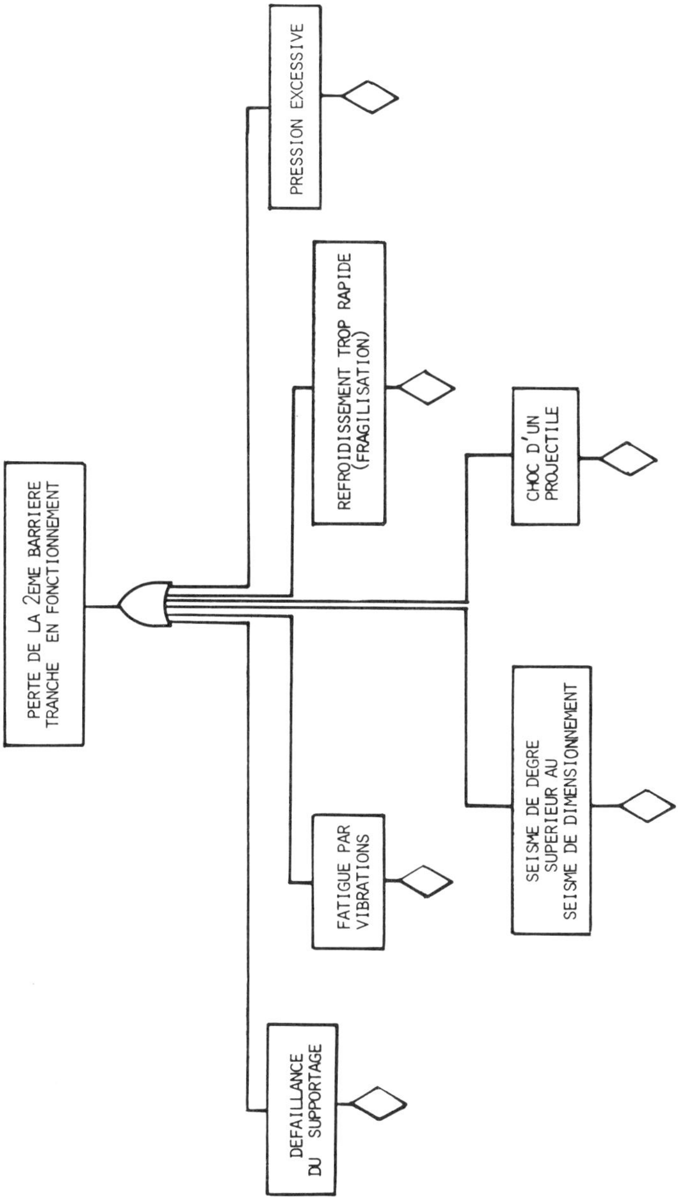

FIG.3. Arbre de défaillance 3A.

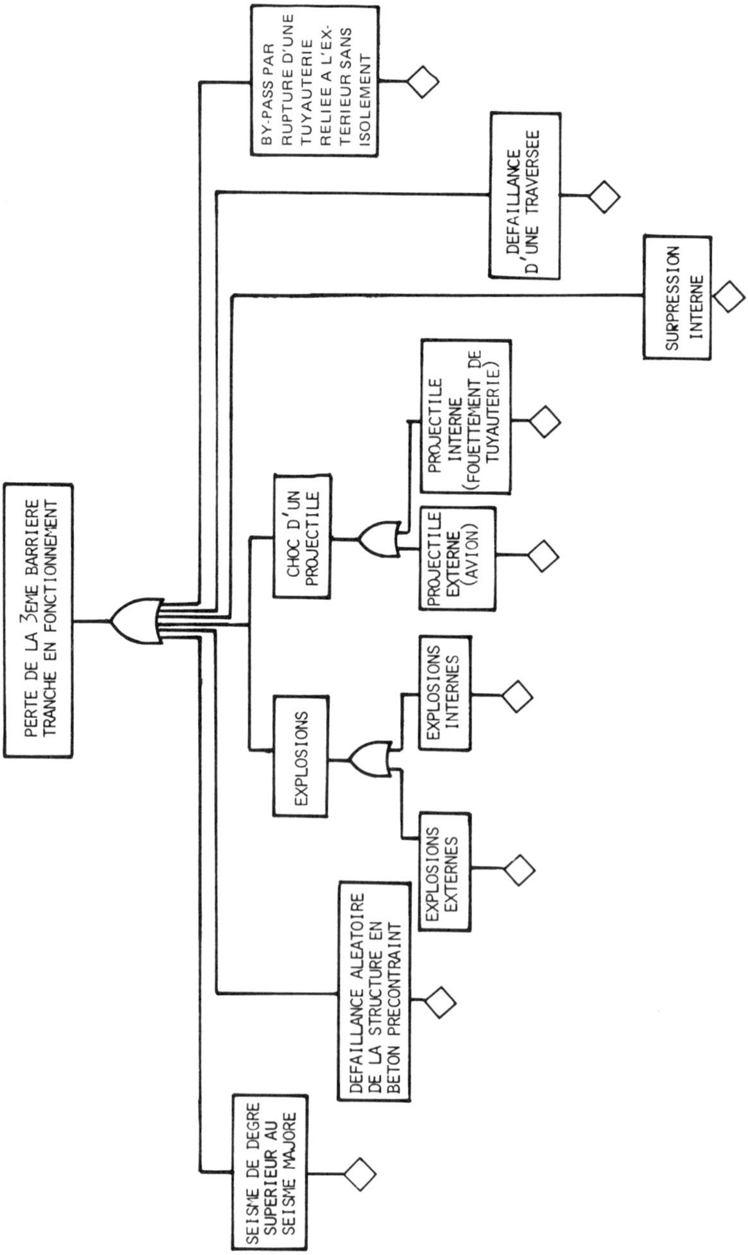

FIG.4. Arbre de défaillance 4A.

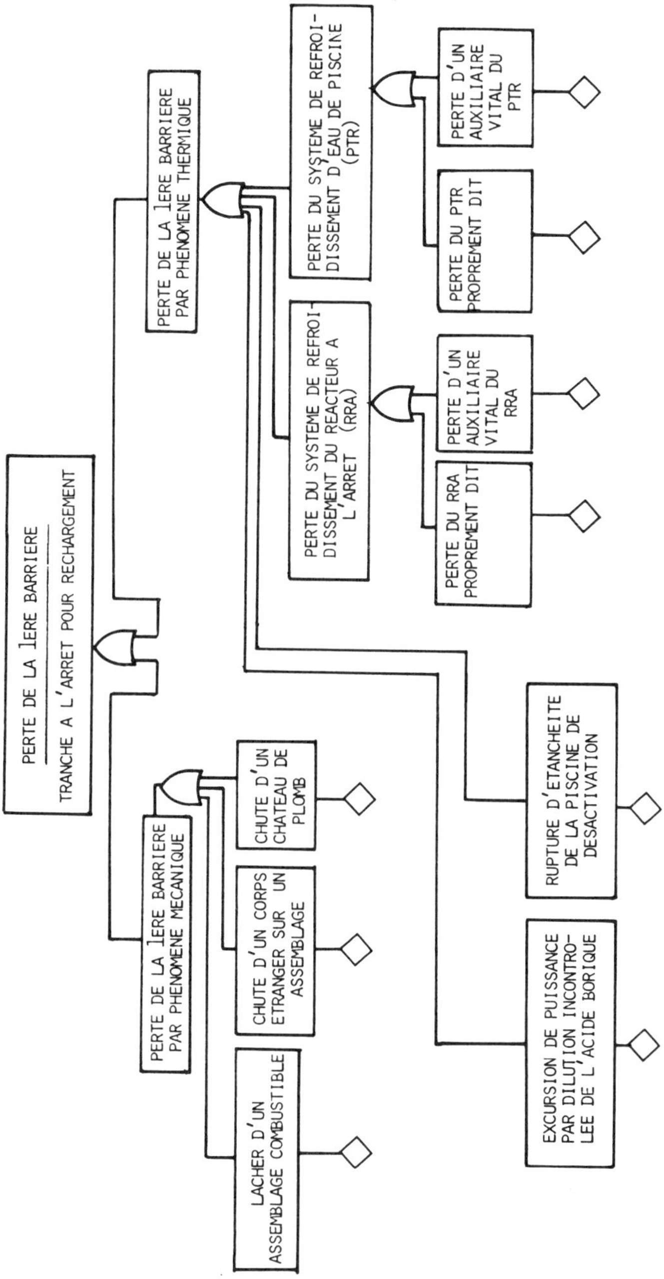

FIG.5. Arbre de défaillance 1B.

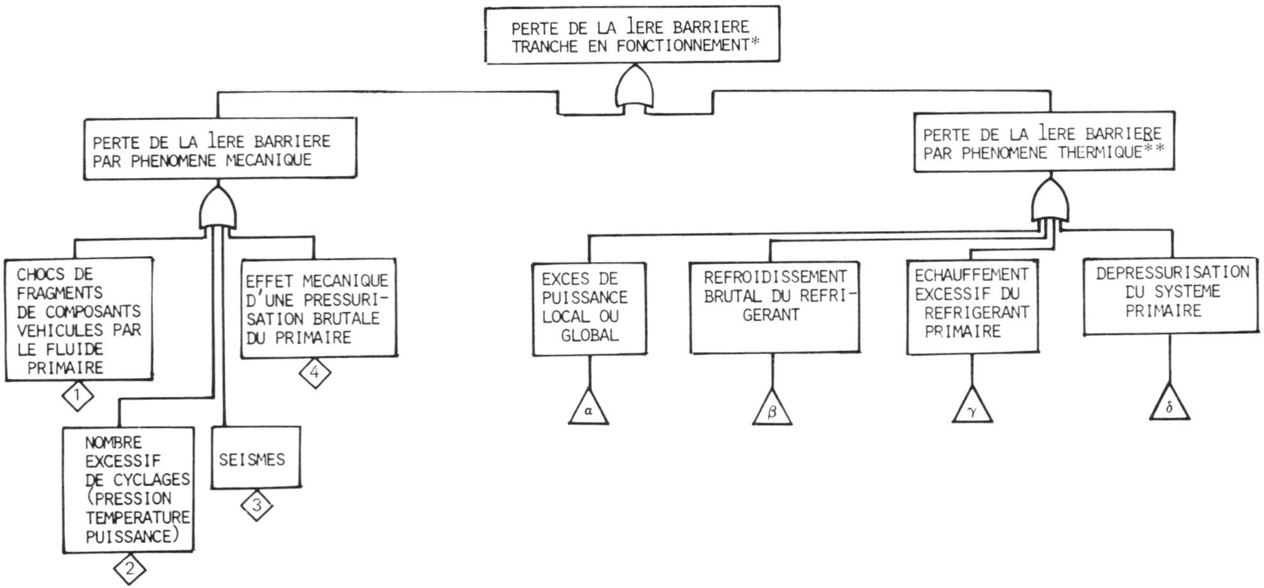

* Cette expression recouvre tous les états de la tranche excepté l'état cuve ouverte.

** Généralement crise d'ébullition ou caléfaction.

FIG.6. Arbre de défaillance 2B.

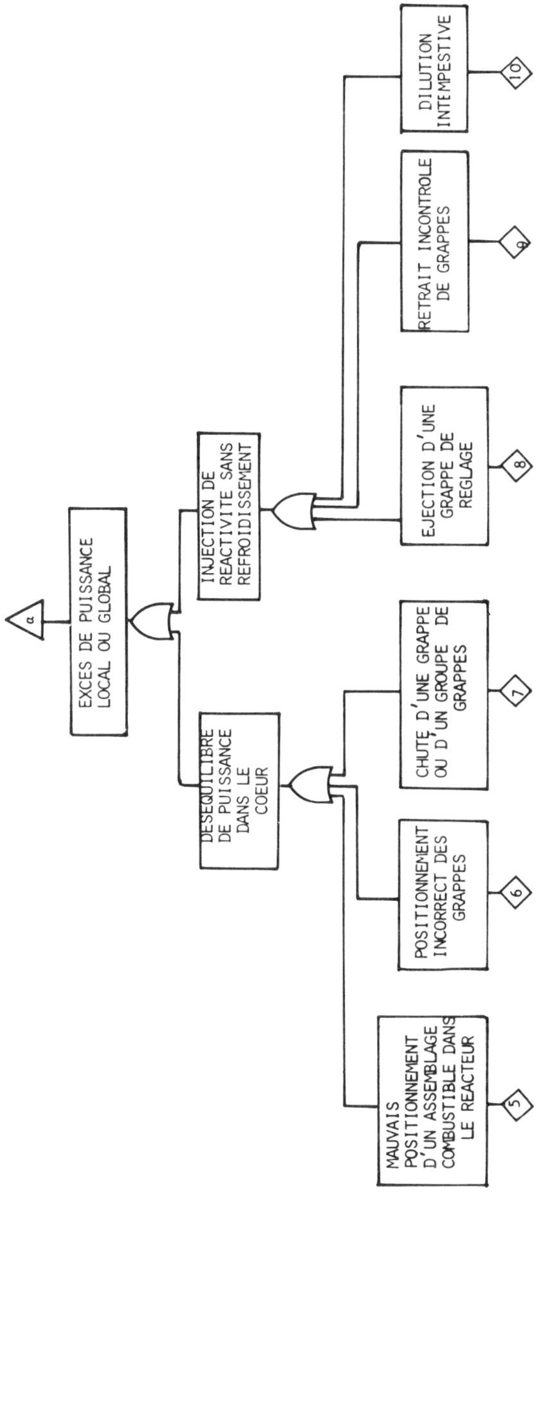
a
EXCES DE PUISSANCE LOCAL OU GLOBAL
DESEQUILIBRE DE PUISSANCE DANS LE COEUR
INJECTION DE REACTIVITE SANS REFROIDISSEMENT
MAUVAIS POSITIONNEMENT D'UN ASSEMBLAGE COMBUSTIBLE DANS LE REACTEUR
5
POSITIONNEMENT INCORRECT DES GRAPPES
6
CHUTE D'UNE GRAPPE OU D'UN GROUPE DE GRAPPES
7
EJECTION D'UNE GRAPPE DE REGLAGE
8
RETRAIT INCONTROLE DE GRAPPES
9
DILUTION INTEMPESTIVE
10

FIG. 6, suite

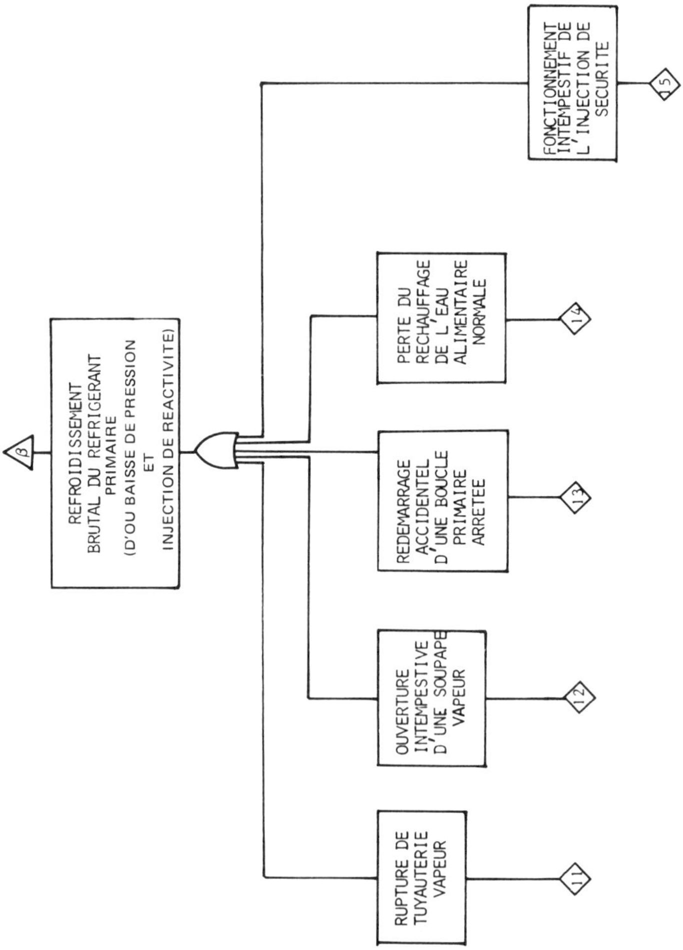
β
REFROIDISSEMENT
BRUTAL DU REFRIGERANT
PRIMAIRE
(D'OU BAISSE DE PRESSION
ET
INJECTION DE REACTIVITE)
RUPTURE DE
TUYAUTERIE
VAPEUR
11
OUVERTURE
INTEMPESTIVE
D'UNE SOUPAPE
VAPEUR
12
REDEMARRAGE
ACCIDENTEL
D'UNE BOUCLE
PRIMAIRE
ARRETEE
13
PERTE DU
RECHAUFFAGE
DE L'EAU
ALIMENTAIRE
NORMALE
14
FONCTIONNEMENT
INTEMPESTIF DE
L'INJECTION DE
SECURITE
15

FIG. 6, suite

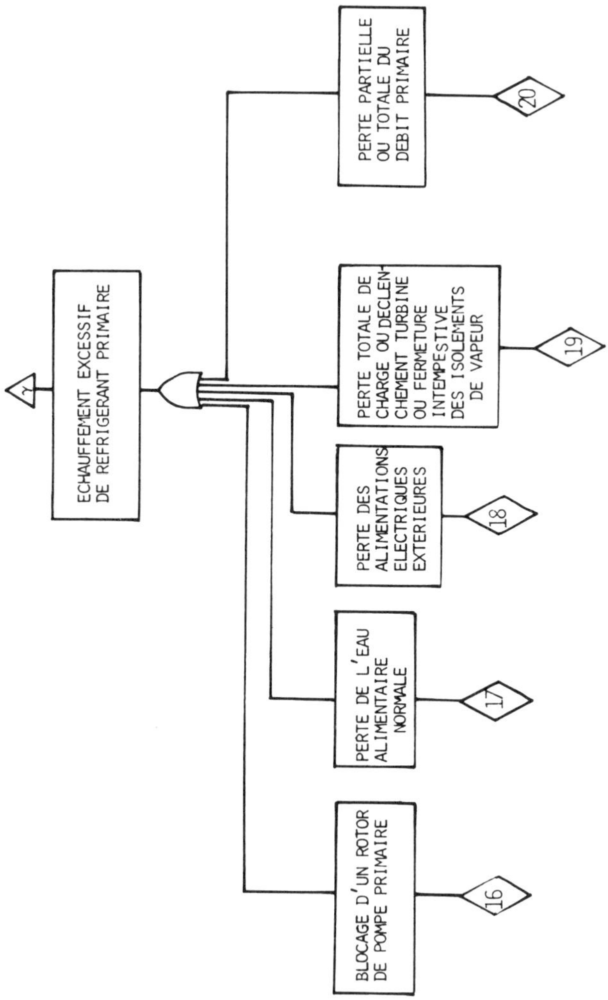
ECHAUFFEMENT EXCESSIF DE REFRIGERANT PRIMAIRE
7
BLOCAGE D'UN ROTOR DE POMPE PRIMAIRE
16
PERTE DE L'EAU ALIMENTAIRE NORMALE
17
PERTE DES ALIMENTATIONS ELECTRIQUES EXTERIEURES
18
PERTE TOTALE DE CHARGE OU DECLENCHEMENT TURBINE OU FERMETURE INTEMPESTIVE DES ISOLEMENTS DE VAPEUR
19
PERTE PARTIELLE OU TOTALE DU DEBIT PRIMAIRE
20

FIG. 6, suite

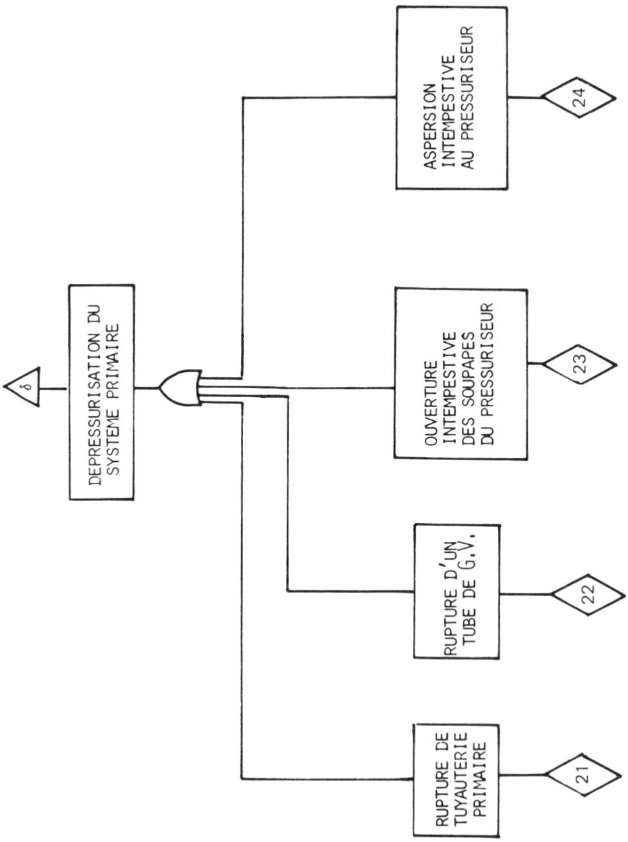
DEPRESSURISATION DU SYSTEME PRIMAIRE
δ
ASPERSION INTEMPESTIVE AU PRESSURISEUR
24
OUVERTURE INTEMPESTIVE DES SOUPAPES DU PRESSURISEUR
23
RUPTURE D'UN TUBE DE G.V.
22
RUPTURE DE TUYAUTERIE PRIMAIRE
21

FIG. 6, suite

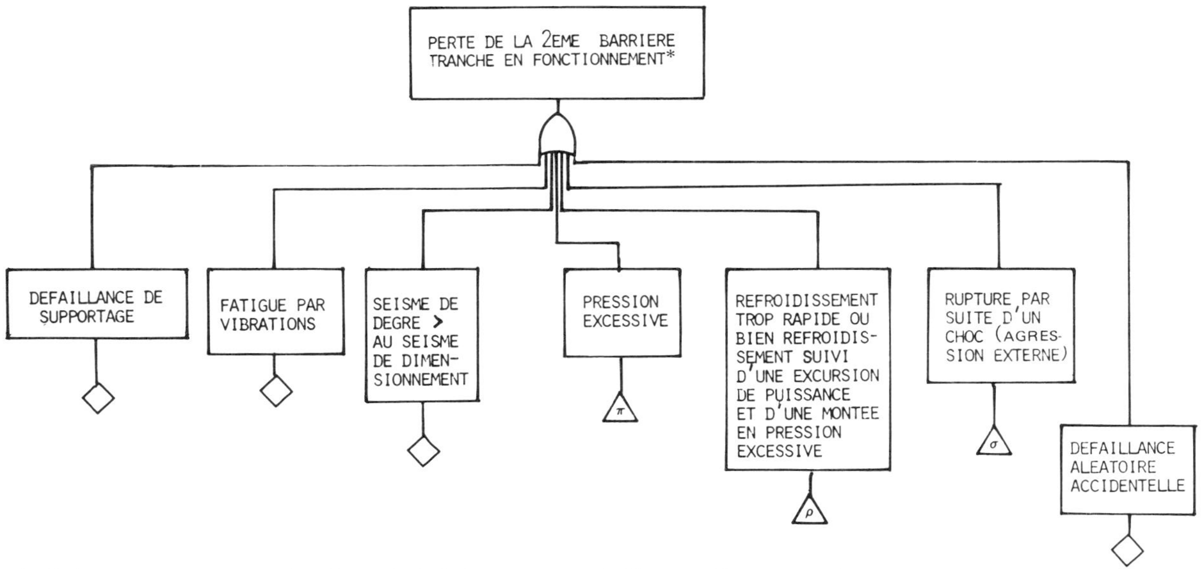

* Cette expression recouvre tous les états de la tranche excepté l'état d'arrêt cuve ouverte.

FIG.7. Arbre de défaillance 3B.

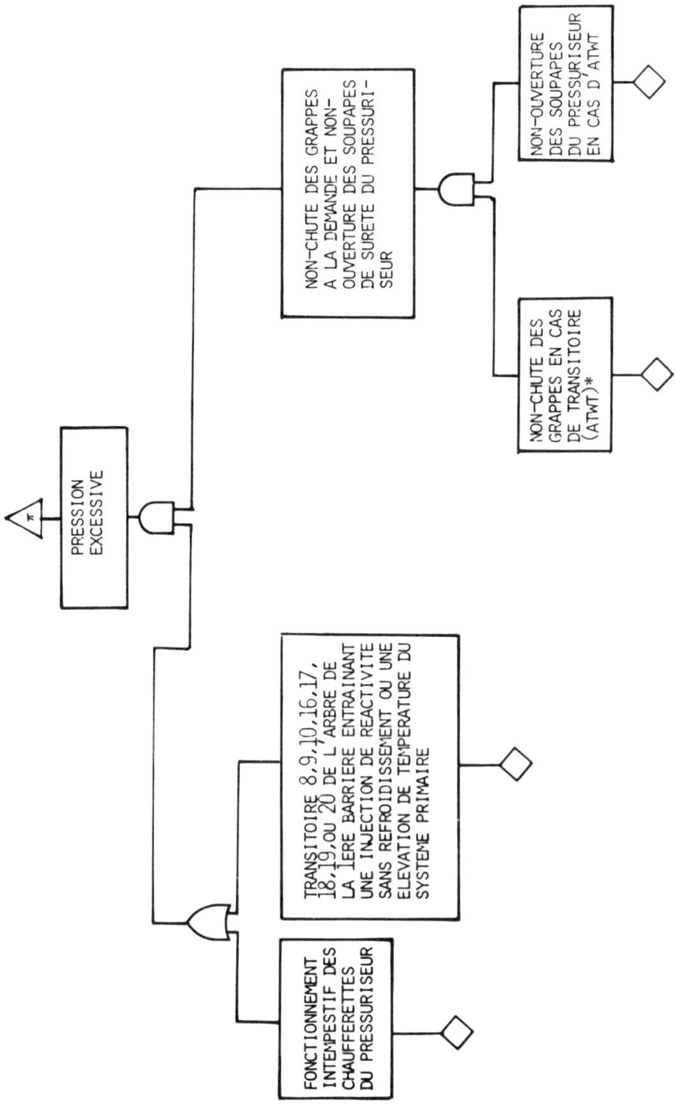

* Anticipated Transient Without Trip.

FIG. 7, suite

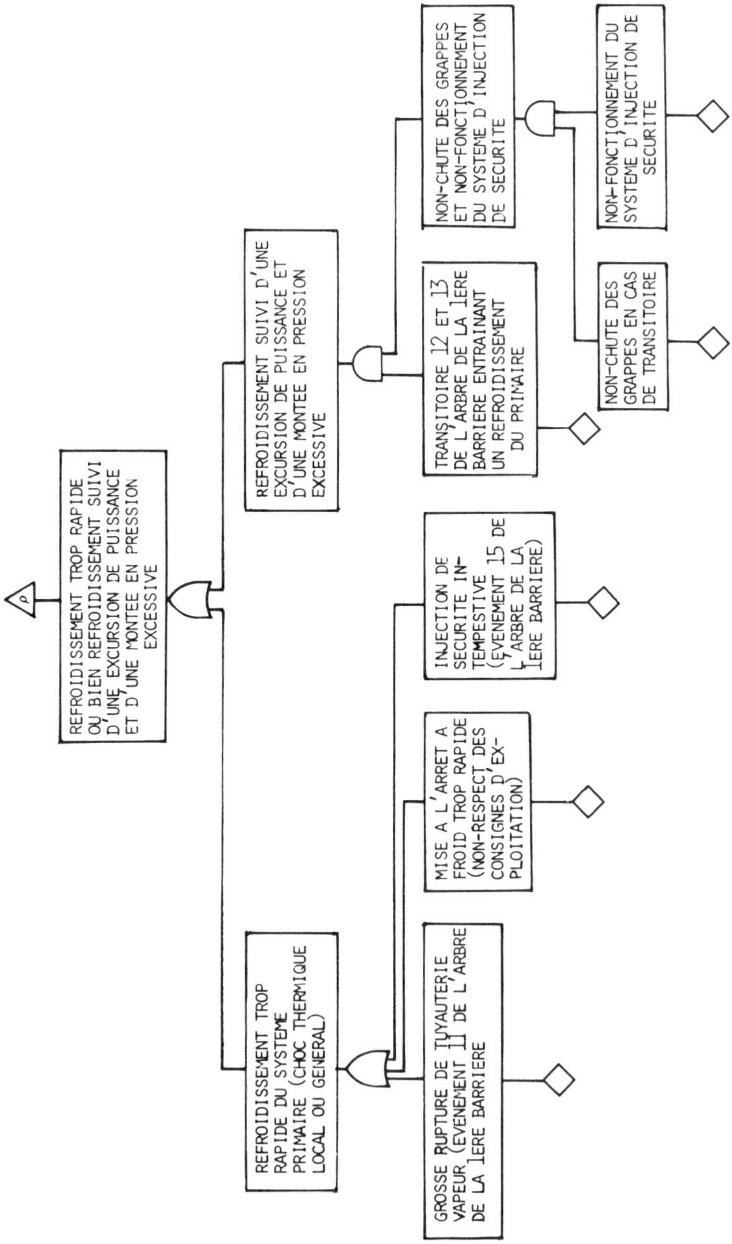
P
REFROIDISSEMENT TROP RAPIDE OU BIEN REFROIDISSEMENT SUIVI D'UNE EXCURSION DE PUISSANCE ET D'UNE MONTEE EN PRESSION EXCESSIVE
REFROIDISSEMENT TROP RAPIDE DU SYSTEME PRIMAIRE (CHOC THERMIQUE LOCAL OU GENERAL)
REFROIDISSEMENT SUIVI D'UNE EXCURSION DE PUISSANCE ET D'UNE MONTEE EN PRESSION EXCESSIVE
GROSSE RUPTURE DE TUYAUTERIE VAPEUR (EVENEMENT 11 DE L'ARBRE DE LA 1ERE BARRIERE
MISE A L'ARRET A FROID TROP RAPIDE (NON-RESPECT DES CONSIGNES D'EXPLOITATION)
INJECTION DE SECURITE INTEMPESTIVE (EVENEMENT 15 DE L'ARBRE DE LA 1ERE BARRIERE)
TRANSITOIRE 12 ET 13 DE L'ARBRE DE LA 1ERE BARRIERE ENTRAINANT UN REFROIDISSEMENT DU PRIMAIRE
NON-CHUTE DES GRAPPES ET NON-FONCTIONNEMENT DU SYSTEME D'INJECTION DE SECURITE
NON-CHUTE DES GRAPPES EN CAS DE TRANSITOIRE
NON-FONCTIONNEMENT DU SYSTEME D'INJECTION DE SECURITE

FIG. 7, suite

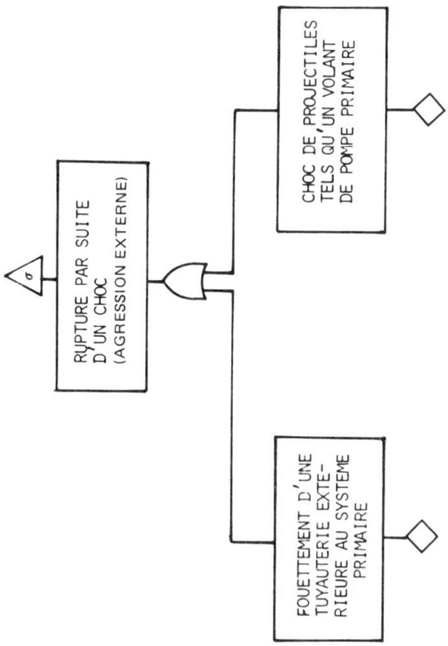
RUPTURE PAR SUITE
D'UN CHOC
(AGRESSION EXTERNE)
CHOC DE PROJECTILES
TELS QU'UN VOLANT
DE POMPE PRIMAIRE
FOUETTEMENT D'UNE
TUYAUTERIE EXTE-
RIEURE AU SYSTEME
PRIMAIRE

FIG. 7, suite

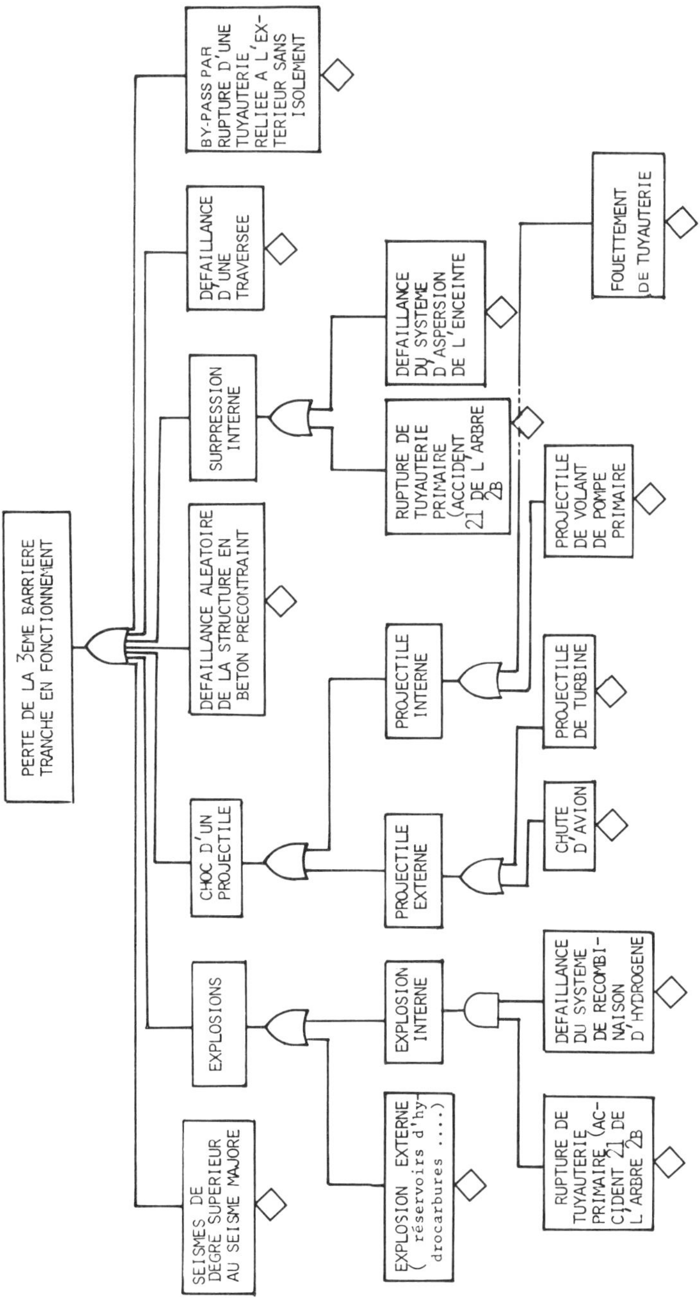

FIG.8. Arbre de défaillance 4B.

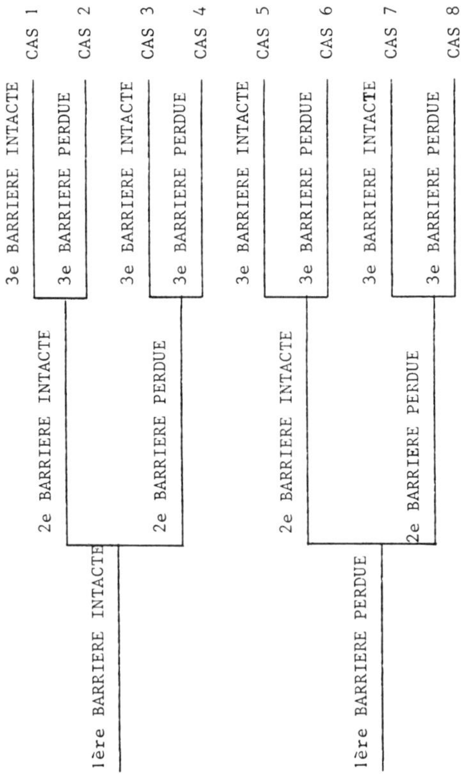

FIG.9. Huit combinaisons possibles de perte des barrières.

DISCUSSION

(on papers IAEA-SM-195/46 and IAEA-SM-195/10)

S. HARTWIG: I should like to ask Mr. Bogani whether the failure rates quoted in Table I are assumed rates or whether they are based on industrial standars for every component used.

G. BOGANI: They are assumptions, but industrial standards are taken into account in deciding on the figures to use (see Section 3 of the paper, which deals among other things with the initial causes of failure). The assumptions are checked at cut-set level to verify whether they meet the overall requirements. If they do not, changes are introduced.

R. GONNOT: I have a question for Mr. Gachot. It appears that you consider only thermal and mechanical attack. Do you disregard chemical attack? If not, how do you class it — with thermal or mechanical attack?

B. GACHOT: We do not disregard chemical attack, but it appears in a more complete fault tree which was not included in this paper. We class it as a form of attack external to the barrier.

R. ZAOUI: What are the mechanical and other causes that lead to destruction of the first barrier?

B. GACHOT: The most important mechanical causes are specified in the fault trees reproduced in our paper. However, these causes have not yet been analysed with sufficient thoroughness and the analysis is continuing.

R. ZAOUI: By what criterion do you judge whether the first barrier has been destroyed?

B. GACHOT: The degree of contamination of the primary fluid by fission products. When a certain maximum permissible value has been reached, the operator may have to shut down the reactor for unloading of one or more burst fuel assemblies.

R. ZAOUI: Do we have at present any measured failure rate data for the first barrier of a PWR?

B. GACHOT: No, there are no experimental results.

N.R. McDONALD: Have you obtained any preliminary results from your study which would enable you to comment on the conclusions of the Rasmussen Report?

B. GACHOT: No, our study is not yet advanced enough to allow any meaningful comparison with the conclusions of the Rasmussen Report.

DETERMINATION, PAR UNE APPROCHE PROBABILISTE, D'UNE REGLE D'EXPLOITATION DES ALIMENTATIONS DE 6,6 kV DES REACTEURS A EAU SOUS PRESSION (TRANCHES DE 900 MW(e))

A. BLIN, Annick CARNINO
CEA, Centre d'études nucléaires de Saclay,
Gif-sur-Yvette

M. BOURSIER, J.-F. GREPPO
Electricité de France,
Paris,
France

Abstract–Résumé

PROBABILISTIC DETERMINATION OF OPERATING RULES FOR 6.6 kV POWER SUPPLIES TO 900 MW(e) PRESSURIZED-WATER UNITS.

The 6.6 kV power supplies to the safety systems of the 900 MW PWR units being built by Electricité de France (EDF) are ensured by four redundant sources within or outside the site. Because of this redundancy, non-availability of one or more of the sources does not jeopardize the electricity supply – it merely means that the unit must be shut down deliberately for safety reasons at some time. The proposed method is a probabilistic approach which makes it possible to determine how long the unit can reasonably continue on power in such circumstances. The study described in the paper is based on operating results recorded by EDF for equipment similar to that envisaged for future nuclear power stations and in service in existing ones. The amount of information gathered and the method employed for gathering it permit a realistic assessment of the reliability parameters of the equipment (failure and repair rates). To allow for the fact that the equipment can be repaired, the method proper involves use of the Markov model, with which one can find, for each configuration of the system, the change over time of the probability P_0 of a simultaneous failure of all power sources. Use of this method requires prior conversion of the actual concept to an equivalent concept. The sensitivity of P_0 to the parameters is studied for each case, a reasonable uncertainty range being obtained for P_0. Common mode failures for external or internal sources are introduced in parametric form so that the parameter value beyond which they can be neglected is determined. On the basis of the results of the study, the authors propose the adoption of a comparative criterion whereby any mode of operation in degraded conditions (one or more sources not available) is permitted provided that the corresponding value of P_0 is always lower than the maximum value attained by it in the reference situation (all sources available). Applying this criterion, one can determine with the help of graphs the operating periods which must not be exceeded. Thus, in the case of non-availability of an external source (the grid) one obtains a maximum period of 12 to 36 hours; in the case of non-availability of an internal source (diesel generator) the maximum period is 150 to 240 hours.

DETERMINATION, PAR UNE APPROCHE PROBABILISTE, D'UNE REGLE D'EXPLOITATION DES ALIMENTATIONS DE 6,6 kV DES REACTEURS A EAU SOUS PRESSION (TRANCHES DE 900 MW(e)).

L'alimentation électrique de 6,6 kV des auxiliaires de sûreté des tranches de réacteur à eau sous pression de 900 MW construites par Electricité de France (EDF) est assurée par quatre sources redondantes externes ou internes au site. Du fait de cette redondance, l'indisponibilité d'une ou plusieurs de ces sources ne compromet pas la fonction alimentation mais nécessite, à terme, un arrêt volontaire de la tranche pour des raisons de sûreté. La méthode proposée permet de déterminer, par une approche probabiliste, un délai acceptable de maintien en puissance de la tranche dans de telles circonstances. L'étude est basée sur les résultats d'exploitation observés par EDF sur des matériels analogues à ceux prévus dans les schémas des futures centrales et en service dans les centrales actuelles. Le nombre de données recueillies et la méthode utilisée pour les obtenir permettent une évaluation réaliste des paramètres de fiabilité de ces matériels (taux de défaillance et de réparation). La méthode proprement dite fait appel, pour tenir compte du caractère réparable des matériels, au modèle de Markov, qui permet de connaître pour chaque configuration du système l'évolution en fonction du temps de la probabilité P_0 de perte simultanée de toutes les sources d'alimentation. L'application de cette méthode nécessite une transposition préalable du schéma réel en un schéma équivalent. Dans chaque cas on effectue une étude de la sensibilité de P_0 aux différents paramètres. Elle permet de déterminer une plage d'incertitude raisonnable pour la probabilité P_0. Les défaillances de mode commun entre sources externes ou internes sont introduites sous forme paramétrique de façon à déterminer la valeur de ces paramètres au-delà de laquelle elles peuvent être négligées. A partir de l'étude précédente, on propose l'adoption d'un critère comparatif conduisant à admettre tout type de fonctionnement en configuration dégradée (une ou plusieurs sources indisponibles) sous réserve que la probabilité P_0 correspondante reste à tout moment inférieure à la valeur maximale atteinte par cette grandeur dans la configuration de référence (toutes sources disponibles). L'application de ce critère permet de déterminer graphiquement des durées maximales de fonctionnement à ne pas dépasser. On obtient ainsi en cas d'indisponibilité d'une source externe (réseau) une durée limite comprise entre 12 et 36 heures et dans le cas de l'indisponibilité d'une source interne (diésel électrogène) une durée limite comprise entre 150 et 240 heures.

INTRODUCTION :

L'étude que nous présentons ici est destinée à étayer par un calcul probabiliste l'élaboration de règles d'exploitation d'un système lié à la sûreté d'une Centrale Nucléaire. Elle comporte essentiellement deux parties : la première concerne la recherche de données opérationnelles, la deuxième expose la méthode mathématique utilisée et la conclusion donne les règles d'exploitation proposées au vu des résultats de l'étude.

La résolution du problème est basée sur la comparaison des probabilités de perte totale des sources selon les différentes configurations possibles du système électrique : perte d'un diesel, perte des deux diesels, perte d'un ou des deux réseaux etc... Ceci permet, avec les données issues du fonctionnement réel du matériel, de déterminer un niveau de probabilité de perte totale lié au schéma donné initialement (deux sources externes et deux sources internes) et, pour ne pas augmenter le risque, il est possible de limiter à ce même niveau les probabilités de perte totale dans toutes les autres configurations dégradées.

1 - RECUEIL DES RESULTATS D'EXPLOITATION RELATIFS AUX MATERIELS CONCERNES :

1.1 Schéma :

Le schéma d'alimentation électrique qui a servi de base à l'étude est indiqué figure 1. Sa conception, conforme au critère 17 de l'USAEC et à la norme IEEE 308, est tel que chaque ensemble d'auxiliaires de sécurité et de sauvegarde peut être alimenté par 3 sources différentes :

.source externe principale (transformateur de soutirage).
.source externe auxiliaire (transformateur auxiliaire).
.source interne (groupe diesel électrogène).

De façon à faciliter la collecte des résultats d'exploitation, ce schéma a été transposé en un schéma équivalent (figure 2) dans lequel chaque zone correspond à un ensemble homogène de matériels pour lequel il est possible de recueillir des informations.

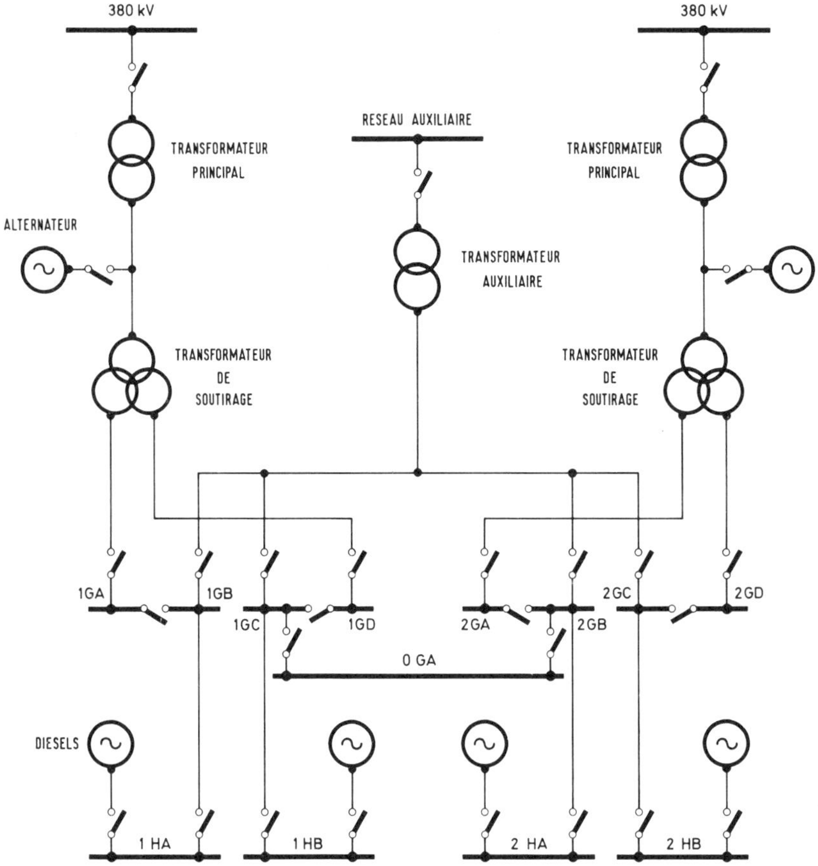

FIG.1. Schéma de principe de l'alimentation électrique des auxiliaires pour les tranches EDF du type à eau sous pression de 900 MW(e) (schéma valable pour une paire de tranches).

1.2 Mode d'acquisition des résultats d'exploitation :

E.D.F. ne disposant pas encore à l'heure actuelle d'un système général d'acquisition de données de fiabilité, nous avons recherché dans les divers documents disponibles au Service de la Production Thermique ou publiés par des organismes extérieurs les renseignements qui nous intéressaient.

Source externe principale :

Ces résultats sont tirés des fiches :

- modèle 221 rédigées par les centrales à l'occasion de tout incident affectant la production d'une tranche (déclenchement, séparation du réseau,...).
- modèle 222 rédigées par ces mêmes centrales lors de tout incident entraînant une avarie importante de matériel.

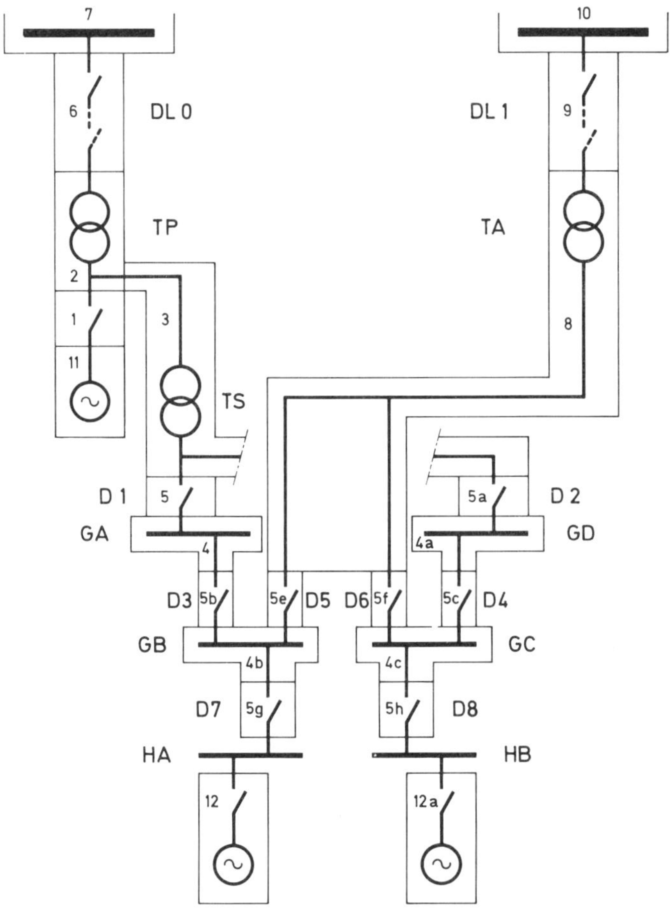

FIG. 2. Définition des zones.

Les incidents affectant l'alimentation des auxiliaires par la source externe principale ont une répercussion quasi-certaine sur la production de la tranche et font donc systématiquement l'objet d'une fiche 221, une fiche 222 venant compléter la précédente si la présence d'une avarie en justifie l'établissement.

On a estimé suffisant, pour cette étude, de faire porter les recherches sur trois années (1970 à 1972 inclus), ce qui correspond à un ensemble de 205 années x tranches.

Tous les matériels dont le comportement est ainsi rapporté ne sont pas parfaitement identiques. Mais on peut remarquer que cet inconvénient est compensé par le fait que la technologie des transformateurs, appareils de coupure et tableaux n'a que faiblement évolué pendant ces dernières années et que cette évolution devrait rester peu sensible dans les prochaines années.

Compte tenu de ces remarques, on peut attribuer aux résultats proposés une valeur statistique suffisante pour ce mode d'approche du problème.

Source externe auxiliaire :

Les incidents affectant cette source n'ayant généralement pas d'influence immédiate sur la production de la tranche ne font pas l'objet d'une fiche 221. Ne possédant pas une statistique analogue à la précédente, il a paru raisonnable d'admettre la transposition suivante :

Source externe principale	Source externe auxiliaire
transfo-soutirage	transfo-auxiliaire
liaison HT principale	liaison HT auxiliaire

Tableaux 6,6 kV :

Les incidents intéressants sont ceux qui nécessitent une mise hors tension d'un tableau. On a retenu tous les incidents nécessitant la mise hors tension des tableaux auxiliaires de tranche et provoquant ainsi l'arrêt de la tranche. (Ils font systématiquement l'objet d'une fiche 221).

Sources internes :

Aucun des documents 221 ou 222 n'étant établi systématiquement lors d'un incident affectant ce type de matériels, nous avons :

.rassemblé toutes les informations disponibles (fiches, comptes rendus,etc...)

.recueilli les résultats publiés par l'USAEC,

.effectué une enquête directe auprès des Unités E.D.F. ou de certains organismes extérieurs exploitant ce type de matériel.

Les données ainsi obtenues sont :

.soit qualitatives (nature des avaries, temps moyen d'indisponibilité),

.soit quantitatives (pourcentage de refus de démarrage, etc...).

Ces derniers résultats comportent, compte tenu de leur mode d'acquisition, une incertitude assez importante, mais regroupent néanmoins assez bien les résultats étrangers publiés dans ce domaine.

Résultats obtenus :

Pour chaque zone, on a retenu les seuls incidents dont le mode de défaillance conduisait à une perte de l'alimentation auxiliaire de secours par la voie considérée. Ce tri a nécessité une analyse incident par incident.

Les résultats portent sur les deux paramètres suivants :

.taux de défaillance (λ_i)

.temps moyen de réparation ($1/\mu_i$)

de chaque zone.

Ils sont indiqués dans le tableau I.

2 - ALIMENTATION ELECTRIQUE D'UNE TRANCHE :

Lorsque le système considéré comporte une redondance formée d'éléments non identiques, on peut appliquer un processus de Markov généralisé.

TABLEAU I. TAUX DE DEFAILLANCE PAR ZONE

Zone	Taux de défaillance λ_i		Temps moyen de répara- tions/zone (heures) ($\overline{T}_i$)	$\lambda_i \overline{T}_i . 10^{-6}$	Temps mini. de répara- tions (heures)	Temps max de répara tions (heures
	(Zone x an)$^{-1}$	(Zone x heure)				
1	0,100	11,4 .10^{-6}	12	137	1	24,75
2	0,140	16,0 "	120	1 920	1,5	1344
3	0,010	1,14 "	8,75	10,0	0,2	17
4,4a,4b,4c,	0,045	5,13 "	89	456	1,5	624
5,5a,5b,5c, 5e,5f,5g	0,025	2,85 "	7,6	21,7	0,2	18,6
6	0,140	16,0 "	21	336	0,2	168
7	0,425	48,5 "	0,2	9,7	ε	1,5
8	0,010	1,14 "	8,75	10,0	0,2	17
9	0,140	16,0 "	21	336	0,2	168
10	0,760	86,8 "	0,2	17,4	ε	1,5
11	0,120	13,7 "	0,03	0,41	-	-
12	Au lancement : 1,5 % Au démarrage : 1,5 % En marche continue : 0,7.10^{-3}/groupexheure		173		1440	0,1

2.1 Position du problème et méthode utilisée :

Rappelons brièvement que les auxiliaires de sûreté peuvent être alimentés :

- par l'une ou l'autre de deux sources externes.
- en cas de défaillance des sources précédentes, par deux sources internes.

Une seule des quatre sources précédentes suffit à fournir la puissance nécessaire à l'alimentation du nombre minimal d'auxiliaires de sûreté requis.

Du point de vue de la fiabilité d'alimentation des auxiliaires de sûreté, le système d'alimentation électrique offre donc une redondance d'ordre 4.

On cherche à évaluer la probabilité de défaillance simultanée[1] de toutes les sources d'alimentation P_o, à partir des différentes configurations imposées par l'indisponibilité d'une ou plusieurs sources.

Tous les éléments du système d'alimentation étant réparables, les quatre sources peuvent être considérées comme telles et il est alors possible de décrire le comportement global du système par un processus de "naissance et de

[1] La probabilité de défaillance simultanée d'un système de n éléments réparables est la probabilité que le n^{ème} élément tombe en panne au temps t, alors que les (n-1) autres éléments sont déjà tombés en panne et ne sont pas encore réparés à l'instant considéré.

mort". Le modèle dit de "MARKOV" peut permettre de représenter un tel processus sous réserve d'un certain nombre d'hypothèses qui seront exposées dans la suite.

Chacune des sources peut présenter plusieurs états : disponible ou indisponible pour les sources externes ; disponible à l'arrêt, disponible en marche ou indisponible pour les sources internes.

Il en résulte pour le système un nombre élevé de combinaisons d'états élémentaires qui peut, toutefois, être réduit en considérant que, sur le plan fiabilité, les deux sources internes et les deux sources externes sont respectivement identiques (mêmes valeurs attribuées à leurs paramètres de fiabilité).

2.2 Schéma équivalent :

Le schéma équivalent proposé est indiqué à la figure 3.

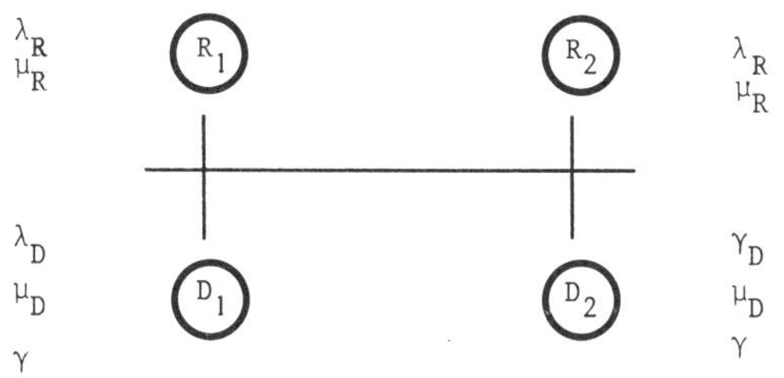

FIG.3. Schéma équivalent.

R_1: source externe principale; R_2: source externe auxiliaire; D_1, D_2: sources internes; λ: taux de défaillance du système; μ: taux moyen de réparation; γ: probabilité de refus de démarrage des sources internes.

2.3 Fonction étudiée :

La fonction étudiée étant l'alimentation des auxiliaires de sûreté vue des jeux de barres secourus, on peut observer que cette fonction est remplie (du fait de la redondance des jeux de barres et des auxiliaires raccordés) dès qu'un jeu de barres est alimenté par une des sources externes ou par sa source interne.

Du point de vue de la fonction envisagée, il n'est donc pas utile de personnaliser chaque jeu de barres et celle-ci est remplie de façon identique par un système comportant un jeu de barres unique alimenté par deux sources externes et deux sources internes redondantes.

2.4 Valeur des paramètres de fiabilité utilisés :

D'après les statistiques opérationnelles de la première partie et en utilisant la technique des arbres de défaillances pour évaluer les paramètres de fiabilité résultants liés à la distribution sur le site, nous avons adopté les valeurs moyennes :

Sources externes :

$\lambda_R = 1,2.10^{-4}/h.$

$\mu_R = 0,06/h.$

Sources internes :

$\lambda_D = 0,7.10^{-3}/h.$

$\mu_D = 0,006\ /h.$

$\gamma = 0,03.$

TABLEAU II. DEFINITION DES DIFFERENTES CONFIGURATIONS ETUDIEES

Configuration	Définition
2 R 2 D_a	- 2 SOURCES EXTERNES DISPONIBLES - 2 SOURCES INTERNES DISPONIBLES A L'ARRET (CONFIGURATION DE REFERENCE)
2 R D_a $\overline{D}$	- 2 SOURCES EXTERNES DISPONIBLES - 1 SOURCE INTERNE DISPONIBLE A L'ARRET - 1 SOURCE INTERNE INDISPONIBLE
2 R D_t $\overline{D}$	- 2 SOURCES EXTERNES DISPONIBLES - 1 SOURCE INTERNE DISPONIBLE EN SERVICE PERMANENT - 1 SOURCE INTERNE INDISPONIBLE
2 R $\overline{2\ D}$	- 2 SOURCES EXTERNES DISPONIBLES - 2 SOURCES INTERNES INDISPONIBLES
R $\overline{R}$ 2 D_a	- 1 SOURCE EXTERNE DISPONIBLE - 1 SOURCE EXTERNE INDISPONIBLE - 2 SOURCES INTERNES INDISPONIBLES A L'ARRET
R $\overline{R}$ 2 D_t	- 1 SOURCE EXTERNE DISPONIBLE - 1 SOURCE EXTERNE INDISPONIBLE - 2 SOURCES INTERNES DISPONIBLES EN SERVICE PERMANENT
R $\overline{R}$ D_t - D_a	- 1 SOURCE EXTERNE DISPONIBLE - 1 SOURCE EXTERNE INDISPONIBLE - 1 SOURCE INTERNE DISPONIBLE A L'ARRET - 1 SOURCE INTERNE DISPONIBLE EN SERVICE PERMANENT
R $\overline{R}$ D_t - $\overline{D}$	- 1 SOURCE EXTERNE DISPONIBLE - 1 SOURCE EXTERNE INDISPONIBLE - 1 SOURCE INTERNE DISPONIBLE EN SERVICE PERMANENT - 1 SOURCE INTERNE INDISPONIBLE
R $\overline{R}$ D_a - $\overline{D}$	- 1 SOURCE EXTERNE DISPONIBLE - 1 SOURCE EXTERNE INDISPONIBLE - 1 SOURCE INTERNE DISPONIBLE A L'ARRET - 1 SOURCE INTERNE INDISPONIBLE

2.5 Configurations étudiées :

Selon la disponibilité des sources externes à l'instant origine le système peut se trouver dans l'un des états décrits dans le tableau II. Nous calculerons pour chaque configuration la probabilité de se trouver dans l'état absorbant (0 source).

2.6 Méthode de Markov et graphes :

Le système se trouvant à l'instant t=0 dans un état initial donné peut à un instant ultérieur t quelconque changer d'état pour se trouver dans un état intermédiaire plus dégradé. La transition se caractérise alors par le taux de défaillance λ de la source dont l'indisponibilité a provoqué le changement d'état.

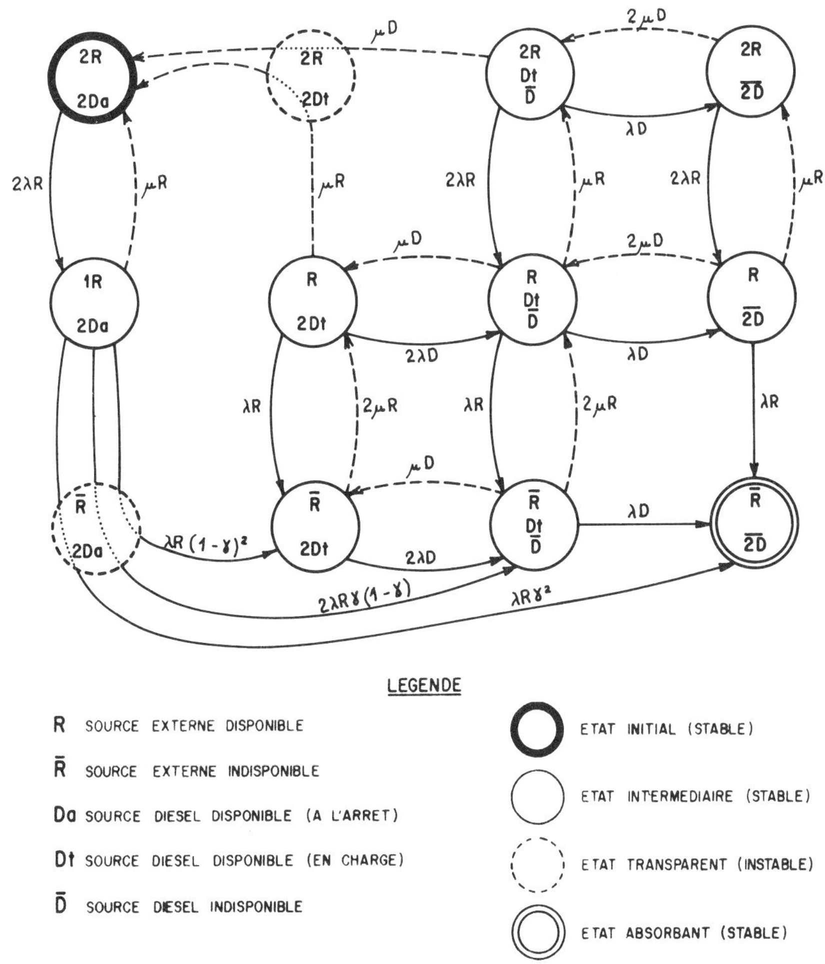

FIG.4. Définition des états et transitions entre états.
Cas de référence: deux sources externes disponibles, deux sources diesel disponibles (à l'arrêt).

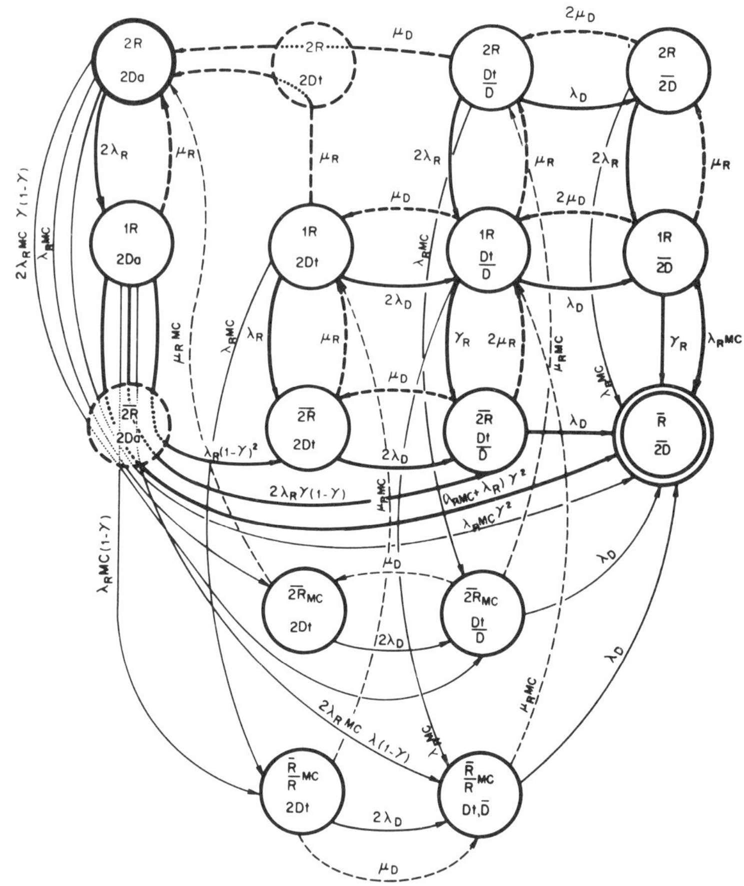

LEGENDE

R SOURCE EXTERNE DISPONIBLE

$\bar{R}$ SOURCE EXTERNE INDISPONIBLE

Da SOURCE DIESEL DISPONIBLE (A L'ARRET)

Dt SOURCE DIESEL DISPONIBLE (EN CHARGE)

$\bar{D}$ SOURCE DIESEL INDISPONIBLE

ETAT INITIAL (STABLE)

ETAT INTERMEDIAIRE (STABLE)

ETAT TRANSPARENT (INSTABLE)

ETAT ABSORBANT (STABLE)

FIG.5. Définition des états et transitions entre états.
Configuration 2R-2Da. Prise en compte des défaillances de mode commun entre sources externes.

D'une façon plus générale, le système étant dans un état intermédiaire quelconque à l'instant t, une transition peut se produire qui se caractérise par :

- un taux de défaillance λ s'il s'agit d'une dégradation.
- un taux de réparation μ s'il s'agit d'une réparation.

On est donc amené à examiner dans chaque configuration toutes les transitions possibles entre états intermédiaires, en leur affectant les taux λ ou μ correspondants.

Ceci conduit à de nombreux graphes dont nous donnons (fig. 4 et 5) des exemples. Ils sont établis avec les hypothèses suivantes :

- dans l'intervalle de temps Δt on ne peut perdre ou retrouver la disponibilité que d'une seule source à la fois
- les possibilités de réparation (pièces détachées, personnel) sont telles que quel que soit le nombre de sources indisponibles les travaux puissent être entrepris sans délai sur l'ensemble des sources.

Cette dernière hypothèse permet d'adopter pour les configurations des taux de réparation μ identiques pour un type donné de sources.

Nous ne rappellerons pas ici le détail du principe de la méthode de Markov. Elle permet à partir des graphes précédemment tracés d'établir pour chaque cas étudié un système d'équations différentielles dont la résolution permet d'obtenir l'expression de la probabilité $P_o(t)$ de perdre toutes les sources à un instant donné, ce qui est l'état absorbant final du graphe.

Les conditions initiales sont telles que pour t=0, $P_o(0)=0$ quelle que soit la configuration étudiée

La résolution de chaque système comporte l'établissement de la matrice de probabilité correspondante.

Les calculs ont été effectués par le Code MARK-GE mis au point par le DSN/CEA à SACLAY.

Lorsque nous avons évalué l'influence des modes communs sur le schéma, nous avons utilisé encore la même méthode avec des états de transition supplémentaires, séparés des précédents et générés par les modes communs seuls (figure n° 4).

2.7 Rappel des hypothèses :

Avant de proposer une interprétation des résultats obtenus, nous récapitulerons les diverses hypothèses faites au cours de l'étude. Elles permettent d'en fixer le domaine de validité.

1-Le schéma de distribution électrique étudié est celui de la figure 1 et les taux de défaillances et de réparation élémentaires ont les valeurs de base indiquées au paragraphe 2.4 .

2-Les valeurs des taux de défaillances et de réparation sont constantes (distribution avec loi exponentielle) ; cette hypothèse est acceptable si le matériel est révisé[2] au moins une fois par an (lors de l'arrêt annuel) et si des échanges préventifs des composants sont effectués avant qu'ils aient atteint la limite de leur durée de vie (ceci afin d'éviter la période d'usure).

[2] Par révision on entend inspection, réparation éventuelle ou échange d'un composant (si l'arrêt programmé coïncide avec la fin de durée de vie du composant) et essai final de vérification de bon fonctionnement.

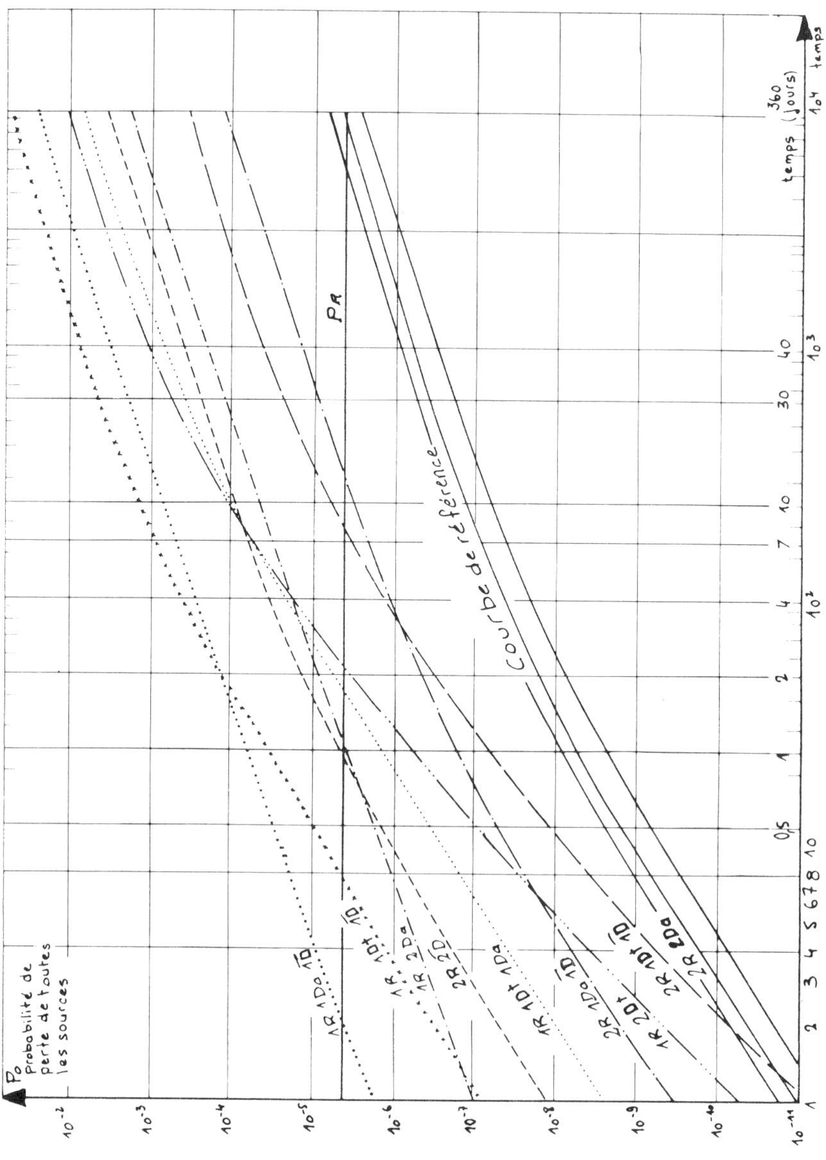

FIG.6. Evolution en fonction du temps de la probabilité de perte totale pour différentes configurations initiales.

3-L'essai périodique d'un diesel n'apporte que la révélation de pannes survenues dans l'intervalle entre essais sans garantir un succès absolu de démarrage à la sollicitation suivante. Il permet de maintenir constante la valeur de la probabilité de refus de démarrage.

4-Pour chaque configuration étudiée, l'instant t=0 pris comme origine est celui où on constate la présence de cette configuration. La probabilité de défaillance P_o de toutes les sources est évidemment nulle à cet instant.

$$P_o(0)=0 \text{ pour toute configuration}$$

5-On dispose du personnel et des pièces de rechange nécessaires pour pouvoir entreprendre les travaux de réparation sans délais quel que soit le nombre de sources indisponibles.

6-Le schéma réel est ramené à un schéma équivalent où les deux sources réseaux sont considérées comme identiques (mêmes caractéristiques de fiabilité).

7-L'étude est faite en se plaçant au niveau des jeux de barres secourus (ceux-ci étant exclus des sources), sachant qu'il suffit que l'un au moins des deux jeux de barres soit alimenté par l'une ou l'autre des quatre sources pour que l'on puisse mettre le réacteur en arrêt sûr.

8-Les défaillances de mode commun affectant les sources internes et externes sont négligées, ce qui suppose que leurs taux respectifs ont des valeurs inférieures à celles établies dans l'étude, c'est-à-dire $\lambda_{RMC} \leq 10^{-7}/h$ pour les sources externes et $\gamma \leq 5 \cdot 10^{-3}$ pour les sources internes.

Interprétation des résultats obtenus (fig. 6 à 9):

Le système retenu, dans sa configuration normale (2R-2Da), satisfait aux conditions minimales imposées par le critère 17 de l'U.S.A.E.C. et la norme I.E.E.E. 308. Cette configuration peut donc être considérée comme la configuration de référence.

Lors du premier démarrage de la tranche, la vérification de chaque élément du système permet de conclure que ce dernier se trouve dans la configuration de référence.[3]

Pendant la période ultérieure de fonctionnement et quelle que soit la durée de cette période, l'évolution prévisible du système peut être décrite de la façon suivante :

-Tant qu'aucun événement ne provoque une modification constatable de l'état des sources (indisponibilité d'une source quelconque, mise en service permanent d'une source interne), la probabilité P_o sur un intervalle de durée t est donnée par la courbe de référence. Les essais périodiques pratiqués pendant cette période n'entraînent, dans la mesure où ils sont réussis, aucune modification constatable du système et ne peuvent de ce fait modifier la valeur de P_o.

-Dès l'instant où l'on constate une modification de la situation précédente - indisponibilité d'une ou des sources(s) externe(s) ou échec d'un essai de source interne - la configuration devient dégradée et la probabilité P_o n'est plus donnée par la courbe de référence.

---

[3] Bien que le non respect de cette condition ne modifie pas la validité de l'étude, on admet que le redémarrage n'a lieu que si les 4 sources sont disponibles.

Une nouvelle configuration est initialisée (t=0, P_O=0) et l'évolution prévisible du système sur une nouvelle période de durée t est décrite par la courbe représentative de la configuration rencontrée.

-Cette situation subsiste jusqu'à l'instant du retour à l'état disponible de la ou des sources(s) indisponible(s). Dès cet instant la configuration de référence est réinitialisée (t=0 , P_O=0) et la probabilité P_O sur une nouvelle période de durée t est à nouveau donnée par la courbe de référence. Si la réparation n'est que partielle, la configuration retrouvée est intermédiaire et la probabilité P_O est donnée par la courbe correspondante.

-Chaque arrêt de tranche provoque une modification constatable de l'état des sources (indisponibilité pour entretien ...). Les vérifications effectuées à cette occasion sont analogues à celles réalisées avant le premier démarrage. Lors du redémarrage qui suit cet arrêt, la situation du système est donc identique à la situation initiale et son évolution prévisible sur une durée t semblable à la précédente.
La période de durée t à prendre en compte est donc celle qui s'écoule entre deux arrêts programmés consécutifs. Cette période est en principe de 1 an sur les tranches PWR.

-La sensibilité des paramètres de fiabilité a été étudiée et donne des marges sur le résultat final qui permettent toujours de déduire les courbes résultantes lorsque les paramètres sont trouvés différents des moyennes indiquées.

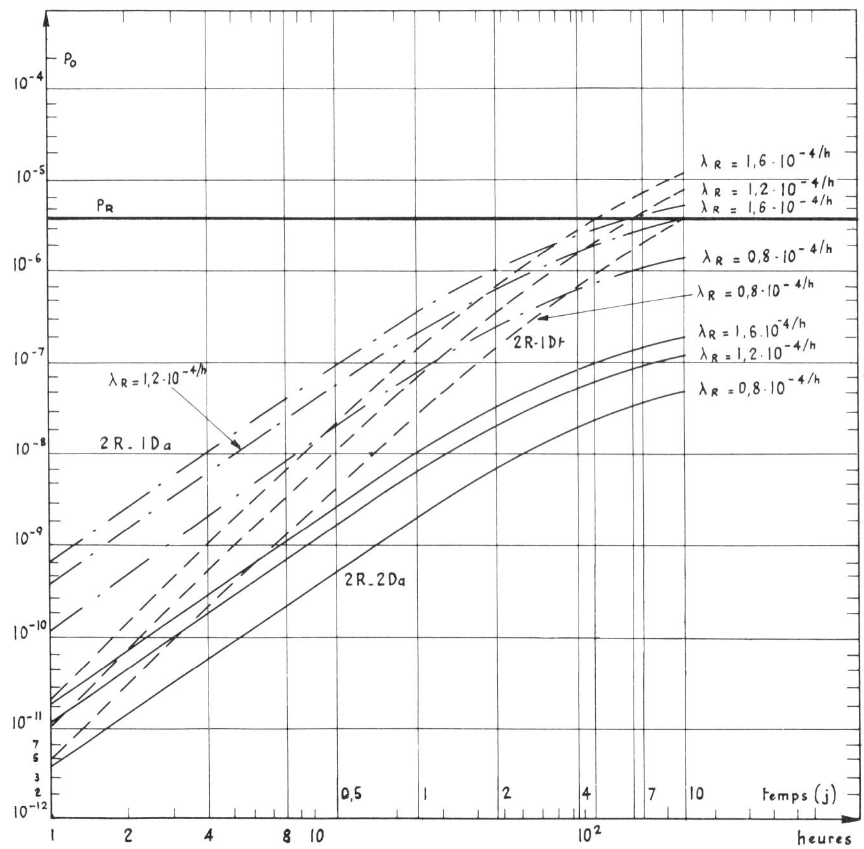

FIG.7. Détermination du délai maximal autorisé en cas de perte d'une source interne.

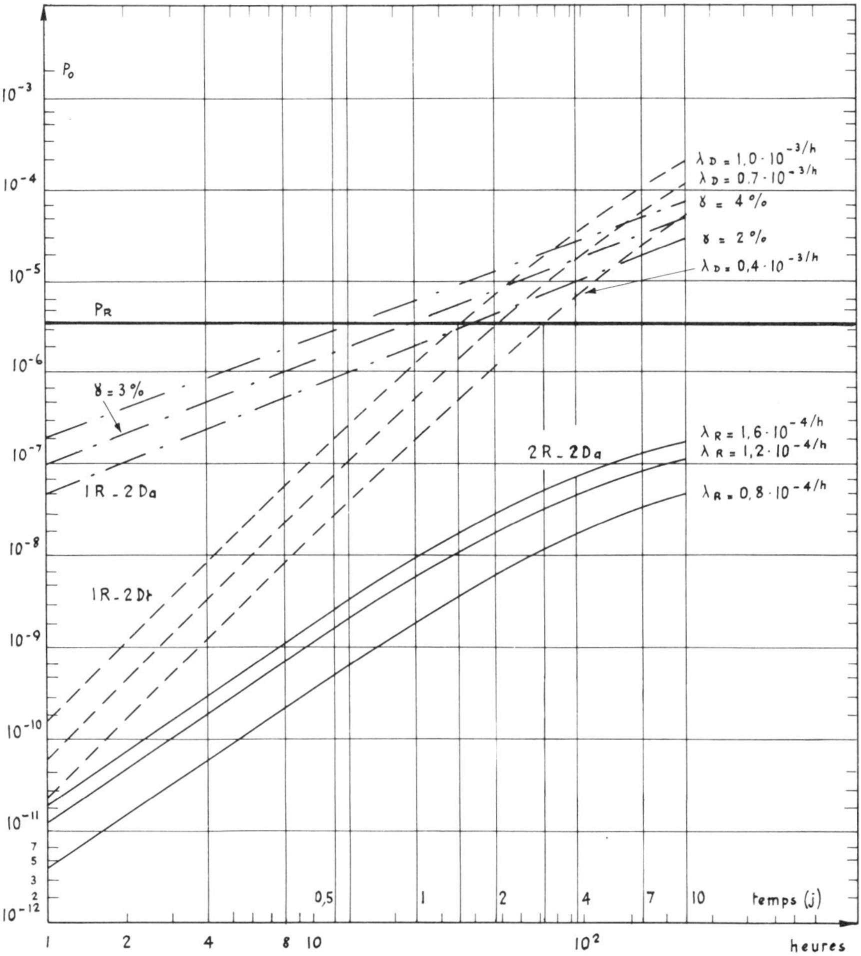

FIG.8. Détermination du délai maximal autorisé en cas de perte d'une source externe.

3 - UTILISATION DU RESEAU DE COURBES POUR DETERMINER DES DUREES LIMITES DE FONCTIONNEMENT EN CONFIGURATION DEGRADEE :

Le risque encouru par perte simultanée des sources est déterminé par le produit du dommage dû à cette perte (quelle que soit l'unité choisie pour l'exprimer : irradiation, durée d'indisponibilité, coût ...) par la probabilité d'occurrence de cet évènement. Dans notre cas le dommage peut être considéré comme constant et l'évolution du risque au cours du temps est représenté, à un facteur constant près, par l'évolution de la probabilité P_0 de perte simultanée de toutes les sources.

La courbe ($2R-2D_a$) représente, par conséquent, l'évolution sur un an d'un risque accepté implicitement par l'application des critères et normes admis.

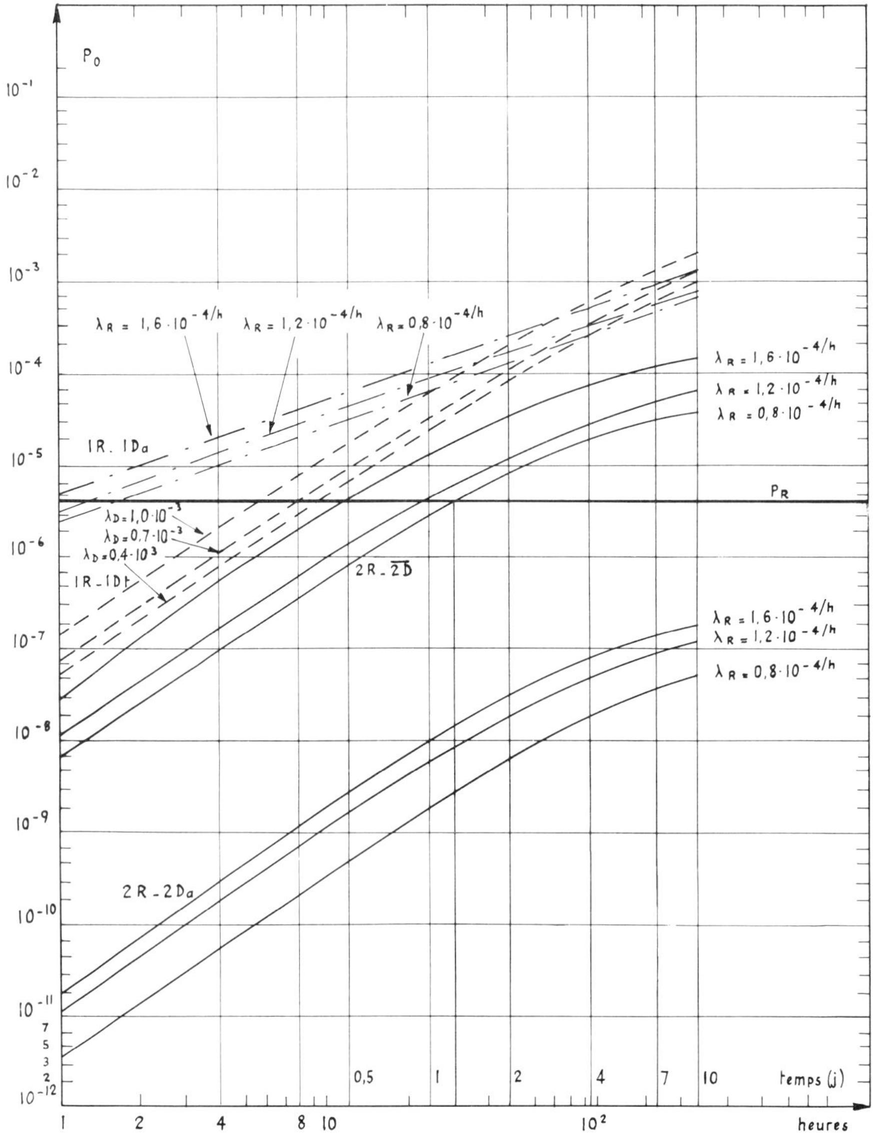

FIG.9. Détermination du délai maximal autorisé en cas de perte de deux sources.

TABLEAU III. CONDUITE A TENIR EN CAS D'INDISPONIBILITE D'UNE OU PLUSIEURS SOURCES

NOMBRE SOURCES DISPO.	DISPONIBILITE DES SOURCES			DUREE LIMITE DE MAINTIEN EN PUISSANCE	CONDUITE A TENIR
	SOURCE EXTERNE PRINCIPALE	SOURCE EXTERNE AUXILIAIRE	SOURCES INTERNES		
4	DISPO	DISPO	DISPO	ILLIMITEE	- PAS D'ACTION
3	DISPO	INDISPO	DISPO	36 HEURES	- DEMARRER LES DEUX DIESELS ET LES FAIRE FONCTIONNER EN CONTINU ET EN CHARGE - ARRETER IMMEDIATEMENT LA TRANCHE SI LA DISPONIBILITE DE LA SOURCE ABSENTE N'EST PAS RETROUVEE AU BOUT DE LA PERIODE INDIQUEE
	INDISPO	DISPO	DISPO	36 HEURES	- MEME CONDUITE A TENIR QUE PRECEDEMMENT
	DISPO	DISPO	1 SOURCE DISPO 1 SOURCE INDISPO	150 HEURES	- VERIFIER LA DISPONIBILITE DES SOURCES RESTANTES - ARRETER IMMEDIATEMENT LA TRANCHE SI LA DISPONIBILITE DE LA SOURCE ABSENTE N'EST PAS RETROUVEE AU BOUT DE LA PERIODE INDIQUEE
2	DISPO	DISPO	INDISPO	12 HEURES	- VERIFIER LA DISPONIBILITE DES SOURCES RESTANTES - ARRETER IMMEDIATEMENT LA TRANCHE SI LA DISPONIBILITE D'AU MOINS UNE DES DEUX SOURCES ABSENTES N'EST PAS RETROUVEE AU BOUT DE LA PERIODE INDIQUEE.
	DISPO	INDISPO	1 SOURCE DISPO 1 " INDISPO	0 HEURE	- ARRETER IMMEDIATEMENT LA TRANCHE
	INDISPO	DISPO	1 SOURCE DISPO 1 " INDISPO		
	INDISPO	INDISPO	DISPO		
1	1 SOURCE QUELCONQUE DISPONIBLE LES AUTRES SOURCES INDISPONIBLES				

Il en résulte que l'on peut considérer le niveau de probabilité P_R atteint au bout d'un an dans cette configuration comme le niveau maximal à ne pas dépasser dans toute configuration dégradée.

Il est possible alors de déterminer une durée maximale de fonctionnement en configuration dégradée par l'abscisse du point d'intersection de la courbe représentative correspondant à la configuration avec la droite horizontale représentant le niveau P_R.

Lors de cette détermination il est souhaitable de prendre, a priori, une marge de sécurité. Pour cela on choisit les abscisses des points d'intersection de la borne supérieure des plages d'incertitude correspondant aux configurations dégradées. Cette borne supérieure est liée à la valeur la plus élevée du paramètre le plus influent d'après l'étude de sensibilité.

Les résultats d'exploitation réels permettront ultérieurement de réduire, le cas échéant, ces marges.

Le tableau III donne les temps de fonctionnement déduits de l'étude d'après le niveau $P_R=5.10^{-6}$, ainsi que les règles proposées pour la méthode d'arrêt.

BIBLIOGRAPHIE

BILLINGTON, R., LEE, S.Y., «A reliability analysis of a heat transfer pump configuration using Markov models», Congrès de Liverpool, avril 1974, à paraître.

BLIN, A., CARNINO, A., JUBAULT, G., JEGU, J., «Programme de calcul de la fiabilité de systèmes complexes PATREC-DE, basé sur la reconnaissance de formes», Congrès de fiabilité, Perros-Guirec, septembre 1974 (édité par le Centre de fiabilité/CNET, Paris).

CARNINO, A., Terminologie et méthodologie utilisées en analyses de fiabilité, Rapport SETS n°11 (février 1973).

GNEDENKO, B.V., BELYAYEV, Y.K., SOLOWYEV, A.D., Mathematical Methods of Reliability Theory, Academic Press, New York and London (1969).

GUYOT, C., Initiation à la maintenabilité, Dunod, Paris (1969).

SCHWOB, M., PEYRACHE, G., Traité de fiabilité, Masson, Paris (1969).

PROBABILITE DE DEFAILLANCE DES CIRCUITS DE REFROIDISSEMENT NORMAUX ET DES CIRCUITS DE REFROIDISSEMENT DE SECOURS DES CENTRALES NUCLEAIRES

H. PROCACCIA
Electricité de France,
Département Etudes et essais thermiques,
Saint-Denis,
France

Abstract–Résumé

PROBABILITY OF FAILURE OF NORMAL AND EMERGENCY COOLANT CIRCUITS IN NUCLEAR POWER STATIONS.

The main difficulty encountered in any evaluation of the reliability of complex systems is that of obtaining statistical data which justify sufficient confidence. It is often enough, however, to assume that the failure rates of equipment of similar design are identical in whatever areas of industry the equipment is used (provided that the operating loads are similar) in order to derive laws for extrapolating from the failure rates determined for conventional equipment at Electricité de France's non-nuclear power stations or taken from the literature. On the basis of this assumption the authors have estimated failure rates for equipment such as the steam generators associated with water- or sodium-cooled systems, large pumps for water or sodium and valves as a function of their unit capacity or of the operating stresses to which they are subjected. By way of example, the probability of failure of circuits for the removal of residual energy from a 1200 MW(e) fast reactor was evaluated by the fault tree method with repairable redundant systems; the average probability is of the order of 0.6×10^{-5} per year per reactor for all the circuits if the availability of the power station during each shut-down has to be ensured and 0.4×10^{-6} per year if one is considering only the safety of the cooling system.

PROBABILITE DE DEFAILLANCE DES CIRCUITS DE REFROIDISSEMENT NORMAUX ET DES CIRCUITS DE REFROIDISSEMENT DE SECOURS DES CENTRALES NUCLEAIRES.

La principale difficulté rencontrée au cours de toute évaluation de fiabilité de systèmes complexes consiste à recueillir des données statistiques présentant un degré de confiance suffisant. Néanmoins il suffit souvent d'admettre l'hypothèse que les taux de défaillance de matériels de conception analogue sont identiques, quelles que soient les industries où ils sont utilisés, à la condition que les charges d'exploitation restent similaires, pour estimer des lois d'extrapolation à partir de taux de défaillance relevés sur les matériels conventionnels des centrales classiques d'Electricité de France ou recueillis dans la littérature. A partir de cette hypothèse, on a évalué les taux de défaillance estimés de matériels tels que générateurs de vapeur, à eau ou à sodium, pompes de grandes dimensions, en eau ou en sodium, et vannes en fonction de leur puissance unitaire ou de leurs contraintes d'exploitation. A titre d'exemple d'application la probabilité de défaillance des circuits d'évacuation de l'énergie résiduelle d'un réacteur à neutrons rapides de 1200 MW(e) a été ensuite évaluée par la méthode des arbres de défauts avec systèmes redondants réparables: cette probabilité est de l'ordre de $0,6 \cdot 10^{-5}$ par année moyenne réacteur pour l'ensemble des circuits si l'on doit assurer la disponibilité de la centrale à chaque arrêt. Elle atteint $0,4 \cdot 10^{-6}$ par an si l'on considère uniquement la sûreté du refroidissement.

INTRODUCTION

La principale difficulté rencontrée au cours de toute évaluation de fiabilité de systèmes complexes consiste à recueillir des données statistiques présentant un degré de confiance suffisant.

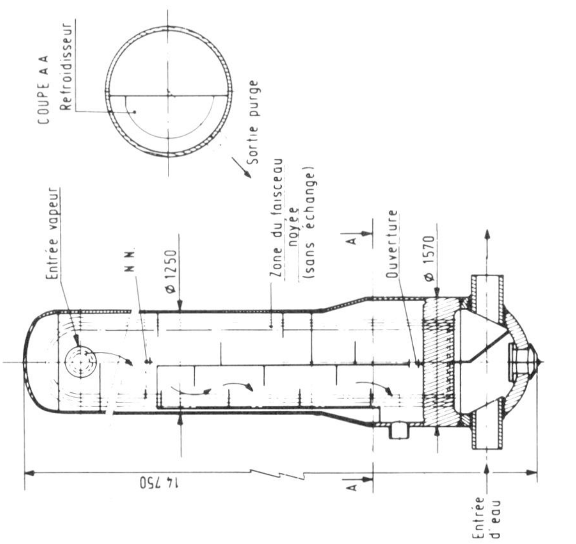

RECHAUFFEUR D'EAU H.P

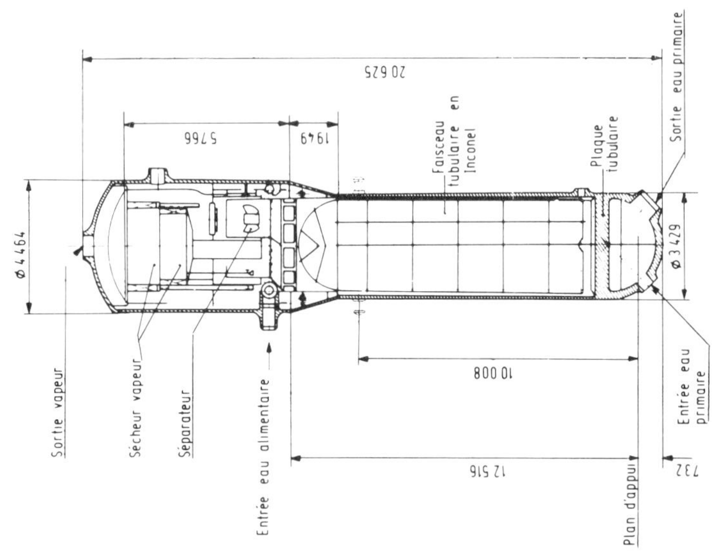

GENERATEUR DE VAPEUR DE CENTRALE P.W.R

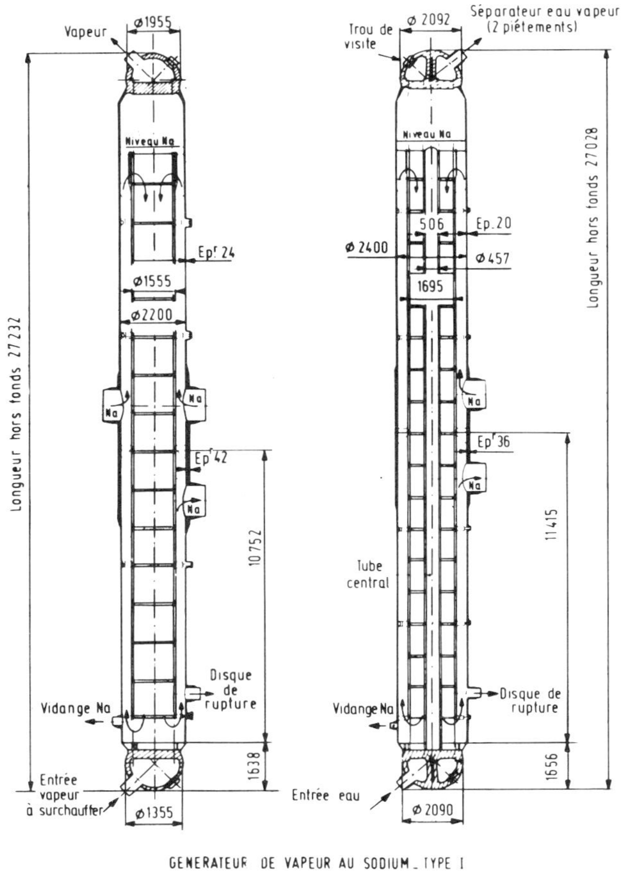

FIG. 1. Analogie entre générateurs de vapeur nucléaires et réchauffeurs (cotes en mm).

En fait, l'expérience statistique sur les principaux matériels d'une centrale de puissance est encore bien modeste ou porte sur des matériels dont la conception est actuellement dépassée, surtout lorsqu'il s'agit d'une technologie nouvelle, telle celle employée en énergie nucléaire.

Or, tant pour des études d'optimisation de filières que pour des études de sûreté de fonctionnement d'installations, l'analyse probabiliste reste une des méthodes les plus objectives de choix de conception.

Dans ces conditions il suffit d'admettre l'hypothèse que les taux de défaillance de matériels de conception analogue sont identiques, quelles que soient les industries où ils sont utilisés, à la condition que les charges d'exploitation restent similaires, pour estimer des lois d'extrapolation à partir de taux de défaillance relevés sur les matériels conventionnels des centrales classiques d'Electricité de France (EDF) ou recueillis dans la littérature.

La marge d'incertitude de toute étude prospective est difficile à évaluer, mais il est vraisemblable que les ordres de grandeur des taux de défaillance définis par cette méthode seront globalement respectés, et que, rapportés à l'ensemble d'un système complexe, les calculs de fiabilité ou de disponibilité resteront valables.

Ainsi, nous avons évalué les taux de défaillance estimés de matériels tels que générateurs de vapeur, à eau ou à sodium, pompes de grandes dimensions, en eau ou en sodium, et vannes en fonction de leur puissance unitaire ou de leurs contraintes d'exploitation.

A partir de ces données élémentaires, nous avons déterminé, à titre d'exemple d'application, la probabilité de défaillance des circuits normaux et des circuits de secours d'évacuation de l'énergie résiduelle d'un réacteur à neutrons rapides de 1200 MW(e) par la méthode des arbres de défauts avec systèmes redondants réparables.

1. ESTIMATION DES TAUX DE DEFAILLANCE DES PRINCIPAUX MATERIELS D'UNE CENTRALE NUCLEAIRE

1.1. Générateurs de vapeur

1.1.1. Critères analogiques

La conception des générateurs de vapeur nucléaires et particulièrement celle des générateurs de vapeur au sodium est trop éloignée de celle des générateurs de vapeur de centrales conventionnelles pour permettre une comparaison directe.

Par contre, cette conception est très proche de celle des échangeurs de chaleur du poste d'eau de toute centrale conventionnelle ou nucléaire (fig.1): boîtes à eau, plaques tubulaires, faisceaux de tubes, chicanages, etc.

Une étude statistique comparative du comportement respectif de ces matériels a donc été effectuée.

1.1.2. Evaluation des facteurs «effet de taille et pression de service»

1.1.2.1. Examen du comportement en exploitation des générateurs de vapeur des centrales à eau légère. L'Atomic Energy of Canada Limited (AEC) [1] a réalisé un recensement de tous les incidents de tubes survenus sur les générateurs de vapeur des centrales nucléaires dans le monde depuis leur mise en service jusqu'en 1971/1972.

Si l'on compare entre eux les différents types de générateurs de vapeur en éliminant les cas singuliers: générateurs de vapeur ayant subi un nombre anormal de pannes, ou ayant subi des pannes d'origine extérieure au fonctionnement propre du générateur de vapeur, ou générateur de vapeur n'ayant subi aucune panne (jeunesse), on constate:

- que leur température de fonctionnement est sensiblement homogène (280 à 300°C)
- que leur pression de service P peut être ramenée à deux niveaux:

$P < 100$ bar (généralement 80 bar)

$P \geqq 140$ bar

- enfin que la nuance des matériaux des tubes ne semble pas être un facteur important d'indisponibilité.

On en déduit qu'il est possible de déterminer les taux de défaillance élémentaires des générateurs de vapeur, d'une part en fonction de leur taille ou plus concrètement en fonction du nombre de tubes constitutifs (fig.2) ou de la surface totale d'échange (fig.3), d'autre part en fonction des deux niveaux de pression définis ci-dessus.

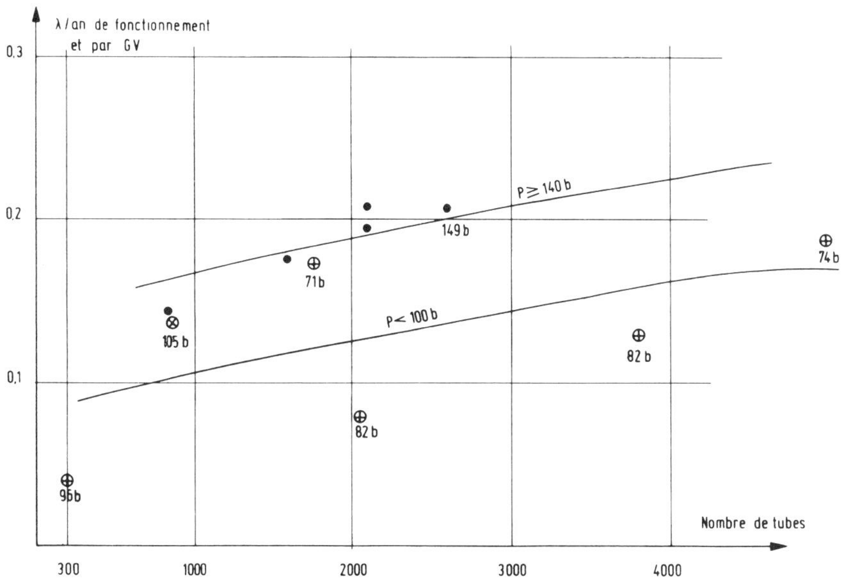

FIG. 2. Taux de défaillance des tubes de générateur de vapeur de centrales à eau légère en fonction du nombre de tubes.

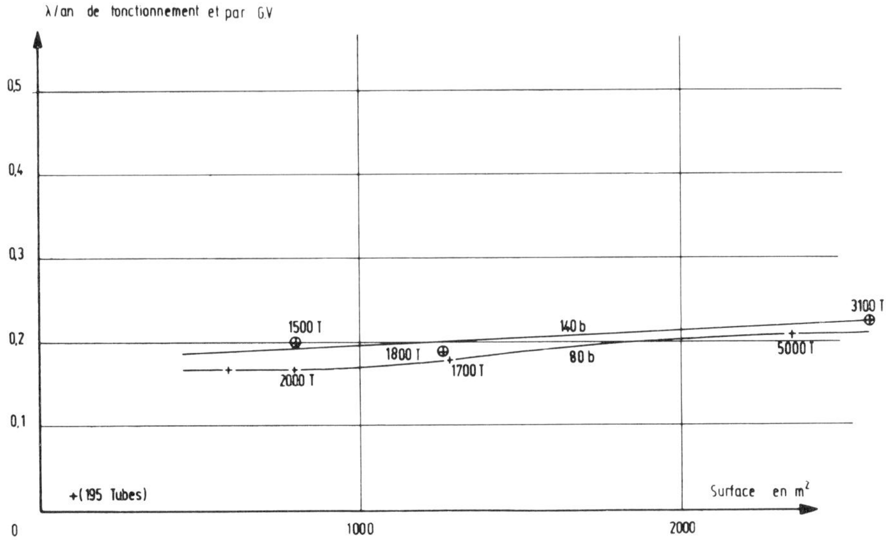

FIG. 3. Taux de défaillance des tubes de générateur de vapeur de centrales à eau légère en fonction de la surface d'échange.

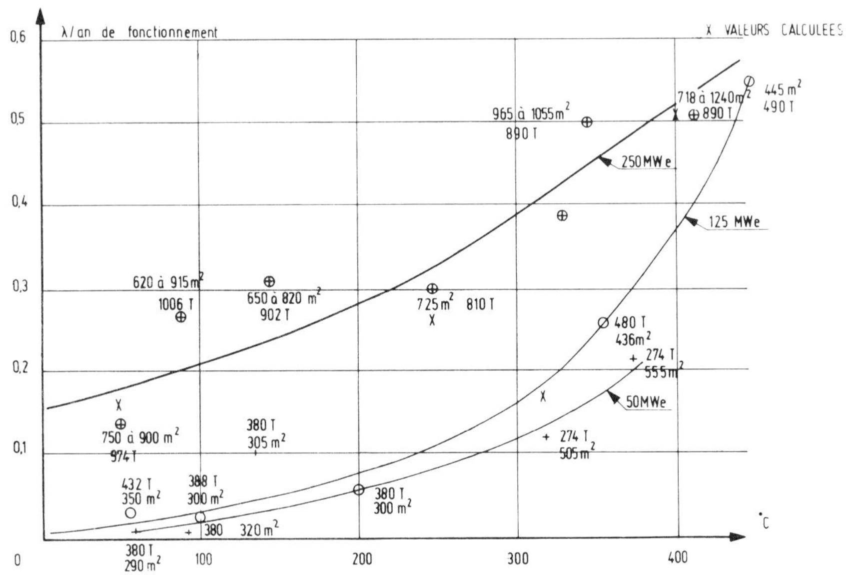

FIG. 4. Taux de défaillance des réchauffeurs en fonction de la température de la vapeur de soutirage.

1.1.2.2. Examen du comportement en exploitation des réchauffeurs de l'eau d'alimentation des centrales conventionnelles EDF. Une analyse détaillée du comportement en exploitation des réchauffeurs de l'eau d'alimentation des centrales conventionnelles a été réalisée dans l'étude [2].

Si l'on veut distinguer le poids relatif des défaillances de tubes par rapport à l'ensemble des matériels que comporte un réchauffeur, on constate dans cette étude que:

- 94% des défaillances sont dues à des défauts de tubes dans les réchauffeurs de petites dimensions (palier 50 MW(e), surfaces d'échange inférieures à 500 m^2)
- cette valeur n'atteint plus que 72% lorsque la taille des réchauffeurs croît (paliers 125 et 250 MW(e)).

1.1.3. Evaluation du facteur «effet de température»

L'étude précédente montre que le comportement en exploitation des réchauffeurs d'eau est fortement conditionné par la température haute du fluide chaud, ainsi que le montre la figure 4.

On déduit de cette figure l'importance relative du facteur «effet de température» sur l'indisponibilité moyenne d'un réchauffeur.

1.1.4. Formule de corrélation entre les statistiques relevées sur le comportement des générateurs de vapeur des centrales à eau légère et les statistiques relevées sur le comportement des réchauffeurs de l'eau d'alimentation

Westinghouse [3], à la suite de l'observation du comportement en exploitation des générateurs de vapeur fabriqués sous sa licence, a établi une formule empirique donnant le nombre estimé de défaillances annuelles d'un type de générateur en fonction de sa surface et du nombre de tubes (donc en fonction de la taille des appareils):

$$\lambda = \frac{5\ N\ t}{10^9} + \frac{S\ t}{10^8} \qquad (1)$$

avec:

λ = taux de défaillance
N = nombre de tubes
S = surface d'échange
t = temps.

Compte tenu des remarques précédentes, il y a lieu de corriger cette formule, d'une part de façon à assurer l'homogénéité avec les statistiques AEC (Canada) et EDF et, d'autre part, de manière à pouvoir l'extrapoler pour des conditions de fonctionnement différentes.

Nous avons ainsi établi la formule suivante:

$$\lambda(t) = A\ t(N+S)\ (1+\alpha\sqrt{P})\ (1+\beta\theta^2) \qquad (2)$$

Dans cette relation:

λ est le taux de défaillance unitaire

N le nombre de tubes

S la surface d'échange en m^2

t le temps (= 8760 heures si l'on veut connaître le taux de défaillance annuel)

θ la température haute de fonctionnement

P la pression de service

A un coefficient égal à $1,28 \cdot 10^{-8}$ pour des échangeurs dont la surface d'échange est supérieure à 500 m^2, ou à $1,06 \cdot 10^{-8}$ dans le cas opposé

α un coefficient égal à 10^{-2}

β un coefficient égal à $0,7 \cdot 10^{-5}$.

1.1.5. Validation de la formule: calage des valeurs calculées par rapport aux valeurs déterminées statistiquement pour les réchauffeurs d'eau

Nous avons représenté sur la figure 4 les valeurs des taux de défaillance estimés par la relation ci-dessus pour différents réchauffeurs.

On peut constater la validité du recoupement avec les valeurs déterminées statistiquement.

1.1.6. Estimation du taux de défaillance d'un générateur de vapeur de centrale à eau légère de 900 MW(th) (type Fessenheim, trois boucles)

Le taux de défaillance d'un générateur de vapeur de 900 MW(th) est estimé à environ 1 à 1,2 par an (fuite de tube et incidents sur les auxiliaires du générateur de vapeur compris) en adoptant les hypothèses de calcul (taux d'utilisation 78%) précédentes.

1.1.7. Estimation du taux de défaillance d'un générateur de vapeur au sodium de 750 MW(th) (type Babcock Atlantique)

Le module de générateur de vapeur à tubes hélicoïdaux proposé par Babcock Atlantique pour la centrale à neutrons rapides de 1200 MW(e) de Creys-Malville est du type à passage unique intégral. La puissance thermique de 750 MW est transmise par l'intermédiaire de 387 tubes de 25 mm de diamètre extérieur, en Incoloy 800 [4].

Le taux de défaillance estimé par la relation ci-dessus est de λ = 0,82 par an de fonctionnement, soit de l'ordre de 0,64 par année calendaire (taux d'utilisation 78%).

Cette valeur correspond à celle prise en compte par Westinghouse pour les études de disponibilité de son projet de centrale à neutrons rapides de 1000 MW(e): 0,64 par an [5] (par contre pour un générateur de vapeur de 1300 MW(th)), valeur estimée à partir du résultat d'exploitation des générateurs de vapeur des centrales à eau légère.

Elle reste en accord avec le résultat de l'évaluation probabiliste de General Electric [6]: 0,7 par an.

Elle est plus élevée, mais toujours du même ordre de grandeur que celle évaluée par Atomics International [7] pour son projet de 1000 MW(e): 0,14 par an.

Enfin, elle est légèrement plus faible que celle constatée par le Liquid Metal Engineering Center [8] sur les installations expérimentales en exploitation: 1 par an.

Si l'on applique la même formule pour évaluer le taux de défaillance d'un générateur de vapeur du type Stein on trouve un résultat équivalent: 1 par an.

1.1.8. Estimation du temps moyen de réparation

La comparaison directe entre générateurs de vapeur de centrales conventionnelles et de centrales nucléaires est sujette à caution. Toutefois, les seules sources statistiques réelles en notre possession proviennent de l'exploitation des centrales conventionnelles EDF [9]: les taux moyens de réparation des générateurs de vapeur sont de l'ordre de 270 à 300 heures équivalentes pleine puissance. Les analyses théoriques formulées par les constructeurs américains sur les générateurs de vapeur de centrales nucléaires donnent des valeurs équivalentes: 10 jours, ce qui laisse supposer que la difficulté d'accessibilité à l'élément défaillant reste du même ordre de grandeur quel que soit le type de générateur de vapeur.

Un temps moyen de réparation de l'ordre de 10 à 12 jours peut donc être adopté pour les générateurs de vapeur nucléaires.

1.2. Evaluation du taux de défaillance des pompes de grandes dimensions

1.2.1. Pompes à eau

1.2.1.1. Formulation du taux de défaillance: effet de taille et complexité de la pompe. L'éventail de pompes à eau utilisées dans une centrale EDF est très large.

D'autre part aux différents paliers techniques des tranches EDF correspond un vaste échantillonnage dimensionnel de pompes (fig.5).

On peut donc estimer (fig.6) l'importance de l'effet de taille au niveau des taux de défaillance élémentaire des pompes, par une analyse statistique du comportement en exploitation de toutes les catégories de pompes utilisées à EDF: pompes d'extraction, pompes de reprise, pompes alimentaires, pompes de circulation, etc. [10].

On peut remarquer que le taux de défaillance élémentaire λ_0 d'une pompe à eau peut s'exprimer relativement simplement en fonction de sa puissance unitaire W_0 (effet de taille) et de sa complexité technologique par la relation:

$$\lambda_0 = 2 \cdot 10^{-5} W_0^{1,9} t \frac{\Delta P}{180} \qquad (3)$$

avec $0 < W_0 < 3{,}5$, exprimé en MW, et t le temps, exprimé en heures. ΔP représente la hauteur manométrique de la pompe en bars.

1.2.1.2. Application aux pompes primaires des centrales à eau légère (type Fessenheim). Pour une puissance unitaire de 5325 kW le taux de défaillance estimé d'une pompe primaire d'une centrale à eau légère du

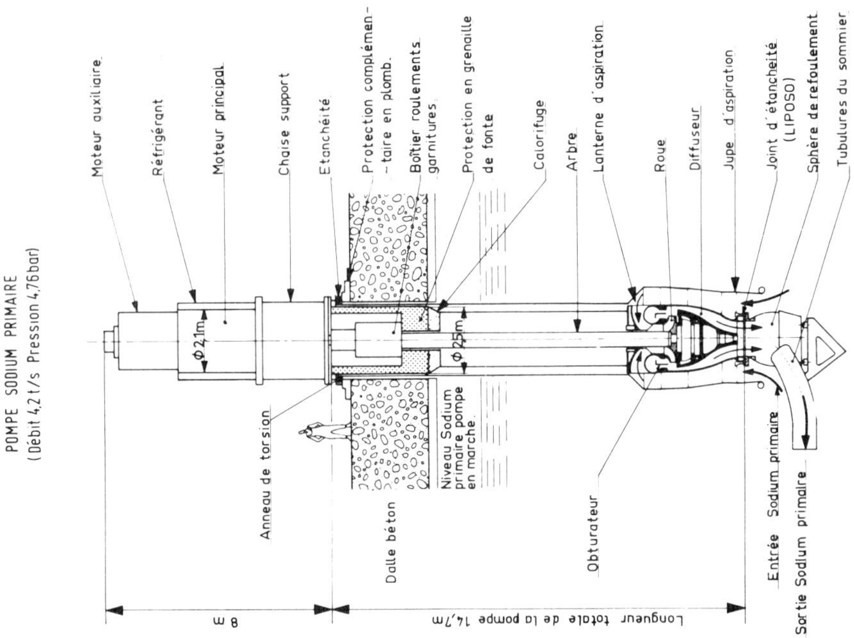
POMPE SODIUM PRIMAIRE
(Débit 4,2 t/s Pression 4,76 bar)
Moteur auxiliaire
Réfrigérant
Moteur principal
Chaise support
Etanchéité
Protection complémen-
-taire en plomb.
Boîtier roulements
garnitures
Protection en grenaille
de fonte
Calorifuge
Arbre
Lanterne d'aspiration
Roue
Diffuseur
Jupe d'aspiration
Joint d'étancheité
(LIPOSO)
Sphère de refoulement
Tubulures du sommier
Φ 2,1 m
Φ 2,5 m
Anneau de torsion
Dalle béton
Niveau Sodium
primaire pompe
en marche
Obturateur
Entrée Sodium primaire
Sortie Sodium primaire
8 m
Longueur totale de la pompe 14,7 m

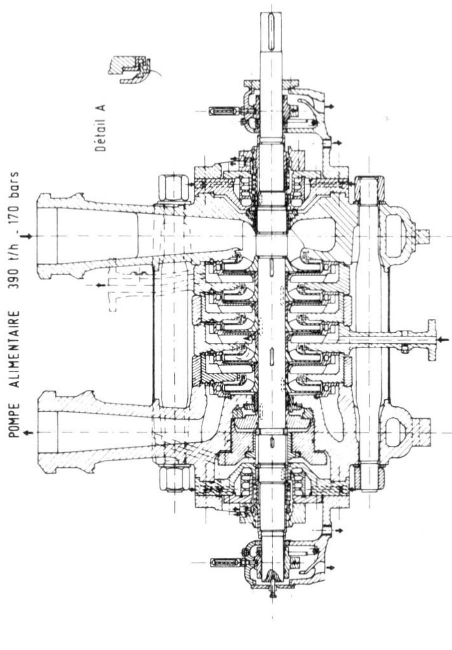
POMPE ALIMENTAIRE 390 t/h 170 bars
Détail A

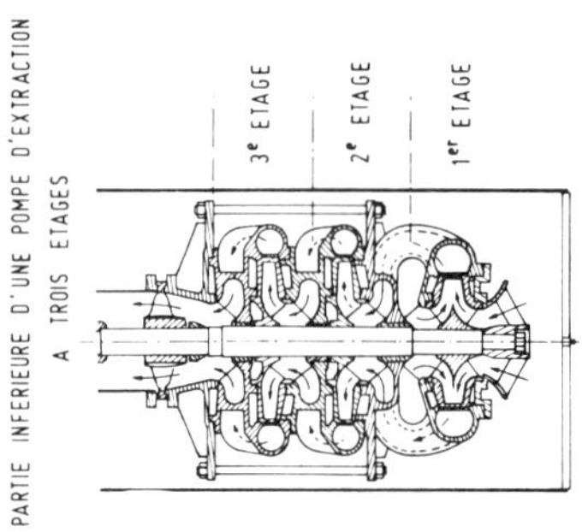
PARTIE INFERIEURE D'UNE POMPE D'EXTRACTION
A TROIS ETAGES
3e ETAGE
2e ETAGE
1er ETAGE

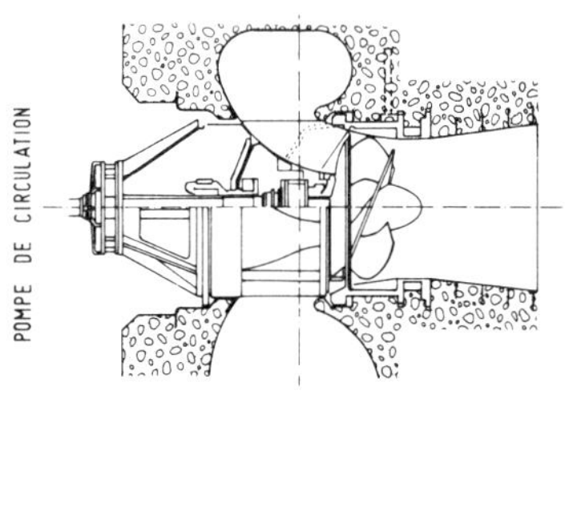
POMPE DE CIRCULATION

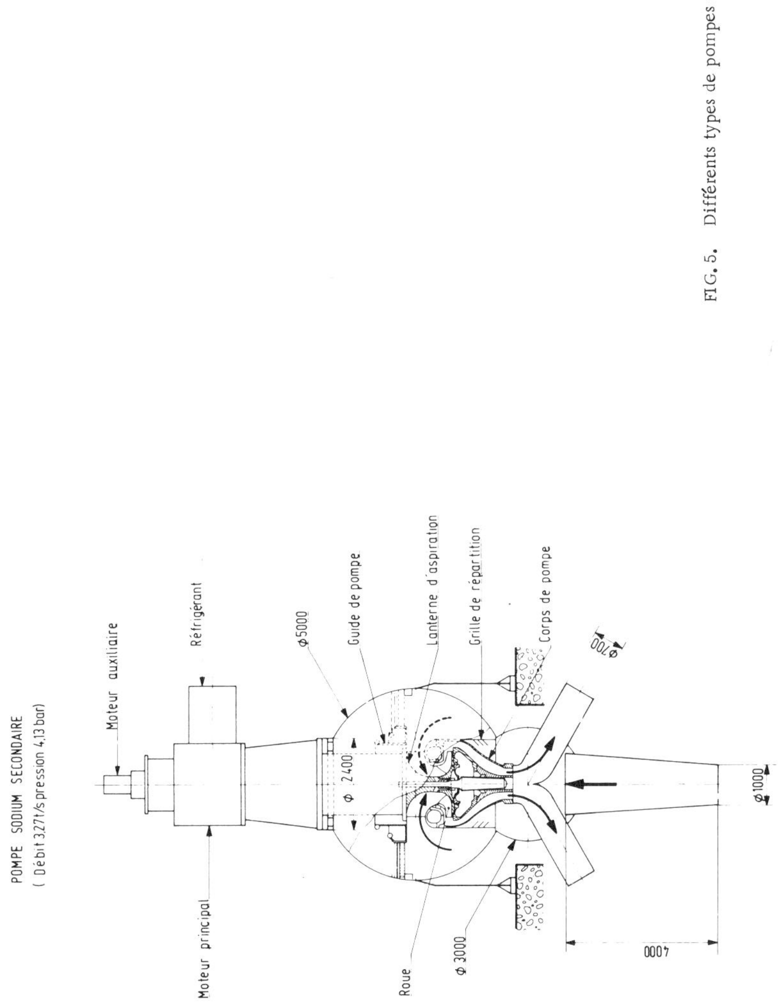

FIG. 5. Différents types de pompes utilisées à EDF.

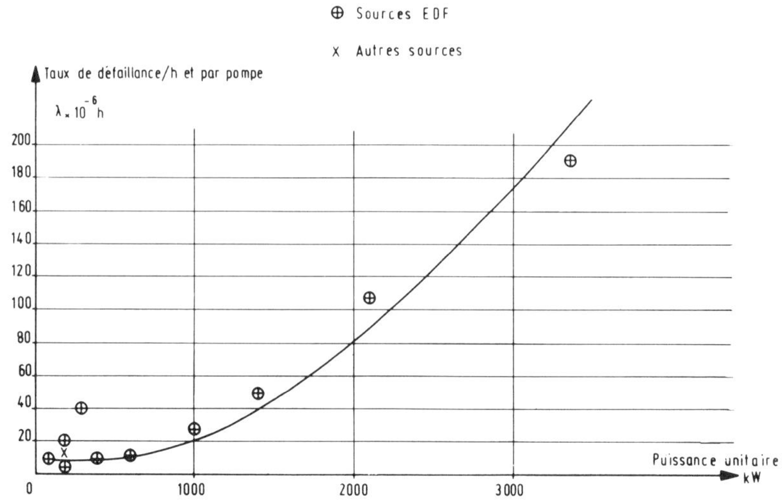

FIG. 6. Pompes à eau: effet de taille.

type Fessenheim sera de l'ordre de 0,23 par an de plein fonctionnement, soit de 0,18 par année calendaire.

Si l'on tient compte du taux de défaillance du moteur d'entraînement ces taux de défaillance seront respectivement de 0,45 et de 0,35 par an.

1.2.2. Pompes à sodium

Bien que le sodium à haute température puisse être assimilé à de l'eau (nombre de Prandlt, viscosité équivalents), on ne peut comparer directement les pompes de puissance à sodium aux pompes de puissance à eau, les premières présentant l'avantage d'être plus simples du fait de la faible hauteur manométrique nécessaire à la circulation du sodium: elles sont généralement centrifuges à roue simple et la plupart des problèmes spécifiques au pompage du sodium sont résolus et ne posent pas de difficultés nouvelles si la taille des pompes augmente, sauf en ce qui concerne la taille des pièces moulées.

Si l'on se réfère aux expériences de démarrage des nouvelles centrales à neutrons rapides Phénix et PFR (Prototype Fast Reactor) (remplacement des pompes primaires), on risque d'être pessimiste lorsqu'il s'agit de définir un taux de défaillance prévisible sur ces matériels.

En fait, on peut encore considérer les pompes incriminées comme des prototypes et estimer que les incidents qu'elles ont subi ne sont que des maladies de jeunesse, eu égard au fonctionnement des pompes des autres installations d'essai: Rapsodie, CGVS (circuit générateur de vapeur sodium), etc.

Pour cette raison on a représenté sur la figure 7 les estimations de taux de défaillance de la partie mécanique des pompes à sodium, estimations données par différents instituts.

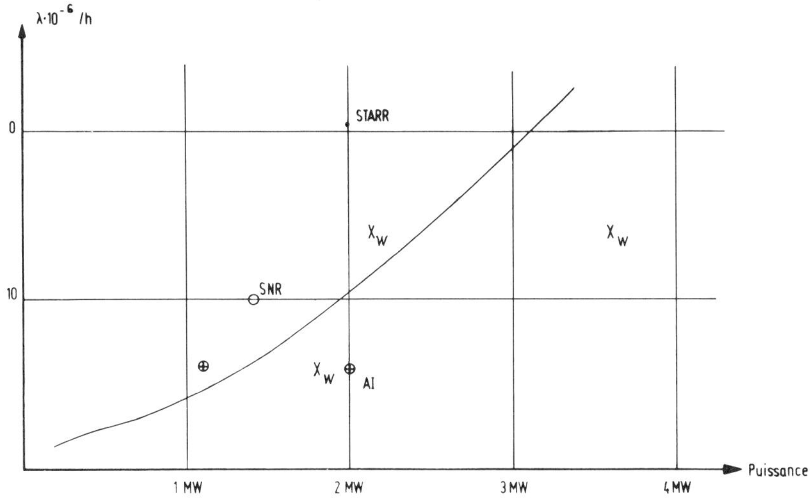

FIG. 7. Taux de défaillance estimé des pompes à sodium (uniquement parties mécaniques propres à la pompe). W = Westinghouse (X_W); AI = Atomics International (⊕); STARR = étude de l'Université de Californie, Los Angeles (UCLA) (●); SNR = réacteur rapide allemand (○).

On constate sur cette figure que les taux de défaillance estimés sont très faibles par rapport à ceux relevés sur les pompes à eau, et qu'une corrélation est possible à partir de la connaissance du taux de défaillance de ces dernières à une puissance donnée λ_{eau}:

$$\lambda_{Na} = \lambda_{eau} \cdot \frac{\Delta P_{na}}{P_{eau}} \cdot \frac{T_{Na}}{T_{eau}} \tag{4}$$

pour des puissances comprises entre 1 et 3,5 MW

λ_{Na}	étant le taux de défaillance de la pompe à sodium pour la puissance W
ΔP_{Na} et T_{Na}	sont respectivement la hauteur manométrique en bar et la température en °C du sodium alimentant la pompe à sodium
P_{eau} et T_{eau}	sont respectivement la pression et la température de l'eau alimentant la pompe à eau, soit P de l'ordre de 180 bars et T de l'ordre de 180°C pour le niveau de puissance considéré.

Cette relation peut s'écrire finalement (λ_{Na} à la puissance W_0):

$$\lambda_{Na} = W_0^{1,9} \cdot t \cdot \Delta P_{Na} \cdot T_{Na} \cdot 10^{-7} \tag{5}$$

La figure 7 permet aussi de constater que la dispersion des résultats est assez vaste entre les instituts, mais que les ordres de grandeur sont respectés si l'on excepte les valeurs très élevées données par le LMEC.

Remarque. La relation (5) ne donne le taux de défaillance que de la partie mécanique de la pompe elle-même.

Si l'on veut connaître la défaillance de la fonction de pompage il y a lieu de rajouter à cette valeur les taux de défaillance:

- de l'entraînement de la pompe: moteur (voir ci-après), variateur de vitesse, sources électriques
- des accessoires de la pompe: arbre d'entraînement, dont la longueur est bien supérieure à celle d'une pompe à eau (taux de défaillance estimé entre 8 et $20 \cdot 10^{-6}$ h^{-1}, clapet anti-retour, liaison pompe sommier)
- de l'instrumentation, de la régulation et du contrôle commande.

C'est ainsi que des statistiques effectuées sur la fonction de pompage de Rapsodie et de Fortissimo [11] font apparaître un taux moyen de défaillance global de $155 \cdot 10^{-6}$ h^{-1} par pompe pendant la vie utile de la centrale.

1.2.3. Turbine d'entraînement

Le taux de défaillance d'une turbine d'entraînement est légèrement supérieur à celui d'un moteur d'entraînement, ainsi que le montrent les diverses statistiques [12] ($25 \cdot 10^{-6}$ h^{-1} pour 3 MW).

1.2.4. Moteurs d'entraînement

Différents auteurs donnent des estimations de taux de défaillance pour les moteurs de puissance. Ces estimations sont résumées dans la note [12].

On constate sur la figure 8 que les estimations de Westinghouse sont intermédiaires par rapport aux estimations des autres instituts, d'où nous en déduisons le tracé d'une courbe moyenne:

$$\lambda_e = 12 \cdot 10^{-6} \, W^{0,6} \tag{6}$$

W étant la puissance unitaire de la pompe en MW.

La disparité des estimations fait que l'on doit accorder une confiance toute relative à cette courbe.

1.2.5. Application à Super Phénix

Le taux de défaillance d'une pompe primaire de Super Phénix (3,3 MW), estimé à partir des valeurs précédentes, sera de l'ordre de 0,57 a^{-1} (0,22 pour la pompe, 0,175 pour l'arbre de transmission, 0,175 pour le moteur d'entraînement), soit 0,45 par année calendaire.

2. EVALUATION DU TAUX DE DEFAILLANCE DES VANNES DE GRANDES DIMENSIONS

Un tour d'horizon des estimations de taux de défaillance de vannes relevées dans la littérature ou établies statistiquement à partir du comportement en exploitation des vannes EDF montre une grande disparité des résultats [12].

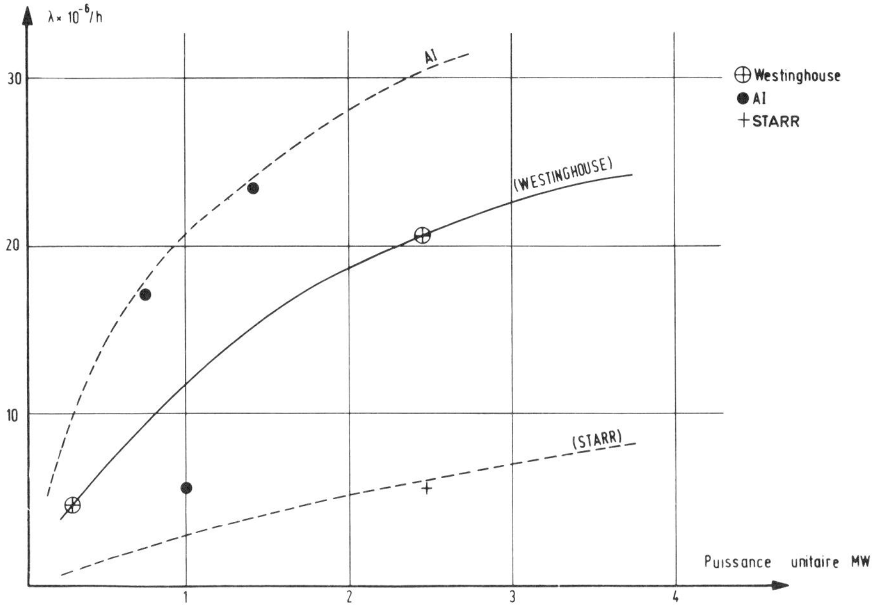

FIG. 8. Taux de défaillance estimé des moteurs électriques.

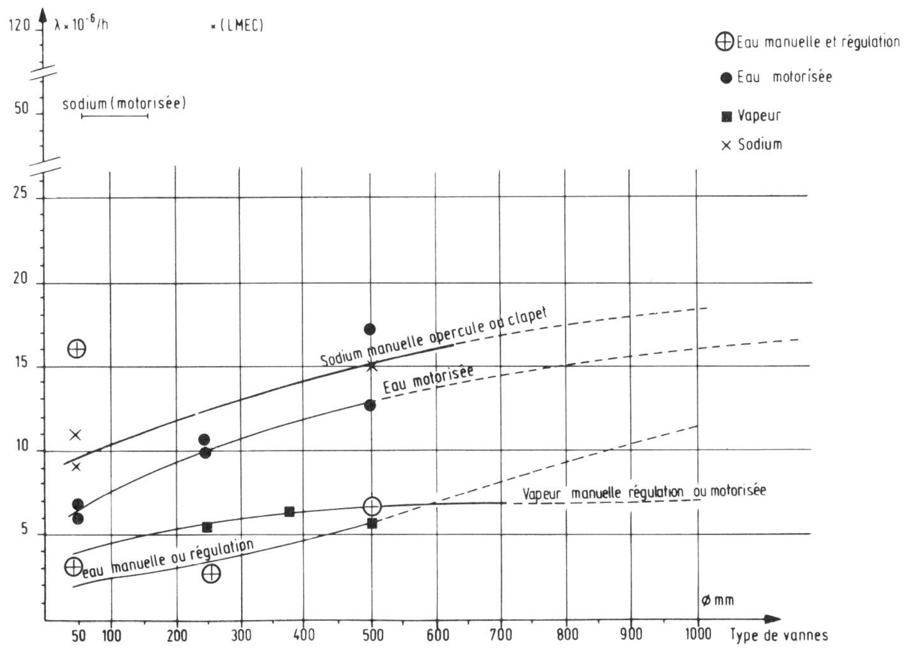

FIG. 9. Taux de défaillance des vannes.

Si l'on excepte les valeurs excessivement élevées ou excessivement faibles, on constate sur la figure 9 que l'on peut distinguer un comportement différent entre:

- les vannes manuelles et les vannes de régulation, dont le taux de défaillance croît rapidement avec les dimensions de la vanne et les caractéristiques thermodynamiques de l'eau
- les vannes motorisées, dont le taux de défaillance unitaire est bien plus élevé que celui constaté sur la première catégorie de vannes, du fait de leur fonctionnement non permanent
- d'autre part que cette distinction n'apparaît pas pour les vannes à vapeur, pour lesquelles le taux de défaillance reste sensiblement constant quel que soit le type de vanne ou ses dimensions; on retrouve, par contre, cette distinction pour les vannes à sodium, mais l'échantillonnage statistique est très faible.

3. APPLICATION PRATIQUE: ESTIMATION DE LA PROBABILITE DE DEFAILLANCE DES CIRCUITS DE REFROIDISSEMENT D'UN REACTEUR RAPIDE DE 1200 MW(e)

Dans une centrale à neutrons rapides l'acquisition des données de base nécessaires à tout calcul probabiliste est d'autant plus difficile que l'expérience technique est faible et que l'extrapolation à des niveaux de puissance élevés conduit bien souvent à des paris technologiques.

Plus que pour n'importe quelle autre filière, la méthode d'extrapolation proposée ci-dessus reste donc la seule source de données possible.

On remarque toutefois que la dispersion des taux de défaillance du matériel entre différentes références bibliographiques reste généralement homogène, et que les extrapolations basées sur la sanction de l'expérience, ou les extrapolations théoriques, sont sensiblement du même ordre de grandeur.

Par contre, il reste très difficile de prendre en compte l'aspect prototype de certains matériels et l'arbre de défauts donné ci-après pour les circuits d'évacuation de l'énergie résiduelle de la centrale de Creys-Malville correspond plus à la centrale du même modèle qui suivra Creys-Malville.

3.1. Description des circuits de refroidissement de Creys-Malville (Super Phénix)

Les circuits de refroidissement de la centrale de Creys-Malville comprennent (fig.10):

a) Les circuits de refroidissement principaux, soit:

- le circuit primaire intégré, avec quatre pompes primaires alimentées par un alternateur à vitesse variable, équipé d'un volant d'inertie, et secourues par diesels, et quatre échangeurs intermédiaires
- quatre boucles secondaires indépendantes, avec quatre pompes exemptes de volants d'inertie et secourues par diesel
- quatre générateurs de vapeur
- deux groupes turbo-alternateurs avec leur poste d'eau associé: chaque poste d'eau comprend une turbo-pompe alimentaire secourue

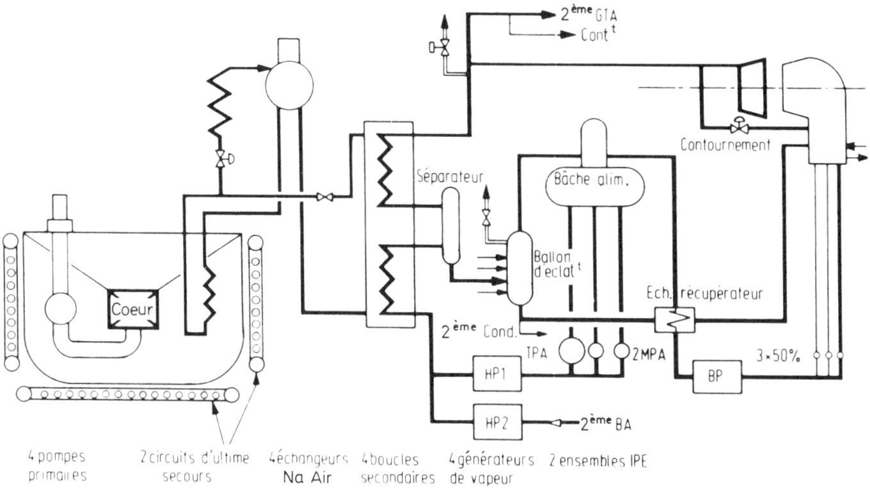

FIG. 10. Circuits de refroidissement de la centrale de Creys-Malville.

par deux moto-pompes à débit réduit (12,5% charge nominale chaudière); deux chaudières auxiliaires assurent le démarrage des groupes et l'alimentation en vapeur des turbo-pompes alimentaires.

Les circuits de démarrage-arrêt, un par poste d'eau, sont constitués de dispositifs de contournement des turbines, et des circuits de retour d'eau au condenseur: quatre ballons séparateurs associés aux générateurs de vapeur, un ballon d'éclatement unique pour les deux groupes, un échangeur-récupérateur par groupe.

Enfin des soupapes de rejet à l'atmosphère assurent l'évacuation transitoire de l'excès de vapeur, en particulier lors de l'arrêt d'urgence après chute de barres avec condenseurs indisponibles.

b) Les moyens chaudières, constitués de quatre échangeurs sodium-air disposés sur les circuits secondaires en parallèle avec les générateurs de vapeur. Ils assurent l'évacuation de la puissance résiduelle du cœur, soit lors d'un manque de tension réseau avec échec de l'ilotage, soit lors d'une perte d'alimentation en eau des générateurs de vapeur par suite d'un incident sur l'installation de production d'électricité (IPE).

La puissance évacuée par l'air et le sodium en convection naturelle est de 0,7% de la puissance nominale (5 MW(th) par échangeur). La mise en service manuelle des ventilateurs auxiliaires permet un doublement de la puissance extraite (11 MW(th) par échangeur).

Notons que ces échangeurs permettent des arrêts autonomes chaudière autorisant les révisions de l'IPE, ou les opérations d'entretien des générateurs de vapeur.

c) Enfin, en cas de perte totale des boucles secondaires ou de fusion partielle du cœur (défaut de mode commun le plus grave), le circuit d'ultime secours permet d'évacuer l'énergie résiduelle.

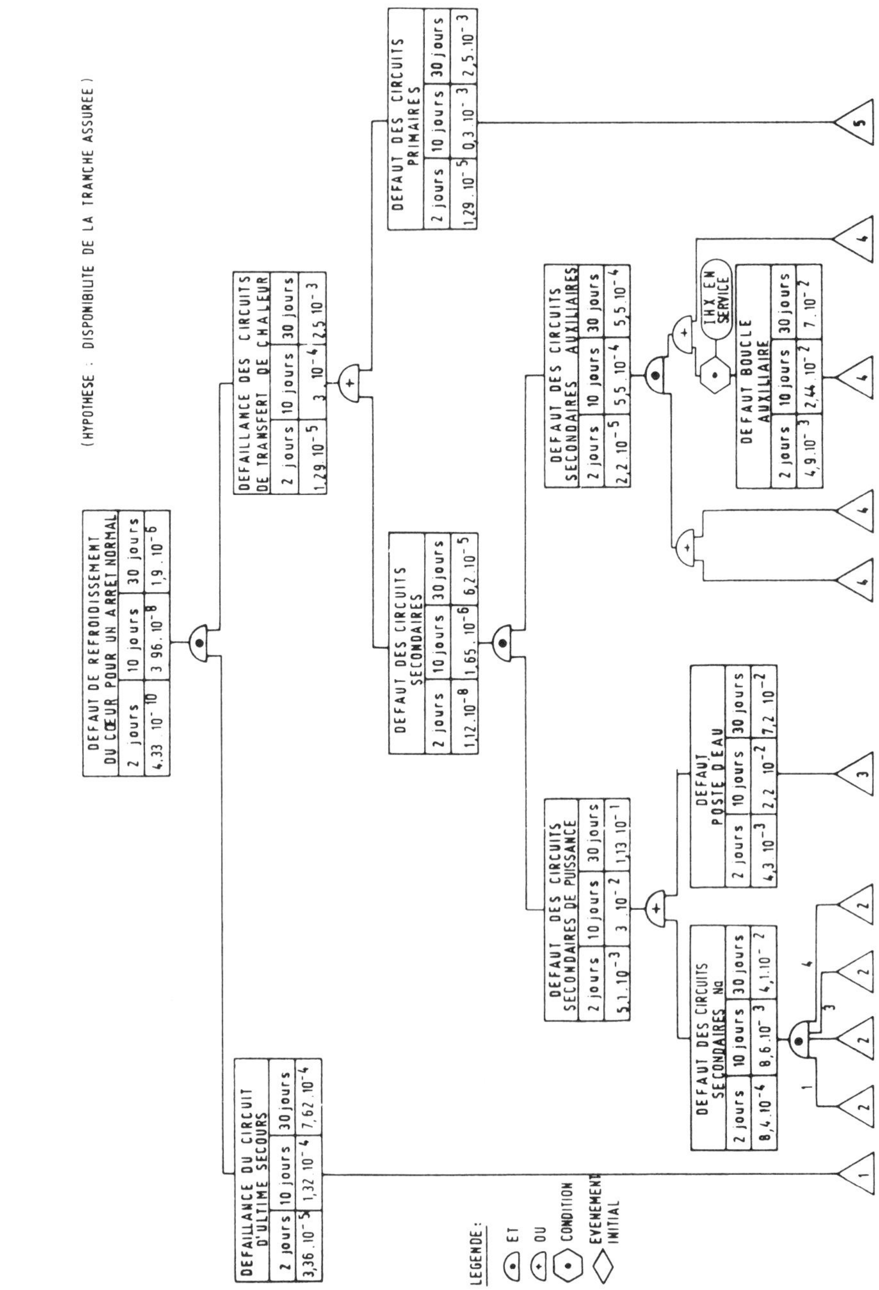

FIG. 11. Arbre de défauts des circuits de refroidissement de Super Phénix.

Ce circuit d'ultime secours est constitué de deux boucles d'eau indépendantes capables chacune d'assurer à elle seule l'évacuation de l'énergie résiduelle (si l'on excepte le cas très improbable de la fusion généralisée du cœur).

Dans l'hypothèse de panneaux de tubes et dans les cas exceptionnels où la température de la cuve primaire atteindrait 700°C, la puissance extraite par un circuit en service serait de l'ordre de 11 MW. Elle atteindrait 13 MW pour deux circuits en service.

En fonctionnement normal, la puissance thermique évacuée est de 2,5 à 3 MW suivant que un ou deux circuits sont en service.

d) La permanence de la circulation de l'eau dans chaque boucle est assurée par:
- le doublement des pompes et des échangeurs
- la possibilité de reprise en secours diesel de ces pompes
- la possibilité de raccorder des moto-pompes de secours alimentées par l'eau du Rhône.

3.2. Etude de la fiabilité des circuits de refroidissement

3.2.1. Méthode de calcul

Le nombre d'éléments à considérer dans une telle étude étant très élevé (de l'ordre de 500), il est nécessaire d'aborder le problème sous forme simplifiée. La méthode des arbres de défauts a été utilisée: elle permet de déterminer la probabilité de défaillance d'un ensemble complexe à partir d'un état initial de cet ensemble. Cette méthode a l'avantage de mettre en évidence les éléments critiques du système, c'est-à-dire ceux dont la probabilité de défaillance est élevée, et pour lesquels les conséquences de la défaillance entraînent une situation dangereuse du système; il suffit ensuite d'élaborer des arbres de défauts pour différents états initiaux possibles de ces seuls éléments critiques, ce qui limite le nombre de cas à considérer, sans oublier de prendre en compte les défauts de mode commun.

Ces arbres de défauts ont été établis pour différentes durées d'arrêts caractéristiques.

3.2.2. Analyse des résultats

Le sommet des arbres de défauts des circuits de refroidissement de la centrale de Creys-Malville est donné à la figure 11 pour les arrêts normaux de différentes durées. Ces arbres de défauts ont été établis pour deux hypothèses d'état de la centrale. D'une part la probabilité de défaillance des circuits de refroidissement a été calculée de manière que la disponibilité de la centrale soit assurée à son redémarrage: l'arrêt n'entraîne, en principe, aucune rupture de gaine, cet arrêt étant réalisé par l'utilisation d'au moins deux boucles de refroidissement (symétrie des écoulements de sodium dans le cœur); dans ces conditions la probabilité de défaillance des circuits est de $1,8 \cdot 10^{-4}$ pendant la durée de vie de la centrale, soit $0,6 \cdot 10^{-5}$ par année moyenne réacteur. D'autre part, cette probabilité a été établie en évaluant la sûreté du refroidissement de l'installation, quelques ruptures de gaines pouvant résulter d'un

ARRETS NORMAUX ET INCIDENTS INTEMPESTIFS		
NOMBRE		
CHAUDS	SEMI - CHAUDS	FROIDS
640	400	80
PROBABILITE TOTALE DE DEFAILLANCE DU REFROIDISSEMENT		
2.10^{-9}	$1,28.10^{-9}$	$1,2.10^{-6}$

ARRET SUR INCIDENTS CIRCUITS PRIMAIRES		
NOMBRE		
CHAUDS	SEMI - CHAUDS	FROIDS
90	90	10
PROBABILITE DE DEFAILLANCE		
$1.4.10^{-9}$	$2,2.10^{-8}$	$1,6.10^{-8}$

ARRET SUR INCIDENTS CIRCUITS SECONDAIRES		
NOMBRE		
CHAUDS	SEMI - CHAUDS	FROIDS
130	90	10
PROBABILITE DE DEFAILLANCE		
$1,3.10^{-7}$	$2,0.10^{-6}$	$1,6.10^{-8}$

ARRET SUR INCIDENTS IPE		
NOMBRE		
CHAUDS	SEMI - CHAUDS	FROIDS
42	58	20
PROBABILITE DE DEFAILLANCE		
$1,5.10^{-10}$	$3,2.10^{-7}$	$1,46.10^{-6}$

ARRET SUR INCIDENT RESEAU		
NOMBRE		
CHAUDS	SEMI - CHAUDS	FROIDS
45	4	1
PROBABILITE DE DEFAILLANCE		
$3,4.10^{-8}$	$2.8.10^{-8}$	$4.9.10^{-6}$

PERTE 1 FIL DU CIRCUIT D'ULTIME SECOURS	
NOMBRE D'ARRETS	PROBABILITE DE DEFAILLANCE
1	2.10^{-6}

PROBABILITE MOYENNE DE DEFAILLANCE DE LA FONCTION DE REFROIDISSEMENT DU CŒUR AU COURS DES ARRETS DE LA CENTRALE		
PENDANT LA VIE DE LA CENTRALE	PAR ARRET	PAR AN
$1,21.10^{-5}$	$0,7.10^{-8}$	$0 4.10^{-6}$

FIG. 12. Probabilité de défaillance de la fonction de refroidissement du cœur compte tenu des différents types d'arrêt (hypothèse: sûreté de l'installation assurée).

refroidissement non symétrique du cœur (arrêt sur une boucle de refroidissement au moins); la probabilité de défaillance des circuits est alors de $1,25 \cdot 10^{-5}$ pendant la durée de vie de la centrale, soit $0,4 \cdot 10^{-6}$ par année moyenne réacteur (fig.12).

On constate d'une manière générale que les systèmes redondants ont une fiabilité bien supérieure à certains circuits passifs de fiabilité unitaire en principe élevée, que l'on songe rarement à doubler.

On constate ainsi:

- que sur les circuits d'ultime secours, les jeux de vannes et les tuyauteries présentent des taux de défaillance plus élevés que les ensembles pompes et échangeurs (doublés)
- que sur les circuits principaux sodium, la défaillance potentielle des tuyauteries, des vannes, des réservoirs d'expansion et des générateurs de vapeur est du même ordre de grandeur que les défaillances de pompes, compte tenu des redondances électriques de ces dernières
- et qu'enfin au niveau du poste d'eau la fiabilité globale des circuits de démarrage-arrêt dépend uniquement de la fiabilité de l'unique ballon d'éclatement.

CONCLUSIONS

Les lois d'estimation des taux de défaillance des gros composants des centrales nucléaires données dans ce rapport peuvent servir de banques de données primaires, lorsqu'il est difficile d'évaluer le taux de défaillance d'un matériel nouveau ou récemment utilisé en pratique.

Ces lois tiennent compte de l'effet de taille des matériels et aussi des différences de conditions d'exploitation.

L'application pratique effectuée à partir de ces lois montre que la probabilité moyenne annuelle de perte totale des circuits de refroidissement de la centrale de Creys-Malville, à la suite d'un arrêt normal ou accidentel, est estimée à environ $0,4 \cdot 10^{-6}$ par an ($1,25 \cdot 10^{-5}$ pendant la vie du réacteur).

Si l'on considère qu'il faille conserver au moins deux circuits de refroidissement en service au cours de cet arrêt pour éviter des dissymétries d'écoulement du sodium dans le cœur, susceptibles d'entraîner quelques ruptures de gaines (disponibilité de la centrale assurée), cette probabilité n'est plus que de $0,6 \cdot 10^{-5}$ par année moyenne.

Il faut noter que ces calculs ont été effectués en prenant en compte des événements de probabilité très faible: perte au moment de l'arrêt de tous les circuits de transfert de puissance (probabilité $2,7 \cdot 10^{-4}\ a^{-1}$), perte de tous les circuits de transfert de puissance et d'un circuit d'ultime secours (probabilité $10^{-4}\ a^{-1}$), perte du réseau d'une durée supérieure à 10 jours, etc.; on peut donc les considérer comme pessimistes.

REFERENCES

[1] STEVENS, P.D., GUILLE, Steam Generator Tube Failure: A World Survey of Water Cooled Nuclear Power Reactors to the End of 1971, Atomic Energy of Canada Ltd (1973) 21 p.

[2] CRAVERO, M., PROCACCIA, H., Etude générale de la fiabilité des tranches thermiques, Etude des incidents réchauffeurs à E.D.F. pendant la dernière décennie, Rapport EDF n°HF 0329/71/02 (1971) 33.

[3] CORNU, G., Westinghouse: Rapport de mission interne EDF (1971) 7.

[4] ROBIN, G., DUCHATELLE, L., BIRAULT, J., et al., Réacteur à neutrons rapides de 1200 MWe, Générateurs de vapeur et composants du circuit secondaire, Bull. Inf. Sci. Tech. (Paris) n°182 (juin 1973) 14.

[5] WESTINGHOUSE, 1000 MWe LMFBR, Conceptual Design, Task II and III, WARD 2000 97, Report prepared for Argonne National Laboratory (1969) 1.B.1 - 1.B.11.

[6] GENERAL ELECTRIC, Advanced Technology Department (BROWN, STITT, WILSON), An Estimate of LMFBR Steam Generator System Reliability and Availability.

[7] ATOMICS INTERNATIONAL, Conceptual Design Report Engineering Studies, vol.II, AI - AEC 12792, Report prepared for Argonne National Laboratory (1970).

[8] LIQUID METAL ENGINEERING CENTER, Failure Data Handbook for Nuclear Power Facilities, LMEC Memo 69-7, vol.I (1969).

[9] PROCACCIA, H., CRAVERO, M., DAVID, J., MATHIOT, M., Etude générale de la disponibilité des tranches thermiques conventionnelles, Analyse des incidents générateur de vapeur pendant la dernière décennie, Rapport EDF n°HF 0329/73/003 (1973) 16.

[10] FERRIOLE, G., Statistiques d'indisponibilité, Fiabilité d'auxiliaires, Note Interne EDF - SPT - EG 225/625 (1974).

[11] ARLAUD, J.C., «Application des méthodes d'analyse de la fiabilité au cas du réacteur Rapsodie», CSNI, Liverpool, 8-10 avril 1974.

[12] PROCACCIA, H., Taux de défaillance élémentaire des matériels mécaniques des centrales nucléaires à eau légère et à neutrons rapides, Lois d'extrapolation en fonction de l'effet de taille et des caractéristiques techniques d'utilisation des matériels, Rapport EDF n°F 329/74/029 (1974) 9.

A RELIABILITY-RISK MODELLING OF NUCLEAR RAD-WASTE FACILITIES

P.H. LEHMANN, A.A. EL-BASSIONI
Carnegie-Mellon University,
Pittsburgh, Pa.,
United States of America

Abstract

A RELIABILITY-RISK MODELLING OF NUCLEAR RAD-WASTE FACILITIES.

Rad-waste disposal systems of nuclear power sites are designed and operated to collect, delay, contain, and concentrate radioactive wastes from reactor plant processes such that on-site and off-site exposures to radiation are well below permissible limits. To assist the designer in achieving minimum release/exposure goals, a computerized reliability-risk model has been developed to simulate the rad-waste system. The objectives of the model are to furnish a practical tool for quantifying the effects of changes in system configuration, operation, and equipment, and for the identification of weak segments in the system design. Primarily, the model comprises a marriage of system analysis, reliability analysis, and release-risk assessment. Provisions have been included in the model to permit the optimization of the system design subject to constraints on cost and rad-releases. The system analysis phase involves the preparation of a physical and functional description of the rad-waste facility accompanied by the formation of a system tree diagram. The reliability analysis phase embodies the formulation of appropriate reliability models and the collection of model parameters. Release-risk assessment constitutes the analytical basis whereupon further system and reliability analyses may be warranted. Release-risk represents the potential for release of radioactivity and is defined as the product of an element's unreliability at time, t, and the radioactivity available for release in time interval, Δt. A computer code (RARISK) has been written to simulate the tree diagram of the rad-waste system. Reliability and release-risk results have been generated for cases which examined the process flow paths of typical rad-waste systems, the effects of repair and standby, the variations of equipment failure and repair rates, and changes in system configurations. The essential feature of this model is that a complex system like the rad-waste facility can be easily decomposed into its basic components and the probabilistic characteristics of these elements furnished with sufficient precision to permit a practical and efficient evaluation of the significant aspects of the rad-waste system's design, operation, release-risks, and cost.

I. INTRODUCTION

With the current energy crisis upon us, it is anticipated that we will experience a significant expansion in the development of nuclear power plants. Associated with this expansion we expect that large quantities of radioactive wastes will need to be processed and stored. These developments will increase the level of normal releases, as well as the risks associated with abnormal or accidental releases of radioactivity to the environment. Therefore, it is desirable that the risks of releases be minimized and more than compensated by the benefits derived from nuclear power.

In the United States as in other nuclear power developing nations, there are regulations governing the quantity of rad-waste discharges to the environs. In most cases, radionuclide concentrations of the discharges have been only a few percent of permissible limits. Hence, little emphasis has ever been placed on establishing mechanistic models for rad-waste system design and performance. Although probabilistic approaches have been applied to the design and safety analyses of the nuclear plants(1-4), quantitative methods for assessing reliability and the risks of radioactivity releases from rad-waste systems have received little or no attention. Therefore, to aid the rad-waste system designer in achieving maximum reliability and minimum release-risk goals, a reliability-risk methodology and computerized model have been developed to simulate the rad-waste system. The objectives of the methodology are to:

(1) provide a practical design and accident-prevention tool for quantifying the effect of changes in rad-waste system configurations, equipment, operations, and additional safeguards;
(2) furnish a computerized methodology which facilitates the identification of weak links in the system;
(3) provide the framework for incorporating optimization algorithms with constraints on system cost and release-risks; and
(4) introduce a measure of safety or risk margin by gauging the probabilistic characteristics of the system against predetermined minimum-release objectives.

II. THE RAD-WASTE SYSTEM

Radioactive waste processing systems at nuclear power sites are designed and operated to collect, delay, contain, and concentrate radioactive waste fluids from normal or accidental reactor plant processes. This is to insure that on-site and off-site exposures to radioactive releases are below permissible limits.

The rad-waste processing system is comprised principally of thousands of feet of piping and many types of valves, pumps, compressors, filters, and decontamination equipment. Present waste systems rely on radioactive decay-holdup and filtration for gaseous wastes, and decay-holdup, filtration, demineralization, evaporation and centrifugation for liquid wastes. Solid wastes are collected from evaporator bottoms, demineralizers, etc., for ultimate disposal.

A schematic of a typical rad-waste processing system for a pressurized water reactor is shown in Fig. 1. Other conceptual rad-waste treatment systems for light water reactor plants and other reactor concepts have been described and analyzed by the United States Energy Research and Development Administration(5). Although it has been observed that the character of rad-waste discharges varies from one reactor plant to another, the operating practices and waste treatment configurations for all rad-waste systems are similar.

Design Bases

The bases for the design and evaluation of the rad-waste system depends on a number of system characteristics. These characteristics include radioactive waste burdens, equipment design specifications, operational controls, and system lay-out. The specific information required involves:

(1) radioactivity source-terms for each input waste stream,
(2) detailed piping and process equipment diagrams,
(3) categorization of each equipment type as to capacity, flow rate, decontamination capability and cost, and
(4) performance tests and equipment maintenance history.

Design release objectives which represent an upper bound on permissible releases are established to set the preliminary design requirements for the rad-waste system. The basic criterion commonly employed for sizing the waste processing system is based on continuous operation assuming that fission products escape into the reactor coolant from fuel/clad elements experiencing 1% defects(6).

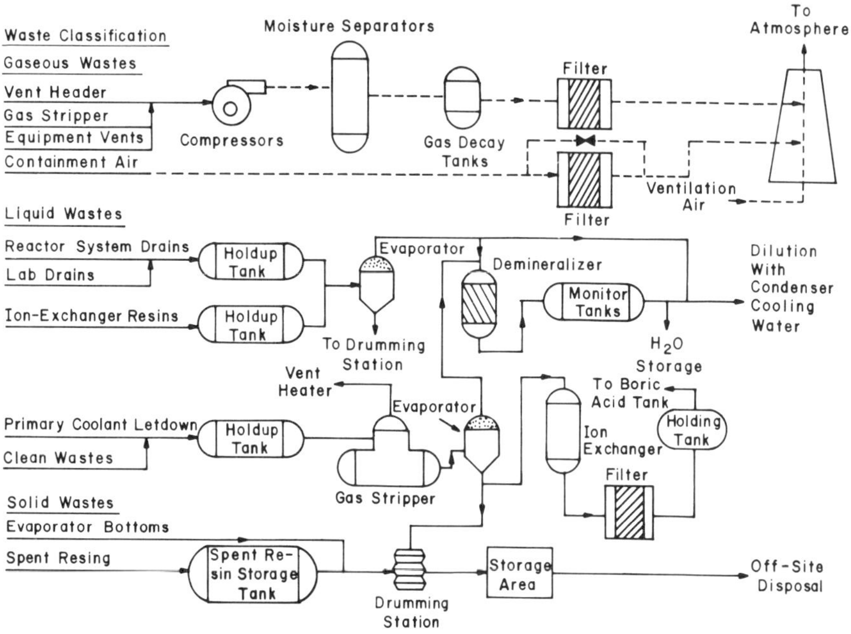

FIG. 1. Waste flow diagram for a typical PWR.

III. THE RELIABILITY-RISK MODEL

The development of the reliability-risk model is a multidisciplinary effort that involves systems analysis, reliability analysis, and release-risk assessment.

System Analysis

The "systems analysis" phase involves the preparation of a physical and functional description of the rad-waste facility accompanied by the formation of a system tree diagram. The formation of system diagrams, flow diagrams or general system layouts is a procedure that a designer normally undertakes to evaluate alternate designs. However, tree diagrams visually enhance the physical and functional relationships shared by the system components and subsystems. Moreover, these techniques facilitate the computer simulation of the system. Tree diagram techniques have been popularized by the fault-tree methods but seldom have been reported for mapping complex systems like the rad-waste facility.

Reliability Analysis

The "reliability analysis" phase involves the formulation of appropriate reliability models and the collection of model parameters. The reliability model employed depends on the equipment states and system elements configuration. Typical states in which any component may exist at time, t, are: operation, failure, repair, or standby; and in practice, the rad-waste system configurations are chains of series and parallel elements.

TABLE I

RELIABILITY MODELS IN COMPUTER PROGRAM

Designation	Equation	Description
RACTIV	$R(t) = e^{-\lambda t}$; λ = failure rate	Reliability of single unit
RPAIR2	$R(t) = e^{-2\lambda t(\lambda/\mu)}$; μ=repair rate	Reliability of 2-element redundant set with single-repair capability(7)
RPAIR3	$R(t) = e^{-6\lambda t(\lambda/\mu)^2}$	Reliability of 3-element redundant set with single-repair capability(7)
RPAIR4	$R(t) = e^{-24\lambda t(\lambda/\mu)^3}$	Reliability of 4-element redundant set with single-repair capability(7)
RSTAN1	$R(t) = e^{-\lambda t}(1+\lambda t)$	Reliability of 2-element redundant standby set with equal failure rates
RSTAN2	$R(t) = \frac{\lambda_A}{\lambda_A-\lambda_S}\left[e^{-\lambda_S t}-e^{-\lambda_A t}\right]+e^{-\lambda_A t}$ λ_S = Failure rate of standby λ_A = Failure rate of active unit	Reliability of 2-element redundant standby set with unequal failure rates
SERIES	$R(t) = \prod_i^N R_i(t)$	Reliability of a system of series elements
PARAL	$R(t) = 1 - \prod_i^N [1-R_i(t)]$	Reliability of a system of parallel elements

The reliability modeling of components or component sets exhibiting the above mentioned states can be undertaken in a dynamic mode or in a quasi-static time frame.

In a dynamic mode, the system can be characterized by a stochastic model in which the behavior of many time-dependent sequential processes are displayed. The evaluation of a system's reliability in this manner is appropriately termed a state calculation which is equivalent to formulating the Markovian model. But because of the compounding complexity of the state calculation as the number of system components exhibiting dependencies increases, it becomes essential to reduce the number of calculations of system states to make solution computationally feasible.

In a static mode, we sort components into independent sets displaying a minimal number of dependencies. Reliability expressions are derived retaining the general approach of the state calculation. Although the models are not dynamic, the effects of failure, repair, and standby are nonetheless treated, but in a quasi-static time frame. For example, at any time, t, the effects of repair or standby for parallel redundant components are treated only in the reduction induced in the set's unreliability.

Table I presents a selection of reliability models used in the computer program. The number of models that have been presented in Table I are by no means exhaustive. Any number of reliability models can be formulated, depending on the complexity of component configurations. Nevertheless, the assumptions associated with these models are:

(1) Components are independent of each other; i.e., the failure of an independent unit does not cause the failure of another unit;
(2) The failure and repair rates of the components are constant; and
(3) The "bad-as-old" concept(8) is used for describing the reliability behavior of components following repair.

The constant failure-rate assumption is justified on the basis of the well-known bathtub curve. Also, component independence is justified assuming components are positioned in different locations to minimize environmental stress variations. And finally, use of the bad-as-old concept is more conservative for design purposes than the good-as-new concept. The basic assumption here is that the system failure distribution following repair depends only on the age of the system and not on its repair history.

Release-Risk Assessment

The "release-risk analysis" phase represents the basis for determining if additional system and reliability analyses are warranted. Release-risk(9) is defined as the product of the failure probability at time, t, and the radioactivity available for release in time interval, Δt. Mathematically, the release-risk can be expressed as:

$$W_k(T) = \int_{t=0}^{T} f_k(t) \cdot \overline{Q}_k(t)dt \tag{1}$$

where $f_k(t)$ = the failure density function for element k; and

$\overline{Q}_k(t)$ = radioactivity available for release through element k at time t.

From Eq. (1), the unreliability at time T for element k is expressed as:

$$F_k(T) = \int_{t=0}^{T} f_k(t)dt \tag{2}$$

Therefore, if we integrate Eq. (1), an upper limit on release risk becomes

$$W_k(T) = F_k(T) \cdot [Q_k(T)] \tag{3}$$

where $Q_k(T)$ is the amount of radioactivity available for release in a period T.

However, the $\overline{Q}_k(t)$ term of Eq. (1) must be expanded to include the different type of functions performed by the rad-waste system components. These functions are segregated into two basic cases which are:

(1) Radioactivity flow through a component with no storage; e.g., components in this category are pumps, valves and short pipes;
(2) Radioactivity flow into and/or through a component with storage; e.g., components in this category are filters, tanks, long pipes, demineralizers, and evaporators.

For case (1), in which we have radioactivity flow through a component with no storage,

$$\overline{Q}_k(t) = C_k(t)\ g_k(t)\ \Delta t \quad , \qquad \text{curies} \tag{4}$$

where $C_k(t)$ = radioactivity concentrations, curies/liter; and
$g_k(t)$ = fluid flow rate, liters/hour.

Since there is no activity holdup in element k during time interval Δt, then radioactive decay need not be considered if the transit time is much less than the decay half-life of the activity.

For case (2) in which we have radioactivity flow into and/or through a component with storage or holdup, the radioactivity in component k is broken down into two categories; i.e., (a) radioactivity initially stored in component k; and (b) radioactivity flowing into component k and heldup or stored.

In category (a), the radioactivity stored in component k and decaying in period t is expressed as

$$[\overline{Q}_{eff}(t)]_k = (Q_0)_k\ e^{-\lambda_{eq} t} \tag{5}$$

where Q_0 = the initial stored activity;
λ_{eq} = the equivalent decay constant for a mixture of isotopes having similar half-lives.

The equivalent decay constant can be approximated by

$$\lambda_{eq} = \sum_i a_i \lambda_i \tag{6}$$

where a_i = the fraction of the ith isotope;
λ_i = the decay constant of the ith isotope.

This expression will serve well for describing the decay process for radioactivities exhibiting similar half-lives; that is, half-lives of days, or months, or years. If the mixture contains a wide range of half-lives and the analysis calls for the added detail, then a compound λ_{eq} can be determined according to each set of activities exhibiting similar half-lives.

For category (b), we define the radioactivity stored or retained in component k due to activity flowing in and decaying in time interval $\Delta t'$ as

$$[\overline{Q}_{in}(t)]_k = \int_{t=0}^{t} [\dot{Q}_{in}(t')]_k e^{-\lambda_{eq}(\Delta t')} dt' \tag{7}$$

where $[\dot{Q}_{in}(t')]_k$ = the rate of radioactivity flowing into component k at time t';
$\Delta t'$ = (t-t') or the time interval over which the activity flows in and decays.

Thus, the total time rate of change of radioactivity initially stored and flowing into component k, including radioactive decay, becomes

$$[\overline{Q}_{total}(t)]_k = [Q_0]_k e^{-\lambda_{eq} t} + \int_{t=0}^{t} [\dot{Q}_{in}(t')]_k e^{-\lambda_{eq}(t-t')} dt' \tag{8}$$

However, to complete the activity balance through the rad-waste system, the decontamination capabilities or factors (DF) for the process equipment should be considered. Characteristically, the radioactivity into and out of decontamination equipment can be expressed as,

$$Q_{in} = DF \cdot Q_{out} \quad ; \qquad DF = \text{decontamination factor} > 1 \tag{9}$$

Here, we make the assumption that the decontamination factor is time invariant, which is justifiable on the basis that replacement or cleaning of the decon-equipment is performed prior to any loss in decontamination capability.

Thus, the net radioactivity trapped in component k is expressed as

$$(\overline{Q}_{net})_k = (Q_{in})_k - \left(\frac{Q_{in}}{DF}\right)_k = \left(\frac{DF-1}{DF}\right)_k (Q_{in})_k \tag{10}$$

Note that the radioactivity flowing out of component k represents the input to component (k+1) or $(Q_{in})_{k+1} = DF_k \cdot (Q_{in})_k$. Therefore, we can quantify the radioactivities in component-wise fashion as the rad-waste fluid is transported from point to point in the system.

As we noted earlier, $W_k(T)$ represents an upper estimate of release-risk and is to be interpreted as the maximum potential activity release over the time interval, not to exceed the total radioactivity available. This interpretation is conservative since we do not take credit for the physical form in which the radioactivity is trapped or stored; e.g., radioactivity entrapped in demineralizer resins versus activity stored as liquid wastes in tanks.

Therefore, Eqs. (1) and (3) provide the bases for the calculation of release-risk with dependence on component type and process function. But for our purposes, we have made several assumptions to make computerization of the methodology more convenient but yet serve to achieve our objectives. These assumptions are that:

(1) Radioactive decay is neglected; and
(2) The release-risk is calculated on the basis of the total radioactivity processed through the system in time steps of one year.

However, the time steps can be altered whenever radioactivity measurements for specific processes are available. This will permit more realistic estimation of the release-risks.

Release-Risk Factor, F_{RR}

To assess the rad-waste system effectiveness, we can best utilize the release-risk results from the foregoing calculations by forming a new quantity called the release-risk factor, F_{RR}. The release risk of the system or a subsystem can be compared to the maximum annual permissible average release, Q_{mpr}, or it can be compared to some other design release objective. The Q_{mpr} value must be determined for each reactor plant in conformance with pertinent regulations. It is normally calculated on the basis of the radioactivity concentration limits and the rates of effluent discharges.

A measure of safety or acceptability of the design can be demonstrated if the calculated release-risk remains less than or equivalent to the Q_{mpr}. That is,

$$W_{sys}(t) \leq Q_{mpr}, \qquad \text{curies/year}$$

Therefore, the release-risk factor becomes,

$$F_{RR}(T) = \frac{W_{sys}(T)}{Q_{mpr}} = \frac{F_{sys}(T) \cdot Q_{sys}(T)}{Q_{mpr}} \tag{11}$$

In Eq. (11), $Q_{sys}(T)$ represents the aggregate of all radioactivities collected and processed in the waste facility from normal and accidental reactor plant operations; and the $F_{sys}(T)$ or unreliability term represents an indicator of how poorly the system is capable of preventing releases.

However, if we wish to compare release-risks for waste facilities of different power plants, another measure can be introduced. The basis for comparison is simply the normalization of the system release-risk, $W_{sys}(T)$ by the electric energy delivered (MWhr) from the power plant in the period of one year. This can be expressed as

$$Z_{sys}(T) = \frac{W_{sys}(T)}{E(T)} \tag{12}$$

where $Z_{sys}(T)$ = power normalized release risk, curies/MWhr;
$W_{sys}(T)$ = release-risk associated with the handling and processing of liquid wastes, curies/year; and
$E(T)$ = plant electric energy delivered, MWhr/year.

Since the generation of fission and activation product activities in the reactor is power dependent, then the $Z_{sys}(T)$ term provides a suitable measure of evaluating the rad-waste system release-risk potential per unit energy delivered by the plant. In addition, this measure furnishes a common basis for comparing different reactor plants and their corresponding rad-waste sources, system design, and operations.

As previously mentioned, an important goal of this methodology is risk reduction. If we utilize the foregoing measures of system effectiveness, we can proceed through a schedule of system design and operation changes which will increase system reliability and/or reduce the quantity of activity available for release such that a release-risk reduction can be realized. If further analysis furnishes no additional risk-reduction, then the system is said to be designed under "minimum risk".

IV. COMPUTER SIMULATION OF THE MODEL

Typically, a system tree diagram, as shown in Fig. 2, can be constructed by identifying each system component or group of components with a unique name. However, the programming and input data effort becomes cumbersome and time consuming for large systems. A more efficient method is to classify and sort the input numerically to minimize data handling and errors.

A computer program was written to calculate reliability and release-risk of nuclear rad-waste systems. A model of the rad-waste system is computer-simulated via a program input package which transforms the tree diagram of Fig. 2 by numerical classification of the types of tree elements. Each block of the tree diagram is classified as a single component, a dependent-component set, a subsystem, or as the total system.

The computational procedure is initiated by first sweeping down the tree diagram for stages $I=1,2,\ldots I_{max}$ and calculating the reliability, unreliability, and release-risk for all basic components and component sets exhibiting standby or repair. Then the procedure is to sweep up the tree from $I=I_{max}$, $I_{max}-1,\ldots 1$ and stage-by-stage re-order elements and apply the combinational reliability models for elements in series or parallel. The results for elements combined at stage I are placed in the next higher stage (I-1). The re-ordering of elements at each stage is necessary such that the correct type and number of elements are combined. This is also important in that the computed results for stage (I) and block (J) must correspond one-to-one to the stage and block symbol of the reference tree diagram. Upon reaching the top of the tree, the

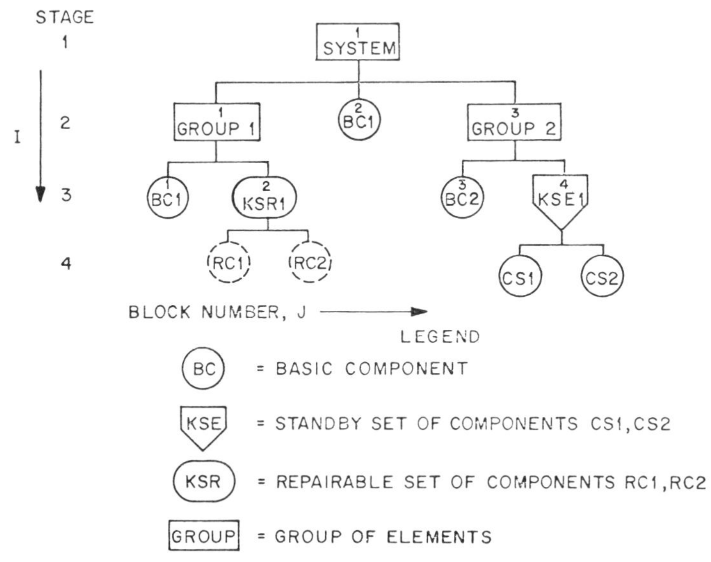

FIG. 2. Tree diagram for a 7-component system.

output is printed and consists of the reliability, R(I,J); unreliability, UR(I,J); radioactivity input, Q(I,J); and the release-risk, W(I,J) for each stage I and block J at time, T. This procedure continues until all time steps are completed.

The program is practical and best suited for direct evaluation of variations in:

system design configurations
operational functions
component modifications and replacements
rad-waste quantities
other system parameters

Also, the code is versatile in that it need not be restricted to the computation of release-risks for rad-waste systems. The program is useful for the calculation of the reliability of any functional system including the calculation of release-risks for fluid systems other than rad-waste facilities.

V. APPLICATION TO A TYPICAL RAD-WASTE SYSTEM

A schematic of a typical liquid rad-waste system is selected to demonstrate the utility of the developed reliability-risk methodology. As shown in Fig. 3, this schematic is similar to some of the existing rad-waste facilities to allow the use of typical experience data. Table II lists the design estimates for normal and maximum activity concentrations of the major waste collection locations for a typical rad-waste facility, while Table III presents the failure and repair rates for several waste treatment components. Although most of the failure data has generally been reported for "failure to function", sources were identified which recorded component leakage and rupture type failures. In cases where no such data existed, estimates were made based on data for components of similar construction and usage.

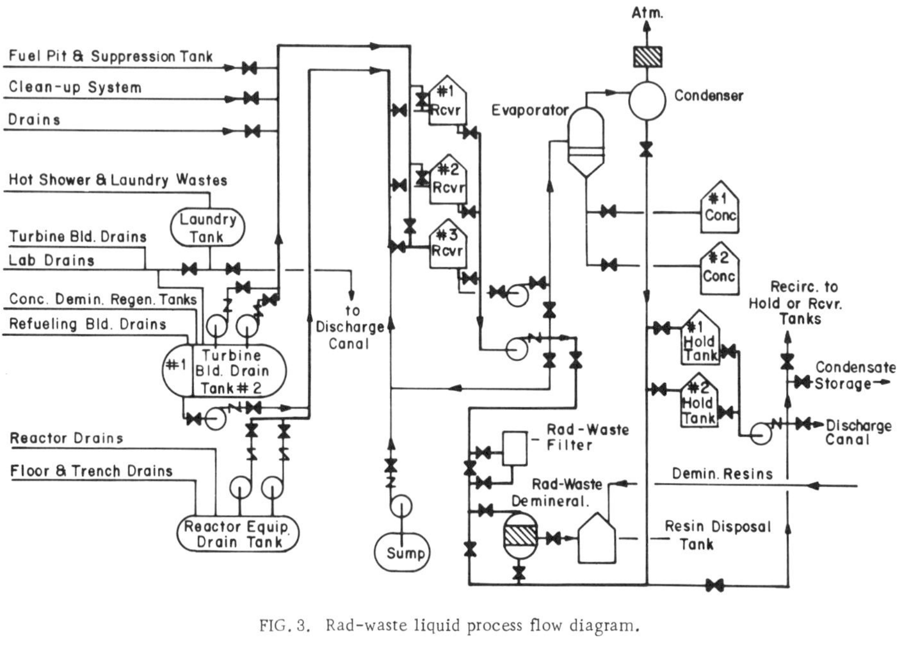

FIG. 3. Rad-waste liquid process flow diagram.

Figure 4 shows a block diagram version of the rad-waste system of Fig. 3. This block diagram is used to construct a tree diagram associated with any segment, operational path or the entire system. In such a tree diagram (Fig. 5), the block elements represent subsets of components or grouping of elements depending on their relationship or operational configuration (series, parallel, standby, etc.).

Several case problems and data sets were prepared to evaluate the system design for a variety of operational modes and for a reasonable range of parameter variations. A tabulation of these cases is shown in Table IV.

Category I represents the main process flow which is taken as the reference problem in this work.

Category II represents analyses of the effects of the modification of the system configuration. In Cases 1 and 2, effects of repair and standby redundancy capabilities were examined. Since high reliabilities were associated with the parallel configurations to which the repair and standby options are added, no dramatic effect on the overall release risk was noted. Nevertheless, repair and standby options provide significant reductions in the local unreliability and release-risk results. Table V shows a comparison between this case and the reference problem of Category I. The tree diagram of Fig. 5 represents the system of Category II, Case 1, which demonstrates the configurational changes of repair/standby to the basic system.

In Case 3, a configuration change in which a bypass line was added in redundancy to segments of the main process flow was examined. Normally very low activities would be processed through the bypass such that excessive operational demands on the main process equipment are not necessary. Therefore, from the design and release-risk point of view, the overall release-risk is significantly reduced. However, it should be noted that bypass lines notoriously represent alternate paths through which high activity releases could occur. Inadvertent releases through bypass lines may occur as a result of operational errors in valve line-up or due to other valve failures such that the redundant features of the configuration are defeated. Nevertheless, these aspects can also be evaluated with the present program by screening every operational path through which potentially high activity releases could develop. Recycling waste prior to release was also examined in Case 3. Recycle paths are considered redundant or alternate to the discharge paths. Addition of a recycle path does enhance locally the reliability associated with the discharge path configuration. However, the decrease in the overall system unreliability and release risk from the reference problem is not dramatic.

Case 4 revealed that the addition of an evaporator and corresponding process equipment has a dramatic impact on the system reliability and release-risk. The significant decrease in the release risk is achieved primarily due to the fact that the largest portion of the processed activity has been diverted to the evaporator segment which has the highest decontamination capability. In other words, low activities are processed through equipment which either exhibits less decontamination capability or exhibits less reliability.

In the last case of Category II, the influence of double containment on the system characteristics was investigated. Double containment provides added physical redundancy. It significantly reduces the potential of releases to the environment, especially in cases of sudden large releases of activity into the waste system as a result of a reactor plant accident.

Category III involves sensitivity analysis. Variations in the failure rates (≈10% increase) induced a comparable change in the system unreliability

TABLE II

LIQUID RAD-WASTE SOURCE ACTIVITIES FOR A TYPICAL RAD-WASTE FACILITY

Collection Location	Volume/Year	Normal Conc.	Max. Conc.	Normal Activity
Laundry Waste Tank	4×10^{4} gal/yr	$< 10^{-6}$ μCi/ml	10^{-4} μCi/ml	1.5×10^{-4} Ci/yr
Turbine Building Drain Tank 1	1.2×10^{5} gal/yr	$< 10^{-6}$ μCi/ml	10^{-4} μCi/ml	4.6×10^{-4} Ci/yr
Turbine Building Drain Tank 2	4×10^{5} gal/yr	6×10^{-5} μCi/ml	10^{-2} μCi/ml	.09 Ci/yr
Rad-Waste Sump and Tanks	1.78×10^{5} gal/yr	$< 10^{-4}$ μCi/ml	10^{-2} μCi/ml	.064 Ci/yr
Reactor Equipment Drain Tank	4×10^{4} gal/yr	< .2 μCi/ml	10 μCi/ml	30 Ci/yr
Spent Waste Demineralizer Resin Tank	5015 ft^3/yr	< 1 Ci/ft^3	10 Ci/ft^3	5-15 Ci/yr

TOTAL NORMAL ACTIVITY ENTERING WASTE SYSTEM = 30.09 Ci/yr

TABLE III

TYPICAL EQUIPMENT FAILURE AND REPAIR RATES

Component	Failure Rate (Leakage)	Repair Rate	Decontamination Factor (DF)
Valve*	2.0×10^{-6}/hr	.2/hr	1
Pump*	5.0×10^{-7}/hr	.2/hr	1
Tank	1.0×10^{-9}/hr	.05/hr	1
Cartridge Filter	1.0×10^{-6}/hr	.17/hr	> 1
Demineralizer	5.0×10^{-7}/hr (estimate)	.05/hr	> 10
Pipe Joints	5.0×10^{-7}/hr	.2/hr	1
Pipe	2.0×10^{-7}/hr	.2/hr	1
Evaporator	1.0×10^{-9}/hr (estimate)	.05/hr	> 1000
Concrete Container	2.0×10^{-6}/hr	-	1

* All type valves or pumps are lumped into one failure-rate estimate.

and release risk. These parameters showed insensitivity to reasonable variations in the repair time. Unless the repair rate remotely approaches the failure rate, high reliability will be maintained over the life cycle of the system. The reliability/repair model in this code characteristically displays this behavior.

In the last three categories, several combinations of the above cases were conducted to obtain further insight of the model's usage. An optimization study was conducted to up-grade the system reliability subject to constraints on the total rad-waste system cost, as well as stagewise constraints on the release-risk. The procedure begins by generating a number of feasible system configurations; the reliability of each configuration is evaluated. The configuration with lowest reliability is rejected and a new improved configuration is generated. This iterative process continues until convergence to an optimal configuration is achieved or until a satisfactory suboptimal configuration is favored.

VI. CONCLUSIONS

A reliability-risk methodology has been presented which is directly applicable to the design and evaluation of nuclear rad-waste facilities. The reliability-risk model has been synthesized by a marriage of system analysis,

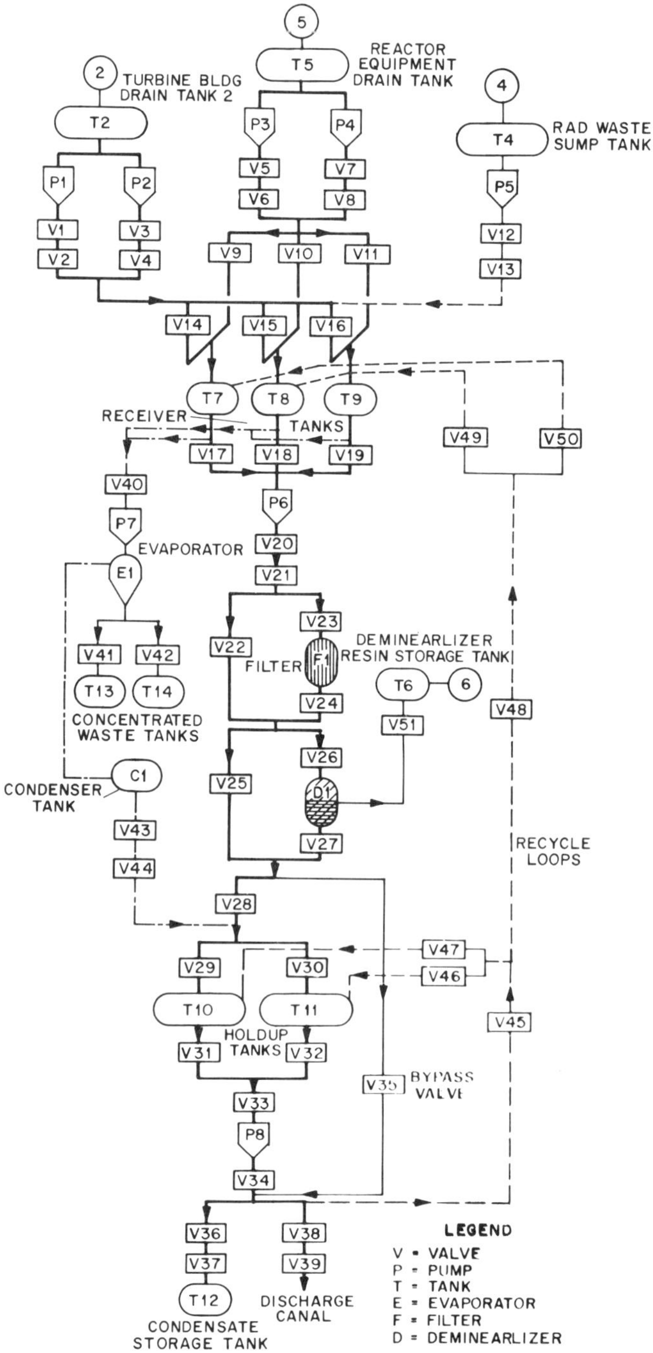

FIG. 4. Typical liquid rad-waste system.

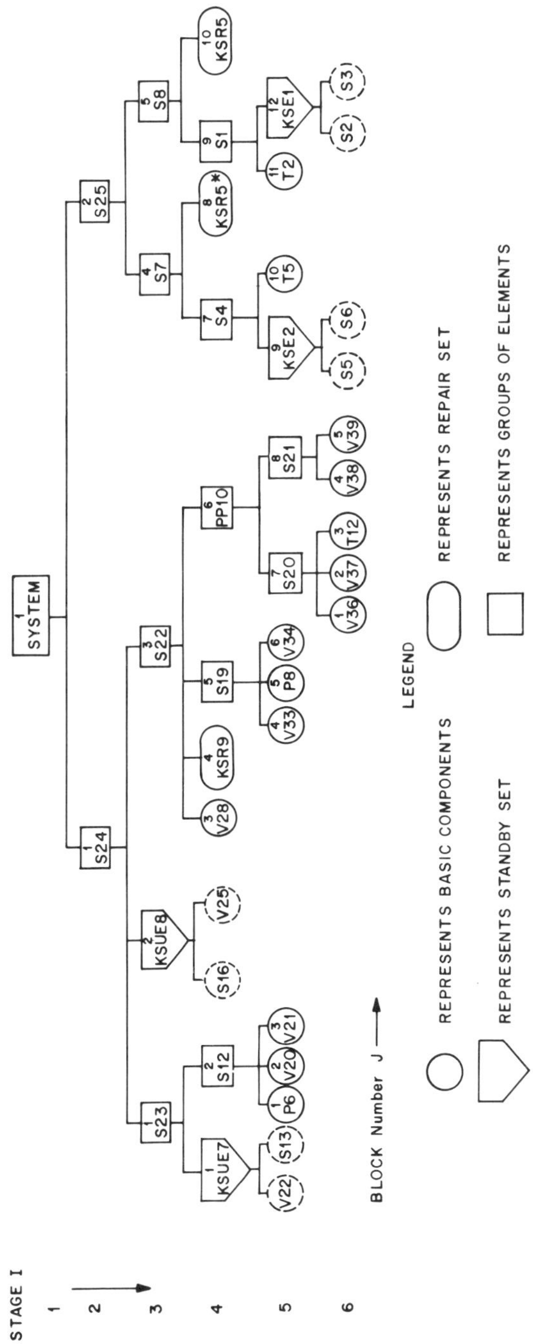

FIG. 5. Typical tree diagram.

TABLE IV

GENERAL PROBLEM DESCRIPTION

Category	Case	Description
I: Basic System		Main process (heavy line (Fig. 4)), normal activities are processed. No repair/standby capability.
II: Configurational Changes	1	Repair/standby capabilities (Fig. 5).
	2	Effect of redundancy (addition of another filter...etc.).
	3	Effect of bypass lines and recycle paths.
	4	Effect of an extra evaporator section.
	5	Effect of double containment.
III: Parameter Changes	1	Effect of changes in the component failure rates.
	2	Effect of changes in the repair time.
IV: Operational Changes		Processing changes required to treat normal vs. high abnormal radioactivity wastes.
V: Parameter/ Configurational Changes		System reliability optimization subject to constraints on releases, overall system cost, and degree of redundancy.
VI: Operation/ Configuration Changes		Combination of the above categories. (Composite case.)

reliability analysis, and release-risk assessment. The methodology has been structured to utilize many types of reliability models, including an optimization algorithm for design of a rad-waste system subject to constraints on system cost and radioactivity releases. Also, utilization of tree diagram techniques and computer simulation of the system has upgraded the designer's capability for bringing into sharp focus any design inadequacies or weak links in the system.

In the past, the overall quantitative assessment of the rad-waste system has been largely a matter of engineering judgement and experience aided by an acknowledgement of the role and reliability of the basic system components. In the present work the prime concern is with the reduction of risks associated with releases from failed components. The case studies demonstrate the model's capability in systematically dissecting a system and weighing its individual segments, operations and parameters. The present model shows that the process of design/evaluation of a rad-waste system is iterative. The procedure is one of initially fixing a reference design and selection of the failure and repair

TABLE V

COMPARISON OF BASIC SYSTEM TO SYSTEM WITH STANDBY AND REPAIR

Category I (Reference System					Category II, Case 1 (Repair/Standby)					
I	J	UR	Q(Ci/yr)	W(Ci/yr)	Type	I	J	UR	Q(Ci/yr)	W(Ci/yr)
3	2	$.96\ 10^{-3}$	30.09	.029	Standby	3	2	$.49\ 10^{-3}$	30.09	.015
4	1	$.74\ 10^{-3}$	30.09	.022	Standby	4	1	$.38\ 10^{-3}$	30.09	.011
4	6	$.12\ 10^{-2}$	3.0	.0036	Repair	4	4	$.28\ 10^{-5}$	3.0	.84-5
4	10	$.41\ 10^{-4}$	30.0	.0012	Rapair	4	8	0.0*	30.0	0.0
4	12	$.41\ 10^{-4}$	.09	.37-5	Rapair	4	10	0.0*	.09	0.0
5	15	$.15\ 10^{-2}$	30.0	.045	Standby	5	9	$.76\ 10^{-3}$	30.0	.023
5	21	$.15\ 10^{-2}$	.09	.14-3	Standby	5	12	$.76\ 10^{-3}$	.09	.68-4

* These computed results occurred because of roundoff.

data. Then the procedure continues through successive computer simulated cases to assess design changes and improvements in failure data, design and/or operations required to minimize system and/or subsystem unreliability and release-risks.

As a design tool, the application of this model for comparative purposes does not demand a substantial body of accurate failure and repair data. Consistency in data use and the execution of sensitivity studies will serve as sufficient bases for making appropriate design/operation decisions. Also, the model need not be restricted to the design and evaluation of rad-waste system. It can be effectively employed to calculate the release-risks associated with other chemical fluid systems. Moreover, the program can be used solely to calculate the reliability and unreliability of any functional system which can be mapped by a tree diagram.

In conclusion, this reliability-risk methodology supplemented by the traditional engineering analyses presently offers the best means for relating the significant aspects of rad-waste system design, operation, release-risks and costs.

REFERENCES

[1] MULVIHILL, R.S., et.al., "A Probabilistic Methodology for the Safety Analysis of Nuclear Power Reactors", SAN-570-2 (1966).

[2] Staff of the Safeguards Division, Authority Health and Safety Branch, UKAEA, "Quantitative Safety Analysis", Nuc. Eng. Design 13 (1970) 183-244.

[3] GARRICK, B.J., et.al., "Reliability Analysis of Nuclear Power Plant Protective Systems", HN-190 (1967).

[4] LINDACKERS, K.H., STOEBEL, W., "Probability Analysis Applied to Light Water Reactors: Loss of Coolant Accidents", Nuc. Safety 14 1 (1973) 14-20.

[5] WASH-1258, "Numerical Guides for Design Objectives and Limiting Conditions for Operation to Meet the Criterion 'As Low As Practicable' for Radioactive Material in Light Water Cooled Nuclear Power Reactor Effluents", Directorate of Regulatory Standards, USAEC 1,2,3 (1973).

[6] BLOMEKE, J.O., et.al., "Management of Radioactive Wastes at Nuclear Power Stations", ORNL-4070 (1968).

[7] McGREGOR, M., "Approximation Formulas for Reliability with Repair", IEEE Trans. on Reliability R-12 4 (1963).

[8] ASCHER, H.E., "Evaluation of Repairable System Reliability Using the 'Bad-as-Old' Concept", IEEE Trans. on Reliability R-17 2 (1968).

[9] LEHMANN, P.H. and A. A. EL-BASSIONI, "A Probabilistic-Risk Modeling of Nuclear Rad-Waste Facilities", Trans. Am. Nuc. Soc. 19 1 (1974) 236.

DISCUSSION

(on papers IAEA-SM-195/11, IAEA-SM-195/13 and IAEA-SM-195/39)

G. BREILING: A question for Mr. Greppo. Your Fig. 5 shows 16 different states of the system with the transitions between states, which suggests that you have taken components of only four different kinds into account. How does this square with the system shown in Fig. 2, which appears to have about 25 separate components?

J.-F. GREPPO: The application of the Markov method requires a transposition from the real scheme shown in Fig. 2 to an equivalent scheme, indicated in Section 2.2. With the fault-tree method it is then possible to calculate the reliability parameters of the equivalent sources.

G. BREILING: We read in the caption to Fig. 5 that common mode failures have been taken into account, and I gather that this was done by adding four supplementary system states to the diagram. How can you, in this way, arrive at a sufficiently complete common mode failure analysis?

J.-F. GREPPO: The choice of additional system states and of the transitions attributable to common mode failures is based on operating experience. Since the study of common mode failures was carried out in parametric form with a view to establishing limiting values that must not be exceeded, it appeared pointless to do a highly refined analysis of each mode and apply statistical data to them.

W.A. WOLFE: As a result of the survey on tube failures carried out by Atomic Energy of Canada Limited, a combined effort to effect tube repairs was mounted with the participation of the utilities, the suppliers and AECL. The purpose of the programme was to develop methods of locating leaking tubes, surveying the surrounding tubes for possible leakage, and performing explosive plugging of defective tubes. Full-scale mock-ups were used to develop these repair procedures, and an actual tube repair has been carried out within the predicted time.

I have a question in this connection for Mr. Procaccia. Tube failures appear to depend on engineering details; areas of cross-flow and the types of tube support used are, for example, significant. Have you considered such factors in estimating failure rates?

H. PROCACCIA: We tried to make it clear in our study that we were seeking orders of magnitude for failure rates, with a view to quantifying the probability of failure in emergency systems which are by nature very complex. Even a large error in the failure rate for a given component within a complex system will not substantially affect the failure rate for the system as a whole. Hence we evaluated the failure rate for steam generators in the best way we could, but without going into design details whose impact on the operational behaviour of the steam generators we could hardly hope to predict in a meaningful numerical form.

J.-C. ARLAUD: Are the incidents you took into account only those which occurred when the installations in question were operating at power?

H. PROCACCIA: Every incident that resulted in a reduction of power output from a station, however slight, was taken into account in our statistics.

G.H. LABAT: You refer to an equipment scale effect which influences reliability, but there is at least one other parameter which varies more or less in parallel with the scale, namely the time of design. In point of fact we witness an almost systematic increase in the scale of our equipment as time goes on. This being so, have you tried to determine whether the so-called 'scale effect' on the reliability or availability of equipment is really due to scale? One could check this point by determining, for example, whether equipment which has remained more or less constant in size over a long period of time has also shown a more or less constant reliability or availability. If no such study has been made, I think it should be; we can then satisfy ourselves whether the 'scale effect' is not sometimes attributable to other parameters, such as new developments in equipment design, modified technological precautions and the like.

H. PROCACCIA: This scale or series effect was illustrated in our paper by the operational behaviour of feed-water reheaters in conventional power stations. Because there is no adequate population for statistical purposes it has not been possible to do a similar study on nuclear steam generators. You are quite right of course: the scale effect as defined in our analysis does encompass new developments in equipment design, and we made it clear that the results were pessimistic because they took account of all the 'teething periods' of the nuclear equipment considered.

P. VERRIERE: I believe you said that the working environment did not modify the failure rates — that pumps, valves and exchangers exhibited the same failure rates whether they were operating in water or in sodium. What sort of statistical base did you use for equipment operating in sodium? Do you really think two exchangers, one operating in water and the other in sodium, would have the same failure modes?

H. PROCACCIA: No, you can see from Fig. 9 that the failure rates for valves in sodium are higher than for valves in water. Conversely, the failure rates for sodium steam generators are lower than those 'expected' for steam generators in water. The failure modes are not of course exactly the same, the consequences of failure even less so; but we find that in general the same processes occur — corrosion, erosion, vibrations and so on. Our source of information for sodium valves was the high-power sodium steam generator test rig (CGVS - 50 MW(th)), an installation of Electricité de France located at Renardières south of Paris.

J.M. BUTTERFIELD: Would coolant flow through the core be maintained by natural circulation during the first few minutes after the loss of electrical supplies and a reactor trip, or is forced circulation required?

H. PROCACCIA: Let me remind you that the primary pumps are assisted by a flywheel which makes it possible to reach half-speed in 50 s. These pumps are also assisted by auxiliary motors, as is the final emergency cooling circuit. Present calculations indicate, however, that natural convection through the core should provide sufficient emergency cooling.

A.V. KARPOV: Did you consider leakage of sodium in your calculations, and, if so, what probability do you attribute to a large escape of sodium?

H. PROCACCIA: Yes, we did allow for possible leakage of sodium in the secondary circuits; one can even regard such losses as appreciable causes of failure. The probability of such leaks occurring varies, depending on the component considered, from say 10^{-5}/h for a small valve to several times 10^{-5}/h for piping; but these are modest leaks that can be brought under control quickly.

A large escape of sodium accompanied by a violent sodium-water interaction in the steam generators would not significantly reduce the probability of adequate cooling in the remaining residual energy removal circuits.

Bear in mind that the four steam generators are supposed to be isolated from each other in four independent structures on different sides of the main reactor building, and that this strictly limits the possibility of a generalized accident involving all the steam generators or all feed sources. Thus one or more cooling circuits would always remain serviceable (main and auxiliary secondary loops) in addition to the two final emergency circuits.

Nevertheless, out of pessimistic conservatism we have accepted in our calculations that a violent sodium-water interaction entailing the failure of three out of four cooling circuits may occur once during the life of the station.

R. GONNOT: You quote a failure rate of 0.18 per calendar year for a Fessenheim-type primary pump. Is this not a huge figure? After all, it would mean 65 primary pump failures in an installation of this kind. Is this acceptable?

H. PROCACCIA: If we consider three loops per reactor and a plant lifetime of 30 years, it would mean 16 failures during the lifetime of one station unit. You are therefore considering all four units, which means 360 pump-years.

The statistical study carried out by EDF's Thermal Production Service covered more than 1000 pump-years, so it is fair to say that we had representative sampling. One could of course question the practice of transposing these results to nuclear pumps, but in fact nuclear pumps differ from conventional feed pumps mainly in the number of stages; and most failures as it happens have been in the equilibration or seal lubrication cooling circuits, rarely in the rotating parts. The transposition that we carried out probably gives a reasonable order of magnitude, and that was the purpose of the exercise.

On the other hand, we cannot take into account technological progress in pump design and construction. Our values are therefore pessimistic and should be recognized as such; but this is of course true of all safety analyses.

J. KRASNODEBSKI: I still believe that Mr. Procaccia's estimate of failure rates for large pumps and motors are very pessimistic. At Ontario Hydro's next nuclear station — the Bruce station, which is to consist of four 750 MW units — the primary heat pumps are rated for 8000 kW, and in order to reach unit availability targets the failure rates for pumps and motors will have to be at least an order of magnitude lower than the rates that could be extrapolated from Mr. Procaccia's Figs 6 and 8. To achieve these failure rates, Ontario Hydro is building a pump test facility that will permit testing and development work on the future primary pumps.

In the United States of America, the American Power Company is using only one 100% turbine-driven boiler feed pump for its large fossil units, which include 1300 MW units. I believe that the failure rates for these very large pumps are much lower than 100×10^{-6}. This proves that with careful specification, choice of supplier and testing, the effect of increase in size need not be very great.

H. PROCACCIA: The extrapolations in Figs 6 and 8 do not go beyond a power rating of 3 MW, and we have made it clear in our paper that the formulae proposed there are intended to give estimates of reliability for power ratings from a few kW up to 3.5 MW. We have, however, done a provisional calculation for a 5 MW pump, but have not gone as high as 8 MW.

Moreover, we have uttered the warning that, particularly where motors of large capacity are concerned, the disparity of the estimations obliges us to treat Fig. 8 with caution.

On the other hand, it does not seem to us unreasonable to anticipate a pumpe incident every five or six years on a 5 MW pump; on the contrary, it seems only realistic to do so. Moreover, the incident may be one entailing only very minor consequences.

B. GACHOT: The primary pump which is the prototype of the pumps that are to be used in the French PWR stations has been tested for several months on a testing rig which makes it possible to reproduce most of the conditions that will be encountered in subsequent operation. This is one reason why one might think that the failure rates determined in Mr.Procaccia's study, influenced as they perhaps are by teething troubles, are a little bit pessimistic where the PWR pumps are concerned. I should add that the precautions taken in the manufacture and quality control of the PWR primary pumps are definitely more stringent than those applied to other pumps for nuclear or conventional power stations.

W.A. WOLFE: The rad-waste system shown here has considerable redundancy; this would increase the maintenance effort and hence the exposure of operating and maintenance personnel. Have you considered a trade-off between safety of the public and total exposure to the population?

P.H. LEHMANN: The purpose of our investigation was to help system designers who require a methodology for the optimization of rad-waste system design and operation, taking into account reliability and constraints associated with costs and radioactive release risks. The results obtained in these design calculations serve for assessing the maintenance effort required and also help to minimize the exposure to radiation of operating and maintenance personnel. Therefore, we are not treating explicitly in these studies a trade-off between safety of the public and population exposure. However, an additional goal implicit in the objectives of the methodology is to assess, via quantitative risk evaluations, the rad-waste system containment capability, with a view to ensuring that radioactive releases to the environment are kept as low as practicable.

SAFETY ASPECTS OF PRESTRESSED CONCRETE PRESSURE VESSELS

F. BREMER
Fried. Krupp G.m.b.H., Krupp Universalbau,
Essen,
Federal Republic of Germany

Abstract

SAFETY ASPECTS OF PRESTRESSED CONCRETE PRESSURE VESSELS.

The increasing demands made on the reduction of the damage probability and higher safety against external influences have focused interest on the favourable load-bearing behaviour of prestressed concrete vessels. Methods used for spatial stress analysis and to furnishing proof of the maximum load-bearing behaviour of the pressure vessel and its liner and insulation have now reached a stage where the service life of the liner can be reliably analysed. The elastic behaviour under maximum stress, for instance, ranges around a factor of $>2 \times p_0$ if p_0 is defined as design pressure. Thermal creeping of the concrete has been sufficiently explored for the temperature ranges considered today. So far the liner has been regarded as a component that is particularly critical from the safety aspect. Meanwhile, engineers have succeeded in developing a leakage detector and a pressure relief system for the transition between liner and concrete. The provision of a prestressed concrete enclosure to protect a steel pressure vessel from bursting may only be regarded as a first step towards optimized safety. The zero-path approach would have to be applied to preclude damage caused to the prestressed concrete enclosure by undefinable forces due to radiation and dynamic loads. However, if a properly designed antiburst enclosure is dimensioned to meet requirements fully, this would almost constitute the first step towards a fully fledged prestressed concrete pressure vessel.

1. GENERAL

As early as 1957, a reinforced concrete pressure vessel was designed and built in France for a Magnox reactor. It must be acknowledged that the purchaser and the engineers who designed this structure showed a remarkably progressive spirit, considering that, at that time, neither the engineering formulas for a clearcut stress analysis of space continuums were available, nor was there any knowledge of the strength and rheological behaviour of concrete in a spatial stress field.

From these first beginnings a method for the construction of prestressed pressure vessels has now evolved which can be regarded as fully developed and absolutely reliable. This is evidenced by the fact that about 20 structures of this type are now in operation or under construction in the western hemisphere. The majority was built in France and the United Kingdom, one in the Federal Republic of Germany and one in the United States of America.

Much research and development work was done to achieve the high engineering standard of these structures. Characteristic of the trend is the considerable research that was done in the United Kingdom to determine the rheological characteristics of concrete, while in the Federal Republic of Germany much was done in the area of three-dimensional stress analysis and the behaviour of the concrete in multiaxial stress and deformation fields. The impetus to solve the spatial stress problems numerically came largely from the United States. Workers in France are known to have done research on the multiaxial mechanical behaviour of concrete.

In contrast to normal reinforced concrete construction, prestressed concrete cannot be fully regarded as a composite material since the size of the tendons and their theoretical action do not ensure bonding between prestressing steel and concrete. This new problem also induced engineers to develop novel fracture hypotheses relating to the structure of such vessels.

According to the results of research work carried out so far, a method was evolved, at least in the Federal Republic of Germany, which makes it possible to determine the probability of failure with a high degree of certainty.

2. RESEARCH INTO CONCRETE AS A BUILDING MATERIAL

The visco-elastic behaviour of concrete has been known for some considerable time. The quantitative knowledge available early in the 'sixties, however, was not adequate to meet the demands imposed on the safety calculation of a prestressed concrete pressure vessel for nuclear purposes. It was undisputably Dr. Brown of the United Kingdom who took the first decisive steps to investigate this phenomenon. The research work done by other workers fully confirmed the results he achieved.

Apart from time as the most important parameter of the viscous effect, the factors governing viscosity are moisture and temperature, the quantity of cement-water paste and the load duration. If the loading is uniaxial, it may be assumed that the visco-elastic properties of concrete at temperatures of up to about + 50°C and under the action of the above parameters are so well known that they have a degree of reliability similar to that of modern numerical computing methods commonly used today for stress analysis. In respect of the viscous properties of concrete in a multiaxial state of stress, the situation is less satisfactory and a great deal of research work has yet to be done in this field. Our own studies and those of other workers have proved, however, that the amount of multiaxial creep is smaller than that of uniaxial creep. In determining the safety of a vessel, one will not always describe the actual condition but may assume to be on the safe side in assessing the vessel's safety.

In all those countries where large-scale research and development work on prestressed concrete pressure vessels is being done, much work has been done in the area of the multiaxial stress and deformation behaviour of concrete.

It was found that the strength of concrete subjected to more than uniaxial compressive stress increases very quickly; it becomes indestructible if compressive stresses of equal magnitude exist in three directions. In such a case, the crystalline strength of the solids is decisive for the failure of the material. Conversely, however, a rapid drop in strength can be noted if, in addition to one or two compressive stresses, a tensile stress acts at the same time in one of the three directions [1]. Research work in this area was very complicated because experimental technologies and the necessary equipment had first to be developed from scratch.

I do not intend to deal with experimental techniques in this paper. Let me state, however, that research workers all agree on one point:

the application of the load must under no circumstances impede transverse strain. It is interesting to note that concrete subject to a compressive load does not fail owing to compressive stress but to transverse strain. The realization that fractures are due to transverse strain is very valuable in assessing the safety of a design. From the results which have now become known, the conclusion can be drawn that research in this area is not only useful for economic reasons; it is imperative for safety's sake. Data on the utilization of the multiaxial compressive strength have already been given in Ref.[2].

Similar pointers are included in the rules for the design of the prestressed concrete pressure vessel for the THTR nuclear power station in the Federal Republic of Germany. In this connection, let me mention the comprehensive research programme now being sponsored by the Federal German government. The results can be expected to be published in 1976.

All other physical properties affecting the stress anlysis, such as shrinkage, linear thermal expansion, modulus of elasticity, specific heat and heat conductance are not so difficult to determine experimentally. It can be expected that these values are so well known to all those engaged in the design and construction of prestressed concrete pressure vessels that they can be regarded as being just as reliable as those obtained by numerical stress analysis.

Considerable research efforts were made to determine whether radiation causes a change in the properties of concrete. It was found, however, that the silicatic or calcitic aggregates in particular lose their properties only above an integrated dosage which known or planned reactors will never reach during their service life of about 40 years.

3. RESEARCH INTO PRESTRESSING STEEL

Research in this area could be restricted to developing prestressing wire of low relaxation and to determining the influence of radiation and of radiolysis (e.g. hydrogen embrittlement), and to the development of suitable tendons having ample load-bearing capacity.

High-tensile prestressing steels, with such a low rate of relaxation that they are of hardly any economic significance, are today available on the world market in sufficient quantities.

For reasons of stress theory, it will always be necessary to arrange the tendons as close to the external surface as possible in order to prevent exposure to intense radiation. The results of research carried out so far showed no adverse effect of residual activity on the properties of prestressing steel.

As a rule, the tendons commonly used in general civil engineering do not exceed an admissible prestressing force of about 250 Mp. Technically and economically speaking, however, this size of tendon is completely inadequate for the construction of prestressed concrete pressure vessels. Prestressing methods are available today for failure loads of up to 1000 Mp. What is known as wrapping prestressing methods can be expected to be introduced soon, representing a further advance in producing tangential and radial prestress.

4. LINER RESEARCH

It is well known that concrete cannot be made impermeable to gas or water under pressure with any means. The reason is that concrete in the requisite workable condition contains a certain amount of water that can neither be combined chemically nor fixed physically. This free moisture diffuses in the course of time into the environment and thus produces a system of open pores and capillary tubes which are permeable to the coolants of the reactor.

Every pressure vessel must therefore be provided with an internal liner (sealing skin). So far, the liner has been regarded as being a safety component of the pressure vessel. Many proposals have been made to furnish sufficient a priori proof of the liner's safety. To this end, among the means available let me mention the proof of liner integrity and the safety of cracks subject to pressure. While the inventor of the method mentioned first and those compelled to use it were anything but happy with it, the method of proving that a crack subject to pressure is safe furnishes hypothetical evidence that a disastrous failure of the vessel is theoretically not possible. It does not furnish direct proof of the liner safety.

Regardless of whether the reactor is water-cooled or gas-cooled, a very ductile material must always be used for the liner. While a simple ferritic steel grade may be adequate for a gas-cooled liner, an austenitic steel must be used with water-cooled reactors for reasons of corrosion prevention. A considerable amount of research work has meanwhile been done concerning the biaxial fracture behaviour of these steels. As far as the austenitic liner is concerned, mention is made of the research project promoted by the Federal German government.

The so-called 'hot-liner' as used by Krupp [3] poses a major object of this research effort.

At least in the Federal Republic of Germany, the authorities will no longer require proof of the limiting condition of the liner integrity for the construction of the next prestressed concrete pressure vessels and will demand other evidence such as a service-life analysis. Certainly such a service-life analysis must be based on intensive research into the biaxial strain behaviour under cyclic loading.

The possibility of helium penetrating the welds of the liner cannot be absolutely excluded, especially in the case of gas-cooled reactors today using helium as coolant. This might cause pressure to build up in the system of pores in the concrete or in the interfacial zone between concrete and liner. If pressure inside the reactor should suddenly drop, there would then be a pressure differential from outside to inside which would destroy the liner.

Research carried out at Krupp has shown that helium movement in the interfacial zones described above is so small as not to permit any damaging pressure build-up. The fact must be admitted, however, that the approval authorities are not satisfied with such safety considerations. Systems have therefore been developed in Germany and, in a somewhat simpler form, in Sweden, which are arranged behind the liner and absolutely prevent any build-up of pressure. The liner is thus even deprived of a large part of its function as a safety component [4].

5. COMPUTATIONAL METHOD

If the Marcoule power station, the first commercial nuclear power plant equipped with a prestressed concrete reactor pressure vessel, is compared with the prestressed concrete pressure vessel for the THTR nuclear power plant in the Federal Republic of Germany, the latest one now under construction, it will be found that the methods of stress theory used for the investigations and the considerations on maximum load-bearing capacity have drastically changed and have become more advanced.

Owing to a lack of analytical or numerical computing methods and suitable electronic data processing equipment, the theory of thin-walled shell-type structures was still used in the design of the Marcoule station and additional studies were only made for rigid corners [5]. The viscosity of the concrete was disregarded. However, since the designers were aware of the qualitative inadequacy of the stress analyses, a factor of 4 was selected to prevent failure. In addition, the assumption was made that the modulus of elasticity and transverse strain remained constant, i.e. were independent of time and temperature, that the concrete behaved elastically, homogeneously and isotropically, and that tensile cracks did not occur in the concrete.

For planning the prestressed concrete pressure vessel of the Oldbury power station, a computer program had already been used which operated on the basis of elasticity equations and was designated Mystic. Here, too, an elastic, isotropic and homogeneous behaviour was still assumed. For the first time, the continuum was divided into elements for the purpose of numerical solution. The storage capacity of the computer, however, did not permit more than 42 elements in the radial direction and more than 100 in axial direction [6].

Here, too, drastic changes have meanwhile occurred. Model tests were carried out for all prestressed concrete pressure vessels built to date or under construction in order to confirm the results of the theoretical strength investigations. In these tests, care was taken to see that physical similarity was as perfect as possible. A conspicuous finding of all the overload tests carried out was that the failure loads determined experimentally were greater than could be expected from the theoretical results. The reasons for this deviation are recognizable today. With the exception of the pressure vessels of Fort St. Vrain and Schmehausen in the Federal Republic of Germany, all stress anlyses were carried out on a two-dimensional basis.

The influence of the third direction on the stress and deformation behaviour was thus disregarded although the integral load-bearing capacity is favourably influenced by the third direction. In addition, knowledge of the material properties under the action of multiaxial stress and deformation fields has increased to the point where a more accurate computation is possible. For the design of the prestressed concrete pressure vessel of the Federal German THTR reactor, different computer programs and computational methods were used to verify the stresses occurring and the maximum load-bearing capacity.

A two-dimensional computer program using the method of dynamic relaxation was employed to investigate the integral behaviour and thus to furnish proof of the structural feasibility of the pressure vessel. Detailed investigations applying particularly to the upper half of the vessel were carried out on a truly three-dimensional basis using a modified safe-3-D

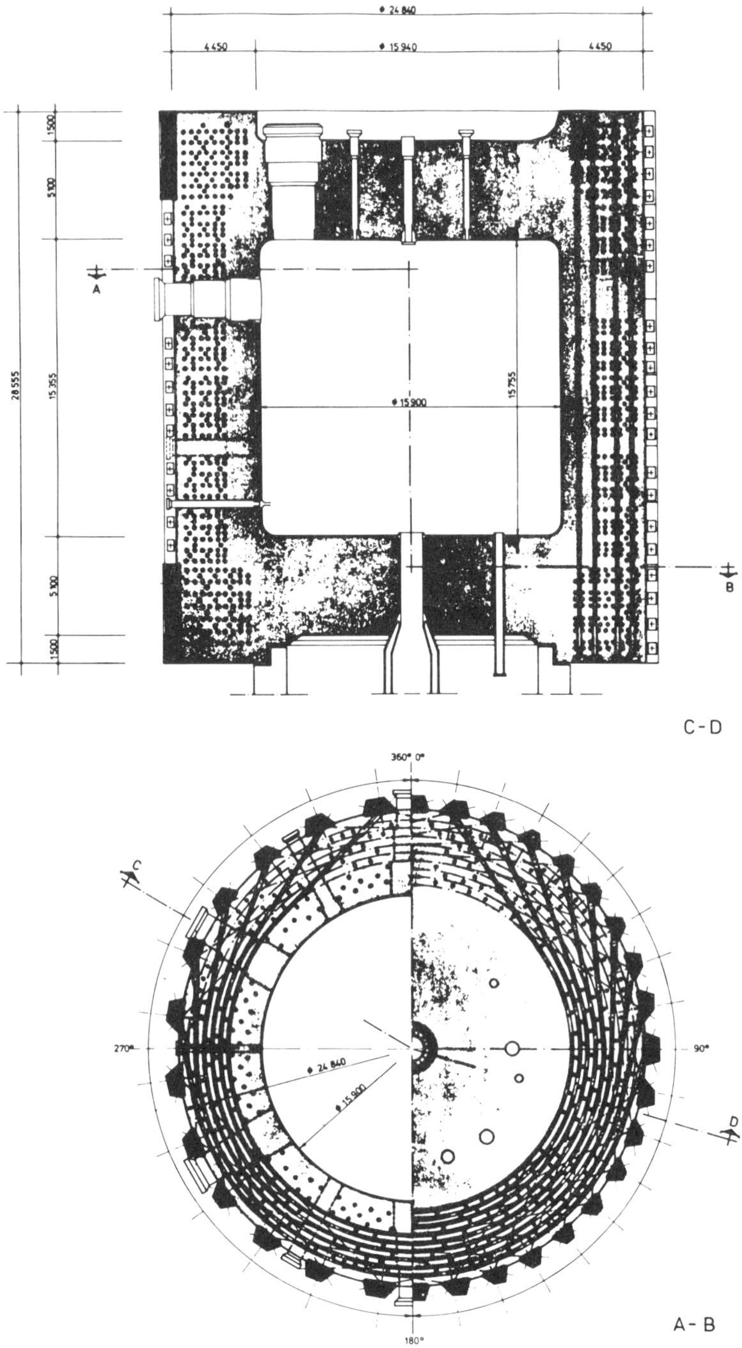

FIG. 1 Cross-section and ground view of the THTR vessel (measurements in mm).

program. The multiaxial strength of the concrete was thus taken into account.

The concrete, which had been specially formulated for this structure, was subjected to thorough tests of all relevant properties long enough before construction started. One of the more important tests dealt with the viscous behaviour of the concrete as a function of temperature and moisture. For various reasons it was assumed that, depending on load combination, 47.5 and 57% of the 90-day values of the prism strength of the concrete would be utilized theoretically. In addition, the tests were required to furnish proof of the maximum load-bearing capacity for which the integral safety factor for the main portions of the pressure vessel was fixed at 2.5, and for the zones containing openings at 2.8.

The pressure vessel designed for the above nuclear power station is characterized in its geometry by the presence of large-diameter openings in the upper half for the blowers and heat exchangers (Fig. 1).

No limitation of the tensile stresses was assumed for the highly stressed internal corners of the pressure vessel. Based on the results obtained for the uncracked condition over the entire volume of the vessel, zones were fixed in which cracking of the concrete had to be anticipated. In a subsequent computation, these zones were then included with a modulus of elasticity that had been substantially changed and had been determined from the amount of reinforcement by a method developed by Rüsch. A kinematic model was used to verify the maximum load-bearing capacity. In determining the mechanism of fracture, the theoretical results of the three-dimensional stress analysis and the results of tests carried out on models were used. It was found that the predetermined safety factors were attained.

Although the computational burden for the THTR vessel has assumed unprecedented proportions, a physically similar model was nevertheless built to a scale of 1:5. In view of the peculiar features and uncommon geometry of the vessel, this appeared to be necessary to ensure the requisite structural safety.

The main items of the test programme were contour measurements and excess pressure tests. The contour displacement measurements are a good yardstick for assessing the correctness of the calculated load-bearing capacity and thus add to the feeling of safety. The overload test was carried out to determine whether all deformations showed a reversible behaviour under a hypothetical excess pressure of 2.1 times the design pressure. The test programme is shown in Fig. 2.

As expected, it proved possible to verify reversibility. The contour displacements were found to agree surprisingly well with the computed results. A superimposition of the deformations is shown in Fig. 3.

Verification of the limiting condition of the liner integrity also involved a high computational burden. The condition was that at no time during the life of the reactor would deformations greater than 6‰ be allowed to occur in the liner. In the case of this reactor, the liner is made of fine-grained steel having a uniaxial elongation at fracture of 24%. In biaxial cyclic tests using strain amplitude of ± 2.5‰, it was found that elongation at fracture decreased to 8% after 100 000 reversals. Compared with the calculated maximum permissible deformation, this would leave a safety factor of 13 against failure of the liner. The hypothetical excess pressure of 1.6 means that the working pressure is calculated within the visco-elastic range

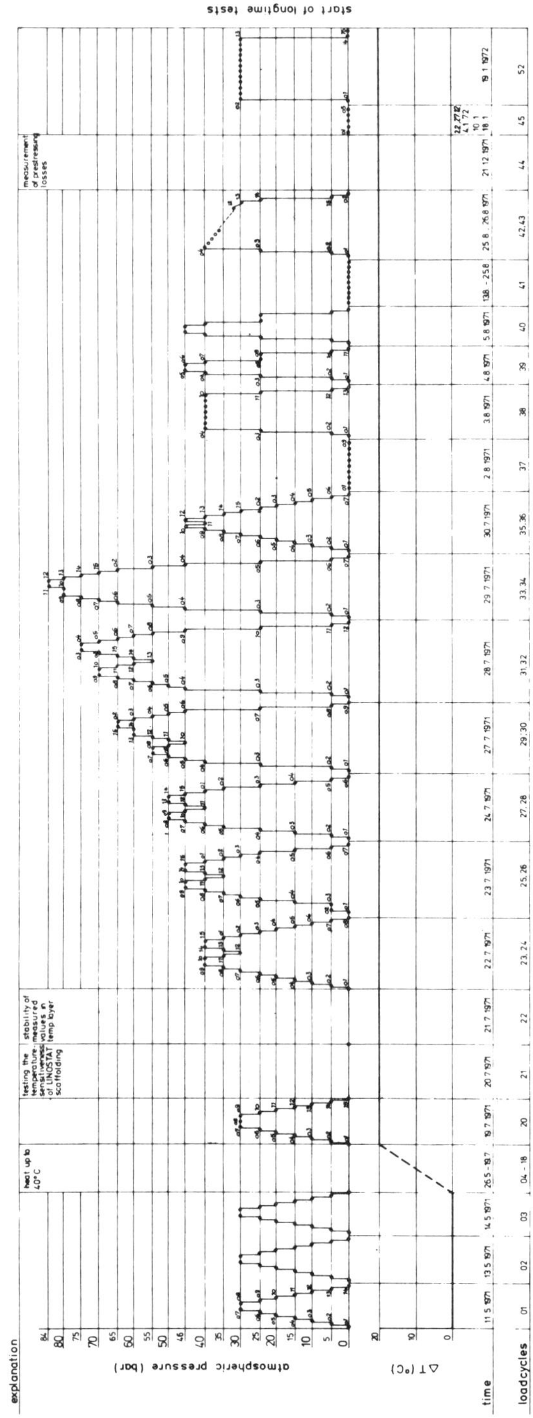

FIG. 2. Test programme.

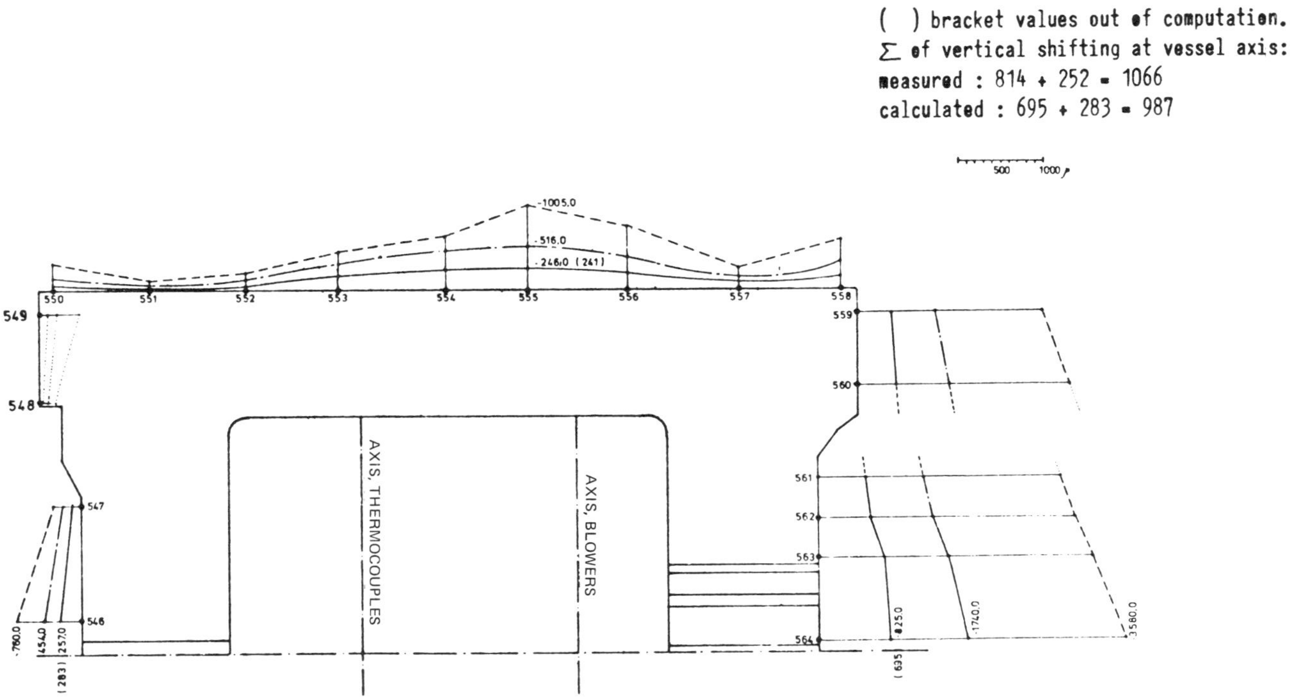

FIG. 3 Contour shifting of the THTR 1:5 model (profile at 90°)
———— : Pi = 47 bar; —·—·—· : Pi = 66 bar; – – – – – : Pi = 85 bar.

and that a sudden hypothetical excess pressure of 0.6 times the working pressure is added within the elastic range. It must be admitted, however, that this method of verifying the maximum load-bearing capacity of the liner is relatively arbitrary and not particularly reliable.

6. CONCLUSIONS

Computational methods are available today for stress analysis which function on a genuine three-dimensional basis and handle problems outside the elastic range.

Also, it is possible to introduce different material characteristics in the three spatial directions and to regard whole regions as non-existent. These methods not only permit more accurate knowledge to be gained of stresses actually occurring in the continuum, but also permit zones in which cracks cannot be excluded or allowed to be included with their actual deformation behaviour. While this adds to safety, it also affects the economy of such a pressure vessel.

For one or two types of concrete, knowledge of the behaviour under load has, as far as reliability is concerned, kept pace with present-day computational possibilities. It would merely be desirable to obtain more knowledge on creep at temperatures of up to about + 100°C and, in particular, on the influence which untensioned bar reinforcement exerts on the modulus of deformation of the concrete. Only initial steps have so far been taken in this direction.

Today, it may be regarded as certain that a prestressed concrete pressure vessel, properly designed and provided with correctly dimensioned prestressing elements, can never fail spontaneously. The main reason for this is that in the region above about 2/3 of its uniaxial ultimate strength the stress-strain curve of concrete flattens out so much that, under an assumed hypothetical overload, the deformations become large enough to be recognized early enough for the reactor to be shut down without this having any adverse effect on heat removal or other criteria. Communications are increasingly being received to the effect that, with the increasing size and complexity of such structures, the probability of a steel pressure vessel failing becomes so great that the question of siting becomes a real problem. A prestressed concrete pressure vessel is a genuine alternative for all types of reactor, so that the question of siting can be viewed from an entirely new angle.

The possibility of surrounding steel pressure vessels for light-water reactors with a containment of prestressed concrete to provide protection if the steel vessel should burst, is now being seriously considered in the Federal Republic of Germany. If, however, the basic design data to be expected are considered, then an anti-burst containment of prestressed concrete can only be a first step towards a fully fledged pressure vessel of prestressed concrete. If, in addition, it should be decided to introduce, as safety criterion, a limiting condition of non-linear elastic maximum load-bearing capacity, as proposed by the author in 1968 on the occasion of a EURATOM meeting, and if the use of partial prestress should continue to be permissible, then we shall not only succeed in making the pressure vessel even more safe but shall be able to make it much more attractive from the aspect of costs.

REFERENCES

[1] BREMER, F., Festigkeits- und Verformungsverhalten des Betons bei mehraxialer Beanspruchung, FIP Congress, Prague (1970).

[2] Proposed Standard Code for Concrete Reactor Vessels and Containments (Proposed Section III, Divison 2), ASME Boiler and Pressure Vessel Code, Am Soc. Mech. Eng. (1972).

[3] BREMER, F., Research work and development of a PCPV for a light water reactor, Br. Nucl. Energy Soc. 1 (1975) 49 - 55.

[4] Federal German Patent No. 1 683 771; and Federal German Patent Apl. No. P 22 27 394.5.

[5] GILL, S., Structures for Nuclear Power (1964).

[6] KELLER, A.Z., CRUNGS, A.R., "Mathematical and computer techniques used in the Oldburry vessels", Prestressed Concrete Pressure Vessels, The Institution of Civil Engineering, McGraw Hill, New York (1968).

RELIABILITY ANALYSIS OF ELECTRICAL POWER SUPPLY SYSTEMS

A. PALAMIDESSI
CNEN-PRV,
Bologna

A.G. COLOMBO, T. LUISI, G. VOLTA
EURATOM-CCR,
Ispra, Varese,
Italy

Abstract

RELIABILITY ANALYSIS OF ELECTRICAL POWER SUPPLY SYSTEMS.

The electrical power supply systems of reactors are required to perform a number of different functions and under various conditions. It is of the utmost practical importance to evaluate the relative importance of component failure for each situation. The paper discusses this problem, starting from a global model of the system. The approach is based on the fault-tree technique and sensitivity analysis. An example of the application of the proposed approach to the electrical power system of a nuclear reactor is also given.

1. INTRODUCTION

In the reliability analysis of electrical supply systems for nuclear power plants two typical problems arise. The first is to assess the availability of the system at various points, more precisely, at various bus-bars. The second is to assess the availability at a given point (a given bus-bar) under various conditions of normal operation and emergency. In both cases it is of the most practical importance to know the relative weight of component failures on the assessed availability.

The approach we have used and present here has the peculiar advantage of permitting the solution of both problems, starting from a global model of the system. By simple variation of the input data the model gives an answer to various problems. The technique used is that of fault-tree analysis and sensitivity analysis described in detail elsewhere [1-3]. The possible use of this technique with a single fault-tree as a global model to give answers to alternative situations is mainly emphasized in this paper.

2. THE MODEL

2.1. Representation

The system to be analysed is represented by means of a fault-tree (Fig.1), the final event being the malfunction of interest at the lowest level; in our case it is the non-availability at the lower point (G 1).

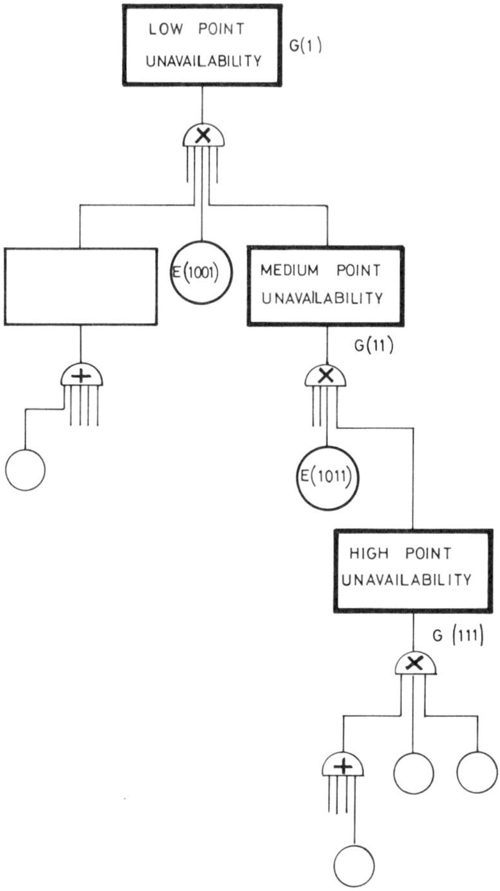

FIG. 1. Diagrammatic representation of a multi-point and multi-situation analysis.

The fault-tree also shows:

(1) The other derived malfunctions of interest for the analysis; for example, medium point non-availability and high point non-availability, G(11) and G(111) in Fig.1.

(2) The primary events that represent 'conditions'; for example, normal operation or emergency condition of a particular subsystem, E(1001) and E(1011) in Fig.1.

2.2. Analysis

For each malfunction of interest the more significant situations are taken into account, each situation being characterized by a combination of conditions. A sensitivity analysis is computed for each situation. M × N analyses are obtained (N is the number of top-events (system malfunctions) taken into account, and M is the number of situations considered).

2.3. Design evaluation and optimization

The sensitivity analysis technique allows the more important primary events in each situation to be identified. Given a weight to each top-event and to each situation, the primary events can be classified according to a prefixed criterion. This classification is the base for a design evaluation and may be used as the starting point for a design optimization.

3. APPLICATION

3.1. Description of the PEC electrical supply system

Figure 2 shows a simplified diagram of a preliminary design for the main electrical system of the PEC reactor. Power supply is taken from two 130 kV separated transmission lines (Suviana and Calenzano) connected in parallel by isolating switch S. Two 100% duty 130/6/6 kV transformers feed the 6 kV bus-bar that supplies the motor generator sets. In the event of loss of supply to this bus-bar, diesel engines are started to maintain the supply to the motor generator sets, which in turn supply the 6 kV no-break bus-bars. Two 100% duty 6 kV/380 V transformers supply the 380 V bus-bars that provide supplies to the sodium pump motors, the instrumentation and the computer. Table I shows the performance of the electrical equipment of the system.

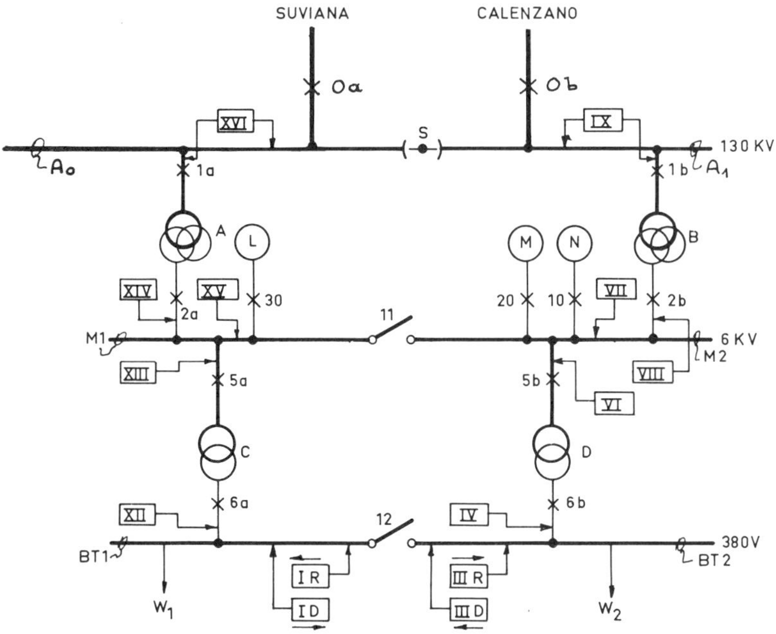

FIG. 2. Simplified scheme of the PEC reactor electrical supply system.

TABLE I. PERFORMANCES OF ELECTRICAL EQUIPMENT

130 kV equipment

Circuit breakers Oa, Ab, 1a, 1b
Type: oil-break switch
Rated voltage: 132 kV
Rated current: 1250 A
Breaking capacity: 7000 MVA

Isolating switch S
Type: rotary
Rated voltage: 132 kV
Rated current: 800 A

Voltage transformers A, B
Type: oil-cooled 3-phase
Ratio of conversion: 130000/V3/100/V3/100/V3 V
Accuracy rating: 0.2/0.5

Current transformers
Type: oil-cooled
Ratio of conversion: 500/1/1 A
Class I/II second. : 0.5/10 P 30

6 kV equipment

Circuit breakers 2a, 2b, 5a, 5b, 10, 20, 30, 11
Type: Dejon
Rated current: 1250 A
Breaking capacity: 250 MVA at 6 kV

No-break diesel-generators L, M, N
Components: Diesel, clutch, flywheel generator,
synchronous motor
Rated output: 100 kVA
Number of poles: 4/6

Voltage transformers C, D
Rated output: 1600 kVA
Ratio of conversion: 6/0.38 kV
Phases connection: Δ/λ
Cooling: Natural

380 V equipment

Panels "Power Centre"
Type: Metal clad equipped with Dejon circuit breakers, extractable, magneto-thermically protected

Panels "Motor Control Centre"
Type: Metal clad

3.2. Fault-tree representation

As stated in Sections 1 and 2, the starting point of the analysis is the representation of the system by means of a fault-tree. In this representation (Fig.3) we consider as final malfunction 'top-event' the loss of supply at bus-bars BT1 and BT2, these bars being hierarchically the lowest level of the system. The top-event is linked by a causal relationship with many events concerning the various parts of the system, i.e. the loss of supply at BT1 and BT2 bus-bars is linked through some intermediate events to related primary events.

The following section shows how the higher level malfunctions have been analysed, using the structure function of the fault-tree of the system.

3.3. Top-events considered

Figure 4 shows the structure function of the system. The solution of problems outlined in Section 3.2 consists in modifying some input data of the fault-tree of Fig.3 as a global model.

The analysed top-events of interest are the following:

TOP 1: corresponding to G(73)
Loss of supply at both BT1 and BT2 bus-bars

TOP 2: corresponding to G(61)
Loss of supply at M1 bus-bar

TOP 3: corresponding to G(100)
Loss of supply at both M1 and M2 bus-bars

TOP 4: corresponding to G(51)
Loss of supply at M2 bus-bar

For each of these top-events (malfunctions of interest) three more significant situations have been taken into account. Various situations related to the various top-events considered are shown in Table II.

3.4. Results

To carry out the analysis described above we have utilized the AWE computer program [3]. Given the structure function of the system, the primary event data and the mission time, the program performs, for each top-event of interest, the following operations:

Computes the primary event probabilities
Finds the more important minimal cut sets
Computes the probability of each minimal cut set
Computes the top-event probability
Classes the cut sets and the primary events according to their 'weight'.

The results of the analysis are summarized in Table III. In the table the primary events are arranged in non-increasing order according to their weight. Each column refers to a particular situation. The weight of the 20th primary event is in any case $\leq 10^{-7}$. The ratio between the weight of the event with relative importance 1 and the weight of the event with relative importance 20 is $\leq 10^{-4}$ for all the situations considered. These values provide very important information for the design evaluation.

FIG. 3. Fault-tree for the PEC reactor supply system.

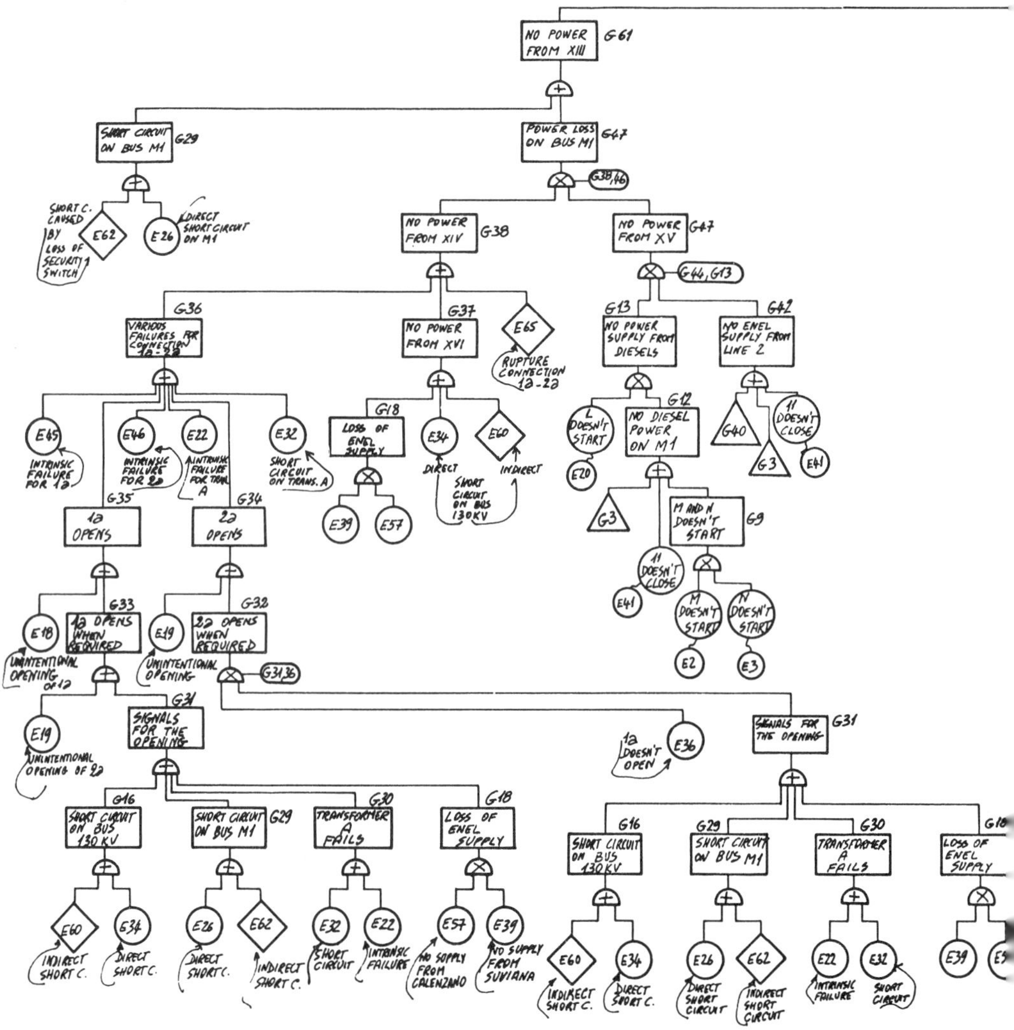

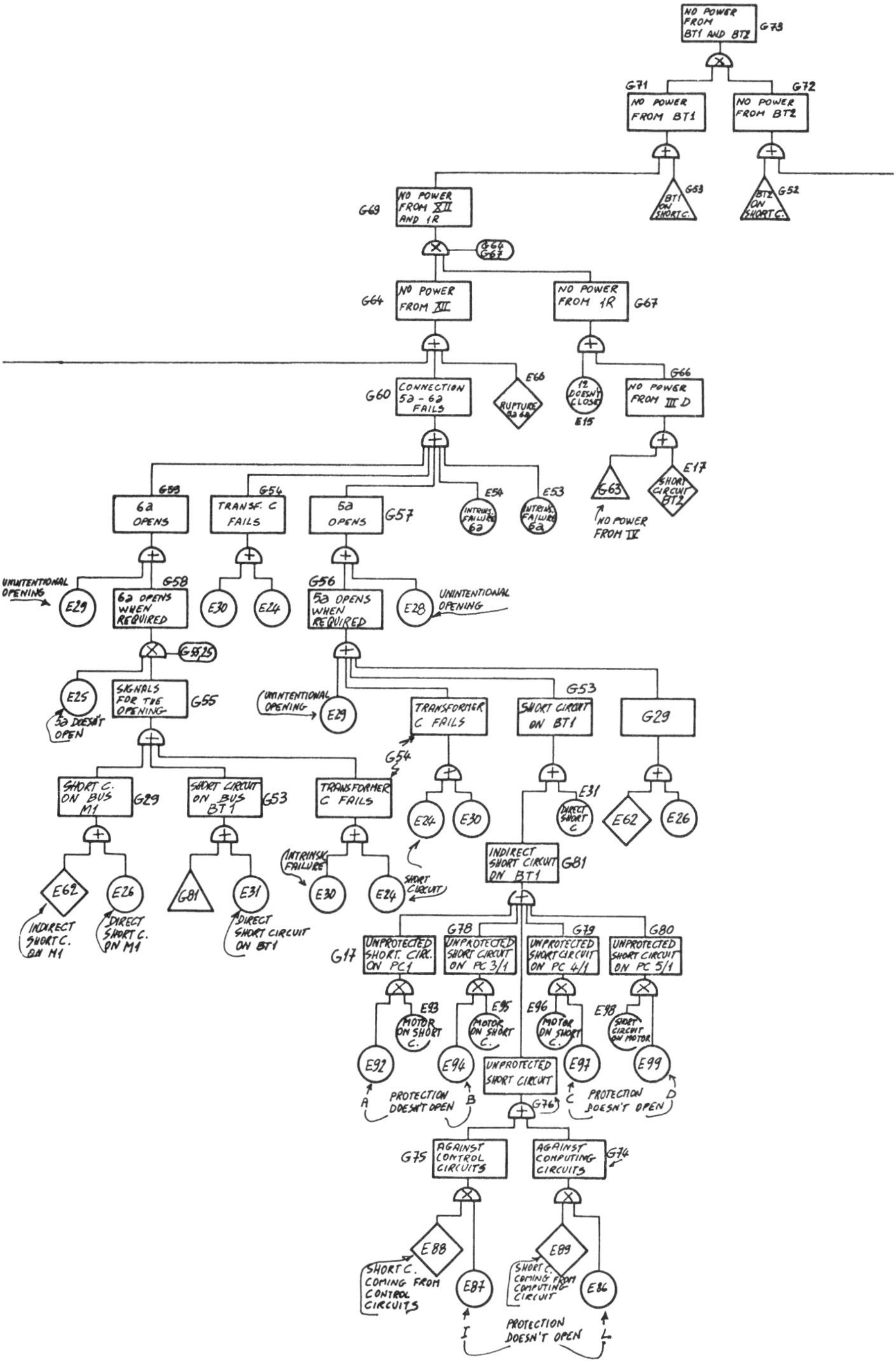
NO POWER FROM BT1 AND BT2
G73
G71
NO POWER FROM BT1
G72
NO POWER FROM BT2
BT1 ON SHORT C.
G53
BT2 ON SHORT C.
G52
G69
NO POWER FROM XII AND 1R
G64 G67
G64
NO POWER FROM XII
NO POWER FROM 1R
G67
G60
CONNECTION 5a - 6a FAILS
E66
RUPTURE 5a 6a
12 DOESN'T CLOSE
E15
G66
NO POWER FROM III D
G63
NO POWER FROM IV
E17
SHORT CIRCUIT BT2
G59
6a OPENS
G54
TRANSF. C FAILS
5a OPENS
G57
E54
INTRINS. FAILURE 6a
E53
INTRINS. FAILURE 5a
UNINTENTIONAL OPENING
E29
G58
6a OPENS WHEN REQUIRED
E30
E24
G56
5a OPENS WHEN REQUIRED
E28
UNINTENTIONAL OPENING
G55 E25
E25
5a DOESN'T OPEN
SIGNALS FOR THE OPENING
G55
UNINTENTIONAL OPENING
E29
TRANSFORMER C FAILS
G53
SHORT CIRCUIT ON BT1
G29
G54
SHORT C. ON BUS M1
G29
SHORT CIRCUIT ON BUS BT1
G53
TRANSFORMER C FAILS
E24
E30
E31
DIRECT SHORT C
E62
E26
INDIRECT SHORT CIRCUIT ON BT1
G81
E62
E26
INDIRECT SHORT C. ON M1
DIRECT SHORT C. ON M1
G81
E31
DIRECT SHORT CIRCUIT ON BT1
INTRINSIC FAILURE
E30
E24
SHORT CIRCUIT
G78
G79
G80
G17
UNPROTECTED SHORT. CIRC. ON PC1
UNPROTECTED SHORT CIRCUIT ON PC 3/1
UNPROTECTED SHORT CIRCUIT ON PC 4/1
UNPROTECTED SHORT CIRCUIT ON PC 5/1
E93
MOTOR ON SHORT C.
E95
MOTOR ON SHORT C.
E96
MOTOR ON SHORT C.
E98
SHORT CIRCUIT ON MOTOR
E92
E94
UNPROTECTED SHORT CIRCUIT
E97
E99
A
PROTECTION DOESN'T OPEN
B
G76
C
PROTECTION DOESN'T OPEN
D
G75
AGAINST CONTROL CIRCUITS
AGAINST COMPUTING CIRCUITS
G74
E88
SHORT C. COMING FROM CONTROL CIRCUITS
E87
E89
SHORT C. COMING FROM COMPUTING CIRCUIT
E86
I
PROTECTION DOESN'T OPEN
L

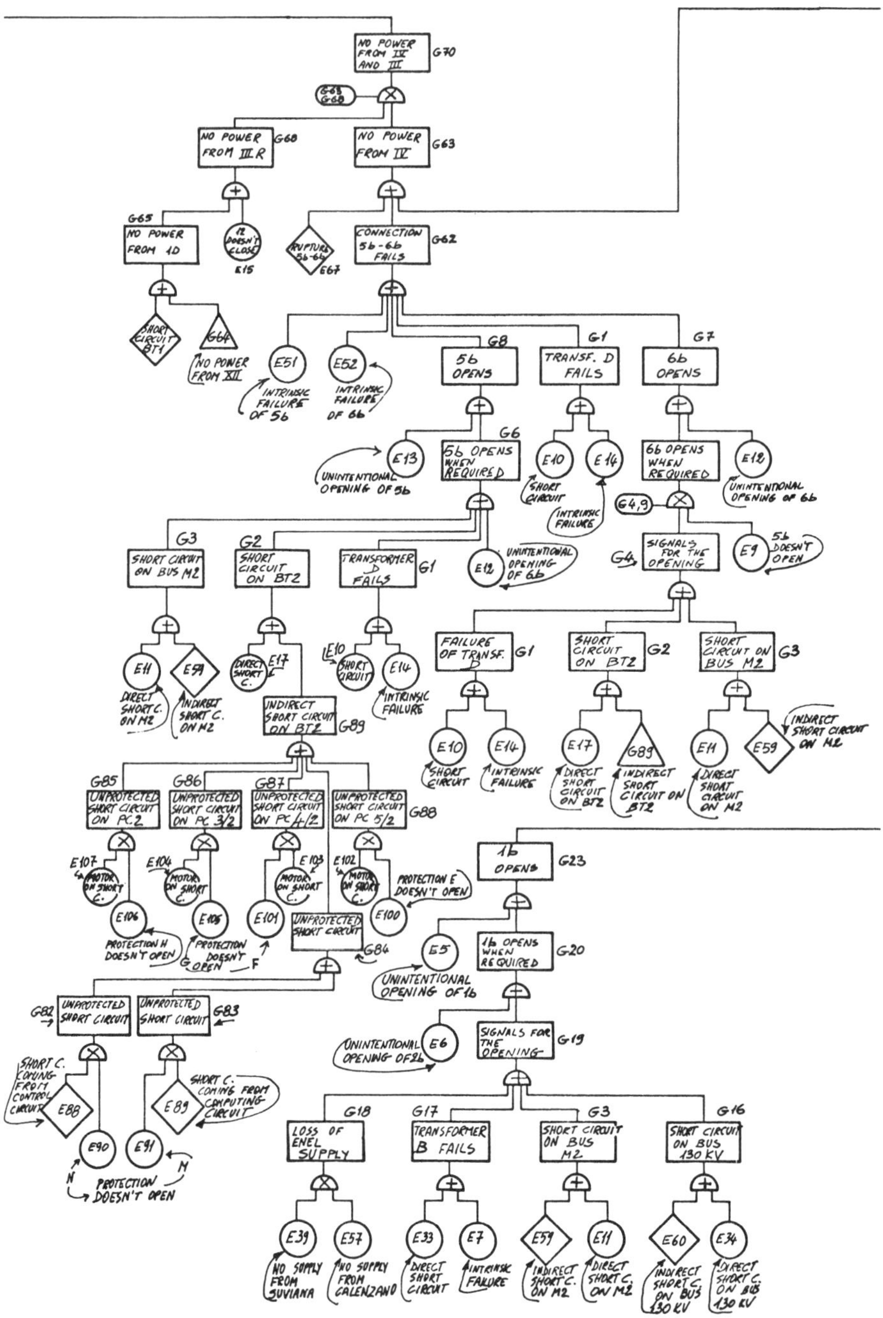
NO POWER FROM IV AND III
G70
NO POWER FROM III R
G68
NO POWER FROM IV
G63
NO POWER FROM 1D
G65
CONNECTION 5b-6b FAILS
G62
5b OPENS
G8
TRANSF. D FAILS
G1
6b OPENS
G7
5b OPENS WHEN REQUIRED
G6
6b OPENS WHEN REQUIRED
UNINTENTIONAL OPENING OF 5b
UNINTENTIONAL OPENING OF 6b
INTRINSIC FAILURE OF 5b
INTRINSIC FAILURE OF 6b
SHORT CIRCUIT ON BUS M2
SHORT CIRCUIT ON BT2
TRANSFORMER D FAILS
SIGNALS FOR THE OPENING
FAILURE OF TRANSF. D
INDIRECT SHORT CIRCUIT ON BT2
G89
UNPROTECTED SHORT CIRCUIT ON PC 2
UNPROTECTED SHORT CIRCUIT ON PC 3/2
UNPROTECTED SHORT CIRCUIT ON PC 4/2
UNPROTECTED SHORT CIRCUIT ON PC 5/2
PROTECTION E DOESN'T OPEN
PROTECTION H DOESN'T OPEN
1b OPENS
G23
1b OPENS WHEN REQUIRED
G20
UNINTENTIONAL OPENING OF 1b
SIGNALS FOR THE OPENING
G19
UNINTENTIONAL OPENING OF 2b
LOSS OF ENEL SUPPLY
G18
TRANSFORMER B FAILS
G17
SHORT CIRCUIT ON BUS M2
G3
SHORT CIRCUIT ON BUS 130 KV
G16
NO SUPPLY FROM SUVIANA
NO SUPPLY FROM CALENZANO
DIRECT SHORT CIRCUIT
INTRINSIC FAILURE
INDIRECT SHORT C. ON M2
DIRECT SHORT C. ON M2
INDIRECT SHORT C. ON BUS 130 KV
DIRECT SHORT C. ON BUS 130 KV
UNPROTECTED SHORT CIRCUIT
G84
G82
G83
SHORT C. COMING FROM CONTROL CIRCUIT
SHORT C. COMING FROM COMPUTING CIRCUIT
PROTECTION DOESN'T OPEN

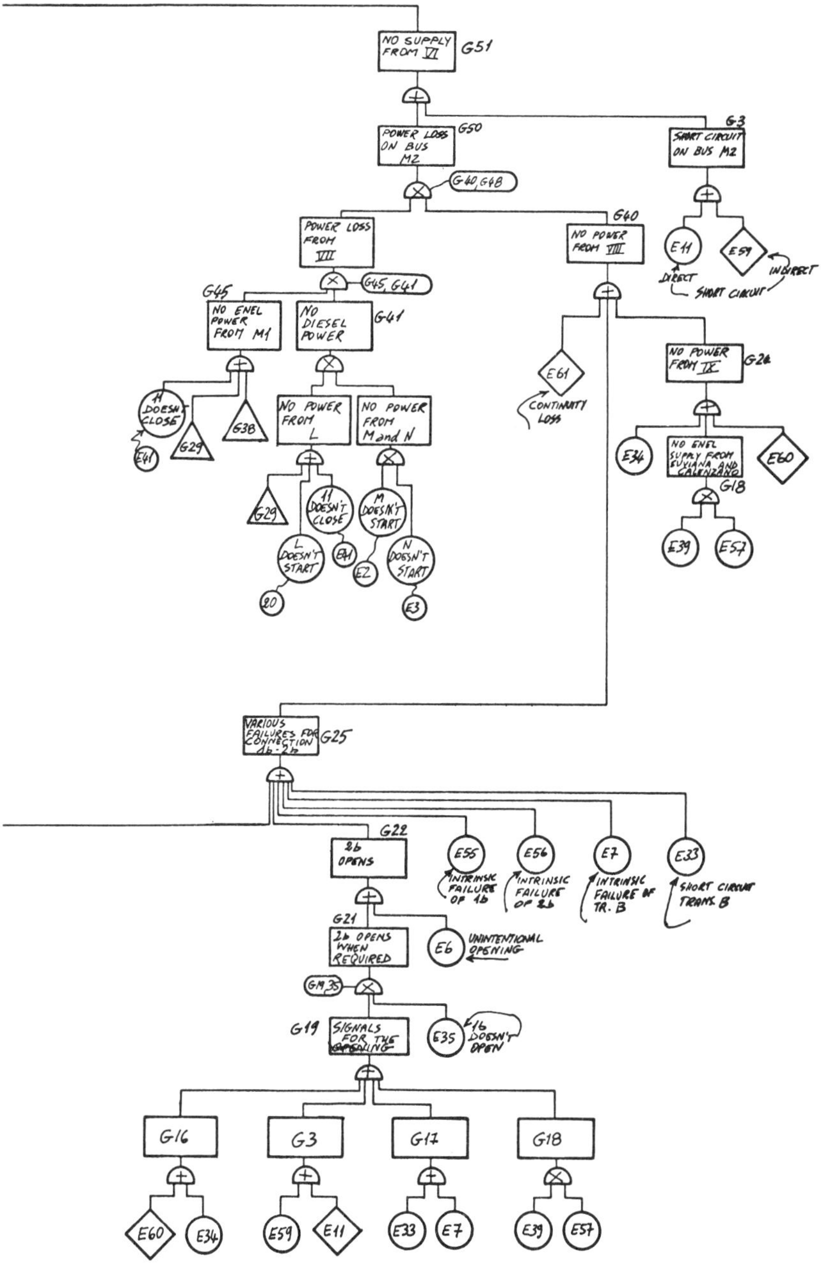
NO SUPPLY FROM VI
G51
POWER LOSS ON BUS M2
G50
G40, G48
SHORT CIRCUIT ON BUS M2
G3
E11
E59
DIRECT
INDIRECT
SHORT CIRCUIT
POWER LOSS FROM VII
G45, G41
NO POWER FROM VIII
G40
G45
NO ENEL POWER FROM M1
NO DIESEL POWER
G41
H DOESN'T CLOSE
E41
G29
G38
NO POWER FROM L
NO POWER FROM M and N
G29
H DOESN'T CLOSE
M DOESN'T START
L DOESN'T START
N DOESN'T START
20
E41
E2
E3
E61
CONTINUITY LOSS
NO POWER FROM IX
G26
E34
NO ENEL SUPPLY FROM SUVIANA AND CALENZANO
E60
G18
E39
E57
VARIOUS FAILURES FOR CONNECTION 1b-2b
G25
G22
2b OPENS
E55
INTRINSIC FAILURE OF 1b
E56
INTRINSIC FAILURE OF 2b
E7
INTRINSIC FAILURE OF TR. B
E33
SHORT CIRCUIT TRANS. B
G21
2b OPENS WHEN REQUIRED
E6
UNINTENTIONAL OPENING
G19, 35
G19
SIGNALS FOR THE OPENING
E35
1b DOESN'T OPEN
G16
G3
G17
G18
E60
E34
E59
E11
E33
E7
E39
E57

```
G(1)=F(10).OR.F(14)
G(82)=F(88).AND.E(90)
G(83)=F(89).AND.E(91)
G(84)=G(82).OR.G(83)
G(85)=F(106).AND.E(107)
G(86)=F(104).AND.E(105)
G(87)=F(101).AND.E(103)
G(88)=F(100).AND.E(102)
G(89)=G(84).OR.G(85).OR.G(86).OR.G(87).OR.G(88)
G(2)=F(17).OR.G(89)
G(3)=F(11).OR.F(59)
G(4)=G(1).OR.G(2).OR.G(3)
G(5)=F(9).AND.G(4)
G(6)=F(12).OR.G(1).OR.G(2).OR.G(3)
G(7)=F(12).OR.G(5)
G(8)=F(13).OR.G(6)
G(9)=F(2).AND.F(3)
G(12)=F(41).OR.G(3).OR.G(9)
G(13)=F(20).AND.G(12)
G(16)=F(34).OR.E(60)
G(17)=F(7).OR.F(33)
G(18)=F(39).AND.E(57)
G(19)=G(3).OR.G(16).OR.G(17).OR.G(18)
G(20)=F(6).OR.G(19)
G(21)=F(35).AND.G(19)
G(22)=F(6).OR.G(21)
G(23)=F(5).OR.G(20)
G(24)=F(34).OR.E(60).OR.G(18)
G(25)=F(7).OR.F(33).OR.E(55).OR.F(56).OR.G(22).OR.G(23)
G(29)=F(26).OR.E(62)
G(30)=F(22).OR.E(32)
G(31)=G(16).OR.G(18).OR.G(29).OR.G(30)
G(32)=F(36).AND.G(31)
G(33)=F(19).OR.G(31)
G(34)=F(19).OR.G(32)
G(35)=F(18).OR.G(33)
G(36)=F(22).OR.F(32).OR.E(45).OR.F(46).OR.G(34).OR.G(35)
G(37)=F(34).OR.E(60).OR.G(18)
G(38)=F(65).OR.G(36).OR.G(37)
G(39)=F(20).OR.F(41).OR.G(29)
G(40)=F(61).OR.G(24).OR.G(25)
G(41)=G(9).AND.G(39)
G(44)=F(41).OR.G(3).OR.G(40)
G(45)=F(41).OR.G(29).OR.G(38)
G(46)=G(13).AND.G(44)
G(47)=G(38).AND.G(46)
G(48)=G(41).AND.G(45)
G(50)=G(40).AND.G(48)
G(51)=G(3).OR.G(50)        TOP 4
G(74)=F(86).AND.E(89)
G(75)=F(87).AND.E(88)
G(76)=G(74).OR.G(75)
G(77)=F(92).AND.E(93)
G(78)=F(94).AND.E(95)
G(79)=E(96).AND.E(97)
G(80)=F(98).AND.E(99)
G(81)=G(76).OR.G(77).OR.G(78).OR.G(79).OR.G(80)
G(53)=F(31).OR.G(81)
G(54)=E(24).OR.E(30)
G(55)=G(29).OR.G(53).OR.G(54)
G(56)=F(29).OR.G(29).OR.G(53).OR.G(54)
G(57)=F(28).OR.G(56)
G(58)=F(25).AND.G(55)
G(59)=F(29).OR.G(58)
G(60)=E(53).OR.E(54).OR.G(54).OR.G(57).OR.G(59)
G(61)=G(29).OR.G(47)       TOP 2
G(62)=F(51).OR.E(52).OR.G(1).OR.G(7).OR.G(8)
G(63)=F(67).OR.G(51).OR.G(62)
G(64)=F(66).OR.G(60).OR.G(61)
G(65)=F(31).OR.G(64)
G(66)=F(17).OR.G(63)
G(67)=F(15).OR.G(66)
G(68)=F(15).OR.G(65)
G(69)=G(64).AND.G(67)
G(70)=G(63).AND.G(68)
G(71)=G(53).OR.G(69)
G(72)=G(2).OR.G(70)
G(73)=G(71).AND.G(72)      TOP 1
G(100)=G(61).AND.G(51)     TOP 3
```

FIG. 4. Structure function of the system.

TABLE II. TABLE OF SITUATIONS FOR TOP-EVENTS ANALYSED

Top-event number and description	Situations	Plant operating conditions				
		Diesel L	Diesel M	Diesel N	Line from Suviana	Line from Calenzano
Top-event 1, G(73) Loss of supply at both BT1 and BT2 bus-bars	1/A	Available	Available	Available	Available	Available
	1/B	Available	Available	Available	Not available	Not available
	1/C	Available	Not available	Not available	Not available	Not available
Top-event 2, G(61) Loss of supply at M1 bus-bar	2/A	Available	Available	Available	Available	Available
	2/B	Available	Available	Available	Not available	Not available
	2/C	Available	Not available	Not available	Not available	Not available
Top-event 3, G(100) Loss of supply at both M1 and M2 bus-bars	3/A	Available	Available	Available	Available	Available
	3/B	Available	Available	Available	Not available	Not available
	3/C	Available	Not available	Not available	Not available	Not available
Top-event 4, G(51) Loss of supply at M2 bus-bar	4/A	Available	Available	Available	Available	Available
	4/B	Available	Available	Available	Not available	Not available
	4/C	Available	Not available	Not available	Not available	Not available

TABLE III. RELATIVE IMPORTANCE OF PRIMARY EVENTS IN VARIOUS SITUATIONS

Relative importance of primary events	Situations											
	1/A	1/B	1/C	2/A	2/B	2/C	3/A	3/B	3/C	4/A	4/B	4/C
1	10	39	2	26	20	20	20	39	2	11	2	2
2	24	57	3	62	39	39	2	57	3	59	3	3
3	14	20	39	20	57	57	3	20	39	2	39	39
4	30	2	57	41	2	2	34	2	57	3	57	57
5	51	3	20	22	3	3	22	3	20	41	20	20
6	52	10	26	32	41	41	11	11	26	7	11	41
7	53	24	10	2	26	26	32	26	11	20	41	26
8	54	11	24	3	11	11	7	59	34	33	59	11
9	12	14	11	34	62	34	33	34	62	34	26	7
10	13	30	34	45	22	62	60	62	7	22	7	33
11	28	41	62	46	32	22		22	22	32	33	34
12	29	26	7	18	59	7		32	33	55	34	62
13	11	51	22	19	34	32		7	32	56	62	59
14	17	52	33	11	45	33		33	59	60	22	22
15	26	53	32	7	46	59		60	60		32	32
16	31	54	14	33	18	60			55		55	55
17	20	12	30	60	19	45			56		56	56
18		13	41	65	7	46			45		60	60
19		28	51		33	55			46			5
20		29	52		60	56			5			6

Finally, Table IV gives a description of the most important primary events. It also gives their index number, failure and restoration parameters for the AWE program.

4. CONCLUSIONS

The most important conclusions are the following:

(1) As indicated in recent papers [4], the main disadvantage of fault-tree techniques in systems description is the amount of time consumed. This is because normally each system situation must be characterized by an ad hoc fault-tree. The proposed model based on a single fault-tree used for all situations of interest allows the time consumed in the system description to be reduced drastically.

(2) The various situations considered can refer both to the safety and the availability of the plant, so the relevance of the primary events can be evaluated with reference to both aspects. This allows a weighted optimization of the system design to be performed.

TABLE IV. TABLE OF MOST IMPORTANT PRIMARY EVENTS

Description	Index	Type of failure distribution	Failure parameters		Type of restoration distribution	Restoration parameters	
	1	1	1.00E-03	0.0	0	0.0	0.0
Diesel M does not start	2	1		0.0	0	0.0	0.0
Diesel N does not start	3	1		0.0	0	0.0	0.0
Unintentional opening of 1b	4	1	1.00E-03	0.0	0	0.0	0.0
Unintentional opening of 1b	5	2	1.00E-06	0.0	2	5.00E-02	0.0
Unintentional opening of 2b	6	2	2.00E-06	0.0	2	5.00E-02	0.0
Intrinsic failure of transformer B	7	2	1.00E-05	0.0	2	8.33E-03	0.0
	8	1	1.00E-03	0.0	0	0.0	0.0
	9	1	1.00E-03	0.0	0	0.0	0.0
Direct short circuit of transformer D	10	2	5.00E-07	0.0	2	6.67E-04	0.0
Direct short circuit on bus-bar M2	11	2	1.00E-07	0.0	2	1.00E-02	0.0
Unintentional opening of 6b	12	2	2.00E-06	0.0	2	5.00E-02	0.0
Unintentional opening of 5b	13	2	2.00E-06	0.0	2	5.00E-02	0.0
Intrinsic failure of transformer D	14	2	6.00E-07	0.0	2	8.33E-03	0.0
	15	1	1.00E-03	0.0	0	0.0	0.0
	16	1	1.00E-03	0.0	0	0.0	0.0
Direct short circuit on bus-bar BT2	17	2	1.00E-07	0.0	2	1.00E-02	0.0
Unintentional opening of 1a	18	2	1.00E-06	0.0	2	5.00E-02	0.0
Unintentional opening of 2a	19	2	2.00E-06	0.0	2	5.00E-02	0.0
Diesel L does not start	20	1	7.00E-02	0.0	0	0.0	0.0
	21	1	1.00E-03	0.0	0	0.0	0.0
Intrinsic failure of transformer A	22	2	1.00E-05	0.0	2	8.33E-03	0.0
	23	1	1.00E-03	0.0	0	0.0	0.0
Direct short circuit of transformer C	24	2	5.00E-07	0.0	2	6.67E-04	0.0
	25	1	1.00E-03	0.0	0	0.0	0.0
Direct short circuit on bus-bar M1	26	2	1.00E-07	0.0	2	1.00E-02	0.0
	27	1	1.00E-03	0.0	0	0.0	0.0
Unintentional opening of 5a	28	2	2.00E-06	0.0	2	5.00E-02	0.0
Unintentional opening of 6a	29	2	2.00E-06	0.0	2	5.00E-02	0.0
Intrinsic failure of transformer C	30	2	6.00E-07	0.0	2	8.33E-03	0.0
Direct short circuit on bus-bar BT1	31	2	1.00E-07	0.0	2	1.00E-02	0.0

Direct short circuit of transformer A	32	2	5.00E-07	0.0	2	6.67E-04	0.0
Direct short circuit of transformer B	33	2	5.00E-07	0.0	2	6.67E-04	0.0
Direct short circuit on 130 kV bus-bars	34	2	1.00E-07	0.0	2	1.00E-02	0.0
	35	1	1.00E-02	0.0	0	0.0	0.0
	36	1	1.00E-02	0.0	0	0.0	0.0
	37	2	5.00E-07	0.0	2	6.67E-04	0.0
No supply from Suviana	39	1		0.0	0	0.0	0.0
	40	2	1.00E-06	0.0	2	2.08E-02	0.0
Switch 11 does not close	41	1	1.00E-03	0.0	0	0.0	0.0
Intrinsic failure of switch 1a	45	2	1.00E-06	0.0	2	2.08E-02	0.0
Intrinsic failure of switch 2a	46	2	1.00E-06	0.0	2	2.08E-02	0.0
Intrinsic failure of switch 5b	51	2	1.00E-06	0.0	2	2.08E-02	0.0
Intrinsic failure of switch 6b	52	2	1.00E-06	0.0	2	2.08E-02	0.0
Intrinsic failure of switch 5a	53	2	1.00E-06	0.0	2	2.08E-02	0.0
Intrinsic failure of switch 6a	54	2	1.00E-06	0.0	2	2.08E-02	0.0
Intrinsic failure of switch 1b	55	2	1.00E-06	0.0	2	2.08E-02	0.0
Intrinsic failure of switch 2b	56	2	1.00E-06	0.0	2	2.08E-02	0.0
No supply from Calenzano	57	1		0.0	0	0.0	0.0
Indirect short circuit on bus-bar M2	59	2	5.00E-07	0.0	2	1.00E-00	0.0
Indirect short circuit on 130 kV bus-bars	60	2	5.00E-07	0.0	2	1.00E-00	0.0
Continuity loss on 1b-2b interconnection	61	2	1.00E-07	0.0	2	5.00E-02	0.0
Indirect short circuit on bus-bar M1	62	2	5.00E-07	0.0	2	1.00E-00	0.0
Continuity loss on 1a-2a interconnection	65	2	1.00E-07	0.0	2	5.00E-02	0.0

REFERENCES

[1] COLOMBO, A.G., VOLTA, G., "Multistep reliability analysis and optimisation of complex systems", OECD-NEA Specialist Meeting on Reliability, Liverpool, 1974.

[2] COLOMBO, A.G., CADI. A Computer Code for System Availability and Reliability Evaluation, Rep. EUR 4940 (1973).

[3] COLOMBO, A.G., VOLTA, G., Sensitivity Analysis in Systems Reliability Evaluation, Joint Research Centre Ispra, Ann. Rep. 1973.

[4] USAEC, Reactor Safety Study. An assessment of accident risks in U.S. commercial nuclear power plants, Rep. WASH-1400.

DISCUSSION

(on papers IAEA-SM-195/2 and IAEA-SM-195/48)

R. GONNOT: I am interested in hearing comments on the papers of Mr. Colombo (IAEA-SM-195/48) and Mr. Greppo (IAEA-SM-195/11) on the relationship which their methods bear to the approach taken in the Rasmussen Report. Rasmussen, in analysing the electrical supply of a power plant, has taken accident conditions as his point of departure. Mr. Greppo assumes normal operating conditions. Mr. Colombo, seeking relative weights to be assigned to the failures of different components, can use both the hypotheses mentioned above. Would anyone care to say what the best approach is?

B. GACHOT: I don't think it is appropriate to speak of the 'best approach', because the three studies mentioned by Mr. Gonnot, which use analytical methods based on fault trees either wholly or in part, pursue rather different objectives which one might regard as complementary. The purpose of the Rasmussen Report was to evaluate the probability of failure of electrical supply systems after an accident involving the rupture of primary piping. Mr. Greppo's aim was to establish operating rules for the event of partial unavailability of electric power distribution systems, a task which required the combined use of a fault-tree analysis and a representation of the evolution of system states with time based on the Markov method. Mr. Colombo, if I understood him properly, was trying to evaluate the probability of electric power distribution systems being partially or totally unavailable in any situation of interest from the standpoint of safety analyses or nuclear power station availability studies, and the situation in question could involve either accident conditions or normal operation. The method is based on the use of a single fault tree, whatever the situation of interest, an approach which doubtless saves much time and effort.

A.G. COLOMBO: I would like to add just two comments to what Mr. Gachot has said. Apart from enabling us to evaluate probabilities for any situation of interest from the standpoint of safety or availability — at different system levels, and with an appreciable saving of time — the method I described also allows an evaluation of the relative importance of each component in all the situations of interest. Secondly, the method can be applied to any type of system, provided the functions demanded of the system are ordered hierarchically.

R. SARTORI: I have a question for Mr. Colombo. Since the supply to the 380 V bus-bars that serve the sodium pump motors depends on the two transformers C and D, you would depend, after an accident at transformer C, on the reliability of D for the lag time of after-heat removal if you need the 380 V bus-bar for after-heat removal.

A.G. COLOMBO: The main aim of our paper was to present a method. In fact, in the case of the scheme we used as an example, analysis shows that the availability of the 380 V bus-bar is strongly linked to the availability of the medium-voltage transformers. The designer will take that feature properly into account in the final design of the system, bearing in mind the actual requirements of the plant.

L. CAVE: I have a comment on Mr. Bremer's presentation. It seems to me that he is unduly optimistic about the risk of failure of concrete pressure vessels. I agree that it may be possible to detect deterioration of the vessel in service in time to prevent failure, though even here there is some probability that the operator will fail to make the necessary measurements. However, there is always a chance that, through an error on the part of the reactor designer, the vessel may be subjected to a sudden accidental overload greater than any specified originally to the vessel designer. As there must always be some finite risk of such an error, it is not possible in my view to claim that concrete pressure vessels are 'absolutely safe'.

F. BREMER: Tests on small-scale models indicate large safety factors for concrete pressure vessels; in this respect they are at least as good as steel vessels.

L. CAVE: Yes, but it is generally accepted that steel vessels are subject to this risk of a sudden catastrophic overload and may fail.

F. BREMER: I think the most important assumption is that with a prestressed concrete vessel the operator will have a good picture at all times of the vessel's load-bearing capability and will be able to take remedial action should a threatening situation develop.

L. CAVE: In my opinion it is by no means certain that the operator would be able to take action quickly enough to prevent the development of a disastrous overload.

D. OKRENT: If one accepts the conclusion of the Rasmussen Report that the probability of core melt-down exceeds the probability of gross failure of a steel pressure vessel by two to three orders of magnitude, I think there is at least a serious question whether prestressed concrete vessels really change the siting problem at all.

It would be interesting to know whether a systematic event- and fault-tree analysis of prestressed concrete vessel failure has been made, covering such factors as design error, corrosion, faulty materials and the like. Would it be possible to quantify such a study?

F. BREMER: Such investigations have been made, or are under way. I have no details of them at present, but I should be glad to get in touch with you later on.

APPLICATION OF RELIABILITY ANALYSIS IN NUCLEAR POWER PLANT DESIGN

Chairman:

A.V. KARPOV, Union of Soviet Socialist Republics

AVAILABILITY: THE OTHER ASPECT OF NUCLEAR POWER PLANT RELIABILITY

L.A. HOPKINS, L.L. CONRADI, J. YOUNG
Boeing Engineering and Construction,
Seattle, Wash.,
United States of America

Abstract

AVAILABILITY: THE OTHER ASPECT OF NUCLEAR POWER PLANT RELIABILITY.

Safety has been and will continue to be a prime factor in nuclear power plant design, licensing and operation. Plant reliability design effort has, to date, been almost exclusively dedicated to developing highly reliable safety systems. There is another aspect to system reliability that deserves greater recognition: plant availability. Plant availability is generally defined as the fraction of time that power is being generated. Availability is a function of plant/equipment reliability, system restoration time, and administrative downtime. The predisposition to apply reliability techniques exclusively to solving the plant safety problem has, in some instances, resulted in decreased plant availability. Conscious design effort to imporve plant operating reliability — as opposed to designing only for high safety system reliability — can result in substantial improvement in the in-use plant availabilities as well as maintaining the required level of plant safety. This paper develops the concept of optimizing plant operating economics as a function of availability within the context of existing safety criteria/requirements. Specific reliability techniques that have application to nuclear power plant design are identified. The economic impact of increased plant availability is explored. This impact is examined within the framework of present plant availability experience and operating costs. The potential for improvement in nuclear power cost is identified and presented graphically. In summary, the authors examine the need for greater effort toward increased plant availability/reliability in the design process and present some measures of the economic potential that can be gained.

INTRODUCTION

Historically, safety has been the forcing function which has driven the design of nuclear power plants. Government regulatory agencies exist whose sole function is to assure minimal risk to the public from these facilities. Reliability techniques have been applied to the design of nuclear plants, primarily to assure the availability of critical safety systems. Features such as fail-safe design, redundancy, physical separation, and functional diversity have been incorporated and, in many cases, are required for nuclear systems which perform a safety function.

Maturation of plant safety designs and current economic factors dictate that proven reliability techniques need to be applied to the balance of plant features as well as to the nuclear steam supply system. This application can aid in the optimization of availability of each nuclear unit. Benefits from increased plant availability will accrue to utilities and consumers alike.

A review of the operating history of United States' commercial light water reactors shows the need for increased emphasis on plant availability. In the calendar year 1972, these plants averaged 73.4% availability [1]. Last year, 1974, proved to be even more disappointing with all commercial plants exhibiting an average of 68.5% availability [2]. This experienced availability is lower than fossil-fueled electrical plants for the same time period.

There are many features of nuclear plants that affect unit availability. These include (1) the complexity of nuclear systems, (2) the requirement for multiple safeguard systems, (3) scheduled test and maintenance activities, (4) refueling outages, and (5) strict operating requirements imposed by regulatory agencies. Inclusion of availability as an identifiable concern throughout all phases of design, development, and utilization is necessary to obtain the most economical operation within the constraints listed above.

DEFINITIONS

Availability can be defined in a number of specific ways; but it is generally interpreted as the fraction of time that a system--in this case, a nuclear power plant--is ready for use or is in use. The time that the system has less than the maximum demand placed on it is still counted as "up-time" as long as the system is in a condition to produce the maximum output. Thus, nuclear power plant availability is one major element in plant capacity factor; the other being the demand, or loads, on the plant.

To assure common understanding of the terminology used in this paper, the following basic definitions are provided:

Unit - a power-generating station consisting of a single steam supply system and associated electrical generating equipment. The term plant is used, in this paper, to refer to a single unit.

Reliability - the probability that an item accomplishes its intended function when operated within specified conditions.

Unit Up-Time - the time that the unit is actually operated with the turbine generator on line.

Unit Down-Time - any time other than unit up-time.

Availability - the ratio of the total unit up-time in the period, divided by the total time in the reporting period, expressed as a per cent.

Maximum Dependable Capacity (MDC) - the gross electrical output as measured at the output terminals of the turbine-generator during the most restrictive seasonal conditions, less the normal service loads.

Unit Capacity Factor - the quotient of the actual net electrical energy output produced by the unit in a given period, divided by the maximum net electrical energy output the unit would have produced if operated at the maximum dependable capacity for the entire period, expressed as a per cent.

PRESENT SAFETY CONSTRAINTS

In the United States, nuclear power plants are built and operated only after detailed reviews show that plant design and proposed method of operation meet prescribed safety standards and criteria. The Nuclear Regulatory Commission (NRC) imposes these safety constraints to assure the health and safety of the public.

The standards and criteria imposed by the NRC require that nuclear power plant design be conservative. A multi-layered safety design concept, referred to as "defense-in-depth," is applied for prevention of and protection against potential nuclear plant accidents.

The initial level of design in this approach is to assure that the plant is designed to conservative standards so that in all phases of operation there is a large tolerance for operator errors, off-normal operation, and component malfunctions. The objective is to provide plant systems such that no single equipment failure will deprive the plant of any needed safety function.

The next design safety level addresses the fact that regardless of the prudence exercised in the design and development process, certain incidents or malfunctions can be expected to occur during the life of the plant. Protective measures are provided to contend or cope with these anticipated occurrences. An example of this type of measure is the reactor shutdown system which is provided to reduce core reactivity quickly if plant conditions require such action.

A final level of safety is achieved by incorporating additional systems and design margins into the plant to protect the public in the event that certain unlikely accidents occur. Major failures of plant components and systems are postulated and safety systems designed to mitigate the effects of such failures. These systems, called engineered safety features, include such systems as emergency core cooling systems, various containment designs, containment sprays, and atmosphere control systems. The purpose of these features is to provide a series of overlapping safety systems and physical barriers to reduce the potential consequences of the postulated accidents to acceptable levels.

Some of the plant safety features have an inverse relationship to plant availability. The reactor shutdown system, vital to the safety of the plant, can adversely affect plant availability if inadvertent or spurious reactor trip signals are received. The engineered safety features are tested and monitored to assure their availability should the need arise. Technical specifications and procedures restrict plant operation in the event of conditions which might affect operation of these engineered safety features.

Unit availability is directly affected by the potential for unavailability or inadvertent operation of these safety related systems, most of which play no direct role in the generation of electric power. These systems require the specific application of reliability analysis techniques to assure that they accomplish their intended functions without compromising plant availability.

RELIABILITY TECHNIQUES

The required emphasis on plant safety quite naturally results in application of reliability analysis techniques to the design of plant safety systems. Both qualitative and quantitative techniques are applied to the design activities associated with these systems. These techniques have been effectively applied to electrical/electronic and mechanical systems alike.

The qualitative reliability techniques primarily identify which component failures can result in system failure and/or the causes of component failures. These qualitative analysis methods provide the basis for development of the system failure logic relationships necessary for the application of the quantitative reliability analysis methods. The quantitative techniques determine the frequency of component and/or system failures.

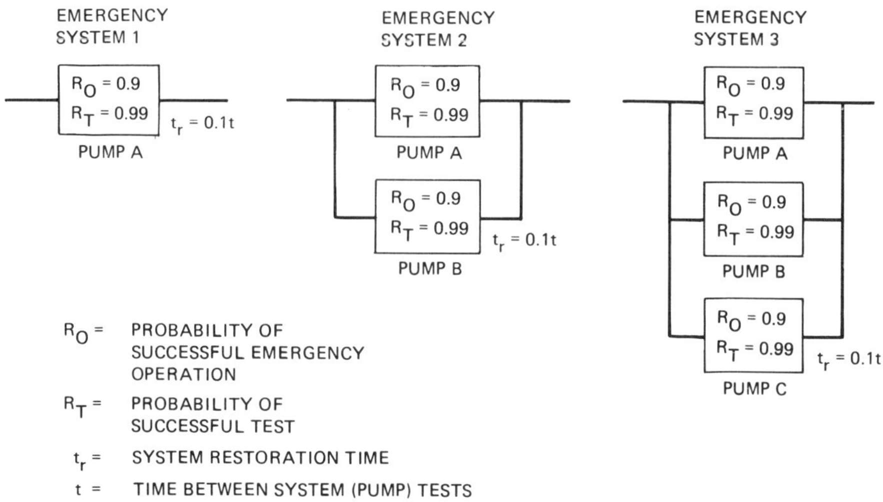

	SYSTEM 1	SYSTEM 2	SYSTEM 3
REQUIRED FOR SAFETY OPERATION	1 OF 1	1 OF 2	1 OF 3
REQUIRED TEST FOR NORMAL OPERATION	1 OF 1	2 OF 2	3 OF 3
SYSTEM RELIABILITY	0.90	0.99	0.999
SYSTEM TEST RELIABILITY	0.99	0.98	0.97
PLANT AVAIL-ABILITY LIMIT	0.999	0.998	0.997

FIG. 1. Comparative analysis of redundant pumps.

The quantitative analyses used to predict the reliability of safety systems also have application to the estimation of availability of nuclear power plants. In fact, the analysis of the safety systems with short-term missions is primarily an availability problem. If the system is available at the time of the emergency situation, there is a high probability that it will operate over the time period required.

The interrelationship between component reliability, system reliability, safety system availability, and plant availability can be illustrated with the aid of a simple example. Figure 1 portrays these relationships for three design approaches to an emergency system whose sole functional element is a pump(s). The basic scenario for this exercise is: the emergency system remains in a standby state (awaiting the occurrence of an emergency condition) for some designated time t; at time t, the pump(s) is tested; if the test(s) is passed, the scenario is repeated; if the test is failed, the plant is shut down and a repair is effected, then the scenario is repeated. The following assumptions were made for the example:

1) The failures of each pump are independent. The pump(s) can only fail during operation or test (i.e., the pump reliability is 1.0 for the standby period).

2) The repair time is $t_r = 0.1\ t$, and is independent of the number of failed pumps.

3) The plant procedures require all pumps to successfully pass a periodic operational test or the plant must be shut down.

4) The output of a single pump is required for successful safety operation (i.e., emergency shutdown).

5) System reliability for safety operation $= 1-(1-R_0)^n = R_S$

where n is the number of pumps, R_0 is the reliability of a pump during shutdown time period.

6) System test reliability $\approx 1-n(1-R_T) = R_{ST}$

where n is the number of pumps, R_T is the probability of a successful pump test.

7) During testing, each pump is available (unless failed) for emergency operation.

8) The plant availability limit associated with this emergency system is:

$$A_p \leq \frac{t-(1-R_{ST})t_r}{t} = 1-0.1(1-R_{ST})$$

It is obvious that as the safety system probability of successful operation is increased by added redundancy, the test reliability of the system decreases. Because the plant can only be operated with a successful system test, the increased redundancy could have a negative effect on plant availability. Determination of the exact nature of this effect would, in more realistic cases, require consideration of such factors as actual length and frequency of test, type of pump failure density function, distribution of repair times, the fact that pumps could fail during standby, and the time at which these failures occur.

The situation on the plant level is far more complex than that in the example. Systems required for power generation must be considered as well as the safety systems. In addition to forced outages due to equipment failures, there are scheduled downtimes for activities such as refueling which also impact plant availability. However, these additional factors can be included in availability analyses of nuclear power plants during the design phase as well as examined after-the-fact in plant experience reports.

Major contributors to plant unavailability, such as the primary coolant system and the secondary system [1], along with the safety systems and the balance of plant equipment, can be analyzed using qualitative and, more importantly, quantitative reliability techniques. The quantitative analysis results (predictions of plant availability/unavailability) then allow studies to be made which relate the change in plant availability and, if applicable, level of safety to change in design. These trade studies can include cost data so the economic impact of the design change, including plant design, equipment test and maintenance, and plant operation, can be measured in relation to plant revenue.

A variety of techniques, developed in various industries, are applicable to this problem of examining plant availability during the design phase. Reliability techniques can be used to examine the impact of hardware failure and repair on availability, the impact of scheduled down-time, and the impact of the total combination. Application of these techniques can vary from performance of a simple single failure examination of the plant systems to development and use of a very complex model representing all conceivable plant operating and failed states. Details of these techniques are available in a number of reliability texts [3], [4], [5].

Appropriate reliability analysis techniques can and should be applied during the design phase to optimize plant availability in relation to plant economics. Some combination of these techniques can provide the basis for development of a plant level availability model. This model will portray the system failure logic that accounts for plant down-time and incorporates the necessary statistical failure models for the constituent elements (components, subsystems, operator interfaces, test and maintenance activities). A side benefit of this activity should be increased plant safety because a plant experiencing increased availability within the context of existing safety requirements will experience fewer transients associated with shutdown and startup and thus will be stressed less.

ECONOMIC IMPACT OF INCREASED AVAILABILITY

Economic forecasts and pricing exercises generally assume a nuclear power plant availability of 75% to 80%. As indicated previously, current experience has shown an average plant availability on the order of 68% to 73%. Figure 2 is a histogram of individual plant availability experience for the year 1974 [2]. The thirty-four plants included in this survey were all 450 MW(e) or larger and were operational on or before September 1974. A striking feature of this diagram is the variance that it portrays. It is this variation, as well as the relatively low average availability, that provides the stimulus for this paper. It is of interest, therefore, to examine the relationship between availability and plant economics. Also of concern is the potential impact of increased availability which can, in part, be achieved by application of reliability analysis to support design.

The costs associated with electrical power production from nuclear power plants can be divided into three categories: capital, fuel, and operation and maintenance (O&M). As an aid to this discussion, the concept of cost factor is introduced. Cost factor is that portion of the total plant cost per kilowatt hour (kwh) of production associated with a given category of cost. These costs and/or cost factors can be written as functions of plant availability. Consequently, once these relationships are understood in detail, the benefits to be gained from consciously designing for high availability can be maximized with respect to cost. Experience in the aerospace industry has indicated that the maximum economic gain can be expected from expenditure of effort during the design phase of a program.

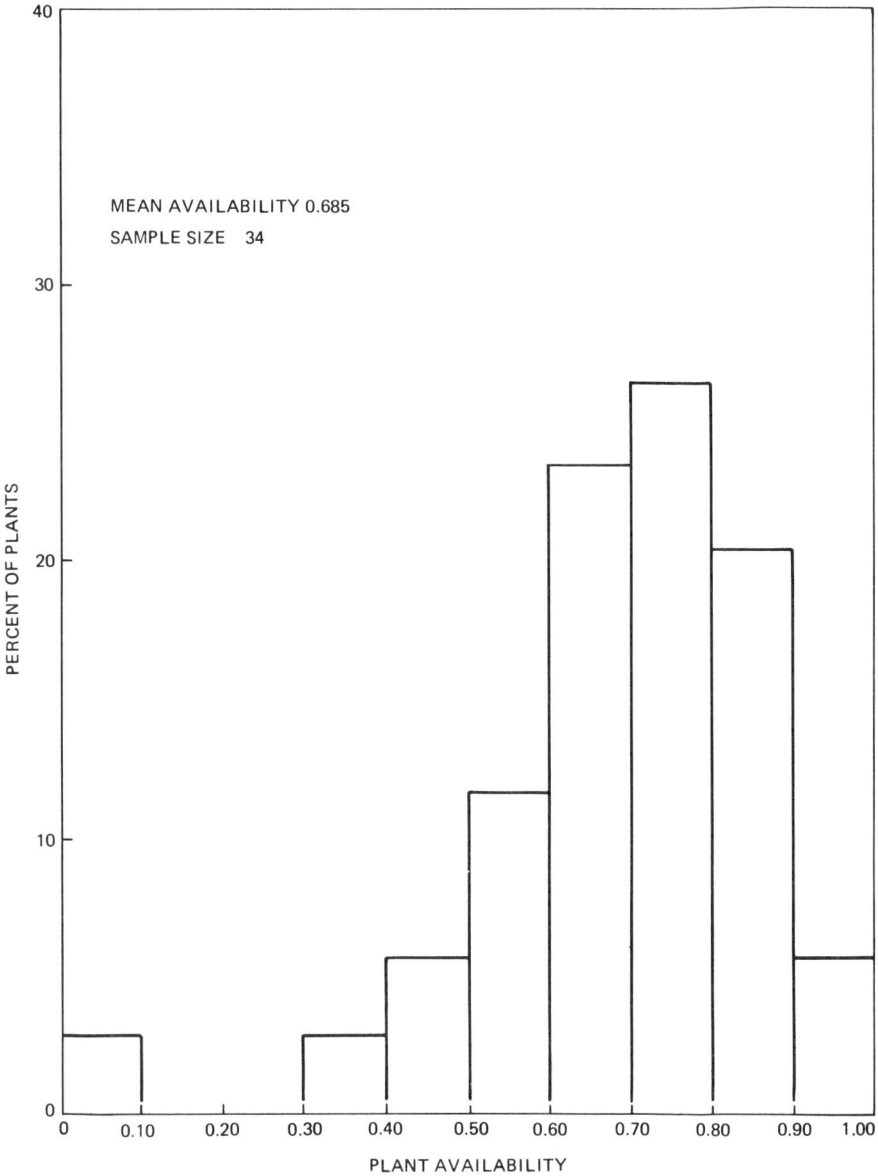

FIG. 2. Histogram of nuclear plant availability experience – 1974.

The impact of improved nuclear power plant availability can be illustrated with the aid of an example. In this example, the capital and operating costs associated with a large light-water nuclear power plant are examined. Relationships between the major plant costs and plant availability are developed based on reasonable physical and economic assumptions. These relationships are defined at the plant level. The primary source of data for this exercise is a USAEC forecast of generation costs for a 1000 MW(e) nuclear power plant for 1982 [6]. These data are based on the following assumptions:

1) Annual generation at 75% capacity factor
2) Annual fixed charge of 15% to translate unit capital costs to annual capital costs
3) Capital costs are $680 million including 8%/year escalation to 1982
4) O&M costs are escalated from 1974 costs at 8%/year
5) The resultant plant cost estimates are:

Cost Component	Cost (Mills/kwh)
Capital	15.5
Fuel	5.6
O&M	1.5

(Mill = 0.001 U.S. dollars)

Parametric extrapolations of these data were made based on the following considerations:

Plant Capacity Factor/Plant Availability - In this example, the plant is assumed to be generating at MDC whenever it is up. This example, though simplistic, does result in definition of an availability/cost envelope and is consequently meaningful.

Fuel Cost/Fuel Cost Factor - Within the constraints of this example, the fuel is assumed to be used at a constant rate when the plant is up. Thus, the fuel cost factor is: c_f = 5.6 mills/kwh. The total fuel cost is directly proportional to plant availability.

Capital Cost/Capital Cost Factor - For a given design, the total capital costs are assumed to remain constant. Thus the capital cost factor is: $C_c = \frac{11.63}{A}$ mills/kwh where A is the plant availability.

O&M Cost/O&M Cost Factor - The total O&M cost is assumed to increase linearly with (1-A). Thus, since the total plant output decreases linearly with (1-A), the O&M cost factor can be expressed:

$C_o = [1+(1-A)k]/A$ mills/kwh. In this example, $k = 0.5$ based on interpretation of the data.

Figure 3 presents the results of these extrapolations. All three cost factors are shown as functions of plant availability. The sum of these three cost factor curves is also plotted as the "total" plant cost per kilowatt-hour as a function of plant availability. It is quite apparent that the capital cost factor has the dominant influence on the "total" cost factor curve.

Added perspective on the impact of increased availability can be gained by considering lost revenue. As previously stated, the parametric data of Figure 3 provided the basic plant cost data for estimating

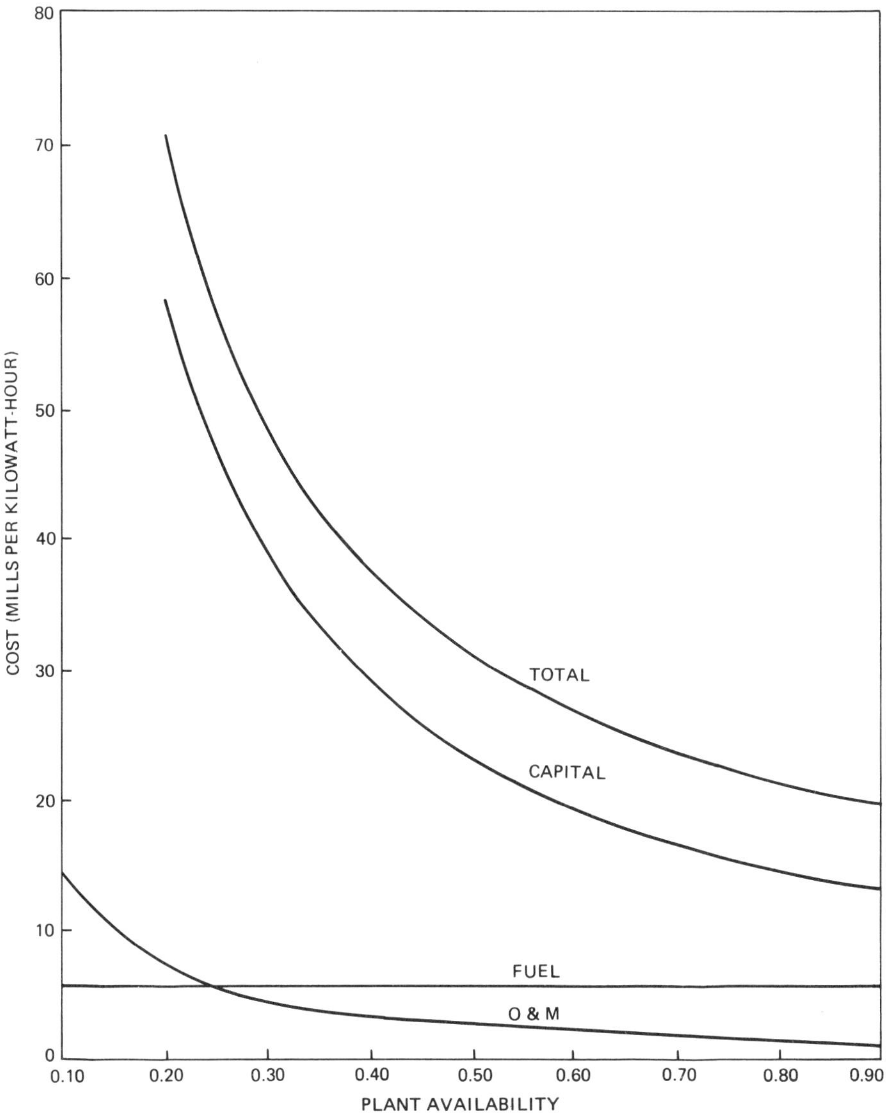

FIG.3. Nuclear power plant cost factors.

plant revenue as a function of availability. Utility pricing information was derived from published industry data in [7] and [8]. The same assumptions regarding escalation were used in this exercise as were used in the determination of plant cost factors. An average 1982 rate for the largest U.S. utilities is projected at 40.3 mills/kwh. Figure 4 depicts this relationship between increasing availability and increasing revenue. Net return per kilowatt-hour is plotted against plant availability. In this example, a minimum plant availability of

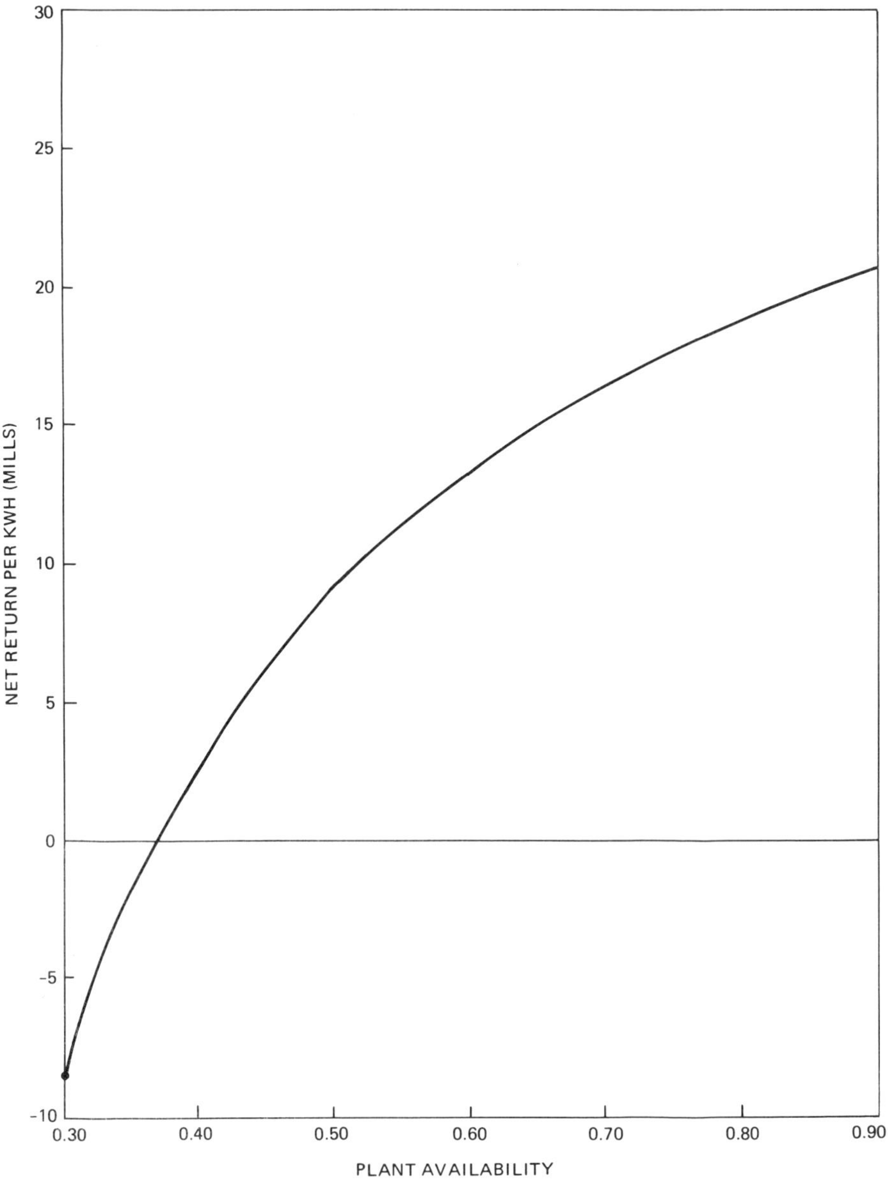

FIG. 4. Cost impact of nuclear power plant availability.

0.37 is required before the net return per kwh becomes positive. It is interesting to note that independent researchers [6] have estimated that nuclear plants are not competitive with coal plants below a capacity factor of 53 per cent. This implies that the break-even point is even lower. In interpreting Figure 4, one should bear in mind that total utility operating costs include several items that are not included in the example plant costs, such as costs associated with maintaining transmission lines, customer service, and business and management office overhead. Thus, the actual break-even point is at a higher availability than that illustrated in Figure 4.

Availability/cost relationships can be directly related to the plant availability models discussed under RELIABILITY TECHNIQUES. These combined models will allow optimization of plant availability with respect to plant cost during design and operational phases.

CONCLUSIONS

Several conclusions relative to nuclear power plant availability can be drawn. First, there appears to be a genuine need for increased nuclear plant availability. This can be accomplished in part by addressing system/plant reliability as an element of design. In this context the system includes software, maintenance/test operating procedures, logistics, and human interfaces as well as the traditional hardware considerations. Second, specific reliability techniques applicable to this problem exist. These techniques have been developed for aerospace and other industry applications over the last 20 years. However, a general framework for application of this methodology and the concept of general plant level availability models must be promoted and developed. Once this latter step is taken, the most pressing needs appear to be: (1) development of detailed, lower-level cost/availability relationships; (2) definition and analysis of plant reliability/availability as a function of system and subsystem interfaces (including operator and maintenance personnel interfaces).

The need for increased plant availability is apparent; the basic methodology exists; and the reward is substantial.

REFERENCES

[1] WILSON, T.R., HILDRETH, M.S.,Jr., GOWER, G.C., Evaluation of Nuclear Power Plant Availability, USAEC Rep. OOE-ES-1 (1974).

[2] Operating Units Status Report, Status Summary Report, United States Nuclear Regulatory Commission (January 1975).

[3] LLOYD, D.K., LIPOW, M., Reliability: Management, Methods and Mathematics, Prentice-Hall, Inc., Englewood Cliffs, New Jersey (1964).

[4] ROBERTS, N.H., Mathematical Methods in Reliability Engineering, McGraw-Hill Book Company (1964).

[5] Reliability Engineering, ARINC Corporation, Prentice-Hall, Inc., Englewood Cliffs, New Jersey (1965).

[6] The Nuclear Industry 1974, USAEC WASH 1174-74 (1974).

[7] "Why utility rate hikes vary so much," Business Week, Industrial Edition No. 2351 (1974) 90d.

[8] Electrical World Directory of Electric Utilities, 1974-1975, 83rd Edition, McGraw-Hill, Inc. (1974).

DISCUSSION

D. HAUCK: One of the reliability analyses performed by your group shows that two turbine generator units may be more reliable than one large one. Could you please elaborate on the design factors underlying this analysis?

L.A. HOPKINS: Please do not misunderstand the context within which the 'redundant turbine example' was presented. It was not a design solution based on an analysis that we did at Boeing; consequently, we do not have details of all the design factors underlying the analysis. However, the following observations can be made:

(1) Some variety of cost-benefit analysis was performed to justify this type of deviation from conventional nuclear plant design. Whether, in fact, the actual analysis is equivalent to what we would recommend cannot be said at present;

(2) Some recent experience in the United States of America with large turbines has been unfavourable. This is particularly true with regard to the amount of time required to effect repairs.

In the light of these considerations, we made no judgement as to whether the example was a good or bad design solution to an availability problem. But we did identify it as one attempt at improvement of plant availability/capacity factor and plant operating economics.

H.V. CRAMER: In the Federal Republic of Germany only single-unit turbines have been installed, and these have been operated with high availability, as high as 98 or 99%, for 20 years. In view of this experience, I wonder whether the results to which Mr. Hopkins has referred are based on practical experience or only on theoretical analyses.

L.A. HOPKINS: I think I have already answered your question in part. An availability of 98% still leaves some room for improvement, though admittedly not much. However, the single turbine in our example is a >1200 MW(e) unit, and I doubt whether the experience of 98% availability would be directly applicable for purposes of comparison.

A. RAMANATH: We know that safety and availability depend in a very direct way on reliability. I wonder whether, with the sophisticated techniques now available, it would be possible for a developing country to obtain an order of merit for the selection of an appropriate system — a system which would give maximum reliability in sizes suitable for that country.

L.A. HOPKINS: In theory, yes, but the techniques available to us would have to be specifically adapted to a problem of that kind. All other system constraints would have to be identified and quantified or related to 'reliability/availability'. Then 'figures of merit' could be derived for available systems.

K. EFFAT: I wonder whether the type of analysis which suggested the possible superiority of two turbine generators over a single larger unit has been applied to other major components such as primary coolant pumps.

L.A. HOPKINS: I think there probably have been a few studies of that kind in the United States, though I cannot quote any specific examples at the moment. But this, precisely, is one of the points we were trying to make in our paper: there have been relatively few applications of available reliability technology to the design of nuclear power plants for the purpose of optimizing plant availability and economics.

P.M. VIBERT: Availability as defined by Mr. Hopkins and as used in the United States is different from availability as defined by UNIPEDE, whose concept is in use in Europe. That is why a joint committee set up by UNIPEDE and the World Energy Conference decided, with the agreement of the representative of the United States, to define non-availability for purposes of international usage as the ratio between unavailable energy and the energy that could have been produced during the period of interest had the plant operated consistently at maximum power. We should be quite clear that non-availability defined in this way is different from the complement to unit availability as defined in the United States.

I have been asked by the members of the joint UNIPEDE-WEC committee on the availability of thermal power stations to draw the attention of participants at this Symposium to the following point. When availability is expressed as a fraction of authorized power rather than as a fraction of the maximum possible power foreseen at the time of ordering, the difference between these two reference values should be specified. It will be necessary to take account of it in construction programmes.

D. NORMAN: I would like to support Mr. Hopkin's approach. In the CEGB we have successfully applied cost/benefit analysis in making decisions on how many feed pumps, auxiliary boilers, fuel-oil pumps and so on should be installed in new power stations. Our work utilizes past operating experience and involves designers and operators. We feel little concern about the question of availability definitions, because we convert outages into costs, and we find no need to define money!

Can I ask Mr. Hopkins whether he, like us, has encountered the problem of random cash flows which, because they arise from rare and costly failures, are difficult to evaluate in a discounted cash flow analysis? We have derived a statistical solution, but would welcome a description of any other work for comparison (the CEGB work, performed in our Marchwood laboratories, is to be published).

L.A. HOPKINS: So far there has been no consideration of a 'discounted cash flow' analysis. We have conducted our studies on the basis of average (lifetime) costs.

W.A. WOLFE: I feel you have increased the complexity of the plant to obtain redundancy. For example, the reactor must be able to follow the decrease in load with a turbine outage. Further, you have lost the cost advantage of scale: two turbines cost more than one. Have you in fact performed a cost-benefit analysis which shows a gain?

L.A. HOPKINS: I agree that increased complexity (both in equipment and in functions) is an element that must be considered in the overall cost benefit analysis. However, it is not immediately obvious that the 'cost advantage of scale' has been lost; two turbines may cost more than one large turbine, but the cost factor is not two, and any measurement of the actual increase or decrease in cost must take into account a possible increase in generated electric power due to the combined effects of differences in turbine reliabilities and repair times.

APPLICATION DES METHODES D'ANALYSE DE LA FIABILITE AU CAS DU REACTEUR RAPSODIE

J.-C. ARLAUD
CEA, Centre d'études
nucléaires de Cadarache,
Saint-Paul-lez-Durance,
France

Abstract–Résumé

APPLICATION OF RELIABILITY ANALYSIS METHODS TO THE RAPSODIE REACTOR.

A study has been made of the variation in the failure rate as a function of the total operating time of the Rapsodie fast reactor, for operation at 24 MW(th) from May 1967 to February 1970 and for operation at 37 MW(th) subsequently (Fortissimo version). In both cases, classical bathtub curves were obtained – without the rise characteristic of wear and tear. The failure rates for both Rapsodie and Fortissimo are very close to 4.3×10^{-3}, giving a mean time between failures of 230 hours. On the other hand, the duration of teething troubles is considerably greater in the case of Rapsodie than in that of Fortissimo. These results relate only to incidents which occurred during reactor operation and which entailed a variation in power. For the classification of incidents, a degree of acuteness is assigned to failures of different components on the basis of a dimensionless number, A, defined for that purpose.

APPLICATION DES METHODES D'ANALYSE DE LA FIABILITE AU CAS DU REACTEUR RAPSODIE.

On a étudié la variation du taux de pannes en fonction du temps total de marche du réacteur rapide Rapsodie et ceci pour chacune des périodes caractéristiques: fonctionnement à 24 MW(th) de mai 1967 à février 1970, puis à 37 MW(th) ultérieurement (version Fortissimo). Dans les deux cas, on a obtenu des courbes classiques en baignoire, sans la remontée caractéristique de l'usure. Les taux de pannes pendant les périodes de vie utile sont pratiquement égaux à $4,3 \cdot 10^{-3}$, donnant une moyenne des temps de bon fonctionnement (MTBF) de l'ordre de 230 heures. Par contre, la durée des pannes infantiles de Rapsodie est nettement supérieure à celle de Fortissimo. Ces résultats ont été obtenus en ne comptabilisant que les incidents survenus pendant la marche du réacteur sous réserve qu'ils entraînent une variation du niveau de puissance. Pour classer ces incidents, on a imaginé et défini un nombre sans dimension, A, permettant d'attribuer une acuité aux pannes des divers composants.

INTRODUCTION

Rapsodie est un réacteur rapide refroidi au sodium dont la première divergence date du 27 janvier 1967. La puissance thermique de 20 MW, initialement prévue, a été atteinte le 17 mars de la même année. Le réacteur a ensuite fonctionné à 24 MW jusqu'au 22 février 1970. Cette période de trois années de fonctionnement (13 300 heures de marche effective) a été suivie d'un arrêt de plusieurs mois décidé pour permettre une adaptation de certains composants à la nouvelle mission du réacteur: dégagement d'une puissance pratiquement doublée. Dans sa nouvelle version, Rapsodie, devenu Fortissimo, a redivergé le 4 mai 1970 pour atteindre, conformément au programme prévu, 37 MW(th) en décembre puis pratiquement 40 MW(th) en 1974.

Au cours de ces huit premières années, dans l'une ou l'autre de ses versions, le réacteur a fonctionné environ 40 000 heures (39 367 heures exactement), perturbées par 190 incidents. Ce temps d'essai et ce nombre d'incidents sont suffisamment élevés pour que l'on tente une étude statistique. Il faut toutefois dès maintenant en préciser les limites.

D'abord, il ne s'agit pas, à partir des résultats obtenus, de prévoir le comportement accidentel d'une série de réacteurs semblables. Rapsodie, comme Fortissimo, sont des prototypes utiles en eux-mêmes et non des têtes de séries.

Ensuite, il n'est pas question d'extrapoler à des engins plus puissants et géométriquement plus grands les résultats de cette étude.

Enfin, nous sommes parfaitement conscients du fait que les modifications apportées au réacteur interdisent, en toute rigueur, l'application directe des méthodes fiabilistes. Nous avons tourné cette dernière difficulté en nous plaçant d'un point de vue global. Compte tenu des modifications intervenues, des changements de consignes d'exploitation, des variations de puissance dégagée, peut-on trouver une loi simple qui donne, par exemple, une bonne approximation pour le taux de pannes de la machine?

Encore faut-il définir la notion de panne. Dans le cadre de cette étude, nous ne retiendrons que les incidents qui ont un effet sur le niveau de puissance désiré, sous réserve qu'ils ne soient pas directement imputables aux exploitants (erreur de pilotage, mauvaise application d'une consigne judicieuse) ou au combustible (présence d'assemblages expérimentaux), incidents d'ailleurs très peu fréquents.

Il s'agit d'étudier la fiabilité intrinsèque de la machine, à partir du moment où elle a divergé. Sont donc exclus de cette étude les incidents de manutention ou les incidents survenus pendant les manutentions ou les périodes de maintenance. Pour se faire une idée plus juste des qualités de Rapsodie, il faut ainsi fournir, en plus des résultats obtenus par une méthode fiabiliste, le facteur de charge. Pendant ces huit dernières années, soit 70 000 heures, le réacteur a fonctionné 40 000 heures, donc environ 6 jours sur 10.

1. DETERMINATION DES TAUX DE PANNES

Après avoir relevé les dates des incidents successifs, nous avons effectué un regroupement par classes, pour Rapsodie (en excluant les 23 premiers incidents très rapprochés représentant manifestement des pannes infantiles) et pour Fortissimo (en excluant les 12 premiers incidents pour la même raison). Nous avons ensuite regroupé les données de Rapsodie et de Fortissimo, au vu des premiers résultats, la variable étant dans ces trois cas le temps qui sépare deux incidents successifs si le premier n'a pas entraîné l'arrêt total du réacteur, ou le temps écoulé entre une mise en route de l'engin et l'instant d'apparition de l'incident.

En portant, sur un graphique de Weibull, en abscisse les bornes supérieures des intervalles de classes et en ordonnée les fréquences cumulées, on obtient les résultats figurant respectivement sur les figures 1, 2 et 3. On voit que les taux de pannes de Rapsodie et de Fortissimo sont constants, respectivement égaux, à la précision de lecture près, à 1/220 et 1/240. Il y aurait donc une légère amélioration des moyennes des temps de bon fonctionnement (MTBF) sur Fortissimo.

Intervalles	Effectifs	Fréquences →	Cumulées
0-200 h	40	0,645	0,645
200-400 h	12	0,193	0,838
400-600 h	7	0,113	0,951
600-800 h	2	0,032	0,983
800-1000	1	0,016	1,000
>1000	0	0,000	
	62		

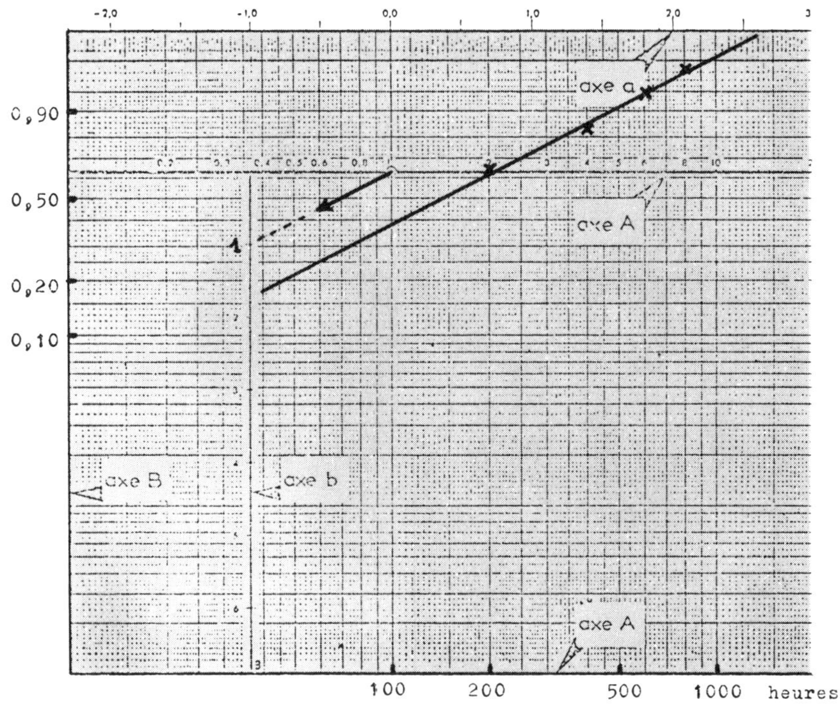

FIG.1. Utilisation du graphique de Weibull: Rapsodie.

Intervalles	Effectifs	Fréquences	Cumulées
0-200 h	51	0,548	0,548
200-400 h	25	0,268	0,816
400-600 h	8	0,086	0,902
600-800 h	4	0,043	0,945
800-1000	0	0,000	0,945
>1000	5	0,053	1,000
	93		

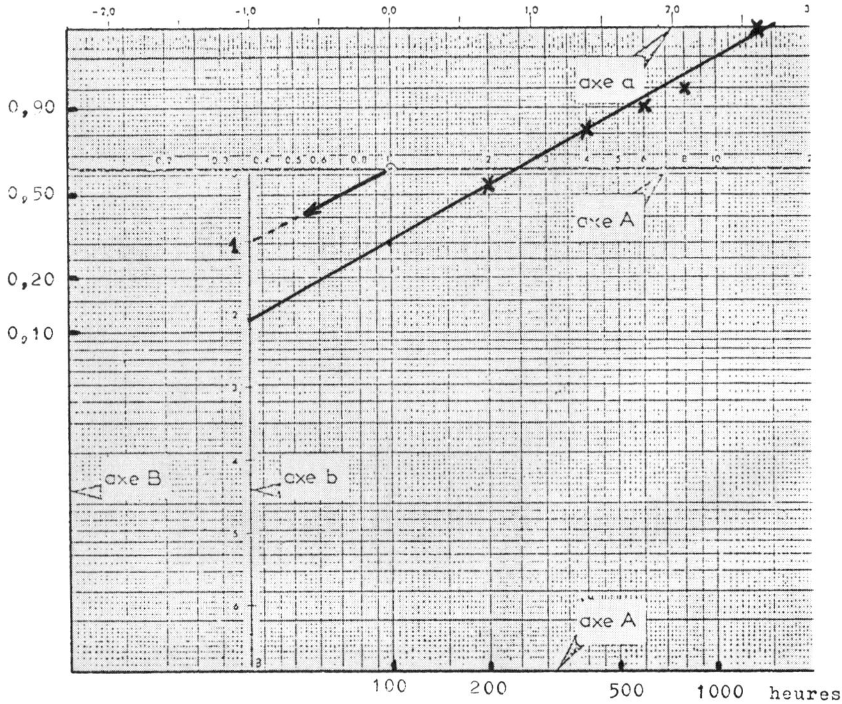

FIG.2. Utilisation du graphique de Weibull: Fortissimo.

Intervalles	Effectifs	Fréquences →	Cumulées
0-200 h	91	0,587	0,587
200-400 h	37	0,238	0,825
400-600 h	15	0,096	0,921
600-800 h	6	0,038	0,959
800-1000	1	0,006	0,965
>1000	5	0,035	1,000
	155		

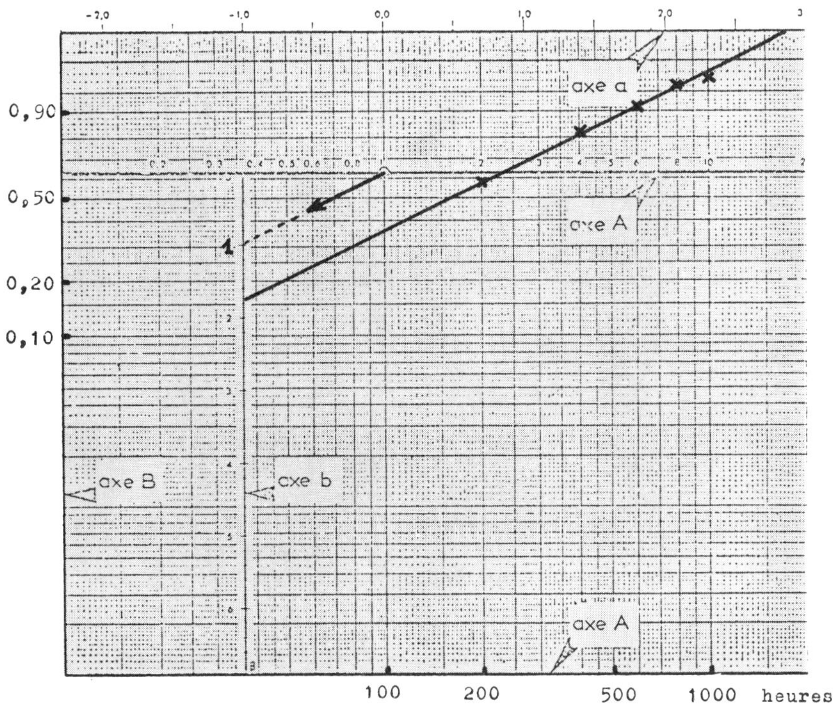

FIG.3. Utilisation du graphique de Weibull: Rapsodie + Fortissimo.

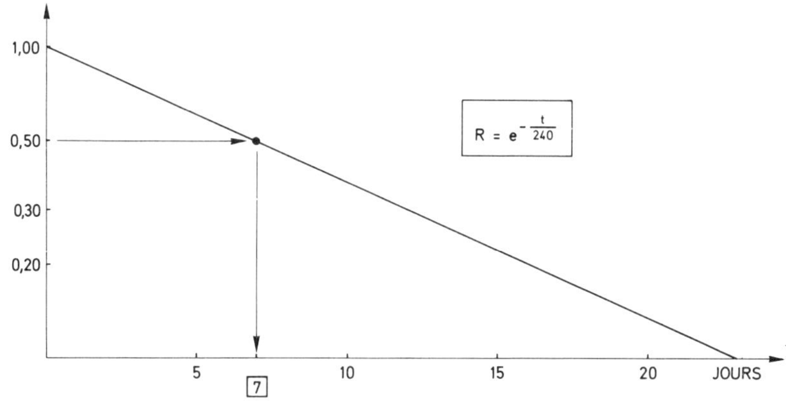

FIG.4. Courbe de fiabilité du réacteur en fonction du temps de marche.

En fait la courbe de fiabilité de la figure 4 donne une bonne idée du fonctionnement du réacteur; cette courbe couvre intégralement huit années de fonctionnement, en excluant les pannes infantiles des deux périodes. On voit sur cette courbe, tracée sur papier semi-logarithmique, qu'après sept jours de marche, la probabilité de panne devient supérieure à la probabilité de non-panne.

2. TAUX DE PANNES DES COMPOSANTS

Il ne s'agit pas de donner le taux de pannes réel des principaux composants mais le taux de pannes apparent, c'est-à-dire celui qui affecte le réacteur en marche.

Dans le tableau I sont rassemblées les grandeurs utiles pour cette étude: nombre d'incidents par composant en fonction du temps de marche du réacteur, et non des composants. Les composants sont: les chaînes neutroniques, les ventilateurs terminaux, les barres de commande, les diésels de secours, le centralisateur de mesures en sodium (TCMS), les pompes à sodium (primaires et secondaires), les circuits d'argon et les thermocouples placés dans le sodium à la sortie des assemblages. Suit une rubrique des incidents divers. Nous donnons de même le temps perdu (nombre d'heures) pour chaque catégorie d'incident. L'étude statistique porte sur cette variable. En utilisant des graphiques de Weibull, nous avons trouvé des relations simples caractérisant les temps perdus pour chaque catégorie d'incidents. Cette étude ne pouvait évidemment se faire que pour les composants ayant eu des incidents fréquents. Nous l'avons donc limitée aux pompes (26 incidents), aux ventilateurs (33) et au TCMS (68). Nous avons ensuite établi par la même méthode une fonction de répartition des temps perdus pour la totalité des incidents (190). Tous ces résultats sont donnés dans les figures 5 à 8.

TABLEAU I. DONNEES DE BASE SUR LES INCIDENTS: TABLEAU RECAPITULATIF

Ce tableau donne, en fonction du temps de marche du réacteur, le nombre d'incidents par composant et le temps perdu, ces deux grandeurs étant totalisées.
Exemple: après 9262 heures de marche il y a eu 13 incidents provenant des ventilateurs terminaux (temps perdu 101 h).

Heures de marche	Chaînes neutroniques	Ventilateurs terminaux	Barres de commande	Diésels de secours	TCMS	Pompes à sodium	Circuits d'argon	Thermo-couples	Divers
(A) 750	11 (63)	5 (20)	1 (1)	5 (42)	1 (4)				
4102	13 (67)	10 (92)	2 (7)	5 (42)	10 (66)	2 (6)	1 (24)		
(B) 9262	14 (69)	13 (101)	2 (7)	5 (42)	21 (140)	6 (138)	2 (122)		
13358	18 (80)	15 (139)	5 (26)	7 (42)	26 (164)	10 (173)	2 (122)	1 (12)	
(C) 13837	20 (84)	15 (139)	6 (27)	8 (42)	28 (169)	16 (189)	2 (122)	1 (12)	1 (0)
16077	22 (88)	18 (142)	6 (27)	8 (42)	34 (242)	16 (189)	2 (122)	3 (13)	1 (0)
22072	25 (97)	25 (365)	7 (32)	8 (42)	40 (316)	19 (204)	2 (122)	3 (13)	1 (0)
(D) 27125	25 (97)	29 (371)	7 (32)	8 (42)	49 (531)	23 (366)	2 (122)	3 (13)	4 (16)
30795	25 (97)	31 (395)	7 (32)	9 (50)	62 (744)	24 (366)	2 (122)	3 (13)	7 (41)
38526	25 (97)	33 (395)	8 (36)	11 (52)	68 (853)	26 (384)	3 (135)	3 (13)	13 (90)

Les séparations horizontales correspondent aux différentes périodes du réacteur: (A) Rapsodie, pannes infantiles; (B) Rapsodie, vie utile; (C) et (D) Fortissimo.

Intervalles	00–04	04–08	08–12	12–16	>16
Effectifs	11	6	3	2	4
Fréquences →	0,423	0,230	0,115	0,077	0,153
↓ Cumulées	0,423	0,653	0,768	0,845	1,000

Fonction de répartition : $F = 1 - e^{-\frac{t}{8}}$

Densité de probabilité : $f = \frac{1}{8}\, e^{-\frac{t}{8}}$

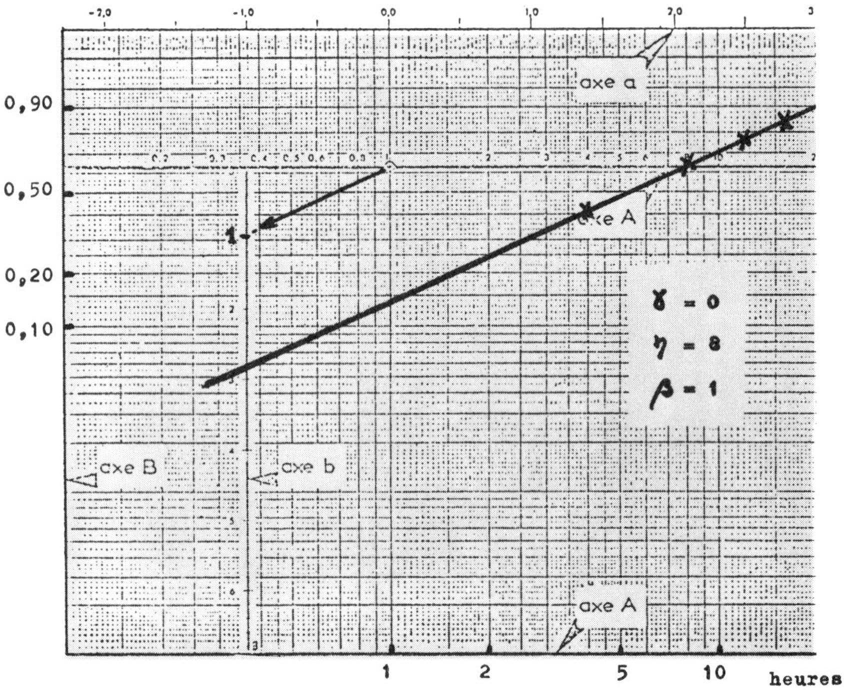

FIG.5. Etude statistique des 26 incidents sur les pompes (Rapsodie + Fortissimo).

Intervalles	00–04	04–08	08–12	12–16	16–20
Effectifs	20	4	2	2	1
Fréquences →	0,606	0,121	0,060	0,060	0,030
↓ Cumulées	0,606	0,727	0,787	0,847	0,877

<u>Fonction de répartition</u> : $F = 1 - e^{-\frac{t+6}{11}}$

<u>Densité de probabilité</u> : $f = \frac{1}{11}\, e^{-\frac{t+6}{11}}$

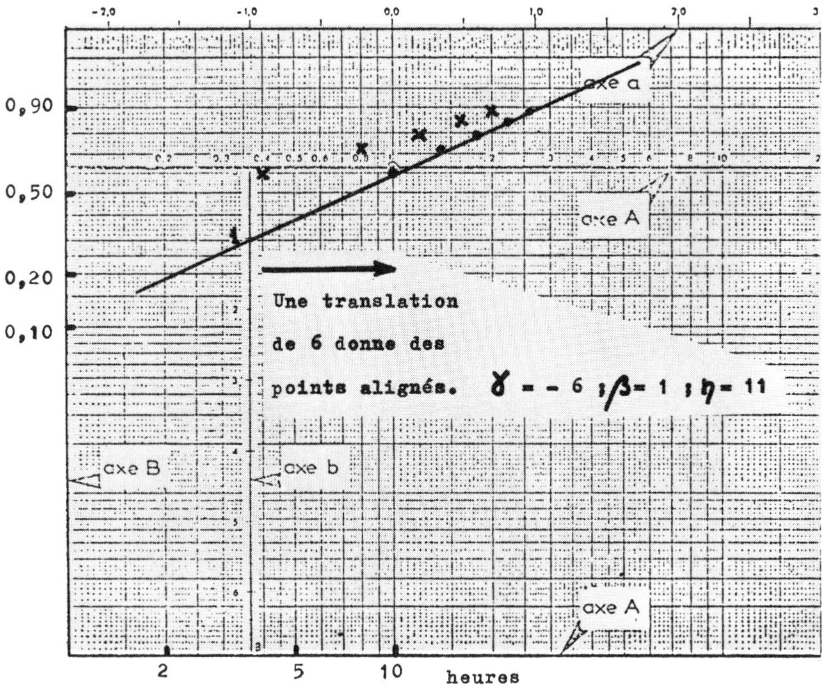

FIG.6. Etude statistique des 33 incidents sur les ventilateurs (Rapsodie + Fortissimo).

Intervalles	00–08	08–16	16–24	24–32	32–40
Effectifs	37	16	5	3	4
Fréquences →	0,544	0,235	0,073	0,044	0,058
↓ Cumulées	0,544	0,779	0,852	0,896	0,954

<u>Fonction de répartition</u> : $F = 1 - e^{-\frac{t}{11}}$

<u>Densité de probabilité</u> : $f = \frac{1}{11} e^{-\frac{t}{11}}$

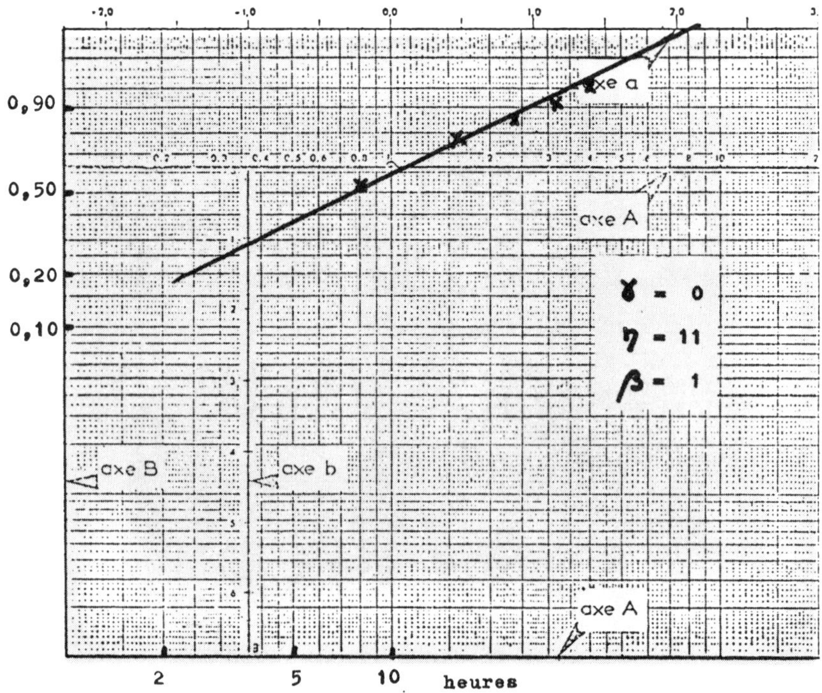

FIG.7. Etude statistique des 68 incidents sur le TCMS (Rapsodie + Fortissimo).

Intervalles	00–10	10–20	20–30	30–40	40–50	50–60
Effectifs	139	27	9	7	2	1
Fréquences →	0,731	0,142	0,047	0,037	0,011	0,005
↓ Cumulées	0,731	0,873	0,920	0,957	0,968	0,973

Fonction de répartition : $F = 1 - e^{-(t/7)^{3/4}}$

Densité de probabilité : $f = \frac{3}{28} \cdot (t/7)^{-\frac{1}{4}} \cdot e^{-(t/7)^{3/4}}$

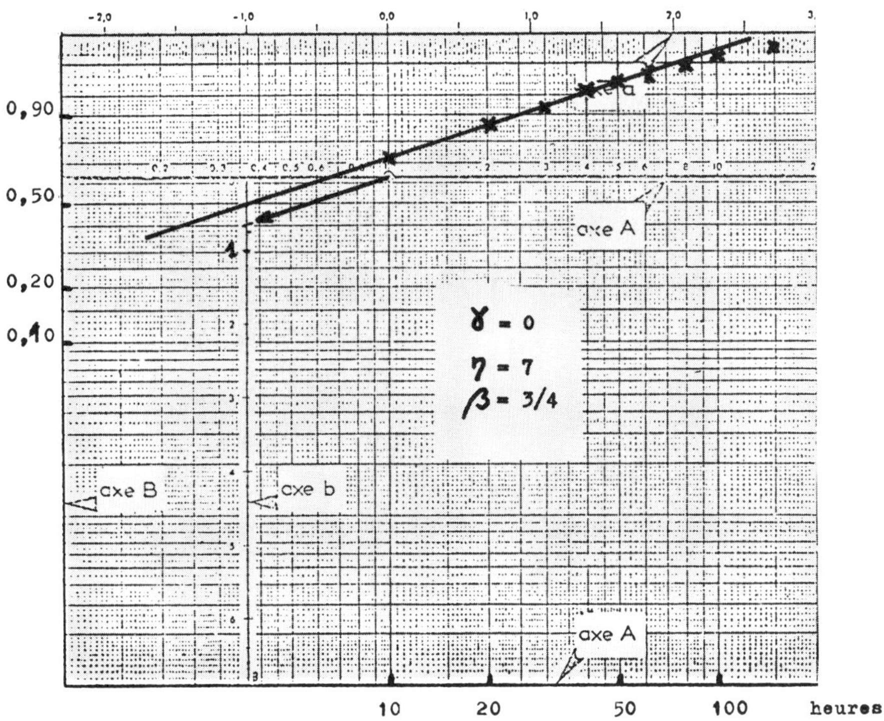

FIG.8. Etude statistique de la totalité (190) des incidents (Rapsodie + Fortissimo).

3. NOTION D'ACUITE ET APPLICATION A RAPSODIE

Afin de relier le nombre d'incidents d'un composant déterminé et le temps perdu qui en découle nous avons imaginé une série de nombres sans dimensions, baptisés acuités.

3.1. Acuité relative A_r

N étant le nombre total d'incidents survenus après T heures de marche d'un réacteur, chaque composant sera responsable de n_i pannes. Si t_i représente le temps total perdu à cause du composant i, on pourra dire qu'en moyenne un incident i coûte t_i/n_i heures. Relativement au temps perdu sur l'ensemble des incidents provenant de la totalité des composants, i sera responsable de

$$\frac{t_i/n_i}{\sum_i t_i/n_i}$$

Multiplions cette grandeur sans dimension par la fréquence d'apparition de l'incident de type i, c'est-à-dire pratiquement, au bout d'un certain temps de fonctionnement du réacteur, par sa probabilité d'apparition. Il vient

$$A_r = \frac{n_i}{N} \cdot \frac{t_i/n_i}{\sum_i t_i/n_i} = \frac{t_i}{N \sum_i t_i/n_i}$$

3.2. Acuité globale A_g

Si λ désigne le taux de pannes du réacteur après T heures de marche et τ le temps moyen perdu par incident ($\tau = T_p/N$), l'acuité globale s'écrit

$$A_g = \lambda \cdot \frac{T_p}{N}$$

Si le taux de pannes est constant, il vient

$$A_g = \frac{N \cdot T_p}{T \cdot N} = \frac{T_p}{T}$$

Ce nombre caractérise le fonctionnement global du réacteur.

3.3. Acuité absolue A_a

C'est le produit des deux acuités précédentes. L'acuité absolue permet des comparaisons chiffrées entre les divers composants de la même machine ou d'engins différents.

3.4. Acuité comparative A_c

C'est le quotient de deux acuités absolues.

TABLEAU II. VARIATION DES ACUITES EN FONCTION DU TEMPS DE MARCHE[a]

Heures de marche	Chaînes neutroniques	Ventilateurs terminaux	Barres de commande	Diésels de secours	TCMS	Pompes à sodium	Circuits d'argon	Thermo-couples	Divers
(A) 750	196,0	62,0	3,1	130,8	12,5				
4102	19,5	26,8	2,0	12,2	19,2	1,7	7,0		
(B) 9262	6,3	9,2	0,6	3,8	12,7	12,5	11,1		
13358	4,3	7,5	1,4	2,2	8,9	9,4	6,6	0,6	
(C) 13837	4,2	7,0	1,3	2,1	8,5	9,5	6,1	0,6	
16077	4,0	6,5	1,2	1,9	11,0	8,6	5,5	0,6	
22072	3,5	13,1	1,2	1,5	11,4	7,3	4,4	0,5	
(D) 27125	3,0	11,6	1,0	1,3	16,6	11,5	3,8	0,4	0,5
30795	2,7	11,2	0,9	1,4	21,0	10,3	3,4	0,4	1,1
38526	2,4	10,0	0,9	1,3	21,7	9,8	3,4	0,3	2,3

[a] L'interprétation de ce tableau est faite à la figure 9, pour chaque composant du réacteur. Les valeurs indiquées sont à multiplier par 10^{-4}.

Les séparations horizontales correspondent aux différentes périodes du réacteur: (A) Rapsodie, pannes infantiles; (B) Rapsodie, vie utile; (C) et (D) Fortissimo.

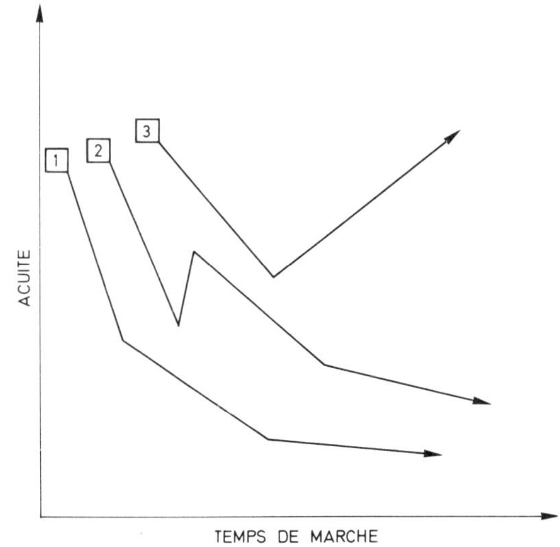

FIG.9. Interprétation des acuités: comportement des composants.

En excluant la première valeur des acuités, qui constitue un point singulier de par sa définition, on constate:

- une amélioration continue du comportement de quatre composants (acuités monotones et décroissantes): chaînes neutroniques, diésels de secours, circuits d'argon et thermocouples du cœur; cette tendance est représentée par la courbe 1
- une amélioration, malgré un incident «grave» isolé, pour trois composants (acuités décroissantes avec maximum relatif): ventilateurs, barres et pompes; tendance: courbe 2
- une aggravation du TCMS: courbe 3.

3.5. Acuités économiques

Les acuités définies précédemment font intervenir les temps perdus. Ces temps perdus peuvent se traduire en termes économiques de façon simple:

- perte liée à l'arrêt du réacteur
- perte liée aux interventions (réparations, changement d'un composant), cette dernière perte pouvant s'exprimer en «heures-équivalentes de non-fonctionnement», c'est-à-dire, connaissant le coût de non-fonctionnement d'un réacteur, en heures fictives d'arrêt.

Supposons un incident i (arrêt de t_i) entraînant le changement du composant (coût C). Si le coût d'une heure de non-fonctionnement est c, les heures fictives d'arrêt seront de C/c, soit t_i'.

Pour obtenir les acuités économiques, on introduira $t_i + t_i'$ dans les relations définies ci-dessus, à la place de t_i.

3.6. Application à Rapsodie

Les résultats du calcul des acuités absolues des composants de Rapsodie (et de Fortissimo) sont donnés dans le tableau II. L'interprétation générale des résultats est faite à la figure 9. Remarquons que pour la période de Rapsodie, les composants les plus défectueux étaient dans l'ordre: les pompes, le TCMS, les ventilateurs.

Pour Fortissimo, on retrouve ces mêmes composants dans un ordre différent: le TCMS, les ventilateurs, les pompes, le TCMS se détachant nettement des autres composants.

CONCLUSIONS

On sait par expérience que plus une machine électro-mécanique est complexe, plus ses chances d'avoir un taux de pannes constant sont grandes. Si Rapsodie n'avait subi aucune modification les résultats que nous obtenons seraient donc parfaitement classiques. En fait, quelles que soient les améliorations apportées au réacteur, le fonctionnement global s'est maintenu au rythme moyen d'un incident tous les 10 jours de marche. Brutalement, on pourrait trouver ce résultat décevant. En fait, en examinant le tableau II (tableau des acuités) on s'aperçoit que tous les composants ont un comportement qui s'améliore sauf un: le TCMS. L'aggravation très sensible du fonctionnement accidentel de cet appareil compense exactement les améliorations de tous les autres composants. Ainsi est mis en évidence (chiffrée) le rôle de ce composant qui fonctionne 20 fois plus mal que les barres de commande, 10 fois plus mal que les chaînes neutroniques et 2 fois plus mal que les ventilateurs terminaux.

RELIABILITY ANALYSIS AND ITS APPLICATIONS IN THE DESIGN OF THE LOVIISA NUCLEAR POWER PLANT

J. ERVAMAA
Technical Research Centre of Finland,
Helsinki

J.K. VAURIO
Imatran Voima Osakeythiö,
Helsinki,
Finland

Abstract

RELIABILITY ANALYSIS AND ITS APPLICATIONS IN THE DESIGN OF THE LOVIISA NUCLEAR POWER PLANT.

The construction of Loviisa Nuclear Power Plant was started in 1970 as a cooperation between the Finnish utility company Imatran Voima Osakeyhtiö and the USSR export organization V/O Atomenergoexport (formerly V/O Technopromexport). Many domestic and foreign companies are participating in the project as subcontractors and consultants. The ordering of the plant gave rise to an extensive investigation on reliability engineering, which was mainly due to the following reasons. (1) The plant will provide, in due course, about 10% of Finland's electric power. (2) It is a requirement of the safety authorities that the conventional safety analysis be supplemented with a probabilistic analysis. (3) The plant is in some respect the first of its kind and the reliability analysis is considered to have a reviewing, unifying and coordinating influence on the non-turnkey delivery. In the report, the aims and organization of the reliability work are discussed in brief, while the methods and realization are discussed in greater detail. The practical tool of the reliability work consists of a systematic reliability analysis, made up of both qualitative and quantitative stages. In quantitative analysis, fault-tree techniques and Monte Carlo simulation are primarily used. Analytical calculation patterns are used for performing sensitivity analyses and, in particular, for the analysis of stand-by type safety systems. Applications have at first concentrated on safety-related systems. The analysis of the reactor protection system, radiation monitoring system, residual heat removal system and a consequence diagram related to a loss of coolant accident are described as examples. Principal problems such as data collection and common mode failure analysis are also discussed.

1. INTRODUCTION

The contract on the construction of the first nuclear power plant unit in Loviisa was signed between the Finnish utility company Imatran Voima Osakeyhtiö (IVO) and the USSR export organization Atomenergoexport (AEE) (formerly V/O Technopromexport) in June 1970. The responsibility is divided between IVO and AEE, but both of them have a great number of subcontractors and consultants both in Finland and other countries, as shown in Fig. 1. IVO coordinates the work, acts as an architect-engineer and designs electrical systems and instrumentation as well as makes preparations for the operation of the plant. AEE's share of the delivery is concentrated on the plant's mechanical components, the most important being a Novovoronesh-type pressurized water reactor and two turbogenerators. The thermal power of the reactor is 1375 MW and the electrical capacity of the turbines is 2 × 220 MW.

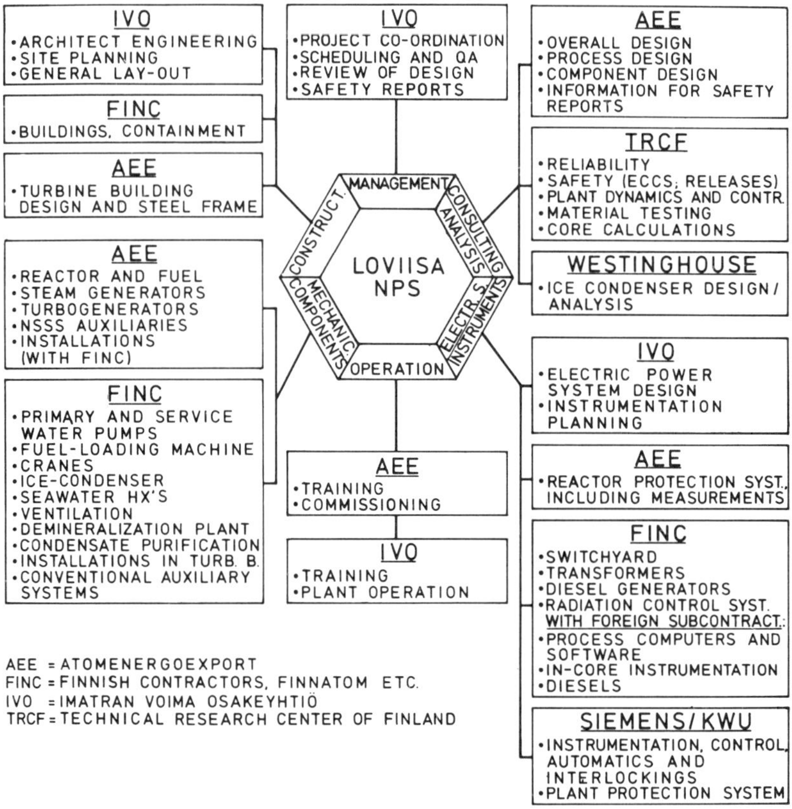

FIG.1. Distribution of deliveries for Loviisa NPS.

The commercial operation of the 1st unit is scheduled to begin in summer 1976. A contract for the second unit was signed in August 1971, and the unit will be started 22 months after the first unit.

The technical structure and realization of the Loviisa plants are described in greater detail in Ref. [1]. Table I includes the most important numerical data on the technical aspects of the plant. The following special features, interesting from the point of view of reliability and safety, are mentioned:

(a) Fuel ratings are conservative. The properties of the cladding material are not exactly the same as those of Zircaloy.

(b) The reactor vessel is large owing to the fuel follower control rods. In contrast to the high thermal capacity of the primary circuit the thermal capacity in each steam generator is small, inasmuch as there are six of them.

(c) The plant is furnished with the following engineered safety systems which have not been standard practice in the USSR:

(1) Steel containment inside the concrete building;
(2) Ice-condenser;
(3) Containment spray-cooling system;
(4) Emergency core-cooling system.

(d) The emergency water pipes are headed directly into the reactor pressure vessel above and below the core, instead of to the primary loops.

(e) The secondary circuit has a separate residual heat removal system, which is used for the long-term shutdown cooling.

TABLE I. TECHNICAL DATA

Reactor type	PWR (VVER-440)
Reactor power	1375 MW(th)
Plant power (net)	420 MW(e)
Plant power (gross)	440 MW(e)
Reactor pressure	125 bar
Outlet temperature	300°C
Core diameter	2.88 m
Power density	87 MW/m^3
No. of primary loops	6
Steam generators	6 horizontal
No. of fuel elements (hexagonal)	349
No. of rods per element	126
Fuel loading (UO_2)	47 t
Average enrichment	3.5%
Fuel rod diameter	9.1 mm
Fuel rod length	2.42 m
Cladding material	Zr + 1% Nb
Rod pitch in assembly	12.2 mm
Average heat rate	129 W/cm
Maximum heat rate	362 W/cm
Maximum heat flux	125 W/cm^2
Average burnup	28600 MW·d/t
Coarse reactivity control	burnable poison, chem-shim
Control rods (with fuel followers)	37
Turbines	2 × 220 MW (3000 rev/min)

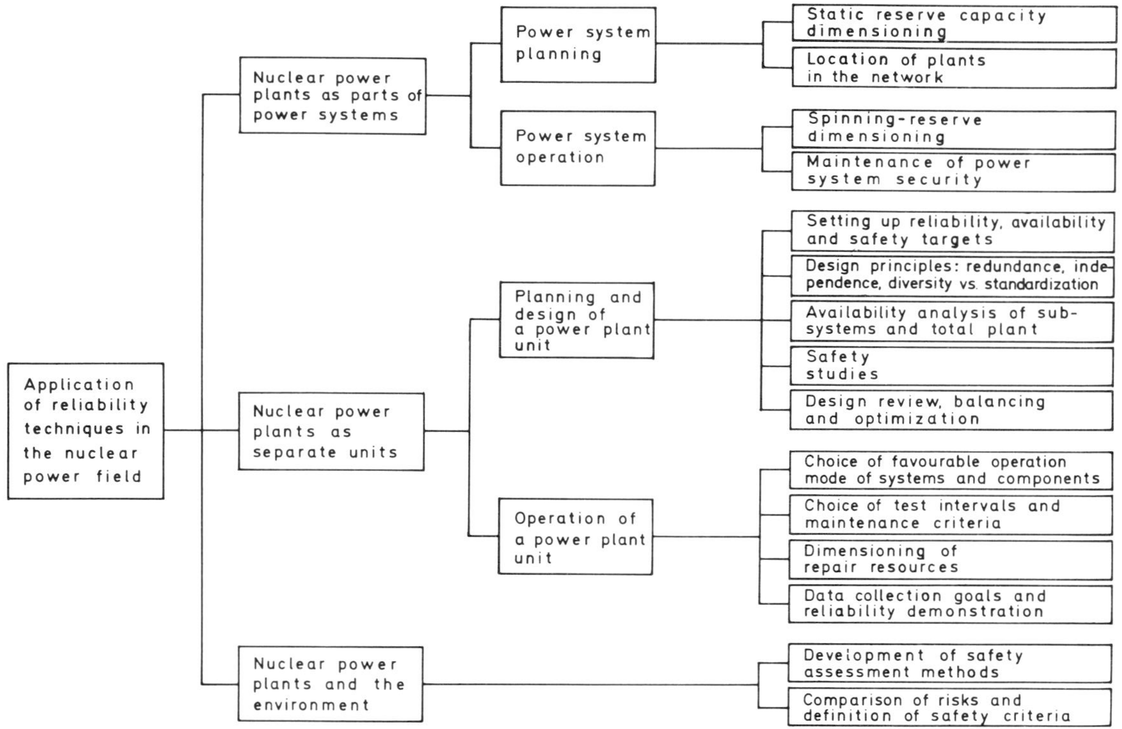

FIG.2. Applications of reliability engineering in the nuclear power field.

Owing to the dispersed way of delivery and the novelty of certain components and systems, IVO decided to apply reliability analysis for eliminating weak points and for achieving a balanced systems design. The review made by the reliability personnel also forms a part of the quality assurance activities.

Since the Loviisa plant is the first nuclear power plant in Finland, the licensing procedure has also been developed in the course of the project. The step-by-step procedure is roughly similar to that of most other countries. In addition, the conventional safety analysis of the plants will be supplemented with a probabilistic safety analysis. This was considered necessary because the plant is the first of its kind, the deliveries are dispersed and previous nuclear experience was relatively scarce.

The high availability of the plant brings a saving of more expensive imported fossil fuels, and the plant's reliability also affects the dimensioning of the reserve capacity [2]. The entire field of application in the long term can be presented with a diagram,as shown in Fig. 2.

To develop a readiness for reliability engineering in the field of nuclear energy, a special Reliability Group was established in 1970 within the Finnish Atomic Energy Commission. Nowadays the group is housed in the Technical Research Centre of Finland, and it includes 12 research engineers.

2. METHODS OF THE RELIABILITY ANALYSIS AND THEIR APPLICATION IN PRACTICE

2.1. Steps of the reliability analysis

In practice it has proved appropriate to carry out the reliability analysis in the following stages:

(1) Description of the system
(2) Determination of the functions to be analysed
(3) Failure analysis
(4) Compilation of the reliability pattern
(5) Acquisition of reliability data
(6) Numerical solution of the problem
(7) Analysis and assessment of the results.

In the first stage of the analysis, detailed information is provided on the system. This includes normal operation, operation in abnormal situations, alarms and indirect detection of failures, testings and test intervals as well as repair data. According to experience, this takes up about 40-70% of the working time needed for the analysis. Smooth collaboration between the designer and the reliability engineer is important especially in this non-turnkey project.

Iterative striving for the ultimate result, taking place as the design proceeds and more data accumulate, is characteristic of the execution of the analysis.

In large projects, the reliability analyses have the effect of co-ordinating the work done in different areas, because it calls for analysing the whole operation as a function of the operation of sub-systems and for centralized collection of the technical data describing the systems.

TRCF RELIABILITY GROUP			System:				Diagram:			
FAILURE MODE AND EFFECTS ANALYSIS			Subsystem:							
Nr. Done by:	Date:	Page:	Equipment:							
Component, function, technical data	Cause of failure	Failure mode	Conse- quences	Effect on the system	Failure detection	Failure code	Failure rate and distribution		Repair time and distr.	Remarks
							$1/10^6$h	$1/10^6$	h	

FIG.3. Failure mode and effects analysis form.

2.2. Failure mode and effects analysis

A form as shown in Fig. 3 has been used when executing the failure mode and effects analysis (FMEA). The form has been developed on the basis of Ref. [3]. A special column has been reserved for the cause of the faults. The purpose is to reveal possibilities for common mode failures. The data section on the right-hand side of the form is not used in the case of FMEA.

A complete FMEA has been carried out in connection with the reliability analyses of the reactor protection system and plant radiation monitoring system. It has become apparent that a large amount of work will be needed if FMEA is realized systematically.

2.3. Common mode failure analysis

Common mode failure involves multiple failures of two or more units due to a common cause so that the failures are not independent of each other. One needs to consider common mode failures when making quantitative assessment of the loss of coolant accident of the plant as well as in connection with the reactor protection system. In both cases one has to examine systems that are extremely reliable in regard to random hardware failures. In such cases the common mode failures often have a dominating role.

Owing to the fact that the many common mode failure possibilities are different in their basic nature, no complete identifying methods exist. In the FMEA one has sought to learn possibilities for a common mode failure by the following measures [4]:

(a) Mutual functional dependencies are considered as potential causes for the failure.

(b) Any exceptional behaviour of environmental factors is identified as a potential common cause of multiple failures.

(c) As a cause of coupled failures, attention is paid to testing, calibration and maintenance errors as well as, to systematic errors in the design and manufacture, especially in the case of several identical items.

(d) The consequences of every failure are examined to see whether that failure could be a common source of other failures through functional dependence or indirectly by generating missiles or by causing fire, flood, pressure impact or some other interaction.

In Ref. [5] important causes of common mode failures have been included in construction and quantification of event trees and fault trees.

The above methods are suitable for the analysis of common mode failures of a certain type. Owing to the multiplicity of common mode failures, the overall results of the analyses often depend on the intuition of the compiler and on his experience and skill.

2.4. Reliability models and the execution of numerical calculations

The fault tree approach has been in use as a reliability model almost exclusively. Its principle of compilation is clearly logical and it is suitable for the analysis of very complicated systems.

The numerical solution of the fault trees is mainly realized with Monte Carlo simulation. The possibility for detailed system description has been considered to be such an important advantage that there has been no desire for any reductions in order to make the calculation times shorter. The calculation times have, however, been shortened by means of the importance sampling method.

The computer program in use, SAFTE-1LR, was originally based on the well-known US program SAFTE-1, which has been published in the literature [6]. Many corrections, improvements and simplifications have, however, been made to the original program, and the functional logic has been so changed that the SAFTE-1LR program can be regarded as an independent program [7]. The program has been written in FORTRAN V for UNIVAC 1108.

The exponential distribution or the constant time can be used as failure time distributions. The exponential, normal, lognormal and uniform distributions or the constant time plus any of those mentioned above can be used as repair time distributions. The fault tree of the system to be analysed is written as a subroutine using logical variables, operators and clauses.

In addition to the ordinary reliability characteristics, the output also includes failure proportions of components, components that have caused a system failure, and components involved in system failures. The output can also be made to include the cut-sets of the system for the first 200 system failures. Two alternative importance sampling methods can be used for reducing the standard deviation of the random results. Unfortunately, importance sampling tends to generate bias in the reliability characteristics if too high an acceleration is used. Acceleration factors of 3 to 5 can be reached without excessive bias.

The following steps will be taken in further development of the program [7]:

(a) Direct or importance sampling availability simulation.
(b) Fast importance sampling reliability simulation of unrepairable systems.
(c) Extremely fast importance sampling simulation of the availability of the stationary state of repairable systems.

For calculation of reliability of unrepairable systems and stationary availability of repairable systems, the analytical approach has also been used. Fault trees have been solved by making use of Boolean algebra and basic rules of probability calculation. The procedure is suitable for small and medium-sized systems. The solution is simple when the gates of the fault tree are independent of one another, but the problem becomes complicated if several gates have common entries. The numerical calculations have been performed with a time-sharing computer using a BASIC program according to a certain procedure. This has proved practicable when one wants to examine the effect of certain parameters or parameter groups on the reliability properties of a system by sensitivity analysis.

2.5. Reliability data and the utilization of quantitative results

The scarcity of reliability data is a well-known fact. The significance of this point to the usefulness of the reliability techniques has, however, perhaps been overemphasized. The accuracy requirements are relatively low, e.g. in the following cases: (1) comparison of system alternatives; (2) localization of weak or overdimensioned points in a system; (3) ascertaining that at least a certain degree of reliability has been achieved.

In practice the following classification of data has proved expedient:

(1) Generic data in terms of representative failure-rate ranges for typical parts, components, equipment and systems.
(2) Average failure rates for different failure modes of certain types of components and equipment, derived from past experience in a large number of applications.
(3) Failure rate data on different failure modes of specified components in relevant operational and environmental conditions.

Often, the source of data in category 1 is in practice the literature. The data in category 2 can be based on operation experiences and the data files of data banks often contain mainly these data. The requirements of category 3 are extremely tight. In a general case this kind of data is not available.

In a practical analysis, rough data of categories 1 or 2 have been used to begin with. Adjustments have been made when necessary in the areas important from the viewpoint of the results. These areas have been identified by utilizing sensitivity studies.

In Table II there are some typical input data that have been used in analyses concerning the Loviisa plant. The validity of failure probabilities expressed "per demand" is uncertain. Dealing in this way with the failures to start does not credit test intervals shorter than approximately four weeks. The repair times in Table II relate to freely accessible equipment. It would be advisable to compare data used in different countries and to analyse common benchmark problems.

3. RELIABILITY ANALYSES RELATED TO THE LOVIISA PROJECT

3.1. Research programme

Table III contains a summary of the analyses carried out up to the end of 1974. The analyses have been performed in accordance with the procedure described in Section 2.1. Safety has been an essential factor when selecting

TABLE II. TYPICAL FAILURE DATA

Component	Failure rate per demand	Failure rate per 10^4 h	Repair time (h)
Diesel generator			
Fail to start	0.02	0.5	20
Fail to operate		1.0	30
Pump aggregate			
Fail to start	0.01	0.1	10
Fail to operate		0.2	15
Motor-operated valves			
Fail to function	0.01	0.1	10
Unintentional function		0.01	15
Non-return valves			
Stick open (fail to close)		0.02	5
Stick closed (fail to open)		0.005	5
Heat exchanger leakage		0.1	35
Valve leakage		0.02	
Relays (50% dangerous)		0.02	
Thyristor switches		0.01	
Temperature channel (10% dangerous)		0.6	
Pressure channel (100% dangerous)		0.15	
Ionization chamber (10% dangerous)		0.15	
d.c. amplifier (50% dangerous)		0.1	
Circuit-breakers, switches		0.01	

the subjects of the analysis, as has been the reviewing of the design, the identification of weak points, and the comparison of alternatives.

Among the clarifications which are being performed at present are a reliability analysis on the containment isolation system, estimation of the probability of leakages in the primary circuit, and the analysis of the consequences of the main pipe break in the primary circuit. The last-mentioned analysis has been in progress for nearly two years.

The analysis has consisted of the compilation of a consequence diagram for a loss of coolant accident and the determination of probabilities of events

TABLE III. RELIABILITY ANALYSES FOR LOVIISA NUCLEAR POWER PLANT

System	Methods used	Results and applications
Feedwater pump system	Availability analysis, fault-tree analysis, crude Monte Carlo simulation and analytical methods	Economic choice between alternatives, comparison of calculation methods
Service water pump scheme	Reliability analysis, analytical method	Choice between alternatives
Primary circuit chemical and volume control system	Reliability analysis, fault-tree analysis, crude Monte Carlo simulation and analytical methods	Reliability of emergency boron control and leakage compensation, system improvement and balancing
ECCS Emergency makeup system Hydroaccumulator system Low-pressure system	Reliability analysis, partial FMEA, fault-tree analysis, importance sampling Monte Carlo simulation, analytical methods, sensitivity analysis	Choice between alternative system configurations, identification of weak points of the system and dominant reliability factors, quantitative reliability estimates, test intervals
Containment spray and cooling system	ditto	ditto
Cooling water system	Reliability analysis, fault-tree analysis, importance sampling Monte Carlo simulation, analytical methods	Quantitative assessment of startup probability and reliability of operation, critical phases of operation, identification of weak points of the system, planning of maintenance and spare part services
Residual heat removal system	ditto	ditto
Reactor protection system	Reliability and availability analysis, partial FMEA, fault-tree analysis, crude and importance sampling Monte Carlo simulation, analytical methods, sensitivity analysis, common mode failure considerations	Identification of weak points, test procedures, quantitative availability estimate, safety assessment
Radiation monitoring system	Reliability and availability analysis; complete FMEA, analytical methods	Quantitative estimates of system reliability and availability, critical components, distribution of failures, compensation of failures

connected with it on the basis of fault-tree analysis. The methods needed in the overall analysis of the safety of a nuclear power plant has been developed at the same time. The idea of the work has been similar to that of Ref. [5]. In this connection it has been necessary to analyse and assess the reliability properties of systems other than those listed in Table III (e.g. the control rod system, electrical systems and the different function modes of the containment). The Systems Reliability Service of UKAEA has taken part in the analysis. The results of Ref. [5] are also to some extent applicable.

In the near future the research programme will include reliability analyses of the spent fuel cooling and handling system, active waste treatment system and the active ventilation system. Increasing attention will be paid to analyses bearing upon the availability of the plants.

3.2. Reliability analysis of the reactor protection system

The reactor protection function consists of four different stages. In the scram proper the thyristor couplings of the control rod drives are opened by a tripping signal, and the control rods drop down. The other functions are fast shutdown, slow shutdown and interlocking.

The system has mainly been realized with relays. It includes two sets of alarm relays, which are connected in a 1-out-of-2 configuration. Each one of the sets of alarm relays receives information on the state of the process through logic connections separately from the neutron flux density channels and from the so-called technological signal channels. To form the technological signal, the pressure and temperature at the reactor outlet, the flow of coolant, the starting of the emergency cooling and the pressure of the containment are controlled. The measuring signals are based on the 2-out-of-3 principle.

Until now most of the analyses have involved the reliability of the scram proper, and not so much effort has been made to analyse the reliability of a fast shutdown and the probability of unnecessary scrams.

The goals and methods have been as shown in Table III. FMEA has been performed in full for the scram and fast shutdown. Fault-tree technique has been used and numerical considerations have been carried out with simulation using the SAFTE-1LR program. The effect of different operational parameters, such as the interval and duration of tests, has been examined with sensitivity analyses, for which the analytical solution was also carried out.

The reliability data employed were first taken from the literature [8]. The aim of the sensitivity analyses was also to clarify the sensitivity of the results to the variations of data. In the event that the dependency proved strong, the data were made more accurate by discussion with experts and by utilizing data banks [9]. In uncertain cases the reliability data were chosen conservatively.

According to the FMEA the scram system fulfils the single failure criterion. Among the dangerous failures are faulty stickings of relays, which may be caused by the adherence of contacts or a mechanical failure in a relay. This kind of fault, preventing a relay from opening in the event of an alarm, is only revealed in connection with testing. A suitable combination of latent safe faults will bring about an unnecessary scram.

TABLE IV. REACTOR PROTECTION SYSTEM UNAVAILABILITY

System	Unavailability	
	Original scheme	New scheme
Instrument channel	$(5\text{-}15) \times 10^{-3}$	
2-out-of-3 configuration of instrument channels	$(0.07\text{-}1.0) \times 10^{-3}$	
Technological signal		
Trip signal from one instrument channel	$(2.9\text{-}3.9) \times 10^{-3}$	$(0.07\text{-}1.0) \times 10^{-3}$
Trip signals from instrument channels of pressure and temperature	1.4×10^{-3}	0.03×10^{-6}
Total system		
Trip signal from neutron flux	1.0×10^{-5}	0.04×10^{-6}
Trip signals from instrument channels of pressure and temperature	2.2×10^{-6}	0.007×10^{-13}
Trip signals from instrument channels of neutron flux and temperature	0.04×10^{-6}	0.002×10^{-13}

Test interval is one month; zero testing time and common mode failures not taken into account.

This concerns in particular the contactors and isolation diodes of power supplies connected in parallel as well as voltage relays with their pre-resistors.

In the numerical considerations of the scram operation the results mentioned in Table IV were obtained. The reliability of the technological signals proved dominant, and the reliability of the system was greatly dependent on the selection of the failure rate of the relays forming the signal.

The fault-tree considerations and sensitivity analyses led to the proposal that changes should be made in the original protection system scheme. The reliability of the technological signals, as well as of the entire system, could be improved considerably (Table IV) through simplifying the scheme by removing certain relays. The sensitivity of numerical results to the selection of failure rate values decreased at the same time.

The length of the testing time has a significant effect in the original scheme and a dominant effect in the changed system. In the case where the neutron flux is the only trip signal, the unavailability of the protection system can be approximated by formula (1) with effective failure rate $\lambda = 8 \times 10^{-6}\ h^{-1}$.

$$u = \frac{5}{24} (\lambda T)^2 + \lambda d \tag{1}$$

$$u = \lambda T \left[\frac{3}{8} (\lambda T)^2 + \frac{11}{4} \lambda d \right] \tag{2}$$

where

T = test interval (h)

d = test duration (h)

In the new scheme the unavailability would be decreased by a factor of 0.05. When both temperature and neutron flux are trip signals, the unavailability in the original scheme is expressed by formula (2) with $\lambda = 8 \times 10^{-6}\ h^{-1}$. In the new scheme the unavailability would decrease by a factor of less than 0.01. The results are comparable to those of Ref. [10].

The random stickings of the rods or random faults in the control rod drives are of minor significance for the reliability of the scram system.

Common mode failures have been investigated by means of FMEA and by visual inspection of layout drawings etc.

3.3. Residual heat removal system

The special feature in the residual heat removal system of the Loviisa plant is that steam generators and the secondary circuit take part in the heat removal at all stages. In the secondary circuit there is a separate residual heat removal plant, which includes two heat exchangers, two emergency feedwater pumps and two residual heat removal pumps as well as a makeup water system. The supply of electricity to the pumps is secured with diesel generators. The discharged amounts of steam remain very small also during a break in the electricity supply, since the steam can be conducted to the residual heat removal plant. When off-site power is available, the turbine bypass can also be used for the first phase of cooling and the normal feedwater pumps can be used instead of the emergency feedwater pumps. At a later stage of the residual heat removal there is water circulating on the secondary side of the steam generators and in the residual heat removal plant. One steam generator is then enough to convey the residual energy, the primary flow taking place through natural circulation.

The reliability block diagram in Fig. 4 shows the alternatives for realizing the residual heat removal.

The following failure possibilities were taken into consideration in the system reliability analysis: (1) latent failure in advance after previous use, (2) disclosing failure in advance, at which the dead time equals the repair time of the equipment, (3) probability of pumps and valves not starting on demand, as well as (4) reliability of the equipment in continuous use.

When it was pessimistically assumed that also the off-site power is lost in connection with the shutdown once a year, the reliability of the supply of electricity to the residual heat removal system proved critical when using the original scheme and diesel generators. It is possible to consider the gas turbines located in the plant area as a standby power system for the nuclear power plant. Then the failure probability of 10^{-3} /year is reduced to 2×10^{-5} /year. Diesel generators with higher capacity or a different scheme can also be used. Then the reliability of the steam condensing will be determined by the reliability of the emergency feedwater system. The redundancy of the pumps may be increased from two to four. Then the

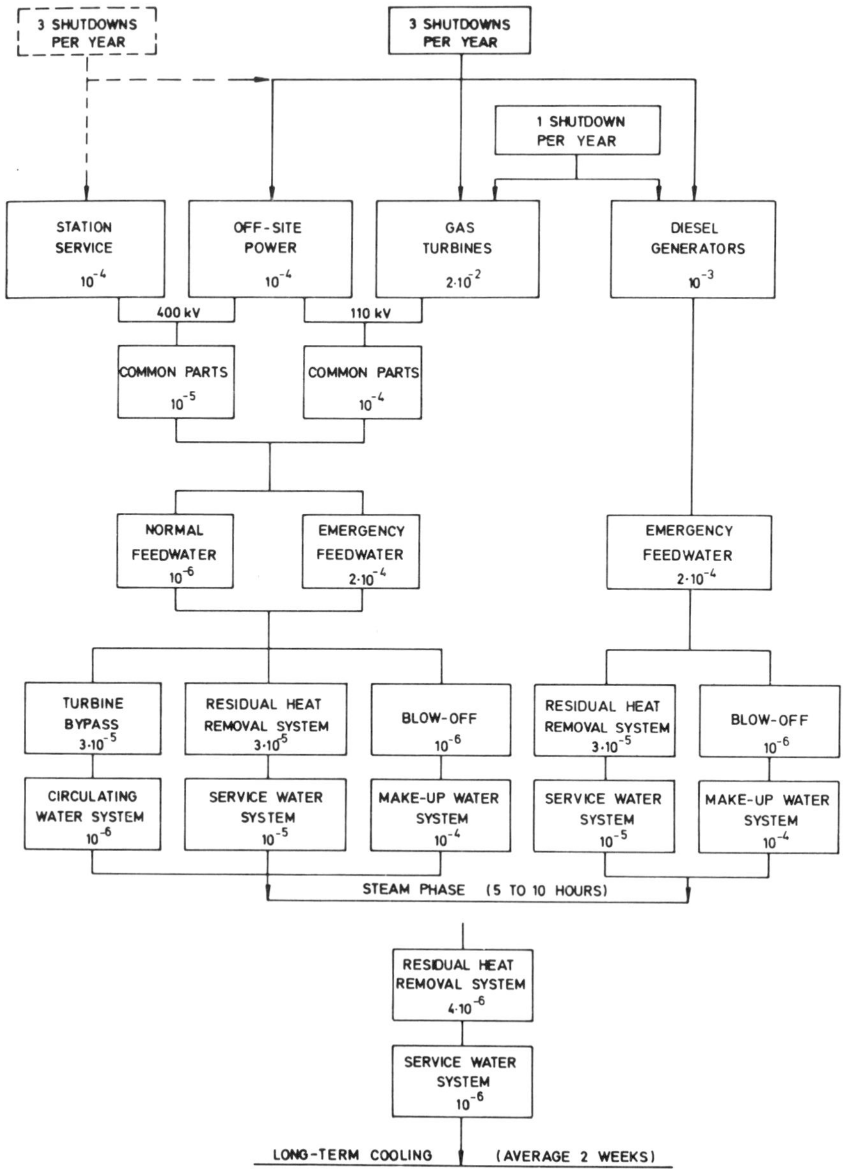

FIG.4. Reliability block diagram of the residual heat removal.

entire system becomes well balanced as to its reliability. The circuit in charge of long-term cooling holds the next place. Its starting is extremely reliable, because the steam phase may, when necessary, be continued until the two pumps have been started. The operation of one pump is enough for residual heat removal.

3.4. Reliability analysis of the radiation monitoring system

The radiation monitoring system is divided into three parts [11]: the area monitoring system, the process radiation monitoring system and the gas and aerosol monitoring system. The measuring channels are either Geiger or scintillation channels. The system is connected to a process computer.

Failure mode and effects analysis was performed on two sample channels, one Geiger channel and one scintillation channel. The FMEA was based on theoretical considerations. All component failures were analysed, but most of the effects were estimated very briefly. Two failure modes were considered for resistors, capacitors, diodes and inductors. All other components were treated assuming one failure mode.

The FMEA showed that only very few component failures which are revealed only by yearly inspection may be important and only few component failures cause failures of the system.

The complete FMEA proved to be quite laborious. Some effects were very difficult to estimate theoretically, e.g. different persons could come to opposite conclusions as to the effect of the same component failure. It is considered practicable to use FMEA on only the most important systems or subsystems.

The numerical calculations were performed for one Geiger channel and one scintillation channel. The literature was used as the source of electronic reliability data. Failure rates based on the mean values of data were 0.2 failures/year for the Geiger channel and 0.3 failures/year for the scintillation channel. Optimistic and worst case estimates of the failure rate were also assessed on the basis of dispersion of data. Quantitative prediction of relative failure percentages indicated that about 40% of the failures of both channels take place in semiconductors, dominating over potentiometers, solderings, connectors etc. As to the modules, more than 70% of the failures take place in measuring units or power sources.

4. CONCLUSIONS

The Loviisa Nuclear Power Plant is being built on a non-turnkey basis coordinated by IVO. The deliveries are divided between many subcontractors and a number of consultants are performing special tasks. The plant will comprise components and systems with minor earlier operation experience. For these reasons IVO has applied reliability methods as a part of the systems design and review procedure. Probabilistic assessments are also used to supplement the conventional safety studies.

To develop capability and experience in reliability analyses in the field of nuclear power, a special project was started in 1970. The trained staff, computer programs, manuals and library concerned with these analyses are now located at the Technical Research Centre of Finland, which is consulting IVO on the reliability work.

The targets for reliability analyses have first been chosen on the basis of their importance to safety. Important protection and monitoring systems, engineered safety systems and parts of the main process have been analysed. In the future more attention will be paid to the availability analyses of nuclear power plants.

As to the methods of reliability technology, fault-tree analysis has played a central role. In numerical studies both Monte Carlo simulation and analytical methods have been used. Reliability data of a rather generic nature have been sufficient in most cases.

One important result of the reliability studies has been an independent, objective checking of the suitability of the design. The analyses have given an idea of the most important reliability factors and possible weak points of the systems, and it has been possible to use the results as a basis for choices between alternative designs, redundancies etc.

In most cases an absolute reliability estimate or a conservative boundary value has been calculated. In general, the emphasis has been on comparative studies. Absolute results have been used cautiously, in particular if relevant data have not been available or when dealing with small probabilities which can be strongly influenced by common mode failures.

REFERENCES

[1] PALMGREN, A., Loviisa nuclear power station: a review at the halfway mark, J. Br. Nucl. Energy Soc. 13 1 (1974) 115.

[2] VIHERIÄVAARA, H., ERVAMAA, J., "Nuclear power plant reliability as a factor influencing power system planning in Finland", 4th Int. Conf. Peaceful Uses At. Energy (Proc. Conf. Geneva, 1971) 4, IAEA, Vienna (1972) 73.

[3] IEEE, Trial Use-Guide: General Principles for Reliability Analysis of Nuclear Power Generating Station Protection Systems, IEEE Std 352, New York (1972).

[4] MANKAMO, T., Common Mode Failures, Technical Research Centre of Finland, Electrical Engineering Laboratory, Report 13, Otaniemi (1975).

[5] USAEC, Reactor Safety Study, an Assessment of Accident Risks in U.S. Commercial Nuclear Power Plants, Rep. WASH-1400 (Draft), (1974).

[6] GARRIC, B.J., et al., Reliability Analysis of Nuclear Power Plant Protective Systems, Holmes and Narver Inc., Los Angeles, HN-190 (1967).

[7] HARJU, T., SAFTE-LR, A Monte Carlo Simulation Program for Reliability and Availability Analysis of Complex Systems, Technical Research Centre of Finland, Electrical Engineering Laboratory, Report 9, Otaniemi (1975).

[8] BALFANZ, H.-P., Ausfallratensammlung, Institut für Reaktorsicherheit der TÜV e.V., IRS-W-8, Köln (1973).

[9] UKAEA, Systems Reliability Service and SYREL-Reliability Data Bank, Personal communication and data from the data bank.

[10] HOWARD, R.S., SCHULTZ, M.A., Comparative reliability analysis of reactor safety systems, Nucl. Saf. 14 6 (1973) 605.

[11] VOIPIO, V.-E., Reliability Analysis of the Radiation Monitoring System of the Loviisa Nuclear Power Plant, Technical Research Centre of Finland, Electrical Engineering Laboratory, Report 8, Otaniemi (1975).

PRACTICAL APPLICATION OF RELIABILITY ENGINEERING IN DETAILED DESIGN AND MAINTENANCE

S.E. BARDEN
Central Electricity Generating Board,
Maidstone, Kent,
United Kingdom

Abstract

PRACTICAL APPLICATION OF RELIABILITY ENGINEERING IN DETAILED DESIGN AND MAINTENANCE.

Modern plant systems are closely coupled combinations of sophisticated and expensive equipment, some important parts of which may be in the development stage (high technology sector), and simpler, crude but not necessarily cheap equipment (low technology sector). Manpower resources involved with such plant systems can also be placed in high and low technology categories (i.e. specialist design and construction staff, and production staff, respectively). Neither can operate effectively without the other, and both are equally important. A sophisticated on-line computer controlling plant or analysing fault symptoms is useless, if not unsafe, if the peripheral sensing and control equipment on plant providing input data is poorly designed and inaccurate, and/or unreliable because of inadequate maintenance. Similarly, the designer can be misled and misinformed, and subsequent design evolution can be wrongly directed, if production records do not accurately reflect what is actually happening on the plant. The application of Reliability Technology can be counter productive if it demands more effort in the collection of data than it saves in facilitating quick, correct engineering decisions, and more accurate assessments of resource requirements. Reliability Engineering techniques must be simplified to make their use widely adopted in the important low technology sector, and established in all financial and contractural procedures associated with design specification and production management. This paper develops this theme with practical examples.

1. INTRODUCTION

Reliability engineering has gained prominence during the last twenty years as a consequence of the United States Space Programme and the existence of sophisticated computers.

It is, however, essentially a simple tool that can be widely exploited by engineers using a hand computer and simple mathematical logic.

This can be illustrated by two very different examples chosen from systems design and maintenance engineering, respectively.

2. RELIABILITY OF INTERCONNECTED GENERATING UNITS

The engineers' instinct for reliability is powerful, and it dominated all developments in electrical power generation throughout the first half of this century.

With small groups of power stations supplying local distribution feeders, ample provision was made in power plant margins and redundancy, particularly where there was any technical uncertainty in design.

The reliability logic of such plant systems is illustrated in the simplified diagram in Fig. 1(a). The extensive use of "ranging" or interconnection of major plant systems at both their input and output connections provided flexibility in operation and made a significant contribution to both reliability and availability of early power stations.

The electrical power engineer adopted this principle with the introduction of the High Voltage Electrical Power Grid System, where load demand was uniformly high over large areas (Fig. 1(b)). This development has played a major role in:

1. bringing electrical power to virtually every citizen of the United Kingdom;
2. keeping operating costs to a minimum by using the most efficient plant to meet the changing power demand;
3. maintaining an extremely reliable service to the consumer by exploiting the flexibility it provided in choice of fuel and plant meeting the power demand.

With this reliable electrical interconnecting system providing operating margins to allow for forced and planned plant downtime, the post-war thermal power stations design developed on the unit principle (Fig. 1(c)), where steam pressures and temperatures are taken near the physical limits of high quality metals to raise thermal efficiency, and interconnections are virtually eliminated in order to reduce construction costs to a minimum.

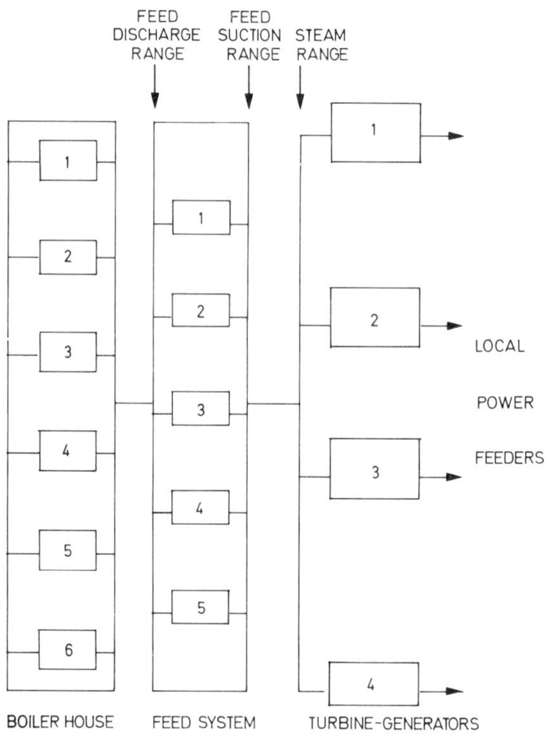

FIG. 1(a). Interconnection for redundancy in early power plant design.

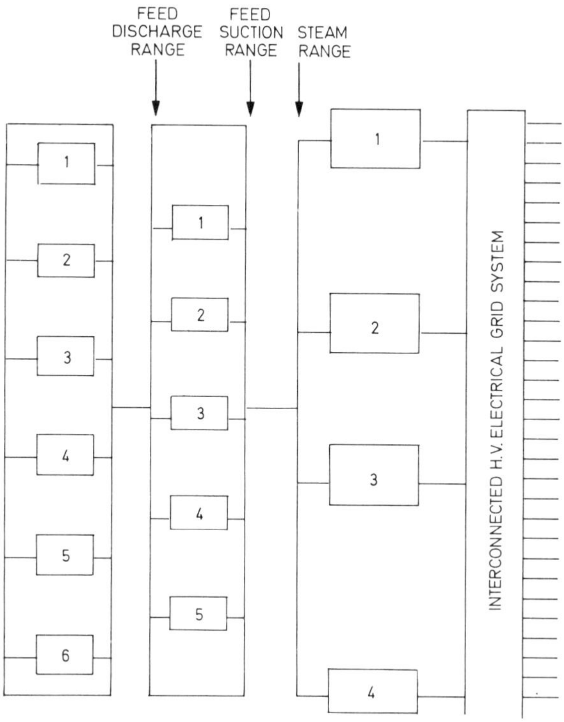

FIG.1(b). Introduction of countrywide H.V. grid systems.

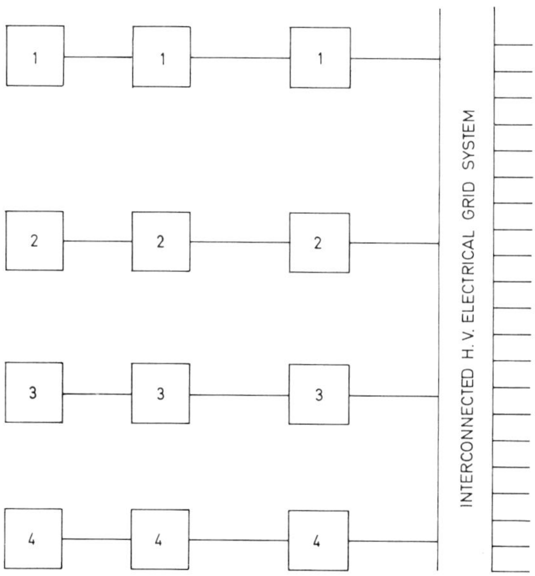

FIG.1(c). Unit generating principle.

The unit principle was, however, not adopted in the early C.E.G.B. "MAGNOX" nuclear plants commissioned about ten years ago. The sixth "Magnox" power station built at Oldbury-on-Severn was the first to have only a single turbine for each nuclear reactor. Each nuclear unit at Oldbury was self contained, with its own essential auxiliaries such as boiler feed pumps, extraction pumps etc.

Three exceptions to the unit principle were, however, built into the station:

1. steam lines from the four boilers on each reactor to the main turbines were interconnected to allow steam transfer;
2. feed pump discharges were interconnected through separate high pressure and low pressure "ranges";
3. cooling water inlet pipework was interconnected.

These interconnections were of limited use in allowing reactor start-up and turbine run-up to low load without the appropriate turbine or reactor respectively being available. It also allowed two reactors to feed into one turbine for load topping up when reactor output had to be reduced when temperatures were restricted to limit corrosion rates within the vessel.

The original design did not, however, provide an interconnection of plant on the feed pump suction side which would allow one reactor to generate its full power through the turbine associated with the other reactor. The effect of this unit design concept on availability of Magnox nuclear power reactors is shown in Fig. 2.

The Oldbury plant was modified in its third year by making the additional interconnections on the feed pump suction side which allowed full reactor output to be obtained when cross-connected, whilst also providing a cooling system for the shut-down reactor.

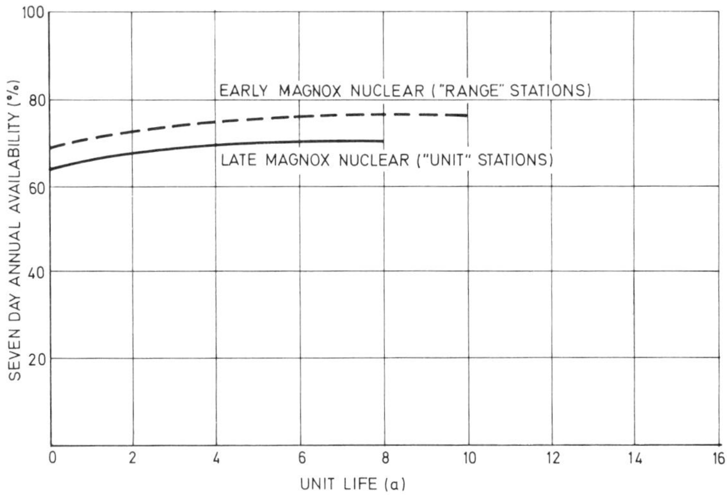

FIG. 2. "Magnox" power plant availability in early operational life.

During the fourth and fifth years of station life, this facility has increased the availability of a single reactor unit by an average of about 20% resulting in a saving in average replacement generation cost of about £2,700,000 over these two years (1).

The engineering instinct that suggests an interconnected system is essential can, however, be examined by a simple analysis in reliability logic (Appendix 1) which considers a cross-connected two reactor unit module.

Take, for example, a unit design with a potential availability of 64%. The maximum increase in availability through full cross connection between reactor and turbine is just over 5%, provided reactor and turbine systems have a comparable availability of about 80%. If either one or other system has a substantially lower availability (say 95% and 67% for instance), the gain in availability is rapidly reduced to about 2% (Fig. 3). This gain in availability is zero, if either reactor or turbine has 100% availability.

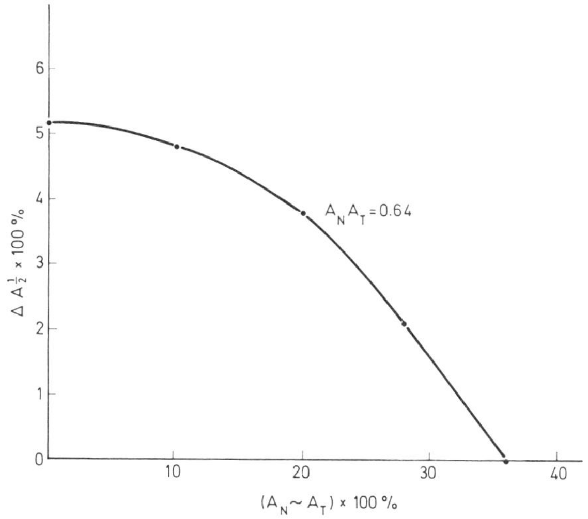

FIG.3. Availability gain from cross-connected two-reactor/turbine unit effect of difference in reactor and turbine availability.

Concern for nuclear plant reliability can lead to an imbalance in the reliability of the reactor systems over the more conventional turbine and feed systems. This must be recognised, and the reliability of all major systems must be given comparable attention in detailed design and maintenance. It is more economic to ensure that the availability and reliability of all major systems are consistently high (above 90%) and that margins and redundancy within each system in each unit are provided to ensure that this is achieved.

Having made the maximum practicable investment in the reliability of all individual plant systems within a unit, it may still remain economic to install interconnections between steam and feedwater pipework, because of the "trouble-shooting" which causes low

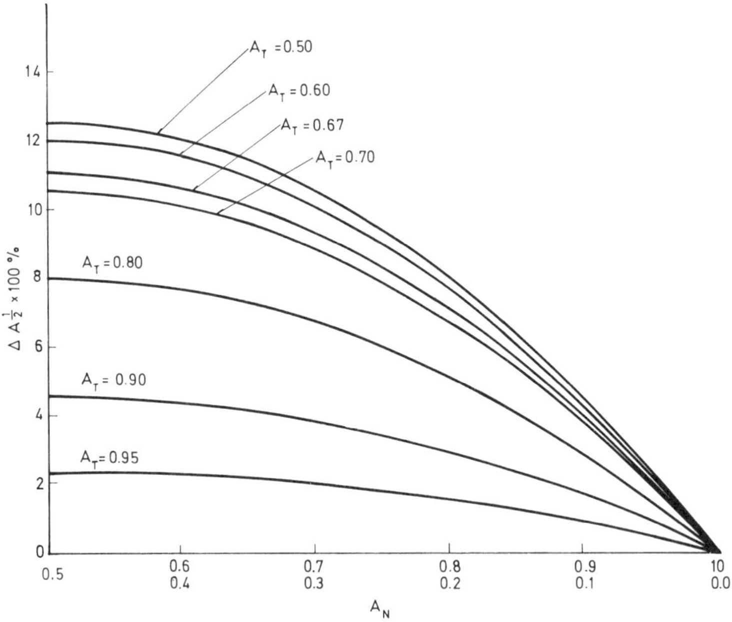

FIG.4. Availability gain from cross-connected two-reactor/turbine unit effect of difference in reactor and turbine availability.

availability in the first few years of plant operating life during and immediately following its commissioning. The potential improvement of availability of over 10% that can be obtained (Fig.4) can help to reduce the period of trouble-shooting, by not allowing construction errors and/or modifications requiring forced downtime in the early life of the plant to restrict the rapid accumulation of operating experience.

However, this limited gain will be readily dissipated if the cross-connection adds significantly to operating and maintenance problems on the plant, by isolation difficulties and cross-contamination.

The flexibility provided by unit interconnections can also be used not only in breakdown situations, but in planned downtime to carry out work intended to anticipate and prevent breakdowns occurring at inconvenient periods, to improve on availability gains based on probability of random breakdown. However, this must not detract from the primary task of ensuring a high consistent reliability for all systems within a generating unit. Cross-connection is no substitute for an inherently reliable generating unit.

3. ECONOMICS OF PLANNED MAINTENANCE ON ROTATING MACHINERY

Planned downtime of nuclear generating plant has developed directly from the post-war development of national High Voltage Electrical grid systems, which allowed opportunities to carry out work on fossil fuelled plant to minimise random plant breakdown at periods where replacement generation cost was near minimum on that particular system.

Work involving non-destructive testing has been expanded with the development of new techniques, but rotating machinery (steam turbines, electric motors, fans and pumps) still have to be periodically stripped for detailed examination of wear or cracking in critical areas not accessible in the assembled state, or for replacement of suspect components.

What evidence is there that detailed maintenance inspection improves the reliability of the plant item being examined?

The United States Navy has reported (2) that there are signs of increased failures in "first flights (of navigating aircraft) after routine maintenance." This is put down to systems which are reliable, but not so easily maintained, which are difficult to reassemble, and to the difficulty in diagnosing the state of components without stripping machinery for detailed examination.

Evidence in this field is not plentiful, perhaps because the facts may be embarrassing. However, in order to improve reliability it is essential to identify how and why failures occur, and the effect of planned preventive maintenance on the subsequent reliability of rotating mechanical plant is important to establish in order to correctly plan time and resources for routine overhauls, and to return the plant to service on schedule in a more reliable condition.

In order to understand the failure mechanisms of complex plant items, it is useful to know how many failures have occurred after various times.

Statistical techniques are being used to determine the patterns of failures. In particular, the technique of cumulative hazard plotting has been used to determine the distribution of times to failure and times to repair (3). The cumulative hazard is defined at time, t, after a repair or maintenance as the instantaneous probability of failure at that time provided that failure has not already occurred. Similarly, we define a cumulative repair rate at time, t, after a failure occurs, to be the instantaneous probability of completion of the repair, provided that the repair is not already complete.

Figures 5 and 6 show these functions plotted from values observed during a reliability study of steam feed pumps. The straight line plots on logarithmic paper indicate Weibull distributions in both cases. All the rotating plant investigated has shown a common feature in that the probability of failure decreases during the period after a repair. Figure 5 shows a typical decreasing failure rate curve for steam-driven feed pumps as a function of running time. It can be seen that while the probability of failure in the first running hour after repair is 1 in 50, the probability is only 1 in 500 in the hundredth running hour after repair. Similarly, decreasing failure rates have been found in all other pumps. The downtime follows a similar distribution which is shown in Figure 6.

An attempt was made to identify a pattern between successive running times and downtimes in order to explain the decreasing failure rate. No clear patterns emerged.

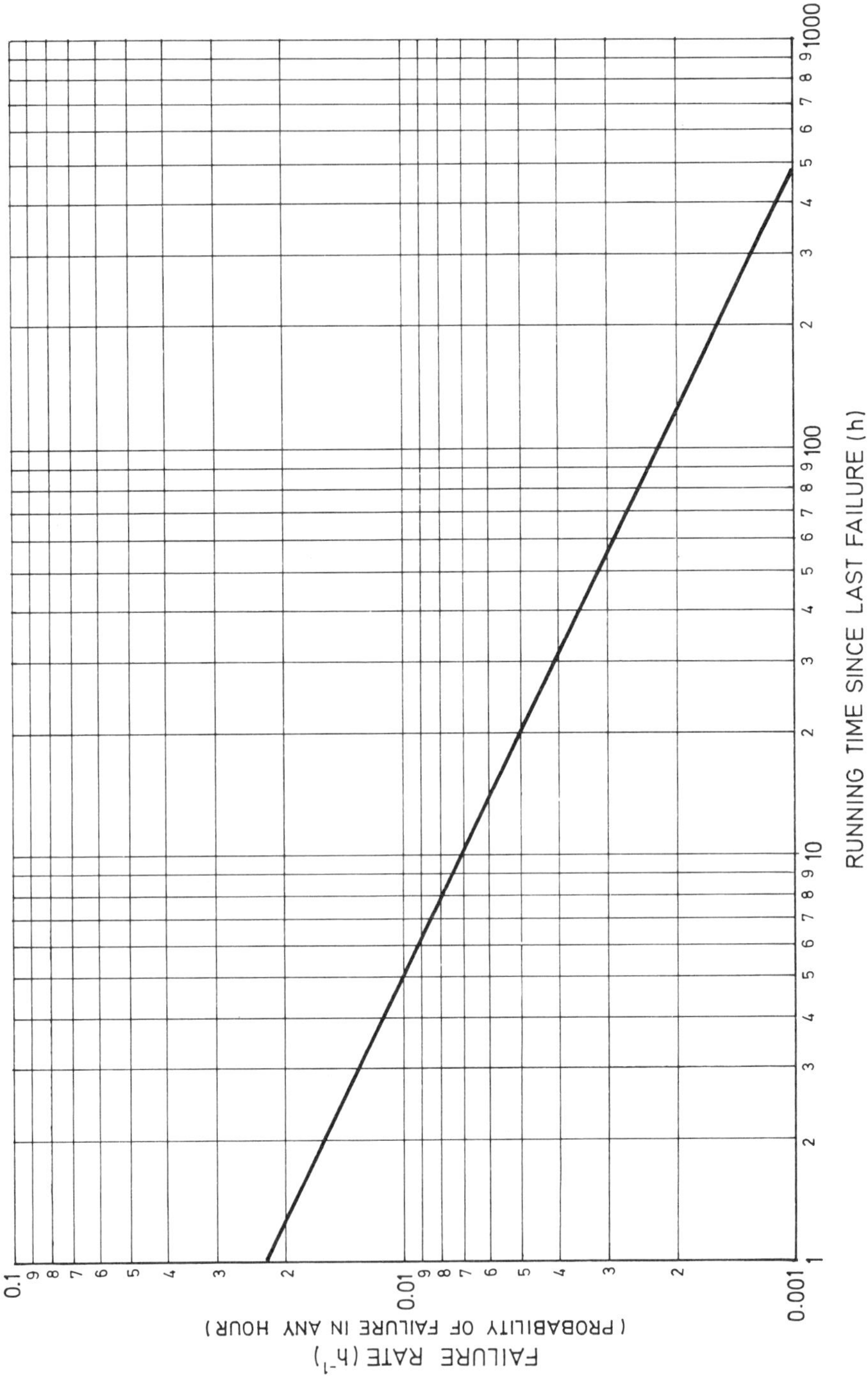

FIG.5. Failure rate of steam-driven feed pumps as a function of time.

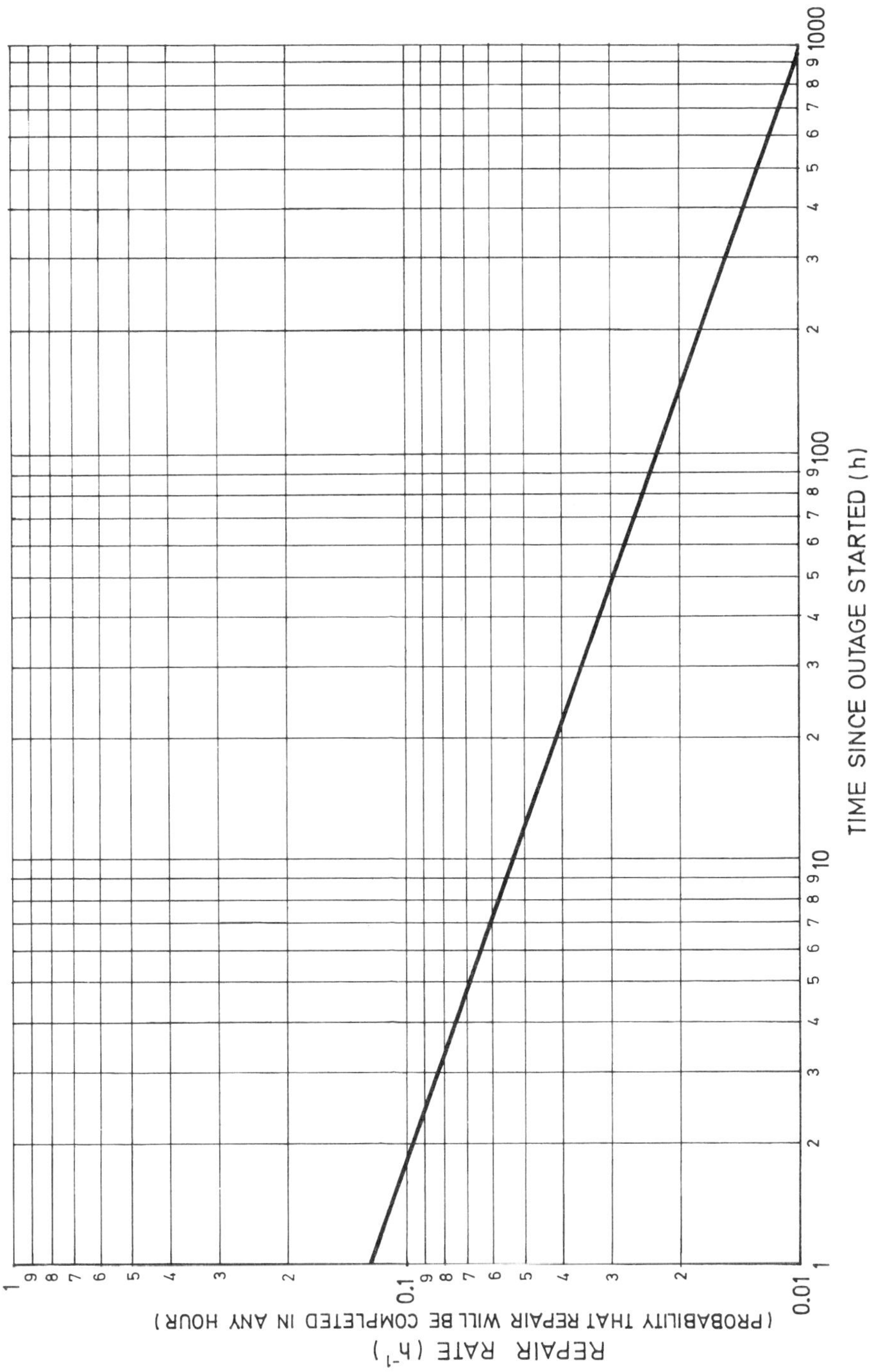

FIG.6. Repair rate of steam-driven feed pumps as a function of time.

The decreasing failure rate with time since outage tends to indicate that unnecessary additional work is carried out. When auxiliary plant becomes available for maintenance due to main plant forced downtime, the natural reaction of maintenance staff is to use the opportunity to inspect the equipment and perform minor overhauls.

If, before this is done, the failure distributions were determined as already described, the effect of this "opportunity" maintenance could be evaluated. It may well be more economic to let certain equipment undergo breakdown before carrying out repair and maintenance.

4. ENGINEERING MANAGEMENT

Modern plant systems are closely coupled combinations of sophisticated and expensive equipment, some important parts of which may be in the development stage (high technology sector), and simpler, but not necessarily cheap, equipment (low technology sector).

Manpower resources involved with such plant systems can also be placed in high and low technology categories (i.e. specialist design and construction staff, and production staff, respectively).

Neither can operate effectively without the other. Both are equally important. A sophisticated on-line computer controlling plant or analysing fault symptoms is useless, if not unsafe, if the peripheral sensing and control equipment on plant providing input data is poorly designed and inaccurate, and/or unreliable because of inadequate maintenance.

Similarly, the designer can be misled and misinformed, and subsequent design evolution can be wrongly directed, if production records do not accurately reflect what is actually happening on the plant.

The application of reliability technology can be counter productive if it demands more effort in the collection of data that it saves in facilitating quick, correct engineering decisions, and more accurate assessments of resource requirements.

Reliability engineering techniques must be simplified to make their use widely adopted in the important production sector by practising engineers who daily make decisions on design and maintenance, and established in financial and contractual procedures associated with design specification and production management.

The examples presented in this paper do not require any more sophisticated equipment than a hand calculator, a standard text book on Reliability Engineering, and an elementary confidence in using quantitative data against which to make and judge engineering decisions.

The competent engineer will not be diverted by a complicated specialised debate on the merits of this or the other calculating technique. He will be concerned about the existence or collection of just enough relevant information from plant to allow him to carry out simple calculations which will sharpen his engineering decisions from "hunches" to "cinches". It will make him more ready to admit mistakes, because he can form personal judgements on what errors were made, and what specific lessons have to be applied in the future.

REFERENCES

(1) KELLY, M., "Modifying Oldbury Power Station to Improve its Availability"
(Electrical Review - 13 October 1973, Page 479)

(2) LAYTON, D.M., Commander United States Navy, "Aviation Accidents"
(Annals of Reliability and Maintainability, Vol. 4 Practical Techniques and Application, Page 317)

(3) RIDPATH, DR. D.L., "Analytical Use of Data for Maintenance and Design"
(Symposium on a Plant Reliability and Data Feedback Pilot Study - Central Electricity Generating Board - South Eastern Region), London, October 1972.

APPENDIX 1

ASSESSMENT OF IMPROVEMENT IN RELIABILITY OBTAINED BY UNIT INTERCONNECTION

Consider a very simple model of a pair of generating units, comprising a Nuclear Reactor with its auxiliary systems, and a TURBINE GENERATOR with its own auxiliary systems.

The availability of each individual unit, A_U, can be expressed in terms of the availabilities A_N, A_T, of Boiler and Turbine systems respectively, as follows:

$$A_U = A_N A_T$$

The availability of any one of these two units is:

$$A_{\frac{1}{2}} = A_N A_T \ (2 - A_N A_T)$$

There is a strong feeling among those who operate generating plant, that the inter-connection (or "ranging") of generating units to increase flexibility will make a substantial contribution to the availability of each unit.

If an inter-connection is installed to allow any nuclear reactor to be operated with the turbine from the adjacent unit, the "ranged" availability, $A_{\frac{1}{2}(R)}$ of a single unit is:

$$A_{\frac{1}{2}(R)} = 2 A_N A_T \ (2 - A_N - A_T) + (A_N A_T)^2$$

The gain in availability of a single unit by having the inter-connection:

$$\Delta A_{\frac{1}{2}} = A_{\frac{1}{2}(R)} - A_{\frac{1}{2}} = 2 A_N A_T \left[1 - A_N - A_T + A_N A_T\right]$$

TABLE A-I.

GAIN IN AVAILABILITY $\Delta A_{\frac{1}{2}}$ FOR DIFFERENT VALUES OF A_N AND A_T

A_N \ A_T	0.95	0.90	0.80	0.70	0.67	0.60	0.50	0.40	0.30
0.95	0.0045 [0.9025]	0.0086 [0.8550]	0.01520 [0.760]	0.0199 [0.6650]	0.021 [0.6365]	0.0228 [0.570]	0.0238 [0.475]	0.0228 [0.380]	0.01995 [0.285]
0.90	0.0086 [0.8550]	0.0162 [0.81]	0.0288 [0.72]	0.0378 [0.63]	0.0398 [0.6030]	0.0432 [0.540]	0.0450 [0.450]	0.0432 [0.360]	[0.270]
0.80	0.01520 [0.760]	0.0288 [0.72]	0.0512 [0.64]	0.0672 [0.56]	0.0704 [0.536]	0.0768 [0.0480]	0.0800 [0.400]	[0.320]	[0.240]
0.70	0.0199 [0.6650]	0.0378 [0.63]	0.0672 [0.56]	0.0882 [0.49]	0.0929 [0.469]	0.1008 [0.420]	0.1050 [0.350]	[0.280]	[0.210]
0.67	0.021 [0.6365]	0.0398 [0.6030]	0.0704 [0.536]	0.0929 [0.469]	0.0980 [0.45]	0.0106 [0.402]	0.1106 [0.335]	[0.268]	[0.201]
0.60	0.0228 [.570]	0.0432 [0.54]	0.0768 [0.480]	0.1008 [0.420]	0.0106 [0.402]	0.1152 [0.36]	0.120 [0.30]	0.1152 [0.24]	0.1008 [0.18]
0.5							0.1250 [0.250]	[0.200]	[0.150]
0.4							0.1200 [0.200]	0.1152 [0.16]	[0.120]
0.3							0.1050 [0.150]	[0.120]	0.0882 [0.09]

Table A-1 gives the gain in availability $\Delta A_{\frac{1}{2}}$ for different values of A_N and A_T :

The figures in [ ] in the table are the individual unit availabilities $\left[A_B \; A_T \right]$.

For example, the availability of any 1 of 2 units <u>without</u> the inter-connection, for $A_N = A_T = 0.8$ is:

$$A_{\frac{1}{2}} = 0.64 \; (2 - 0.64) = 0.87$$

This is increased with an inter-connection by 0.0512 to 0.9212.

Taking the example of the 2 units without inter-connections, with $A_N = 0.95$, $A_T = 0.67$:

$$A_{\frac{1}{2}} = 0.6365 \; (2 - 0.6365) = 0.8680$$

This is increased with an inter-connection by only 0.021 to 0.889. The small direct increase of A_T from 0.67 to 0.70 will give an increased value of $A_{\frac{1}{2}} = 0.665 \; (2 - 0.665) = 0.888$ which is comparable to the effect of the inter-connection.

DISCUSSION

(on papers IAEA-SM-195/26 and IAEA-SM-195/28)

W.A. WOLFE: I have a number of questions for Mr. Ervamaa. Do you set a reliability target for the plant initially, and is that value then sub-divided to set targets for different sub-systems?

J. ERVAMAA: No definite reliability targets have been set in the studies bearing on safety, although the work done thus far has given us some idea what target figures might be reasonable in connection with various accident sequences. On the other hand, there is no quantitative expression of risk levels associated with nuclear power plants which could have been expected to gain general acceptance.

The establishment of a numerical target for availability is subject to cost-benefit analysis, too. If the target is very high, one has to duplicate everything, including the reactor itself, and that would be expensive. We consider it a practicable course to assume realistic availability values for the main components, such as the reactor and the turbogenerators, and to design the rest of the plant in a balanced manner so that weak points are eliminated.

It is our intention to set separate targets for individual components, but we have not done that yet. Obviously the cost of components affects the sub-division: higher availability will be required in systems consisting of cheaper components because the addition of redundancy and hence the increase of availability is cheaper in such cases.

W.A. WOLFE: Is the radiation field at the equipment considered in the failure mode and effects analysis?

J. ERVAMAA: It is, especially in connection with repair characteristics. Of course the radiation field may increase the failure rate of certain components and also lead to common mode failures, but we do not really have enough data to take these effects accurately into account. The main components of the plant and auxiliary systems are protected against radiation by shielding in order to facilitate maintenance, and the actual environmental conditions are specified for the component manufacturers in the invitations to bid; so I think we can safely say that the problem is under control.

W.A. WOLFE: Who actually does your analyses?

J. ERVAMAA: The analyses are carried out by the reliability personnel of the Technical Research Centre of Finland, in close co-operation with the staff of Imatran Voima Osakeyhtiö, the owner and operator of the plant.

G. BREILING: I have a question for Mr. Barden. Mr. Taylor (paper IAEA-SM-195/16) has shown us that about 45% of the design error failures reflected in abnormal occurrence reports had to be attributed to 'unknown causes or effects'. Do you see any possibility of overcoming these problems by, say, equipping the plant engineer with a hand computer? In my opinion the work described by Mr. Ervamaa has to be acknowledged as representing the 'state of the art', where the problem is one of making complex systems more reliable.

S.E. BARDEN: The reason why so many abnormal occurrences remain unexplained is to be sought, in my opinion, in the dearth of detailed and accurate information on the operation and maintenance of the plant. If simple diagnosis and reliability analysis is carried out by engineers working on the plant, the chances of obtaining all the relevant details are greatly improved — particularly if errors or omissions in operation and/or maintenance are seen as opportunities for learning to operate and maintain the plant more efficiently.

Sophisticated techniques should be tested initially on reliable data derived from special testing in laboratories or on test rigs, before being applied to plant data of doubtful reliability. Wherever possible, the reliability expert must simplify his methods for plant engineers using a readily available computing facility, and the hand computer is the most readily available computing facility I know of.

J.R. TAYLOR: In response to Mr. Barden's comments, I would like to emphasize that no operator can have experience of all plants. The value of a system like the American abnormal occurrence reports is that it allows everyone to share the experience which is available. I think it is also important to remember that only a very small number of the failures referred to have a significant disrupting effect on the plant or any consequences for safety. Nevertheless, the failure records are very valuable as a means of introducing more realism into reliability analyses at the design stage.

S.E. BARDEN: Records of serious or abnormal incidents are warning signals that a disciplined quantitative examination is necessary on plants with similar (not only identical) equipment, to establish what the risk is of a repetition of the fault. Such quantitative examinations can be carried out effectively only by engineers working on the systems in question, whose task will involve getting all the engineering details of the fault directly from their colleagues at the plant where the incident occurred, and also examining their own plant records.

My experience is that one does not always need to take action, apart from installing simple equipment to give an early warning of any deterioration in performance and/or physical condition which could lead to failure.

Nominally identical plant need not be, and generally is not, identical in critical details, and it is only plant engineers familiar with the details of the plant design and performance who can make reliable quantitative judgements of the significance of an abnormal incident that has occurred on similar plant elsewhere.

There is one important exception to these general remarks; this concerns the repair times required for large items of plant in modern generating units. Repair time increases rapidly with the size of an item of equipment, and this tendency must be heeded at the design stage. Protection or condition monitoring equipment must be provided to give an early warning of failures that could lead to serious damage.

G. FERRIOLE: Have you already carried out reliability studies on a certain number of components with a view to optimizing the frequency or type of preventive maintenance?

S.E. BARDEN: The most rapid way of converging on the optimum frequency of preventive maintenance is to determine under controlled conditions the effect of cutting out the preventive maintenance already being carried out on selected items of plant — in other words, to increase the period between maintenance inspections to infinity. If this is done on critical plant items which directly affect plant safety, and there is insufficient redundancy to back up possible forced breakdown, then the condition of the plant concerned must be monitored to anticipate breakdown.

What must not be done is to examine or test plant items or systems at every opportunity without quantitative information on the effect such testing or inspection is having on the failure distribution of the equipment.

Unfortunately, there is not enough quantitative information to optimize maintenance frequencies on most plants. I feel we carry out too much maintenance or testing in many cases as an insurance policy, and pay the price of more frequent breakdowns.

D. NORMAN: I have some additional information on maintenance planning. Within the Generation Development and Construction Division of the CEGB we have made some progress on the problem of optimizing maintenance frequencies. We have examined feed pump data and have found a maintenance interval which balances the risk of failures resulting from maintenance against the risk of those which maintenance prevents. However, great care is needed, because the practical consequences of each type of failure are different. We therefore use outage costs as our criterion in finding this balance. Even then, however, the result is not yet of practical value, because each plant item probably has a different optimum maintenance frequency, and all of these would need to be properly integrated to produce a good overall maintenance programme. Our work is therefore only a demonstration, for power station staff, of what might be achieved from a careful analysis of their data, in the hope that they will be motivated to collect and use it (this would provide better data for general use). This hope emphasizes Mr. Barden's point, that such analysis will not be performed by station staff unless it appears both simple and useful. This is why I feel we should make more effort to present our work simply. I accept that complex analysis is often necessary (and personally I enjoy this); however, my job is not just to get analyses performed, but to see that the results are understood and applied in actual decision making.

THE RELATIONSHIP BETWEEN RELIABILITY AND SAFETY IN NUCLEAR POWER PLANTS

L. CAVE
Pollution Prevention (Consultants) Ltd,
Crawley, Sussex,
United Kingdom

Abstract

THE RELATIONSHIP BETWEEN RELIABILITY AND SAFETY IN NUCLEAR POWER PLANTS.

This paper is concerned primarily with the problem of taking account of the relationship between reliability and safety at the conceptual design stage. The relationship between availability and safety is discussed and a composite set of quantitative criteria, in a form suitable for use in the design stage, is derived. This set has been used as a basis for the studies employed for illustrative purposes in this paper. A method is described by which reliability engineering techniques can be applied to reactor design at the conceptual stage in order to clarify the more important safety aspects. By the use of this method it is possible to determine the reliability which will be required from each of the major protection systems (i.e. reactor shutdown, core cooling and containment) and from certain critical structures in order that the reactor should meet a given set of safety criteria, specified in quantitative terms. The application of this method to the case of an SGHWR is described. Possible ways of reducing the reliability required from the reactor protection systems are discussed.

1 INTRODUCTION

In the aero-space and the defence industries it has been generally accepted for many years that in order to obtain high reliability from complex equipment it is necessary to specify at the ordering stage the reliability which is required.

The techniques of reliability engineering have been used to ensure that the design, fabrication and provision for in-service inspection and maintenance are such that the specified standards are achieved throughout the life of the equipment. During the past decade there has been an increasing acceptance of this view in other industries, including nuclear power.

In the case of a nuclear power station the potential hazard to the public which is created by the presence of a large concentration of fission products obviously necessitate a high degree of reliability. Although there is no *a priori* reason why the safety requirements should be more severe than those dictated by economic considerations it is shown below that in practice this is the case for most systems and plant items.

The use of safety objectives defined in quantitative terms has now been adopted by the USAEC, initially with respect to certain aspects of LWR safety (Ref. 1), but the U.K. licensing authority prefer to express their requirements in qualitative terms, although much of the pioneer work was carried out in the U.K. (Ref. 2, 3 and 4).

In this paper the relationship between safety and availability is discussed and a method is described for developing a set of criteria which covers both aspects. The problem of translating these general criteria into design criteria, which define the reliability requirements for particular sub-systems, is then discussed. The method is illustrated by reference to the design of an SGHWR.

2 ECONOMIC CRITERIA IN RELATION TO REACTOR FAULTS

In order to define economic criteria it is necessary to estimate the loss of availability likely to arise from specific reactor faults, so that, having regard to the value of the lost generation, the maximum acceptable frequency of particular faults can be defined. At the conceptual design stage the outages associated with particular faults can only be estimated approximately and they will tend to vary with reactor type. For illustrative purposes the lengths of outages for the more important types of fault and the corresponding values for their maximum acceptable frequency, per site year, which would probably be appropriate for a 660 MW(e) SGHWR station are shown in Table A1.2 of Appendix 1. The maximum acceptable frequencies are based on the assumptions that the average increase in fuel cost per reactor day is £2 x 10^5, if the duty of the nuclear plant has to be taken over by fossil fired plant, and that the maximum acceptable expectation of loss per reactor year is of the same order (see Appendix 1).

One of the most interesting results which is obtained from the simple analysis described in Appendix 1 relates to the proportion of the annual unscheduled outage of the reactor attributable to events which are of importance in connection with both safety and availability. It is shown that the events of this type would contribute less than one per cent of the total unscheduled outage time. This result emphasizes the limited significance which economic considerations have in relation to the definition of design criteria for the major safety-related systems of the reactor.

3 SAFETY CRITERIA

3.1 Acceptable Standards of Risk

Consideration of the hazards inseparable from everyday life and of the effects of severe reactor accidents has led to the suggestion both in U.K. (Ref. 5) and U.S. (Ref. 1), that in a country with a large nuclear power programme an acceptable standard for the over-all risk of a large release to atmosphere (e.g. 10 per cent or more of the gaseous and volatile fission products), is about 10^{-6} per reactor year for typical sites in those countries. In the U.K. this would include semi-urban sites, such as Hartlepool or Heysham.

With the conservative assumptions that radiation effects such as delayed cancers are linearly proportional to dose and that there is no threshold below which damage ceases, a consistent set of safety criteria for less severe faults could be established by reference to the criterion chosen for large releases. However as discussed in Ref. 6, on sites with a high population density (say 10^4 per km^2) within a distance of about 5 km it is necessary to consider also the possibility of immediate effects, due to whole body radiation, which are not linear. This aspect is also discussed in Ref. 7. It has been argued (Ref. 8) that in most, if not all, serious accident conditions radio-active buoyancy effects would cause the plume of gaseous and volatile fission products to rise, thus reducing the whole body dose to a level at which "radiation-sickness" would not be a problem.

It is assumed in the remainder of this paper that the population distribution is such that the immediate effects can be ignored. With this reservation it is convenient to use the modified 'Farmer line' (Ref. 5) as a basis for the safety criteria. However, experience has shown that it is not necessary to consider the detailed distribution of releases, due to individual fault sequences, within the various decades, but to use instead, the total expectation (defined in Appendix 1) implied by the Farmer line as an over-all target. This expectation is 12 curies I^{131} per reactor year but, for the reasons described below, it is preferable to treat separately the regions above and below 10^3 curies I^{131}. It should be noted that in most situations the relative proportions of the various isotopes in the plume are reasonably constant, so that it is permissible to define the targets in terms of any one of the predominant isotopes; as in most safety studies, I^{131} has been chosen.

In the high release region ($10^3 - 10^8$ Ci I^{131}) an expectation of 10 Ci I^{131} per reactor year has been taken as an over-all target. In the lower region (0.1 to 10^3 Ci I^{131}) the remaining 2 curies per reactor year of the total target value derived from the Farmer line are divided between the following groups:

(i) Minor in-reactor faults, for which availability considerations are likely to be more limiting than the requirements of public safety.

(ii) Faults outside the reactor containment (e.g. in the cooling pond), whose frequencies are not likely to be severely limited by availability considerations.

3.2 Criteria for Reactor System Design

The procedure described in the previous Section leads to the set of criteria, for in-reactor faults, shown in Table 1.

Table 1 : Proposed criteria based on safety and availability considerations, for a 2 x 660 MW(e) SGHWR station, in a semi-urban location, for use in deriving reliability requirements for reactor protection systems.

Size of release to atmosphere (curies I^{131})*	0.1 to 1	1 to 10	10 to 100	10^2 to 10^3	10^3 to 10^8
Maximum acceptable frequency (summed over all foreseeable fault sequences) per site year	10^{-2}	10^{-2}	10^{-2}	10^{-3}	Total expectation (summed over all foreseeable fault sequences) 20 Ci per site year
Basis for criterion	Availability				Safety

*The presence of appropriate fractions of other isotopes is implied.

In some countries a further consideration, that of the public's reaction to small but relatively frequent releases (e.g. a few curies of I^{131} every 10 years) from a nuclear site, might prove more restrictive than the availability requirements in the range 1 to 10^3 curies.

The use of criteria of the form shown in Table 1 as a basis for deriving the reliability requirements for the principal safety-related reactor systems is described briefly in the next Section.

4 DERIVATION OF RELIABILITY REQUIREMENTS FOR REACTOR PROTECTION SYSTEMS

4.1 Basis of Method

The systems and plant items considered are those which have a protection role. They are as follows:

(i) Reactor shut-down systems

(ii) Decay heat removal systems (DHRS)

(iii) Systems required to provide protection against local faults

(iv) Containment system

(v) Critical structures

The reliability required from each of the systems (i), (ii) and (iii) is dependent on the magnitude of the release to the environment which could occur if it failed and on the frequency of demands for the system to operate. Similar considerations apply to the "critical structures", which are defined as structures whose failure could render the other protection systems ineffective and, at the same time, create conditions leading to a large release of activity.

In the case of systems (i), (ii) and (iii) it is assumed that their functional performance is adequate, so that their starting and, in (ii), their running reliabilities are the points of interest. In the case of the containment system, however, several types of response are possible, depending on the circumstances, so that it is necessary to define a set of performance criteria. For the critical structures it is necessary only to determine the maximum acceptable rates of failure.

A considerable simplification of the analysis results from adopting the view that if either the reactor shut-down systems or the DHRS fail to operate when required as a result of some fault condition, melt down of the core is usually inevitable. Thus an upper limit to the reliability required from one of these systems can be obtained by assuming that its failure would lead to the release of the full inventory of volatile and gaseous fission products from the fuel.

It is convenient to derive first an expression for the reliability of the reactor protection systems on the assumption that the containment is completely ineffective. As shown in Appendix 2, u_i, the maximum acceptable unreliability (or failure rate per demand) for system S_i is given by the equation

$$u_i = \frac{E_i}{R_i f_i} \qquad \text{...... Equation 1}$$

where E_i is the sub-target expectation allocated to system S_i (see Appendix 2)

R_i is the release to atmosphere resulting from a failure of system S_i

and f_i is the summated frequency of all faults which necessitate the operation of system S_i

As discussed in Appendix 2, the containment system can be characterized by a factor P_i which represents the expectation of release to atmosphere for a unit release of I^{131} from the fuel into the containment, in the conditions associated with failure of system S_i. As described in Appendix 2, P_i represents not only the "decontamination factor" for a particular sequence of events in the containment but also embodies the estimated probabilities for the various alternative sequences. It will be seen therefore that it reduces to a single number the complex event trees for the containment, which are a feature of the method of analysis used in the Rasmussen study (Ref. 7). This simplification facilitates the determination of system design criteria, as u'_i (the maximum acceptable unreliability for system S_i in a contained reactor) is given directly by the equation

$$u_i' = \frac{E_i}{R_i f_i P_i} \qquad \text{...... Equation 2}$$

As the design proceeds it may be found more economical to re-arrange the initial allocation of the target expectation between the various systems.

4.2 Application of the Method to a Pressure-tube Reactor

The method described above has been used to determine the system reliabilities required for a 660 MW(e) SGHWR in order to meet the same over-all criteria as those defined in Table 1.

From consideration of the conceptual design values were chosen for the estimated frequencies of demand and for the initial allocation of the target expectation (i.e. 10 Ci I^{131} per reactor year) between classes of faults. These are shown in Table 2.

Table 2 : Estimated frequencies of demand on reactor protection systems and initial allocation of target expectation between classes of fault.

Serial No.	System	Estimated frequency of demand per r.y.	Target expectation Ci I^{131} per r.y.
1	Reactor shut down (2 systems)	0.2	2
2 (a)	Decay heat removal, pressurised, starting reliability	4	1
(b)	As 2 (a) running reliability	4	1
3 (a)	Decay heat removal, depressurised, starting reliability	10^{-2}	1
(b)	As 3 (a) running reliability	10^{-2}	1
4	Critical structures	Not Applicable	2
5	Local protection systems (e.g. channel blockage detection)	See text	2
		Total	10

The use of the method to define the maximum acceptable failure rates for the critical structures is straight forward but its application to local protection systems presents some difficulties, as it is necessary to take into account random variations in channel rating. Moreover, in a pressure tube reactor it is also necessary to take into account the risk that a fault which initially involves only one channel will escalate to involve neighbouring ones, or even the whole core. The method has been extended to cover such cases but the treatment is too lengthy to describe in this paper.

The results obtained by this method for the principal reactor protection systems are shown in Table 3, both with and without credit for the containment.

Table 3 : Maximum acceptable unreliability for reactor protection systems, with and without taking credit for containment, for a 660 MW(e) SGHWR

Protection System	Maximum acceptable unreliability		Ratio
	With containment	Without containment	
	failures per demand	failures per demand	
Reactor shut-down system	5×10^{-6}	2×10^{-7}	25 : 1
Decay heat removal system (pressurised)- starting	0.25×10^{-6}	0.05×10^{-7}	50 : 1
Decay heat removal system (depressurised)-starting	100×10^{-6}	20×10^{-7}	50 : 1

The assumptions concerning containment performance are described in Appendix 2. It will be seen that unless credit is taken for containment it is doubtful whether the necessary reliability could be achieved for the DHRS (pressurised) without using at least three separate and diverse systems.

5 POSSIBLE MEANS OF REDUCING RELIABILITY REQUIREMENTS FOR REACTOR PROTECTION SYSTEMS IN AN SGHWR

From Equations 1 and 2 it can be seen that these requirements could be reduced by increasing E_i or decreasing one or more of R_i, f_i and P_i. However E_i is outside the designer's control.

Clearly R_i is simply proportional to the thermal power of the reactor. However a less conservative assumption concerning the proportion of I^{131} released into the containment could be justified in some fault sequences. This might lead to a decrease of a factor of about 2 in the reliability requirements.

f_i is dependent on a number of design features but it is unlikely that it could be decreased by a factor of more than 3, in any of the classes of faults, without major changes in current design concepts. In general the faults which make the major contribution to f_i, in any of the classes, do not make a substantial contribution to the total expected outage of the plant as a whole. Consequently there is little economic incentive to make design changes in order to reduce f_i.

Direct substantiation of the assumptions made in deriving values of P_i is difficult and agreement would have to be reached with the licensing authorities concerned on the basis of semi-quantitative arguments. Consequently the assumptions must be conservative but in due course further test work may justify the use of larger values for P_i.

6 CONCLUSIONS

The reliability which would be required from any of the reactor protection systems on economic grounds can be represented by a failure rate of about 10^{-4} per demand.

In the case of the SGHWR design used here as an illustrative example it can be seen by reference to Table 3 that it is only in the case of the DHRS required in depressurisation accidents that a failure rate of order 10^{-4} per demand, in conjunction with an effective containment, would be acceptable on safety grounds. For other cases the safety requirements necessitate a reliability which is greater by at least an order of magnitude. It is believed that this situation exists in most, if not all thermal reactors. Studies, similar to those reported here for SGHWR, which have been made by the present author for fast reactors, show that it is certainly the case for both LMFBR and GCFR (See Refs. 9 and 10).

Nevertheless, for reactor faults which could give only small releases of I^{131} (less than 10^3 curies) to atmosphere, economic considerations are likely to be more restrictive than safety requirements based on public health considerations and about as restrictive as considerations based on the public's acceptance of nuclear power. However, faults of this type are not associated with failures of the main reactor protection systems.

ACKNOWLEDGEMENTS

The permission of the South of Scotland Electricity Board to make use of data relating to reliability requirements for a 660 MW(e) SGHWR is gratefully acknowledged. However, the safety criteria used as a basis for the derivation of system reliabilities do not necessarily reflect the views of S.S.E.B. or of the U.K. licensing authority.

APPENDIX 1

NOTES ON DERIVATION OF RELIABILITY CRITERIA

A1.1 Expectation of Loss

If an event "i" has a frequency of occurrence of f_i and the cost, or loss, associated with the event is c_i then the product $f_i c_i$ is defined as the "expectation of loss" for event "i". In a system subject to numerous possible events such as "i" per year, the annual expectation of loss, E, is defined by $\Sigma f_i c_i$, where the summation extends over all such events.

E represents the average annual loss which would be expected over a large number of years, clearly there could be a substantial variation from year to year.

A1.2 Expectation of Loss as a Guide to Reliability Requirements

Consider firstly a hypothetical reactor in which there is only one system which affects both safety and availability.

Assume that the average outage time following the failure of this system is t days and that the value of the lost generation is £c per day. There will exist some frequency of failure (denoted by f) which, from the economic point of view, is an optimum, since to improve the system would cost more than the value of the generation saved. In this context the value of the generation saved is its present worth at the start of the reactor's life, which is assumed to be 10 times the annual value.

Assuming that during the life of the reactor the average value of c is £10^5 per reactor day, the present worth of the expectation of loss (10 ftc) can be written down, as in Table A1.1, for various combinations of f and t.

Table A1.1 : Present worth (in £ sterling) of the expectation of loss as a function of f and t. (c = £10^5 per reactor day)

Average duration of outage, t (days)	Frequency of failure, f, per reactor year					
	1	10^{-1}	10^{-2}	10^{-3}	10^{-4}	10^{-5}
1	10^6	10^5	10^4	10^3	10^2	10
10	10^7	10^6	10^5	10^4	10^3	10^2
100	10^8	10^7	10^6	10^5	10^4	10^3
1000	–	10^8	10^7	10^6	10^5	10^4

At the present level of plant costs it is unlikely that the reliability of any major reactor sub-system could be improved by a factor of 10 without incurring an additional cost of order £10^5. Reference to Table A 1.1. shows that the optimum values of f for particular values of t are of the following order:

t (days)	1	10	100	1000
f (failures per reactor year)	10^{-1}	10^{-2}	10^{-3}	10^{-4}

It will be seen that 'ft', the "expectation of outage", is the same for all values of t.

In an actual nuclear power station there could be several major items of plant, or sub-systems, whose failure would affect both safety and availability. If there are n such systems the total expectation of loss per reactor year due to outages caused by their failures is $\sum_1^n 10 f_i t_i c$.

The contribution made to the total expectation of loss by any one system is largely independent of the others. Consequently each can be optimised, from the economic point of view, in the same way as the single system examined above. Thus from the result obtained above it follows that in an optimum design f should be inversely proportional to t, for each system, unless the cost of improving the reliability of some systems is markedly less than for the others.

A1.3 Reliability Requirements for a 660 MW(e) SGHWR

In the case of a 2 x 660 MW(e) SGHWR station the average difference in fuel cost between the nuclear and fossil fired stations is estimated to be about 0.6p per u.s.o., which gives an average value for the output of £10^5 per reactor day. However, to provide a margin to cover the approximate nature of the method, the calculations of the maximum acceptable failure frequencies have been based on a value of c of £2 x 10^5, other factors remaining unchanged. Thus the maximum acceptable frequencies per reactor year are halved.

In SGHWR the types of fault sequence which are important from the points of view both of safety and availability are those in which, as a minimum, some breach of the primary coolant circuit occurs, permitting the escape of fission products, notably I^{131} (from defective fuel elements), dissolved in the coolant or liberated by further damage to the fuel. The fault sequences of this type which are believed to be important can be grouped into four general classes. These are shown in Table A1.2, together with the estimated outage times, maximum acceptable frequencies (on economic grounds) and magnitudes of I^{131} releases to atmosphere. The releases have been estimated on the basis that the containment retains only 99 per cent of the I^{131} released into it. This is thought to be a conservative assumption in all the cases considered.

Table A1.2 : Maximum acceptable frequencies, on economic grounds, for particular types of reactor fault, for a 2 x 660 MW(e) SGHWR station.

Serial No.	Class of fault	Estimated outage time (days)	Maximum acceptable frequency, on economic grounds, for type of fault	Estimated release to atmosphere Ci I^{131}*
1	Single channel (in core) faults without gross fuel damage, but requiring replacement of pressure tube	Order 10	10^{-2} per site year	$\ngtr 50$
2	Single channel (in core) faults with gross fuel damage	Order 100	10^{-3} " " "	$\sim 10^3$
3	Major circuit failures outside core, without gross fuel damage (e.g. melt down in only 1 or 2 channels)	Order 100	10^{-3} " " "	$\sim 10^3$
4	Major circuit failure outside core, with gross fuel damage in many channels	Order 1000	10^{-4} " " "	$\sim 10^5$

*These values are based on conservative estimates of containment performance.

From Table A1.2 a further table can be prepared showing the maximum acceptable frequencies, based on economic considerations, for particular sizes of release of I^{131}. These are shown on Table A1.3, which shows also the expectation of non-availability as days per site year, and the expectation of release, as curies I^{131} per site year.

Table A1.3 : Maximum acceptable frequencies for I^{131} releases, based on economic considerations, for a 2 x 660 MW(e) SGHWR station.

Size of release to atmosphere (curies I^{131})	0.1 to 10^2	10^2 to 10^3	10^3 to 10^4	$> 10^4$	Total
Maximum acceptable frequency per site year	10^{-2}	10^{-3}	10^{-3}	10^{-4}	-
Expectation of non-availability (days per site year)	.1	.1	.1	$\not>$ .1	$\not>$ 0.4
Expectation of release, curies I^{131} per site year (maximum)	1	0.9	9	> 10	> 20

It will be seen from Table A1.3 that the expectation of non-availability due to the classes of faults defined in Table A1.2 is of order 1 day per site year (for a 2 reactor site). For nuclear or fossil fired plant an annual availability of 80 per cent is unlikely to be exceeded; allowing 33 days per year per reactor for scheduled maintenance and refuelling, the non-availability due to unscheduled events is 80 days per site year. Thus only some one per cent of the unscheduled non-availability would be due to faults of the types shown in Table A1.2, which emphasises the limited significance of safety related faults in relation to plant availability.

A1.4 Economic criteria derived from consideration of material damage outside the station

The cost to the community resulting from large releases of gaseous fission products (up to 10^6 curies I^{131} and corresponding fractions of other isotopes) has been discussed in Ref. 11. It is shown that the total cost for the largest release considered could be as high as £6 x 10^8; the dominant factor would be the cost of rehabilitation of the people living in the area affected by caesium contamination. This cost is less than an order of magnitude greater than the replacement cost of the station together with the differential cost of power to make good its output during the replacement period.

Thus the wider economic considerations would not require a much higher standard of reliability than those based on the risk to the plant itself. In practice the standard of reliability required on safety grounds would be more than sufficient to satisfy the wider economic requirement, as the frequency of releases greater than 10^6 Ci I^{131} would be of order 10^{-6} per reactor year, or less.

APPENDIX 2

DERIVATION OF MAXIMUM ACCEPTABLE UNRELIABILITIES FOR REACTOR PROTECTION SYSTEMS

A2.1 Containment Assumed to be Completely Ineffective

Denote the maximum acceptable unreliability for a protection system S_i by u_i.

Assume that the total number of demands for the system to operate per reactor year (i.e. the summated frequency) is f_i. The value of u_i can be estimated as described below.

In the limit any one failure to operate would lead to melt down of the fuel and to a potential release from it of the total inventory of I^{131}, although in practice, in some fault sequences at least, there might be some retention within the molten fuel. Denote this potential release by R_i. Melt down of the fuel would be accompanied by failure of the primary circuit, either as a cause or as an effect. If it is assumed that the containment would be ineffective in these conditions, the expectation of release to the atmosphere as a result of failures of the system S_i would be $u_i R_i f_i$ curies of I^{131} per reactor year.

In any reactor system the individual fault sequences which make a significant contribution to the total expectation of release can be grouped into 5 broad classes, according to the type of protection system failure. This grouping is shown in Table 2 of the main text.

In the early stages of developing a design it is usually advisable to divide the total target expectation for large releases equally between the 5 classes. i.e. in terms of the criteria developed in Appendix 1 a sub-target expectation of 2 curies I^{131} per reactor year would be allocated to each class. Further sub-division within each class would be necessary, as indicated in Table 2 of the main text. Denote by E_i the expectation allocated to fault sequences in which failure of system S_i occurs. Thus we have:

$$u_i = \frac{E_i}{R_i f_i} \qquad \text{...... Equation 1}$$

E_i and R_i are known. By a study of the conceptual design in the light of operating experience with other reactors and non-nuclear plant, f_i can be estimated. Obviously such estimates are subject to error but a factor of 3 either way should encompass the 90 per cent confidence limits, except for structural fails (pipes and vessels) where the band of uncertainty may be wider (see Ref. 7).

A2.2 Containment Assumed to be Partially Effective

If the reactor is contained there will be a number of possible event-sequences which describe the behaviour of the containment following a reactor fault. Associated with each sequence there will be some specific fraction of the 'free' I^{131} (i.e. the I^{131} released from the fuel into the containment) which escapes to atmosphere. Let F_j denote the fraction for sequence j and let p_{ji} denote the probability of sequence j occurring after some particular reactor fault, i.

It should be noted that p_{ji} may vary with the type of fault which has resulted in damage to the fuel and the extent of that damage.

Let P_i denote the expectation of release to atmosphere, per curie released into the containment, following a fault sequence embodying failure of protection system S_i. Thus we have $P_i = \Sigma F_j p_{ji}$ curies per curie of 'free' I^{131}. The summation extends over all possible event-sequences.

With the same notation as in Equation 1 we can now write:

$$u_i' = \frac{E_i}{R_i f_i P_i} \qquad \text{...... Equation 2}$$

where u_i' is used instead of u_i to indicate that the effect of the containment has been taken into account.

Clearly, in order to determine a numerical value for P_i the design of the containment system must be defined.

In the illustrative example given in the main text the values of F_j and p_{ji} which were assumed gave values for P_i of 3.4×10^{-2} for fault sequences in which the reactor failed to trip and of 1.3×10^{-2} for fault sequences leading to core melt down as a result of inadequate decay heat removal.

REFERENCES

[1] Regulating Staff, USAEC, Technical report on anticipated transients without scram for water-cooled power reactors, Document WASH-1270, September 1973.

[2] FARMER, F.R., "Siting criteria - A new approach", Containment and Siting of Nuclear Power Plants (Proc. Symp. Vienna, 1967), IAEA, Vienna (1967) 303.

[3] ADAMS, C.A., STONE, G.N., "Safety and siting of nuclear power stations in the United Kingdom", Containment and Siting of Nuclear Power Plants (Proc. Symp. Vienna, 1967), IAEA, Vienna (1967) 129.

[4] CAVE, L., HOLMES, R.E., "Suitability of the Advanced Gas-cooled Reactor for urban siting", Containment and Siting of Nuclear Power Plants (Proc. Symp. Vienna, 1967), IAEA, Vienna (1967) 273.

[5] FARMER, F.R., "Development of adequate risk standards", Principles and Standards of Reactor Safety (Proc. Symp. Jülich, 1973), IAEA, Vienna (1973) 3.

[6] CAVE, L., HALLIDAY, P., "Suitability of gas-cooled reactors for fully urban sites", Paper 10, BNES Symposium on Safety and Siting, London, March 1969.

[7] RASMUSSEN, N., USAEC Reactor Safety Study - An Assessment of Accident Risks in U.S. Commercial NPP, WASH-1400 (Draft).

[8] BEATTIE, J.R., Proceedings, BNES Symposium on Safety and Siting, London, March 1969, and discussion p.24.

[9] CAVE, L., "A comparative study of the safety of liquid-metal cooled and gas-cooled fast reactors", Paper 14/4, ANS Fast Reactor Safety Meeting, Los Angeles, April 1974.

[10] CAVE, L., A comparison between the safety of liquid-metal cooled and gas-cooled fast reactors, Nucl. Eng. Int. (Nov. 1974) 950.

[11] BEATTIE, J.R., BELL, G.D., "A possible standard of risk for large accidental releases", Principles and Standards of Reactor Safety (Proc. Symp. Jülich, 1973), IAEA, Vienna (1973) 11.

RELIABILITY STUDIES IN A DEVELOPING TECHNOLOGY

L.A. MITCHELL, C. OSGOOD, S.J. RADCLIFFE
Central Electricity Generating Board,
Berkeley Nuclear Laboratories,
Berkeley, Glos.,
United Kingdom

Abstract

RELIABILITY STUDIES IN A DEVELOPING TECHNOLOGY.

The standard methods of reliability analysis can only be applied if valid failure statistics are available. In a developing technology the statistics which have been accumulated, over many years of conventional experience, are often rendered useless by environmental effects. Thus new data, which take account of the new environment, are required. This paper discusses the problem of optimizing the acquisition of these data when time-scales and resources are limited. It is concluded that the most fruitful strategy in assessing the reliability of mechanisms is to study the failures of individual joints whilst developing, where necessary, analytical tools to facilitate the use of these data. The approach is illustrated by examples from the field of tribology. Failures of rolling element bearings in moist, high-pressure carbon dioxide illustrate the important effects of apparently minor changes in the environment. New analytical techniques are developed from a study of friction failures in sliding joints.

1. STATEMENT OF THE PROBLEM

The first stage in the assessment of the reliability of a system is to establish the functional relationships between its individual components. If the component reliabilities are known, and are statistically independent, the overall reliability of the system can be computed using existing analytic techniques.

In any established technology many years of operational experience provide a wealth of information on component failures which, if correctly interpreted, may be used as input data for future reliability analyses. The special problem which arises in any developing technology, for example in nuclear applications, is that relevant operating experience is at best limited and at worst non-existant. This is because the basic failure modes can be strongly influenced by changes in scale and apparently trivial differences in environment. Thus predictions of component reliabilities must be based on data acquired from research and development programmes. Such programmes need to be carefully formulated if they are to provide an economical and sufficient estimate of system reliability.

There are three basic constraints within which all research and development programmes must be optimised. Two of them arise from limits on available time and resources; the third, of a slightly different nature, is determined by the level of theoretical understanding. The independent variables in the optimisation process are the number of tests performed and the size of the component tested, where size can be changed by scaling or by testing sub-assemblies. The difference between a research and a development programme lies in which of the constraints are assumed to be fixed. In a development programme the timescale and level of theoretical understanding are fixed but the resources allocated can be variable. In a research programme resources are fixed, timescales are relaxed and the objective is to ease the constraints (arising from a lack of theoretical understanding) in future development programmes.

It is clear that the cheapest and fastest method of assessing the reliability of a system is the one which performs tests on the smallest elements from which the reliability of the system can be calculated theoretically. In the example given in section 2 an estimate of the reliability of a control rod actuator is required. Proving tests demonstrated that the element most likely to fail was a ball-bearing and analysis of the mechanism showed that ball-bearing failures could be considered to be statistically independent. Thus a development programme to determine the parameters describing ball-bearing performance in the appropriate environment was performed.

In many mechanisms the assumption of statistical independence is invalid. This is especially true for mechanisms with sliding joints where friction forces on one joint may alter the loads on others. Thus,for mechanisms with sliding joints there is currently no alternative to testing the complete machine. To improve on this situation requires research in two areas, the first being to consider the parameters controlling the failure of joints and the second to elucidate the methods by which data for individual joints can be combined. Section 3 presents the results of some theoretical work; a solution is given for frictionally induced failures in sliding joints and methods for the combination of data from individual joints are discussed.

2. A DEVELOPMENT PROBLEM

2.1 The Background

The methods adopted in establishing the reliability of the control rod actuators for the Advanced Gas-cooled Reactor (AGR) provide a good example of a successful development programme. The actuator consists essentially of a motor driven gear-train connected to a winch which raises and lowers the control rod. The complete mechanism is contained within the pressure circuit and is therefore exposed to the AGR coolant, but at a low enough temperature to permit the use of grease lubricated bearings. There are several components which might be expected to reduce the reliability of the actuator (e.g. gears, bearings, limit switches etc.). Failures of limit switches are clearly statistically independent of failures in other components but this is not true for gears and ball-bearings. In the later cases the basic failure rates are functions of load so that only if the transmitted forces and torques are large, compared with those arising from friction, can their failures be simply predicted from the reliabilities of individual components.

Conventional engineering components were used throughout the actuator and thus the reliability in <u>air</u> could be predicted from standard data. Experience with similar mechanisms in Magnox reactors gave confidence that such predictions were applicable to CO_2 atmospheres. Thus for the AGR it was considered that a simulated life test would suffice; its main purpose being to detect the presence of gross design errors. During the tests several ball-bearing failures occurred, which suggested that air data was not in fact appropriate.

2.2 Analysis of Failures in'Life' Tests

The distribution of ball-bearing lifetimes is usually described by the Weibull function viz:

$$F = 1 - \exp(-AL^{e})$$

where F is the failure fraction at a given time.
L is the time.
A,e are parameters of the distribution.

The variables are normally plotted using the equation in the form:

$$\ln \ln\left(\frac{1}{1-F}\right) = e \ln(L) + \ln(A)$$

so that the equation has been transformed to a straight line with e as the slope and ln(A) as the intercept. If experiments are performed to determine the distribution functions for various loads, geometries, speeds, lubrications and materials it is found that, provided the failures are attributable to fatigue, a family of approximately parallel lines is generated. Thus the parameter e is only weakly dependent on these variables and to a first approximation may be taken to be constant. Given an understanding of the dependence of the parameter A on the variables listed above, data from all experiments can be reduced to a single, unique distribution. For the purposes of comparison it is convenient to use the non-dimensional variable L/L_{10} where L_{10} is time at which 10% failures would be predicted. Ball-bearing fatigue analysis is based on a model due to Lundberg and Palmgren [1] from which semi-empirical formulae have been developed. Modifications of these formulae are used by manufacturers to make recommendations on the choice of bearings for various loads geometries and speeds. There is no sound theoretical basis relating lifetimes to lubrication conditions but the scientific literature [2-6] indicates that lifetime is reduced if a full elastohydrodynamic film is not formed.

For the bearing failures in the 'life' tests, loads, speeds, test durations were known. A service factor of 2, to allow for uncertainties in the loads (due to friction in the gears, vibrations etc.) was assumed. From an assessment of the scientific literature on lubrication it was decided that, under the test conditions, bearing fatigue life would have been reduced by a factor of the order of 6. Hence the non-dimensional lifetime (L/L_{10}) for each failure was calculated. The information that could be extracted from the 'life' tests was equivalent to the failure of three bearings, at known times, from four tests. To express these results in terms of failure fractions the standard 'median rank correction' was applied to allow for the small sample. These results, together with the expectation, based on air data, are plotted in fig.1.

It is clear from fig.1 that the observed lifetimes were substantially shorter than expected. The statistical significance of these observations was estimated using the binomial distribution, from which it can be shown that the probability that these failures were taken from the expected distribution is less than 1 in 1000. It was concluded, therefore, that there was a systematic effect reducing the lifetimes of the bearings and that the most likely cause lay in the operating environment.

2.3 Laboratory Bearing Tests

The failures observed in the 'life' tests showed all the characteristics of fatigue, including a relatively long period of satisfactory operation followed by a rapid deterioration corresponding to the onset of 'pitting'. Thus the assumption of statistical independence used in the analysis of the mechanism remained valid and the only requirement was for new data appropriate to the AGR operating environment.

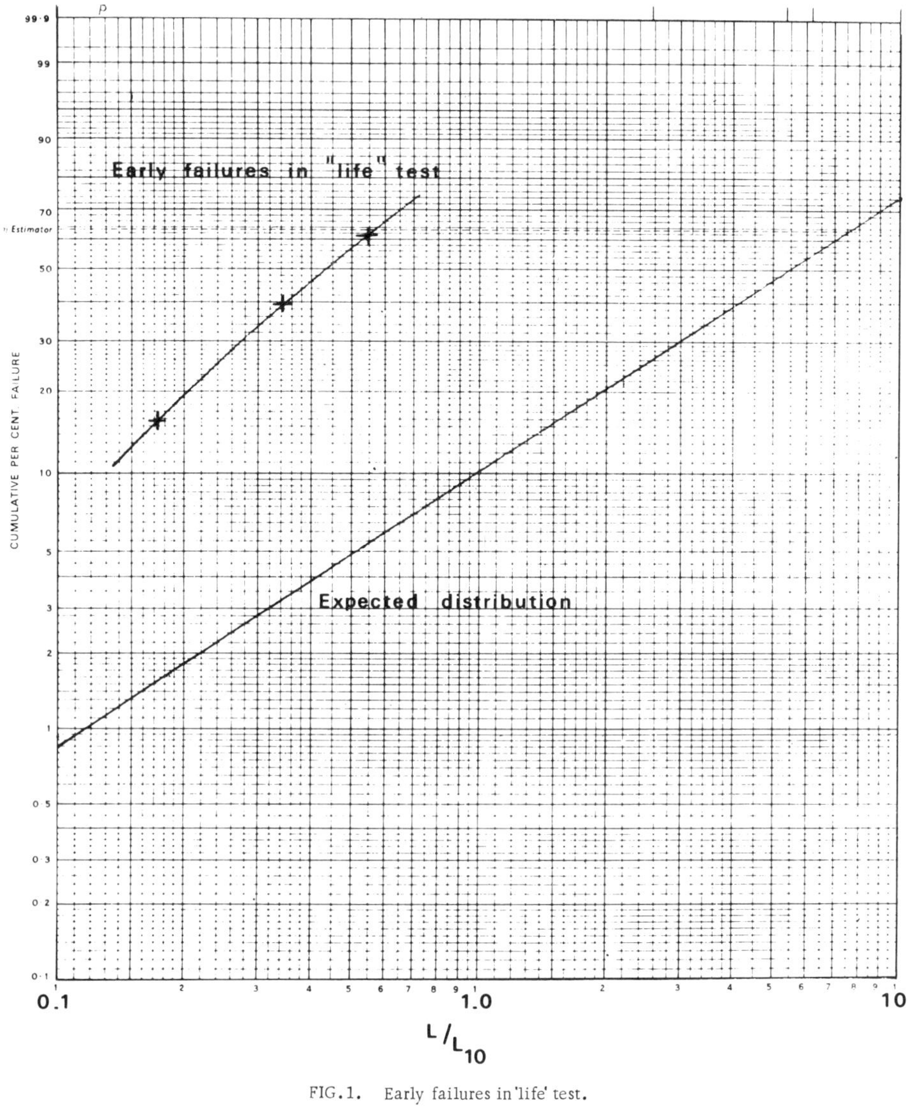

FIG.1. Early failures in 'life' test.

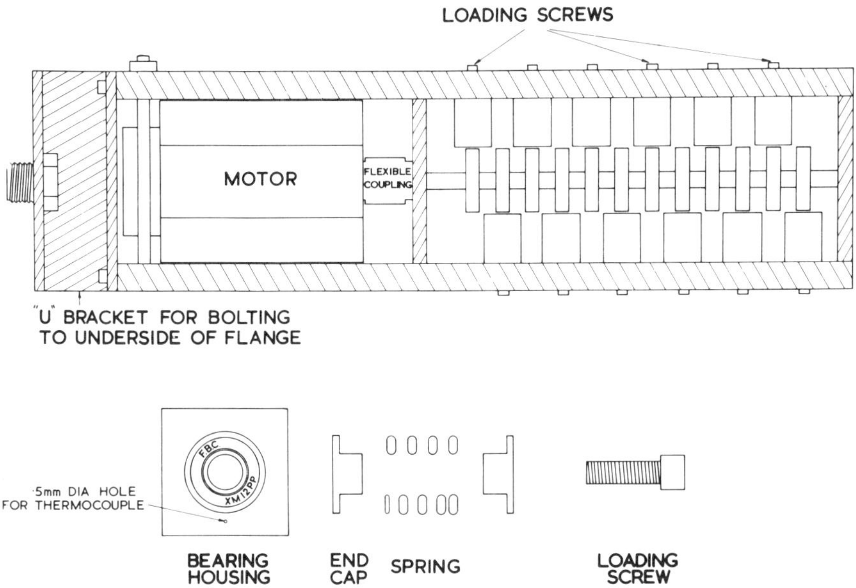

FIG.2. Bearing test rig.

The tests were carried out in CO_2 at (4.14 MN/m^2)(600 psi†) with 5-10 vpm‡ O_2 and 10-20 vpm CO. The moisture level was controlled to ±10% at various values between 50 vpm and 1300 vpm. The bearing test assembly is shown diagrammatically in fig.2. It consists of a motor driven shaft (motor speed was 1450 rpm) on which the bearings (type 6001 double shielded) are mounted. The bearing outer races were held in steel housings which were radially loaded by a spring-screw system. Bearing temperatures and moisture level were continuously monitored. Tests were terminated either when the rig noise exceeded an arbitrary level or when the motor stalled. At the conclusion of each test the number of fatigue failures was determined by stripping each bearing and performing a thorough visual examination. These results were analysed using the methods described above, and are shown in fig.3. It is clear that these results are adequately described by a Weibull distribution of similar slope to that for air data but reduced in life by a factor of 15-20.

To investigate the effect of moisture level, further tests were performed with moisture levels in the range 5-50 vpm. Fig.4 shows the life reduction factor plotted against moisture level. Below ~30 vpm the life reduction factor is of the order of 5 whilst above 100 vpm it is approximately constant and about 20. Between these values there is some uncertainty.

---

† lbf/in^2.

‡ Parts per million by volume.

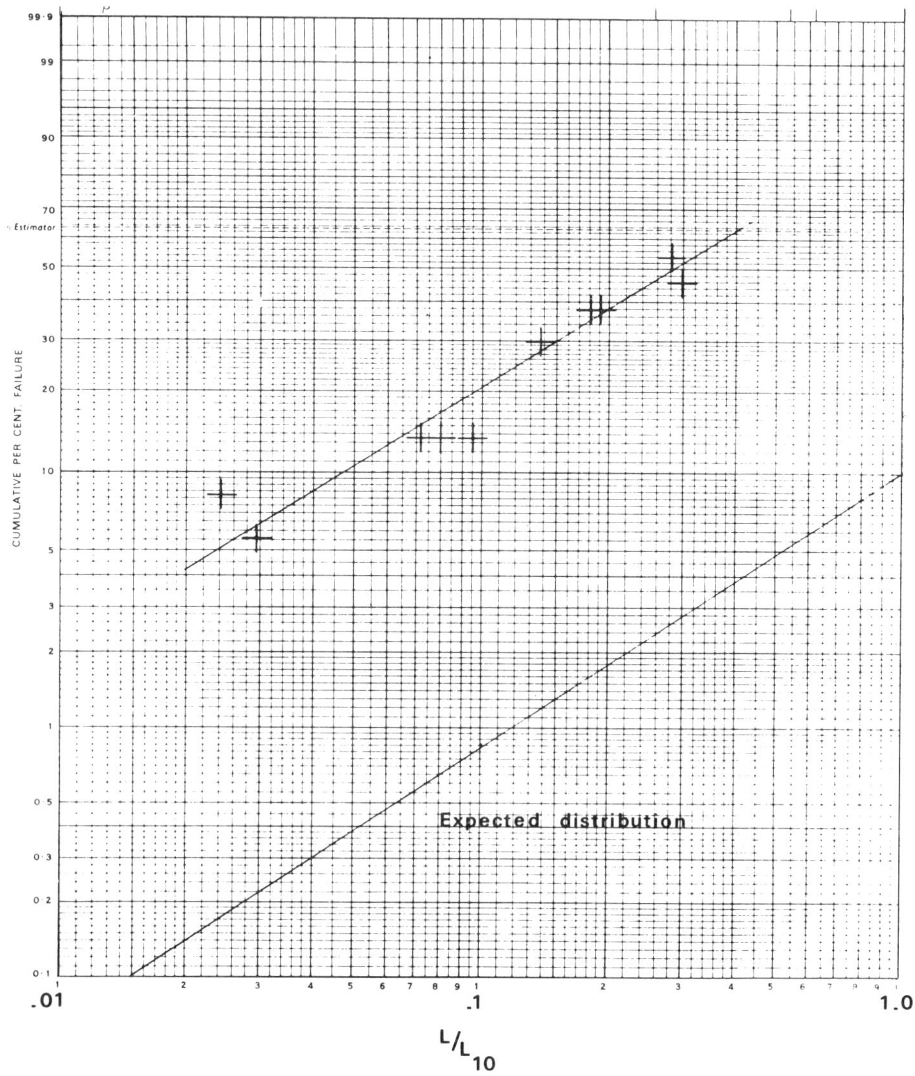

FIG. 3. Results from laboratory bearing tests at high moisture levels.

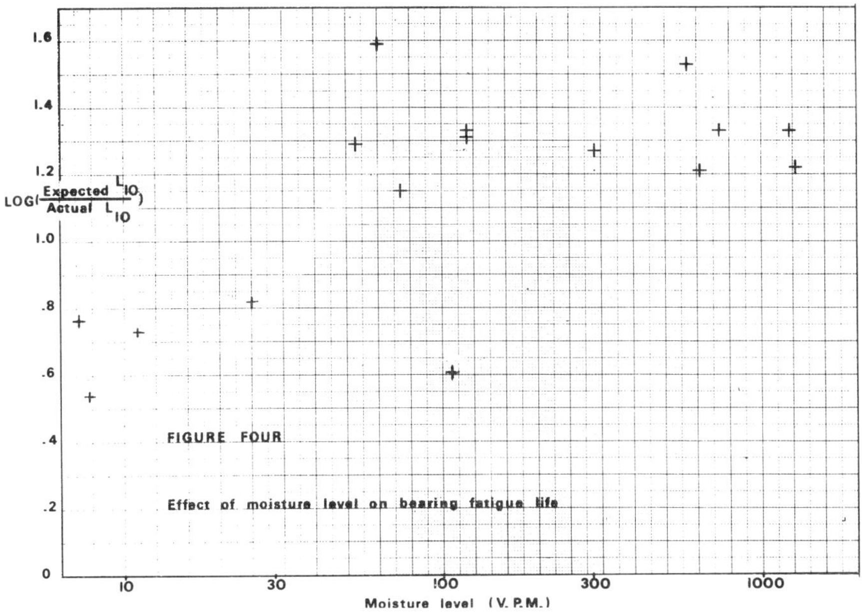

FIG.4. Effect of moisture level on bearing fatigue life.

2.4 Summary

A study of figs.1 and 3 shows that the failure distribution obtained from the laboratory experiments provides an explanation of the observations from the 'life' tests. These results support the view that, provided a mechanism can be analysed, its reliability may be predicted from basic data. However, the data must be relevant to the application, and it is important to note that apparently small changes in environment or other conditions can have a dramatic effect on performance.

3. A RESEARCH APPROACH

The objectives of a research programme were defined in section 1. as being to relax the constraint on development programmes imposed by a lack of understanding. The level of current understanding permitted the problem of section 2. to be reduced to a study of ball-bearings but did not allow reliability predictions to be made from a knowledge of fatigue processes. Thus research here might be directed towards the provision of a link between simple fatigue data and behaviour of ball-bearings as an assembly. However, the problem was reduced to a manageable scale only because the individual bearing failures could be shown to be statistically independent. Thus a fruitful research topic is the study of that class of mechanisms which cannot, as yet, be considered as a system of sub-assemblies. It is clear that we require the ability to analyse complex mechanisms so that the mechanical conditions at a given joint can be determined at any time during the operation of the mechanism. A less obvious but equally important area is the determination of the factors influencing the failure of joints. This is necessary in order to specify which of the variables, met in actual operation, need to be accurately simulated in a meaningful test.

The above discussion specifies a broad research field and the authors have therefore chosen two smaller areas which are described below.

3.1 Frictional Failure of Sliding Joints

Seizure is a term widely used to describe failures of sliding joints and according to the OECD Research Group on Wear of Engineering Materials [7], should be taken to mean:
'the stopping of relative motion as a result of interfacial friction'.
It is precisely this type of failure which will be discussed in this section. It follows from this definition that there are two modes of frictional failure:

(i) Self-jamming failure - relationships between transmission angles and the coefficient of friction becomes such that increasing the driving force simply increases the forces opposing motion.

(ii) Cumulative failure - if a succession of high friction values causes energy to be dissipated faster than it is supplied by the driving forces, the mechanism will lose kinetic energy and may eventually stop.

In an ideal situation, where the coefficient of friction is constant, it is simple to design mechanisms so that these failures will not occur. However, it has been shown [8-11] that the coefficient of friction has inherent statistical fluctuations associated with it, and that these fluctuations can cause the instantaneous coefficient of friction to differ considerably from the mean value. This is not surprising since friction results from a summation of the forces acting at the many discrete points of contact between the two rubbing surfaces. A theoretical model which offers an explanation of these fluctuations has been proposed by Mitchell and Osgood [12]. In the context of this paper the importance of these fluctuations is that they lead to a finite probability of self-jamming or cumulative failure, even when the mean coefficient of friction is well below the critical level at which such failures would be certain.

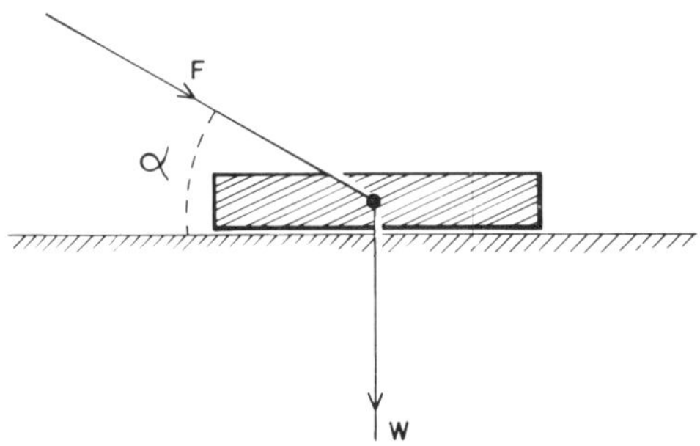

FIG.5. An idealized joint.

3.1.1 Self-Jamming Failure

The reliability implications of self-jamming failure are best understood by the study of the idealised joint shown in fig.5 where F is the driving force, supplied by a pin-jointed link, which moves the weight W (expressed in force units). Thus, in order to drive the weight at constant speed, the component of F parallel to the plane must be equal to the force resulting from friction, i.e.

$$F \cos\alpha = \mu\,[W + F \sin\alpha]$$

$$\therefore \quad \frac{F}{W} = \frac{\mu}{\cos\alpha - \mu \sin\alpha}$$

and clearly the mechanism will self-jam if:

$$\mu \geqslant \cot\alpha$$

The probability that this condition will occur can be predicted if the distribution of friction coefficients is known. However, this information is insufficient since it does not allow the probability of failure to be predicted as a function of sliding distance. Mitchell and Osgood [13] have shown that this problem can be overcome by introducing the parameter

$$\overline{\left|\frac{d\mu}{dx}\right|}$$

which is the mean of the absolute value of the rate of change of friction coefficient with sliding distance. Using this parameter and some theoretical ideas developed by Kimura [14] they derived the following expression for the probability that the mechanism would successfully slide a distance X:

$$\text{Prob}(X) = A \exp\left\{\frac{-P(\mu_{crit}) \cdot \overline{\left|\frac{d\mu}{dx}\right|} \cdot X}{2A}\right\} \qquad \ldots(1)$$

where $P(\mu)$ is the density function describing the distribution of coefficients of friction,

μ_{crit} is the critical coefficient of friction for self-jamming ,

and

$$A = \int_0^{\mu_{crit}} P(\mu)\, d\mu \, .$$

To appreciate the implications of equation (1) assume a Gaussian distribution function:

$$P(\mu) = \frac{1}{\sqrt{2\pi}\sigma_\mu} \, e^{-\frac{(\mu - \bar{\mu})^2}{2\sigma_\mu^2}}$$

where $\bar{\mu}$ is the mean value of the coefficient of friction and σ_μ is its standard deviation. It is then convenient to express the mean slope

$$\left(\overline{\left|\frac{d\mu}{dx}\right|}\right)$$

in terms of the related quantity L which is the half-power wavelength. Following Whitehouse and Archard [17]:

$$\overline{\left|\frac{d\mu}{dx}\right|} = \frac{\sigma_\mu}{\ell}\left[\frac{1 - \rho^2(\ell)}{\pi}\right]^{\frac{1}{2}} \qquad \ldots(2)$$

where $\rho(\ell)$ is the correlation coefficient for sampling interval ℓ. It is clear from equation (2) that, in general, the mean slope will be a function of the sampling interval and that the form of the function will be dependent on the power spectrum of the friction fluctuations. It is reasonable to assume that the fluctuations in friction force will have a white noise power spectrum.

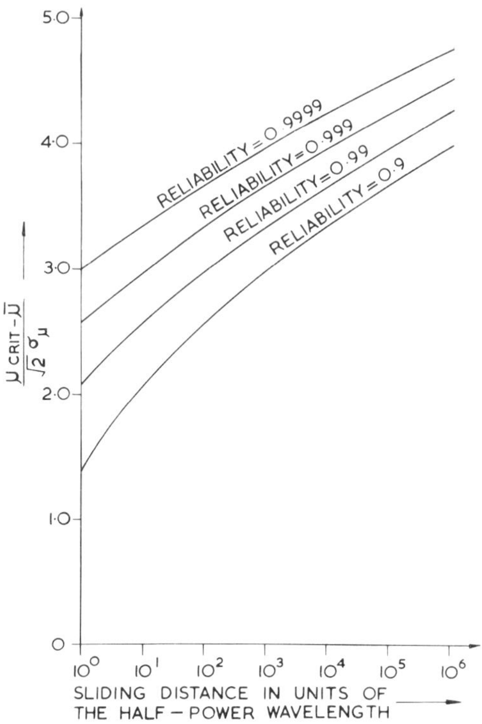

FIG.6. Relations between sliding distance, friction parameters and reliability for an idealized joint.

However, most mechanical joints have a simple spring-mass equivalent and thus the observed power spectrum, after passage through this 'filter', can reasonably be modelled as white noise limited at 12 dB per octave in the upper frequencies. The correlation function derived from this power spectrum is:

$$\rho(\ell) = (1 + \frac{2\pi\ell}{L}) \exp[-2\pi\ell/L]$$

Substituting this expression in equation (2) it can be seen that if $\ell \to 0$ then:

$$\left|\frac{d\mu}{dx}\right| \to \frac{2\sqrt{\pi}\ \sigma_\mu}{L}$$

Using this value and writing $t = \{\frac{\mu_{crit} - \bar{\mu}}{\sqrt{2}\ \sigma_\mu}\}$ in equation (1) we find:

$$\mathrm{Prob}(n) = \tfrac{1}{2}\{1 + \mathrm{erf}(t)\} \exp\{\frac{-\sqrt{2}\ n \exp[-t^2]}{1 + \mathrm{erf}(t)}\}$$

where $n = X/L$ and is the sliding distance expressed in units of the half power wavelength. The probability of failure, in sliding a distance nL, is simply [1 - Prob(n)] and contours of constant failure probability are plotted in fig.6 as a function of t and n. The half power wavelength, L, is seldom longer than about 1 mm so that, from fig.6, it is clear that in order to slide 1 m with a probability of failure less than 0.001 say, the critical friction level must be at least five standard deviations above the mean value.

3.1.2 Cumulative Failure

If there is a limit on the available driving forces, and if the further conditions of a limit on the maximum kinetic energy stored in the machine is imposed, then a cumulative failure can be envisaged. If a succession of high friction values should occur such that the energy absorbed by the combination of output forces and friction forces is greater than the energy input to the machine, then the kinetic energy of the machine would be reduced and could reach zero, at which time the machine would have failed. For a linkage mechanism where it is not possible for the kinetic energy of one link to be zero without the kinetic energy of all links being zero, the condition for failure can be written as:

$$E = 0 \qquad \ldots(3)$$

where E is the total kinetic energy of the mechanism.

In computing the kinetic energy it is necessary to consider the forces acting on the system to be constant over short 'steps'. Such a digital calculation would be carried out using the following rules:

$$\begin{aligned} E_{i+1} &= E_i + \Delta E_i \quad &\text{if } E_i + \Delta E_i < E_{max} \\ E_{i+1} &= E_{max} \quad &\text{if } E_i + \Delta E_i \geq E_{max} \end{aligned} \qquad \ldots(4)$$

where E_i is the kinetic energy at the ith calculation step, E_{max} is the maximum kinetic energy of a governed system and:

$$\Delta E_i = \Delta E_{in} - \Delta E_F - \Delta E_o \qquad \ldots(5)$$

where ΔE_i is the change in kinetic energy during the ith step resulting from the actions of the input, friction and output forces.

Using the above conditions it is necessary to determine the probability of failures to slide a given distance having started the machine from a prescribed value of the kinetic energy. By suitable choice of values for the parameters above it is possible to consider mechanisms of various types. For example, if E_{max} is finite the mechanism is governed whereas, for an ungoverned mechanism, $E_{max} \to \infty$. The probability distribution contains enough information to decide both the mean running distance to failure, for a continuously operating mechanism and the frequency of failure for an intermittently operating mechanism.

To illustrate the computational method, consider a block being pushed along a plane by a force F, having a maximum value F_{max} (fig.5). Thus in the simple case when $\alpha = 0$, $\Delta E_{in} = F\Delta x$ (where Δx is the computational step length). Given a weight W (the friction force will be μW) and an initial energy E_S, then after I computational steps the kinetic energy is:

$$E_I = E_S + I\Delta x(F - \bar{\mu}W) - W\Delta x \sum_{i=0}^{I} \Delta\mu_i \qquad \ldots(6)$$

$$\text{where } F = F_{max} \quad \text{if } E_I < E_{max}$$

$$F = \mu W \quad \text{if } E_I = E_{max} \qquad \ldots(7)$$

where $\bar{\mu}$ is the mean coefficient of friction and $\Delta\mu$ is the difference between the instantaneous and mean values of the friction coefficient. A value for E_S is

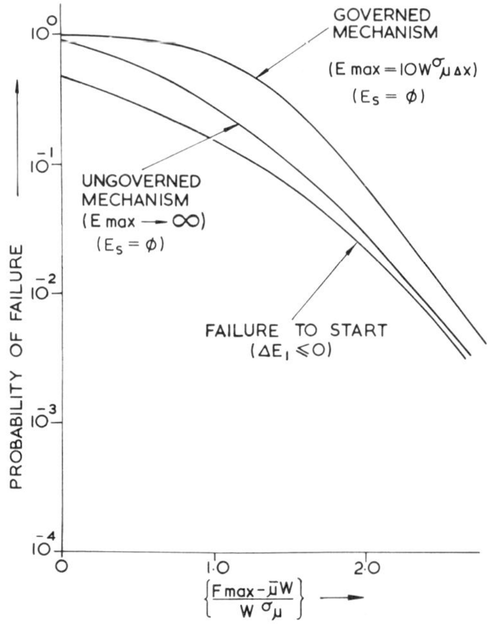

FIG.7. The relation between driving force and failure probability for cumulative failures.

chosen and relevant values for F, W, Δx and $\bar{\mu}$ are assigned. A series of values of $\Delta\mu_i$ are generated so that the changes in E can be followed until E = 0. The generation of sequences of $\Delta\mu_i$ is repeated many times until the probability of failure is determined with sufficient accuracy. For convenience the computer program uses the dimensionless parameters:

$$\frac{F - \mu W}{W\sigma_\mu}, \quad \frac{\Delta\mu}{\sigma_\mu}, \quad \frac{E_S}{W\sigma_\mu \Delta x}, \quad \frac{E_{max}}{W\sigma_\mu \Delta x}$$

so that μ is measured in units of σ_μ, forces in terms of $W\sigma_\mu$ and energies in terms of $W\sigma_\mu \Delta x$. The distribution of friction values $P(\mu)$ is assumed to be Gaussian with mean $\bar{\mu}$ and standard deviation σ_μ. A computer subroutine to generate random numbers with an exponential correlation function was available and for simplicity this was used to generate a sequence of values of:

$$\left\{\frac{\Delta\mu_i}{\sigma_\mu}\right\} .$$

Fig.7 shows the effect of increasing the driving force on the failure of governed (E_{max} is finite) and ungoverned ($E_{max} \to \infty$) mechanisms. From this diagram it is clear that cumulative failure, at low failure probabilities is indistinguishable from failure to start and is much less important, in general, than self-jamming failure.

3.2 The Prediction of the Reliability of Mechanical Systems

As suggested above there is no general analytic method for predicting the reliability of a complex mechanical system from the reliabilities of its constituent joints and links. Thus it is necessary to devise ways in which the work described in 3.1 may be used more effectively. The authors' view is that there are two basic approaches which are likely to prove fruitful:

i. theoretical - in this case the mechanical system to be evaluated is modelled, as accurately as possible and in all its aspects, on a digital computer. The model is then 'tested' to failure repeatedly until sufficient data is obtained. This method will be expensive in computer time but if the model is accurate it will be cheaper and more informative than a full-scale 'life' test.

ii. experimental - for large systems, or where high reliabilities are required, multiple tests to failure are not feasible. Thus methods of failure prediction which do not depend on the occurrence of failures in a 'life' test are necessary. Although this appears to be impossible, it is theoretically feasible by extrapolation of the statistical distribution of forces and energies associated with key components of the system. These may be determined by suitable instrumentation of an operating test rig.

These two items represent the overall objectives of research in this field. To attain these it is necessary to consider where further research is required. It appears that there are four basic areas:

(a) the development of automated methods of mechanisms analysis in the presence of friction and wear.

(b) analysis of critical value problems of random variables.

(c) studies of individual joints and the parameters controlling their behaviour.

(d) basic studies of the processes leading to joint failures, with particular emphasis on scaling problems.

To realise the theoretical methods described in (i) work is required in all the areas (a) to (d) above, but for the experimental approach, (ii), only (a) and (b) need to be actively pursued. The work described in 3.1 represents a useful contribution to the study of critical value problems associated with random variables. Since it is central to the success of the methods (i) and (ii) above the most pressing need is for work in area (a) - automated methods of mechanisms analysis. To the authors' knowledge there are no researchers working in this field; however, the techniques used by Chace [15] and Uicker [16] in their computer programs for dynamic mechanisms analysis appear to be capable of extension to satisfy the needs identified here.

Although many development programmes (such as section 2 of this paper) look in detail at a particular joint there have been few systematic studies leading to a generalised view of joint behaviour. Similarly most of the processes leading to failure have received attention at the basic level but there has been comparatively less work on coupling the basic data to practical problems. To take the particular example of friction: there have been many studies made but few have done more than measure the mean value so that most experiments do not provide the detailed statistical data required for failure analyses.

This study of the techniques of reliability prediction demonstrates the requirement for a more detailed approach to the acquisition of basic experimental data in the field of tribology.

4. DISCUSSION AND CONCLUSIONS

This paper has considered the problem of predicting the reliability of mechanisms in a developing technology. Since the authors' field of experience is centred in tribology all the work described has concerned failures of joints or bearings.

Clearly, the cheapest means of establishing the reliability of a mechanical system is to test the smallest element permitted by the availability of analytic techniques. The application of this criterion to the example, given in section 2, led to the choice of rolling element bearings as the component to be tested since:

(i) the bearing failures could be shown to be statistically independent,

(ii) the analytic techniques for predicting bearing failures from fatigue data were not adequate.

The interpretation of the bearing test results, using the well known methods of Weibull statistics, demonstrated the critical importance of apparently minor changes in atmosphere and operating conditions. Thus whenever a mechanical system can be reduced in this manner the function of the research organisation is to provide failure statistics for the basic component. This type of work will always be relatively short term since it will be coupled to a development timescale.

In general, it is not possible to predict mechanism reliabilities from the failure rates of more basic elements since these failures are not statistically independent. Thus full-scale tests under simulated operating conditions are necessary. In these circumstances three objectives for research work can be identified:

(i) on a short timescale - to define the accuracy required of the simulation by exploring the sensitivity of failure processes to minor changes.

(ii) on an intermediate timescale - to ease the constraints on development timescales by increasing the information obtained from large scale tests.

(iii) on a longer timescale - to develop an analytic capability to increase the proportion of systems which may be assessed in terms of their basic elements.

The automated methods of mechanism analysis and the analysis of critical value problems for random variables, described in section 3, provide a basis for work towards the intermediate and longer term objectives.

Although this paper has been restricted to a discussion of mechanisms the methods of Weibull statistics have much wider application. Similarly it will be found that the analysis of critical value problems for random variables can be applied in situations other than frictional failures. For example the methods could be used in describing force fluctuations in a structure: in this case a strength criterion would be substituted for the critical coefficient of friction and the analysis would predict failure probability from overstressing. It is significant that the possibility of predicting joint failures from friction measurements has required considerable changes in the type of friction measurements which need to be made. If the methods described in this paper are to be applied to other fields then similar changes in the data produced from other relevant basic scientific studies will be required.

REFERENCES

[1] LUNDBERG, G., PALMGREN, A., Dynamic Capacity of Rolling Bearings, Acta Polytechnica 7 (1947).

[2] BAMBERGER, E.N., Life Adjustment Factors for Ball and Roller Bearings. ASME (1971).

[3] HOBBS, R.A., Fatigue Lives of Ball Bearings lubricated with Oils and Fire Resistant Fluids. (Proc. Conf., London, 1972) Instn. Mech. Engnrs., (1972) 1.

[4] VAESSEN, G.H.G., de GEE, A.W.J., Rolling Contact Fatigue of Maraging Steels of different production History: Influence of Film Thickness/ Roughness Ratio. (Proc. Conf., London, 1972) Instn. Mech. Engnrs., (1972) 40.

[5] MOYER, C.A., BIANCHI, A., Relating Lubricant Parameters and Bearing Fatigue Life in the Mixed EHL Regime. (Proc. Conf., London, 1972) Instn. Mech. Engnrs., (1972) 95.

[6] ANDRÉASSON, A., LUND, T., Ball Bearing Endurance Testing considering Elastohydrodynamic Lubrication. (Proc. Conf., London, 1972) Instn. Mech. Engnrs. (1972) 138.

[7] OECD RESEARCH GROUP on Wear of Engineering Materials, 'Friction, Wear and Lubrication - Terms and Definitions' (1966)

[8] RABINOWICZ, E., RIGHTMIRE, B.G., TEDHOLD, C.E., WILLIAMS, R.D., Trans. ASME, 77 (1955) 981.

[9] SIMKINS, T.E., 'Surfaces and Interfaces', The Similarities of Static and Kinetic Friction, Proc. 13th Army Materials Conf., Sagamore, Syracuse Univ. Press (1966).

[10] PODALSECK, S., SHEN, H.K., 'Adhesion or Cold Welding of Materials in Space Environment', ASTM STP 431, Am. Soc. Testing Mats. (1967).

[11] KILBURN, R.F., 'Friction Viewed as a Random Process', J. Lub. Tech. 96F(2) (1974) 291.

[12] MITCHELL, L.A., OSGOOD, C., 'A Theory of Friction and Wear Based on a New Characterisation of Asperity Interactions', CEGB Report RD/B/N2858, (1973).

[13] MITCHELL, L.A., OSGOOD, C., 'Prediction of the Reliability of Mechanisms from Friction Measurements', 1st European Trib. Conf. (Proc. I. Mech. E. Conf., London, 1973) Paper No. G265/73.

[14] KIMURA, Y., 'Estimation of the Number and the Mean Area of Real Contact Points on the Basis of Surface Profiles', Wear, 15, (1970) 47.

[15] CHACE, M.A., SMITH, D.A., 'DAMN - Digital Computer Program for the Dynamic Analysis of Generalised Mechanical Systems', SAE Transactions 80 (1971) 969.

[16] SHETH, P.N., UICKER, J.J. Jr., 'IMP - A Computer-Aided Design Analysis System for Mechanisms and Linkages', ASME Paper 71-VIBR-80.

[17] WHITEHOUSE, D. J., ARCHARD, J. F., "The Properties of Random Surfaces of Significance in their Contact" Proc. Roy. Soc. Lond A, 316, 1970, 97 - 121.

DISCUSSION

W.A. WOLFE: I should like to compliment Mr. Osgood on a clear exposition of the interplay between reliability assessments, scientific understanding and research and development. With regard to reliability calculations, I wonder whether Mr. Osgood has any particular reason for preferring a mean value calculation to a median rank estimate.

C. OSGOOD: As tribologists rather than reliability experts we were not in a good position to judge the merits of the various estimation methods. However, it was clear that our basic conclusions would be unaffected by the choice of method, so we used the first method that came to hand.

J.O. TATTERSALL: Is there any other work going on at Berkeley in which the problem of engineering reliability is being attacked on the basis of the fundamental properties of materials, thus avoiding the need for historical data? This is a question which I would like to address not just to Mr. Osgood but to any participant who may care to answer. Is anyone, for example, approaching the problem of vessel failure by considering the fundamental properties of the materials and the quality assurance data associated with particular vessels?

C. OSGOOD: I do not know of any other work of this type at Berkeley. It is perhaps difficult to start such work taking reliability as one's point of departure; in our case it came about because we saw the reliability implications of our work in tribology. You might say that we were fortunate in having a solution looking for a problem!

H.E. KONGSØ: At Risø in Denmark they are trying to evaluate the probability of failure of pressure vessels on the basis of probabilistic fracture mechanics. Anyone interested in this work can refer to Mr. P.E. Becher of the Department of Reactor Technology for further information.

D. NORMAN: In answer to Mr. Tattersall's question, there has been other work, based on the physics of failure, similar to Mr. Osgood's. The most notable is the work of Becher and Pederson on the probability of steel pressure vessel failure, which was published about 15 months ago in Nuclear Engineering and Design. In the electrical field, several CIGRE papers were published about four years ago on insulation life distributions. Similar work on capacitors was published in the Electronics World about 10 years ago, including one paper by B. Epstein. I believe Mr. Volta of Ispra has also done work on pressure vessels.

References to early work on fatigue failures can be found in Mr. Buckland's book on the statistical analysis of the life characteristic (a bibliographic guide). Some later work is covered in 'Mechanical Survival', a book by Bompas-Smith. Finally, a recent paper in Technometrics dealt with the problem of inspecting for cracks and subsequent crack growth.

IAEA-SM-195/55

ИССЛЕДОВАНИЕ АТОМНЫХ ЭЛЕКТРОСТАНЦИЙ НА НАДЕЖНОСТЬ

Я.НЕМЕЦ и Я.СЕДЛАЧЕК
Технический университет,
Прага,
Чехословацкая Социалистическая Республика

Abstract–Аннотация

RELIABILITY STUDY OF NUCLEAR POWER STATIONS.

The paper considers the physical and mathematical bases underlying nuclear power station reliability from the standpoints of material fatigue, thermal yield and ageing. The risk of failure of nuclear power station components is determined by means of the Markov stochastic process.

ИССЛЕДОВАНИЕ АТОМНЫХ ЭЛЕКТРОСТАНЦИЙ НА НАДЕЖНОСТЬ.

В докладе рассматриваются физические и математические основы надежности атомных электростанций с точки зрения процессов усталости, теплотекучести и старения материала. С помощью стохастического процесса Маркова определялся риск неполадок в компонентах АЭС.

Постоянно возрастающий недостаток источников энергии и необходимость защиты окружающей среды от загрязнения выбросами из классических электростанций говорит о необходимости ускоренного использования атомных энергетических источников. В частности, реакторы с водой под давлением в последние годы оказались – с экономической точки зрения – вполне конкурентоспособными с обычными источниками электроэнергии. В ближайшее десятилетие энергетика, вероятно, перейдет от типов станций, в настоящее время ставших уже классическими, на тепловых нейтронах, к электростанциям с реакторами на быстрых нейтронах. При обсуждении преимуществ отдельных типов и размеров электростанций с точки зрения коэффициентов полезного действия, безопасности, капиталовложений и проблем распределения электроэнергии наилучшими оказываются атомные электростанции (АЭС) средних размеров мощностью около 1000 МВт. В будущем предполагается строительство сравнительно большого количества таких блоков, создание густой сети атомных электростанций вблизи крупных городов и в густонаселенных частях земного шара.

Решающим фактором становится надежность электростанций, представляющая комплексную проблему, которую следует решать с точки зрения теории систем. Очевидно, что физика излома материалов не достигла до сих пор настолько объективного определения законов накопления повреждений в случае сочетания усталости, теплотекучести, коррозии под напряжением и нарушения вследствие радиоактивного излучения, чтобы на их основе мы могли высказать теоретические предположения устойчивости первого контура этих электростанций. Кроме того, повышение рабочей температуры и показатели размеров усложняют проблемы переноса лабораторных исследований на работу действующих сложных механических систем, которыми являются первые контуры указанных электростанций. Нет других таких установок, которые подвергались бы столь комплексному повреждению при эксплуатации. Отсутствие достаточной информации затрудняет разработку положений о риске АЭС.

Опыт эксплуатации АЭС свидетельствует о том, что радиоактивные отходы и выбросы в атмосферу из этих станций не являются значительными, как это предполагалось вначале. Основной опасностью остается авария с радиационным повреждением контура под давлением и подсоединенного к реактору трубопровода высокого давления. Такая авария может действительно произойти только в случае расплавления активной зоны вследствие нескольких нарушений работы установки. Авария может произойти в первом контуре в случае появления трещины, вызывающей потерю теплоносителя и сопровождаемой частичным или полным отказом системы аварийного расхолаживания. Кроме того, при отключении реактора может также произойти – в случае аварии системы – отвод остаточного тепла, испарение охлаждающей воды или перегрев теплоносителя и расплавление активной зоны. Имея ввиду производственные причины, которые вызывают повышение давления или нарушение расхолаживания после отключения реактора, необходимо в первую очередь учитывать умелые действия оператора при управлении этим чувствительным и сложным оборудованием в переходных или аварийных ситуациях. Специально созданной для этой цели группой специалистов при КАЭ США, под руководством проф.Рассмуссена, испытывались тысячи возможностей при отдельных вышеуказанных обстоятельствах и определялась вероятность отказа оборудования. Самый большой риск заключается в возможности расплавления активной зоны. КАЭ оценила его в своем отчете WASH 1400 как величину $6 \cdot 10^{-5}$ в год на реактор; одновременно Комиссия определила, что опаснее всего может быть разрыв трубопроводов приблизительно до диаметра 100 мм. Бесспорным является вывод, приведенный в указанном отчете, что техническая деятельность человека в воздушном и автомобильном транспорте, аварии плотин гидроэлектростанций и природные бедствия, как, например, тайфун или землетрясение, на много раз превышают риск эксплуатации реакторов с водой под давлением на АЭС. Из отчета следует также, что количество возможных жертв и экономические потери ниже, чем при других массовых авариях. Данные о вероятности отказа АЭС указанного типа, по которым – в случае отказа установки – отходы радиоактивного горючего электростанции составляли бы 50%, являются в вышеприведенном отчете несомненно более реальными, чем предостерегающие данные из отчета WASH 740.

Здесь мы не ставим своей целью рассмотреть все предположения, из которых исходят вышеуказанные заключения. Частично их можно отнести к более безопасным, частично к менее. Необходимо иметь ввиду, что вследствие широкого использования АЭС в серийном производстве компонентов, работающих под давлением, будут принимать участие также заводы и цехи с меньшей технической базой. Применяемая до сих пор автоматизация производства корпусов большой толщины, изготовленных из более прочных материалов, не достигла еще такого уровня, чтобы мы могли в достаточной мере исключить влияние субъективного фактора, так как физические параметры сварки, требуемые с точки зрения окончательного качества сварных швов, до сих пор еще объективно не определены. Далее коротко рассматриваются предельные состояния корпусов в сложных эксплуатационных условиях, в которых имеют место и прогрессируют дефекты материалов сварных швов.

В действительности это выглядит следующим образом:

а) До сихпорне разработана теория надежности сложных механических систем, которая учитывала бы нелинейные законы накопления повреждений и взаимодействия, определяемые изменениями силового потока в локальных зонах и переход к дезинтеграции разрушения; не известны также начальные параметры процессов.

б) Пока еще не все контрольные испытания, проведенные в лабораторных условиях, можно в полной мере использовать в процессе эксплуатации АЭС.

в) Производственная диагностика еще не в такой мере комплексная, чтобы можно было надежно определять тонкие места и степень их повреждения в процессе эксплуатации оборудования электростанции; теория и практика предупредительных сигналов пока находится только в самом начале своего развития, особенно это касается их применения.

г) Численное выражение риска возможного разрушения часто имеет только статистический характер, т.е. оценивается только наличие разрушения независимо от времени эксплуатации оборудования. Динамическая схема риска разрушений характеризует непосредственное соединение с их механикой, что с теоретической точки зрения еще окончательно не решено, в частности, вопрос комбинации процессов разрушения.

Дополнительно к первой проблеме можно отметить следующее. Вероятность неисправности корпуса под давлением исследуется с точки зрения распределения внешних влияний, подвергающих корпус нагрузке и вызывающих его разрушение. Из величин вариационных коэффициентов для устойчивости материала V_M и для распределения нагрузки V_S можно найти интенсивность повреждений корпусов при условии, что будет достигнут определенный уровень решения конструкций. Фройдентал пользовался этим методом при решении вопроса комбинации текучести материала под влиянием тепла и усталости по упрощенным законам о накоплении повреждений на корпусах реакторов с водой под давлением и сделал вывод, что интенсивность разрушений будет менее 10^{-7} /год на корпус. Если бы мы сделали аналогичные расчеты для корпусов реакторов на других неактивных рабочих участках, например в химических установках, и учитывали бы при этом различие в уровне качества производства и в контроле, то получили бы величину приблизительно на половину порядка меньше. При этом давления и температуры являются сопоставимыми. Однако здесь не возникает разрушение корпусов реакторов вследствие радиоактивного облучения. В работе трубопроводов, имеющих самый большой диаметр около 500 мм и подвергающихся различным побочным нагрузкам (расширительные эффекты, аэроупругие воздействия, большие температурные градиенты и т.д.), можно ожидать, при условии соответствующего уровня производства и регулярного контроля, интенсивность разрушений порядка 10^{-6} /год на систему. С этим можно сравнить трубопроводы высокого давления в классических электростанциях или на важных химических производственных участках. Опыт показывает, что при вышеуказанном, менее трудоемком и более верном сопоставлении аварии встречаются чаще, чем это вытекает из приведенных предположений. Причину этого надо искать в следующем:

а) Фрактография разрушений корпусов реакторов и частей трубопроводов в классических производствах показывает, что почти ни в одном случае здесь не встречается тип разрушения, предполагаемый в соответствии с линейной механикой разрушения. Возникшая в случае начального поверхностного дефекта одна магистральная трещина, как правило, не увеличивается, причем как барьер действуют укрепление материала в пластической зоне на корне трещины, а также напряжения, образованные окружающей квазиэластичной средой. Величина удельной энергии G_I на 1 см^2 поверхности трещины здесь сравнительно большая и зависит от графика нагрузки. Общая затраченная работа при этом процессе разрушения составляет $L_I = G_I \cdot F_I$. На предельную величину коэффициента интенсивности напряжения при вышеуказанной схеме процесса разрушения оказывает влияние система нагрузки. Однако более часто встречается второй тип разрушений, когда энергия G_{II}, затраченная на единицу новой поверхности в корпусе, сверхнизкая. Она мало зависит от деформационного упрочнения и барьеров напряжения вокруг трещины. Мы говорим о дезинтеграционном типе разрушения, сопровождаемом ветвлением и соединением системы трещин, тормозным эффектом которой является увеличение поверхности трещин. Общая затраченная энергия $L_{II} = G_{II} \cdot F_{II}$; здесь она ниже, чем в первом случае, несмотря на то, что поверхность излома в несколько раз больше в связи с отсутствием образования барьеров на корне трещины. Таким типом разрушений является ветвистая трещина, которая при высоких температурах, коррозии под напряжением и в случае разрушения материала излучением тянется по границам зерен. Если мы у этого типа разрушений определим надежность корпуса, то она будет тем больше, чем большей будет интенсивность возрастания общей поверхности трещин. Скорость движения трещины будет высокая, потому что энергия G_{II} является низкой. Здесь происходят многочисленные взаимодействия между тонкими поверхностями и дефектами в материале. Трещина ищет самые тонкие поверхности и путь к своему дальнейшему развитию. Это последовательный ряд квазихрупких скачков. В действительности здесь всегда появляются изломы смешанного характера и один тип разрушений часто переходит в другой. Так например, трещина начинается как "пластическая", магистрального типа, с незначительными взаимодействиями, но после достижения определенной длины она переходит в дезинтеграционный тип, а именно под влиянием коррозии при напряжении и старения материала. На характер этих изломов оказывает влияние также местная (локальная) напряженность, которая связана с формой корпуса и с характером сварного соединения. В теоретическом анализе процесса разрушения, протекающего в авиационных конструкциях, показано, что трещина переходит из одного типа в другой, а именно в местах, где существенно изменяется форма конструкции: далее трещина под определенными углами продвигается через массивные профили. Эти условия нельзя имитировать на тонких образцах малых размеров. Сложным способом будут проявляться области вокруг фланцев, горловин и цоколей на корпусах реакторов, некоторые места в арматуре первичного трубопровода и т.д.

б) Очень сложным является вопрос правильного выбора моделей при испытании предельных состояний. Здесь оказывает свое влияние фактор размеров. Этот фактор может переменить процесс первого типа на второй и, в частности, может сократить этап развития разрушения. Если мы выводим вариационные коэффициенты V_M из лабораторных испытаний малых корпусов, то потом следует ожидать сокращения жизнеспособности корпусов на практике по сравнению с лабораторными испытаниями, а именно под влиянием коэффициентов размеров и геометрии. Коэффициенты размеров мы можем вывести из двух моделей, разработанных во второй половине двадцатого века: частично из модели Ирвина по механике разрушения, а также из статистической модели Веибулла. Несмотря на то, что их окончательные результаты существенно не отличаются, разность проявляется прежде всего в физическом характере процесса разрушения, где первая модель показывает развитие единственной решающей трещины, а вторая – отвечает случайному соединению системы слабых мест. В первом случае встречается эффект изменений пластической зоны при различном расстоянии краев корпусов и напряженности, во втором случае вероятно проявляется самый серьезный дефект из всей системы дефектов, который зависит от размера корпуса с технической точки зрения. Вследствие этого весьма важно, какой этап сыграет решающую роль во всем развитии катастрофического излома и будет ли это этап расширения или образования усталостной трещины на дефекте материала. В большинстве случаев на практике – в отличие от лабораторных испытаний с малыми корпусами – в процессе эксплуатации встречается еще один этап увеличения трещин, вследствие чего эффект размеров оказывает здесь значительное влияние. Из этого вытекает необходимость экспериментального свидетельства надежности, важность стратегии экспериментальных испытаний, выбор размеров испытательных образцов и затрудняющих условий.

в) Высказанные до сих пор соображения были сделаны исходя из оценки надежности АЭС при использовании материалов, основой которых являлись квазистатические типы повреждения корпусов. Вполне ясно, что решающую роль при докритическом нарастании дефектов играет низкоциклическая механическая и термическая усталость в сочетании с дальнейшими эффектами вырождения материалов. Однако, какое число изменений учитывается у этого сложного (комплексного) напряжения ? Было подчеркнуто квазистатическое понимание этого вопроса. Если допустить дезинтеграционный тип разрушения, тогда можно говорить о скачках трещин и о росте повреждения по этапам. В таком случае высшие эластичные деформации будут интегрироваться в целую систему и в случае одного изменения во внешнем состоянии электростанции будут действовать целые системы динамических воздействий на поврежденные и слабые места. Деформация развивается в квантах и повторяемый проход окажет влияние на корни трещин и дефектов. В случае накопления повреждений, вместо одного изменения проявится ряд изменений, если они будут выше своей начальной величины. Эксплуатационная диагностика должна решить, какие воздействия являются активными и какие не являются активными.

г) Основным новым подходом к решению этой проблемы является математическая формулировка динамического понимания риска при разрушении. Следует рассматривать другой способ определения риска разрушений, а именно способ, тесно связанный с характеристиками последовательной деградации деталей, причина которой заключается в усталостном разрушении $X_M(t)$, текучести $X_c(t)$, старении $X_s(t)$ или в сочтании этих процессов, действующих одновременно. Значит, необходимо будет подготовить материальную базу для оценки надежности, выполняемой при использовании условных вероятностей, которые исходят из кинетики процессов повреждения корпусов по многим причинам, представляемым вектором показателей времени.

$$\{\underline{X(t)}\} \equiv \{X_м(t),\ X_c(t),\ X_s(t)\} \tag{1}$$

так что риск разрушения будет определен вероятностью $r_{мcs}(t)$ во времени t,

$$r_{мcs}(t) = \Pr\{отказ\ (АЭС)_K / X_м(t),\ X_c(t),\ X_s(t)\} \tag{2}$$

$$(0 \le r_{мcs}(t) \le 1,\ t > 0)$$

На основе вышеизложенного мы указали на необходимость привести в соответствие риск разрушения $r(t)$ и те процессы разрушения тел, которые в течение времени использования $\langle 0, t \rangle$ приводят исходную надежность компонентов E_m, $m = 1, \ldots, K$ к такому уровню, когда необходимо будет учитывать возможность более серьезных аварий; то есть, мы хотим вычислить динамический риск r_d как функцию времени, для которой действительна формула

$$0 \le r_s < r(t_1) \le r(t_2), \qquad (t_2 > t_1 > 0) \tag{3}$$

причем поведение функции $r(t)$ более точно определено стохастическими процессами:

процесс усталостного разрушения в пункте E_i

$$\{X_i(t)\} \equiv \{X_м(t)\} \equiv \{X_{E_i}(t)\} \tag{4}$$

процесс течения в пункте E_j

$$\{X_j(t)\} \equiv \{X_c(t)\} \equiv \{Y_{E_j}(t)\} \tag{5}$$

процесс старения в пункте E_k

$$\{X_k(t)\} \equiv \{X_s(t)\} \equiv \{ZE_k(t)\} \tag{6}$$

где E_i, E_j, E_k обозначают составные части системы или ее критические места.

С процессами номеров 4, 5 и 6 мы связываем также следующие случаи:

$U_i(t)$ . . . процесс (4) во времени $<0, t>$ абсорбирован (т.е. происходит усталостное разрушение), (7)

$C_j(t)$ . . . процесс (5) во времени $<0, t>$ абсорбирован (т.е. происходит разрушение по поводу текучести), (8)

$S_k(t)$ . . . в промежутке времени $<0, t>$ произойдет разрушешение вследствие старения (6). (9)

Из рассматриваемых процессов: (5) и (4) мы учитываем как процессы вероятностные (стохастические), процесс (6) в первой апроксимации – как процесс детерминистический. Кроме процессов (4), (5) и (6) в дальнейшем можно учитывать их взаимные сочетания, так, например, сочетание симультанной усталости и разрушения по поводу текучести, или же лучше сказать, композиция указанных процессов определяет новый процесс

$$\{\xi_{M\times c}(t)\} \equiv \{X_{E_i}(t) \times Y_{E_i}(t)\} \tag{10}$$

Мы можем учитывать также суперпозицию процесса $\{X_{E_i}(t)\}$ с детерминистической составляющей $S_d(t)$, которая обозначает регулярное старение, и вследствие этого получаем процесс

$$\{\eta_{M+d}(t)\} \equiv \{X_{E_i}(t) + S_d(t)\} \tag{11}$$

и т.п.

Общий процесс действия всех разрушений (включая также мелкие, они не действуют в вышеуказанных процессах и их сочетаниях), который мы обозначаем $\{N(t)\}$ и который выражает количество n разрушений ($n = 0,1,2,\ldots$), предполагаемых в течение промежутка времени $<0, t>$, это диференцируемый вероятностный (стохастический) процесс Маркова ; его можно в общем предположить неоднородным (тип подобный процессу Поиссона). Он зависит в общем от состояний n и от времени t; для вероятностей перехода в этом процессе действительна формула

$$\Pr\{n_t \xrightarrow{\Delta t} n_t + 1\} = P_{n,n+1}(t, t+\Delta t) = \Delta t \cdot \lambda(n,t) + 0(\Delta t) \tag{12}$$

$$(n_t = 0, 1, 2, \ldots, \qquad 0 \le t < +\infty)$$

Другими словами, отношение (12) уточняет условную вероятность

$$\Pr\{N(t+\Delta t) = n+1/N(t) = n\} = \lambda(n, t) \cdot \Delta t + 0(\Delta t) \quad (13)$$

$$(n = 0, 1, 2, \ldots, \qquad 0 \leq t < \infty)$$

или же для $n = 0$ эта условная вероятность выражается отношением

$$\Pr\{N(t+\Delta t) = 1 = \text{первый отказ}/N(t) = 0\} = \lambda(t) \cdot \Delta t + 0(\Delta t) \quad (14)$$

что значит

$$r(t) = \lambda(t) \cdot \Delta t$$

Указанное вероятностное отношение (14) будет нами использовано как самое пригодное для вычисления риска в случае серьезного постепенного разрушения компонента АЭС. Однако первое серьезное разрушение системы с компонентами (установками) – (K), которое возникает по истечении времени t в течение промежутка времени $<0, t>$, является каким-то общим событием $X_{ijk}(t)$ и ограничивается одним из случаев вышеприведенных типов разрушений (7), (8), (9) частей E_i, E_j или E_k системы $(АЭС)_K$. Это мы запишем как операцию логического сложения возможных событий

$$X_{ijk}(t) = [U_i(t) \cup C_j(t) \cup S_k(t)] \quad (15)$$

$$(i, j, k = 1, 2, \ldots, K; \qquad 0 < t < +\infty)$$

Безусловная вероятность этого события $P[X_{ijk}(t)]$, следовательно, определяется известным отношением при сложных событиях и для нее действует формула

$$P[X_{ijk}(t)] = \Pr\left\{\begin{matrix}\text{в течение } <0, t> \text{ произойдет событие} \\ U_i(t) \text{ или } C_j(t) \text{ или } S_k(t)\end{matrix}\right\} = \quad (16)$$

$$= \Pr\left\{\begin{matrix}\text{первое усталостное нарушение или разру-} \\ \text{шение в связи с текучестью или старением}\end{matrix}\right\} =$$

$$= P[U_i(t)] + P[C_j(t)] + P[S_k(t)] - P[U_i(t) \cup C_j(t)] -$$

$$- P[U_i(t) \cap S_k(t)] - P[C_j(t) \cap S_k(t)] +$$

$$+ P[U_i(t) \cap C_j(t) \cap S_k(t)]$$

Исходя в дальнейшем из предположения, что события (15) независимы друг от друга, мы получаем в случае первого разрушения системы (в течение промежутка времени $<0, t>$) вероятность

$$P[X_{ijk}(t)] = P[U_i(t)] + P[C_j(t)] + P[S_k(t)] - P[U_i(t)] \cdot P[C_j(t)] + \\ + P[U_i(t)] \cdot P[S_k(t)] - P[C_j(t)] \cdot P[S_k(t)] + \\ + P[U_i(t)] \cdot P[C_j(t)] \cdot P[S_k(t)] \qquad (17)$$

Так как $U_i(t)$, $C_j(t)$, $S_k(t)$ являются случайными событиями, связанными с усталостной жизнеспособностью τ_M, или же, другими словами, жизнеспособностью, на которую влияет разрушение по поводу текучести τ_c, или же жизнеспособностью τ_s в процессе старения, можно более точно определить неусловную вероятность (17) для первого разрушения, а именно подстановкой отдельных дистрибутивных функций случайных переменных τ_M, τ_c, τ_s, т.е.

$F_{M_i}[t, X_{krit}; \mu_M(t)]$ = вероятность усталостного излома компонента E_i во времени $<0, t>$ при интенсивности $\mu_M(t)$ процесса усталости; (18а)

$F_{c_j}[t, Y_{krit}; \mu_c(t)]$ = вероятность разрушения в связи с текучестью компонента E_j во времени $<0, t>$ при интенсивности $\mu_c(t)$ процесса текучести; (18б)

$F_{s_k}[t, Z_{krit}; \mu_s(t)]$ = вероятность разрушения компонента E_k старением во времени $<0, t>$ при интенсивности процесса $\mu_s(t)$ (18в)

Из чего мы получаем для события $X_{ijk}(t)$ (т.е. для первого разрушения системы $(АЭС)_K$, отношение

$$\Pr\{X_{ijk}(t)\} = F_1(t) = F_{M_i}[t, X_{krit}; \mu_M(t)] + F_{c_j}[t, Y_{krit}; \mu_c(t)] \\ + F_{s_k}[t, Z_{krit}; \mu_s(t)] - F_{M_i}[t, X_{krit}; \mu_M(t)] \\ \times F_{c_j}[t, Y_{krit}; \mu_c(t)] - F_{M_i}[t, X_{krit}; \mu_M(t)] \\ \times F_{s_k}[t, Z_{krit}; \mu_s(t)] - F_{c_j}[t, Y_{krit}; \mu_c(t)] \\ \times F_{s_k}[t, Z_{krit}; \mu_s(t)] + F_{M_i}(t) \cdot F_{c_j}(t) \cdot F_{s_k}(t) \qquad (19)$$

где

$$F_1(t) = Pr\{X_{ijk}(t)\} = Pr\{\text{первые разрушения во времени } \tau \leq t\}$$
$$= \Psi\,[t : X_{krit},\ Y_{krit},\ Z_{krit} : \mu_M(t),\ \mu_c(t),\ \mu_s(t)] \tag{20}$$

Надежность $R_{total}(t)$ всей системы $(АЭС)_K$ будет в таком случае представлять при помощи (20) дополнение до величины 1 или же для нее действительна формула:

$$R_{total}(t) = R_{(АЭС)_K}(t) = 1 - F_1(t) \qquad (0 < t < +\infty) \tag{21}$$

При помощи приведенной характеристики надежности (21) можно определить риск разрушения $r_d = r(t)$ на основе известного отношения для интенсивности $\lambda(t)$, что определяется следующим

$$r(t) = \lambda(t)\,dt = -\frac{R'_{total}(t)}{R_{total}(t)} = -d\,[\ln R_{total}(t)] \tag{22}$$

Вследствие этого можно установить соответствующую окончательную полную интенсивность $\lambda_{Mcs}(t)$ в процессе $\{N_{Mcs}(t)\}$ серьезных разрушений, которая выражает каусальное отношение нарушений и дает возможность выразить при помощи числового значения риск разрушения в связи с усталостью, текучестью или же старением, а также в сочетании указанных причин и их последствий. Так, например, для усталостных или текучестных разрушений, или для сочетания этих процессов (т.е. сочетание "усталость – текучесть") действительна формула, в которой интенсивность $\lambda_{Mc}(t)$ дана отношением

$$\lambda_{Mc}(t) = \frac{f_{M_i}(t) + f_{c_j}(t) - f_{M_i}(t)\cdot F_{c_j}(t) - f_{c_j}(t)\cdot F_{M_i}(t)}{1 - F_{M_i}(t) - F_{c_j}(t) + F_{M_i}(t)\cdot F_{c_j}(t)} \tag{23}$$

так что риск такой аварии определяется отношением $r_d = \lambda_{Mc}(t)\,dt$.
Таким же способом можно учитывать сочетание процессов "усталость и старение" или "текучесть и старение", или все три процесса вместе взятые и определить для них соответствующие интенсивности $\lambda_{Ms}(t)$, $\lambda_{cs}(t)$, $\lambda_{Mcs}(t)$, иными словами, выразить при помощи числового значения соответствующие риски разрушений в любом моменте $\tau \in \langle t, t+\Delta t\rangle$, т.е. по истечении любого времени эксплуатации t.
В случае, если нам удастся достичь удовлетворительного согласования теории механики разрушения с точным объяснением тех процессов, которые представляют собой настоящую причину разрушений, т.е. с про-

цессами усталости, текучести и старения, и определить соответствующие им дистрибутивные функции (18а), (18б), (18в), то можно исходя из общих отношений (приведенных под номерами от 19 до 23), получить синтез механики разрушения с теорией надежности и таким образом выразить при помощи числового значения возрастающие со временем динамические риски разрушений.

Здесь была кратко показана математическая формулировка проблемы надежной эксплуатации составных частей оборудования АЭС.

Кроме того, был дан перечень некоторых аспектов (точек зрения), которые необходимо подвергнуть дальнейшему исследованию, чтобы достичь соответствующего уровня при оценке надежности атомных электростанций и обеспечить прогресс в этом направлении. Вполне понятно, что нам предстоит еще долго проверять предположения, отображенные в наших вычислениях, и сопоставлять их с практическим опытом. Для этого чрезвычайно целесообразно провести на международном уровне отбор нескольких реакторов с водой под давлением и частей первичных трубопроводов и на них более подробно исследовать кинетику развития дефектов, проверить законы накопления повреждений и старения материала. Прежде всего необходимо экспериментально удостовериться в остаточной жизнеспособности выбранных частей первого контура по истечении определенного этапа эксплуатации АЭС. Следует изучить не только потенциальные источники разрушений, но также на поврежденных корпусах искать пути к их уменьшению, к искусственному образованию барьеров против расширения усталостных трещин. Необходимо внедрить комплексную диагностику эксплуатации и, в частности, уделить особое внимание обучению обслуживающего персонала. Действующие до сих пор инструкции по эксплуатации не подвергаются систематическому корректированию и учету приобретаемого опыта; они никак не отвечают условиям наиболее эффективной защиты оборудования. Ввиду этого необходимо в максимально возможном объеме осуществлять международный обмен опытом. Целый ряд средств для повышения жизнеспособности и уменьшения последствий разрушений не были до сих пор проверены и они основаны иногда на довольно фиктивных предположениях. Несмотря на то, что вся наука в области надежности механических систем находится на начальной стадии развития, полученные результаты объективно говорят о том, что АЭС являются гораздо более безопасными, чем это ожидалось ранее. Основой дальнейших успехов по повышению надежности АЭС является прежде всего специализированное обучение персонала, который мог бы не только принимать участие в научно-исследовательской работе, но прежде всего на практике использовать приобретенный опыт.

TESTING AND INSPECTION OF NUCLEAR POWER PLANT SYSTEMS AND COMPONENTS

Chairmen:

M.S.R. SARMA, India
K. EFFAT, Egypt

CONTROLES NON DESTRUCTIFS ET EXAMENS PERIODIQUES DES CIRCUITS PRIMAIRES DE REACTEURS

A.C. PROT*, R. SAGLIO**
CEA, Centre d'études nucléaires de Saclay,
Gif-sur-Yvette,
France

Abstract–Résumé

NON-DESTRUCTIVE TESTING AND PERIODIC INSPECTION OF THE PRIMARY CIRCUITS OF REACTORS.

A knowledge of failures in the different components of a nuclear power plant is important for ensuring the reliability of the plant. The primary coolant circuit is particularly important from the safety point of view, which is why so many countries are working on regulations concerning periodic inspection of this system. The authors start by describing the parallel development of such regulations and of the non-destructive testing techniques designed to meet their requirements. After stating these requirements, they discuss the progress made in France during the past few years in different areas: ultrasonic non-destructive testing, especially through the stainless steel lining of the reactor vessel; extension of the results to external testing (advantage inherent in the method), to the testing of the mixed welds of safe-ends and of austenitic stainless steel welds, and to the testing of the reactor cover bolts; improvement of an acoustic holography process (which is compared with the ultrasonic method); improvement of testing methods based on Foucault currents (especially for the inspection of steam generators); and improvement of detection and location by acoustic emission during hydraulic trials and continuous operation. In conclusion, the authors suggest the main lines of work that should be followed to achieve better non-destructive testing during the construction, entry into service and operation of nuclear power plants.

CONTROLES NON DESTRUCTIFS ET EXAMENS PERIODIQUES DES CIRCUITS PRIMAIRES DE REACTEURS.

La connaissance des défaillances des divers composants d'une centrale est importante pour assurer la fiabilité de celle-ci. Parmi ces composants, le circuit primaire principal joue un rôle très particulier, par suite de son importance pour la sûreté de l'installation. C'est pourquoi chaque pays s'efforce de mettre au point une réglementation relative à l'examen périodique de ce circuit. Dans un premier chapitre les auteurs montrent l'évolution parallèle de cette réglementation et des études de contrôles non destructifs permettant d'en satisfaire les exigences. Ensuite, après avoir rappelé ces exigences, ils exposent les progrès effectués en France depuis plusieurs années dans différents domaines: contrôles non destructifs par ultrasons, en particulier au travers du revêtement en acier inoxydable des cuves; extension des résultats obtenus au contrôle par l'extérieur (avantages propres à la méthode développée), au contrôle des soudures mixtes des embouts de sécurité (safe-ends) et des soudures d'acier inoxydable austénitique, et au contrôle des goujons de fermeture des couvercles; améliorations apportées à un procédé d'holographie acoustique et comparaison avec la méthode précédente; amélioration des méthodes de contrôle par courants de Foucault particulièrement dans l'optique des inspections des générateurs de vapeur; amélioration de la détection et de la localisation par émission acoustique au cours des épreuves hydrauliques et en fonctionnement continu. En conclusion les auteurs indiquent quels sont les axes d'effort privilégiés à envisager pour l'avenir afin d'améliorer les contrôles non destructifs au cours de la fabrication, de la mise en service et de l'exploitation des centrales nucléaires.

---

* Département de sûreté nucléaire.
** Département de technologie, Section des techniques avancées.

INTRODUCTION

L'irruption rapide et massive de l'énergie nucléaire a vu se développer de manière parallèle technique et réglementation, la seconde étant généralement sensiblement décalée par rapport à la première. En effet, l'expérience acquise sur les installations réalisées, la solution des problèmes imparfaitement résolus au moment des études et bien mis en évidence au cours de l'exploitation, servent de guide pour les extrapolations ultérieures.

En ce qui concerne les centrales de puissance, le problème de fiabilité s'est immédiatement posé, car c'est lui qui conditionnait la réussite du pari nucléaire. Aussi une attention toute particulière lui a-t-elle été consacrée, consciemment ou non, et les problèmes de fiabilité à tous les stades, de la conception à la fabrication et à l'exploitation, sont maintenant bien connus de tous les experts. La réglementation fait largement appel aux notions de qualité et de fiabilité, et c'est ainsi que l'on a vu se développer partout dans le monde des textes de nature et de portée très diverses, mais qui tous ont pour objectif d'assurer la fiabilité des contrôles. Cette fiabilité à laquelle ce colloque est entièrement consacré, est naturellement étroitement liée à celle des divers composants de la centrale, et contribue pour une grande part à la Sûreté.

Dans le cas des réacteurs à eau, le circuit primaire principal constitue un élément privilégié par l'importance qu'il a sur le plan de la Sûreté. Aussi a-t-il fait l'objet de réglements particuliers :

— Code ASME Section XI aux USA, dont s'inspirent de nombreux pays,

— Arrêté et circulaire du 26 Février 1974 pour la France.

Pour ne pas alourdir cet exposé, nous noterons seulement la très grande différence d'esprit qui a présidé à l'élaboration de ces textes :

— La Section XI du Code ASME, conforme à l'esprit des autres sections, est un document précis et détaillé, ne laissant que peu de place à l'interprétation.

— L'arrêté français, beaucoup plus général, impose un cadre moins précis à l'intérieur duquel le constructeur ou l'exploitant dispose d'une autonomie assez grande. En retour, cette généralité volontaire est beaucoup plus contraignante.

Rappelons deux articles de l'arrêté français qui ont guidé très étroitement le développement des études de contrôles non destructifs associées au développement des méthodes d'inspection périodique, et qui en précisent les exigences :

Art. 42 — L'utilisateur mettra en oeuvre les moyens nécessaires pour connaître l'évolution des défauts de l'appareil constatés tant à l'issue de la construction qu'au cours de l'exploitation.

En particulier, il procédera, avant la mise en service de l'appareil, à une visite de celui-ci au moins aussi approfondie que celle qui est prescrite à l'article 44 (§ 2).

Art. 43 — Sans préjudice des dispositions des articles 41 et 42, l'utilisateur devra s'assurer, par une surveillance constante durant le fonctionnement et par des vérifications et un entretien appropriés, que l'appareil et ses accessoires, notamment les dispositifs de contrôle, de régulation et de sécurité demeurent constamment en bon état.

PROBLEMES POSES PAR LES INSPECTIONS PERIODIQUES

Ils sont clairement exprimés par les deux articles précédents. Pour mieux les situer, disons seulement que l'évolution éventuelle des défauts s'étend sur toute la vie de l'installation, c'est-à-dire que **pendant au moins 30 ans**, les méthodes utilisées doivent être reproductibles, et que toute évolution constatée doit, sans ambiguïté, être associée à un défaut de la structure contrôlée, et non à une dérive ou une modification de la chaîne de mesure.

Cet aspect du problème, fondamental à notre avis, diffère profondément de ce que l'on rencontre plus généralement dans l'industrie : une fois le contrôle effectué en usine, il est rare que l'on ait besoin de le reprendre aussi souvent et pendant une aussi longue période (mises à part les techniques aérospatiales et pétrochimiques).

La conséquence presque immédiate de cette exigence, est la nécessité d'une inspection repère : c'est la préinspection qui permet de connaître l'état des structures contrôlées, juste avant mise en service. Il est alors évident que cette préinspection doit être réalisée avec les mêmes appareillages que les suivantes.

Une autre conséquence, tout aussi immédiate, est que les procédés de contrôle utilisés, doivent posséder un très bon pouvoir de résolution afin de permettre la détection aussi précoce que possible des évolutions éventuelles.

Nous ne parlerons pas des problèmes d'accessibilité qui se posent toujours au contrôleur. Ils sont essentiels, mais font partie des paramètres qui échappent à sa compétence, au moins dans une très large mesure.

Par contre, un autre aspect doit être souligné : les méthodes utilisées doivent être exploitables, c'est-à-dire donner des informations précises, d'interprétation facile, et, dans la mesure du possible, exemptes d'ambiguïté. C'est donc dire que les défauts doivent être non seulement tous décelés, mais en outre suffisamment bien caractérisés pour que leur inocuité ou leur danger puissent être appréciés par des méthodes adéquates (par exemple mécanique de la rupture) avec le minimum d'incertitude.

D'autre part, la continuité des objectifs de fiabilité des contrôles non destructifs, du stade de la conception à l'exploitation, requiert une homogénéité dans les méthodes qui n'est pas toujours réalisée. Nous pensons qu'à l'avenir, il sera nécessaire que les contrôles en fabrication soient adaptés aux méthodes élaborées requises pour les inspections périodiques.

CONTROLES PAR ULTRASONS

Compte tenu des exigences propres aux inspections périodiques :

— accessibilité généralement sur une seule face,

— examen à distance par suite de l'environnement radioactif,

c'est la méthode la plus couramment utilisée.

Les paramètres importants qui lui sont liés sont en particulier :

— nécessité d'un milieu de couplage entre traducteur et pièce à contrôler pour permettre le transfert de l'énergie ultrasonore dans le matériau,

— sensibilité à la structure métallurgique considérée : en particulier influence très importante du rechargement en acier inoxydable de la paroi interne des cuves, et difficultés pour les soudures d'acier inoxydable,

— absorption non négligeable due aux épaisseurs traversées,

— absence de résolution angulaire propre aux traducteurs classiques,

— absence de reproductibilité à long terme due à l'impossibilité de réaliser des traducteurs de mêmes caractéristiques.

— absence de mémoire des procédés habituels.

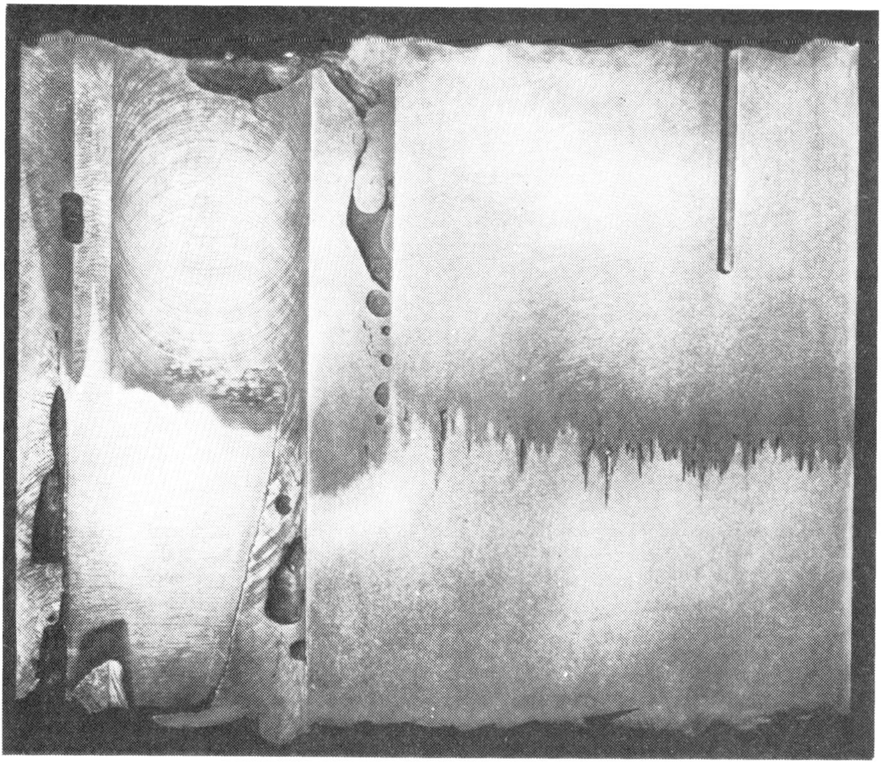

FIG.1. Coupe macrographique d'une soudure défectueuse (faisceau d'électrons).

Chacun de ces points a fait l'objet d'études poussées au Commissariat à l'Energie Atomique ; elles ont abouti à la définition d'un matériel basé essentiellement sur l'utilisation de traducteurs focalisés. Les résultats les plus marquants ont été les suivants :

— La traversée du revêtement inoxydable est rendue possible en ondes longitudinales par utilisation de traducteurs spéciaux. Le contrôle d'une épaisseur de 200 mm nécessite par exemple trois traducteurs contrôlant chacun un tiers de l'épaisseur, avec la même sensibilité et le même pouvoir de résolution. Ceci est obtenu par un dimensionnement correct des traducteurs, qui permet en particulier d'ajuster indépendamment la longueur de la focale et le diamètre du foyer. Toutefois, la dimension minimale du foyer que l'on peut envisager dans ce cas, est limitée par les caractéristiques géométriques de la surface du revêtement, et de la zone de transition acier inoxydable - acier ferritique.

— Il est désormais possible de réaliser des traducteurs focalisés en ondes transversales, avec le même pouvoir de résolution et les mêmes caractéristiques qu'en ondes longitudinales en évitant les aberrations introduites par l'interface milieu de couplage - matériau.

Les figures 1, 2 et 3 montrent ce que l'on peut attendre de l'utilisation de tels traducteurs. Elles montrent également que les cartographies ainsi réalisées (C.Scan) constituent une excellente mémoire au même titre que les films en radiographie. Ces résultats

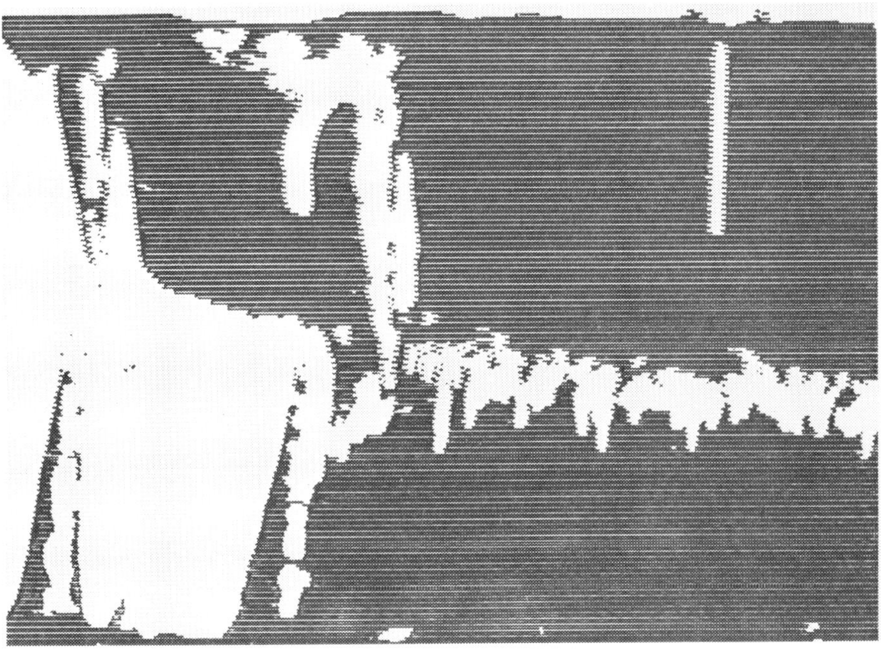

FIG.2. Enregistrement cartographique du contrôle en ondes longitudinales.

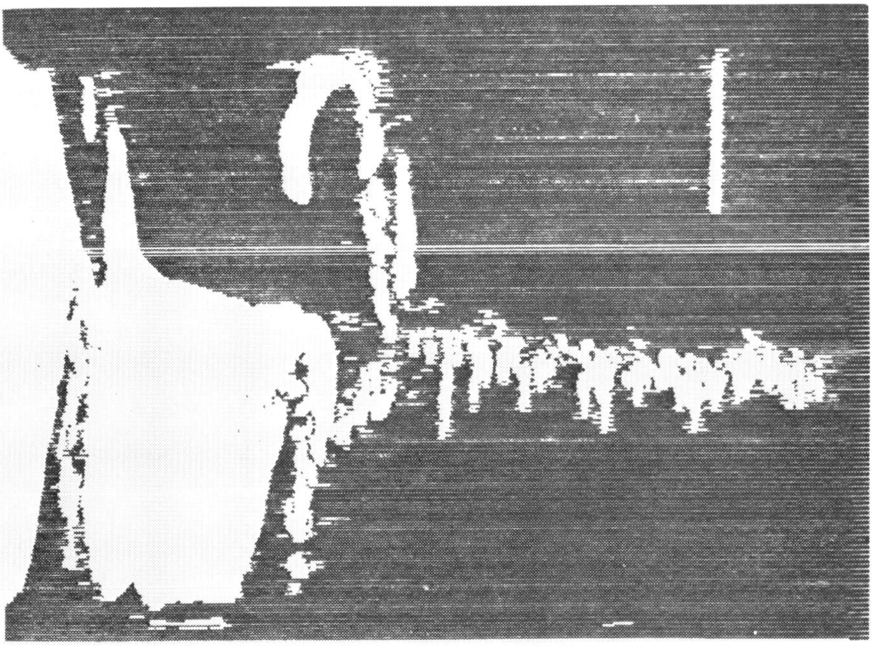

FIG.3. Enregistrement cartographique du contrôle en ondes transversales.

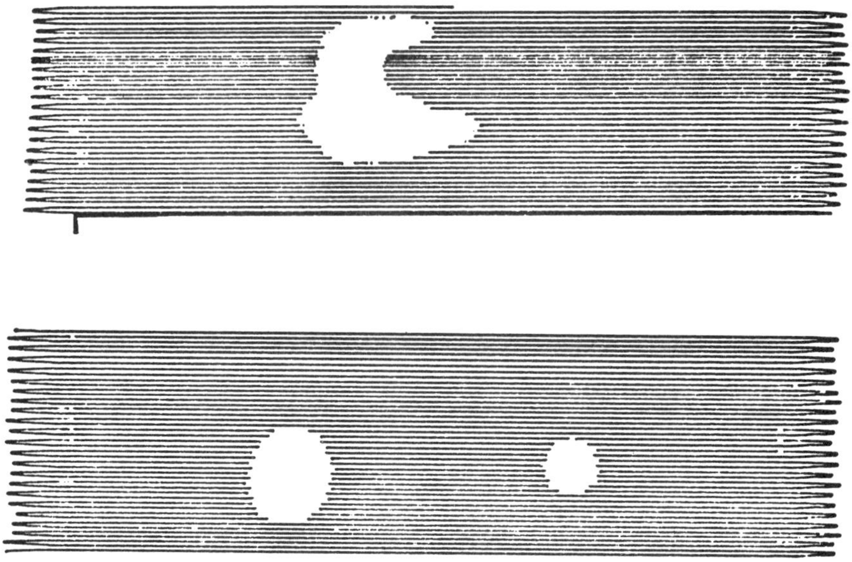

FIG.4. Cartographies de deux trous à fond plat dans une tôle de 200 mm d'épaisseur, rechargée en acier inoxydable (9 mm).
En haut: Traducteur plan 2 MHz. Seul le trou de 3 mm de diamètre est visible.
En bas: Traducteur focalisé. Trous de 3 et 2 mm de diamètre.

(qui correspondent à des soudures par faisceau d'électrons), ont été obtenus avec des dimensions de tache focale adaptées à la résolution souhaitée. Dans le cas du contrôle des cuves de réacteur PWR, qui est généralement réalisé par l'intérieur, c'est-à-dire au travers du revêtement inoxydable, la résolution obtenue est moins bonne, mais cependant largement supérieure à ce que l'on obtient avec les traducteurs standard, et c'est là un résultat très important. Les figures 4 et 5 le montrent clairement.

— La **reproductibilité dans le temps** est également assurée, et ce, quels que soient les progrès qui pourront ultérieurement être réalisés dans le domaine de l'électronique du contrôle ou du traitement du signal. En effet, au niveau de la tache focale, les hétérogénéités de champ ultrasonore que l'on rencontre dans des traducteurs plans (ou à plus forte raison avec des traducteurs d'angle à semelle plastique) disparaissent. Seules les sensibilités entre deux traducteurs peuvent être différentes. Les figures 6 et 7 témoignent de cette qualité, essentielle pour répondre aux exigences de l'inspection en service.

En corollaire de cet ensemble de propriétés, le choix de la méthode par immersion est une condition essentielle. Elle permet d'envisager tous les types de contrôle: par l'intérieur (couplage par l'eau de la cuve), par l'extérieur (couplage local). Dans ce cas toutefois, le couplage dépend éventuellement de la qualité d'adhérence des peintures utilisées, et de leur évolution possible au cours de la vie du réacteur (température, rayonnement).

Un autre avantage de l'immersion est la possibilité de pouvoir effectuer toutes les corrections du faisceau ultrasonore, nécessitées par la géométrie de la pièce : concavité ou convexité. Ces paramètres ne sont pratiquement jamais pris en compte ; mais l'on peut constater qu'ils ont beaucoup d'influence dès qu'il s'agit de déceler correctement un défaut, et mieux encore de le dimensionner.

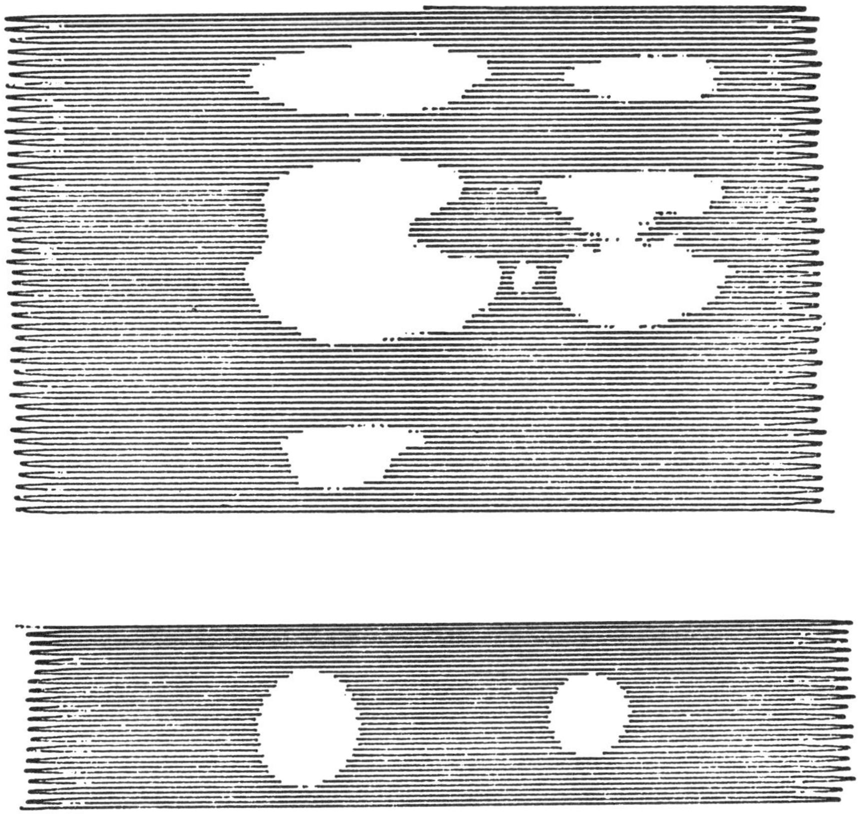

FIG.5. Même figure que la figure 4, le gain étant augmenté.
En haut, les trous sont «vus» plusieurs fois.

Un exemple des résultats d'une telle correction est donné par le contrôle des goujons de cuve de réacteur. Les figures 8, 9 et 10 montrent d'une part les étalons et la machine utilisés, d'autre part l'enregistrement effectué à l'aide d'un traducteur adapté à la géométrie de l'alésage. La détection du défaut de 0,3 mm en fond de filet est obtenue sans ambiguïté. On verra plus loin les résultats du même contrôle par courants de Foucault.

En ce qui concerne les aciers inoxydables, et plus particulièrement la soudure bimétallique (safe-end) de raccordement des tubulures de cuve aux tuyauteries primaires, les mêmes traducteurs permettent de reculer sensiblement les limites actuelles de sensibilité. La figure 11 est un exemple de ce qui peut être obtenu dans ce cas. Force est cependant de reconnaître que ces résultats ne peuvent être considérés comme parfaitement satisfaisants.

Avant d'en terminer avec le contrôle par ultrasons, signalons une importante propriété des traducteurs focalisés. C'est leur aptitude à déceler des défauts mal orientés. La figure 12 indique clairement la différence avec des traducteurs plans. Cette propriété est particulièrement utile, car des défauts plans de grande dimension, légèrement désorientés

suite du texte p.545

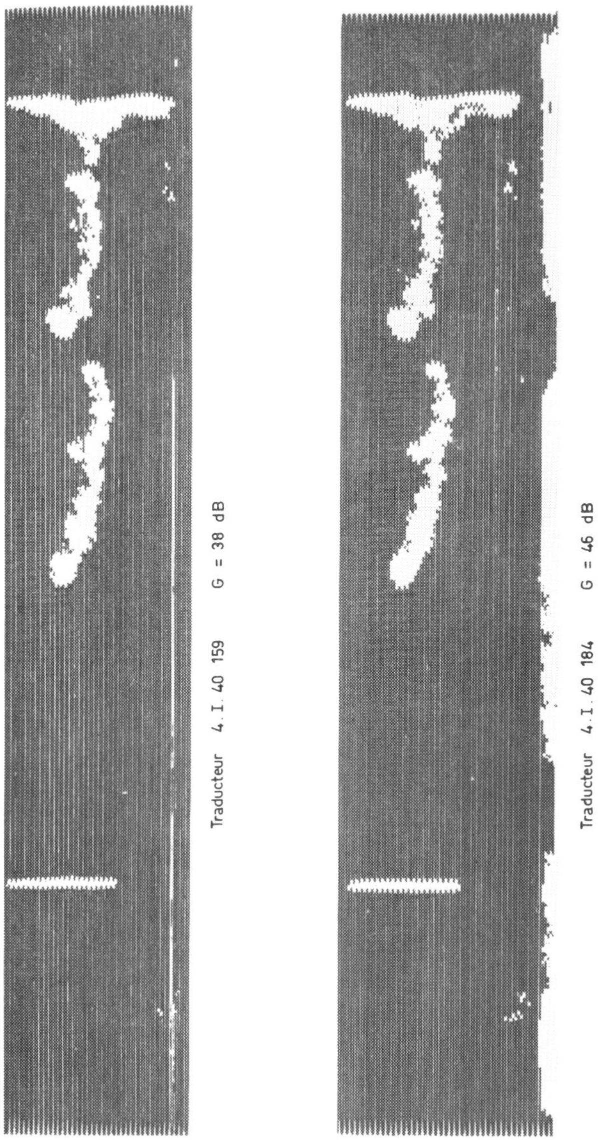

FIG.6. Mise en évidence de la reproductibilité due à l'usage de traducteurs focalisés.

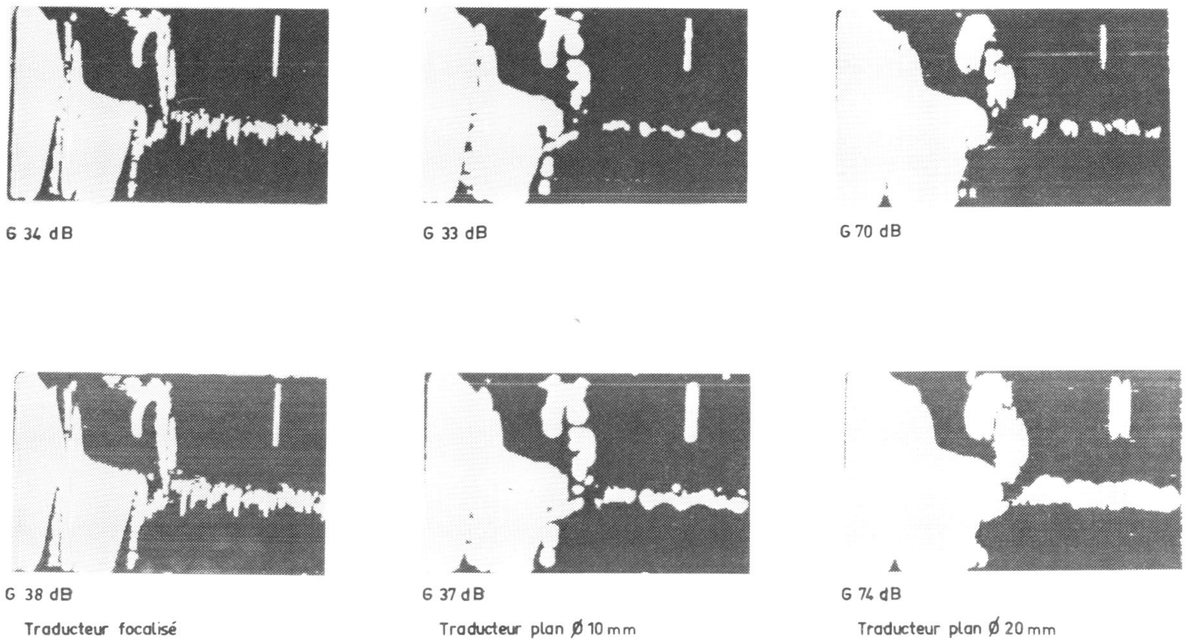

FIG.7. Résultats comparatifs en fonction du gain: traducteur plan et traducteur focalisé.

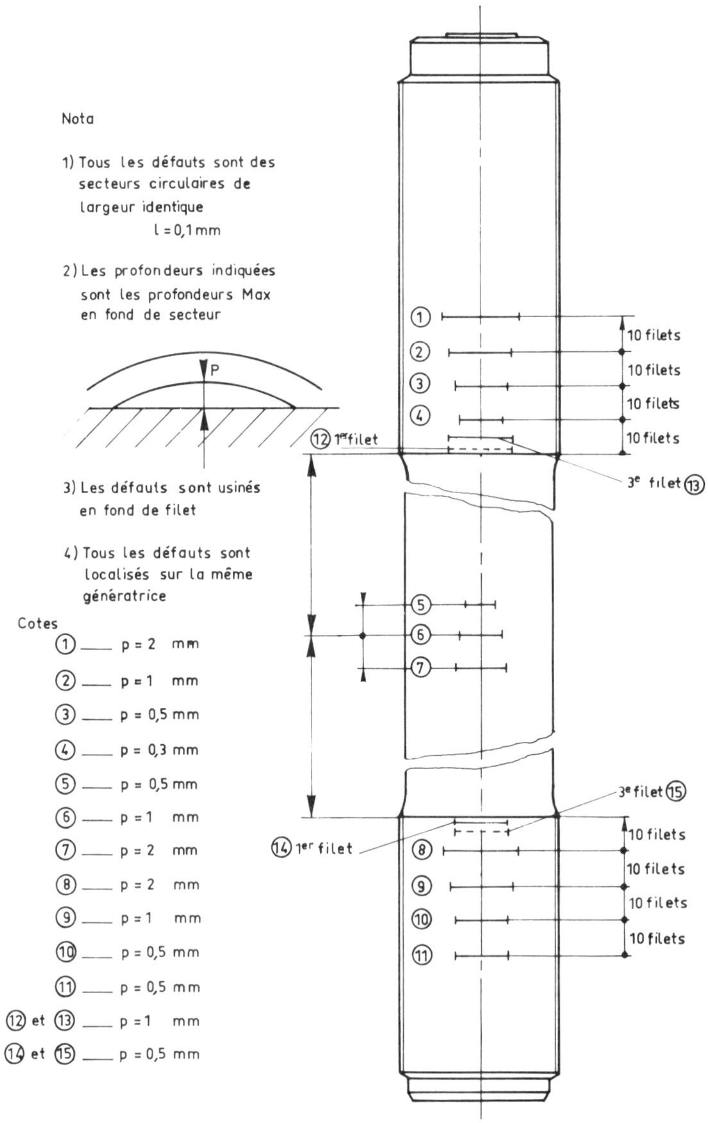

FIG.8. Goujon de fermeture de couvercle de cuve: défauts étalons.

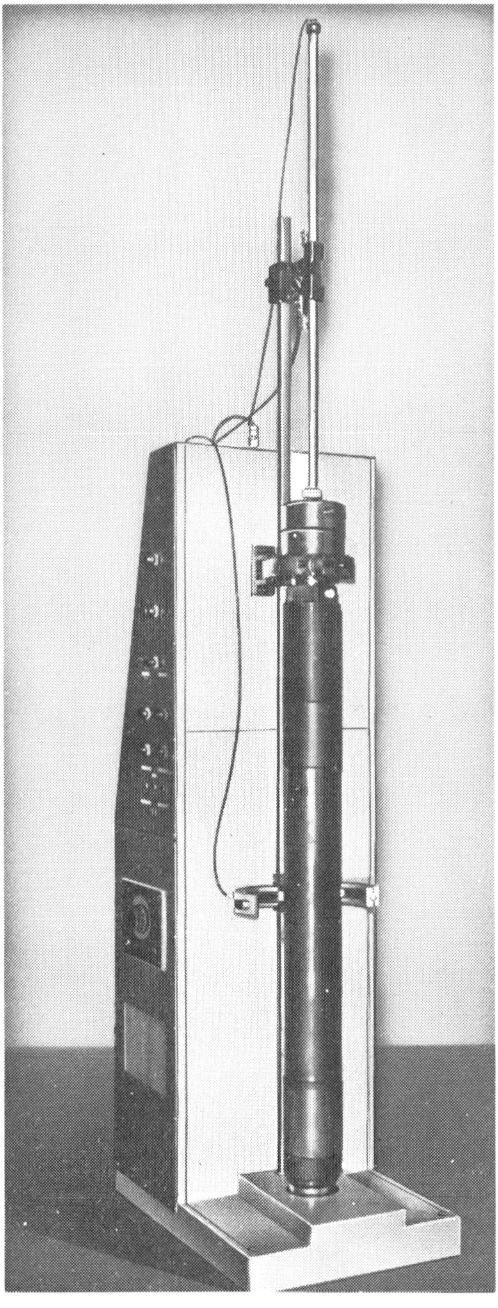

FIG.9. Machine de contrôle des goujons.

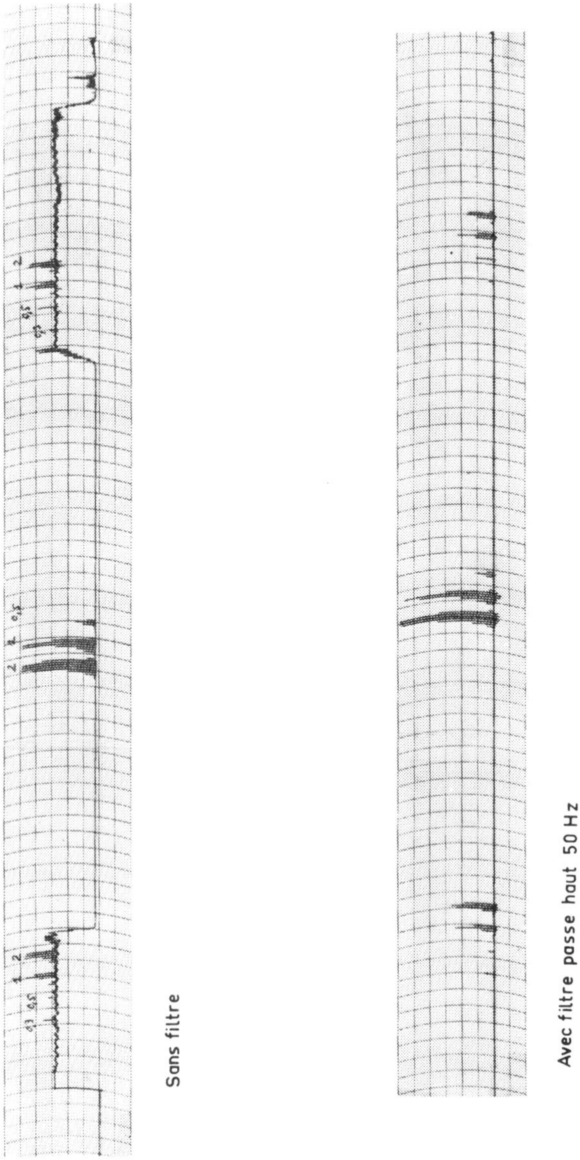

FIG.10. Enregistrement du contrôle par ultrasons sur goujon: défauts étalons.

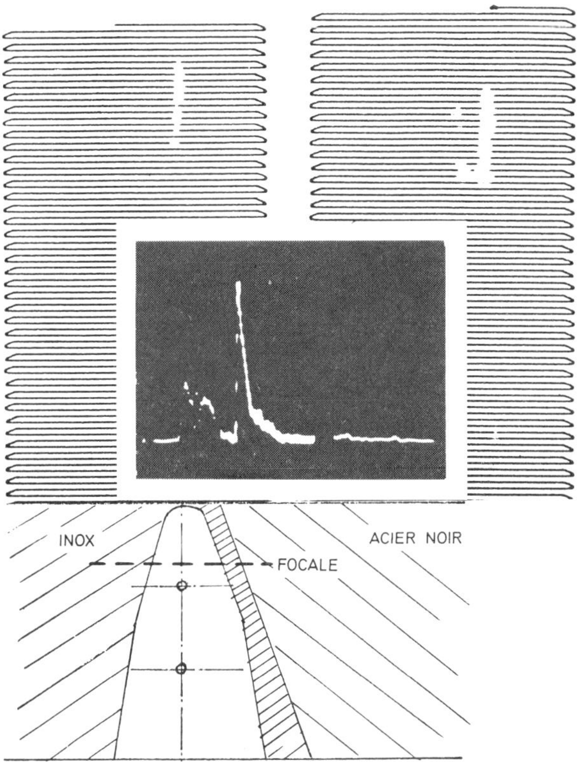

FIG.11. Contrôle d'un anneau intermédiaire: cartographie d'un trou de 3 mm de diamètre.

peuvent échapper totalement à un examen classique, même dans le cas d'une méthode tandem. Une étude analytique de cette propriété sera faite en cours d'année et permettra d'optimiser les contrôles.

Signalons enfin une étude des possibilités de l'holographie acoustique. Elle a été menée avec un appareillage «HOLOSONICS» utilisé avec des traducteurs focalisés, de conception nouvelle, particulièrement adaptés au contrôle en ondes transversales.

La figure 13 est un exemple de ce qui peut être obtenu sur des défauts réels. On notera la similitude de forme avec l'enregistrement cartographique (C.Scan) réalisé avec un seul traducteur focalisé. La comparaison avec la macrographie de la figure 14 montre clairement que la cartographie, tout en donnant une allure de la forme du défaut aussi exacte que l'holographie, permet de mieux en apprécier la nature continue. Il s'agit là d'un défaut inhérent à l'holographie dont la «richesse» en information est d'autant plus grande que l'objet diffractant se rapproche d'une forme sphérique ou cylindrique (réflexion spéculaire).

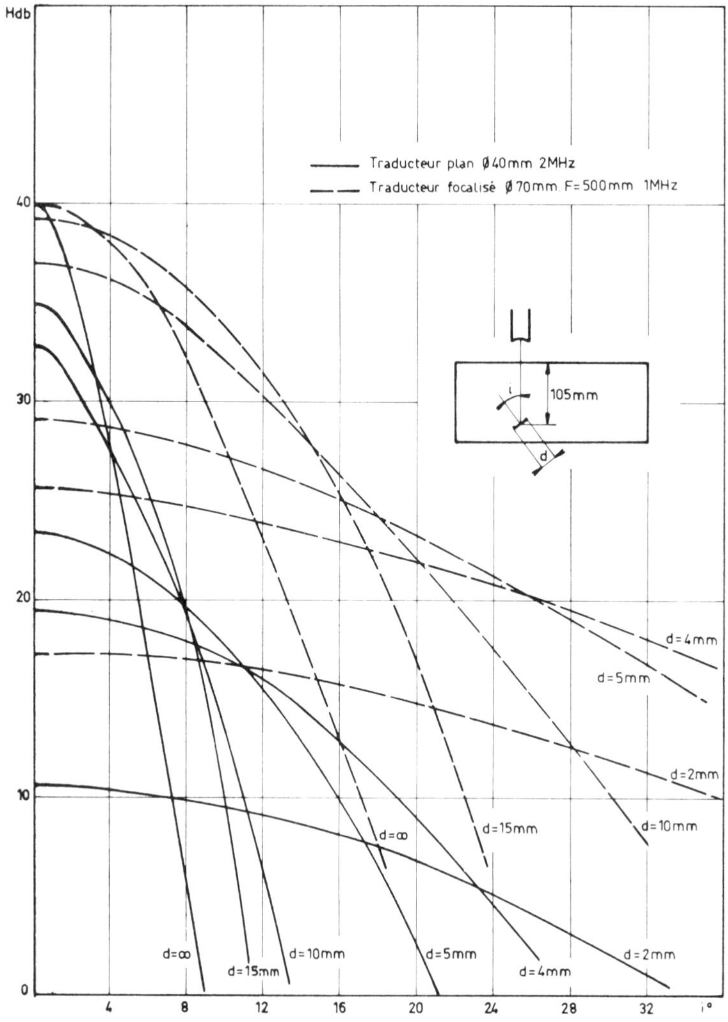

FIG.12. Détection de défauts mal orientés: comparaison traducteur plan — traducteur focalisé.

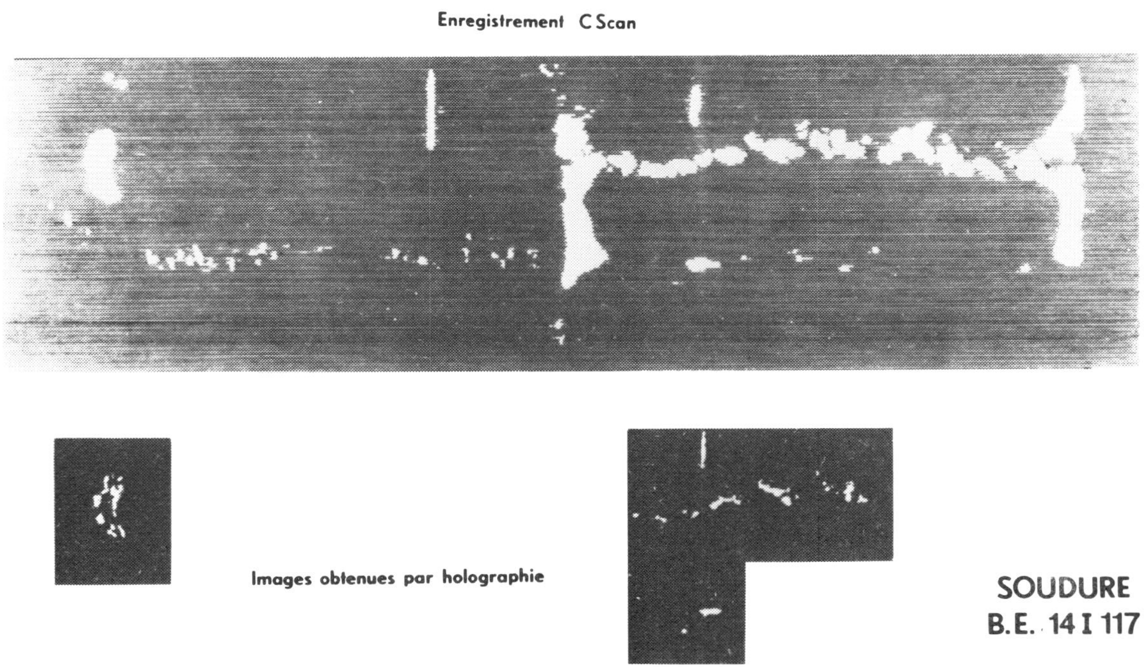

FIG.13. Comparaison cartographie ultrasonore – holographie ultrasonore.

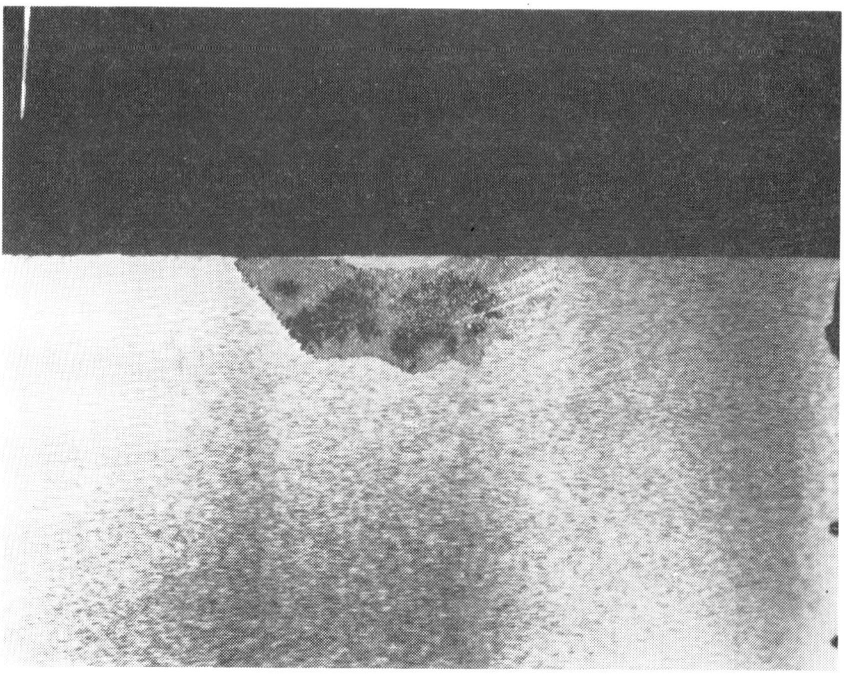

FIG.14. Macrographie du défaut mis en évidence sur la figure 13.

CONTROLES PAR COURANTS DE FOUCAULT

Les études effectuées dans ce domaine à Saclay ont essentiellement porté sur deux points :

— contrôle des goujons de cuve. La figure 15 montre le résultat sur défauts étalons (voir plus haut, contrôle par ultrasons).

— contrôle des tubes d'échangeurs (générateurs de vapeur). A ce propos, il faut signaler que dans un PWR c'est à ce niveau que l'on trouve la plus grande surface d'échange avec le milieu ambiant, et par conséquent, cette partie du circuit primaire doit être surveillée du point de vue Sûreté et fiabilité avec une attention toute particulière. Les résultats acquis à ce jour permettent :

- de contrôler les tubes avec le minimum d'interférences dues aux plaques tubulaires ou aux entretoises,
- de déceler des défauts au niveau du dudgeonnage.

Le type de sonde utilisé permet le contrôle des tubes droits, en U ou spiralés.

Des études sont en cours pour réaliser des sondes permettant une bonne sensibilité aux défauts transversaux.

D'autre part, un appareillage basse fréquence permet de résoudre le même type de problèmes sur des tubes en acier ferritique.

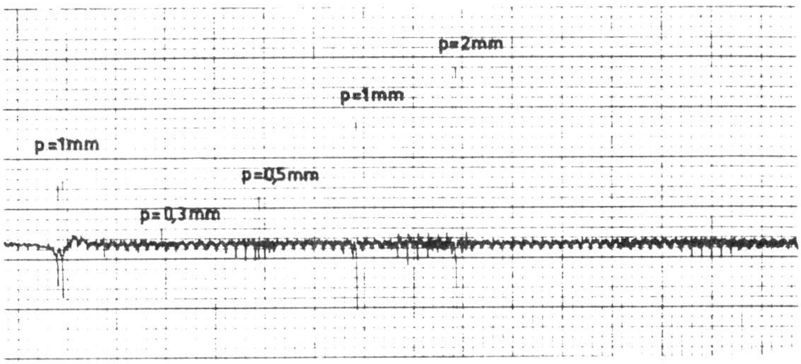

a) Filetage supérieur

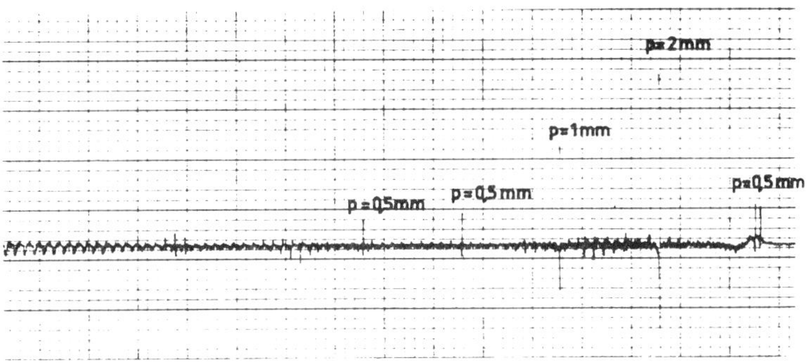

b) Filetage inférieur

FIG.15. Enregistrement du contrôle par courants de Foucault d'un goujon: défauts étalons.

EMISSION ACOUSTIQUE

Ainsi qu'on l'a vu plus haut, l'article 43 de l'Arrêté du 26 Février 1974 recommande «une surveillance constante durant le fonctionnement».
Toutes les méthodes que nous venons de signaler sont applicables uniquement réacteur à l'arrêt. La solution envisagée pour répondre à la fois aux exigences des articles 42 et 43, peut se trouver dans l'utilisation des méthodes d'écoute de l'émission acoustique.

Rappelons en effet que ces méthodes ont pour avantage principal de ne déceler que des défauts en cours d'évolution, ce qui correspond précisément au problème posé par les inspections périodiques et par la surveillance en continu.

Nos études ont porté sur les divers aspects du problème :

– étude en laboratoire sur éprouvettes de diverses natures,

– définition d'une chaîne complète : détecteur, préamplificateur, filtre, dispositif de traitement du signal, localisation par triangulation.

Là encore, des traducteurs de type nouveau ont été mis au point, qui améliorent notablement la sensibilité. Actuellement, nous disposons d'une chaîne opérationnelle de 26 capteurs et de l'électronique associée. Ce chiffre doit être porté à 32. Cette chaîne a été essayée aussi bien en laboratoire qu'en ambiance industrielle et permet d'envisager dès maintenant, la surveillance d'une cuve nucléaire au cours de la première épreuve hydraulique avant préinspection. La surveillance en continu exige de prendre en compte l'environnement sévère d'un réacteur en fonctionnement, l'une des solutions possibles étant le transfert du signal dans une zone hors rayonnement et température, par l'intermédiaire de guides d'ondes. Cette solution est l'une de celles en cours d'étude.

CONCLUSION

L'exposé quelque peu aride et didactique des diverses études réalisées au Commissariat à l'Energie Atomique, a pu laisser supposer que nous avions oublié la préoccupation majeure de ce colloque : la fiabilité.

Bien au contraire, nous pensons que les contrôles non destructifs constituent un jalon essentiel dans l'obtention de cette caractéristique des installations nucléaires. Notre exposé a été limité aux inspections périodiques : elles jouent en effet un rôle essentiel dans la sûreté, en ce qu'elles permettent d'améliorer la fiabilité du circuit primaire.

Tous les problèmes ne sont pas résolus. Il reste à améliorer la qualité des contrôles des soudures d'acier inoxydable austénitique, à rechercher davantage de précision dans le dimensionnement des défauts, à assurer leur détection dans toutes les conditions de forme, de nature d'orientation ; il reste à confirmer les espoirs que suscite l'émission acoustique pour la surveillance en continu, et à rechercher des méthodes de plus en plus élaborées de traitement de signal tant pour les ultrasons que pour les courants de Foucault et l'émission acoustique. C'est à ce prix seulement que la fiabilité des circuits primaires pourra réellement être assurée.

BIBLIOGRAPHIE

BARRACHIN, B., CHRETIEN, Nicole, PROT, A.C., ROCHE, R., SAGLIO, R., TOMACHEVSKI, E., «Application des contrôles non destructifs à la sûreté des réacteurs: Programme d'études du Commissariat à l'énergie atomique», Principles and Standards of Reactor Safety (C.R. Coll. Juliers, 1973), AIEA, Vienne (1973) 383.

PROT, A.C., SAGLIO, R., «Les contrôles non destructifs dans l'inspection en service des réacteurs nucléaires», Utilisation des contrôles non destructifs pour la surveillance des installations en service, Journée d'information du COFREND (Comité français pour les essais non destructifs), 27 avril 1973, à paraître.

RICHTER, H.U., Zur Ultraschallprüfung austenitischen Schweissverbindungen, Technik 23 10, 11 (1968).

SAGLIO, R., PROT, A.C., «Improvements in ultrasonic testing methods of welds, specially in the presence of austenitic stainless steel cladding», Proc. 2nd Int. Conf. on Pressure Vessel Technology, San Antonio, 1973, II, Materials, Fabrication and Inspection, ASME, United Engineering Center, New York (1973) 701-14.

SAGLIO, R., «Routine check-out, French Atomic Energy Commission research report», Proc. Conf. on Periodic Inspection of Pressurized Components, London, June 1974, Inst. Mech. Engineers (1974) 139-45.

WÜSTENBERG, H., «Limiting influences on the reliability of ultrasonic in-service inspection methods», Ibid. p. 108-13.

THE EFFECT OF TEST AND REPAIR STRATEGIES ON REACTOR SAFETY

E. DRESSLER, H. SPINDLER
Laboratorium für Reaktorregelung und Anlagensicherung Garching,
Technische Universität München,
Garching,
Federal Republic of Germany

Abstract

THE EFFECT OF TEST AND REPAIR STRATEGIES ON REACTOR SAFETY.

This paper deals with the effect of test and repair strategies of cold-standby systems, especially of engineered safeguards, on reactor safety. The unavailability of these systems is of high importance when the risk of nuclear power plants is to be discussed. Component failures in cold-standby systems normally can only be revealed when the systems are tested. The unavailability of such a system is therefore dependent on the length of test intervals. Of course it is also dependent on the redundancy of the system and on the component failure rates. As for repair times, our investigations show that their effect on the unavailability usually is negligible for cold-standby systems as they are installed in nuclear power plants. On the other hand, the authors find that the unavailability of such systems is strongly dependent on the method of testing. It is shown that consecutive tests are more favourable than simultaneous ones. As the redundant lines of engineered safeguards in modern German nuclear power plants are only rarely intermeshed, these systems can be tested consecutively and their unavailability can thus be lowered without additional effort.

The risk to the population caused by nuclear power plants can be sub-divided into the risk connected with the normal operation of the plants and the risk connected with accidents. Though accidents are unlikely, engineered safeguards are installed to cope with them. One example of an engineered safeguard is the emergency-core-cooling system which is installed to rule the consequences of a loss-of-coolant accident. Normally, engineered safeguards are cold-standby systems. They have to be available when an accident occurs which they have to control. Therefore the risk connected with an accident is proportional to the unavailability of the corresponding engineered safeguard.

If a cold-standby system fails, this can only be revealed when it is tested. Such tests are performed periodically, e.g. in a four-week period. They usually only answer the question: Is the system intact or not ? and do not apply to future failures. The aim of these tests is to find failures of components of the engineered safeguard and so to lower its unavailability. Beyond this, engineered safeguards are m-out-of-n systems, which means that only m of n existing lines are necessary to cope with an accident. For the sake of simplicity we assume that the existing lines are equal and consist of N components which are all necessary for the operation of the line. If there are redundant components in a line, their part in the unavailability is usually negligible. If we assume the failure rates of the components to be constant, we are allowed to define a resulting failure rate λ for a whole line, which is given by

$$\lambda = \sum_{i=1}^{N} \lambda_i \tag{1}$$

Based on the assumptions made above, it is quite easy to calculate the unavailability of an m-out-of-n system within one test period T if we assume that all components are intact at the beginning of the period and all lines are tested simultaneously. The probability that the system is not available within this period is identical with its failure probability. So the unavailability $U_{m/n}$ of an m-out-of-n system is given by the mean failure probability $P_{m/n}$ within T.

$$U_{m/n} = \frac{1}{T} \int_0^T p_{m/n}(t)\, dt \tag{2}$$

For example, we get the following result for a 2-out-of-4 system with p_k as the failure probability of line k:

$$p_{2/4}(t) = p_1p_2p_3 + p_1p_2p_4 + p_1p_3p_4 + p_2p_3p_4 - 3p_1p_2p_3p_4 \tag{3}$$

where

$$p_1 = p_2 = p_3 = p_4 = 1 - e^{-\lambda t} \tag{4}$$

$$U_{2/4} = 1 - \frac{1}{\lambda T}\left[3(1 - e^{-2\lambda T}) - \frac{8}{3}(1 - e^{-3\lambda T}) + \frac{3}{4}(1 - e^{-4\lambda T})\right] \tag{5}$$

Formula (5) for the mean unavailability is valid if the system is intact at $t = 0$ and if no component failure is repaired before T. It is also valid for all succeeding periods, if the repair times of the components are negligibly small. This can be seen from Fig.1, which shows the time-dependent unavailability.

To see the effect of various parameters formula (5) can be approximated:

$$U_{2/4} \simeq (\lambda T)^3 - 1.8\,(\lambda T)^4 \tag{6}$$

From this example the unavailability of other system configurations was calculated and can be seen in Table I. It is obvious from Table I that the unavailability of engineered safeguards can be reduced by lowering the failure rates of components or the length of the test period or by raising the redundancy. Reducing the failure rates of components might be possible by means of most accurate quality control, etc. However, this is quite difficult because components of nuclear power plants are of a considerably high standard. Reducing the length of the test period is easier, but it should be avoided that components fail due to too frequent testing. Raising redundancy means additional lines and is therefore rather expensive.

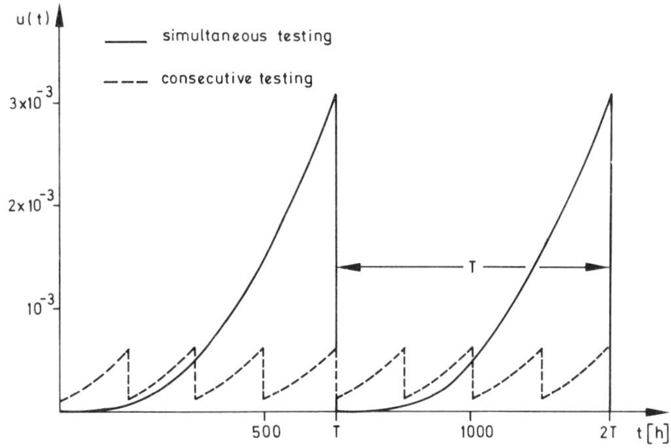

FIG. 1. Time-dependent unavailability u(t) of a 2-out-of-4 system.

TABLE I. UNAVAILABILITIES OF STANDBY SYSTEMS

System	Simultaneous testing	Consecutive testing
1/2	$0.333(\lambda T)^2 - 0.250(\lambda T)^3$	$0.208(\lambda T)^2 - 0.125(\lambda T)^3$
1/3	$0.250(\lambda T)^3 - 0.300(\lambda T)^4$	$0.083(\lambda T)^3 - 0.072(\lambda T)^4$
1/4	$0.200(\lambda T)^4 - 0.333(\lambda T)^5$	$0.033(\lambda T)^4 - 0.036(\lambda T)^5$
2/3	$1.000(\lambda T)^2 - 1.250(\lambda T)^3$	$0.667(\lambda T)^2 - 0.583(\lambda T)^3$
2/4	$1.000(\lambda T)^3 - 1.800(\lambda T)^4$	$0.375(\lambda T)^3 - 0.436(\lambda T)^4$
3/4	$2.000(\lambda T)^2 - 3.500(\lambda T)^3$	$1.375(\lambda T)^2 - 1.625(\lambda T)^3$

Besides, raising the redundancy more and more might lead to the effect of common mode failures dominating the system's unavailability. So it is quite important to look for other possibilities of lowering the unavailability of the cold-standby systems under consideration.

For this purpose, one has to check whether the repair times, which were neglected in the above formulae, could be of importance. In our investigations [1] we assumed that a mean repair time τ for all components of one line can be found, and that the failure rate of a component which is found to have failed is negligibly small compared with the resulting failure rate of the whole line. Under these assumptions we get for the simultaneous testing of all lines:

$$U_{m/n} = \frac{1}{T} \int_{\tau}^{T+\tau} p_{m/n}(t)\, dt \tag{7}$$

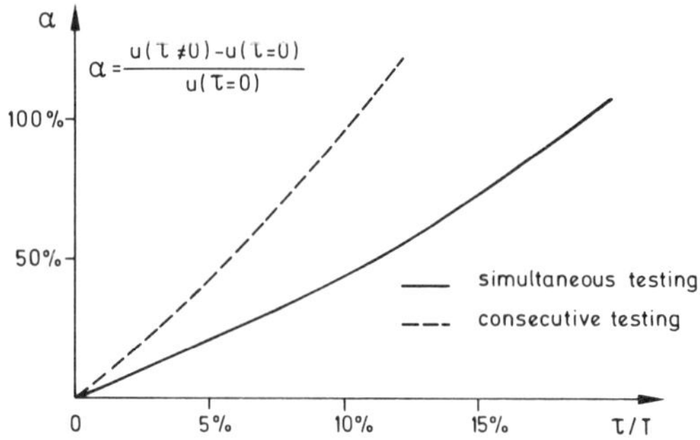

FIG. 2. Effect of repair time τ for a 2-out-of-4 system.

By means of this formula we arrive at the following approximation, e.g. for a 2-out-of-4 system:

$$U_{2/4} \simeq \lambda^3 T^3 (1+\tau/T)^4 - \frac{9}{5}\lambda^4 T^4 (1+\tau/T)^5 \tag{8}$$

which is valid with an accuracy of better than 10% for $\lambda T < 0.3$ and $\tau/T < 0.5$. For $\lambda T < 0.1$ the effect of the repair time on the unavailability is almost independent of the failure rate. Figure 2 shows this effect for a 2-out-of-4 system. We learn from Fig.2 that up to about $\tau/T = 2\%$ the increase of the system unavailability remains below 10%. The repair is of such negligible significance so long as the mean repair time of a line is smaller than about 14 hours for a four week testing period. This is normally applicable for engineered safeguards that are in use.

There is one method to lower the unavailability of cold-standby systems which has not yet been mentioned. We have always mentioned simultaneous tests here because tests are normally performed in this manner. It is, of course, possible to test in a different way, e.g. with the same frequency as before, but not periodically. As long as all lines of the system are still tested simultaneously, this would lead to an even greater unavailability. This does not apply to systems in which the lines are tested consecutively. Consecutive periodical tests seem to be especially favourable. In the example shown in Fig.3, there is an overlapping of three lines if all lines are tested together, whereas this overlapping is no longer possible if the lines are tested consecutively. This thought is confirmed by our calculations, which prove that when testing consecutively we get a considerable improvement in the system's unavailability. If we consider a 2-out-of-4 system within the time period $0 \leq t \leq T/4$, we have to take into account the following failure probabilities for the four lines:

$$p_1 = 1 - e^{-\lambda(t+3T/4)} \qquad p_3 = 1 - e^{-\lambda(t+T/4)}$$

$$p_2 = 1 - e^{-\lambda(t+T/2)} \qquad p_4 = 1 - e^{-\lambda t} \tag{9}$$

The resulting failure probability $p_{2/4}$ can be calculated with formula (3). The time-dependent unavailability which results from this is periodic with

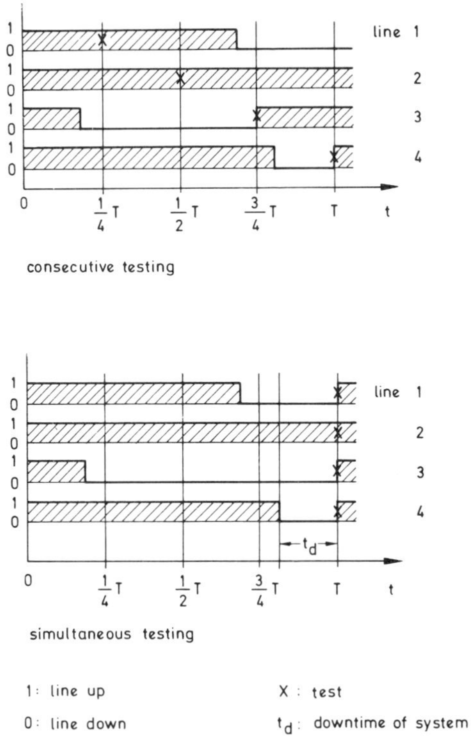

FIG. 3. Consecutive and simultaneous testing of a 2-out-of-4 system.

T/4 and can be seen from Fig.1. Consequently the mean unavailability for consecutive testing is:

$$U_{2/4} = \frac{4}{T} \int_0^{T/4} p_{2/4}(t)\, dt \simeq 0.375\ \lambda^3 T^3 - 0.436\ \lambda^4 T^4 \tag{10}$$

The unavailabilities for consecutive testing of other system configurations can be calculated in the same manner. For some system configurations their unavailabilities are listed in Table I. The amount of improvement of system unavailability by consecutive testing is dependent on the redundancy of the system configuration under consideration. The unavailability of a 2-out-of-4 system is about a factor of 2.6 lower for consecutive testing than for simultaneous testing. For a 2-out-of-3 system this factor is about 1.5 and for a 1-out-of-4 system about 6. The influence of repair times for consecutive testing can be calculated in the same way as for simultaneous testing. It can be seen from Fig.2 that the effect of repair times for consecutive testing is somewhat higher, but normally its influence is also negligible for this kind of testing.

We conclude that with practically no additional effort it is possible to improve the availability of most of the engineered safeguards by testing their lines consecutively. Stress should be laid on the fact that this improvement cannot be influenced by the error spreads of the failure or repair rates as long as the failure rates remain within reasonable limits. It should also be mentioned that consecutive tests can be easily performed in systems which are not intermeshed, which is the case, for instance, in modern German nuclear power plants.

REFERENCE

[1] DRESSLER, E., SPINDLER, H., Die Nichtverfügbarkeit von Bereitschaftssystemen in Abhängigkeit von Teststrategie und Reparaturzeit, MRR-144, Munich (March 1975).

LIMITED TEST DATA: THE CHOICE BETWEEN CONFIDENCE LIMITS AND INVERSE PROBABILITY

P. NICHOLS
Nuclear Health and Safety Department,
Central Electricity Generating Board,
London,
United Kingdom

Abstract

LIMITED TEST DATA: THE CHOICE BETWEEN CONFIDENCE LIMITS AND INVERSE PROBABILITY.

For a unit which has been successfully designed to a high standard of reliability, any test programme of reasonable size will result in only a small number of failures. In these circumstances the failure rate estimated from the tests will depend on the statistical treatment applied. When a large number of units is to be manufactured, an unexpected high failure rate will certainly result in a large number of failures, so it is necessary to guard against optimistic unrepresentative test results by using a confidence limit approach. If only a small number of production units is involved, failures may not occur even with a higher than expected failure rate, and so one may be able to accept a method which allows for the possibility of either optimistic or pessimistic test results, and in this case an inverse probability approach, based on Bayes' theorem, might be used. The paper first draws attention to an apparently significant difference in the numerical results from the two methods, particularly for the overall probability of several units arranged in redundant logic. It then discusses a possible objection to the inverse method, followed by a demonstration that, for a large population and a very reasonable choice of prior probability, the inverse probability and confidence limit methods give the same numerical result. Finally, it is argued that a confidence limit approach is overpessimistic when a small number of production units is involved, and that both methods give the same answer for a large population.

1. INTRODUCTION

1.1 For any population consisting of failed and unfailed units, it is relatively easy to predict the chance of any particular number of failures occurring in a sample of given size, provided that the probability of failure, p, for individual units is known. Thus, for a sample size of m + n units, the probability of exactly m failures is

$$P(m,n,p) = \frac{(m+n)!}{m!\,n!}\,p^m(1-p)^n \tag{1}$$

and the probability of not more than m failures is

$$Q(m,n,p) = \frac{(m+n)!}{m!\,n!}\,p^m(1-p)^n + \frac{(m+n)!}{(m-1)!(n+1)!}\,p^{m-1}(1-p)^{n+1} + \frac{(m+n)!}{(m-2)!(n+2)!}\,p^{m-2}(1-p)^{n+2} + \ldots + (1-p)^{m+n} \tag{2}$$

1.2 The converse problem, namely to predict the failure rate of the population when m failures have been observed in a sample of m + n, is not so simple, and, indeed, there is no unique solution. There are two approaches in common use.

2. CONFIDENCE LIMITS

2.1 For any chosen values of m and n (with $n \geqslant 1$), equation (2) shows that the probability of m failures or less in m + n units becomes progressively smaller as the value of p increases. Thus, for example, one can choose a value p* of p for which

$$Q(m,n,p^*) = 0.01 \tag{3}$$

Suppose now that m failures have been found in m + n units. Then one immediately feels that it is rather unlikely that the unit failure probability is greater than p*, and one can be more precise by saying that there is a 99% confidence of this.

2.2 More generally, one can say that there is a confidence 1 - Q(m,n,p) that the failure rate for the population is less than p.

3. INVERSE PROBABILITY

3.1 Equation (1) expresses the probability of a particular test result T when given the individual unit failure probability p. This can be written Prob(T|p). Similarly, for a particular combination C of failures of untested units, an expression for Prob(C|p) can also be written down. In order to relate this to the test result one must derive Prob(C|T) from Prob(T|p) and Prob(C|p). This can only be done if additional information is available on the relative likelihood of the various values of p in the range from zero to unity, which can be expressed as a frequency distribution function, Freq(p). The following relation can then be derived by the use of Bayes' Theorem:

$$\text{Prob}(C|T) = \frac{\int_0^1 \text{Prob}(C|p)\ \text{Prob}(T|p)\ \text{Freq}(p)\ dp}{\int_0^1 \text{Prob}(T|p)\ \text{Freq}(p)\ dp} \tag{4}$$

3.2 Prob(T|p) is equal to P(m,n,p). For a large sample (large m) this has a sharp maximum and the form of Freq(p) is almost immaterial. It is often assumed that all values of p are equally likely, that is

$$\text{Freq}(p) = 1 \qquad 0 \leqslant p \leqslant 1 \tag{5}$$

3.3 The failure probability for a single additional untested unit can be obtained by putting

$$\text{Prob}(C|p) = p \tag{6}$$

Making the substitutions in equation (4) then gives for the failure probability of this single unit

$$\frac{m + 1}{m + n + 2} \tag{7}$$

3.4 Equation (4) can also be used to show that the skew distribution

$$\text{Freq}(p) = 2p \qquad 0 \leqslant p \leqslant 1 \tag{8}$$

is equivalent to an additional failure in the test results, and that

$$\text{Freq}(p) = 2(1 - p) \qquad 0 \leqslant p \leqslant 1 \tag{9}$$

is equivalent to an additional unfailed unit.

Table I

Estimates of Failure Rate

Number of Failures	0	1	2
99% Confidence Limit	$\frac{4.61}{N}$	$\frac{6.64}{N}$	$\frac{8.41}{N}$
95% Confidence Limit	$\frac{3.00}{N}$	$\frac{4.74}{N}$	$\frac{6.30}{N}$
Inverse Probability	$\frac{1}{N + 2}$	$\frac{2}{N + 2}$	$\frac{3}{N + 2}$
Best Estimate	0	$\frac{1}{N}$	$\frac{2}{N}$

N is the number of tests

4. COMPARISON OF THE TWO METHODS

4.1 For any unit having high reliability, the number of failures even in a large sample will be small. Equation (2) can then be represented by a Poisson approximation and Table I constructed. This shows that the confidence limit and inverse probability methods can lead to significantly different answers, particularly when an unsafe situation can only result from the simultaneous failure of several redundant units. That is, when the square, cube or higher power of the unit failure probability has to be taken.

4.2 Quite clearly the use of inverse probability can lead to economy in plant provision and test programmes, and this raises three important questions.

(i) Is the confidence limit method unnecessarily pessimistic?

(ii) Is the inverse probability method logically sound?

(iii) Are the two methods actually solving the same problem, or is there only a superficial resemblance?

4.3 Taking the first question first, the confidence limit method is concerned with predicting the failure rate for the population. If a large number of units is to be manufactured, an unexpected high population failure rate will certainly result in an excessive number of failures, so that the confidence limit method is formulated to guard against incorrect conclusions from optimistic unrepresentative test results. The analyst can select a confidence appropriate to the economic risk involved, so that the method cannot be described as over pessimistic. However, with any reasonable choice of confidence it will usually result in an overestimate of failure rate.

4.4 If only a small number of production units is involved, failures may not occur even with a higher than expected population failure rate. The inverse probability method, which allows for the possibility of either optimistic or pessimistic test results, may then be acceptable. In practice the number of production units is unlikely to be less than the test sample size, and so the method is more applicable to the situation where independent data can be obtained by repeat tests on a small number of units, for example, start tests on diesel generators.

4.5 In answer to the second question, it has to be admitted that the application of the inverse probability method can never be completely rigorous, because of the need to assume a prior probability distribution, Freq(p). On the other hand, paragraph 3.4 indicates that reasonable alternative choices for this distribution do not produce intolerable changes in the results obtained. From the foregoing, it appears that the two methods have been used to treat logically distinct problems, so that one now comes to the key question: can a reasonable prior probability distribution be chosen so that the two methods give the same answer when solving the same problem?

4.6 It appears very difficult to derive anything other than the population failure rate by the confidence limit method. Fortunately the inverse probability method can also be used to predict the population failure rate, but before doing this it is of interest to use it to examine the simultaneous failure of two units, or more.

5. REDUNDANT UNITS

5.1 When an unsafe condition can only result from the failure of more than one similar unit, the units are said to be redundant. Thus, if there have been m failures in (m + n) units, equation (4) can be used to predict the simultaneous failure probability of two untested units by putting

$$\text{Prob}(C|p) = p^2 \tag{10}$$

which leads to the result

$$\frac{m+1}{m+n+2} \cdot \frac{m+2}{m+n+3} \tag{11}$$

This is not the result that might at first be expected, namely

$$\left\{\frac{m+1}{m+n+2}\right\}^2 \tag{12}$$

Expression (11) arises because the failure probabilities for the two units are not independent, both being conditional on the same test results. In expression (11) the first factor is the probability of the first unit failing, given the test result. The second factor is the probability of the second unit failing, given the failure of the first as an addition to the available test information. This is reasonable, since one is only interested in failure probability for the second unit when the first has already failed. This effect will increase with the number of units, and so there is hope that the two methods may "come together" for a very large number.

6. FAILURE PROBABILITY FOR A LARGE NUMBER OF UNITS

6.1 If the event C in equation (4) is taken to be the failure of not more than r units out of a total of r + s untested units, then

$$\text{Prob}(C|p) = Q(r,s,p) \tag{13}$$

The prior probability distribution chosen is

$$\text{Freq}(p) \text{ proportional to } \frac{1}{1 - p}$$

and it can be inferred from paragraph 3.4 that this is equivalent to the deletion of one sound unit from the test results. This, then, is not an unreasonable choice. The appropriate substitutions can now be made in equation (4), giving

$$\text{Prob}(C|T) = \frac{\int_0^1 Q(r,s,p)\, P(m,n-1,p)\, dp}{\int_0^1 P(m,n-1,p)\, dp} \tag{14}$$

The manipulation of this expression is given in the Appendix, which shows that, for large r and s

$$\text{Prob}(C|T) = 1 - Q(m,n,\frac{r}{r+s}) \tag{15}$$

Although r and s must both be large, this relation holds for any ratio of r to s.

6.2 Thus, if m + n units have been tested, and m failures have occurred, then

(i) the confidence limit method says that the population failure rate is less than $\frac{r}{r+s}$ with confidence $1 - Q(m,n,\frac{r}{r+s})$

(ii) with the chosen prior probability function, the inverse probability method says that, provided r and s are large, the number of failures in r + s units will not exceed r with probability $1 - Q(m,n,\frac{r}{r+s})$

Thus the two methods give essentially the same answer.

7. CONCLUSIONS

7.1 Although the confidence limit and inverse probability methods may appear to give significantly different answers, this may be due to the fact that different problems, with only superficial resemblance, are being solved.

7.2 With a reasonable prior probability distribution, equivalent to the deletion of one success from the test results, the inverse probability method gives the same numerical results for a large population as the confidence limit method.

7.3 The inverse probability method will show significant advantage when the population is small, and so care must be taken to make an honest assessment of population size. Thus, for gross failure of a reactor pressure envelope, the population must be the number of reactors of that type. However, if it could be argued that only one such failure would occur without legislative action, then the relevant population for considering the reliability of the trip and post trip systems would be the equipment appropriate to one nuclear plant. Consequently the inverse probability approach would show the main advantage in this latter part of the probability assessment.

Appendix

Supporting Analysis

A.1 Two functions have been defined in paragraph 1.1 of the main text, namely,

$$P(m,n,p) = \frac{(m+n)!}{m!\,n!}\,p^{m}(1-p)^{n} \tag{1}$$

$$Q(m,n,p) = \frac{(m+n)!}{m!\,n!}\,p^{m}(1-p)^{n} + \frac{(m+n)!}{(m-1)!(n+1)!}\,p^{m-1}(1-p)^{n+1} + \frac{(m+n)!}{(m-2)!(n+2)!}\,p^{m-2}(1-p)^{n+2} + \ldots + (1-p)^{m+n} \tag{2}$$

A.2 By repeated application of the formula for integration by parts, it follows that,

$$\int_{o}^{p} P(m,n,p)\,dp = \frac{(m+n)!}{m!\,n!}\int_{o}^{p} p^{m}(1-p)^{n}\,dp = \frac{1}{m+n+1}\Big[1 - Q(m,n+1,p)\Big] \tag{16}$$

A.3 From equation (2),

$$Q(m,n,0) = 1 \tag{17}$$

$$Q(m,n,1) = 0, \text{ for } n \neq 0 \tag{18}$$

Using (16) and (18),

$$\int_{o}^{1} P(m,n,p)\,dp = \frac{1}{m+n+1} \tag{19}$$

Also, from (16),

$$\frac{d}{dp}Q(m,n+1,p) = -(m+n+1)P(m,n,p) \tag{20}$$

A.4 Using the above expressions, and integrating by parts,

$$\int_{o}^{1} Q(r,s,p)P(m,n-1,p)\,dp = \frac{1}{m+n}\Big[(1-Q(m,n,p))Q(r,s,p)\Big]_{o}^{1} + \frac{r+s}{m+n}\int_{o}^{1} P(r,s-1,p)(1-Q(m,n,p))\,dp$$

$$= \frac{1}{m+n}\left[1 - (r+s)\int_{o}^{1} P(r,s-1,p)Q(m,n,p)\,dp\right]$$

since the first term vanishes. Now, for very large r and s, $P(r,s-1,p)$ has a very sharp maximum at $p = \frac{r}{r+s-1}$, which is essentially the same as

$p = \frac{r}{r + s}$. Thus, the only contribution to the integral comes from the region of this point and

$$(r + s) \int_0^1 P(r,s - 1,p)Q(m,n,p)\, dp = Q(m,n,\frac{r}{r + s})(r + s) \int_0^1 P(r,s - 1,p)\, dp$$

$$= Q(m,n,\frac{r}{r + s}) \qquad (21)$$

since r and s are large compared with m and n and Q is well behaved in the region of the maximum.

A.5 Finally, using these results and equation (19),

$$\frac{\int_0^1 Q(r,s,p)P(m,n - 1,p)\, dp}{\int_0^1 P(m,n - 1,p)\, dp} = 1 - Q(m,n,\frac{r}{r + s}) \qquad (22)$$

which is the result quoted in paragraph 6.1.

DISCUSSION

(on papers IAEA-SM-195/6, IAEA-SM-195/20 and IAEA-SM-195/29)

L. ALVAREZ DE BUERGO: I believe that the non-destructive testing methods discussed by Mr. Prot will be usable if power plants are specifically designed and built with a view to making their application possible — in other words, with a view to allowing in-service inspections. However, it would be interesting to know whether Mr. Prot has any experience of the practical application of these methods in existing stations, or knows of any results obtained with them. In Spain I might say we have had serious difficulties in carrying out inspections on pressure vessels and safe ends.

A.C. PROT: No, we have no experience with stations that are not designed for periodic inspection. What one could do, in situations where no incident has yet occurred, is to determine in successive inspections whether any defects registered earlier have developed further. It is in this sort of thing that the value of high-resolution control methods lies.

With regard to safe ends, I must admit that the results obtained so far are not absolutely satisfactory. A solution that one can fall back on is gammagraphy, although this presents other problems. The difficulty is not only to detect defects but to estimate their dimensions accurately and in a reproducible manner.

N.R. McDONALD: Perhaps Mr. Prot could describe any experience that has been obtained in France with the application of acoustic emission testing in practical industrial situations. Are you optimistic about the chances of this technique ultimately achieving the objectives outlined in your paper, particularly of its overcoming problems associated with other sources of noise and emission?

A.C. PROT: For the moment our experience is limited to hydraulic testing of pressure vessels, either in the factory or on site, but we are familiar with other work, particularly the American work done on EBOR at Idaho Falls. We feel that the technique has now reached a point where large-scale testing will be necessary to determine the real influence of operating noise and, perhaps, to find ways of overcoming the problems associated with it.

A.V. KARPOV: What is the maximum sensitivity of your ultrasonic and acoustic emission defect control equipment, and what is the maximum temperature at which it can operate?

A.C. PROT: Where the ultrasonic methods are concerned, it is fair to say that the use of focussing probes significantly reduces the gain required in the instruments, so that an improved signal-to-noise ratio is obtained. (With conventional equipment one is working in the neighbourhood of 40 db at minimum power.) Since the probes are selected as a function of the problem to be dealt with, there is no real limitation; for example we succeeded in mapping accurately a flat-bottomed hole 2 mm in diameter in an aluminium plate with a thickness of 600 mm. As to the acoustic emission technique, it is more appropriate to speak of detectable signal level. Unfortunately I can't recall the order of magnitude off hand.

The question of temperature does not arise with the ultrasound technique because it is used when the reactor is shut down, at 50°C. For in-service inspections using the acoustic emission method, probes withstanding temperatures of the order of 350°C will be needed, or else the signal will have to be transmitted by means of an inert wave-guide to a zone not exposed to high temperature and radiation.

H. WÜSTENBERG: I have a question concerning the influence of surface irregularities on the probability of detecting defects. Can you somehow allow for the fact that strongly focussed acoustic beams are often deflected by such irregularities and have a 'shadow', as it were, precisely where a defect is located?

A.C. PROT: The use of focussing probes makes it possible to avoid problems associated with the state of the surface, at the price of some reduction in resolution.

H. WÜSTENBERG: What is the total time required for inspection of a pressure vessel with focussing probes?

A.C. PROT: Since the work is carried out with the vessel immersed, and since in the case of PWRs the focusses are of the order of 10 to 15 mm, the inspection time is roughly the same as — if not less than — that required by conventional techniques.

H.E. KONGSØ: Have you tried to estimate the probability of overlooking a defect of critical size in a reactor pressure vessel during an ultrasonic test?

A.C. PROT: No, all we have done is to seek methods that would enable us to detect large but unfavourably oriented defects (defects of critical size) more reliably. The methods utilizing focussing probes belong in this category.

J.-C. ARLAUD: I would like to put to Mr. Prot a general question on the inspection of welds. Are there any statistical tables concerning the quality of welds as a function of their nature (done under argon, for example, or by electron bombardment), the type of material welded, the shape of the pieces welded, or the skill of the individual welder? If such statistics do not exist, could you from personal experience suggest any values?

A.C. PROT: That is a good question. Unfortunately I have no statistics on these various problems. What we have tried to do is to ensure detection of defects in all situations — that is, all types of material, dimensions, orientation and so on — and you could see from the curves I presented that the use of focussing probes very appreciably improved the detection of large but unfavourably oriented defects. This procedure is now being optimized by an analytical study using computers.

Where heavy plate is concerned, statistical material is difficult to obtain; one has to make cuts and prepare micrographs or macrographs. On the other hand, we do have experience of electron bombardment welding methods.

B. CAFFREY: I should like to ask Mr. Nichols whether he could elaborate on how he deduces that use of the inverse probability method improves the reliability of gas turbines at a particular nuclear plant by a factor of 30. Could he also clarify why he feels that the inverse probability method is adequate for a small set of gas turbines installed at a particular station? These are surely a sub-set of a larger population of a particular type of gas turbine and their reliability may be largely a function of their design and construction.

P. NICHOLS: CEGB nuclear stations have four gas turbines, any one of which can carry the essential supplies system load. With one out for maintenance, we are interested in the probability that no unit out of three will start. The factor of 3 must therefore be cubed, that is to about 30.

In the event of an incident we are interested only in the performance of the gas turbines at the station concerned. We are not interested in predicting the performance of turbines at other stations, although their past performance may provide useful statistics.

R.A. WALLER: Is the limit you determine by the inverse probability method and which corresponds to the confidence limit in reality a percentile of the posterior distribution on failure rates?

P. NICHOLS: The mathematical formulae appear to be the same, although the starting definitions are different.

R.A. WALLER: Are your inverse probability method and Bayesian analyses the same?

P. NICHOLS: The inverse probability method is an application of Bayes' theorem with a particular assumption for the prior probability.

R.A. WALLER: Is the gain provided for small samples by the inverse probability method simply a consequence of our assumptions of knowledge expressed in the assumed prior model?

P. NICHOLS: No. The confidence limit method essentially predicts the probability that the fraction of failed units in a large population will not exceed a chosen value. The individual unit failure probabilities are not independent, being conditional on the test results. In allowing for this interdependence, higher failure probabilities are produced than those emerging from the inverse probability method applied to a few units.

J. KRASNODEBSKI: I would like to make a general observation which represents, I think, the attitude of the plant designer and operator: the primary objective of reliability analysis and testing is failure prevention, not the production of sophisticated statistics.

P. NICHOLS: In many cases the plant at one station may be different from that at other stations, although in many respects similar. More conservative methods of analysis can then be applied to the general mass of data, the more elegant types of analysis being applied within a given station to give additional reassurance.

G.H. LABAT: I have a question for Mr. Dressler. Your analysis implies periodic annulling of the failure probability through testing, on the hypothesis that the failure rate, λ, is constant. In making this assumption one decides not to take into account, in particular, defects associated with duration of use. However, to the extent that such defects can exist at all, one must be aware that annulling of unavailability through testing can only be partially valid. After all, one may legitimately ask whether, after a simultaneous test of all lines in the emergency core cooling system of a reactor, it is in fact possible to guarantee zero unavailability. Can one guarantee that a demand following immediately on the test will be met with 100% certainty? I think one needs to be cautious in replying to this question, and that one should consider, in addition to tests, such things as inspections to reveal ageing.

E. DRESSLER: For simplicity's sake we assumed that a failed component is completely intact at the end of repair, and that repaired components are as good as new. But, even if we took imperfect repair into account, that would not change the qualitative results presented in our paper. One might perhaps overcome some problems by testing a line once more immediately after the completion of a repair.

R. GONNOT: Do you take the duration of a test into account in your evaluation? Also, do you take into account the errors that might be made in the course of the test?

E. DRESSLER: In modern German nuclear power plants engineered safeguards have to be operable during tests, so the length of the test need not be taken into account. As to the second part of your question, the answer is no, because allowance for errors made during testing would not influence the qualitative results of our investigation.

J.R. TAYLOR: I would like to pursue the previous question further by asking Mr. Dressler whether he feels that the extra protection against common mode failure in testing given by consecutive testing is in fact valuable.

E. DRESSLER: That there is extra protection seems certain. The value of the improvement depends, however, on the specific kind of common mode failure.

D. NORMAN: I would like to reiterate the difference between the 'bad-as-old' and 'good-as-new' concepts of availability. It is important to recognize that, for many plant items, a repair only improves the condition of the component which failed, the other components remaining in a degraded state so that the reliability of the plant is not as 'good as new'. The work of H. Ascher in the United States of America has covered this situation well. His results have implications for much of our analysis of inspection and maintenance intervals. For example, the actual variability in gas turbine unit starting reliability between major overhauls may well be such as to swamp differences between estimates of that starting reliability using inverse probability methods and those using confidence intervals. Therefore, in this particular example it may be more important to establish trends in gas turbine starting reliability between overhauls than to worry about which type of estimate to make. However, in situations which are not time-dependent, of course, the distinction is important.

P. NICHOLS: Although the analysis given in my paper is not time-dependent, it can readily be extended to cover the time-dependent situation. Perhaps it was a little unfortunate that I chose to present a time-independent analysis and a time-dependent example.

ULTRASONIC TESTING OF AUSTENITIC STEEL WELD JOINTS

B. KUHLOW, E. NEUMANN, H. WÜSTENBERG,
E. NABEL, E. MUNDRY
Bundesanstalt für Materialprüfung,
Berlin,
Federal Republic of Germany

Abstract

ULTRASONIC TESTING OF AUSTENITIC STEEL WELD JOINTS.

The inspection of austenitic welds by the well-known ultrasonic angle beam methods is difficult. The formation of grains and dendrites of large volume in the weld metal results in so much ultrasonic scattering that normal ultrasonic testing methods only yield results with limited reliability. To improve the reliability of the inspection results, an ultrasonic probe system operating as a transmitter-receiver with obliquely incident waves has been developed. The beams of the two probes cross in a limited 'focal area' in a distance corresponding to the near-field length of the probes. The natural bundling of the sound field at this point leads to a constriction of the focal area, resulting in a substantial improvement of the signal-to-noise ratio. By frequency-dependent measurements of the ultrasonic attenuation in the frequency range 0.5 - 5 MHz for longitudinal and shear waves in the base material and in the weld material, it was found that longitudinal as well as shear waves are suitable with the exception of the uppermost layer of the weld specimen. The disturbance of the results of measurements by surface waves can be avoided in this case only by use of longitudinal waves. The transceiver angle probes have been tested in an actual inspection of austenitic welds in reactor components.

INTRODUCTION

In addition to X-ray inspection, ultrasonic technique is a common method for testing materials for defects.

However, the inspection of austenitic welds by means of the well-known ultrasonic angle beam method is difficult, owing to the strong ultrasonic scattering which occurs in this material.

By using special testing techniques only, which are reported in this paper, one can get reliable results. The γ-α-transformation — i. e. the transformation of a cubic face-centred lattice into a cubic body-centred lattice — did not take place in austenitic steels. If the rate of cooling in the weld metal is too low, it leads to the formation of coarse grains in the austenitic weld metal which at first solidifies with a fine crystalline structure.

Figure 1 shows the polished section of such an austenitic double U-butt weld. Starting from the side walls of the weld, the weld metal crystallites grow predominantly in the direction [100], and this leads to the so-called dendrite formation. The comparatively large extension of the dendrites leads to a higher scattering in the area of a weld joint [1-4].

FIG. 1. Austenitic weld joint (double U-butt weld in 50-mm-thick sheets) in the steel X6 CrNi 1811.

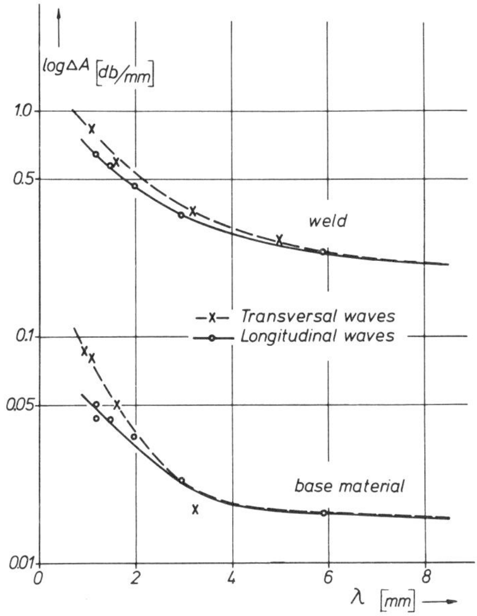

FIG. 2. Ultrasonic attenuation (ΔA) in austenitic base material and in the weld material.

MEASUREMENTS OF ULTRASONIC ATTENUATION

Investigations were made to determine whether longitudinal or transverse waves are better suited for testing austenitic weld joints. Therefore the ultrasonic attenuation of the parent metal and of the weld material with sound penetration in the plate thickness direction was studied for transverse and longitudinal waves with a spectral scan of the ultrasonic echoes in the frequency range 0.5 - 5 MHz (see Fig.2).

The attenuation of such waves, both for the same wavelength in the parent metal and for an integral measurement across the weld region, differs markedly only below a wavelength of 2 mm ($\geq$ 2.9 MHz for longitudinal waves and $\geq$ 1.6 MHz for transverse waves). However, the attenuation of the transverse waves in comparison to that of the longitudinal waves in the frequency range examined was never greater than a factor of 2. This difference cannot be considered to be significant [5, 6]. For this reason, both wave types were employed for the detection of defects in austenitic steel weld joints.

It must be noted that the integral attenuation in the weld seam differs from the attenuation in the parent metal, which in its turn is almost equal to the attenuation in normal ferritic material, by a factor of 10.

ULTRASONIC TEST PROCEDURE WITH TRANSMITTER-RECEIVER ANGLE PROBES

In the pulse echo technique normally used, all scatter echoes from the entire material region penetrated by the sound beam are registered.

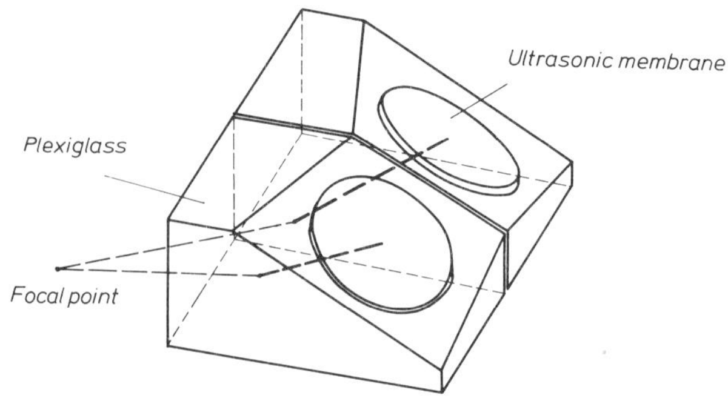

FIG. 3. Transmitter-receiver angle beam probe.

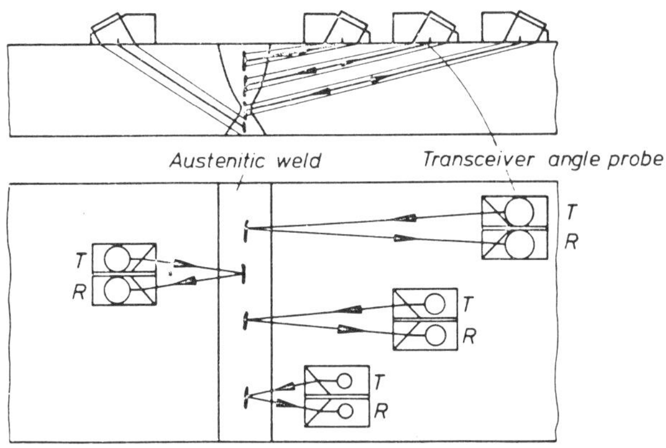

FIG. 4. Ultrasonic testing of austenitic steel weld joints with transceiver angle probes.

To improve the ratio between the flaw and scatter echoes originating from the austenitic weld structure, transceiver angle probes were constructed (Fig. 3). The directional lobes of the two transducers (transmitter and receiver) are inclined towards one another and cross in the focal area [7-10]. It is suitable to choose the focus area at a distance corresponding to the near-field length, because the sound beam has its narrowest section there. However, if one works with a broadband ultrasonic pulse, the near-field length is not so sharply distinguished. The advantage of this method over the normal 'pulse echo' technique is that the structural scatter echoes from the entire material region penetrated by the sound beam are not registered, but only those coming from the focal area, whereas the echo from a defect in the focal area remains constant.

For testing a thick specimen, some different probes had to be calculated and built because of the limited focal area of one probe.

For complete coverage of a 60-mm-thick austenitic weld in an ultrasonic test, the seam was divided into four 15-mm-thick layers (Fig. 4).

The upper region of the specimen near the probe had to be tested with longitudinal waves to suppress interfering echoes from surface waves.

The choice of the angle of incidence depends on the orientation of the flaws to be detected. Assuming that the flaw orientation corresponds to that of the weld side walls — and that these flaws are the dangerous ones — one has to choose large angles of incidence because of the directional effect of the flaws in the reflection of the ultrasonic pulse.

Angles of incidence in the steel to be tested were in the range of 70° for the uppermost layer to 50° for the bottom layer. This choice of angles is not compulsory. It was determined by the requirement to keep the echo path for all four layers more or less equal so that, when comparing the probes, the attenuation and divergence losses could be kept constant.

INFLUENCE OF THE COUPLING LAYER

The echo transmittance at the wedge-material/steel interface also has to be considered [11]. The transducer produces longitudinal waves. They travel through the wedge material. Between probe and test piece lies a coupling layer, mostly a liquid, in order to achieve good and uniform coupling for rough or uneven surfaces as well. Interferences within the coupling layer lead to a sound transmission which depends on the thickness of the layer (Fig. 5).

Besides depending on the thickness d of the layer, the transmission factor D also depends on the wave resistances W and velocities of sound c in the wedge material ①, the coupling layer ②, and the test piece ③, as well as on the angle α and on the absorption a in the coupling layer.

The calculated transmission for the sound pressure refers to monochromatic sound and a plane parallel coupling layer. For a broadband ultrasonic pulse, the transmission results in the superposition of the D-values for all frequencies which appear. The most effective sound transmission is present

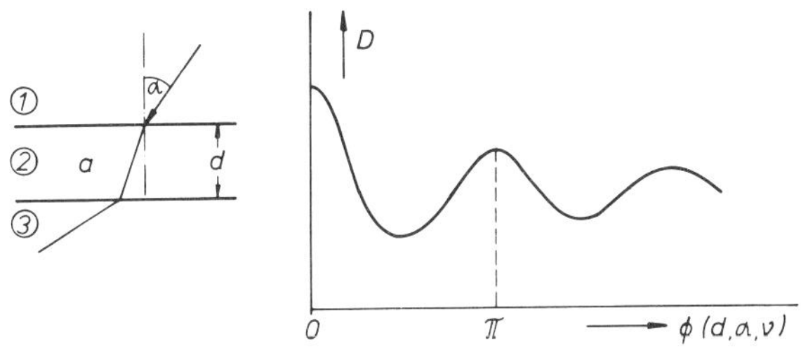

$$D = \frac{t \cdot e^{-al}}{\sqrt{1 - 2re^{-2al} \cdot \cos\phi + r^2 e^{-4al}}}$$

$$l = \frac{d}{\sqrt{1-(c_2/c_1)^2 \sin^2\alpha}} , \qquad \phi = \frac{2\pi\nu}{c_2} \cdot 2d \sqrt{1-(c_2/c_1)^2 \sin^2\alpha}$$

$t = t_{12}(W_1, W_2) \cdot t_{23}(W_2, W_3)$ Transmission

$r = r_{21}(W_2, W_1) \cdot r_{23}(W_2, W_3)$ Reflection

FIG. 5. Transmission factor D of a plane parallel coupling layer.

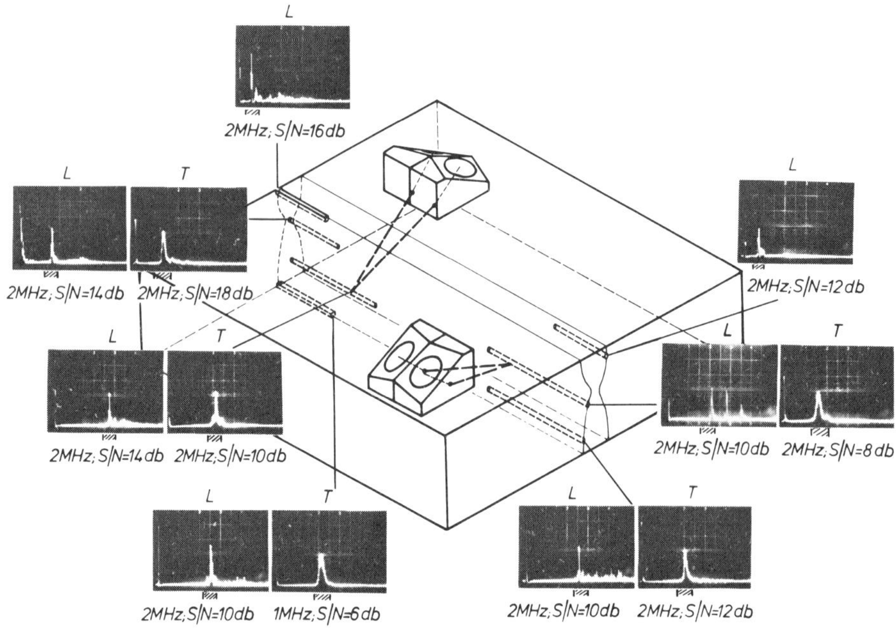

FIG. 6. Longitudinal (L) and transverse (T) wave echoes from artificial flaws in austenitic steel weld joints.

in all cases for a disappearing thickness of the coupling layer, particularly if one works with short ultrasonic pulses. For the ultrasonic testing of strongly scattering materials, it is favourable to use broadband ultrasonic pulses. The signal-to-noise ratio of such short-time pulses is better because of their shorter coherence length. Moreover, their local resolution is better, so that the ultrasonic signals from reflectors lying very close together — which can be considered macroscopic compared with the grain boundary dimensions — can be separated better [12, 13].

The second favourable thickness of the coupling layer, which corresponds to the value $\Phi = \Pi$ — for normal incidence of the sound this is a half-λ-layer — acts as a narrow-band filter for the ultrasonic pulses. Hence the signal-to-noise ratio and the local resolution is lowered.

The best sound transmission from the probe into the test piece and vice versa is given by a large value of the transmission on the interfaces. To avoid the transmission oscillations depending on the thickness of the coupling layer, one has to choose a small value of r. This can be achieved by fitting the wave resistances at one interface, for example at the interface wedge material/coupling layer. In that case, interferences in the coupling layer and therefore the transmission oscillations vanish.

RESULTS

The probes were tested with a commercially available ultrasonic instrument on artificial test flaws in the steel X6 CrNi 1811 which is used for reactor vessel components. These flaws were placed in the base material, in the transition zone and in the weld joint in the form of cylindrical boreholes with a diameter of 2 mm and 3 mm and flat bottom holes perpendicular to the testing surface (Fig. 6). The results obtained for the signal-to-noise ratio were 6-10 db for flat bottom holes and were always >10 db for cylindrical boreholes.

Echoes coming from the weld side-walls caused by the jump of the wave resistance there disturb the interpretation of the flaw echoes. By suitable signal processing — for example, by the use of auto-correlation — an immediate improvement of the signal-to-noise ratio of about 6 db is achieved. Difficulties only arise in the region of the test piece surface which is not facing the probe. The side-wall scatter reflections are additionally reflected by the surface and reach the probe nearly at the same time as the effective signal from this region.

REFERENCES

[1] MIDECKE, P., Über die Ultraschallschwächung in austenitischen Chrom-Nickel-Stählen, Materialprüfung 3 (1961) 1.

[2] RICHTER, H.-U., Zur Ultraschallprüfung austenitischer Schweissverbindungen, Die Technik 23 (1968) 610; 23 (1968) 692, with an extensive citation index.

[3] MARINI, J., ROULE, M., Contrôle automatique par ultrasons d'assemblages d'angle en acier austénitique, Rapport STA-NT 28, Direction de la métallurgie, Département de technologie, Section des techniques avancées (Nov. 1970).

[4] LINKE, D., Gefügeeinflüsse bei der Ultraschallprüfung von ES-Schweissverbindungen, Schweisstechnik 23 (1973) 538.

[5] MATSUMOTO, S., KIMURA, K., The relation between grain size and ultrasonic attenuation coefficient in austenitic stainless steel and iron, Trans. Natl. Res. Inst. Met. 14 (1972) 155.

[6] GOEBBELS, K., GRETER, S.-E., Zur Ultraschallschwächung von Longitudinal- und Transversalwellen in Eisen und Stahl, Vortrag auf der Frühjahrstagung der Deutschen Physikalischen Gesellschaft, Münster, 1975.

[7] WÜSTENBERG, H., SCHULZ, E., Versuche zur Feststellung plattierungsnaher Reflexionsstellen an Reaktorteilen mit Ultraschall, Vortragstagung der Deutschen Gesellschaft für Zerstörungsfreie Prüfverfahren, Saarbrücken, 1972.

[8] RÖNTGEN TECHNISCHER DIENST NV, private communication, Rotterdam (1972).

[9] PELSENEER, J.P., LOUIS, G., Ultrasonic testing of austenitic steel castings and welds, Br. J. Non-Destr. Test. 16 4 (1974) 107.

[10] KOLB, K., WÖLFEL, M., Ultraschallprüfung auf Unterplattierungsrisse an Reaktorteilen, Materialpruefung 16 (1974) 74.

[11] CANELLA, G., The effect of couplant thickness in ultrasonic contact testing, Br. J. Non-Destr. Test. 16 6 (1974) 179.

[12] SPROULE, D. O., Ultrasonic testing of welds, Br. Weld. J. 6 (1959) 470.

[13] BRADFIELD, G., "Ultrasonic flaw detection in metals", The Physical Examination of Metals (CHALMERS, B., Ed.), A.G. Quarrel, London (1960).

INVESTIGATION WITH AUTOMATIC ULTRASONIC EQUIPMENT TO TRACE FLAWS IN A LARGE TEST PIECE, AND EXPERIENCE GAINED IN CARRYING OUT INSPECTIONS

J.P. LINDNER
Maschinenfabrik Augsburg-Nürnberg AG,
Nürnberg,
Federal Republic of Germany

Abstract

INVESTIGATION WITH AUTOMATIC ULTRASONIC EQUIPMENT TO TRACE FLAWS IN A LARGE TEST PIECE, AND EXPERIENCE GAINED IN CARRYING OUT INSPECTIONS.

Based on the FRG codes providing guide lines for the Reactor Safety Commission regarding the size and location of flaws to be detected during in-service inspections, investigations were carried out into the possibility of detecting defects in thick-walled reactor pressure vessel components with the aid of ultrasonic inspection systems. A large test rig was used and, in a similar manner to the in-service inspections on a reactor, the tests were carried out with remote-controlled, automatically guided inspection equipment. For this purpose, a test specimen weighing about 10 tons was produced and provided with two weld seams having a large number of artificial defects. Essential parameters for the various reflectors in the test specimen were the size, location, angle and roughness or structure of the reflecting surfaces. As it is known that austenitic cladding has a considerable influence on flaw detection, the tests were undertaken first without cladding and then with cladding. A manipulator was designed for automatic remote-controlled inspection with which the inspection system travels on a meandering route over the area to be inspected. The inspection system employed was of the same type as the one used for baseline tests during external inspections of reactor vessel walls with parallel surfaces. Digital data collection was by a magnetic tape recorder designed to store both the data of the ultrasonic inspection system as well as the allied position data. The data stored on the tape are evaluated with electronic data processing programmes specially developed for this purpose. These programmes allow locally coherent indication patterns to be prepared, thus simplifying the interpretation of the data obtained. The author initially describes the equipment with the aid of which the studies were undertaken. A detailed discussion is then presented on the design of the test specimen and the inspection systems employed. Following this, the results obtained are explained and evaluated for the individual parameters of the inspection. Finally, the possibilities are given for employing such inspection equipment for ultrasonic in-service inspections on reactor pressure vessels. A report is given on the experience gained in this respect both for internal and external inspections.

Based on the Codes of the Federal Republic of Germany and the Guidelines of the Reactor Safety Commission for the detection of flaws by in-service inspection tests of pressurized components, various tests have been conducted to establish the ability of ultrasonic testing systems to detect flaws, by size and location, in thick-walled reactor pressure vessels. An extensive test facility was developed to carry out these tests. The work was carried out at the Maschinenfabrik Augsburg-Nürnberg (M.A.N) plant by various institutions participating in a research programme sponsored by the Minister of the Federal Republic of Germany for Research and Technology. The objective of the tests was to establish the flaw detection capabilities of present inspection equipment. On the basis of the experience accumulated during these tests, the systems were improved and refined and then used for performing base-line tests on reactor pressure vessels.

I should like to give you a brief outline on the test setups, on how the tests were performed and on the major test results on a large pressure vessel specimen. I shall then deal briefly with experience gained in applying such equipment in the field.

Regarding beam angle and sensitivity setting, the Safety Commission's guidelines contain the following provisions:

(a) The angles of incidence shall be so selected that cracks and other separations with the most probable orientations are indicated. Most probable flaw orientations are predominantly such surfaces which are perpendicular to the main stressing directions.

(b) To check for flaws inside the vessel, general procedure shall be to combine tandem and single-probe techniques for scanning the most probable orientations.

(c) The sensitivity setting of the ultrasonic inspection equipment depends on the test technique applied. For some important techniques the rules are as follows:

(1) When using a technique where the sound beam is directed perpendicular to a crack surface of anticipated orientation or when using a tandem technique lined up for this orientation, the recording limit must be 6 dB lower than the echo amplitude of a flat-bottom hole reflector of 10-mm diameter with the same orientation. This condition must be satisfied at every point of the sound beam cross-section. Tests are required to be carried out to substantiate that normal fluctuations in the orientation of flaw surfaces can be scanned at these limits.

(2) When a single-probe technique with a single crystal is applied and advantage is taken only of the diffuse scatter reflection of crack surfaces for the indication, a flat-bottom hole reflector of 3-mm diameter with an orientation perpendicular to the sound beam must be used as the recording limit. Here, too, this condition must be satisfied at every point of the sound beam cross-section.

The inspection equipment and the test specimen were devised for these specific objectives.

In previous work with an automated in-service inspection setup, it was found that certain parameters, such as size, location, angle and roughness of the reflecting surfaces have an influence on detection and interpretation by ultrasonics. It was also found that the geometric conditions of the surface ot the test specimen and the cladding were liable to impose additional complications on the ultrasonic test. In view of this, individual theoretical and experimental tests were carried out beforehand to establish the quantitative influence of these parameters. Most of these tests were carried out by the Federal Institute for Materials Testing. Although the tests were very informative, it is beyond the scope of this paper to deal with the results. However, on the basis of the test results, the Kraftwerksunion developed a test specimen containing test flaws with all parameters likely to have an influence on the probability of flaw detection.

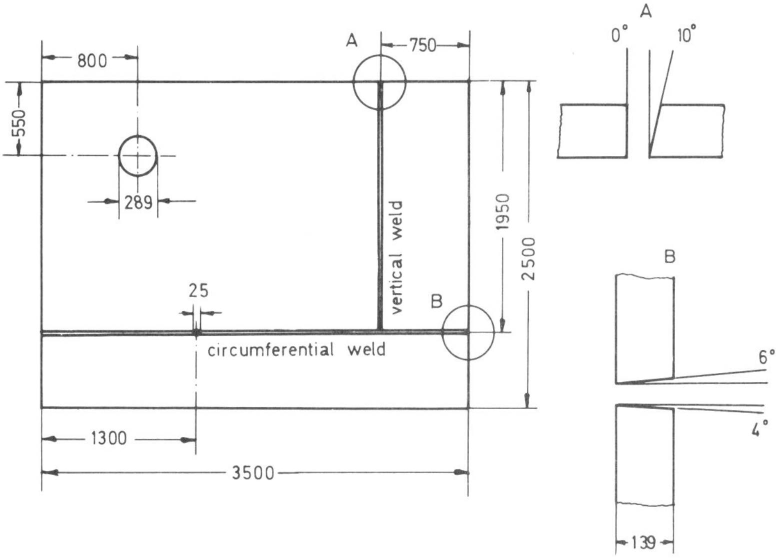

FIG. 1. Schematic diagram of a pressure vessel specimen.

FIG. 2. Test piece on the test rig.

The outer dimensions of the specimen wall are shown in Fig. 1. It is 3.5 m long and 2.5 m high. Longitudinally, the wall is bent to a radius of 2.8 m. The wall thickness is 139 mm without cladding. The material corresponds to that used in the manufacture of reactor pressure vessels. The total weight of the non-clad test specimen is approximately 9 t. The specimen was cut into three sections with a view to producing a longitudinal weld and a circumferential weld on welding the sections together again. To achieve different test flaw angles, the weld flanks have differing angles between 0° and 10°. To simulate a coolant connection, the test body has a circular cutout of 289-mm diameter at the top left. On these weld flanks a number of test pads with diameters ranging from 5 to 28 mm, corresponding to reflecting surfaces between 19 and 600 mm^2, were tacked in place. The small test pads with surface roughnesses over 0.5 mm were taken from tensile test specimens. The large test pads were taken from compact tension specimens with a surface roughness of about 0.2 mm. A total of approximately 90 such test pads were placed on the weld flanks and worked into the weld. Figure 2 shows the final test specimen on the test rig. The tests were carried out in two phases, first without cladding, then with cladding, in order to clearly establish the influence of the cladding on testability.

A test specimen, which in principle satisfies all conditions, was thus developed to examine the capabilities of ultrasonic inspection systems on the basis of the requirements set by the authorities.

The tests were carried out on a remote-controlled automatic inspection system scanning the test area. The test system consisted of the probes, the manipulating equipment, the ultrasonic electronic gear and the data logging system.

A multi-probe system was used, combining the tandem and the single-probe techniques. Figure 3 shows the layout of the system. In all, a full test cycle embraces 14 test functions. On devising the test system, the 139-mm-thick wall was sub-divided into four more or less identical test zones. One complete test cycle consists of 14 individual test functions, which are

(a) Four for monitoring the coupling of the four probes. This was done via the back echo.
(b) Two for tandem-probe inspection of the middle zones.
(c) Four for inspection of the two outer zones from the left and from the right by the single-probe technique.
(d) Three for inspecting the middle zones by the single-probe technique.
(e) One for inspection by the pitch-and-catch tandem technique.

Thus, each volume increment is subjected to at least two inspection processes. Coupling was effected by flowing water introduced during the test. The water gap between the probe and the surface of the test specimen was approximately 1.5 mm. The test was carried out with 2-MHz transversal waves. The beam angle was 45°. The probes were adjusted by means of an adjusting block especially made for this purpose. Without going into the details of the adjusting procedure, it may be said that the adjusting block was a simplified form of the reference block V 1 adapted to the specific problem.

Mechanical guidance of the probes was, in principle, handled by a manipulator of the type used in actual practice for external inspection tests on reactor pressure vessels of boiling-water reactors. Fundamentally, the

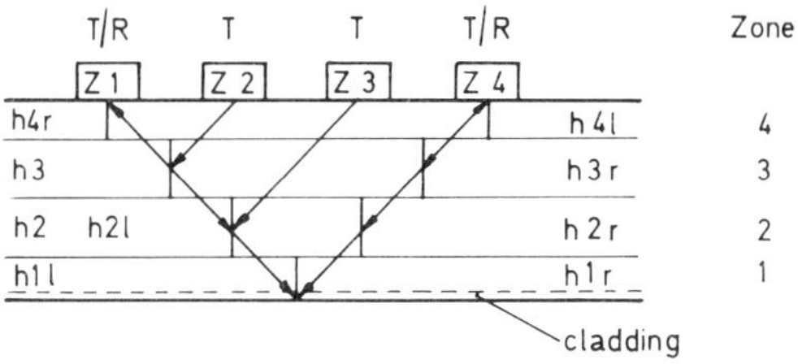

T Transmitter
R Reciever
r Single-Probe Technique from the Right
l Single-Probe Technique from the Left
2,3 Tandem-Probe Technique

FIG. 3. Schematic diagram of a multi-probe system.

manipulator consists of a vertical rail on which a carriage is driven by a rope winch. The carriage has a manipulating arm on which the probe systems are mounted. With the aid of this manipulating system it is possible to scan all points of the vessel's surfaces in the area of the rail and the manipulating arm. An electric control system enables the probes to scan the desired test areas on a meander-shaped pattern. Figure 4 shows the vertical rail and the test carriage which is designed for a gap width of 350 mm. At the rear there is the control desk for the individual manipulator motions, and next to it is the ultrasonic testing gear which I shall now deal with. The testing speed was infinitely variable between 5 and 50 mm/s. The measurements were made at a speed of about 30 mm/s. The actual test tracks are vertical, the space between one test track and the next is 10 mm. For both motions the respective position is indicated via digital position transmitters on the control desk. Simultaneously with this indication, the position data is signalled to the data logging system and stored on magnetic tape together with the ultrasonic inspection data.

The electronic gear of the ultrasonic inspection equipment was built by the firm of Krautkrämer. It is a 15-channel system. Fourteen channels were used for testing the 14 test functions of the probe system, and one channel for transmitting the data to the data logging system. The cycle frequency was 1600 Hz, which at a testing speed of 30 mm/s means that every 0.3 mm a defined test zone was scanned with a definite test function, for example by the tandem technique. The signals received by the system were pre-amplified near the probe and then transmitted into the electronic gear. The electronic gear amplified the signals again and transmitted them to gates and suppression controls for further evaluation. In the gates, the timing of the echoes was evaluated. Adjustable suppression controls enabled the echo amplitude to be evaluated. In the present case, four threshold values differing by 6 dB were set and related to indication patterns of 1, 2, 4 and 8. For checking the coupling and the beam, only one threshold value was set and any failure to reach this was recorded.

The digitalized values were stored on magnetic tape. The position data, the ultrasonic inspection data and a code input through the electronic gear were stored within a logic set. A high-speed printer printed the stored data immediately after they had been recorded on tape.

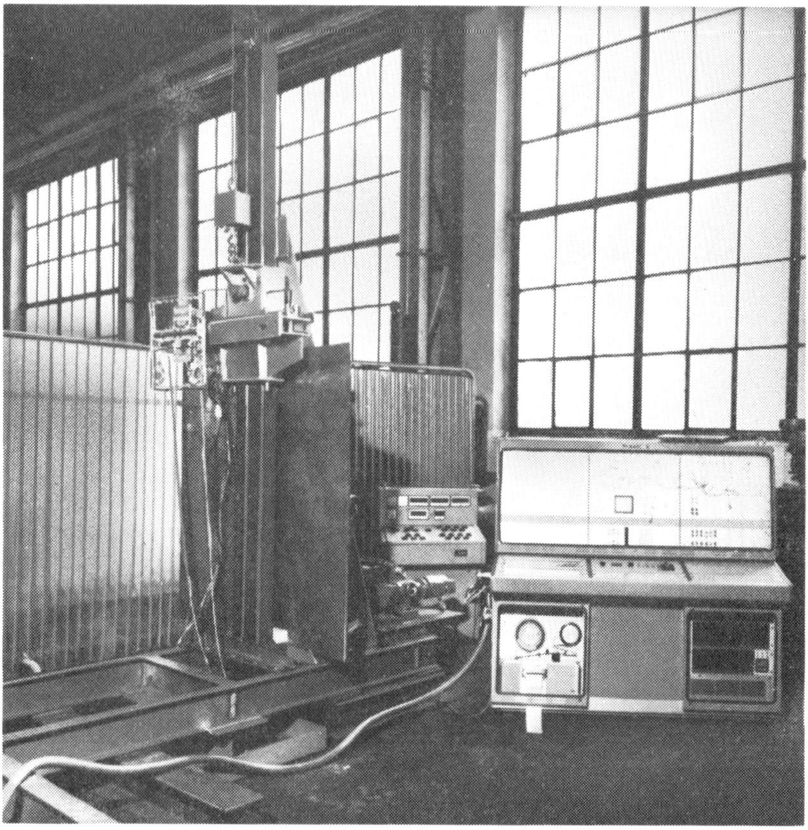

FIG. 4. Test rig without test piece.

With the aid of evaluation programmes which my co-workers and I developed for our own purposes, B- and C-scope presentations were made of the stored data, which enabled an evaluation and also a comparison to be made of two consecutive tests.

The tests were carried out before and after cladding. The two-layer strip cladding was put on by hand. The thickness of the cladding was roughly 8 to 10 mm. In the area of the weld, the surface of the cladding was ground over.

In addition to actual flaw detectability and interpretation of the indications, a thorough examination was made to establish within what tolerances the indications are reproducible.

The scope of this paper enables only the results, which were very extensive, to be shown in summarized form and by way of examples. In summarizing, it can be said that with the aid of the equipment we used in this experiment, we were successful in detecting the reflectors which were introduced.

Figure 5 shows the pattern of indications in the form of a C-scope presentation for a 10-mm large reflector located perpendicular to the surface about 50 mm deep in the material. From the figure it can be seen how the indication pattern rises to the highest point at the centre where the beam

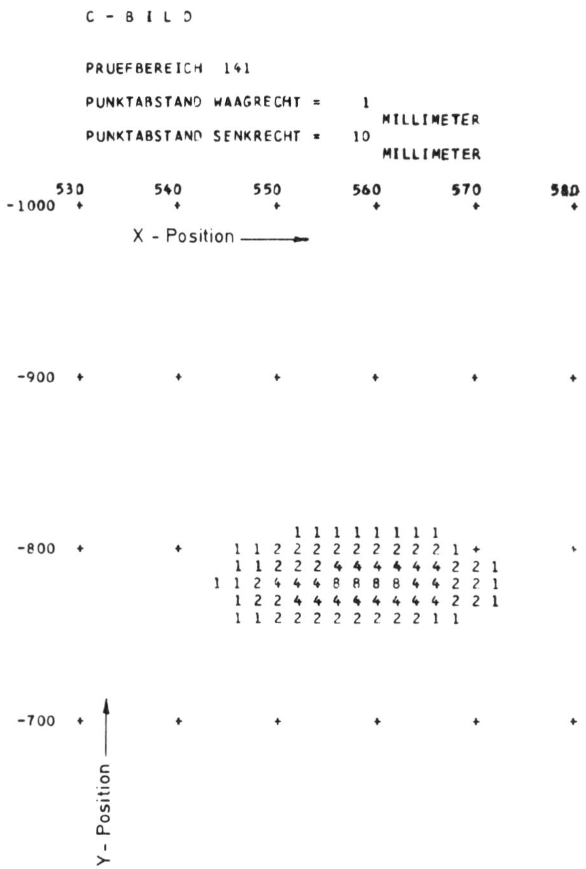

FIG. 5. Digital C-picture of a reflector as printed out by a computer.

hits the reflector best and then drops outwards according to the digital step setting. The indication pattern corresponds to that obtained in a test with the tandem probe technique.

Figure 6 shows a comparison between two reproduction runs after a revision of the starting positions. Where the equal sign is seen, the indication has not changed in its position on the grid; at a "one" the indication is one step higher than the original test, at an "A" it is one step lower. Symbols are also available for larger deviations, but we can see from the figure that, except for small deviations, the indication pattern is practically identical.

Summarizing the results, it can be said that the system used guarantees a high probability of flaw detection. The tests revealed that an inclination of the reflector of about 10° results in a reduction of the indication by about 10 dB. This is only an average figure since, as is known, the angular dependence is stronger for large reflectors than for small ones. The figure was established for a test specimen without cladding. As the investigations have not yet been concluded on the test specimen with cladding, a corresponding comparison figure is not yet available.

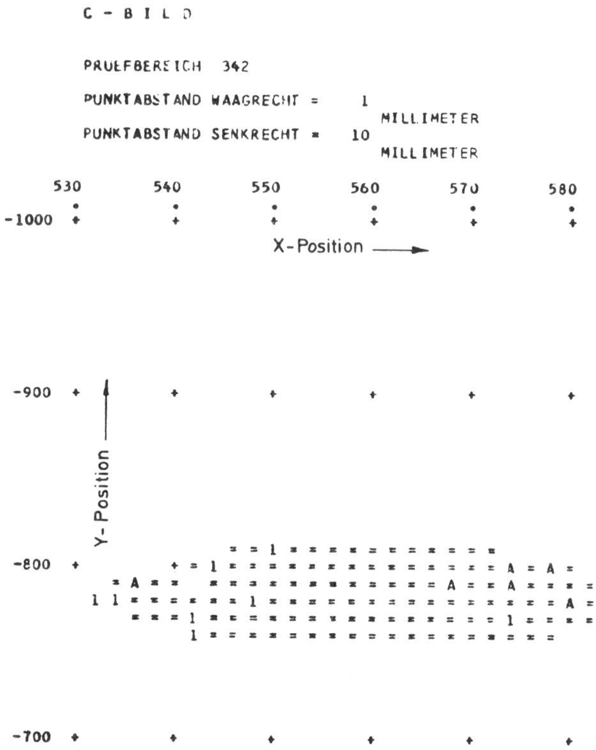

FIG. 6. Comparison of a reflector's C-picture.

The cladding results in a sensitivity reduction of about 15-20 dB which, however, can be compensated by a correspondingly higher gain setting. With 2-MHz waves, it was not found that the roughness of reflecting surfaces had any influence on the height of the indication at these roughness depths. This also coincides with the findings of the Federal German Institute for Materials Testing (Bundesanstalt für Materialprüfung, Berlin) in tests on their own specimens.

Surface corrugations resulting in a changed beam angle have generally the same influence as a deviation of the angular position of the reflector from the ideal position. Here, too, small angles can result in a pronounced change in the result.

The inspection system which was tried out on the test specimen has been used for base-line tests on reactor pressure vessels. Disregarding the coupling conditions, no major differences were anticipated for external and internal tests. Consequently, this system was adopted for internal and external tests of vessel sections with parallel walls. In other areas, for instance at the nozzles, suitably adapted systems were used.

Figure 7 shows the complete testing equipment for the bottom of the reactor pressure vessel from inside, adapted for a base-line test. The probe systems are mounted on the arm of a central mast, about 18 m long, extending into the pressure vessel.

FIG. 7. Central mast manipulator.

The bottom part of the inspection system with the arms inside the pressure vessel is shown in Fig. 8.

An extensive internal test with the corresponding inspection equipment has also been carried out on two reactors which are already in operation. An external test of the reactor pressure vessel is carried out in similar fashion, except that the manipulators are adapted to the specific conditions. Here, too, the capability of such equipment has been verified in the field.

FIG. 8. Bottom part of the central mast manipulator in the reactor pressure vessel.

In addition, special equipment has been developed for inspecting the cover and the ligaments at the bottom of boiling-water reactors. Ultrasonic testing technology has thus reached a stage where in-service inspection tests can be carried out on a reactor pressure vessel to provide extensive information with a high degree of reliability.

A GENERAL APPROACH TO COMPUTERIZED INSPECTION OF ENGINEERED SAFETY SYSTEMS

W. HAWICKHORST
Laboratorium für Reaktorregelung und Anlagensicherung Garching,
Technische Universität München,
Garching,
Federal Republic of Germany

Abstract

A GENERAL APPROACH TO COMPUTERIZED INSPECTION OF ENGINEERED SAFETY SYSTEMS.

Engineered safety systems are in a standby state during the normal plant operation. Therefore their ability to fulfil their safeguarding functions must be proven by periodic test-start-ups. This paper describes work done on the development of computerized inspection techniques. Generally, engineered safety systems consist of a number of mechanical components and the instrumentation equipment, which is required for supervision. It is assumed that all disturbances are caused by a finite number of component and/or instrumentation faults which are stochastic in nature. To inspect a system, two steps must be carried out. First, an overall performance check, based on the verification of fixed normal operating points is made. If deviations are indicated, disturbance analysis is the next step. Assuming that a system might contain a number of n possible basic faults, theoretically 2^n different disturbance patterns could occur. However, the analysis of large systems can be made easier by decomposition into sub-systems, each of which contains fewer basic faults than the original system. When the failed sub-systems are detected through the process of verification, using the normal operating characteristics for all sub-systems, the prime cause analysis can be restricted to them. The prime cause analysis is based on the substitution of the predefined basic faults. Disturbance patterns of higher complexity can be ignored, since their probability of occurrence has been considerably reduced by the decomposition. Using this concept, simulation experiments are being carried out. An idea of the achievable reliability of computerized inspection techniques can be obtained from the results discussed in this paper.

1. INTRODUCTION

Engineered safety systems are in a standby state during the normal plant operation. In contrast to the electronic part of a safety system, they have no dynamic self-checking features; and their ability to operate must be proven by periodical test-start-ups [1]. The test procedure is presently carried out by the operational staff, according to the specified plant operating instructions. This means that current inspection techniques have a limited efficiency with respect to the test run frequency and accuracy, which in turn depends upon the qualification and the experience of the employed operator. This is particularly true if system irregularities occur, which require a disturbance analysis.

The aim of present investigations is to establish possible improvements in the availability of engineered safety systems by replacing the manual procedures with process computer controls [2]. Computerized inspection techniques should meet the following requirements.

(i) Deviations of the system characteristics from the designed state, which are serious enough to affect a system's safeguarding functions should be reliably detected.

(ii) Misinterpretations of detected irregularities should not cause safety risks.

(iii) To enable useful repair instructions, the disturbance analysis should be broken down to the component level.

2. THE BLOCK STRUCTURE OF ENGINEERED SAFETY SYSTEMS; POSSIBLE DISTURBANCE PATTERNS

Generally engineered safety systems consist of a number of mechanical components such as pumps, valves, heat exchangers etc., as well as instrumentation equipment, which is needed for supervision (Fig. 1). For standby systems, like engineered safety systems, it is assumed that every disturbance is caused by basic faults which are stochastic in nature. Basic faults can be component faults and also instrumentation faults, if the instrumentation equipment is handled as an integral part of the system. The simplest disturbance pattern which might occur contains one basic fault; but more complex ones might occur which contain more. Since the number of components and instruments which build a system is finite, there exists only a finite number of different basic faults and of possible disturbance patterns. Let n be the total number of basic faults for a given system with a uniform probability p. The total number of different disturbance patterns is therefore given by 2^n, and the probability that a disturbance might occur which contains $m(m \leqq n)$ basic faults is given by the binomial distribution.

$$p_m = \binom{n}{m} \times p^m \times (1 - p)^{n-m}$$

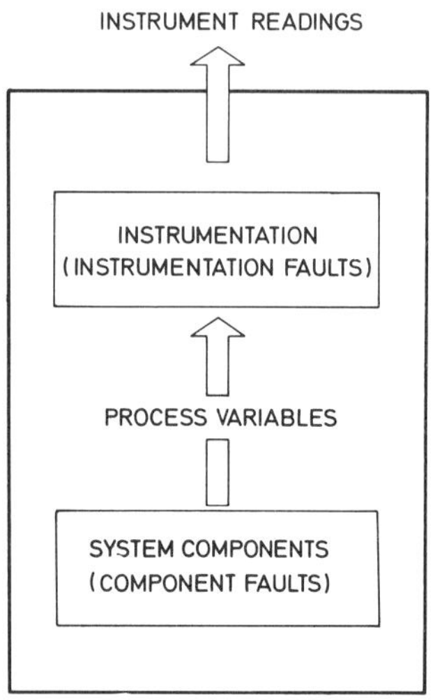

FIG.1. Block diagram of engineered safety systems: information flow and failure sources.

3. THE DECOMPOSITION CONCEPT

If n is again the total number of basic faults then, theoretically, 2^n different disturbance patterns must be taken into consideration for the analysis of an actual disturbance situation. Therefore the analysis becomes more unmanageable as system complexity increases. For practical applications, however, one can make use of the fact that with increasing complexity of disturbances, their probability of occurrence decreases considerably. If the basic faults show a uniform probability p, the probability that a disturbance might contain more than m ($m \leq n$) basic faults is given by

$$p_c = \sum_{i=m+1}^{n} \binom{n}{i} \times p^i \times (1-p)^{n-i}$$

That means, for any given m the accumulated probability of disturbances with a complexity greater than m can be reduced if the total number of basic faults to be considered simultaneously is also reduced. If n can be reduced so far that the probability of disturbance patterns with complexities greater than m becomes negligible compared with those of lower complexities, disturbance analysis becomes very simple. The analysis can, therefore, be restricted to disturbance patterns with a complexity less than or equal to m without deteriorating the performance of system status identification significantly.

An example based on failure rates and inspection intervals, typical for engineered safety system of light water reactors, is given in Fig. 2.

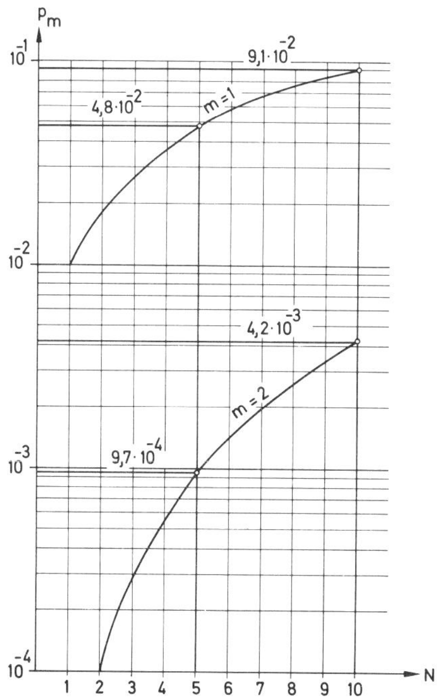

FIG.2. Probability of disturbance patterns as a function of complexity and total number of basic faults, $p = 10^{-2}$.

It is assumed that the basic faults show a uniform probability p with failure rates of $\lambda = 10 \times 10^{-6}\ h^{-1}$ and inspection intervals of $\tau = 10^3$ h. Then p is approximately given by

$$p = \lambda \times \tau = 10 \times 10^{-6}\ h^{-1} \times 10^3\ h = 10^{-2}$$

The diagram shows the probability of disturbance patterns which contain one and two basic faults as a function of the total number of possible basic faults n. Two systems are compared containing 10 and 5 possible basic faults. The absolute probability of disturbances is greater for the larger system. However, of more importance is the fact that the ratio of the probabilities p_2 and p_1 is reduced from

$$\frac{p_2}{p_1} = \frac{4.2 \times 10^{-3}}{9.1 \times 10^{-2}} = 4.6 \times 10^{-2}; \quad n = 10$$

for the larger system to

$$\frac{p_2}{p_1} = \frac{9.7 \times 10^{-4}}{4.8 \times 10^{-2}} = 2.0 \times 10^{-2}; \quad n = 5$$

for the small one. That confirms the common knowledge that in smaller systems a higher ratio of disturbances is of less complexity than in larger systems.

For the analysis of a large system, n can be reduced by decomposition into sub-systems, each of which contains fewer basic faults than the original system.

Definition: A sub-system contains a part of a given system, which can be described completely and definitely by use of the corresponding sub-set of the measured process variables. Each process variable can reside in more than one sub-system.

Figure 3 shows a simplified scheme of a BWR core flooding loop, operating in the recirculation mode. A possible decomposition of this system leads to the sub-systems 'flooding pump' and 'pipe network', which are connected by two process variables: the pumping pressure and the pump flow. However, the decomposition of a given system generally requires installation of additional instrumentation equipment.

The influence of decomposition on the total number of basic faults to be considered simultaneously is shown in Fig. 4. There are two sub-systems given, SS_i and SS_k, each of which contains sets of component faults CF and instrumentation faults IF. Actual disturbances are assumed to be located inside the shaded areas. If only one sub-system has failed, for example SS_i, the disturbance analysis can be restricted to that sub-system only, without considering failures of the instrumentation which corresponds to both sub-systems. For this case the total number of basic faults to be considered is given by the set notation

$$n = |CF_i \cup IF_i|$$

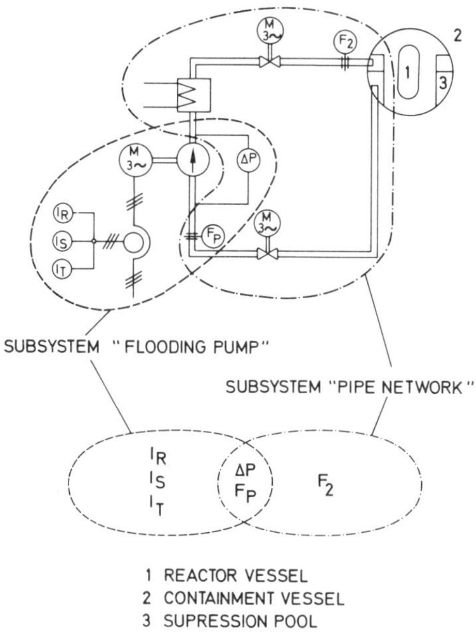

FIG.3. Decomposition of a system into sub-systems.

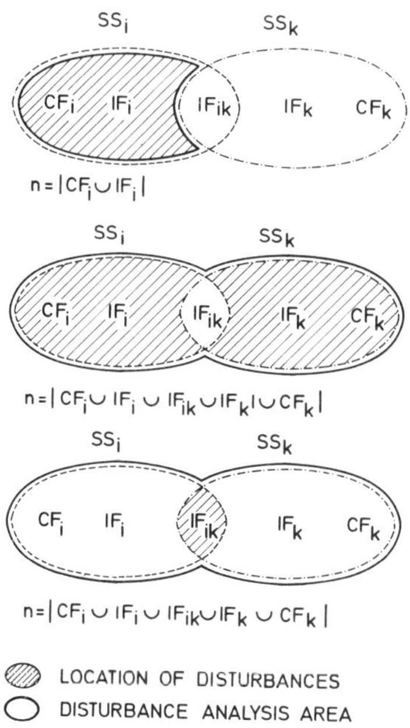

FIG.4. Number of basic faults, to be considered for various locations of disturbances.

If both sub-systems show irregularities or if the common instrumentation has failed, disturbance analysis has to be carried out for both sub-systems and n is given by the set notation

$$n = \left| CF_i \cup IF_i \cup IF_{i,k} \cup IF_k \cup CF_k \right|$$

That means, for these types of disturbances the decomposition concept seems to be of no advantage. But this problem can be solved by installation of duplicate instruments for the measurement of the common process variables.

Definition: Two or more instruments, which are used for measuring the same process variable, are called a set of duplicate or multiple instruments if and only if their readings are identical, independent of system disturbances, except in case of failures of the instruments themselves. An example will make this definition clearer; Through a pipe without branches the flow remains constant. Therefore two flowmeters which are installed at different positions will normally show identical readings. These instruments are called duplicate instruments, if no leakage of the pipe can occur between the two measuring positions, which would lead to different readings.

Multiple instruments build up special types of sub-systems, called DUMMY sub-systems.

Definition: DUMMY sub-systems contain only duplicate or multiple instruments.

From the definition given, it is clear that DUMMIES can contain only instrument faults but not component faults. The influence of DUMMIES on the total number of basic faults to be considered for disturbance analysis purposes is shown in Fig. 5. If again SS_i and SS_k show irregularities, the total number of basic faults to be considered simultaneously for analysing SS_i and SS_k is given by

$$n_i = \left| CF_i \cup IF_i \right|$$

and

$$n_k = \left| CF_k \cup IF_k \right|$$

That means, even if both sub-systems SS_i and SS_k show irregularities, n_i and n_k are not greater than they would be if only one of the sub-systems fails. Even for a failure of one of the duplicate instruments, for example an instrument which corresponds to SS_i and the DUMMY-sub-system D_{ik}, n is still less than it would be without the DUMMY sub-system:

$$n = \left| CF_i \cup IF_i \cup IF_{i,k} \right|$$

Although the installation of additional instrumentation equipment yields additional failure sources, decomposition has been found to be a useful

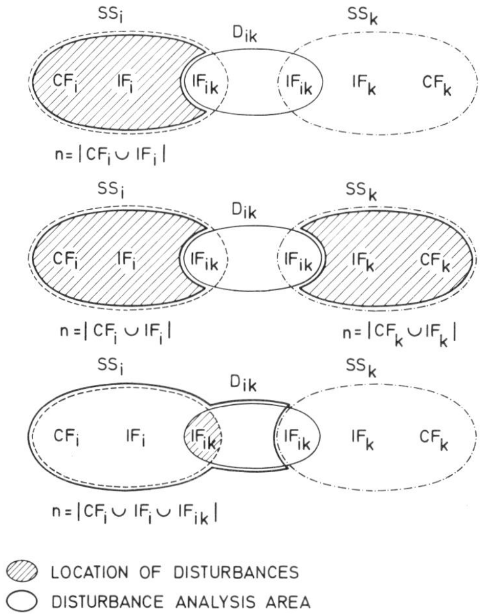

FIG.5. The influence of DUMMIES.

method for simplifying the disturbance analysis of large systems by reducing the number of basic faults to be considered simultaneously. The decomposition concept is also advantageous for the implementation of computerized inspection techniques with process computers. Since process computers have limited capacity in core memory, the inspection routine should be segmented into sub-routines which correspond to the sub-systems created by decomposition.

4. SYSTEM STATUS IDENTIFICATION BY COMPUTERS

Figure 6 shows the general structure of an inspection routine. During a test start-up the system runs under predefined operating conditions. The different test operating points should contain all possible operating modes of the system; and, for convenience, the test data should be measured during steady state operation. When the test is completed, system status identification starts.

4.1. Overall performance check

The first step for system status identification is an overall performance check, which can be executed very simply. The normal system operating characteristics can be verified by comparison of the actual system data, measured during the test start-up, with stored nominal values, for all different operating points. If irregularities are indicated, disturbance

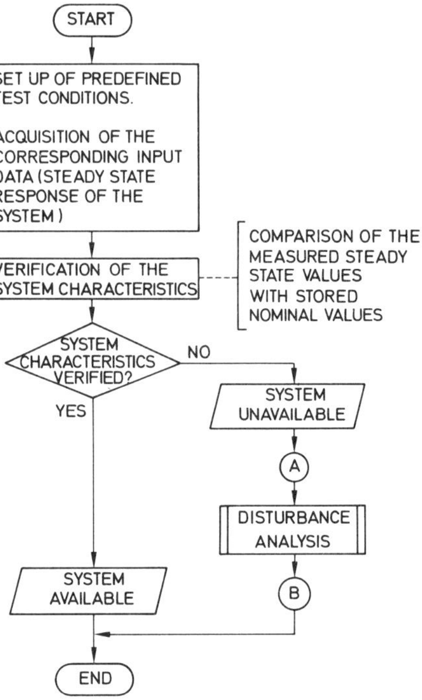

FIG.6. General structure of an inspection routine.

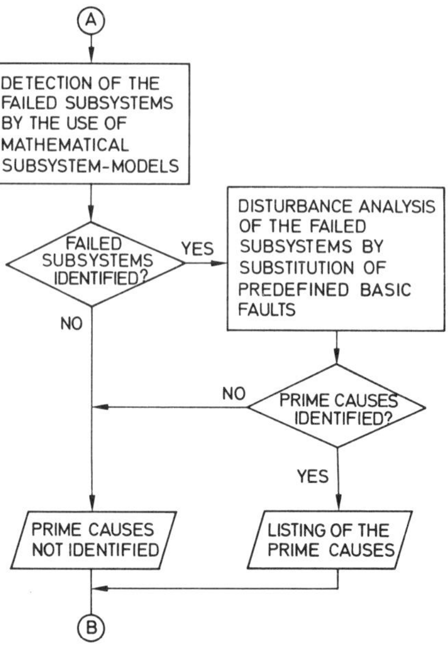

FIG.7. Disturbance analysis routine.

analysis follows. It has been found that component faults as well as instrumentation faults can be detected with very high reliability by using the described overall performance check. Although there is theoretically the possibility that failures could occur which compensate other failures, creating an illusion of normal overall performance, it is extremely unlikely.

To prevent safety risks caused by any misinterpretations of detected irregularities, the system is stated to be unavailable whenever the overall performance check indicates irregularities. The type of disturbance that has occurred is not considered.

4.2. Disturbance analysis

Disturbance analysis contains the detection of failed sub-systems and the prime cause analysis (Fig. 7).

4.2.1. Sub-systems check

A sub-systems check is necessary if the prime cause analysis is to be restricted to the failed sub-systems. Failed sub-systems can be detected by the verification of the system's test data with mathematical sub-system models which represent the normal operating characteristics. Verification of the test data by fixed nominal values, similar to the overall performance check, is not useful. Because of the interactions of the sub-systems, failures in one sub-system can lead to abnormal operation of others. As seen in Fig. 8 a failure of the sub-system 'pump network' changes the operating point of both sub-systems; therefore both would be found to have failed if a sub-systems check based on a fixed normal operating point was made. However, a check based on the normal operating characteristics of both sub-systems shows that the pump is still in normal operation.

Independently from actual disturbance patterns, the sub-systems check must be made for all sub-systems.

4.2.2. Prime cause analysis

For the failed sub-systems a prime cause analysis is undertaken to find the faulty components, and the faulty instruments. It is assumed that a given system can be separated into sub-systems which are small enough that, in making an analysis, all disturbance patterns containing more than one basic fault can be ignored, without deteriorating significantly the analysis.

The component fault analysis can be based on the verification of normal characteristics or on the substitution of predefined basic faults. As seen in Fig. 9, these two methods are in principle different. In Fig. 9 it is assumed that a failure of a pump leads to a pumping pressure drop. If pump failures should be detected by verification of the normal pump characteristic, all operating points which differ from the normal characteristic by more than an acceptable error fulfil the failure criterion. If the pump failure analysis is based on the pumping pressure drop, the failure criterion is fulfilled only by operating points below the normal pump characteristic. Since faulty instrument readings can range the full scale, the reliability

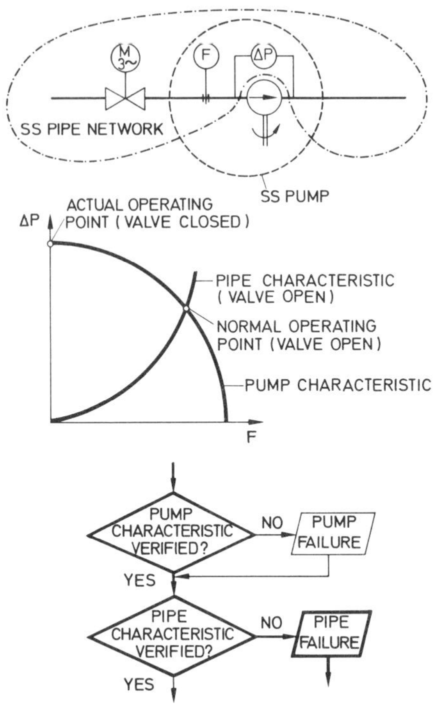

FIG.8. Detection of failed sub-systems; actual disturbance: valve fails to open.

of a component fault analysis, based on the verification of normal characteristics, is therefore much more deteriorated by instrument faults than the reliability of an analysis based on the substitution of predefined basic faults. However, the latter method requires knowledge of the relative consequences of all possible component faults. This problem must be solved by a detailed system analysis.

The instrument fault analysis is somewhat simpler. It is assumed that the behaviour of faulty instruments is unknown; therefore no use is made of the readings of instruments assumed to have failed. Their readings are substituted by readings of redundant instruments or by values computed by the sub-system models which represent the normal operating characteristics; these are still valid because no further faults are to be considered.

A prime cause analysis routine for one sub-system is shown in Fig. 10. The analysis is carried out separately for component faults and instrumentation faults. The results are compared, and an interpretation is given. If only one component or one instrumentation fault has been identified, the result is clear and unambiguous. If a component fault and an instrumentation fault have been identified, the alternatives are printed out. Separate analysis for component faults and instrumentation faults have been found to be a useful method of presenting misinterpretations.

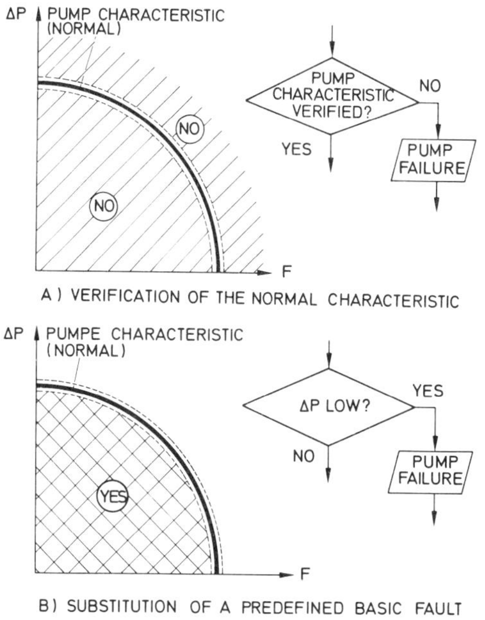

FIG.9. Prime cause analysis methods.

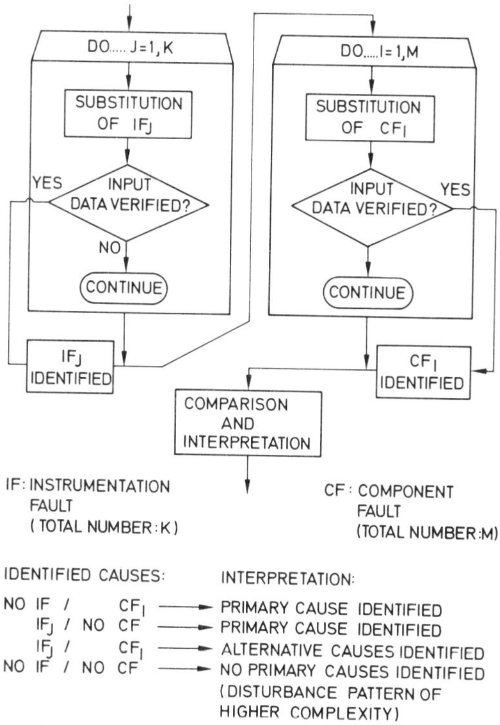

FIG.10. Prime causes analysis routine for one sub-system.

5. SIMULATION TESTS

With the concept discussed in this paper, experimental investigations are being carried out, using a hybrid simulation as described in the block diagram of Fig. 11. The system which has to be supervised is represented by a simplified analog model of one strand of a boiling water reactor core flooding loop [3]. In the analog model, stochastic failures of the loop components, as well as instrumentation faults, can be generated by a digital fault generation routine. The subsequently started checkout routine, which has to be tested, should recognize the initiated disturbances. An estimation of the routine performance is derived through comparison of the generated faulty status with the identified loop status over a large number of tests.

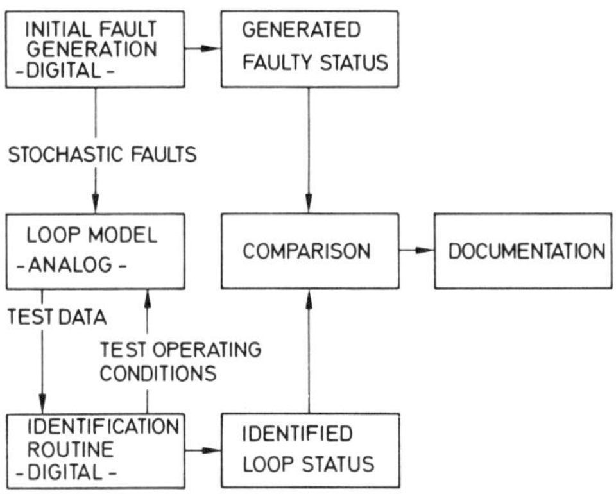

FIG.11. Block diagram of the hybrid simulation used for testing of the inspection routine.

Figure 3 illustrates the simulated loop configuration together with the sub-system boundaries. Shown are the containment, housing the reactor vessel and the suppression pool. The flooding system is operating in the recirculation mode. This mode of operation can be used for testing of the loop during the normal plant operation. Measured variables are the pumping pressure, pump flow and the current of the drive motor. The inaccuracy of measurement is simulated by white noise. The actual states of the valves are signalized by position indicators. Possible component faults are damage of the drive motor or flooding pump, failures of the valves in the closed or opened state and internal leakages of the heat exchanger. Instrumentation faults can falsify the instrument readings. The generated component faults (total number 6) as well as the instrumentation faults (total number 6) show a uniform distribution of $p = 10^{-2}$ per run.

The first simulation experiments were undertaken to estimate the modelling error. In practical application, the modelling error covers the inevitable deviations of the operating characteristics of the system model, which is used for disturbance analysis purposes, from the real

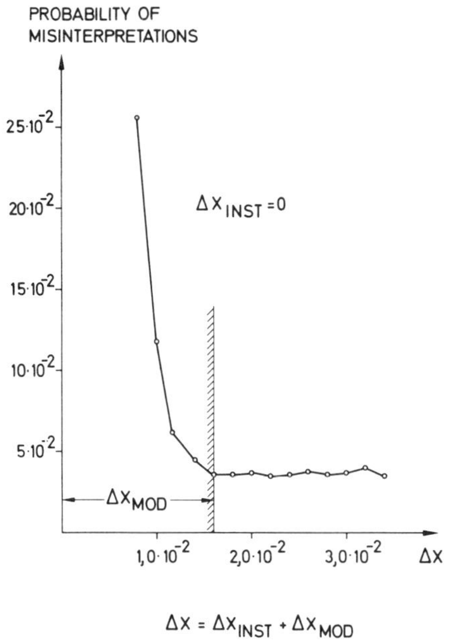

FIG.12. Estimation of the modelling error, based on 3000 disturbance analyses.

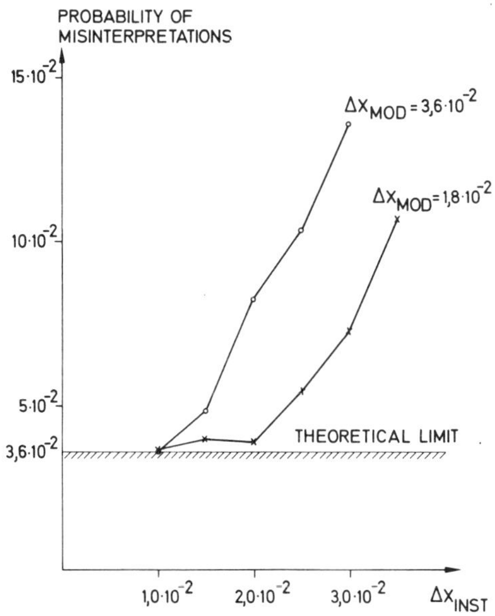

FIG.13. Analysis reliability as a function of the modelling error considered and the inaccuracy of measurement, based on 3000 disturbance analyses.

system. The modelling error is handled as a contribution to the full scale inaccuracy of measurement:

$$\Delta x = \Delta x_{inst.} + \Delta x_{mod.}$$

In the hybrid simulation, the modelling error covers deviations between the analog and the digital loop models. For the simulation experiment the modelling error can be easily estimated by setting the inaccuracy of the instruments to zero; then computing the probability of misinterpretations as a function of Δx. Figure 12 shows an estimation. The modelling error is given the value of Δx which corresponds to the end of the significant probability drop. To prevent misinterpretations, the modelling error, considered for the disturbance analysis, should be somewhat greater.

An idea of the achievable reliability of the disturbance analysis can be obtained from Fig. 13. The probability of misinterpretations of detected disturbances depends on the inaccuracy of the installed instruments and on the modelling error considered, and is less than 4×10^{-2} per analysis job for measuring errors of 10^{-2}. It should be noted that the restriction of the analysis to disturbance patterns which contain only one basic fault limits the probability of misinterpretations to 3.6×10^{-2}.

The probability of undetected disturbances (failures of the overall performance check) was found to be less than 10^{-3} per analysed disturbance situation.

REFERENCES

[1] KRAFTWERK UNION AG, 770 MWe-Kernkraftwerk mit Siedewasserreaktor für Standort Brunsbüttel, Sicherheitsbericht Band I und II, Nov. 1969.

[2] HAWICKHORST, W., Voruntersuchungen zur Funktionsprüfung von Sicherheitseinrichtungen mit Prozessrechnern, MRR 126, Laboratorium für Reaktorregelung und Anlagensicherung der Technischen Universität München, Munich, Jan.1973.

[3] HAWICKHORST, W., Simulationsuntersuchungen zur Funktionsprüfung technischer Sicherheitseinrichtungen mit Prozessrechnern, Teil 1: Analoges Modell zur Simulation von Störungen am Nachkühlsystem eines Siedewasserreaktors, MRR 133, Laboratorium für Reaktorregelung und Anlagensicherung der Technischen Universität München, Munich, Dec.1973.

DISCUSSION

B. GACHOT: Is your computerized method of carrying out periodic inspections on engineered safety systems being used at present in any German nuclear plant? If it is, what additional equipment beyond what would normally be available does this method require?

W. HAWICKHORST: The answer to the first part of your question is no. We first concentrated on developing the theoretical concept; then we started to run simulation experiments with the simplified loop, which is to be completed in July 1975. We will decide then whether we should continue with a large-scale simulation or go over to practical application.

Although we have no practical experience yet, I will try to answer the second part of your question. Since the possibility of testing engineered safety systems is an absolute requirement, independent of the application of computerized inspection techniques, no additional equipment is required for that; and we have found that we can base our method largely on the instruments that are normally installed. Since process computers are installed in nearly all plants now entering into operation, the additional investment consists mostly of the interface needed to connect the system to the computer.

B. CAFFREY: I would like to pursue Mr. Gachot's question further by asking Mr. Hawickhorst whether his group has already decided where to install the system initially for testing purposes, and whether they have any idea of the additional cost of computer hardware and instrumentation which its installation in a commercial nuclear plant would require at 1975 prices.

W. HAWICKHORST: No, there has not been any decision yet as to where the system should first be installed. The second part of your question highlights what may be a serious problem. As I said, the additional costs will be mainly due to the interface, but I cannot give you exact — or even approximate — figures at the present time.

G. BREILING: Figure 8 suggests separate verification of pump and pipe characteristics. I would like to know what special devices are needed in the emergency core cooling system to measure those characteristics — throttles, pump motor speed controls and the like.

W. HAWICKHORST: The design of engineered safeguards systems requires that the normal operating characteristics of all components should be exactly known. Pump characteristics, for example, could be obtained from the manufacturer or measured during pre-operational tests on the plants, after which the information is stored in the computer. Verification of speed characteristics require only that measurements should be made at the actual operating points of the components or sub-systems during a test start-up. No additional devices are required.

IN-SERVICE INSPECTION OF NUCLEAR POWER PLANT COMPONENTS
An opportunity to perfect ultrasonic non-destructive test methods

H. WÜSTENBERG, J. KUTZNER,
E. MUNDRY, E. NABEL, E. NEUMANN
Federal Institute for Materials Testing,
Berlin,
Federal Republic of Germany

Abstract

IN-SERVICE INSPECTION OF NUCLEAR POWER PLANT COMPONENTS: AN OPPORTUNITY TO PERFECT ULTRASONIC NON-DESTRUCTIVE TEST METHODS.

The need for specialized ultrasonic non-destructive techniques for in-service inspection has been a strong incentive to development in this field. Not only have new or improved ultrasonic methods and procedures been established, but there have also been changes in some philosophical aspects of the planning and execution of ultrasonic testing inspections and the evaluation of the results. This refers to the basic information sources in use for the evaluation, the sensitivity setting and the detection probability. These aspects are discussed in the paper. A survey of possible ultrasonic testing-methods is given for near and far surface defects and for coarse grain structures. The problems of coupling fluctuations, of the sensitivity setting in connection with the detection probability and the catalogue of possible defects and the evaluation of the results are described.

1. INTRODUCTION

The safety of light water cooled nuclear power plants — today mostly equipped with pressurized or boiling water reactors — is strongly dependent on the integrity of all the pressurized components of the system, particularly on that of the pressure vessel. Much importance is therefore attached to the quality assurance of those components by the application of non-destructive testing methods during fabrication, acceptance and in-service inspection.

As in fabrication control, but more especially for in-service inspection, the ultrasonic method is the most important non-destructive testing method in use, the following contribution has been restricted to this method.

2. SPECIAL ASPECTS OF IN-SERVICE INSPECTION

Whereas in fabrication and acceptance control predominantly manual ultrasonic techniques are used, ultrasonic in-service inspections must be carried out automatically and by remote control in view of the high radiation level of many components in the primary circuit. We have learned from experience in recent years that such requirements have initiated considerable new developments and solutions, sometimes fundamentally different from those of the manual inspection techniques used so far. One of the reasons is that the information provided by an automatic inspection is correlated to the material defects to be detected in quite another manner than is the case in manual inspection. The most important information from manual inspection is the position of a reflector — determined generally from the flight time of the echo — and the echo amplitude. By manually optimizing the probe position, the echo amplitude can be brought to its maximum. Thus the relation between echo amplitude and significance of the corresponding reflector is much better than in automatic inspections, where such an optimization is in general impossible. It is well known that the maximum echo amplitude is used for the evaluation of the results of manual inspections. To do the same for automatic inspection may be a very doubtful procedure. Instead of the echo amplitude, therefore, the dependence of the echo indication on the probe position has to be recorded continuously. For manual inspections this can only be done discontinuously.

Both manual and automatic ultrasonic inspections are very useful, each particularly in its own specific field of applicability, but it should not be assumed that the results of automatic and manual tests can be compared without contradiction. Therefore one of the basic prerequisites of a successful in-service inspection of a reactor pressure vessel is a first basic automatic inspection of the virgin vessel. The results of this basic inspection are the reference values for all in-service inspections to follow, the reliability of which is based to a large extent upon comparison with the results of previous inspections.

Since the test conditions are aggravated considerably by geometry, cladding and other material-dependent complications, the problem of the reliability of the ultrasonic methods applied is a challenge to the specialist in non-destructive testing. It is already clear that this challenge has been a strong incentive to the further development of ultrasonic testing techniques in general, and not only within the field of in-service inspection.

We have also learned to classify some general problems of non-destructive inspections more clearly. For example, in the field of ultrasonic testing in particular the techniques for finding defects must be clearly distinguished from those for analysing the ultrasonic indications detected. The classical pulse-echo techniques for weld inspection, which are also very important for in-service inspection, are examples of procedures solely for finding defects. High reliability here means high probability of finding all defects of all types, positions, orientations, etc., that may exist in the test piece. The question of probability of finding defects can only be answered if the catalogue of defects to be expected — their size, position, orientation etc. — can be defined by the manufacturers and users of the components. The methods for finding the defects must be adapted to this catalogue of defects. Unfortunately, in general ultrasonic methods with a high potential for finding defects are rather poorly qualified for analysing the indications and vice versa.

In the following we describe some examples of investigations and developments that have been induced by in-service inspection techniques.

3. DEVELOPMENTS FOR IN-SERVICE INSPECTION

3.1. Techniques for internal defects far from the surface

The direction of incidence of the ultrasonic beams used for the detection of internal defects must be adapted to the catalogue of possible defects to be detected. The techniques generally used for this purpose are presented in

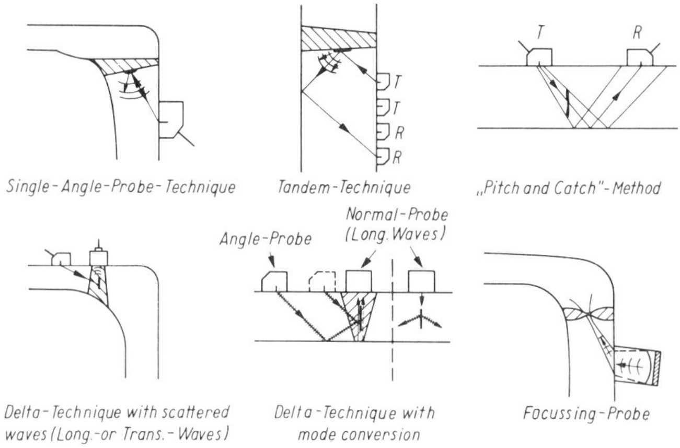

FIG. 1. Methods of ultrasonic testing for internal defects.

Fig. 1. In the single-probe technique the adaptation of the beam direction to defects perpendicular to the surface is poor. This disadvantage can be avoided with the tandem technique. In this case, however, considerable influence from the cladding and the surfaces must be taken into account.

The automatic inspection of thick-walled components by means of the tandem technique was first proposed by de Sterke [1]. The use of focussing probes is described in Ref. [2]. In this case the adaptation to defects of oblique orientation relative to the beam direction is improved by the fact that the generally used low frequency and the excitation of only a small, restricted area of the reflector cause a reflection with poor directivity, thus resulting in an improved probability of finding the defect.

In some cases of special geometry on thick-walled welds with a restricted accessibility the 'Delta technique' may be used.

3.2. Techniques for defects near the surface

The classical way to detect defects near the surface is to use the corner effect, i.e. to use the corner formed by defect and surface as a reflector. Because of the considerable ultrasonic noise from the grain structure of the cladding, the detectability of defects in this area by means of the corner effect is restricted. For both clad and unclad surfaces a transmitter-receiver technique (TR technique) with longitudinal waves of oblique incidence, shown in Fig. 2 on the right-hand side, is a very helpful method. By application of longitudinal waves the disturbing influence of surface waves (Rayleigh waves) can be suppressed and so, even for defects starting directly from the surface a sufficient detectability can be guaranteed where the probe has been suitably constructed adequately for the problem [3, 4]. For this purpose focussing probes can also be used (Fig. 2, bottom).

Under certain circumstances a scattering technique with longitudinal waves of oblique incidence (Fig. 2, bottom left) can be used successfully for the detection of cracks near the surface and for the estimation of the depth of such cracks. Advantage is taken of the fact that the waves scattered by the grain of the material are shaded off more or less by the cracks [5, 6].

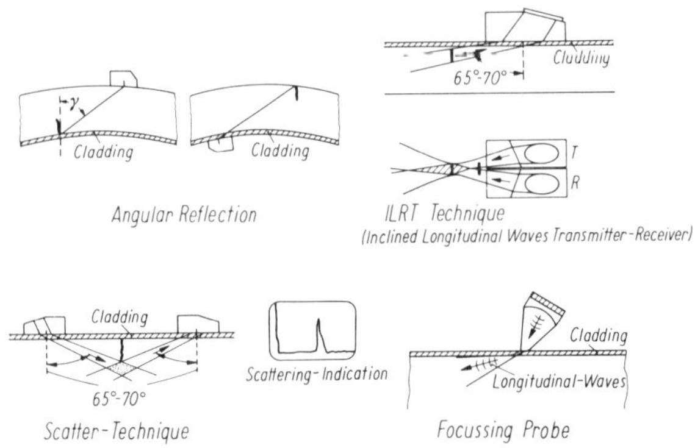

FIG. 2. Methods of ultrasonic testing for defects near the surface.

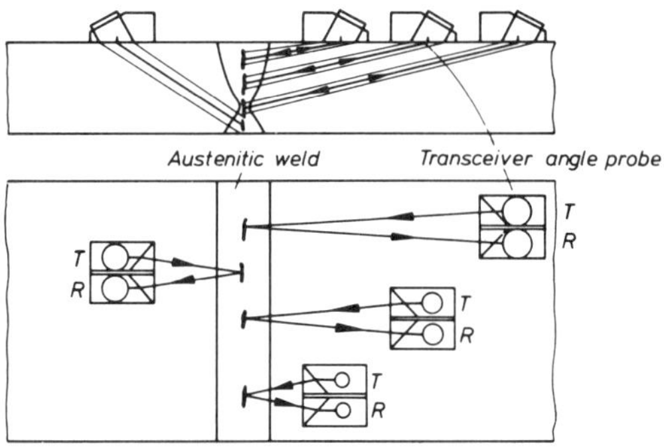

FIG. 3. Ultrasonic testing of austenitic steel weld joints with transceiver angle probes.

3.3. Technique for coarse-grained materials with high ultrasonic grain noise

For the inspection of austenitic welds more than 20 mm thick and with flat surfaces – producing no spurious echoes – a modification of the TR technique with longitudinal waves of oblique incidence, described above and shown in Fig. 2, can be applied.

Figure 3 presents special probes for this technique. The intersection of the transmitted and the received beam – caused by a suitable inclination of the membranes – defines an area of rather high defect detectability, since the signal to noise ratio is increased by the fact that indications can only be received from this limited area. This effect can be further improved if the area of beam intersection is positioned at the end of the near field, where the cross-section of a sound field of a piston source is concentrated to a minimum by interference effects. The signal to noise ratio can be further improved by the use of wide-banded 'shock wave' pulses [7].

3.4. Coupling conditions

In principle, direct coupling and immersion technique can both be used. For application of the immersion technique the probes must be guided parallel

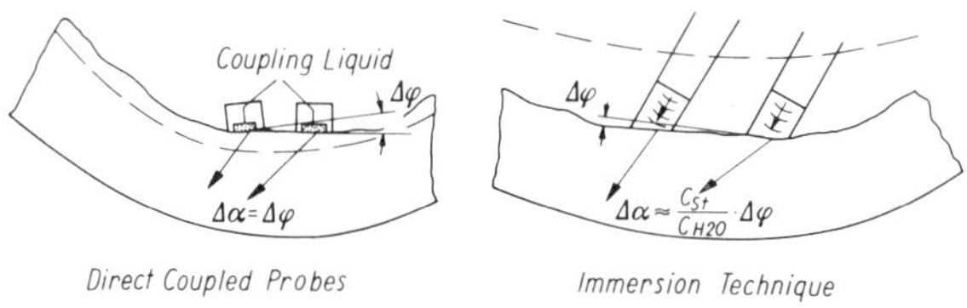

FIG. 4. Probe coupling methods.

to the surface by a special mechanical arrangement without touching the surface. In the case of direct coupling the probes are directed by the surface itself. The thickness of the generally very thin liquid coupling layer, and therefore its acoustic transmission factor, varies from probe position to probe position.

Figure 4 demonstrates that fluctuations of the actual angle of incidence due to unevenness of the surface are larger for the immersion technique than for the direct coupling technique by about a factor 2 for transverse waves and 4 for longitudinal waves. The main disadvantage of direct coupling is a considerable fluctuation of the acoustic coupling due to the variations of the layer thickness mentioned above. To overcome this difficulty the thickness of the coupling layer must be higher than a minimum value, beyond which such fluctuations of the acoustic coupling and therefore of the test sensitivity disappear [8, 9].

3.5. Choice of recording level

According to the adaptation of the selected inspection technique either to the catalogue of defects or to certain types of defects in this catalogue, the recording level must be set in such a way that the probability of finding defects of these types is as high as possible, or at least sufficient. For the sensitivity calibration and the determination of the recording level the DGS (AVG) diagram is a very useful tool, which should be understood here – as we recommend generally – as a 'reflectivity diagram', i.e. a diagram of only the acoustic reflectivity and not of the corresponding 'defect size'. Figure 5 shows such a diagram. In this diagram the noise level due to the grain structure of the material – relative to the back wall echo – is plotted below. Spontaneous coupling fluctuations or fluctuations due to geometric irregularities, which cannot be compensated by automatic readjustment of the sensitivity, must be considered by plotting the range of spread of the sensitivity. To guarantee the required recording level for all probe positions, the actual recording level must be adjusted to the lower limit of this range of spread. This actual recording level is the most important measure for the probability of flaw detection. With the help of this level and of the catalogue of defects mentioned above it is possible to evaluate what defect sizes and orientations are detectable by the method selected [10]. On the other hand, the requirement of certain recording levels must guarantee the probability of finding all defects defined by the catalogue of possible defects.

The range of sensitivity spread, plotted in Fig. 5, entails that for some probe positions there is only a very small distance between signal and noise.

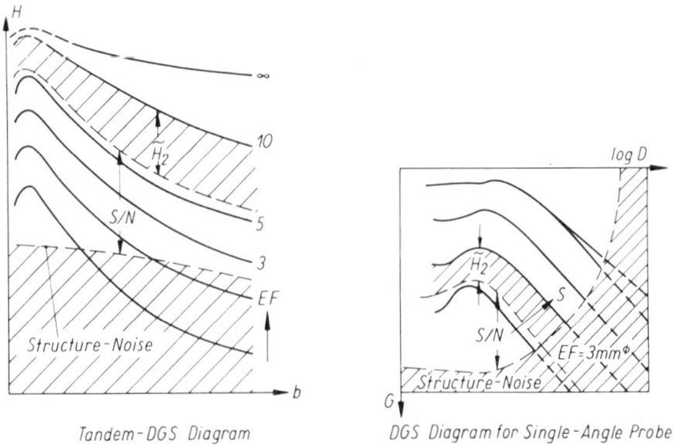

FIG. 5. Required and possible sensitivities in DGS diagrams.

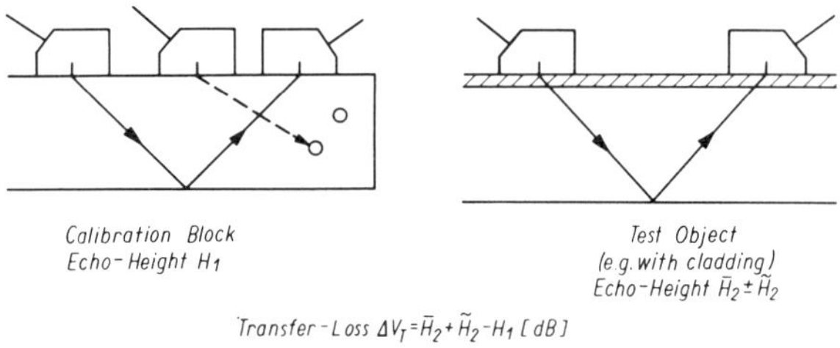

FIG. 6. Transfer loss in ultrasonic testing.

In such cases it would seem to be better to define the recording level relative to the noise level – e.g. 6 dB above it – rather than to use a recording level absolutely defined by the reflectivity of a particular circular disc reflector. However, one principal remark concerning this point must be made: for pure defect detection techniques the recording level should be based upon the well-defined reflectivity of a given test reflector and not upon the noise level, because otherwise the actual capability of the test method cannot be specified. Furthermore, it must be taken into account that all defect detection methods operate with more or less divergent ultrasonic beams, thereby causing considerable fluctuations of the noise level, depending on the formation and geometry of the grain structure.

The use of focussing sound beams completely changes the situation, because grain noise can arise only out of a small volume. In this case it may be reasonable to define the recording level by means of the grain noise.

The fluctuations in sensitivity, taken into account in Fig. 5, must be measured by a V-transmission technique both on a reference block and on the test object (Fig. 6). Therefore it is reasonable to use a reference block without cladding so that no fluctuations of the V-transfer can occur on it. The V-transfer at the object must be measured for a sufficient number of different, representative positions in order to get statistical values [11, 12]. The fluctuations in sensitivity are mainly caused by the cladding [13].

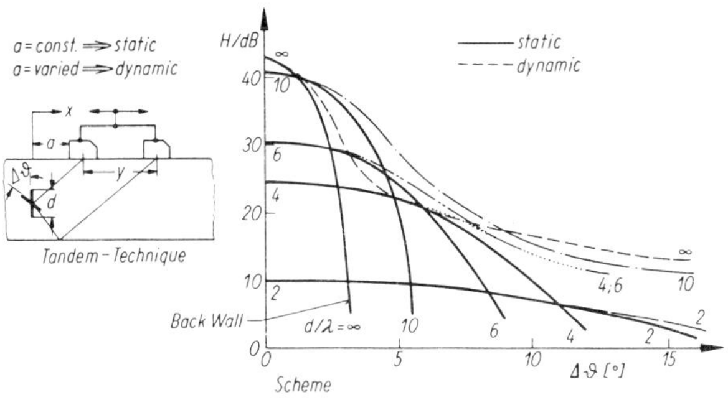

FIG. 7. Dependence of the echo height on the orientation of a reflector.

3.6. Influence of defect orientation

The echo amplitude depends on the angle of inclination between reflector and ultrasonic beam direction. Therefore large reflectors may produce smaller echo indications than small ones do. This is shown in Fig. 7 for a tandem technique. Two cases must be distinguished:

(1) The relative position between probe system and reflector remains constant; only the reflector orientation is varied. The curves for this 'static case' are plotted in Fig. 7. In this case a strong dependence on the angle of inclination is observed.

(2) If the probe system is moved relative to the reflector, the situation becomes more favourable. For this 'dynamic case', however, the results are dependent on details of the probe arrangement and it is impossible to plot a generally valid curve. Only a typical example of this case is shown in Fig. 7. For the determination of the recording level of an inspection technique this behaviour of inclined reflectors relative to their optimal orientation must be considered, possibly by model calculations. But this requires a precise knowledge of possible defect orientations. These considerations are the reason for the application of the tandem technique in the Federal Republic of Germany. This technique is especially sensitive for defects with an orientation perpendicular to the surface and has therefore a very good probability of finding such defects [14].

3.7. Definition of inspection zones for the tandem technique

If the tandem technique is used, the area of potential flaw detection is given by the range of intersection of the two ultrasonic beams. The position in depth of this section depends on the distance between the probes. For automatic inspection several commutable probes must be used. The distances between these probes must be selected in such a way that the extensions in depth of the resulting inspection zones overlap each other. At the periphery of such a zone the drop in sensitivity should not exceed, e.g. 6 dB. In Fig. 8 a possible quantitative subdivision of the test object into such zones is indicated. Once more a distinction must be made between a static and a dynamic

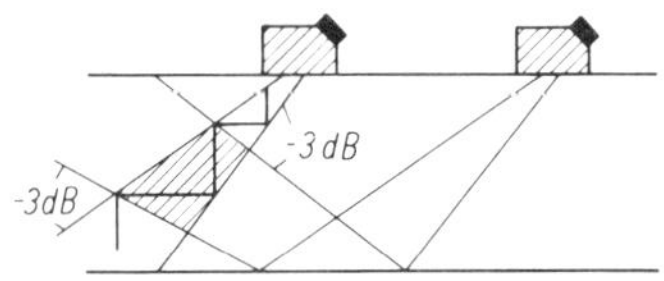

FIG. 8. Zone repartition in the tandem technique.

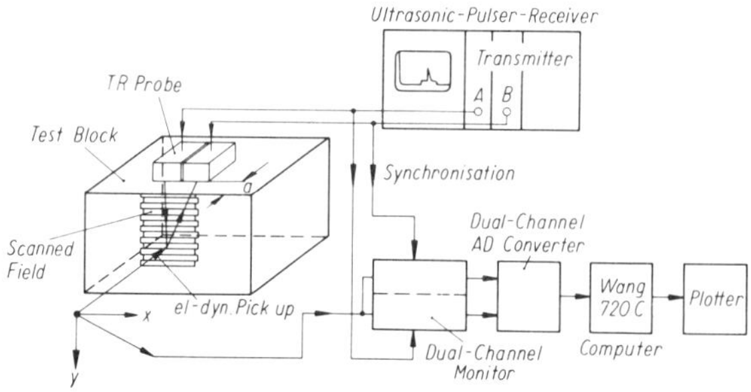

FIG. 9. System of measuring the sensitivity distribution of a probe arrangement.

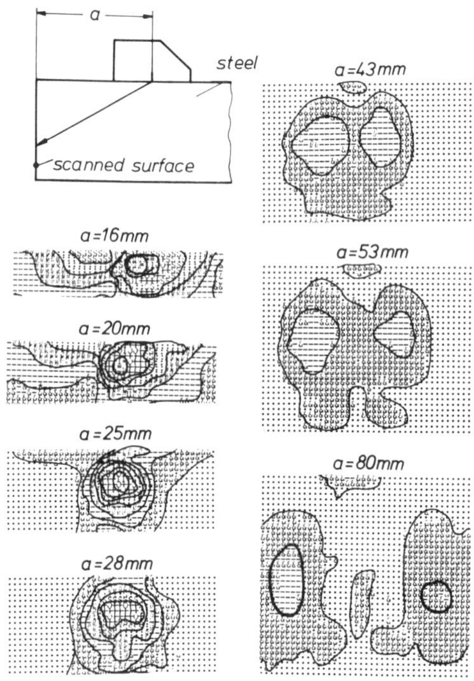

FIG. 10. Sensitivity distribution of a TR probe, measured by an electrodynamic microphone.

subdivision. In the static case only the position in depth of the reflector is varied, the projected distance — parallel to the surface — relative to the probe system remaining constant. In the dynamic case the position in depth of the flaw and the projected distance are both varied. The dynamic subdivision results in zones of somewhat larger extension in depth than does the static one [14].

3.8. Influence of sound fields on in-service inspection

The capacity of an ultrasonic testing technique strongly depends on the sound field of the transducers in the material tested. Therefore the Federal Institute for Materials Testing has developed a device for measuring the sound fields in steel of transducer arrangements, e.g. transmitter-receiver transducers or tandem arrangements. With this device it is possible to determine not only the original sound field of a transducer but also the sensitivity distribution resulting from a superposition of two sound fields [1]. A computer calculates the product of the two sound fields point by point and the results are printed. Figure 9 shows the basic scheme of this device.

The sound field of a transmitter-receiver transducer with inclined longitudinal waves, as measured with the device described above, is presented in Fig. 10. This transducer has been developed by the Röntgen-Technische Dienst and the Federal Institute for Materials Testing for the detection of 'under-clad' cracks [3]. Similar transducers are also used for testing near-surface areas and austenitic welds. Figure 10 shows cross-sections of the sensitivity distribution for various distances between transducer and the surface of measurement. For a distance of about 10 mm — at the crossing of the two sound beams — a focussing is observed, while for larger distances the separation of the two sound beams from each other can be seen from the sensitivity distribution.

3.9. Techniques for the characterization of reflectors

Figure 11 gives a survey of the techniques for characterizing reflectors. With regard to the first group of techniques, it should be borne in mind that the 'classical' ultrasonic-pulse-echo techniques are normally only able to detect and locate flaws. Statements about the kind of flaw detected are only possible with additional knowledge, e.g. about the kind of defects that usually occur in a certain area of a welded seam. The use of the echo amplitude for estimating the defect size is very problematic, as is the determination of defect extension by means of the half-value or 20 dB drop methods [15]. In the field of in-service inspections, where — apart from the few cases where simple comparison between successive inspections is sufficient — quantitative characterization of the detected defects is absolutely necessary, special techniques for characterizing reflectors become most important.

Direct data of a defect are provided by all those methods that give a direct image of the defect, the well-known scanning with focussing probes [2] and the holographic techniques. Two other techniques of an indirect defect characterization are shown in Fig. 11. Frequency analysis of defect indications — ultrasonic pulse spectrometry — makes it possible to relate a measured spectrum to a certain orientation and size of the defect, either by means of the known physical diffraction behaviour of the reflector itself

Information Sources

Conventional techniques used (Flaw detection techniques)

- sound path → Flaw location (kind, cause)
- max. Echo height → Equivalent flaw size
- Flaw scanning → 6 dB or 20 dB drop

Extensions

New techniques – Development not yet finished (Flaw analysis techniques)

- Holography and Imaging Techniques → real flaw representation
- Frequency analysis → Pattern recognition → flaw determination; → direct flaw size estimation
- Echo dynamic → Pattern recognition → flaw determination; → direct flaw size estimation

FIG. 11. Possibilities of interpreting ultrasonic information.

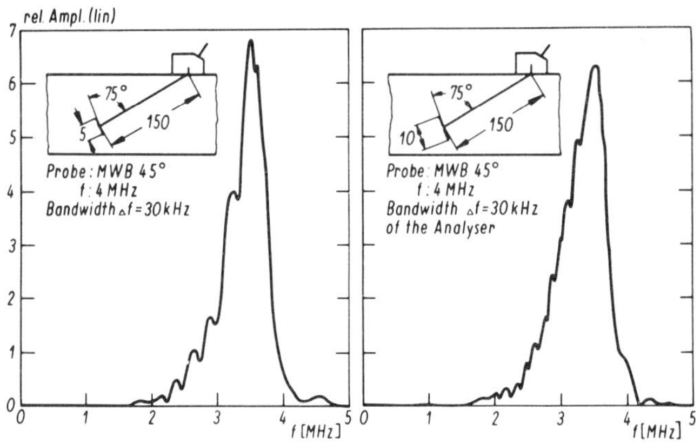

FIG. 12. Frequency spectra of a pulse from a flat reflector.

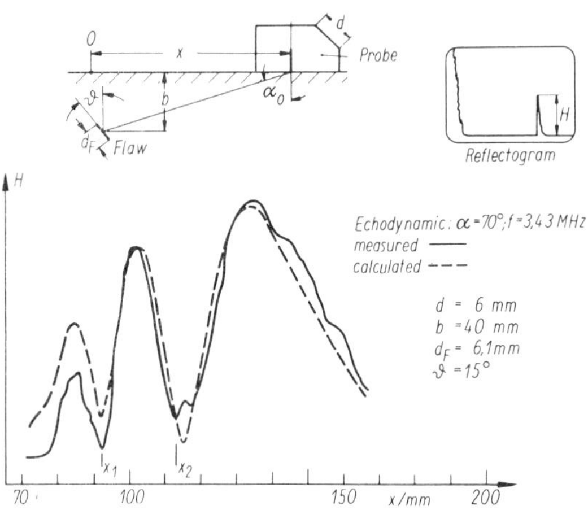

FIG. 13. Echo dynamic for flaw size estimation.

or by comparing the spectrum with a multitude of spectra of defects of different shape, size and orientation stored in a computer, using a technique of pattern recognition. Similar possibilities are provided by the echo-dynamic-technique, that is, the determination of the variation of echo amplitude and position on the screen caused by a movement of the probe relative to the reflector.

These two techniques shall be explained in more detail by two examples:

(1) Characterization of a defect indication by pulse spectrometry. Figure 12 shows two spectra of defects with constant inclination of 15° relative to the beam direction, but of different size. Small defects have a spectrum with large distances between the minima of interference, those distances becoming smaller with growing defect size. From this effect conclusions concerning the extension in depth of a defect can be drawn, and in addition — by measuring spectra for several positions of the probe relative to the defect — also conclusions concerning the orientation are possible [16, 17].

(2) The echo-dynamic method. Figure 13 shows the change of the echo indication caused by movement of a probe with a broad beam relative to a reflector. The echo amplitude as a function of the probe position shows a series of alternating minima and maxima. The minima are caused partially by the sound field of the transducer and partially by the reflector itself. If the divergence of the beam is sufficiently large — small diaphragms — the minima are only caused by the reflector. The relative position of the maxima and minima offers information about the orientation and the extension of the reflector if, for instance, measurements with probes of two different frequencies are performed. The echo-dynamic method is a simplification of the technique of holographic scanning with a reduced information output and therefore also reduced possibilities for statements about the defect. In many cases, however, this simplified technique offers some additional information about flaws.

4. CONCLUDING REMARKS

By describing examples of developments and investigations in connection with in-service inspections, the paper attempts to give an impression of the variety of possible ultrasonic techniques that can and must be used. The necessity of in-service inspection is a challenge to the people working in the field of ultrasonic testing. This challenge has initiated not only the development of new techniques but also an attempt at a systematic analysis of the operation and capacity of the well-known techniques of ultrasonic testing, an attempt that has turned out to be urgently necessary.

REFERENCES

[1] De STERKE, A., "Ultrasonic examination of welded joints in steel plates of thicknesses up to 300 mm; IIW/IIS Colloquium, Delft, 1966, Doc. V 318-66/OE.

[2] SAGLIO, R., "Routine checkout", Conf. Periodic Inspection of Pressurized Components, Institution of Mechanical Engineers, London, 1974, Paper C 116/74.

[3] WÜSTENBERG, H., SCHULZ, E., "Versuche zur Feststellung plattierungsnaher Reflexionsstellen an Reaktorteilen mit Ultraschall", Meeting Zerstörungsfreie Materialprüfung, DGZfP, Saarbrücken, 1972.

[4] BUNDESANSTALT FUR MATERIALPRÜFUNG, Report on Research Project RS 27/1 to the M. A. N. Company, 1. Entwicklung und Erprobung von Ultraschallprüfmethoden zum Auffinden plattierungsnaher Fehler; 2. Untersuchungen zur Transferkorrektur am Testkörper II; 3. Konstruktion von Diagrammen zur Bestimmung der Ersatzfehlergrösse bei der Tandem-Methode; 4. Abhängigkeit der Echohöhe von der Schräglage der Fehler bei der Tandem-Methode, January 1972.

[5] BÖTTCHER, B., SCHULZ, E., WÜSTENBERG, H., "A new method for crack depth determination in ultrasonic materials testing", 7th Int. Conf. Non-destructive Testing, Warsaw, 1973, Paper H-05.

[6] BUNDESANSTALT FÜR MATERIALPRÜFUNG, Report on Research Project RS 27/2, Abhängigkeit der Streuanzeigentechnik vom Oberflächenzustand und von der Frequenz, May 1974 (available from IRS-Cologne).

[7] NEUMANN, E., WÜSTENBERG, H., NABEL, E., LEISNER, W., "Ultraschallprüfung von Schweissverbindungen austenitischer Stähle", Annual Meeting DGZfP, Nijmegen, Holland, 1974.

[8] RATH, W., "Possibilities and requirements for repetitive tests on nuclear reactors", Symp. Mechanization and Automation of Inspection on Pressurized Plants, Tokyo, 1973, Japan Industrial Planning Association.

[9] HEINRICH, D., LINDNER, J., MÜLLER, G., REGNET, K., Untersuchungen über die Fehlererkennbarkeit mit ferngesteuerten Ultraschall-Prüfeinrichtungen an einem grossen Behälterwand-Testkörper mit natürlichen Fehlern. Teil A: Untersuchungen am unplattierten Testkörper, M. A. N. Rep. No. VIII, 1974, Research Project RS 27/2.

[10] BUNDESANSTALT FÜR MATERIALPRÜFUNG, Report on Research Project RS 27/2, Hilfsmittel zur Fehlergrössenabschätzung bei der Wiederholungsprüfung mit Ultraschall: Echodynamik, October 1974.

[11] MAYER, W., NEUMANN, A., STREICHER, V., DANNEHL, G. Ergebnis von Ultraschall-Messungen am Reaktordruckbehälter des Kernkraftwerkes Philippsburg, KWU internal Rep. R 214 - 2887.

[12] SHCHERBINSKII, V. G., Reliability of ultrasonic control in relation to the quality of acoustic contact, Central Scientific-Research Institute of Machine-Building Technology; trans. from Defektoskopiya 2 (1971) 88-93.

[13] BUNDESANSTALT FÜR MATERIALPRÜFUNG, Report on Research Project RS 27/2, Untersuchungen zum Einfluss der Plattierung und der Geometrie auf die Prüfbarkeit von dickwandigen Reaktorkomponenten mit Ultraschall bei der Tandem- und der Einkopftechnik, October 1972 - July 1973.

[14] KUTZNER, J., WÜSTENBERG, H., MÖHRLE, W., SCHULZ, E., "Zonenaufteilung, Empfindlichkeitseinstellung und Prüfkopfhalterung bei der manuellen Ultraschallprüfung mit dem Tandemverfahren", presented at DGZfP Meeting, Berlin, May 1975.

[15] WÜSTENBERG, H., MUNDRY, E., Beitrag zur Halbwertstiefenausdehnung von Reflexionsstellen in der Materialprüfung mit Ultraschall, Materialprüfung 13 10 (1971) 329-35.

[16] ADLER, L., WHALEY, H. L., Interference effect in a multifrequency ultrasonic pulse echo and its application to flaw characterization, IASA 51 (1972) 881-87.

[17] NABEL, E., NEUMANN, E., MUNDRY, E., WÜSTENBERG, H., "Analysis of ultrasonic echoes by means of pulse spectroscopy", presented at V. Konferenz für zeitgemässe Bemessung und VI. Kongress für Materialprüfung, Budapest, 1974.

DEVELOPMENT AND OPERATION OF ONCE-THROUGH STEAM GENERATORS *

U. HENNINGS
Babcock – Brown Boveri Reaktor GmbH,
Mannheim,
Federal Republic of Germany

Abstract

DEVELOPMENT AND OPERATION OF ONCE-THROUGH STEAM GENERATORS.

At the time when the once-through steam generator was developed for nuclear loops, Babcock and Wilcox had already accumulated considerable experience over a period of several decades in the construction of conventional steam boilers.

The experimental program for determining the essential data relating to heat transfer, flow behaviour, controllability, flow stability, mechanical load under operating and transient conditions began over ten years ago. Experimental units were run with full-length tubes for conducting basic research, for testing design data, for ascertaining detailed operating features and characteristics during parallel operation.

Testing of materials was carried out concurrently, various tube materials being exposed to hydrochemical and stress conditions which may develop over a long operating period and also in fault conditions. These tests showed Inconel 600 to have excellent properties as pipe material in the presence of water treated with volatile chemicals and a corresponding stress during operation. In particular it was proved that the problems of stress corrosion can be overcome. Large-scale versions of this type have been in operation since the end of 1972 in pre-operation testing (block I, Oconee power plant of the Duke Power Co.). In the meantime, four years of experience in the operation of steam generators have been accumulated.

In the pre-operation testing it was important to prove that the results from the large-scale version corresponded to those of design calculations and model trials.

In this work the following, inter alia, were determined: the live steam temperature against flow rate; the temperature profile of steam outlet against cross-section; the axial temperature distribution; the shift in the final evaporation point; the primary and secondary pressure drop in the tube bundle; the natural circulation capacity during simulated fault conditions; the vibration behaviour in the range from part load to overload; the thermodynamics during overload and during unequal load of parallel steam generators.

In Europe once-through steam generators of this type are being used for the first time in the Mühlheim-Kärlich nuclear power plant of the RWE. They provide superheated live steam which keeps to a minimum the propagation of contaminating fission products to the secondary loop in case of leakages. They reduce the necessity of a blowdown, since they do not have a defined water level during normal operation.

These properties are being confirmed by existing operating experience and especially by the inspections that are being carried out.

* For technical reasons, only the abstract of this paper is available for publication in these Proceedings.

DISCUSSION

(on papers IAEA-SM-195/25 and IAEA-SM-195/3)

A.C. PROT: Listening to Mr. Wüstenberg's presentation, I think I now understand somewhat better the question he asked after my paper this morning concerning the possibility of distortion of the ultrasonic beam by the surface through which it penetrates the test piece. It is true that the technique using focussing probes is sensitive to such distortion. One has to remember, too, that the results of periodic inspections must be in quantitative form so that a decision can be taken whether to shut the reactor down or not. We must hope, therefore, that designers and manufacturers will do everything in their power to improve the situation and make the method as quantitative as possible.

H. WÜSTENBERG: At the periodic testing stage it is desirable that all non-destructive test methods should yield quantitative results to back up the evidence of defects, so that assessments — for example in terms of fracture criteria — can be made. But precisely with ultrasonic testing methods these quantitative results are particularly hard to obtain. The surface state of the equipment to be tested, particularly pressure vessels, is unsatisfactory in many of the nuclear power stations being set up today. In the Federal Republic of Germany efforts have been made to overcome this difficulty through the development of appropriate procedures applied during tests, for example sensitivity corrections. This means, however, that the test methods used then have to be regarded for the most part as pure defect location techniques. All indications which are judged — in accordance with previously established criteria — worth recording must then be investigated further to determine their causes. Often a decision can be taken only with the help of analytical methods applied to the direct image of the defect. Hence we are thinking, in Germany too, of using focussing probes for this purpose. But we doubt whether it will be possible to scan, say, a whole pressure vessel with such probes at the initial defect-seeking stage, given the likely surface condition of the vessel.

A. SEVČIK: To what extent are the techniques described in your paper used in operating nuclear power stations?

H. WÜSTENBERG: Apart from focussing probes, all the techniques have been used in remotely controlled testing installations — for base measurements at five nuclear power plants and after two years of operation at one nuclear power plant in the Federal Republic of Germany, and also after two years of operation at one plant in Holland.

A. SEVČIK: Has the Federal Materials Testing Institute laid down any rules or instructions for the application of these techniques?

H. WÜSTENBERG: Periodic testing is carried out in the Federal Republic of Germany in accordance with guidelines laid down for that purpose by the Reactor Safety Commission. These guidelines prescribe the general conditions for the use of individual techniques and the efficiency required of them, but without specifying technical details for individual cases. At the moment standards for periodic testing are being drafted, and the Federal Materials Testing Institute is taking part in the preparation of this draft.

A. SEVČIK: Has German industry produced any specialized testing equipment?

H. WÜSTENBERG: Yes, a number of firms are engaged in the development and production of testing equipment. Systems are available both for boiling-water and for pressurized-water reactors.

W.A. WOLFE: How much does one rely on the operator of the inspection apparatus to detect flaws?

H. WÜSTENBERG: The training and reliability of the operator can have a decisive influence on the reliability of the results obtained from the ultrasonic techniques manually applied during manufacture. For periodic testing, on the other hand, fully automatic and remotely controlled testing installations are used, and the handling of this equipment, including computers, must naturally be mastered by the testing staff; but any faulty operation can be recognized as such almost immediately from the data put out by the system, so one's dependence on the individual operator is greatly reduced.

W.A. WOLFE: When a flaw is detected, how is it evaluated?

H. WÜSTENBERG: In ultrasound testing the evaluation of flaw indications proceeds in two stages. First one must decide whether a given indication is worth recording or not, a decision made in accordance with criteria derived from extensive practical experience with the technique used (for example echo height, recording lengths, frequency of indication per metre and so on), and based on specification of values for the recording limits. It is important to bear in mind that the flaw detection probability is decisively influenced by these limits.

Second, all positively recorded indications must be investigated to determine their causes. In many cases it is fairly certain that a minor cause will be found for the indication. But in all cases for which no satisfactory explanation is found, the safety implications of the indication must be determined by means of analytical techniques applied to the direct image of the flaw.

Evaluation of flaw indications purely by comparison with earlier test results seems to me a questionable procedure, because there are so many factors that limit the reproducibility of periodic testing.

A.V. KARPOV: Do you have any practical experience of inspecting reactor pressure vessels after a certain period of operation?

H. WÜSTENBERG: An automatic ultrasonic test was carried out after two years of operation at the Obrigheim nuclear power plant. A similar test was carried out by KWU at Borssele in Holland, in co-operation with the Röntgen Technical Service of Rotterdam and the UKAEA (Risley). In the course of this latter test the growth of sub-lining cracks known to be in the material was monitored. After almost two years of operation the same measurement results were obtained and in no instance was any crack growth observed.

So far complete zero-point inspections have been carried out automatically on five reactors in the Federal Republic of Germany before their entry into service (Würgassen, Stade, Brunnsbüttel, Biblis A and the Neckar power plant).

L. ALVAREZ DE BUERGO: I have a question for Mr. Hennings. It seems obvious that the steam generators designed by Babcock and Wilcox make it possible to increase plant efficiency and at the same time are simpler in design than the conventional tube models. Nevertheless, I believe they have given some problems with leaks, problems which have been explained to some extent by the metallurgists and the chemists. My specific question is whether special precautions have been taken in this respect — whether the chemistry of the primary and secondary circuits has been subject to more restrictive specifications as a result.

U. HENNINGS: It is true that some PWR steam generator tubes have shown defects and leaks which necessitated repairs. This does not apply, however, to steam generators of the OTSG type of B&W design. Straight Inconel 600 tubes are working in 12 OTSG units at present, and have already accumulated an aggregate operational life of four full power OTSG unit years.

The defects on the secondary side of steam generator tubes are due to stress corrosion cracking. For this to occur, three conditions must be fulfilled: (1) the tube material must be of a kind which is susceptible to it in the first place; (2) the level of operational stresses must be high enough; and (3) the environmental conditions must be such that it can occur. An extensive experimental investigation of the stress-corrosion-cracking behaviour of Inconel 600 indicated susceptibility to caustic and/or chloride attack at stresses greater than 0.75 yield strength. Accordingly, a secondary water specification and treatment was worked out that precludes chemical attack; at the same time a stress level of maximum 40% yield — far below the threshold of 75% — was adopted. Laboratory results and operational experience with 180 000 tubes in B&W OTSGs have demonstrated that stress corrosion cracking is no longer a problem.

H.V. CRAMER: In the slide you showed in your presentation, can you explain the dark-brown colour of the tubes just above the tube plate? What is this dark stuff?

U. HENNINGS: Upon inspection, the OTSG secondary side tubes and tube support plates were found to be covered with a greenish-black deposit. This deposit was very light, however, and much of the original reflectiveness of the tubes was still in evidence.

One aim of the inspection programme was to obtain samples of oxides that might have accumulated in the generator. Attempts to collect them with a knife-edge, stainless steel spatula were futile because of the very small amounts of oxide present. The reddish-brown oxide was easily removed for oxide analysis by wiping with filter paper. The results of the

TABLE A. SPECIFICATION OF WATER QUALITY

	Specification	Average value before power operation	Average during operation at power Jun. - Nov.	Nov. - Mar.
ppm B	<2100	600	1200	~800
ppm ^{7}Li	0.2 - 2.0	0.35	0.35	0.25
pH at 25°C	4.6 - 8.5	8.0	6	6.2
ppm O_2	0.1 (max.)	< 0.1	<0.1	~0
ppm Cl	0.1 (max.)	0.07	0.05	< 0.05
ppm F	0.1 (max.)	<0.05	<0.05	< 0.05
$H_2Ncm^3/kg \cdot H_2O$	15 - 40	15	15	~20
Total gas, Ncm^3/kgH_2O	100 (max.)	40	35	38
Suspended matter, ppm	<0.1[a]	0.025	0.015	<0.010
ppm N_2H_4	0.1 - 1.0	0.5	-	-

[a] Fill water specification.

wipe specimen and scale analyses indicated predominantly corrosion product iron with lesser quantities of nickel and chromium. None of these materials will be detrimental to the steam generator in the quantities found. The occurrence of the reddish-brown iron oxide film was due to maloperation during the pre-inspection phase: air was erroneously introduced instead of nitrogen when the water of the OTSG was drained.

J.H. REMEYSEN: Could you tell us how the feed water is treated?

U. HENNINGS: The feed-water treatment involves the use of hydrazine, with some excess of hydrogene. The water quality specified and maintained is given in Table A.

QUALITY ASSURANCE FOR NUCLEAR POWER PLANTS

Chairman:

L. ALVAREZ DE BUERGO, Spain

ASSURANCE DE LA QUALITE ET FIABILITE

J. NORMAND
Technicatome,
CEA, Centre d'études nucléaires de Saclay,
Gif-sur-Yvette

M. CHARON
CIRNA, Groupement pour les activités atomiques
et avancées,
Le Plessis-Robinson,
France

Abstract–Résumé

QUALITY ASSURANCE AND RELIABILITY.

Concern for obtaining high-quality products which will function properly when required to do so is nothing new — it is one manifestation of a conscientious attitude to work. However, the complexity and cost of equipment and the consequences of even temporary immobilization are such that it has become necessary to make special arrangements for obtaining high-quality products and examining what one has obtained. Each unit within an enterprise must examine its own work or arrange for it to be examined; a unit whose specific task is quality assurance is responsible for overall checking, but does not relieve other units of their responsibility. Quality assurance is a form of mutual assistance within an enterprise, designed to remove the causes of faults as far as possible. It begins very early in a project and continues through the ordering stage, construction, start-up trials and operation. Quality and hence reliability are the direct result of what is done at all stages of a project. They depend on constant attention to detail, for even a minor piece of poor workmanship can, in the case of an essential item of equipment, give rise to serious operational difficulties.

ASSURANCE DE LA QUALITE ET FIABILITE.

Chercher à obtenir des produits de qualité qui répondent bien au service qu'on en attend, au moment où leur fonction le requiert, n'est pas un souci nouveau: cela s'appelle conscience du travail bien fait. Mais la complexité des matériels, leur coût, les conséquences de leur immobilisation même temporaire sont tels qu'il est devenu nécessaire de prévoir et d'organiser les éléments permettant d'assurer la qualité, puis d'analyser ce qu'on a obtenu. Chaque unité doit effectuer ou faire effectuer un examen de ses propres travaux; une unité ayant pour souci spécifique la qualité est chargée du contrôle global des fournitures sans pour autant se substituer à ces unités d'exécution. L'assurance de la qualité se présente comme une assistance mutuelle dans les entreprises pour éliminer au maximum les causes de défauts. Elle commence très tôt dans l'élaboration d'un projet et se poursuit à travers la commande, la réalisation, les essais de mise en service et l'exploitation. La qualité et donc la fiabilité sont le résultat direct du travail de ceux qui sont chargés des réalisations, à tous les stades; c'est un souci de tous les instants car une inattention ou une légère malfaçon sur un matériel essentiel peuvent être la source de graves difficultés au cours de l'exploitation.

INTRODUCTION

Chercher à obtenir des produits de qualité qui répondent bien au service qu'on en attend, n'est pas un souci nouveau, même s'il se précise et s'amplifie de façon importante dans le nucléaire à l'heure actuelle. Cela s'appelle en fait conscience du travail bien fait.

N'était-ce pas ce sentiment que tendaient à développer les corporations de métier du Moyen-Age, concrétisé en particulier par la réalisation du Chef-d'Oeuvre nécessaire pour devenir Maître-compagnon et matérialisé par les nombreuses cathédrales par exemple ? Mais à cette époque un cerveau humain pouvait maîtriser l'ensemble de l'ouvrage sans trop de difficultés et de ce fait, il n'existait pas de règles écrites.

Au XXème siècle les matériels mis en jeu sont devenus très complexes et onéreux, les entreprises sont proportionnellement beaucoup plus osées et nécessitent de prendre toutes les précautions pour réussir, car vous n'ignorez pas que les premières applications des règles d'Assurance-Qualité à une échelle industrielle ont concerné l'aéronautique et plus encore l'aérospatiale. En effet, dans une fusée chaque fonction, même si elle n'entre en jeu qu'une fois, doit être assurée sans défaillance exactement au moment précis où l'opérateur en a besoin ; et le moindre défaut peut compromettre tout le programme ou mettre en danger la vie des astronautes.

Les problèmes concernant une Centrale Nucléaire sont de même nature même si les implications et les conséquences des défaillances sont différentes ou d'un autre ordre. Il s'agit ici de réaliser des installations à la fois maîtrisables, robustes et disponibles, c'est-à-dire, en un mot fiable. Il faut également qu'elles soient sûres, c'est-à-dire qu'elles permettent de protéger efficacement les populations et les travailleurs contre les risques de type radioactif ou autre tel que réaction sodium-eau par exemple.

La fiabilité d'une installation productrice d'énergie électrique est une exigence économique absolue pour deux raisons fondamentales : la première est que l'impossibilité de stocker l'électricité conduit à devoir disposer d'installations ayant un taux élevé de disponibilité ; la seconde est que, quel que soit le mode de production, les investissements des installations de base sont très lourds et la durée d'utilisation réelle est un facteur important de rentabilité. Les indisponibilités des Centrales sont très onéreuses. Des défauts de fonctionnement au démarrage se traduisent par une perte moyenne pouvant atteindre 10 % du prix des investissements. Les défaillances au cours de la vie de la Centrale peuvent atteindre 15 % du prix des investissements. Par ailleurs, le progrès technique joue un rôle déterminant dans la course vers les plus grandes puissances unitaires dont le rythme résulte d'un équilibre entre de fortes incitations économiques - diminution des frais d'investissement et d'exploitation - et le recul des limites technologiques.

La fiabilité indispensable au bon fonctionnement d'un réacteur nucléaire durant les 20 ou 30 années de son existence repose en grande partie sur la qualité des principaux équipements le constituant.

Face à cette recherche de matériels de hautes performances, l'ingénieur est confronté à de multiples problèmes. Une Centrale est un assemblage de quelques appareils principaux, entouré d'un nombre impressionnant de matériels petits et moyens, le tout habillé d'un dense réseau d'appareillage. Mais la tenue, au fil des ans, de chaque gros ensemble dépend elle-même d'une infinité de facteurs : le calcul d'une aube de turbine tient compte de contraintes de toute nature ; son processus de fabrication est long et à chaque stade imparfait ; elle est soumise à des régimes de marche variés et parfois difficiles à analyser. Il suffit pourtant d'une corrélation mal établie, d'un mauvais maillon dans cette chaîne, pour déterminer l'incident et l'arrêt de plus ou moins longue durée. Une turbine est, d'autre part, protégée par certains systèmes - régulateurs, systèmes de sécurité, de graissage, etc. - très fiables et organisés de façon redondante, ce qui n'exclut pas totalement les défaillances, dues bien souvent à des enchaînements de circonstances et à des défauts très localisés. Que l'on modifie un tant soit peu les hypothèses techniques ou les choix, et le résultat devient tout autre.

La nécessité d'obtenir du premier coup des installations fiables et sûres, répondant à la demande sans défaillance, possédant un facteur de charge aussi élevé que possible, a conduit à mettre en place un système destiné à veiller à la qualité qui concerne en fait tous ceux qui participent à la réalisation de la Centrale, depuis le Projet jusqu'à l'exploitation. Nous allons essayer d'exposer de façon simple en quoi consiste ce système et prendre quelques exemples montrant l'importance de bien organiser le système d'assurance de la qualité, de bien prévoir les divers aspects techniques, fonctionnels, technologiques, donc de bien choisir en s'entourant de toutes les précautions possibles, enfin de s'assurer que le but recherché est bien obtenu, aussi bien en matière de matériels que de fonctions assurées par les matériels.

La fiabilité est une conséquence directe de la qualité.

1. VUE GLOBALE SUR LE SYSTEME-QUALITE

Il est nécessaire de parler de système-qualité car l'obtention de la qualité est l'oeuvre commune de tous ceux qui participent à la réalisation du matériel aussi bien horizontalement depuis les études jusqu'à l'exploitation, que verticalement depuis le Président Directeur jusqu'au soudeur. On ne saurait limiter l'assurance-qualité à l'action d'une équipe restreinte.

L'élément primordial d'obtention de la qualité est la bonne exécution du travail tant au niveau de l'Ingéniérie que du Constructeur ou de l'Exploitant de la Centrale. Celui qui fait la qualité, c'est celui qui travaille. C'est aussi pourquoi il nous paraît indispensable que les organismes techniques soient responsables des fournitures livrées de même que des coûts et des délais ; ils doivent avoir le souci de la qualité et s'assurer de la validité des travaux qu'ils effectuent ou font effectuer.

La mise en place d'un système d'assurance-qualité comprend :

- d'une part, des mesures d'organisation de la qualité dans l'ensemble de la chaîne comprenant l'Ingéniérie, le Constructeur et les Sous-Traitants,
- d'autre part, des mesures techniques concernant la préparation de la réalisation et l'assurance que les résultats obtenus sont adéquats au but poursuivi.

Une unité particulière, que nous convenons d'appeler Assurance-Qualité, est généralement chargée de mettre en place et animer ce système. L'action de cette unité Assurance-Qualité doit répondre à un certain nombre de principes, qui sont notamment les suivants :

a) Principes généraux

1°) Le travail effectué par une personne ou un groupe doit être vérifié par une autre personne ou un autre groupe ; on évitera cependant les redondances d'ensemble entre les diverses parties concernées par ce travail. En effet, un homme seul ne peut tout savoir, il peut se tromper et ne pas penser à tout : rien n'est aussi dangereux que de laisser un même individu faire un travail d'étude, assurer le contrôle, puis la réalisation comme cela s'est vu parfois.

2°) Les personnes ayant une responsabilité spécifique concernant la qualité et la conformité aux exigences de qualité doivent être indépendantes des personnes ayant pour responsabilité les coûts, les délais, ou l'accomplissement de l'activité contrôlée.

3°) Toute opération doit faire l'objet de documents préalables définissant les conditions de sa réalisation.

b) <u>Principes pratiques</u>

Ces principes concernent la structure de l'organisation des entreprises :

1°) Chaque unité d'exécution ou chaque entreprise doit avoir le souci de la qualité de la tâche qu'elle exécute et/ou de la fourniture qu'elle livre, conjointement avec ses responsabilités des coûts et des délais; chaque unité doit donc assurer la vérification de ses propres travaux (contrôle au 1er degré).

2°) Une unité ayant pour mission spécifique la qualité est rattachée à la Direction de l'Entreprise ; sa mission consiste à assurer la qualité des fournitures à travers tous les éléments pouvant en affecter le niveau, depuis les études jusqu'aux essais de mise en service et à l'inspection en service. Cette unité a un rôle de contrôle au 2ème degré des unités d'exécution. Tout cela signifie que la qualité a pris de nos jours une telle importance que c'est devenu une fonction de la Direction des Entreprises au même titre que la gestion financière par exemple.

<u>NOTA</u>.- L'application de ce 2ème principe à des Sous-Commandiers est adaptée au type de fourniture à livrer et à leur importance.

c) L'organisation de ce système-qualité est conçue de telle façon que les vérifications et les contrôles à tous les stades, la coordination à tous les échelons, le contrôle de l'exécution des exigences contractuelles et leur respect, etc... ne peuvent pas être évités ou omis.

L'organisation de la qualité, les procédures d'élaboration et de diffusion des documents, plans, comptes rendus sont décrits dans un Manuel descriptif de l'Assurance-Qualité.

L'Unité d'Assurance de Qualité n'a pas pour vocation d'assumer les tâches des autres unités de l'Entreprise, mais elle se doit de les coordonner, en évaluant les procédures de travail précises établies par celles-ci (ou en provoquant leur établissement si nécessaire), puis en vérifiant leur application, en plus de la vérification exercée par les unités elles-mêmes.

L'Assurance de Qualité est constituée par <u>l'ensemble des actions</u> destinées à garantir que le niveau de qualité du matériel livré est celui attendu, cette garantie étant matérialisée après la livraison par des documents écrits et archivés démontrant que chaque opération ou acte pouvant influer sur ce niveau a été exécuté comme prévu et requis.

Cet ensemble d'actions pratiques varie évidemment d'un Etablissement à l'autre et, à l'intérieur d'un même Etablissement, d'un type de matériel à un autre. Pour un type de matériel donné, une Entreprise doit donc définir cet ensemble d'actions, dans un <u>Programme d'Assurance de Qualité</u> du matériel considéré.

Il faut définir, prévoir, mais le système d'Assurance-Qualité doit, pour être efficace, ne pas réduire la responsabilité des personnes ou des unités qui font le travail sous une avalanche de procédures, de comptes rendus; il n'y a rien de pire que le désintérêt au travail. C'est pourquoi il est important de rechercher l'efficacité de l'Assurance-Qualité dans un allègement du carcan que certaines méthodes paraissent entraîner, sans pour autant avoir une garantie sur la qualité des fournitures et donc sur la fiabilité.

Ce sont là les grandes lignes de l'Assurance-Qualité. Il est bien certain que les systèmes d'application sont fort divers selon les pays, mais l'examen des faits permet de classer les systèmes en 2 groupes principaux :

- l'un où le Constructeur est responsable aussi bien de la réalisation que du contrôle,
- l'autre où le contrôle est effectué par des organismes extérieurs plus ou moins officiels ou réglementaires.

Par ailleurs, dans certaines entreprises, l'unité Assurance-Qualité est chargée d'assurer également le contrôle des fabrications, dans d'autres entreprises cela est séparé.

2. DOMAINES PRINCIPAUX D'APPLICATION DE L'ASSURANCE DE QUALITE DANS L'ENTREPRISE

L'Assurance de Qualité s'intéresse à toutes les opérations pouvant affecter le niveau de qualité du produit livré. On ne peut donc pas dresser une liste exhaustive des domaines auxquels elle s'intéresse.

Cependant, il y a un certain nombre de domaines particulièrement importants où l'Organisation d'Assurance de Qualité doit se montrer spécialement vigilante, où rien ne doit être laissé au hasard.

L'Organisation de l'Entreprise

Dans une Industrie où les impératifs relatifs à la Qualité (donc à la fiabilité et à la sécurité) et les intérêts commerciaux à court terme (bilans d'affaires et planning) ne sont pas toujours convergents, le système doit donner la priorité à la qualité. L'Organisation de Qualité doit donc avoir toute l'autorité voulue pour faire respecter les impératifs de qualité.

L'Organisation de Qualité ainsi définie doit veiller, non seulement dans sa propre organisation, mais dans l'ensemble de l'Entreprise, à ce que chacun voie son rôle, sa hiérarchie, ses responsabilités, son autorité clairement définis : si la responsabilité d'une tâche n'est pas bien définie entre deux individus ou deux organisations, on constate que souvent aucun des deux ne la remplira complètement, chacun en toute bonne foi se reposant sur l'autre.

. Le choix et la qualification des fournisseurs est un élément essentiel.

. L'établissement, la diffusion, la révision des plans et spécifications utilisés en atelier

Qui déclenche ? Qui rédige ? Qui diffuse ? On doit être certain, pendant toute la vie du matériel, que les bons documents sont parvenus à toutes les personnes intéressées, mais aussi que les mauvais documents (indices périmés par exemple) ont été retirés des circuits et n'ont donc pas pu être utilisés par erreur.

. Le Fabricant doit pouvoir justifier la validité des calculs nécessaires dans le cadre de son Marché. L'Ingénieur d'études doit non seulement élaborer le document mais aussi proposer par écrit des méthodes de vérification telles que :

- références à des réalisations antérieures,

- démonstration expérimentale (modèles ou maquettes),

- calcul selon une méthode simplifiée ou différente, portant sur une partie de l'ensemble objet du calcul principal : test partiel ou recoupements par d'autres méthodes de calcul.

Il s'agit dans la plupart des cas de pouvoir vérifier l'ordre de grandeur des résultats obtenus, par des procédés simples ; le recours à un calcul nouveau et complet, faisant appel éventuellement à un programme, doit rester l'exception.

L'Unité-Qualité du Fabricant doit veiller spécialement à ce que les résultats des calculs approuvés soient appliqués et respectés sur les plans, notes et au cours de la fabrication.

. Particularités concernant les plans et documents connexes

. La vérification des plans et nomenclatures par le Fabricant en matière de qualité porte en particulier sur le choix des matériaux, le niveau de qualité, les repères de soudure, les cotes principales ou fonctionnelles, les possibilités de montage, d'assemblage ou de manutention, le respect des données fournies par les notes de calcul, la correspondance des repères avec les Cahiers de Procédés de Soudage et les programmes de fabrication, les moyens indispensables aux contrôles, épreuves et tests.
Les matériaux en particulier doivent faire l'objet d'études et d'essais préalables, notamment pour ce qui concerne leur tenue en température, leur soudabilité ; l'addition de résiduels tels que Bore ou Niobium pour les aciers inoxydables peut modifier considérablement le comportement des matériaux indiqués dans les normes, en ce qui concerne notamment la soudabilité (fissurations par exemple lors du soudage).

. Opérations de fabrication

Pas d'instructions verbales, de procédures inexistantes parce que "le soudeur sait comment faire", d'électrodes distribuées à quiconque en demande, de traitement thermique pour lequel la position des thermocouples est inconnue. On doit pouvoir retrouver après plusieurs années quel soudeur a exécuté telle portion de joint soudé, quel lot d'électrodes a été utilisé, si ces électrodes ont été étuvées pendant le temps voulu... Tous ces éléments participent au niveau de qualité de la pièce.

. Formation et qualification des soudeurs

L'organisation de Qualité doit s'assurer que les qualifications nécessaires et/ou requises ont été faites.

. Contrôles non-destructifs

Les contrôleurs sont-ils qualifiés ? Qui les convoque ? Suivant quels documents travaillent-ils ? A qui envoient-ils leurs rapports ?

. Etalonnage des moyens de contrôle

Un contrôle exécuté avec un appareil dont l'étalonnage est douteux est lui-même douteux. Toutes les pièces contrôlées avec cet appareil doivent être mises en question.

. Documentation

C'est une partie fondamentale de l'Assurance de Qualité puisque c'est la seule information qui subsiste matériellement sur les conditions d'exécution d'une opération. Chacun dans l'Entreprise doit donc savoir ce qu'il doit noter, à qui il doit diffuser ces notes et où il doit les classer. Une opération quelconque qui ne laisse pas de trace écrite est par définition sujette à caution vis-à-vis de la qualité.

. Anomalies

Que fait le contrôleur qui découvre une pièce non conforme, c'est-à-dire déviant des documents de travail qui la définissent ? Quel est l'enchaînement des opérations qui conduiront à ce qu'une action soit entreprise sur cette pièce ? L'organisme qui prend la décision finale sur cette pièce est-il réellement indépendant des considérations autres que la qualité du produit, et donnera-t-il bien la priorité à celle-ci ?

. Vérifications du Système - Inspections-Qualité

Toutes les questions ci-dessus trouvent réponse dans le Manuel d'Assurance-Qualité. Il appartient alors à l'Organisation de Qualité de vérifier si le Manuel est appliqué, en exécutant des inspections-qualité.

3. LES SERVICES QUE PEUT RENDRE L'ASSURANCE-QUALITE EN VUE DE LA FIABILITE

Outre le rôle d'organisation et de systématisation, l'Assurance-Qualité amène une coordination entre les divers domaines, elle aboutit a ne rien laisser au hasard.

D'abord au niveau du Sous-Traitant, du fabricant de tubes, par exemple : celui-ci fournit un élément de l'ensemble de la Centrale, même s'il s'agit d'un élément important, ce producteur de tubes ne peut pas tout connaître de la qualité à fournir pour ces tubes, notamment en ce qui concerne de nouveaux types de Centrales. En particulier, il ne connaît pas nécessairement les conditions exactes de service, il n'a pas réalisé que pour tel type de Centrale, les tubes ne constituent qu'une paroi fort mince entre sodium et eau, et que certains défauts qui pouvaient être tolérés pour des réacteurs à eau ne peuvent absolument pas être acceptés pour des réacteurs rapides, refroidis au sodium.

Par ailleurs, la recherche d'une meilleure utilisation du matériau, se caractérisant par une analyse fine des contraintes, donc par des taux de travail plus proches des limites admissibles, nécessite que le système qualité veille à définir des exigences complémentaires sur la santé du métal, sur l'absence de défauts dans son épaisseur. Il faut absolument chercher à éliminer tous les paramètres qui peuvent, par la défaillance d'un organe, provoquer l'incident.

Il en est de même pour les matériaux choisis, la soudabilité, etc... Ce n'est pas à l'Assurance-Qualité de débrouiller tous les problèmes, il y a des spécialistes et des Services pour cela, car l'Assurance-Qualité se présente de ce point de vue comme un coordonnateur entre les divers domaines de spécialités ; elle se présente comme une assistance mutuelle pour éliminer au maximum les causes de défauts. Elle a une action particulière à assurer pour ce qui concerne les interfaces : entre les études et la réalisation, entre les procédés de soudage et l'inspection en service. Il y a donc à coordonner les dérogations, centraliser les anomalies, en un mot être le garant de la conformité aux exigences contractuelles et réglementaires.

Le rôle de l'Assurance-Qualité commence très tôt dans l'élaboration d'un Projet, par les essais préliminaires, par le choix et la tenue du matériau dans les conditions de service ; ce rôle se poursuit au stade de la commande, de la réalisation des calculs, des fabrications, des montages sur site, des essais de mise en service ; et ce n'est pas encore là l'ultime étape puisqu'il faut envisager les mesures à prendre en cours d'exploitation, les inspections en service, les entretiens à effectuer et éventuellement les interventions sur matériels défectueux.

Ce dernier point illustre bien que "l'assurance ne coûte cher qu'avant l'accident". Exemples :

- Intervention dans boîte tubulaire de Générateur de Vapeur de Centrale Nucléaire.
  Nettoyage par décapants, difficile en raison de présence dans les circuits aussi bien d'aciers non alliés que d'aciers inoxydables. Travail de soudage dans des conditions difficiles avec soudeurs en scaphandre, opérant dans la boîte à eau du générateur de vapeur. Le réacteur a été immobilisé presque 2 ans.

- Dans un autre réacteur, problème de coudes ailetés. La Société chargée de la réalisation a estimé suffisant d'utiliser des coudes ailetés d'un type déjà utilisé dans un précédent réacteur. Par la suite, des coudes ayant même épaisseur que la paroi ont été montés et mis en essais sans qu'il y ait eu au préalable d'essais hydrauliques ou d'étude de soudage. Il s'agit cependant d'un matériel sensible aux vibrations, et la vitesse réelle du fluide était nettement supérieure à la vitesse dans le réacteur de référence. Des éléments des ailettes se sont retrouvés cassés et entraînés. Cet exemple illustre à la fois le manque d'essais préliminaires hydrauliques, le manque d'études et de références sérieuses pour les études, le manque d'essais de soudage à l'échelle réelle.

- Les défauts constatés sur des tubes de générateurs de vapeur poussent à effectuer désormais des essais sur lot de pré-séries. Nous demandons aux fournisseurs de tubes la fabrication d'un lot préliminaire de 50 tubes, qui subissent tous les essais de recette ordinairement prévus ainsi que des essais particuliers sur la soudabilité du produit, et la production réelle de tubes ne commence qu'après avoir obtenu les accords nécessaires de tous les spécialistes. Des étalons d'état de surface sont définis à cette occasion, ainsi que le seuil de rebut au contrôle ultrasonore défini sur des défauts réels et non en se basant uniquement sur des défauts-étalons. Mais le nombre de défauts rencontrés sur les tubes et les générateurs de vapeur impose des précautions particulières. Ce sont là des exigences importantes pour s'assurer des moyens de fabrication et de contrôle du fournisseur, et de la qualité d'un produit.

Quand on sait que toute réparation en période de démarrage ou de service nécessite non seulement l'arrêt prolongé du réacteur, mais également des interventions et des réparations longues, on s'aperçoit qu'il est nécessaire d'examiner au préalable avec soin toutes les causes possibles de défaillance, et d'être plutôt deux personnes pour les examiner, afin d'éviter toute réparation et toute indisponibilité coûteuses des Centrales.

Un autre point essentiel consiste à conserver des traces précises des divers éléments qui touchent à la qualité, afin de pouvoir à tout moment de la vie de la Centrale s'assurer que tout ce qui était prévisible pour la qualité a bien été fait et en cas de défectuosité de pouvoir retrouver immédiatement tous les éléments nécessaires pour expliquer le phénomène et en retrouver les causes.

4. NORMES D'ASSURANCE-QUALITE

Les prescriptions existantes actuellement sur l'assurance-qualité sont contenues dans les principaux documents suivants :

4.1. Documents américains

Un travail important a été réalisé aux U.S.A. en ce domaine, où il existe notamment :

CODE FOR FEDERAL REGULATIONS (émis par l'ancienne US.AEC)	10 CFR 50 (App B)	Critères d'assurance-qualité (les fameux 18 critères)
ANSI	N 45.2	Prescriptions d'assurance-qualité pour les Centrales Nucléaires.
US.AEC	RDT Standard F2 - 2T	Prescriptions pour les programmes d'Assurance-Qualité
ASME III	§ NA 4000	Assurance de la qualité

Par ailleurs, l'association de normalisation américaine adhérent à l'I.S.O., dénommée AFNOR a édité ou édite actuellement les spécifications suivantes :

N. 18.7	Contrôle de la gestion de la qualité	Administrative control
N. 45.1	Inspections en service	In-service inspections
N. 45.2.6	Qualification du personnel	Qualification of personnel
N. 45.2.9	Rapports	Records
N. 45.2.10	Terminologie d'assurance-qualité	QA terms and definitions
N. 45.2.11	Conception – Etudes	Design
N. 45.2.12	Inspection-qualité	Auditing
N. 45.2.17	Soudage	Welding
N. 402	Assurance-qualité pour les réacteurs de recherche	QA in research reactors

4.2. Autres documents

- En dehors des normes américaines, il n'existe qu'assez peu de spécifications éditées par les associations nationales de normalisation : les British Standards ont édité une norme.

- Ailleurs, les prescriptions sont souvent le fait des exploitants propriétaires de Centrales : c'est le cas notamment en France, d'Electricité de France qui a édité ses prescriptions sous forme d'un document général en 3 pages ; celui-ci s'est donné pour but de demander aux fournisseurs de mettre en place une organisation de la qualité et de la définir dans un état descriptif sans pour autant définir des règles astreignantes.[1]

- L'Agence Atomique Internationale a déjà réuni une équipe d'experts pour mettre au point un texte sur l'Assurance de la qualité. Mais il est évident que l'Assurance-Qualité n'est pas faite uniquement pour assurer la sûreté, mais aussi la fiabilité de la Centrale, son fonctionnement et pas seulement son arrêt. Selon nous, ce document doit être valable pour tous les types de Centrale et pour satisfaire les divers critères concourant à la qualité.

5. CONCLUSION

En conclusion l'Assurance-Qualité est un système essentiel pour assurer la fiabilité d'une Centrale Nucléaire. Ce ne sont pas nécessairement des règles nouvelles, beaucoup pratiquaient l'Assurance-Qualité sans le savoir ou du moins sans lui donner de nom ; c'est alors une systématisation de ce qui existait, un certain formalisme, l'obligation pour les fournisseurs de définir au préalable ce qu'ils font ou vont faire, et pour le propriétaire exploitant la Centrale la nécessité de définir ses exigences. Le système d'Assurance-Qualité est une garantie importante de bon fonctionnement de la Centrale d'autant plus nécessaire que le nombre de Centrales à construire grandit et que l'attention des personnes et des entreprises risque de s'émousser et d'entraîner une moins grande application des règles, donc une baisse de qualité. L'unité Qualité est donc là pour veiller tout au long des réalisations à ce que la qualité ne dérive pas.

L'Assurance-Qualité ne saurait être le travail d'une unité ayant pour souci spécifique la qualité. C'est le travail de tous ceux qui réalisent à quelque niveau ou à quelque stade qu'ils se trouvent. C'est un souci de tous les instants car une inattention ou une légère malfaçon sur un matériel essentiel peuvent être la source de graves difficultés en fonctionnement.

[1] Voir mémoire IAEA-SM-195/45: VAUJOUR, J.-A., «L'assurance de qualité en France: Le système d'Electricité de France» (ces comptes rendus).

THREE STEPS TO A MORE SUCCESSFUL QUALITY ASSURANCE PROGRAM

W.E. FERRISS
Bechtel Power Corporation,
Ann Arbor, Mich.,
United States of America

Abstract

THREE STEPS TO A MORE SUCCESSFUL QUALITY ASSURANCE PROGRAM.

The three steps that will be presented are by no means a cure-all for the variety of problems and challenges that a Quality Assurance (QA) Department is faced with in its role in the design and construction of a nuclear power plant. However, these steps are considered to be three of the most important ones in the realization of an effective and efficient QA program.

Step 1. Awareness. With the multitude of people involved in activities that effect the resultant Quality of Design, Procurement, and Construction of a nuclear power plant, a concerted effort has been put forth at Bechtel to promote 'Quality Awareness'. This effort has resulted in presentations to thousands of engineers, buyers, superintendents, supervisors and many others to make them more aware of their role in the Quality program. These presentations cover the Quality criteria, organizations, manuals, and implementation responsibilities that constitute the Company Quality program. In addition to the above, many specialized courses covering inspection techniques, communications, auditing, problem solving, etc. have been given to the people involved in the Generic Quality Assurance functions.

Step 2. Attitude. Nuclear Power Quality requirements are quite stringent and have presented additional requirements to engineers, buyers, and superintendents who have previously designed and constructed fossil fuel power plants. Logically there was a resistance to these new requirements and a number of attitudes had to be changed. The most effective way that we have found to accomplish this is through communications from top management expressing their support of the Quality Program.

Step 3. Objective and economical compliance. With the increased awareness and a more positive attitude toward Quality requirements, Bechtel has been able to devote considerable effort on finding effective methods to comply with Quality requirements in the most economical way.

The complete presentation will include several examples of the implementation of each of the above steps taken from actual experience.

INTRODUCTION

Bechtel has been involved in the design and construction of nuclear facilities for many years and is one of the most experienced design-construction organizations in the world. One of the primary reasons for our success is our belief in "quality".

One of the elements required by the U.S. AEC (NRC) in their Code of Federal Regulations for QA Criteria for nuclear power plants is a QA program that provides for indoctrination and training of personnel who perform activities affecting quality, to assure that suitable proficiency is achieved and maintained.

I would like to present three steps to a more successful QA program which exemplify Bechtel's commitment to quality, including the indoctrination of personnel in the quality program and the steps Bechtel has taken to realize a more effective and economical quality program for the design and construction of nuclear power plants.

STEP 1: AWARENESS.

With the multitude of people involved in activities that affect the resultant quality of design, procurement, and construction of a nuclear power plant, a concerted effort has been put forth at Bechtel to promote quality awareness. This effort has resulted in presentations to over 5,000 engineers, buyers, superintendents, supervisors and many others in Bechtel's three power divisions, to make them more aware of their role in the quality program. These presentations cover quality criteria, organizations, manuals, and implementation responsibilities that constitute the company's quality program. In addition, many specialized courses covering such topics as inspection techniques, communications, auditing and problem solving have been given to persons involved in the generic quality assurance functions.

Typical examples of the quality awareness activities during 1974 are:

(a) A quality workshop held in the Ann Arbor, Michigan Area Office during the week of 5/13/74-5/17/74, which was called the "Quality Awareness Week". More than 400 employees participated in the workshop that was divided into four courses, the subject matter included: General Indoctrination to Quality; Written and Verbal Communications; Auditor Training; and Quality Problem Solving.

(b) A follow-up to the Quality Workshop, the General Indoctrination to Quality presentation, has been given on a weekly basis, since August 1974, in the San Francisco Power Division Office. Other offices in Los Angeles, Gaithersburg, Maryland, and Houston have been giving similar quality awareness presentations since mid 1974.

(c) A half-day Quality Awareness Seminar was held in San Francisco in December 1974, attended by 100 management personnel. The keynote speaker, a member of the Board of Directors, stressed the need to maintain individual dedication to the living ideals of quality of performance and quality of product. The seminar panel chairman, a Bechtel Vice-President, stated: "It is the responsibility of management to establish the standards of quality and to make sure that employees at every level are aware of them. We want to make quality awareness a part of our work habits".

STEP 2: ATTITUDE.

Nuclear power quality requirements are quite stringent and have presented new challenges to engineers, buyers, and superintendents who have previously designed and constructed fossil fuel power plants. Understandably there was a resistance to these new requirements and a number of attitudes had to be changed. The most effective way that we have found to accomplish this attitude change is through communications from top management expressing their support of the quality program.

Typical examples of management support to the quality program are:

(a) In January 1974 an area office manager issued a letter to be given to all new employees during their indoctrination to the company. The letter emphasized management's support of the quality program with such statements as:

> "As a new employee, Bechtel welcomes you to become a member of our <u>Quality Team</u>."
>
> "One of the primary reasons for our success is our belief in <u>Quality</u>."
>
> "In order for our Quality Assurance Program to be a success, it requires that each individual involved in <u>any</u> activity, no matter how small it may seem, be personally committed to <u>Quality</u>."

(b) As a further endorsement of management's support of quality, a member of the Board of Directors added the following note to the new employee letter and forwarded it to all power division managers:

> "I feel this is an excellent way to begin building <u>attitude</u> from the start, and would hope each of you are doing something similar."

(c) Another letter from the area office manager in January 1974 stated:

> "It is not possible to emphasize sufficiently the importance of quality assurance and quality control in all of our nuclear work. Proper performance of our quality responsibilities is the most critical subject facing us this year. It is important that this message be transmitted to our Bechtel people throughout our projects."

STEP 3: OBJECTIVE AND ECONOMICAL COMPLIANCE.

As a result of increased awareness and a more positive attitude towards quality requirements, Bechtel has been able to devote considerable effort to find effective methods of complying with quality requirements in the most economical way.

Typical examples of steps taken in 1974 to realize more objective, effective, and economical methods of accomplishing the quality requirements are:

(a) The Nuclear Quality Assurance Manual, which was completely revised to provide greater detail and clarity to the requirements of the Quality Assurance Program. This manual provides divisional policy in assigning responsibilities (who) for the many aspects of the Quality Assurance Program (what) for nuclear power plant design and construction. This major revision was the coordinated effort of the Quality Assurance Department and the functional departments of engineering, procurement and construction, and resulted in a much-improved definition and understanding of the QA requirements.

(b) Numerous standardization programs have resulted in a more consistent accomplishment of quality-related activities. These programs include standard Engineering Department procedures, standard Quality Assurance auditor training and standard designs for safety-related systems.

(c) Calibration of measuring and test equipment is being performed at the construction site instead of at off-site facilities, which results in reduced handling and transportation costs in addition to more timely support to the construction activities.

(d) Use of a standard form for specifying required supplier documents, which also contains the Supplier's Statement of Conformance, streamlines the document review and acceptance, and simplifies the accountability and traceability requirements.

SUMMARY

Let's briefly review the three steps:

Awareness - Communicate to as many people as you can the basic elements of your quality program and what their role is in the program.

Attitude - Have top management outwardly express their support of the quality program and ensure this position is known to everyone involved in nuclear power plant activities.

Objective and Economical Compliance -

As people become more aware of the quality program and develop more positive attitudes, considerable effort should be devoted to finding more effective methods of complying with the quality requirements in the most economical way.

L'ASSURANCE DE QUALITE EN FRANCE
Le système d'Electricité de France

J.-A. VAUJOUR
Direction de l'équipement,
Electricité de France,
Paris,
France

Abstract-Résumé

QUALITY ASSURANCE IN FRANCE — THE SYSTEM APPLIED BY ELECTRICITE DE FRANCE.

For the design and construction of nuclear power stations, Electricité de France (EDF) has adopted a method which is based on its experience — taking account of its position as virtually the sole client of an industry in the course of development — and which enables it to fulfil its statutory obligations. Together with the power station constructors, EDF defines the technical objectives to be attained and the technical rules to be complied with, both of which are set forth in the contracts; it examines the designs and technical specifications submitted by the constructors and, if necessary, has them modified so that they conform to the contractual provisions; it maintains a check on the construction work to ensure compliance with the contracts as it has accepted them, examining any requests by the constructors for the waiving of certain provisions, and co-ordinates the activities of the different contractors involved in a single project. EDF requires that the constructors have a 'quality control organization' conforming to specifications which are regularly up-dated and which ensure compliance with the principles set forth in the contracts. In particular, each contracting body must have a system of 'internal controls' for ensuring the required quality in all operations and also a system of 'external controls' for enabling the management to verify the proper function of the system of 'internal controls'.

L'ASSURANCE DE QUALITE EN FRANCE — LE SYSTEME D'ELECTRICITE DE FRANCE.

Electricité de France (EDF) a adopté, pour la conception et la réalisation des centrales nucléaires, une méthode qui dérive de son expérience, tient compte de sa position de client presque unique d'une industrie en cours de structuration et permet de satisfaire aux obligations réglementaires auxquelles cet établissement est soumis. EDF définit en commun avec ses constructeurs les objectifs techniques à atteindre et les règles techniques à respecter, qui figurent les uns et les autres dans les textes des contrats. EDF examine les études et spécifications techniques établies par les constructeurs et, le cas échéant, les fait modifier pour les rendre conformes aux dispositions contractuelles. EDF contrôle le respect de ces documents dans la forme où il les a acceptés, instruisant les demandes de dérogation que les constructeurs peuvent présenter; il assure la coordination entre les différents constructeurs intervenant dans une même réalisation. Il exige de ses constructeurs qu'ils disposent d'une «organisation de la qualité», exposée dans des états descriptifs régulièrement tenus à jour et satisfaisant à des principes fixés dans les contrats. Notamment, chaque organisme qui agit doit disposer d'un mécanisme de «contrôle interne» qui lui permette d'obtenir que ses actions aient la qualité requise; chaque entreprise doit également disposer d'un mécanisme de «contrôle externe» qui permette à sa direction de s'assurer du bon fonctionnement des mécanismes de contrôle interne.

Il n'existe pas en France, à l'heure actuelle, de règlementation concernant l'assurance de qualité pour les installations nucléaires. Le CEA et EDF ont développé, lors de la conception, de la réalisation et de l'exploitation des installations nucléaires qu'ils ont fait construire, des méthodes propres dans ce domaine. Nous décrivons ci-après le système adopté par EDF dont le programme actuel de construction de centrales nucléaires représente de l'ordre de 6000 MW(e) par an.

1. GENERALITES

1.1 - Définition de "l'assurance de qualité"

ELECTRICITE DE FRANCE, les Constructeurs avec lesquels il passe des contrats, les sous-traitants et sous-commandiers de ces Constructeurs prennent des dispositions pour que les ouvrages, matériels ou équipements de la centrale nucléaire aient la qualité requise, c'est-à-dire qu'ils soient aptes à assurer la fonction qui leur est confiée. Cela implique notamment que ces ouvrages, matériels et équipements soient conçus, réalisés et exploités, conformément aux lois, aux règlements et aux règles de l'art et qu'ils aient en service un comportement satisfaisant.

L'ensemble de ces dispositions constitue ce que nous convenons d'appeler un "système d'assurance de qualité".

1.2 - L'expérience d'EDF

En vue d'obtenir la qualité requise pour les ouvrages et installations de production d'électricité qu'il met en service, ELECTRICITE DE FRANCE a adopté, dès sa création, la méthode suivante : EDF effectue lui-même les études d'avant-projet des centrales, confie à des constructeurs l'étude et la réalisation des matériels de ces centrales, contrôle directement l'action de ces constructeurs et assure lui-même la coordination entre ceux-ci.

EDF exerce en effet directement un contrôle sur l'action des constructeurs, tant au niveau de la conception, qu'à ceux de la construction, du montage et de la mise en service des matériels. Cela implique d'ailleurs que soient définis au départ, et de façon précise, par EDF et les constructeurs, les objectifs à atteindre et les règles techniques qui doivent être suivies. Les uns et les autres figurent dans les textes des contrats. Ce travail est essentiel si l'on veut pouvoir effectuer un contrôle qui ait un sens, c'est-à-dire qui soit une comparaison entre ce qui est prévu et ce qui est réalisé. EDF estime que cette façon de procéder permet d'obtenir une assurance complémentaire que le fonctionnement des installations, au-delà des délais de garantie prévus par les contrats, sera sans danger pour les travailleurs ou le public, et assez régulier pour lui épargner des indisponibilités coûteuses.

Pour les centrales nucléaires d'ailleurs, à la préoccupation naturelle d'EDF pour la sécurité des personnes, viennent s'ajouter les dispositions réglementaires qui, à elles seules, l'obligeraient, en tant qu'exploitant, à exercer un tel contrôle et à en rendre compte à l'Administration.

Cette méthode a prouvé son efficacité. EDF a été conduit à l'adapter à la situation nouvelle créée par le développement de son programme nucléaire et l'évolution de la réglementation française. Nous décrivons ci-après la situation actuelle.

1.3 - Les actions d'EDF et des constructeurs

1.3.1 - Etudes générales

EDF procède de façon continue à l'établissement et à la mise à jour de documents de doctrine technique générale qui prendront place dans les contrats et qui précisent les règles techniques générales qui doivent être, de son propre avis et de celui des constructeurs, respectées (règles techniques relatives à la conception, à la fabrication, aux montages, aux essais de mise en service). De la même façon, EDF définit, en accord avec les constructeurs, la nature et le contenu des documents que ceux-ci doivent lui soumettre lors de l'exécution des contrats. Enfin, EDF établit des dossiers de standardisation de systèmes, composants ou matériels. Tous ces documents peuvent faire l'objet d'une communication à l'Administration pour agrément.

1.3.2 - Etudes de tranches-type

EDF établit, pour chaque type de tranche nucléaire, un projet-type. A ce projet-type sont associés des documents techniques généraux qui prendront place dans des contrats, et qui fixent, en accord avec les Constructeurs, la nature et les caractéristiques des matériels, les règles techniques particulières à ces matériels, les documents techniques particuliers que doivent établir les constructeurs. EDF peut rédiger pour un type de centrale donné, un rapport de sûreté de référence qui s'appuie sur les documents ci-dessus et peut être présenté à l'agrément de l'Administration.

1.3.3 - Etudes préliminaires à l'engagement de la réalisation d'une tranche

EDF établit, pour chaque tranche nucléaire, un projet. A ce projet sont associés des documents techniques particuliers à chaque contrat et qui fixent, en accord avec le titulaire du contrat, les conditions particulières auxquelles doit satisfaire le matériel. EDF rédige un rapport de sûreté préliminaire qui s'appuie sur les documents ci-dessus et est présenté à l'Administration avec la demande d'autorisation de création de la centrale.

1.3.4 - Réalisation d'une tranche

Les constructeurs procèdent aux études, à la fabrication, au montage et à la mise en service des matériels. EDF contrôle que les documents techniques fournis par les constructeurs, en exécution des contrats, satisfont aux conditions techniques de ces derniers et que la fabrication, le montage et la mise en service sont effectués suivant les dispositions de ces documents techniques. EDF établit un rapport provisoire de sûreté qui s'appuie sur ces documents techniques, dans lequel il rend compte des conditions réelles de la réalisation et auquel il joint des consignes provisoires d'exploitation. Ces documents sont remis à l'Administration qui donne ou refuse l'autorisation de procéder aux essais préliminaires à la mise en service

1.3.5 - Mise en exploitation normale d'une tranche

EDF rédige un rapport définitif de sûreté qui traduit la situation exacte de l'installation à sa mise en exploitation normale. Il y joint des règles d'exploitation. Ces documents sont remis à l'Administration qui donne ou refuse l'autorisation de mise en service.

1.3.6 - Exploitation d'une tranche

EDF exploite la tranche en se conformant aux règles d'exploitation acceptées. Il procède à l'entretien des matériels. Il soumet à l'Administration ses projets de modifications de l'installation ou des règles d'exploitation.

1.3.7 - "Organisation de la qualité"

Pour que les actions d'EDF et des constructeurs soient convenablement menées et qu'en définitive, la qualité requise pour le matériel soit obtenue, il convient que l'organisation adoptée soit bonne. Cette organisation nécessaire constitue ce que nous convenons d'appeler "l'organisation de la qualité". L'organisation de la qualité adoptée par EDF et celle qu'EDF demande aux constructeurs, sous-traitants et sous-commandiers d'adopter, sont décrites dans la suite du présent chapitre.

2. PRINCIPES RELATIFS A L'ORGANISATION DE QUALITE

ELECTRICITE DE FRANCE adopte, en ce qui concerne son organisation de la qualité, les principes ci-après et prend les dispositions nécessaires pour que les constructeurs avec lesquels il passe des contrats, les sous-traitants et sous-commandiers de ces constructeurs respectent également ces principes dans leur organisation de la qualité.

2.1 - Description par écrit

L'organisation de la qualité doit être décrite de façon précise et détaillée dans des documents écrits tenus à jour.

Ces documents doivent notamment indiquer :

- les effectifs et leur répartition dans les différents services et usines,
- la liste des moyens techniques dont disposent, le cas échéant, chaque service et chaque usine,
- les structures, les missions et les responsabilités de chaque service et de chaque personne dont l'activité touche à l'organisation de la qualité,
- la liste de ces personnes et des indications sur leur compétence,
- les modalités d'élaboration, de correction et de vérification des documents ainsi que celles de leur circulation et de leur diffusion,
- la nature et les modalités d'exercice du contrôle, par le client, de l'action du fournisseur.

2.2 - Contrôle interne

L'organisation de la qualité doit être telle que :

- les différentes activités soient convenablement coordonnées et en particulier que soit mise en place une bonne articulation entre les personnes et les services qui en sont chargés,
- les personnes chargées d'effectuer des études ou d'établir des projets, des plans, des schémas ou des spécifications aient une connaissance suffisante des règles qu'ils doivent respecter et des sujétions réelles de la fabrication,
- les études, projets, plans, schémas ou spécifications établis par une personne soient vérifiés par un supérieur de cette personne d'un niveau hiérarchique suffisant pour juger convenablement de l'intégration de la partie concernée dans l'ensemble, et d'une qualification adaptée à cette tâche de vérification,
- la procédure suivie pour l'établissement et la vérification de ces documents soit conduite suivant des règles écrites et traduites sur un état visé par les personnes qu'elle concerne,
- les opérations d'approvisionnement, de fabrication, de montage, de contrôle, d'essais et de mise en service fassent l'objet de spécifications techniques définissant de façon précise et complète ces opérations,

- ces opérations ne soient entreprises qu'après qu'aient été établies les spécifications bonnes pour exécution et soient réellement effectuées suivant ces spécifications,
- ces opérations donnent lieu à des comptes rendus ou procès-verbaux qui soient tenus à la disposition des personnes qui ont effectué les études ou établi les spécifications techniques.

L'ensemble des vérifications et contrôles visés dans ce paragraphe constitue ce que l'on convient d'appeler le "contrôle interne".

2.3 - Contrôle externe

L'organisation de la qualité suppose notamment de la part de la Direction :

- que celle-ci veille à ce que soient intégralement respectées les dispositions contenues dans les documents de description de l'organisation de la qualité visées en 2.1 ci-dessus, et que ces documents soient tenus à jour,

- qu'elle ait le souci constant de juger de l'efficacité de l'organisation de la qualité et apporte à celle-ci, le cas échéant, les redressements ou améliorations utiles.

Dans ce but, la Direction doit disposer d'agents ayant l'indépendance, la compétence et l'autorité voulues pour l'assister dans cette action que l'on convient d'appeler "contrôle externe".

Ces agents ont pour mission :

- de contrôler que les documents de description de l'organisation de la qualité sont tenus à jour et que leurs dispositions sont intégralement appliquées,
- de contrôler, en procédant le cas échéant à des sondages, que ces dispositions contribuent effectivement à l'obtention de la qualité requise,
- de proposer à la Direction toutes mesures correctives qui s'avéreraient nécessaires au vu des résultats de ces contrôles.

3. APPLICATION DE CES PRINCIPES

3.1 - A l'ELECTRICITE DE FRANCE

Les deux Directions d'EDF chargées des centrales nucléaires sont, d'une part, la Direction de l'Equipement (conception et réalisation), et, d'autre part, la Direction de la Production-Transport - Service de la Production Thermique (exploitation).

La Direction de l'Equipement comporte différentes unités : les Régions d'Equipement, le Service d'Etudes et Projets Thermiques et Nucléaires (S.E.P.T.E.N.), le Service Contrôle des Fabrications (S.C.F.). Le SEPTEN et le SCF sont chargés, le premier dans le domaine de la conception, le second dans celui de la fabrication des matériels, composants et systèmes, d'établir la doctrine technique et de veiller à son application. Les Régions d'Equipement sont chargées de mener à bien la réalisation des centrales nucléaires.

Le Service de la Production Thermique comporte des Groupes Régionaux de Production Thermique dont dépendent les centrales en exploitation et des Services Centraux, notamment un Département Sûreté.

3.1.1. - Organisation de la conception

3.1.1.1. - Définition des activités

Les activités d'EDF concernant la conception sont celles qui sont définies aux paragraphes 1.3.1, 1.3.2, 1.3.3 et 1.3.4, à l'exception de celles qui concernent la construction en usine, le montage sur le site et la mise en service, qui sont définies aux paragraphes 3.1.2 et 3.1.3 ci-après.

3.1.1.2 - Organisation

Les activités décrites en 3.1.1.1. sont réparties entre le SEPTEN et la Région d'Equipement chargée de la réalisation de la façon suivante :

- Le SEPTEN :

  . est chargé (en accord avec le SCF pour ce qui touche à la construction en usine et dans la concertation avec les Régions d'Equipement) d'effectuer les actions définies aux paragraphes 1.3.1 et 1.3.2 relatives à la conception ; les clauses techniques particulières des contrats (§ 1.3.3) relatives à la conception, mises au point avec les constructeurs par la Région d'Equipement, doivent recevoir l'accord du SEPTEN,

  . assiste la Région d'Equipement pour le contrôle des études des constructeurs,

  . donne à la Région d'Equipement ses avis sur tout ce qui touche à cette mission de contrôle,

  . établit des comptes rendus de son action de contrôle,

  . contrôle l'organisation de la qualité des bureaux d'études.

- La Région d'Equipement :

  . est chargée (en liaison avec le SEPTEN et sous réserve des dispositions du § 3.1.2.2.) d'effectuer les actions définies en 1.3.3 et 1.3.4,

  . en règle générale, suit les avis du SEPTEN,

  . notifie aux constructeurs les positions d'EDF,

  . rend compte à l'Administration en se faisant assister par le SEPTEN.

L'organisation interne du SEPTEN est telle que les principes figurant au § 2.2 ci-dessus soient respectés. Le contrôle externe du SEPTEN est assuré pour le compte de la Direction du SEPTEN par un attaché au Chef du SEPTEN.

L'organisation interne des Régions d'Equipement doit être telle qu'elle permette d'effectuer convenablement les tâches décrites ci-dessus et notamment que les avis du SEPTEN soient en règle générale suivis et que les positions d'EDF soient notifiées aux constructeurs en temps voulu. Le SEPTEN exerce le contrôle externe du fonctionnement de cette organisation, pour ce qui concerne la conception, pour le compte du Directeur de l'Equipement.

3.1.2 - Organisation de la qualité de la construction en usine

3.1.2.1 - Définition des activités

Les activités d'EDF concernant la construction en usine sont, parmi celles qui sont définies aux §§ 1.3.1, 1.3.2, 1.3.3 et 1.3.4, les suivantes :

- convenir (pièces contractuelles) avec les constructeurs des règles techniques que ceux-ci doivent respecter dans la construction (choix et mise en oeuvre des matériaux, contrôles associés),
- s'assurer que les constructeurs établissent les spécifications techniques prévues par les pièces contractuelles et que ces spécifications satisfont aux règles techniques visées ci-dessus,
- s'assurer que les constructeurs prennent les dispositions nécessaires pour parvenir à une exécution et un contrôle satisfaisants de leurs fabrications et notamment qu'ils effectuent leurs approvisionnements et exécutent les opérations de fabrication et de contrôle associées en se conformant aux spécifications techniques reconnues bonnes pour exécution.

3.1.2.2 - Organisation

Les activités décrites en 3.1.2.1 sont réparties entre le SCF et la Région d'Equipement, chargée de la réalisation, de la façon suivante :

- Le SCF :
  . met au point avec les constructeurs (dans la concertation avec le SEPTEN et les Régions d'Equipement) les règles techniques générales relatives à la construction (cahier des prescriptions de fabrication et de contrôle, clauses techniques relatives à la fabrication et au contrôle, cahiers de spécifications techniques), les clauses particulières de l'espèce, mises au point par la Région d'Equipement, devant recevoir également l'accord du SCF,
  . contrôle l'action des fournisseurs pendant la construction (examen des spécifications techniques et contrôle de l'action proprement dite),
  . donne à la Région d'Equipement ses avis sur tout ce qui touche à cette mission de contrôle, prend, le cas échéant, des mesures conservatoires à l'égard des constructeurs et en rend compte à la Région d'Equipement,
  . établit des comptes rendus de son action de contrôle,
  . contrôle l'organisation de la qualité des fabricants.

- La Région d'Equipement :
  . en règle générale, suit les avis du SCF,
  . notifie aux constructeurs les positions d'EDF,
  . rend compte à l'Administration en se faisant assister par le SCF.

L'organisation interne du SCF est telle que les principes figurant au § 2.2 ci-dessus soient respectés. Le "Contrôle Externe" est assuré par la Direction du SCF.

L'organisation interne des Régions d'Equipement doit être telle que les avis du SCF soient en règle générale suivis et que les positions d'EDF soient notifiées aux constructeurs en temps voulu. Le SCF exerce le contrôle externe du fonctionnement de cette organisation, pour ce qui concerne la construction en usine, pour le compte du Directeur de l'Equipement.

3.1.3 - Organisation de la qualité durant l'exploitation

3.1.3.1 - Définition des activités

Les activités d'EDF concernant l'exploitation des tranches sont celles qui sont définies au paragraphe 1.3.5 et au paragraphe 1.3.6, complétées par des actions destinées à améliorer la disponibilité des tranches et la sécurité des agents (formation permanente, essais périodiques de matériels, utilisation de procédures écrites, ..).

3.1.3.2 - Organisation

Les activités décrites en 3.1.3.1 sont réparties entre l'unité d'exploitation et les Services Centraux du Service de la Production Thermique (SPT) de la façon suivante :

- . le respect des règles générales d'exploitation relève de la seule responsabilité du Chef de Centrale,
  . un contrôle interne à la Centrale, placé sous la responsabilité du Chef de Centrale Adjoint porte sur les points suivants :
    - formation et entraînement du personnel,
    - préparation et mise à jour des consignes d'exploitation ainsi que des procédures d'essais et d'entretien,
    - dépouillement des essais,
    - préparation, vérification et archivage des documents périodiques ou réglementaires liés à la sûreté,
    - modification en cours d'exploitation.
- Une mission de contrôle externe des actions de sûreté est confiée au Chef du Département de Sûreté du Service de la Production Thermique qui est chargé de s'assurer du respect des dispositions du Rapport de Sûreté et des règles générales d'exploitation propres à chaque Centrale ainsi que la réglementation générale et les prescriptions particulières.

3.2 - Chez les Constructeurs

L'Organisation de la qualité qu'EDF demande d'adopter aux Constructeurs, sous-traitants et sous-commandiers est analogue à celle qu'il s'applique lui-même (cf. annexe à la Convention "Organisation de la Qualité").

Annexe à la Convention

ORGANISATION DE LA QUALITE

La présente annexe à la Convention a pour objet de définir des règles générales que le Constructeur doit respecter en vue d'obtenir que la fourniture objet du présent contrat (études, approvisionnements, fabrication et contrôles associés, montage et contrôles associés, essais, mise en service ...) ait, d'une manière générale, la qualité requise et qu'en particulier les matériels correspondants ou concernés par les prestations du Constructeur soient aptes à satisfaire l'usage auquel ils sont destinés.

1 - Le Constructeur prend les dispositions permettant de contribuer, dans le cadre de ses obligations contractuelles et pour la part de prestations qui lui est confiée, à obtenir que soient mis à la disposition d'E.D.F. des ouvrages, matériels ou équipements réalisés suivant les règles de l'art et dont le comportement en service soit satisfaisant.

Le Constructeur prend aux mêmes fins des dispositions de même nature à l'égard de tous les sous-traitants et sous-commandiers auxquels il fait appel directement, ou qui contribuent indirectement à l'exécution du contrat.

Le Constructeur prend toutes mesures pour que les dispositions qui sont ainsi prises par lui-même et par les sous-traitants et sous-commandiers soient effectivement appliquées. Ces mesures doivent notamment permettre que soient respectés les lois et règlements applicables, les règles et prescriptions techniques contractuelles ainsi que toutes les dispositions contractuelles relatives à l'établissement des études, plans, schémas et spécifications "bon pour exécution".

2 - Le Constructeur et les sous-traitants et sous-commandiers doivent notamment prendre, en ce qui concerne tant leur propre activité que leurs relations mutuelles, les dispositions nécessaires d'organisation qui les mettent en mesure de satisfaire aux obligations du paragraphe 1 ci-dessus. Ces dispositions constituent ce qui est dénommé dans le présent contrat : "organisation de la qualité".

3 - L'organisation de la qualité doit être décrite de façon précise et détaillée dans des documents écrits (appelés "états descriptifs de l'organisation de la qualité") et tenus à jour par le Constructeur et les sous-traitants et sous-commandiers, tant en ce qui les concerne que vis-à-vis des tiers par rapport à eux-mêmes. Ces états descriptifs doivent être tenus, en permanence, à la disposition d'EDF.

Les états descriptifs établis par le Constructeur, les sous-traitants et les sous-commandiers doivent notamment comporter :

- les effectifs et leur répartition dans les différents services et usines,
- la liste des moyens techniques principaux dont dispose chaque service et chaque usine,
- une description précise des structures et une définition détaillée des missions et des responsabilités de chaque service, de chaque usine, et de chaque personne dont l'activité touche à l'organisation de la qualité,
- la liste de ces personnes, accompagnée d'indications sur leur compétence, en renvoyant, le cas échéant, à des documents de caractère plus confidentiel,
- la description détaillée des modalités d'élaboration, de correction et de vérification des documents ainsi que de leur circulation et de leur diffusion.

L'état descriptif du Constructeur doit préciser la nature et les modalités d'exercice du contrôle de l'action des sous-traitants et sous-commandiers.

4 - L'organisation de la qualité doit être telle que :

- les différentes activités soient convenablement coordonnées et, en particulier, que soit mise en place une bonne articulation entre les personnes et services qui en sont chargés,
- les personnes chargées d'effectuer des études ou d'établir des projets, des plans, des schémas ou des spécifications aient une connaissance suffisante des règles qu'ils doivent respecter et des sujétions réelles de la fabrication,
- les études, projets, plans, schémas ou spécifications établis par une personne soient vérifiés par un supérieur de cette personne, d'un niveau hiérarchique suffisant pour juger convenablement de l'intégration de la partie concernée dans l'ensemble, et d'une qualification adaptée à cette tâche de vérification,
- la procédure suivie pour l'établissement et la vérification de ces documents soit conduite suivant des règles écrites et traduites sur un état visés par les personnes qu'elle concerne,
- les opérations d'approvisionnement, de fabrication, de montage de contrôle, d'essais et de mise en service fassent l'objet de spécifications techniques définissant de façon précise et complète ces opérations,
- ces opérations ne soient entreprises qu'après qu'aient été établies les spécifications bonnes pour exécution et soient réellement effectuées suivant ces spécifications,
- ces opérations donnent lieu à des comptes rendus ou procès-verbaux qui doivent être tenus à la disposition des personnes qui ont effectué les études ou établi les spécifications techniques.

L'ensemble des vérifications et contrôles visés dans ce paragraphe constitue ce que l'on convient d'appeler le "contrôle interne".

5 - L'organisation de la qualité suppose notamment de la part de la Direction :

- que celle-ci veille à ce que soient intégralement respectées les dispositions contenues dans l'état descriptif de l'organisation de la qualité et que cet état soit tenu à jour,
- qu'elle ait le souci constant de juger de l'efficacité de l'organisation de la qualité et apporte à celle-ci, le cas échéant, les redressements ou améliorations utiles.

Dans ce but, la Direction doit disposer d'agents ayant l'indépendance, la compétence et l'autorité voulues pour l'assister dans cette action que l'on convient d'appeler "contrôle externe".

Ces agents ont pour mission :

- de contrôler que l'état descriptif de l'organisation de la qualité est à jour et que ses dispositions sont intégralement appliquées,
- de contrôler en procédant le cas échéant à des sondages, que ces dispositions permettent de satisfaire aux prescriptions du paragraphe 1,
- de proposer à la Direction toutes mesures correctives qui s'avèreraient nécessaires au vu des résultats de ces contrôles.

6 - Outre l'application des dispositions de l'article 7 du C.C.A.G., il est précisé qu'aucun sous-traité ni sous-commande ne peut être

passé tant que le Constructeur n'a pas certifié que le sous-traitant ou le sous-commandier dispose d'une organisation de la qualité telle qu'elle permette de satisfaire aux dispositions qui précèdent.

7 - Les dispositions de la présente annexe doivent figurer dans tout sous-traité ou sous-commande.

8 - EDF se réserve le droit d'exercer ou de faire exercer par telles personnes de son choix le contrôle de la mise en oeuvre et de l'application des dispositions qui précèdent, tant par le Constructeur que par les sous-traitants et sous-commandiers. Le Constructeur et les sous-traitants et sous-commandiers doivent assurer aux représentants dûment mandatés d'EDF le libre accès de leurs services et de leurs usines, pendant les heures de travail, et leur donner toutes facilités pour leur permettre de remplir entièrement leur mission.

9 - L'inobservation de l'une quelconque des dispositions qui précèdent constitue un "manquement grave du Constructeur à ses obligations contractuelles" et pourra entraîner l'application de l'article 28-5 du C.C.A.G.

DISCUSSION

(on papers IAEA-SM-195/9, IAEA-SM-195/41 and IAEA-SM-195/45)

A.V. KARPOV: I would like to direct a question to Mr. Ferriss and to our French colleagues. Are there any standards in your countries covering quality assurance measures to be carried out on a reactor during the operational phase?

W.E. FERRISS: The Bechtel Corporation is involved in the design and construction of nuclear power plants; the utilities are responsible for the operation of the plants. However, the Nuclear Regulatory Commission has published a set of quality standards for use during the operational phase: this is called the 'Orange Book'.

J.-A. VAUJOUR: In France there are no standards of the kind Mr.Karpov has in mind.

A.V. KARPOV: What reliability tests are planned for the primary circuit of Phénix, and when are they to be carried out?

J. NORMAND: Phénix is a plant with an integrated primary circuit; the circuit components, pumps and heat exchanger are located within the main vessel. This means that in-service testing will in fact bear on the vessel, and experiments related to in-service testing of the vessel welds of Super-Phénix are in progress at present.

P.G. SMITH: I should like to ask Mr. Ferriss whether the experience of the Bechtel Corporation suggests that the application of quality assurance to the construction of nuclear power stations can be shown to be economically advantageous in terms of the improvement of reliability, irrespective of the enhancement of safety. Does the United States power industry apply the same quality assurance programmes to items destined for nuclear sites which are not defined as 'safety-related' items, or to conventional (fossil-fired) plant?

W.E. FERRISS: The Bechtel quality assurance programme designed to satisfy the requirements of the Nuclear Regulatory Commission is in fact applied to safety-related components, systems and structures; normally it is applied to items which are not directly related to safety only at the request of the client. Bechtel does have a quality assurance programme for fossil fuel power plants.

L. CAVE: This Symposium has shown the need to treat reliability in a quantitative manner if safety and availability requirements are to be met. However, in the nuclear industry at present, the documents specifying quality assurance requirements are all written in very qualitative terms. Do not many of you have the impression, as I do, that something should be done to put these requirements into quantitative terms? In the munitions industry 'quality' has been specified in quantitative terms for many years, and in the aero-space industry, too, this objective has been achieved. I agree that it may be difficult in the nuclear industry, but I feel that a determined effort should be made. Reliance on exhortation is particularly likely to be ineffective in the case of sub-contractors who have both nuclear and non-nuclear products going through their shops at the same time; it is difficult for the man on the shop floor to appreciate that a difference in quality may be required.

B. CAFFREY: Each speaker has emphasized how important an appropriate attitude to quality is, if quality is in fact to be achieved.

I would like to add to their comments and offer a personal opinion as to how that attitude can be implanted. I feel that an 'open' approach is necessary; this means that all parties should feel free to express their ideas on the subject quite openly. It also implies a positive attitude by all parties, for confidence breeds confidence as surely as diffidence is degenerative.

Finally, a question. Does anyone (Mr. Ferriss perhaps) have an idea what fraction of the total nuclear plant cost is attributable to quality?

W.E. FERRISS: I would say that 10 to 12% of the nuclear power plant cost can be attributed to activities associated in one way or another with quality assurance.

W.A. WOLFE: Ten per cent for quality assurance strikes me as a very high figure, because if the calculation includes such items as interest during construction and civil works, it is a much larger fraction of the acquirement cost. In your instructions to bidders, do you specify the cost of quality assurance as a separate item, and, if so, can you state it as a percentage of the acquirement cost of the items?

W.E. FERRISS: The cost of quality assurance is a separate cost item for many sub-contracts, but not at present for procured items.

A.C. PROT: Quality assurance is often defined as 'good management', a term which in fact appears in a recent document of the IAEA. This implies that it bears a relationship not just to safety but to availability as well. I wonder what Mr. Ferriss thinks of this definition.

W.E. FERRISS: I can only repeat what I said before, namely that the quality assurance programmes developed by Bechtel to satisfy the requirements of the Nuclear Regulatory Commission are focussed primarily on safety-related items.

J. KRASNODEBSKI: I would like to ask Mr. Vaujour whether in France availability targets are specified for units, systems and individual items of equipment, and in particular whether standards for reliability and maintenance are specified in quantitative terms to suppliers.

J.-A. VAUJOUR: There are no reliability clauses in our contracts except in special cases. In the present state of our knowledge it is extremely difficult to quantify the probability of failure of equipment, other than for a very large production series of constant quality. This being so, it is difficult to exact guarantees from the suppliers.

S. HARTWIG: Do you think there is a limit of reliability beyond which the effort to secure a further quantitative gain in quality is no longer worth while — beyond which, that is, any residual risk should be covered by insurance?

J.-A. VAUJOUR: Where items of plant which have a direct bearing on nuclear safety are concerned, it is hard to imagine that one would ever be able to establish a level of reliability beyond which further progress would no longer be sought. For items whose failure would affect only availability and not safety, it is theoretically possible to establish a level beyond which the wiser course would be to cover any residual risk by insurance; but in view of what has already been said, you will realize that even this is extremely difficult.

B. CAFFREY: The Canadian approach referred to by Mr. Krasnodebski — that of including reliability requirements in enquiry specification documents — has also been mentioned by Mr. Cave of the United Kingdom and is in fact very attractive for the utilities. It raises questions connected with the commercial aspect of contracts, however, and might, I feel, present certain practical difficulites. Successful implementation probably depends on the commercial terms of the contract, particularly on the type of contract price arrangement selected (cost plus, fixed price or incentive arrangement with a penalty/bonus clause). It would be interesting to hear from those who have experience of handling such reliability requirements how successful they have been, and what sort of price arrangements applied.

L. CAVE: I know that the Tennessee Valley Authority and the South of Scotland Electricity Board both have contracts which impose a penalty on the manufacturer for failure to achieve a certain level of availability during the first few years of plant life.

W.A. WOLFE: Let me add a word about reliability specifications in contracts. For many years we have had such clauses in contracts relating to fuel, and these specifications have been met. Recently Hydro Quebec has included reliability clauses in specification, particularly for pumps, requiring studies and data in support of the manufacturers' claims. I understand, too, from my British colleagues that United Kingdom turbine manufacturers now have identifiable reliability groups as a result of contract requirements.

G. BOGANI: Have you, Mr. Vaujour, established specific quality classes for components?

J.-A. VAUJOUR: In a sense, yes. There are only two true classes: items of plant whose failure would have consequences for safety or availability and, in class two, all other items. In the first of these classes we have however established several levels which correspond to the extent and the stringency of the non-destructive tests imposed on the equipment.

A. J. SCHEPENS: I have a question for Mr. Normand. Is it always possible, under a quality assurance programme, to reconcile the wishes of the operator and the habitual manufacturing practices of the supplier, especially in the case of turnkey contracts?

J. NORMAND: The main objective of course is to build power stations that function properly, using the best and the most advanced manufacturing technology. The engineers must take care to supply plants, even under turnkey arrangements, of the best possible quality. Obviously, traditional practice and technology may not be adequate for the purpose, and in such cases preliminary tests will be needed to improve the technology. Close contact should always be maintained between the supplier's engineers and the operator; we have in fact achieved this kind of co-operation at Technicatome, which is a subsidiary of the CEA, and EDF. A point to be noted is that the stations delivered to EDF in France are not supplied under turnkey arrangements. However, the concept of our quality assurance system does give the owner-operator precisely the scope he needs to define his requirements, and enables the supplier to assume proper responsibility for the quality of his products.

NUCLEAR TECHNOLOGY STANDARDS AND RELIABILITY

Chairman:

L. ALVAREZ DE BUERGO, Spain

О СОТРУДНИЧЕСТВЕ СТРАН-ЧЛЕНОВ СОВЕТА ЭКОНОМИЧЕСКОЙ ВЗАИМОПОМОЩИ В ОБЛАСТИ НАДЕЖНОСТИ АТОМНЫХ ЭЛЕКТРОСТАНЦИЙ

А.П.БАРЧЕНКОВ, М.КРОЙТЦБУРГ
Секретариат Совета Экономической Взаимопомощи*

Abstract–Аннотация

CO-OPERATION AMONG COUNTRIES MEMBERS OF THE COUNCIL FOR MUTUAL ECONOMIC ASSISTANCE IN CONNECTION WITH NUCLEAR POWER PLANT RELIABILITY.

In recent years the countries belonging to the Council for Mutual Economic Assistance have developed bilateral and multilateral co-operation in the field of nuclear power plant reliability on a steadily increasing scale. This paper describes the stages in the evolution of that co-operation, which has helped to accelerate the construction and improve the technical and economic characteristics of nuclear power plants. It gives a number of examples of scientific and technological work, planning and design studies, and work associated with the actual construction and reliable operation of nuclear power plants.

О СОТРУДНИЧЕСТВЕ СТРАН-ЧЛЕНОВ СОВЕТА ЭКОНОМИЧЕСКОЙ ВЗАИМОПОМОЩИ В ОБЛАСТИ НАДЕЖНОСТИ АТОМНЫХ ЭЛЕКТРОСТАНЦИЙ.

Страны-члены Совета Экономической Взаимопомощи (СЭВ) за последние годы все шире развивают двустороннее и многостороннее сотрудничество в области обеспечения надежности атомных электростанций (АЭС). В докладе приводятся этапы развития сотрудничества стран-членов СЭВ по надежности АЭС, что способствует ускорению их строительства и улучшению технико-экономических показателей. В докладе приводятся некоторые примеры работ, проводимых в области научно-технических и проектно-конструкторских разработок, а также в области сооружения и надежной эксплуатации АЭС.

На современном этапе одной из важных проблем топливно-энергетической области, по которой широко развивается сотрудничество стран-членов СЭВ, является атомная энергетика. Это определяется тем, что энергетика в целом, и атомная энергетика в частности, является одной из наиболее сложных, капиталоемких, динамично развивающихся сфер экономики.

Внимание к развитию атомной энергетики в странах-членах СЭВ определяется также тем, что атомные электростанции с отдельными типами реакторов к настоящему времени достигли оправданного уровня экономической эффективности, надежности и способствуют уменьшению загрязнения окружающей среды.

В принятой XXV сессией СЭВ Комплексной программе дальнейшего углубления и совершенствования сотрудничества и развития социалистической экономической интеграции стран-членов определены основные направления сотрудничества в области развития атомной энергетики на длительную перспективу.

Комплексной программой, в частности, предусматривается разработка предложений о создании совместными усилиями заинтересованных стран-членов СЭВ научно-технических, производственных и организационных предпосылок для ускорения и эффективного внедрения в народное хозяйство атомной энергии в промышленных масштабах.

* Адрес: СССР, Москва, Г-205, Проспект Калинина, 56.

Учитывая, что развитие атомной энергетики связано с проведением большого комплекса научно-исследовательских, опытно-конструкторских и экспериментальных работ, а также накоплением опыта по освоению и эксплуатации атомных электростанций, что требует значительных затрат денежных средств и материальных ресурсов, страны-члены СЭВ осуществляют развитие атомной энергетики на основе широкого сотрудничества в рамках Совета Экономической Взаимопомощи.

Прогресс в развитии атомной энергетики, результаты научно-технических, проектно-конструкторских разработок, проведенных на основе сотрудничества, содействовали странам-членам СЭВ в разработке вопросов развития атомной энергетики.

Намечаемое странами-членами СЭВ на ближайший период строительство атомных электростанций базируется на применении разработанных и освоенных энергетических реакторов на тепловых нейтронах, в частности, с водой под давлением, типа ВВЭР единичной электрической мощностью 440 МВт (ВВЭР-440).

В 1974 г. были введены в строй энергоблоки с реакторами типа ВВЭР-440 на атомных электростанциях "Козлодуй" в НРБ, "Норд" в ГДР и на Кольской АЭС в СССР общей электрической мощностью 1320 МВт.

На конец 1974 г. общая мощность атомных электростанций стран-членов СЭВ (НРБ, ГДР, СССР и ЧССР) достигла около 6000 МВт. Осуществляется сотрудничество по подготовке к сооружению атомных электростанций в ВНР, Республике Куба, ПНР и СРР.

Проводятся работы в области прогнозирования развития атомной энергетики в качестве составной части общего прогноза о развитии генерирующих мощностей стран-членов СЭВ на период до 1990 г.

Как следует из прогноза, страны-члены СЭВ намечают к 1980 г. мощность атомных электростанций довести примерно до 30 000 МВт, а к 1990 г. мощности увеличатся в несколько раз.

В разрабатываемых органами СЭВ прогнозах развитие производства энергетического оборудования рассматривается с учетом сооружения в странах-членах СЭВ атомных электростанций с реакторами большой мощности на тепловых нейтронах и возможности применения в ближайшие 10 лет реакторов на быстрых нейтронах. Производство оборудования для атомных электростанций будет базироваться на широкой специализации и кооперировании. Для решения этих задач создано Международное хозяйственное объединение "Интератомэнерго", членами которого являются назначенные странами предприятия европейских стран-членов СЭВ и СФРЮ.

Вопросы надежности и безаварийности работы атомных электростанций, их безопасности для окружающей среды занимают одно из важных мест в сотрудничестве стран-членов СЭВ в области реакторной науки и техники и атомной энергетики. С начала сотрудничества стран-членов СЭВ по развитию атомной энергетики и строительству атомных электростанций систематически проводятся симпозиумы, семинары и другие встречи ученых и специалистов по обмену научным и техническим опытом в области повышения надежности атомных электростанций. Так например, в 1968 г. Советом Экономической Взаимопомощи был проведен в Москве симпозиум о состоянии и перспективах развития атомных электростанций с водо-водяными реакторами, где вопросам надежности атомных электростанций было уделено особое внимание.

На симпозиуме был освещен опыт по исследованиям и расчетам нестационарных теплофизических и гидродинамических характеристик активной зоны реактора, а также виброустойчивости ее отдельных элементов и устранения причин пульсации потока теплоносителя первого контура. Были рассмотрены вопросы критических тепловых потоков в пучках тепловыделяющих элементов в стационарном и нестационарном режимах теплообмена, связанных с расхолаживанием активной зоны реактора при внезапном прекращении циркуляции теплоносителя, вопросы надежности работы систем контроля герметичности тепловыделяющих элементов, системы управления и защиты реактора и по ряду других проблем, связанных с повышением надежности атомных энергетических реакторов.

В 1972 г. на основе имеющегося опыта эксплуатации АЭС был проведен семинар по вопросам надежности работы атомных электростанций с водо-водяными реакторами. На семинаре были рассмотрены вопросы, связанные с обеспечением более глубокого выгорания ядерного горючего и освоением прогрессивных методов регулирования реактивности активной зоны реактора с использованием борной кислоты, сокращением времени между физическим и энергетическим пуском реактора, а также вопросы перехода с принудительной на естественную циркуляцию теплоносителя первого контура, что связано с повышением надежности расхолаживания реактора в аварийных ситуациях.

Вопрос, связанный с анализом возможных аварий реакторов при потере теплоносителя, детально рассмотрен на совещании специалистов стран-членов СЭВ в ноябре 1974 г.

В согласованных программах научно-технического сотрудничества заинтересованных стран-членов СЭВ в области реакторов на тепловых нейтронах большой мощности значительное место занимают вопросы, связанные с исследованиями и разработкой опытных образцов более совершенных конструкций основного реакторного оборудования для атомных электростанций, отвечающего современным требованиям надежности, безопасности и экономичности.

К таким элементам оборудования относятся парогенераторы, насосы большой производительности для первого контура, приводы системы управления и защиты реактора, быстродействующая арматура, датчики для внутриреакторного контроля и т.п. На основе широкого сотрудничества также рассматриваются вопросы, связанные с выбором конструкционных материалов для оборудования первого контура, выработки рекомендации по водным режимам реакторной установки и применению специальных ионообменных смол и т.д.

В рамках сотрудничества по проблемам проектирования, строительства и эксплуатации АЭС рассматриваются, в частности, вопросы:

– анализ и обобщение мирового опыта в области проектирования, сооружения и эксплуатации атомных электростанций и разработка предложений по улучшению технико-экономических показателей и надежности действующих АЭС и совершенствованию проектных решений по атомным электростанциям;

– разработка предложений по сотрудничеству и обмену опытом в подготовке специалистов;

– разработка предложений на основе обобщений опыта стран в части строительства, монтажа и пуско-наладочных работ в целях повышения экономичности строительства АЭС.

Развивая сотрудничество стран-членов СЭВ в этой области, был разработан ряд материалов, относящихся, в частности, к следующим проблемам:

- принципы составления правил технической эксплуатации АЭС;
- принципы надежности и радиационной безопасности АЭС;
- организация и проведение контроля сварных соединений в процессе эксплуатации;
- форма эксплуатационного отчета АЭС;
- вопрос терминологии в области АЭС;
- условия и критерии при выборе площадок для строительства АЭС;
- сотрудничество заинтересованных стран-членов СЭВ в подготовке специалистов в области атомной энергетики;
- основные принципы по водному режиму и методам очистки радиоактивных вод АЭС и др.

Результаты разработки этих нормативных положений используются заинтересованными странами-членами СЭВ при составлении ими национальных документов.

Например,"Основные принципы составления правил технической эксплуатации АЭС" составлены применительно к АЭС с реакторами типа ВВЭР, как наиболее распространенными в странах-членах СЭВ, однако они могут быть использованы и применительно к составлению правил технической эксплуатации с реакторами других типов. В данном материале освещен круг вопросов и требований к эксплуатации АЭС, в частности, вопросы безопасности, в том числе ядерной и радиационной; требования к тепломеханическому оборудованию, в том числе к ядерной паропроизводящей установке, вентиляции, транспортно-технологическим операциям, системе удаления газообразных отходов и т.д.; требования к электротехническому оборудованию, в том числе к генераторам, распределительным устройствам, релейной защите и автоматике и т.д.; вопросы организации эксплуатации, в том числе требования к оперативному персоналу, диспетчерскому управлению, ремонту оборудования, технической документации и т.д.

Другим примером совместно разработанного документа являются "Основные положения по организации и проведению контроля за состоянием основного металла и сварных соединений трубопроводов и оборудования АЭС в процессе эксплуатации". В нем представлены рекомендации по организации и объему контроля оборудования АЭС в процессе монтажа и эксплуатации. Кроме того, даны мероприятия по контролю, которые должны быть выполнены на стадии проектирования трубопроводов и оборудования АЭС.

Направляя таким образом свои усилия, с одной стороны, на вопросы научно-технического сотрудничества и проектно-конструкторских разработок, а с другой стороны, на вопросы, связанные с сооружением и эксплуатацией АЭС, включая и вопросы подготовки эксплуатационного персонала, может быть охвачен весь комплекс вопросов, связанный с обеспечением надежности АЭС.

Осуществляемое странами-членами СЭВ сотрудничество в области ядерной энергетики, оказывает положительное влияние на повышение надежности атомных электростанций.

Развитие и углубление экономического и научно-технического сотрудничества стран-членов в области атомной энергетики будет и дальше занимать важное место в деятельности СЭВ.

CODES, STANDARDS AND SAFETY REQUIREMENTS FOR ANGRA UNIT 1, BRAZIL'S FIRST NUCLEAR POWER PLANT

M.T. MAGALHÃES
Furnas — Centrais Elétricas S.A.,
Rio de Janeiro,
Brazil

Abstract

CODES, STANDARDS AND SAFETY REQUIREMENTS FOR ANGRA UNIT 1, BRAZIL'S FIRST NUCLEAR POWER PLANT.

At the time of the call for bids (June 1970) for Angra Nuclear Power Plant, only one relevant site selection regulation had been formulated. Thus a group of requirements were selected from abroad (basically United States ones) to be included in the bid specification. Since a United States company was the successful bidder, the requirements were imposed without major difficulty. The paper describes the consequent problems, namely the design modifications, cost increase and design and construction schedule slippage that resulted from the subsequent issuance of a great number of regulatory guides by the United States Atomic Energy Commission which were adopted by the Brazilian Regulatory Body. The paper concludes by mentioning some important lessons learned from the experience to date.

The first nuclear power plant in Brazil is in course of construction and will be operated under the responsibility of FURNAS - Centrais Elétricas S.A., a power company operating throughout the South-Eastern region of Brazil, where it produces bulk power for sale to distributors. FURNAS is the largest subsidiary of the federal government holding company ELETROBRÁS

The FURNAS installed capacity is today about 3350 MW hydro power and 650 MW oil-fired power. Notwithstanding the hydraulics resources to be explored, the Brazilian Government has decided to build a nuclear power plant in order to increase the thermal augmentation to the system and to prepare technicians for nuclear technology.

The Angra project will comprise a pressurized water reactor rated at 1876 MW(th), enclosed in a steel containment structure, a 657-MW(e), 1800-rev/min three-casing condensing turbine, a 760-MVA, 0.9-p.f. (power factor), hydrogen-cooled generator and related equipment.

The project is being carried out on the basis of four main contracts: the first one includes the supply and erection of the nuclear steam supply system, the turbo-generator set with all its auxiliaries and the overall design (plant contract); the second comprises the nuclear fuel; the third covers the steel containment design, supply and erection; and the fourth one includes the civil works.

The bid specification of design and equipment was a direct reflection of the policies adopted for the conduct of the overall project. In the Angra case one fundamental policy was to minimize potential problems by utilizing proven design concepts as well as experienced suppliers.

Although National Nuclear Energy Commission (CNEN), which is the Brazilian regulatory body, and the Brazilian Standard Association (ABNT) existed at the time of the call for bids (June 1970), only one relevant CNEN site selection regulation had by then been formulated and no one single ABNT standard relating to the nuclear field was applicable at the time. Thus, a group of codes, standards and regulations were selected from abroad to be included in the bid specification.

In the bid specification the following requirements were imposed, related to the subject:

(i) The bidder should comply with a selected list of mandatory codes and standards, basically U.S. ones (ASME, ANSI, IEEE codes and standards) or, alternatively, conform to national standards in the country of origin of the bidder, if equivalent. Detailed and comprehensive technical justification should be provided for all deviations from the mandatory codes and standards. For specification purposes relating to radiological protection it was decided to adopt standards prepared by international organizations (ICRP and IAEA);

(ii) The design should comply with a design criterion based on the United States Atomic Energy Commission's (USAEC) General Design Criteria;

(iii) The project should comply with quality assurance requirements derived from United States Atomic Energy Commission's (USAEC) Quality Assurance Criteria for Nuclear Power Plants.

Since a United States company, with a pressurized water reactor design, was the successful bidder for the partial turnkey type contract the three items above were imposed without difficulty, based on United States rules.

In the plant contract terms one more requisite was added: equipment, material and design features for the plant should be licensable on the basis of USAEC safety requirements as of 26 January, 1971, the date the bid tenders were submitted.

To complement this requirement the concept of a reference plant was established. The intention was to have a design essentially equal to a previously designed and licensed plant with changes limited to those required by currently known deficiencies. A U.S. nuclear power plant with approximately the same power rating, designed and manufactured by the same companies as the Angra Plant and with a schedule about one year ahead of the Angra schedule, was selected. Unhappily the schedule delay of the selected reference plant has cancelled out the anticipated benefits.

At the time of the signing of the fuel contract there were no specific standards or regulations in existence which were directly applicable to the design and fabrication of nuclear fuel. Therefore, to obtain some protection in the event that the fuel design or fabrication should be defective, a warranty in the fuel contract was imposed which guaranteed the mechanical integrity of each discharged region which should reach a warranted average discharge burnup. For licensing purposes a reference fuel design level was included in the fuel contract, equivalent to the level of design expressed in the construction permit of a selected U.S. nuclear plant.

Another move that was very useful in verifying the quality and reliability of the plant was the contracting of an experienced architect-engineering firm to assist in the performance of independent quality assurance, quality control and design review checks.

Subsequent to the signing of the nuclear steam supply system and nuclear fuel contracts, CNEN has issued some safety regulations and has adopted a number of the Regulatory Guides and new criteria proposed by USAEC. Recently ABNT is adopting a series of standards based on ISO standards; however, this move has not affected the Angra project.

As a consequence of the Angra Plant contract terms and conditions and the actions of CNEN, a sizeable cost increase has been incurred mainly as a result of the extensive additional engineering analysis and design work that had to be done or re-done to comply or prove compliance with the imposed requirements. At the present time this added cost is about 4% of the estimated total plant cost at the time the major contracts were signed.

Design and construction schedule slippage, due to licensing, has been a minor problem.

In addition to the application of a multitude of Regulatory Guides, the Utility and the Regulatory Body are taking the approach of performing a detailed design review. The overall objective of this effort is to assure a uniform design approach throughout the plant. This has been done mainly to avoid the situation where the designer is being overcautious in the design of some areas of the plant as a result of imposed regulations and being lax in the design of other areas which are equally important from the plant reliability or safety standpoint but are not subject to specific safety regulations or criteria.

The steps adopted for the Angra project have been effective, but to minimize the effects of the above problems on the cost and schedule of a plant in a situation similar to the Angra project, the following actions should be considered. A flexible contract with the designers, suppliers and constructors should be made in relation to extra work to facilitate the implementation of changes when a new regulation is imposed; an agreement should be made with the regulatory body to establish a procedure for determining the applicability of codes, standards and regulatory guides; at least a reference nuclear steam supply system design should be selected and incorporated into the contract as the basis for design.

DISCUSSION

(on papers IAEA-SM-195/57 and IAEA-SM-195/53)

J.A. CREMADES: I have a few comments to make in connection with Mr. Magalhães' presentation. You mentioned the problem created by the continuous evolution of safety requirements in the United States of America and their adoption in other countries. As a solution you suggest the introduction of flexible contracts with suppliers which would make allowance for the cost and schedule delay implications of implementing new regulations. A certain degree of flexibility is already included for this purpose in current contracts, so I should be interested to know exactly how much flexibility you are recommending.

A point to bear in mind is that sellers in the United States have the opportunity of discussing new guidelines with the Nuclear Regulatory Commission and experience shows that these discussions can be lengthy. Also, since authorization to construct a nuclear power plant depends in most countries on a certain percentage of participation by local industry, it is not likely that any seller will be prepared to absorb all or a major part of the cost of meeting new safety requirements in a foreign country, particularly in the area of quality assurance.

M.T. MAGALHÃES: The Angra plant contract has no flexibilty relating to 'extra work'. The best solution for a new contract of course is to have a 'cost plus' kind of arrangement in which the cost of equipment and engineering is clearly spelled out. In this way the true costs can be checked more easily. In our case, all the cost of implementing new regulations falls on FURNAS. I agree that it is impossible for sellers to forecast the cost increases which new regulations will entail.

J.R. CABRAL: You mentioned that FURNAS and the National Nuclear Energy Commission (CNEN) are performing a detailed design review. Is this done jointly by the two bodies or separately? And what is the position of CNEN with regard to design approval?

M.T. MAGALHÃES: The design review is being performed independently by FURNAS and CNEN. The latter does not itself approve the design of the plant; it issues to FURNAS a construction permit and subsequently an operation permit, based, respectively, on the Preliminary Safety Analysis Report and the Final Safety Analysis Report submitted by FURNAS.

J. ERVAMAA: Mr. Barchenkov has mentioned many interesting meetings and seminars concerning, among other things, the reliability of nuclear power plants. Have the proceedings of these meetings, or reports on them, been published, and are they generally available?

A.P. BARCHENKOV: Yes, they are published and distributed to Member States of CMEA and also to the IAEA Secretariat under the scientific and technical information exchange programme.

OPERATION AND MAINTENANCE
OF NUCLEAR POWER PLANTS

Chairman:

K. EFFAT, Egypt

AN ENGINEERED MAINTENANCE APPROACH TO IMPROVED NUCLEAR PLANT AVAILABILITY

R.K. ROBINSON, J.T. STRINGER, M.L. FAUGHT
United Nuclear Industries, Inc.,
Richland, Wash.,
United States of America

Abstract

AN ENGINEERED MAINTENANCE APPROACH TO IMPROVED NUCLEAR PLANT AVAILABILITY.

Forced outages resulting from equipment failures have seriously reduced nuclear power plant availability. The current high construction costs of new power plants and world-wide energy shortages provide unusual incentive to improve the operating reliability of nuclear power plants. Since the first plutonium producing reactor at Hanford in the State of Washington, United States of America, went critical in 1944, the need to improve plant availability has been a major objective. It was quickly recognized that one of the most important guarantees for high on-line operating reliability was optimum equipment and system maintenance. Consequently, for the original Hanford production reactors a maintenance program based on a planned and engineered approach was explored and developed to replace the older breakdown and preventive maintenance philosophy. Initial implementation of this program resulted in a 50% reduction in plant equipment forced outages. The dual purpose N Reactor, which achieved full operating capability in 1964, has incorporated a similar engineered maintenance program in its operating philosophy. All equipment essential to assuring nuclear safety and plant availability was included in this program. Acceptable equipment performance limits were established for all programmed equipment together with detailed procedures for routinely verifying equipment performance. Performance data feedback was accumulated by the program and routinely monitored to detect undesirable performance trends for each equipment item included in the program. This program was completed in 1972 with the result that equipment forced outage hours have been significantly reduced. In 1973 forced outage hours for the U.S. nuclear industry averaged 23 per 100 operating hours as compared to 8 per 100 at N Reactor. In 1974 the U.S. nuclear industry average was 20 compared to 6 at N Reactor. This paper presents the scope, development, implementation, and benefits gained from the engineered maintenance program at Hanford's N Reactor.

1. INTRODUCTION

Considerable attention and emphasis has recently been placed on nuclear reactor plant operating availability. The nuclear industry is witnessing an increasing awareness and concern regarding the availability of nuclear plants. Availability and performance comparisons are now being routinely made of one nuclear plant to another, one vendor's design versus another's, and fossil fueled against nuclear powered plants both nationally and internationally. The nuclear critic points to the data and loudly proclaims how bad things are; the nuclear proponent points to the same data and counters that performance is good and improving.

Regardless of how the data is interpreted, the fact remains that whether a plant is fossil or nuclear fueled, each percentage point of unavailability represents a large loss in revenue to a utility plus additional incremental costs for the required purchase of replacement electrical power from higher cost sources. For a 1000 MW(e) nuclear power plant, one percentage point in availability is worth approximately $1,000,000 (U.S.) and for an equivalent sized fossil fueled plant it is worth about \$1,500,000 (U.S.) using average power costs of 10.52 mills/kWh and 17.03 mills/kWh for nuclear and fossil fueled plants respectively.[1] The incentive to improve component and system operating reliability and plant availability is obvious.

Many suggestions, recommendations, and actions have been made and taken with the intent of improving plant reliability. Most of these suggestions, recommendations and actions have been influenced by the more visible and currently significant contributions to plant down times. Typical items include steam generator tube leaks, turbine blade fatigue failures, assorted valve problems, pump failures, and others.

Consequently, emphasis is being placed on improved design, more reliable components, and increased quality assurance activity on the part of the supplier and vendors. It is obvious that improvements can and should be made in these areas of component performance. However, while improvements in these areas will improve plant availability during the early operating periods and make the job easier and less costly to maintain good plant availability in the later years, good, long-term availability is not possible without an effective surveillance and maintenance system judiciously applied and managed by the plant operator.

Figure 1 shows a typical and well demonstrated plant life curve in terms of availability. The first phase is an initial startup and transition period with relatively low availability. This phase of low availability is quite short for plants which are based on well established technology and an established, repetitive design. It is predictably longer for early generation plants, the pioneers of new design and technology. Following the initial shakedown period, plant availability is relatively high, with gradually decreasing performance over the years as equipment and systems age, deteriorate and wear out. The rate of this decrease in availability can be significantly controlled by the

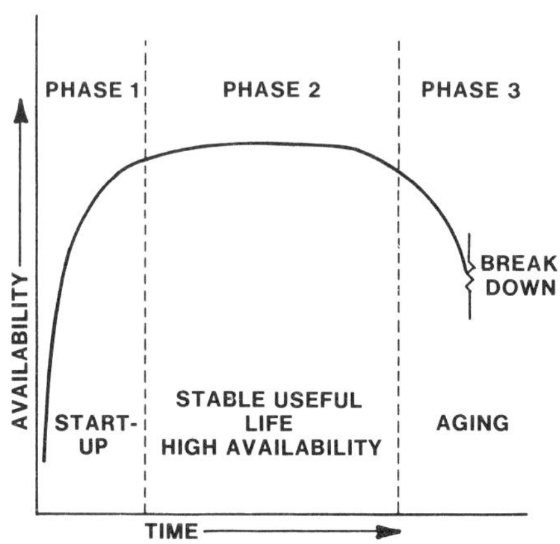

FIG.1. Typical performance life of systems and equipment.

maintenance philosophy applied in the plant, e.g., breakdown maintenance, preventive maintenance, or an engineered maintenance approach consisting of an integrated surveillance and maintenance system.

It has been demonstrated on the dual purpose N Reactor at Hanford, Washington, that the application of an integrated surveillance and maintenance system can continue to improve plant availability and decrease the number of forced outage hours to operating hours, even in a plant which is eleven years old. The N Reactor, in conjunction with the Washington Public Power Supply System electrical generating facilities, currently holds the world's record of 29 billion kilowatt hours generated in a single nuclear reactor plant (as of 1 March, 1975). In 1974 the N Reactor had only 6 forced outage hours per 100 operating hours compared with the nuclear industry average of 20.

2. BACKGROUND

Since the first plutonium producing reactor at Hanford went critical in 1944, the need to improve plant availability has been a major objective. Through the late 1940s and the 1950s, continued improvements were made in the typical breakdown and preventive maintenance systems. In 1961 it was recognized that further improvement in plant availability required a new approach and consequently a maintenance program based on an integrated, engineered approach was explored and developed to replace the standard breakdown and preventive maintenance philosophy. Initial implementation of this program resulted in a reduction of plant equipment forced outages by 50 percent in the original production reactors as shown in Figure 2.

The newest of the Hanford reactors, N Reactor, was designed and built as a dual purpose reactor so that, in addition to plutonium, steam is generated and supplied to an adjacent turbine/generator plant for the production of electrical power. The reactor is pressurized, light water cooled, graphite moderated, metallic uranium fueled with a nominal power rating of 4000 MW(th).

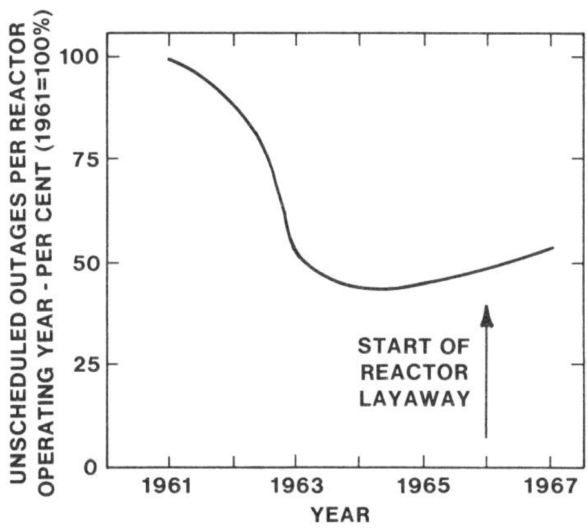

FIG.2. Effect of engineered maintenance program on single pass reactors 1961-1967.

The reactor has been tested at 4800 MW(th). A more detailed description of the N Reactor can be found in reference [2]. United Nuclear Industries operates the N Reactor and fuel fabrication facilities for the U. S. Energy Research and Development Administration. The reactor plant went critical on December 31, 1963 and achieved full power on December 12, 1964, operating in the single purpose, plutonium production mode. The electrical generating facilities were added in April, 1966 with a design rating of 860 MW(e). The electrical generating facilities were built and are operated by the Washington Public Power Supply System.

The size, design features, materials application and operating requirements of the N Reactor constituted a major step forward in engineering and technology. Consequently, the typical growing pains were experienced when plants incorporating advanced concepts are brought on line - experiences currently typified in the nuclear industry today. Initially, N-Reactor downtime was high as major equipment problems such as steam generator leaks, turbine blade fatigue cracks, pump and valve operability problems, etc. were experienced and solved. The plant then settled down to a forced outage level of approximately 25 forced outage hours per 100 operating hours from 1966 to 1969, as shown in Figure 3. This forced outage performance was similar to U. S. nuclear industry experience in 1973 and 1974.[3][4]

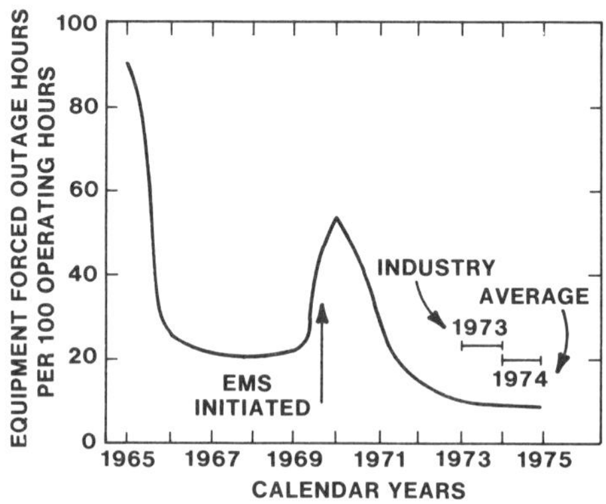

FIG.3. N Reactor equipment forced outage performance.

Unfortunately, many of the lessons learned from maintaining the original 15 to 20 year old production reactors were not initially applied to N Reactor. Conventional maintenance and surveillance programs were adopted: programs that are now in common use by the nuclear industry. These programs are typified by:

- Preventive Maintenance Programs
- Materials Surveillance Programs
- Equipment Calibration and Standards Programs

- Component Testing and
- Technical Specification Surveillance

It was found at N Reactor that these conventional maintenance systems were not adequate to provide and sustain the availability performance desired. The very rapid increase in forced outage hours experienced between 1969 and 1970 as shown on Figure 3 and the resultant severe decrease in plant availability shown on Figure 4 prompted a revaluation of the N Reactor maintenance systems in light of the lessons learned on the older reactors. The increasing forced outage frequency was attributed to aging equipment and a maintenance system which placed too much reliance on breakdown repair rather than anticipation and prevention. In addition, in the original design of N Reactor, many components and critical equipment had been installed such that they were extremely difficult to get at to maintain. Much of this equipment was inaccessible during reactor operation. High radiation levels to personnel during plant shutdown required considerable manpower to maintain the equipment because of limits on personnel radiation exposure. This situation resulted in both forced outage and scheduled outage downtime being extended beyond the time considered reasonable and necessary to perform a normally simple maintenance job. Therefore, as part of the effort to improve plant availability, a system of design changes was initiated to relocate or reposition components and equipment to make them more accessible and easier to maintain.

An integrated surveillance and maintenance system was developed and implemented on N Reactor from 1970 to 1972 which resulted in forced outage hours being reduced by approximately 90 percent. In 1973, forced outage hours were only 8 hours per 100 operating hours compared with an average of about 22 hours per 100

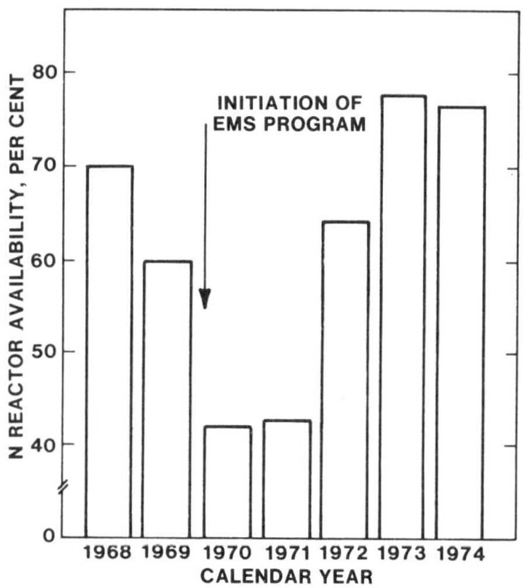

FIG. 4. N Reactor availability.

operating hours for the U. S. nuclear industry. In 1974, the N Reactor celebrated its 10th year of operation by experiencing only 6 forced outage hours per 100 operating hours compared with the U. S. nuclear industry average of about 20.

3. INTEGRATED SURVEILLANCE AND MAINTENANCE PROGRAM

Basically, the Engineered Maintenance System now in use at N Reactor integrates Company resources into a tightly managed program that has the full support and formal backing of top management, and involves a dedicated and close cooperative effort of the operating groups, the maintenance craftsmen, the engineering support of designers and plant engineering personnel, and the quality assurance inspectors. The prime objective of the system is to ensure plant safety, extend equipment and system life, and assure high plant on-line availability with a minimum number of forced equipment outages.

The N Reactor Engineered Maintenance System was developed over a period of approximately three years, i.e., 1970, 1971 and 1972. The task required approximately 30 manyears of effort, only about one-half of which involved experienced maintenance or plant engineering personnel. Because of this and since N Reactor is much more complex [2] than commercial reactor plants due to its dual purpose design, it is felt less time and effort would be required for application of a similar program to typical nuclear electrical power generation stations. The administrative effort required to sustain the program is about 6 manyears per year.

It is interesting to note that at N Reactor the direct maintenance costs to implement the Engineered Maintenance System program were offset by the resulting reduction of breakdown maintenance requirements. Hence, the only incremental costs associated with the development and implementation of the program were for engineering and administration, while maintenance craft resources remained essentially stable.

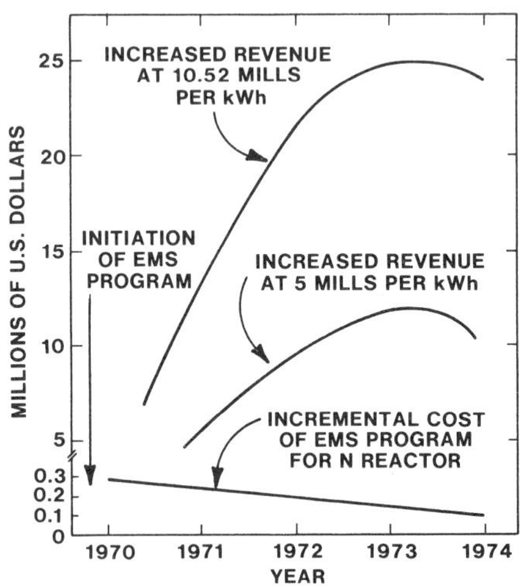

FIG.5. Value of reduced forced outages at N Reactor for two assumed electrical power costs.

Figure 5 shows the development and administrative costs associated with the Engineered Maintenance System program at N Reactor. Also shown is the resultant gains in revenue possible by reducing forced outages. The benefits of an Engineered Maintenance System at the N Reactor are obvious.

4. DEVELOPMENT AND IMPLEMENTATION OF AN ENGINEERED MAINTENANCE SYSTEM

Maintenance programs can be broken down into three broad range categories. These are:

- Breakdown Maintenance which allows equipment to operate to failure before it is repaired or replaced. This type of maintenance requires little planning but is inefficient in the use of manpower and results in excessive downtime. It is characterized by work being performed on an emergency basis.

- Preventive Maintenance utilizes scheduled inspections and periodic equipment overhaul or replacement as a means to prevent breakdowns. This plan is effective, but generally quite expensive and has little predictive value for future performance. It is also not readily adaptive to trend analysis of performance.

- Engineered Maintenance takes the strong points from the first two types and adds predictive testing techniques to aid in determining frequency of overhaul, replacement or repair. The purpose is to maximize operating time and eliminate unnecessary work. Equipment is categorized into levels of importance and a system of predictive maintenance is applied. This maintenance system operates through scheduled function tests and inspections. Overhauls and repairs are performed as indicated from analysis of conditions found during these tests and inspections.

The basic steps in the development and implementation of an Engineered Maintenance System are summarized in Table I. The first step is to characterize all the reactor equipment and systems into degrees of importance relative to their effect on nuclear safety, plant life and operating continuity. This characterization is based on a detailed review of Design Criteria, Operating Bases, Technical Specifications, Operating Safety Limits, etc. Once this characterization is completed, all equipment and systems are placed into an appropriate maintenance philosophy group of Breakdown Maintenance, Preventive Maintenance or Engineered Maintenance depending upon the potential effects of failures. All those systems and associated equipment considered critical to nuclear safety, plant life and operating continuity are placed under the Engineered Maintenance System.

The next step is to define the performance requirements of the critical equipment and establish the limits of acceptable operation. Inspection, testing and maintenance requirements and frequencies are then established to achieve the defined performance requirements.

TABLE I. ENGINEERED MAINTENANCE PROGRAM

- <u>Categorize Equipment</u>

  Categorize equipment into degrees of importance and effect on reactor safety and availability.

- <u>Define Equipment Performance Requirements</u>

  Define critical equipment performance requirements and establish limits of acceptable operation.

- <u>Establish Surveillance & Maintenance Requirements</u>

  Establish inspection, testing and maintenance requirements and frequencies to achieve the defined performance requirements.

- <u>Provide Specific Procedures</u>

  Provide specific procedures for functional testing, periodic calibration, inspection and maintenance of critical equipment and systems.

- <u>Integrate into Planning</u>

  Categorize work as to outage or nonoutage, establish required labor, radiation exposure control, other resource requirements, and integrate work activities into outage and nonoutage periods.

- <u>Close the Loop</u>

  Provide a system of data collection and audit records for trends analysis to provide the necessary feedback to assure surveillance and maintenance frequencies are adequate to meet performance requirements.

It is desirable from an audit and performance viewpoint that the critical equipment be broken down into functional units that can be tested as a complete system, such as power range flux monitor system, emergency cooling system, containment system, etc. Overlapping of the systems is difficult to avoid unless the system interfaces are carefully chosen and clearly defined.

After the system descriptions and interfaces have been tentatively determined, priorities for procedure preparation can be established to guide the work. Assignments are made to the specific person best qualified for a given system. The assigned engineer should have a good understanding of the nuclear process along with considerable maintenance experience as an engineer, supervisor, or craftsman. It is also important that he have the capability to grasp the overall system functions and the ability to detail the meticulous procedures necessary for checking the system's capability for performing its design functions.

Preparation for writing the procedures and instructions is the most difficult phase of the development process since it involves intensive study of the system. Actual work or observation of work on the system and its components is essential at this point.

After the engineer becomes thoroughly familiar with the system, he is in a position to prepare a draft of the scope of the procedure. The scope consists of a brief description of what is included in the specific system, the standard requirements for the specific system, and a basis or justification for the testing and maintenance requirements. The most important content of the scope is the statement of standard requirements for the system. These statements specify the functional requirements of the system in relation to the process. Limits required by the process must be adhered to in these statements. In many instances the limits stated are more restrictive than the process actually requires to allow for reasonable system error. Also included are the minimum system test and inspection frequencies. Inspection frequencies are chosen after considering such factors as (1) consequences of system failure, (2) expected or experienced failure rates, (3) whether inspections can be performed during reactor operation, (4) applicable code requirements, and (5) manufacturer recommendations. Detailed knowledge of the system and modern testing techniques, if properly applied, can greatly reduce the amount of work necessary to provide assurance that system reliability is being maintained at an acceptable level.

The testing and maintenance procedures form the heart of the system and provide the maintenance and operating personnel with the necessary instructions for performing the work. The procedures are divided, where appropriate, into three categories, electrical, instrument, and mechanical. This division is made primarily to facilitate the preparation so that engineers best suited by experience and education can be assigned specific procedures. At the N Reactor it also facilitates the assignment of the work to the craft people in the field.

The special test equipment required for each inspection is listed for each of the three categories. Test equipment, used for calibrating critical process instruments, must be certified as traceable to National Bureau of Standards calibration.

The procedures are written in detail so that the average craftsman and operator can follow them without reference to other documents. This practice makes it necessary for the engineer preparing the procedure to write in a style such that there is no confusion regarding the requirements. Frequently, the first or second drafts are reviewed by people who are not familiar with the system in order to assure it can be followed in logical steps.

After the procedures have been prepared, check sheets can then be designed and cross referenced to fit the instructions. The purposes of the check sheets are to:

- Provide a record of as found conditions that provide the basis for evaluating system performance and analysis of trends.

- Provide a guide for a craftsman or operator performing the inspection.

- Provide a record of repairs and adjustments made to the system.

- Provide an auditable record to prove that the specified inspections were properly performed on time.

If the system is found to be outside of the limits specified in the procedure, the corrective actions required to bring the sub-system into limits are recorded on the check sheet. These sheets provide a record of system performance in relation to failure rate, a basis for redesign to improve performance, and input to the analytical evaluation for trends and overhaul frequencies. This is the basic key to the success of the program.

To be effective the checks, testing and maintenance must be thoroughly integrated into the planning and scheduling activities of reactor operation. The procedures must be separated into those which can be performed during reactor operation and those which must be done during shutdown. Availability of resources, required testing frequency, plant operating schedule, etc. must all be carefully weighed and conflicts resolved. Work packages are then prepared from the procedures and assigned to the maintenance craftsmen and reactor operators including engineering backup where required.

The check sheets are filled out by the maintenance craftsmen, operators and engineers performing the work. Out-of-limit situations are usually handled on-the-spot by the people performing the work and so recorded on the check sheets. Those that cannot be remedied are reported so that further engineering assistance can be obtained. All deviations from acceptable limits or performance must be resolved by correction or an approved waiver obtained.

Mini-computer programs are utilized extensively in performing inspections on large multi-channel systems such as coolant flow, temperature monitors and fuel failure monitors, each of which contain checks on over 1000 channels in the N Reactor.

An important aspect of any maintenance system is the feedback of actual performance data into the planning and scheduling cycle. It must provide the mechanism for data collection and records for analysis of trends to assure the surveillance and maintenance frequencies are adequately meeting the performance requirements.

At N Reactor, check sheets are audited by first-line supervision of the people doing the work, then by analysts who prepare the work packages and schedule the work. The check sheets are then used as input into a computer program which analyzes the data for conformance to specified limits, plots trend information that helps anticipate changes in failure rates, assists scheduling work requirements, and provides the bases for revising testing frequencies to conform with actual performance experience.

Witness audits of scheduled inspections are also performed by engineering personnel. These provide opportunities for education of operators and craftsmen and for feedback to engineers in improving procedures and limits of performance.

The procedures and frequencies of an Engineered Maintenance System must be mandatory to be effective. Therefore, some mechanism for obtaining waivers must be provided to permit the flexibility required to meet possible conflicting objectives. When surveillance and maintenance procedures cannot be performed on the frequency scheduled or a performance limit cannot be met, a waiver is requested by the operating or maintenance personnel and an engineering evaluation is made of the situation. A waiver may be issued provided that nuclear safety and plant life are not compromised. Each waiver has a specific time limit established within which corrective action must be taken.

Preventive maintenance, such as scheduled lubrication, equipment inspections, and overhauls which do not affect nuclear safety, reactor life, or operating continuity, complements the Engineered Maintenance program. Preventive Maintenance is scheduled along with system checks, but on a less formal basis. Records of conformance to schedule are kept.

Whenever any breakdown repair is performed on a sub-system covered by the program, the applicable inspections from the program are made and an appropriate check sheet filled out. This documents the fact that the sub-system will perform its function within the specified limits. This information is also fed into the analysis system to provide a true picture of system performance and to determine if the inspection and maintenance frequency should be changed.

5. CONCLUSION

For N Reactor, the establishment of an integrated surveillance and maintenance system has saved many millions of dollars by increasing on-line availability through decreased forced equipment outages. Such a system can be useful and advantageous to any plant, nuclear or otherwise, that is plagued with the problem of forced equipment outages and unacceptable on-line availability.

REFERENCES

[1] Nucleonics Week 15 48(November 28, 1974) 8, McGraw Hill.
[2] STAFF OF N REACTOR PROJECT, "Hanford's New Production Reactor," Nuclear Engineering and Design 4 (1966) 299-310, North-Holland Publishing Corp., Amsterdam.
[3] Nucleonics Week 15 20(May 16, 1974) 1, McGraw Hill.
[4] DIRECTOR OF REGULATION, "Operating Statics Report," USERDA, January 24, 1975.

THE THROW-AWAY MAINTENANCE PHILOSOPHY VERSUS REPAIR

Nadia KONTOLEON, J.M. KONTOLEON*, N.G. CHRYSOCHOIDES
Reactor Department,
Nuclear Research Center "Democritos",
Athens, Greece

Abstract

THE THROW-AWAY MAINTENANCE PHILOSOPHY VERSUS REPAIR.

The question as to whether it is more advisable to repair or replace a part of a system upon failure is one that has received much attention over the past few years. From an economic point of view, there are cases in which repair seems to be better than replacement while in other cases the opposite holds. However, the problem of increased demand for high reliability and system availability, which directly affects the improvement of protection and engineered safeguard systems in the nuclear industry, has also to be seriously considered. The increased complexity of the systems and the shortage of skilled technicians to deal with them, together with the increased component densities as well as integrated circuit densities, makes the throw-away maintenance philosophy more appealing. In this work, a mathematical analysis is presented to demonstrate the effect of the throw-away maintenance policy on reliability in the case of a system which is subject to both the partial and the catastrophic type of failures. Partial failures are those which are catastrophic to the part, but not to the system. The mathematical formulation uses a Markov process which is continuous in time and has three discrete states, namely working, partial fault present and under repair. Times to failure and repair times are considered to be exponentially distributed. It is assumed that catastrophic failures are discovered as soon as they occur, while partial faults are discovered only at regular times when proof tests are carried out. It is shown that the throw-away maintenance procedure improves reliability, and mathematical expressions are derived which can be used straightforwardly in reliability calculations. The problem of minimizing system unavailability with respect to imposed cost constraints is also discussed and applied to a system which is subject to both the partial and the catastrophic type of failures.

1. INTRODUCTION

The question as to whether it is more advisable to repair or replace a part of a system upon failure is one that has received much attention over the past few years. Besides the economics involved, there is the problem of the increased demand for high reliability and system availability, which has a direct influence on the improvement of reactor protection and engineered safeguard systems. There is no doubt that from an economic point of view there are cases in which repair is superior to replacement while the opposite is true in other cases. However, when one looks into the future and considers the impact of integrated circuits, the increased component densitics and the shortage of skilled technicians to deal with the increased complexity of systems, the throw-away maintenance philosophy becomes even more appealing. This means that a supply facility must be available to accomodate the throw-away parts as well as to be able to provide for the storage of physically larger replacement items. Certainly there are several factors which must be considered in order to arrive at a final

* Present address: Department of Electrical Engineering, Wollongong University, Wollongong, N.S.W., Australia 2500.

decision in favour of either repair or replacement. In this paper, a mathematical analysis is presented to demonstrate how the throw-away maintenance policy may affect reliability in the case of a system which is subject to both the partial and the catastrophic type of failures. Partial failures are those which are catastrophic to the part but not to the system. It is assumed that catastrophic failures are discovered as soon as they occur while partial faults are discovered only at regular times, when proof tests are carried out. It is further assumed that for catastrophic types of failure, repair is really replacement of the failed item by a new one, known to be functioning correctly. It is evident that this repair procedure also rectifies any partial fault existing at the time when the catastrophic fault occurs, and so the probability of success based on the statistical independence assumption between partial and catastrophic failures is certainly pessimistic. This is also valied in cases where, instead of repair or replacement, a repair is carried out, followed by a detailed check to identify any partial faults.

The fact that a single-part failure may not cause an entire-system failure does not necessarily mean that the system remains unaffected; partial failures may change the imposed stresses which in turn affect reliability. Thus, to ensure the desired reliability goal when a system is subject to both partial and catastrophic failures, some inspection policy must be followed which depends on the partial and catastrophic failure rates of the system.

In this paper it is assumed that proof tests are carried out at regular time intervals. First the impact of the throw-away maintenance philosophy upon system reliability is formulated mathematically by a Markov process [1, 2] assuming that times to failure and repair times have an exponential distribution. Then, following the replacement policy outlined above, or following instead a repair procedure which is accompanied by a detailed system checking, the problem of optimizing system availability with respect to imposed cost constraints is discussed. Finally, the impact of testing on system availability and its economic implications are taken into consideration.

2. PROBLEM FORMULATION AND MODEL EQUATIONS

For a system subject to both partial and catastrophic faults, there are three states describing its behaviour at any instant between regular proof tests. These are:

(a) The system is functioning properly. This state is denoted by S_w, and the probability of the system being at this state at time t is denoted by $P_w(t)$, for $0 \leq t \leq T$, where T is the test period.

(b) The system is under repair or replacement. This state is denoted by S_c and the probability of the system being at this state at time t is denoted by $P_c(t)$, for $0 \leq t \leq T$. Here it is assumed that either repair followed by a detailed checking, or replacement, will rectify any partial fault which may exist by the time a catastrophic fault occurred.

(c) A partial fault is present. This state is denoted by S_p and the probability of its occurrence at time t, by $P_p(t)$, for $0 \leq t \leq T$.

It is assumed that at the beginning of each test period the system is functioning correctly. The state will change from the working state to one of the other two states if either a catastrophic or a partial fault occurs. Considering that the transition time δt is sufficiently small, it may be assumed that during this time only one transition may occur, which changes the state of the system. Then, the following four transitions are possible:

(1) $S_w \rightarrow S_c$ with probability equal to $\lambda_c^{\delta t}$

(2) $S_w \rightarrow S_p$ with probability equal to $\lambda_p^{\delta t}$

(3) $S_c \rightarrow S_w$ with probability equal to $K_r^{\delta t}$

(4) $S_p \rightarrow S_c$ with probability equal to $\lambda_c^{\delta t}$

where λ_c and λ_p are the failure rates for catastrophic and partial failures, respectively, and K_r is the repair (or replacement) rate.

At time t between test intervals the system can be in one of three states, i.e. S_w with probability $P_w(t)$, S_c with probability $P_c(t)$ and S_p with probability $P_p(t)$. Therefore,

$$P_w(t) + P_c(t) + P_p(t) = 1 \tag{1}$$

Since it is assumed that at time t = 0 the system is at the working state, we have

$$P_w(0) = 1 \tag{2}$$

The differential equations relating the states of the system can be written in a matrix form as

$$\left[\frac{dP_i(t)}{dt}\right] = [P_i(t)][C] \tag{3}$$

where [C] is a state-transition-coefficient square 3×3 matrix, which can easily be obtained from the reliability transition diagram shown in Fig. 1.

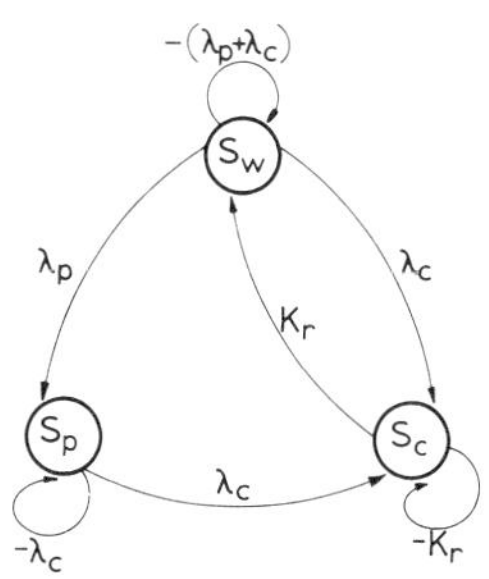

FIG. 1. Reliability transition diagram, S_w, S_p and S_c denote the working state, partial fault present and catastrophic fault present, respectively.

Thus

$$[C] = \begin{bmatrix} -(\lambda_p + \lambda_c) & \lambda_c & \lambda_p \\ K_r & -K_r & 0 \\ 0 & \lambda_c & -\lambda_c \end{bmatrix}$$

Taking the Laplace transform of Eq. (3) and considering the boundary condition, Eq. (1), one obtains

$$[L_i(S)] = [L_i(S)]\,\frac{1}{S}\,[C] + \frac{1}{S}\,[(P_i(0)] \tag{4}$$

where $L_i(S)$ is the Laplace transform of $P_i(t)$. Therefore,

$$\left.\begin{aligned} L_w(S) &= \frac{(S + K_r)(S + \lambda_c)}{S(S + \lambda_c + \lambda_p)(S + \lambda_c + K_r)} \\ L_p(S) &= \frac{\lambda_p(S + K_r)}{S(S + \lambda_c + \lambda_p)(S + \lambda_c + K_r)} \\ L_c(S) &= \frac{\lambda_c}{S(S + \lambda_c + K_r)} \end{aligned}\right\} \tag{5}$$

These transforms can be inverted to give the following expressions for the required probabilities

$$\left.\begin{aligned} P_w(t) &= \frac{\lambda_c K_r}{(\lambda_c + K_r)(\lambda_p + \lambda_c)} + \left(1 - \frac{\lambda_c K_r}{(K_r - \lambda_p)(\lambda_p + \lambda_c)}\right)\exp\left\{-(\lambda_p + \lambda_c)t\right\} \\ &\quad + \frac{\lambda_c K_r}{(K_r - \lambda_p)(\lambda_c + K_r)}\exp\left\{-(\lambda_c + K_r)t\right\} \\ P_p(t) &= \frac{\lambda_p K_r}{(\lambda_p + \lambda_c)(\lambda_c + K_r)} - \left(1 - \frac{\lambda_c K_r}{(K_r - \lambda_p)(\lambda_p + \lambda_c)}\right)\exp\left\{-(\lambda_p + \lambda_c)t\right\} \\ &\quad - \frac{\lambda_p \lambda_c}{(K_r - \lambda_p)(\lambda_c + K_r)}\exp\left\{-(\lambda_c + K_r)t\right\} \\ P_c(t) &= \frac{\lambda_c}{(\lambda_c + K_r)}\left(1 - \exp\left\{-(\lambda_c + K_r)t\right\}\right) \end{aligned}\right\} \tag{6}$$

The above expressions are used in the following to demonstrate the effect of the throw-away maintenance philosophy versus repair.

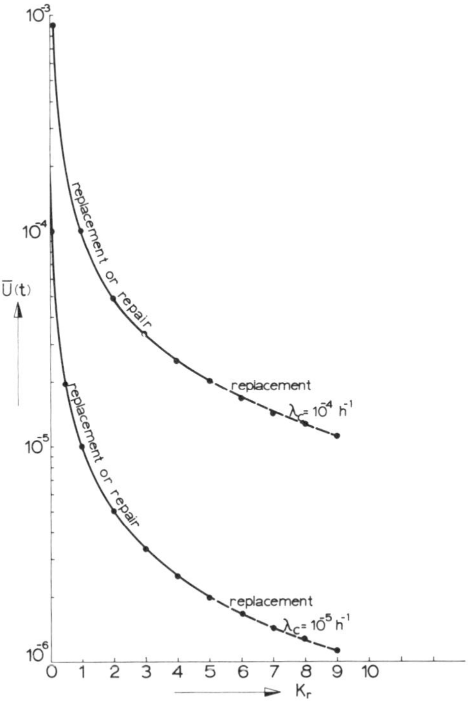

FIG.2. Impact of repair rate on system unavailability for T = 1000 h.

3. REPAIR VERSUS REPLACEMENT

From Eq.(6), the dependence of the mean system unavailability upon the system repair rate is

$$\bar{U}(t) = \frac{\lambda_c}{(\lambda_c + K_r)}\left[1 - \frac{1}{T(\lambda_c + K_r)}\left(1 - \exp\left\{-(\lambda_c + K_r)T\right\}\right)\right] \tag{7}$$

This is represented graphically in Fig.2 for $\lambda_c = 10^{-5}\ h^{-1}$ and $10^{-4}\ h^{-1}$ and T = 1000 h. The dashed lines impose a limit on the maximum system availability which can be achieved following a purely repair action. Normally, following a replacement policy, much higher system repair rates can be assumed, as indicated by the dashed lines of Fig.2.

However, the question as to whether it is more advisable to repair or replace a part of a system upon failure is one which requires an examination of the economic factors involved for the particular system under consideration. In general, the expression which relates the total system-cost, due to replacement or repair policies, with the system repair rate, is a non-linear one. Normally, the price to be paid for an increase of the system repair rate through a purely repair action involves a considerable increase in the total expense. Also, even when very high system repair rates are set, the question of economics involved must be investigated when choosing a replacement policy.

Let us consider the problem of minimizing $\overline{U}(t)$ subject to a given cost constraint C_c (given in cost units per unit time), when (a) a replacement policy is employed and (b) a repair procedure is carried out. For demonstration purposes a simple case will be considered which has been applied to the study of transistorized protective systems used in nuclear reactors. The cost functions associated with such systems may be written in the form

$$C_t = \{C_0 + C_1(K_r) + C_2 D_t\}/T \tag{8}$$

where C_0 is a fixed cost, $C_1(K_r)$ is a non-linear function of K_r, C_2 is a cost per unit of time in the event of a failure, D_t is the system downtime and C_t the total cost per unit of time. We then have

$$D_t = T\,\overline{U}(t) \tag{9}$$

and

$$C_1(K_r) = \begin{cases} C_3 K_r^n & \text{for replacement} \\ C_4 K_r^n & \text{for repair} \end{cases} \tag{10}$$

where C_3, C_4 are expressed in cost units and n is an exponent, which in this particular example assumes values equal to 1.5 and 1.2 for replacement and repair, respectively. In addition, the following constraints are applied:

$$\begin{aligned} &0 \leq K_r \leq 5 && \text{for repair} \\ &0 \leq K_r \leq 10 && \text{for replacement} \end{aligned} \tag{11}$$

As can be seen from Figs 2 and 3, the maximum availability for $\lambda_c = 1 \times 10^{-5}\ h^{-1}$ is obtained through a replacement policy, while for $\lambda_c = 3 \times 10^{-4}\ h^{-1}$ the maximum availability requires a repair procedure. In both cases T = 1000 h, C_c = 0.26 U/h, while C_0 = 100 U, C_2 = 1000 U/h, C_3 = 12 U, C_4 = 15 U. It is obvious that the optimum policy to be followed depends upon many parameters, and among them the system failure rate is of primary importance.

4. OPTIMUM TEST DURATION AND PERIOD

Assuming that a repair action is not carried out during T, system availability is given by

$$\overline{U}(t) = \left[1 - \frac{1}{T\lambda_c}(1 - \exp(-\lambda_c T))\right] \tag{12}$$

which for a test duration equal to r becomes

$$\overline{U}(t) = \left[1 - \frac{1}{T\lambda_c}(1 - \exp\{-\lambda_c(T-r)\})\right] \tag{13}$$

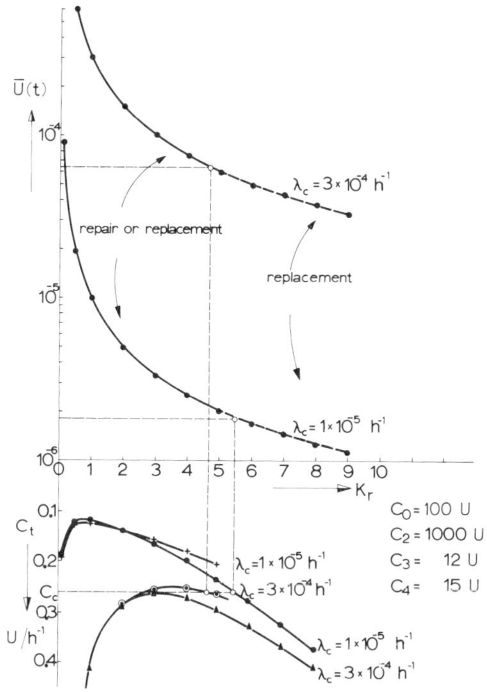

FIG.3. Cost and unavailability versus repair rate for T = 1000 h.

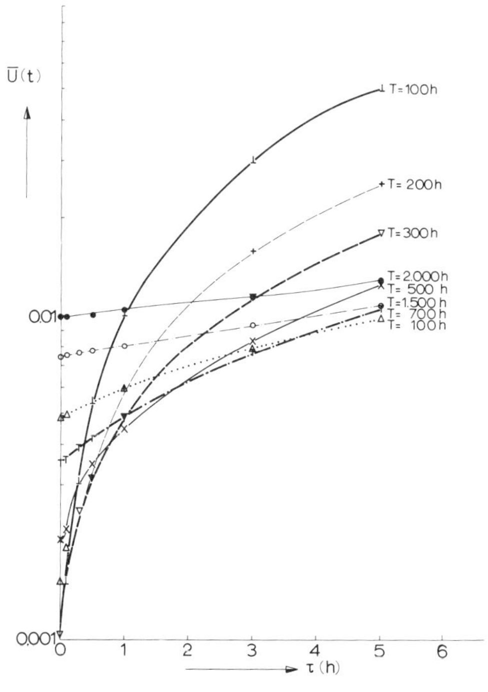

FIG.4. Unavailability versus test duration for various test intervals T and for $\lambda_c = 1 \times 10^{-5}\ h^{-1}$.

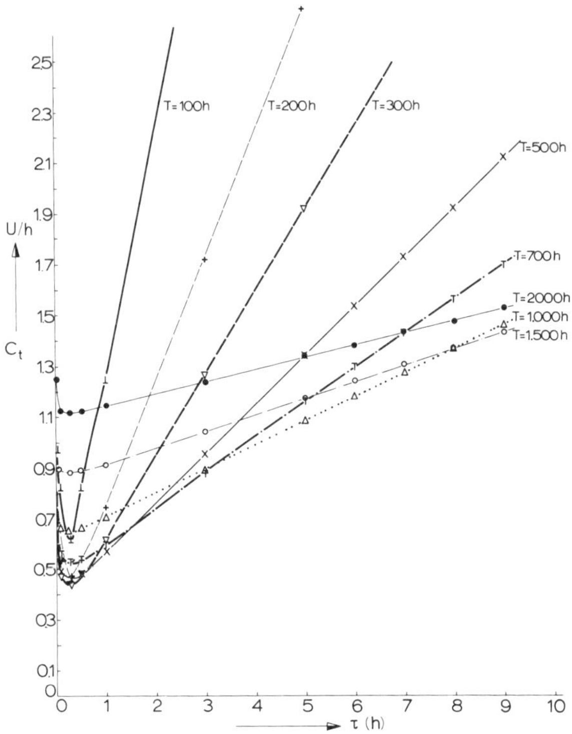

FIG. 5. Cost versus test duration for various test intervals T and for $\lambda_c = 1 \times 10^{-5}\ h^{-1}$.

Then, $C_1(K_r)$ for the repair action is of the form

$$C_1(K_r) = C_4\left(\frac{1}{r}\right)^n \tag{14}$$

and there exists an optimum pair (T, r) for which the cost becomes minimum for a fixed availability. In Fig. 4 the dependence of unavailability on test duration is shown. From Fig. 5 it can be seen that for 0.5 h, the minimum time needed to perform a test, and $\overline{U}(t) = 0.0035$, which is the unavailability value which must not be exceeded, the optimum pair (T, r) is T = 300 h and r = 0.5 h, while if the minimum time required to perform a test is about 2 h the optimum time interval, for which the cost is minimum and unavailability does not exceed the given value $\overline{U}(t) = 0.009$, is T = 700 h. Thus, depending on the minimum time required to perform a test, there is an optimum test interval for which the cost is minimum subject to an unavailability constraint.

5. CONCLUSIONS

An analysis has been formulated for systems subject to both catastrophic and partial faults under the conditions of replacement of the failed part or repair including a check. The formulation was made using a Markov process, continuous in time and with three discrete states, each one being ultimately reached from any other. The times to failure, and repair or replacement times, were considered to follow the exponential law. The replacement or repair procedure with proof test also restores any partial fault present. As was demonstrated, system repair rates, test intervals and test duration influence system availability and cost, which are of the utmost importance in nuclear power plants. To investigate throw-away philosophy versus repair it becomes clear that one must consider the economics associated with the system under consideration as well as the system failure rate.

REFERENCES

[1] COX, R.D., MILLER, D.H., The Theory of Stochastic Processes, Methuen and Co., London (1965).
[2] TA KACS, L., Stochastic Processes, Methuen and Co., London (1960).

MAINTENANCE EXPERIENCE AT RAJASTHAN ATOMIC POWER STATION

M.S.R. SARMA, S. JAYABARATHAN
Rajasthan Atomic Power Station,
Anushakti, Kota, Rajasthan,
India

Abstract

MAINTENANCE EXPERIENCE AT RAJASTHAN ATOMIC POWER STATION.

The Rajasthan Atomic Power Station is a twin-unit station of the CANDU-PHW type with a nominal rating of 200 MW(e) net per unit and on-power refuelling capability. Unit-I became critical in August 1972 and was declared commercial in December 1973. Unit-II is in an advanced stage of construction. Maintenance and repairs of both the nuclear and conventional equipment are done during routine planning, breakdown outages, poison shutdown outages and planned annual shutdown periods. Yearly, a period of three to four weeks may be needed for carrying out major inspection and maintenance activity on large equipment such as the steam turbine, generator, main condenser, station transformer, primary coolant recirculation pumps, moderator pumps and fuelling machines. Preventive maintenance jobs are planned and executed in accessible areas daily as a matter of routine and during poison shutdown outages in inaccessible areas. This requires that the maintenance crews be more or less continually on the alert. In addition to gamma radiation, radiation doses due to tritium are encountered and pose special problems for the maintenance personnel. During the past two years (1972-74) it was found that there is a need, in the maintenance crews, for more specialists for various radioactive and conventional jobs. This paper outlines the significant nuclear and conventional maintenance activities as well as problems in maintenance encountered during the period 1972-74.

1. INTRODUCTION

The Rajasthan Atomic Power Station is the second nuclear power station that was built in India, the first being at Tarapur. The station lies on the bank of the Rana Pratap Sagar Lake in Rajasthan. This station consists of two identical units of the CANDU-PHW type, each having a net generating capacity of 200 MW(e). The first unit achieved criticality in August 1972, and the second unit is expected to go critical by end of 1976 (Fig.1).

The station is owned by the Department of Atomic Energy (DAE). The design of the nuclear part of the station was carried out by Atomic Energy of Canada Ltd, while that of the conventional part was done by the Montreal Engineering Company of Canada. The DAE is in charge of the on-site construction work and the entire project management responsibilities; it also operates the plant after completion. Commissioning services have been provided by the Hydro-Electric Power Commission of Ontario, Canada, for Unit-I; Unit-II will be commissioned by Indian staff.

2. HIGHLIGHTS AND STATION UPKEEP DURING COMMISSIONING AND OPERATION

Unit-I construction activities were completed by the end of 1970. The heavy-water systems were commissioned with light demineralized water. The initial fuel loading was done manually, without the fuelling machines,

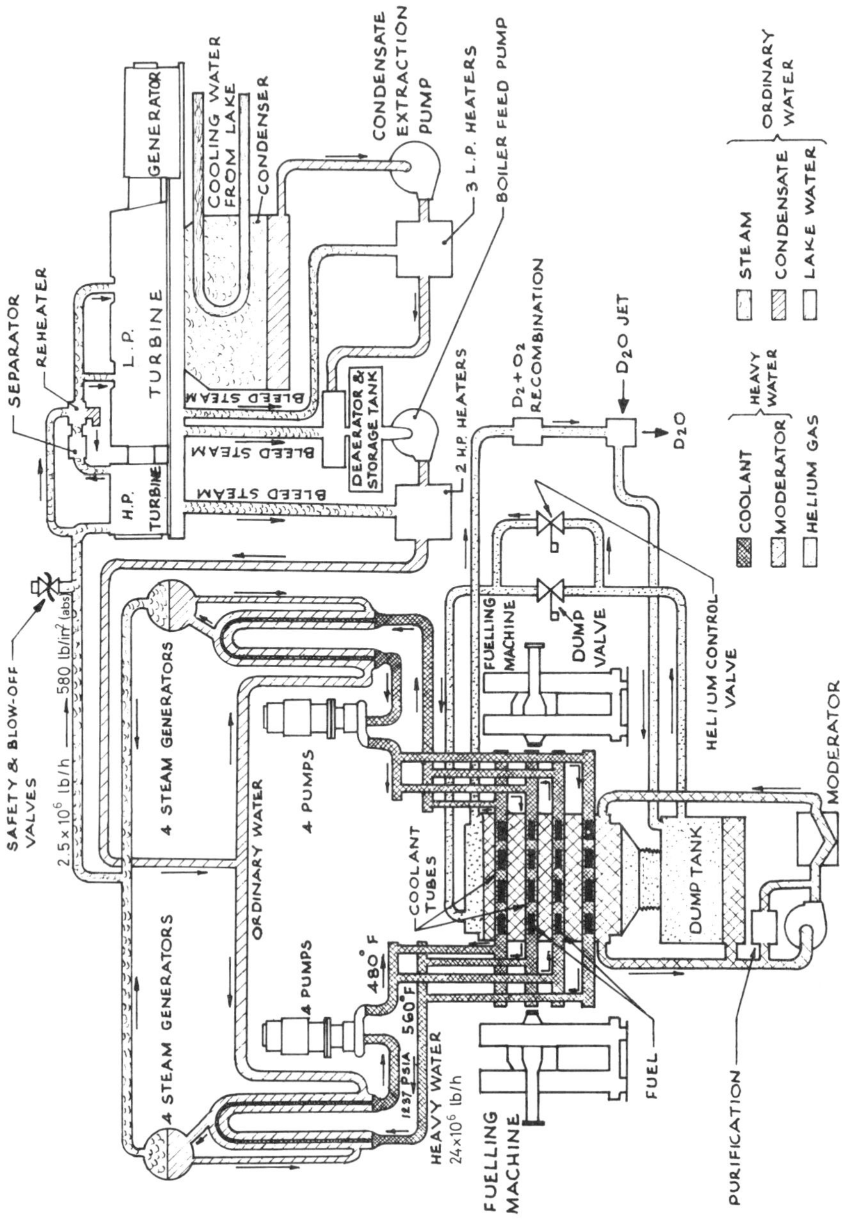

FIG. 1. Simplified flow diagram of Rajasthan Atomic Power Station (H.P. = high pressure, L.P. = low pressure).

in April 1972. Heavy water was filled in the above systems in July 1972 and the first criticality was achieved on 11 August 1972.

The fuelling capability was first established at low power in September 1972, and on-power fuelling capability at high power was demonstrated successfully in September 1973. Since October 1974, we are doing routine refuelling on-power on automatic operations. So far, we have fuelled over 70 channels. To date, the fuel behaviour in Unit-I has been satisfactory, and we have found only two channels containing one defective fuel bundle in each.

The turbine was rolled and Unit-I was synchronized for the first time on 30 November 1972. In the initial stages, the turbine bearings had high temperatures and the station had to be shut down on a few occasions for inspection of the bearings. On inspection, the bearings were found wiped up of the "babbit metal" and they had to be re-machined. We had to flush the lubricating oil system for a long time, as we observed sand and debris in the main strainers and governor valves. We have the problem of load hunting when the power level is between 110 and 130 MW(e). We do not know the reason for this hunting yet. So far, the performance of the turbine generator has been satisfactory except for the high-pressure turbine blade failure which occurred in July 1974 and cost us a long outage of three and a half months for repair/replacement work.

Unit-I has been in commercial operation since 16 December 1973. The outage details for 1973 and 1974 are highlighted below.

The total number of outages in 1973 was 70. Of these, 34 were due to problems of a nuclear nature and 36 were due to conventional problems. Of the nuclear problems, 26 were forced, 4 spurious and 4 due to human error. Of the conventional problems, 30 were forced, 1 spurious and 5 due to human error. The unit operation from December 1973 to June 1974 was satisfactory. The average availability factor for these seven months was 82% and the maximum power level was 200 MW(e). We had 49 outages in 1974, of which 22 were due to nuclear problems and 27 to conventional problems. In the nuclear category 12 were forced, 8 spurious and 2 due to human error. In the conventional category, 17 were forced, 7 spurious and 3 due to human error. The grid caused 14 outages in 1973 and 7 outages in 1974. As can be seen, the number of outages decreased considerably in 1974, thus indicating that the plant is approaching maturity.

3. MAINTENANCE PLANNING AND MAN · REM BUDGETING

For the past two years (1973 and 1974) most of our maintenance activities were concentrated on conventional equipment, for example (1) air compressors, (2) refrigeration compressors, (3) water treatment and chlorination plant, (4) motor generator sets, (5) diesels, (6) boiler feed pumps, (7) reheater drain pumps, (8) deaerator, (9) ventilation fan motors and (10) turbine. On the nuclear side, we had problems with the primary heat transport pumps and motors, the pressurizing pump and the shutdown cooling pump. We had frequent shutdowns of our heavy-water recovery driers, owing to defective thermocouples, burning of heater elements and damage of fan motors. Our experience with heavy-water upgrading plant was extremely good for the past couple of years. We had problems with the auxiliary boilers, owing to defective components and inadequate steam supply to the upgrading plant.

Now, the auxiliary boilers are being replaced by the manufacturer. There were very frequent troubles in our thermal shield cooling system equipment because of abnormal vibrations in the fan, motor and the chamber. Also, the shear pins of the blow-out panels used to break because of vibration; and in almost all the poison outages, thermal shield cooling equipment had one problem or the other. Motor bearings used to get jammed and fan belts used to break. In the 1973 annual shutdown, most of the vibrating structures were reinforced with extra supports and the chamber was grouted properly. Now, the problems have decreased considerably. The latest annual inspection of 1974 revealed that our main condenser tubes are extremely clean, owing to the clear water running in the Chambal river.

There used to be frequent failures of the Resistance Temperature Detectors (RTDs) of the temperature monitoring system of the reactor coolant channels. These RTDs are located in the feeder cabinets which are highly radioactive and hot in congested locations. Quite a lot of man • rem has to be spent in repairing the RTDs to keep the temperature monitoring system functioning all the time.

Generally, maintenance and repair work of nuclear and conventional equipment is planned every week, with daily checking of any system having a deficiency report. Weekly planning includes the major continuous activities of maintenance, preventive maintenance and breakdown maintenance. Periodically, a poison outage plan will be produced listing the maintenance work to be done during the poison outage period, which is 36 h. Advantage is taken in every 36-h poison outage period to do preventive maintenance jobs in the inaccessible and accessible areas in order to cut down the annual outage period. Hence the maintenance task force is always kept alert and ready to handle the poison outage jobs on a round-the-clock basis. The maintenance crews (electrical, mechanical, control and fuel handling) are scheduled to work round-the-clock with engineers and maintainers and their assignment is considered completed only after the Unit is synchronized.

There is no outage for refuelling the reactor, as the CANDU fuelling system is capable of refuelling the reactor on power. However, an annual shutdown period of three to four weeks may be required for carrying out major inspection and maintenance of large equipment such as the steam turbine, generator, main condenser, station transformers, primary heat transport pumps, switchyard electrical equipment, replacement of cobalt in the reactivity control mechanisms and the fuelling machines. The 1973 annual shutdown was planned mainly to incorporate several engineering changes/modifications into the station systems, and in that period most of the maintenance deficiencies were corrected and preventive maintenance jobs were done. The damage to the high-pressure-turbine blade in 1974 caused an outage and the repair period of three and a half months was also used for planned annual maintenance activities.

Our major concern regarding radiation is related to tritium and to gamma radiation in the various areas. The total man·rem utilization was 160 man·rem for 1973, and 350 man·rem for 1974. These figures for 1973 and 1974 are low for the following reasons: the background radiation fields are low because the system chemistry is maintained well and we had only two failed fuel bundles. This reduced the general radiation in various accessible and inaccessible areas. In December 1974, we had two heavy-water spills which increased the tritium in the air activity in the reactor building.

Even accessible areas had 25-60 $(MPC)_a$ tritium activity compared with normal levels of <1 $(MPC)_a$.

Highlights of maintenance experience with various station equipment will now be presented.

3.1. Experience with on-power fuelling

The Rajasthan Atomic Power Station is one of the few nuclear stations in the world in which on-power refuelling is done regularly on automatic operation. To date, more than 70 channels have been refuelled on power. During refuelling, we had some problems with the mechanical components, but the automatic operations were well carried out, smoothly and repeatedly.

3.1.1. Fuelling-Machine Maintenance

The fuel-handling system needs frequent maintenance because it is highly sophisticated. Our objective is to do on-power fuelling on 40 to 50 reactor channels in a month, with eight bundle changes in each channel. This programme gives a maintenance period of 5 to 10 d in a month for maintaining the fuel-handling system and preparing for refuelling work.

We frequently have to change the skinner seals, which are the seals between the machine and the coolant channel. These seals are metallic and can be used for 25 to 50 clampings. Some seal rings leaked even after 10 clampings. The fuelling machine has to be brought out to an accessible area for replacing the skinner seal. The replacement of the skinner seal is essential, as otherwise heavy water will leak through the end fittings during the entire fuelling operations.

We had frequent problems with the closure plugs which are the pressure boundaries of the coolant channels. Our experience is that the closure plugs, installed after completing the fuelling operations, leaked heavy water at normal pressure and temperature. Either the same closure plug has to be reseated by the fuelling machine or a spare plug installed from the machine. It is a time-consuming job and the machine has to be brought out in order to install a new spare plug.

Sometimes the closure plugs used to be tight in the end fittings and the fuelling machine rams could not remove them by normal means. Stronger forces had to be used on the rams, to remove the tight plugs. These tight plugs revealed, during inspection, some corrosion and lithium hydroxide depositions in the plug mechanism. On one occasion, an installed spare closure plug also indicated a severe heavy-water leak in the channel after fuelling operations, and the fuelling machine could not be unclamped from the channel. The reactor had to be shut down and the system was cooled. The feeder connections had to be frozen by means of ice plugs, and the machine was unclamped for replacing the leaky plug.

3.2. Coolant channel creep measurement of Unit-I in 1973

The coolant channels are free to float axially to a limited extent. Similarly, channels are free to contract to a limited extent. The expansion and contraction of the channel is regulated by two mechanical stops and they are adjustable to a limited extent. We have a maximum provision of 2.56 inches for each channel. This clearance is available for thermal expansion and for longitudinal creep due to neutron irradiation. As a preventive maintenance check, once in two years we have to check the creep measurement using optical or theodolite instruments. During the 1973 annual shutdown, about 98 channels were measured by optical and theodolite instruments and the creep was found to be within limits. Considerable man·rem was expended for the creep measurement work, as the personnel had to work very close to the reactor end fittings. In the 1973 annual shutdown, when we did the first checkup of creep measurement, 23 people worked and got a total dose of 12 rem. We have to do this creep measurement check once every two years to verify whether clearance is available for further creep development. The design estimate is that we may have to replace a few coolant channels in the annual shutdown periods after about 16 years of reactor operation.

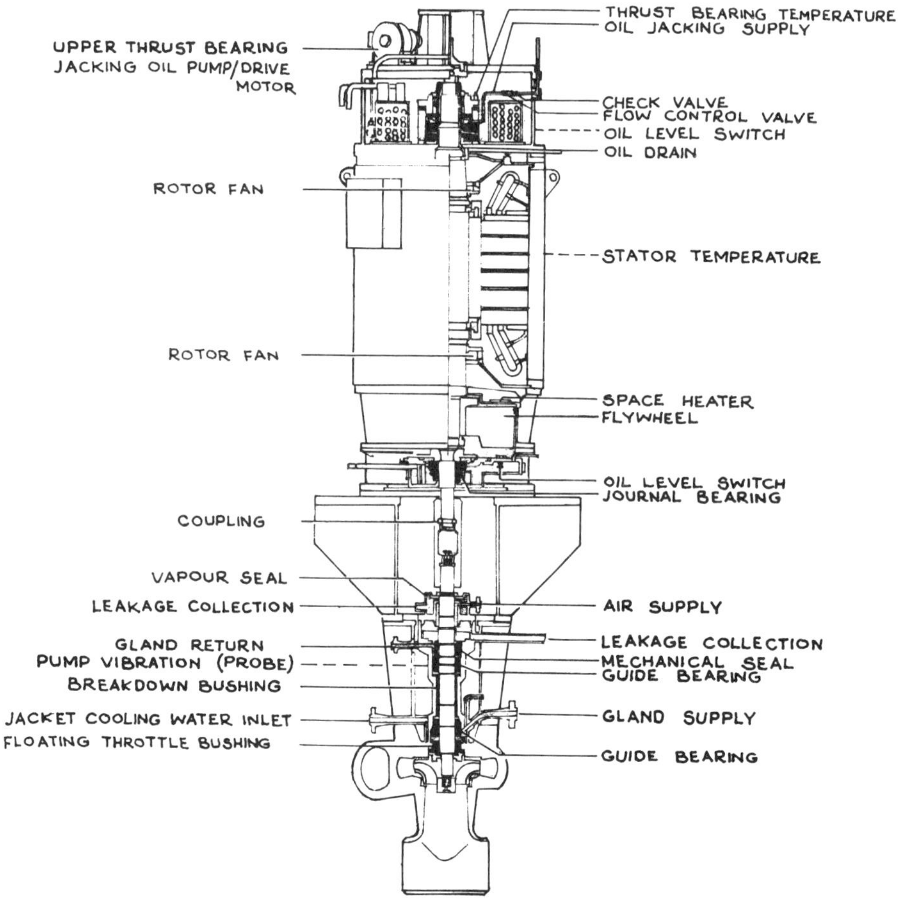

FIG. 2. Schematic diagram of primary heat transport pump/motor.

3.3. Problems with pressurizing pumps (Fig. 2)

The main pressurizing pump of the primary heat transport system is a special multi-stage (13 stages) centrifugal pump of vertical type, running at 3000 rev/min with a discharge pressure of 1800 lb/in^2(g) and a motor with a motive force of 350 h.p.. The suction water enters the first bottom stage and the discharge goes upwards up to the sixth stage. The discharge of the sixth stage enters the top and goes downwards through 7 stages. The main discharge is in the middle. Our experience with this pump is that it is highly sensitive to partial suction and the entire rotor gets jammed owing to gas-locking and cavitation. On four occasions, the rotating elements of the pump were jammed because of momentary loss of suction. During momentary suction loss, the floating rotor with 13-stage impellers loses its balanced position and is pulled upwards owing to the upward thrust exerted by the last seven stages. This upward movement of the rotor causes severe rubbing of the wearing rings, jamming the entire rotating assembly. During inspection, we observed that both faces of the mechanical seal were damaged badly. The wearing rings of 7 out of 13 stages were seized because of the close tolerances between the impeller and the wearing rings. The motor bearing (thrust) was also damaged.

3.4. Problems with primary heat transport pumps and motors (Fig. 3)

There are 8 pumps with 2 spare units. Pump capacity is 5670 gal(imp)/min with a head of 490 ft. The motor has 1170 h.p. at a speed of 3000 rev/min. The pump is a single-stage centrifugal type with a complex gland, cooled by cold high-pressure clean heavy water. The gland coolant is injected at high pressure into the impeller casing to prevent the hot system water from entering into the glands. The motor is a squirrel-cage induction type with a huge fly-wheel of about a ton mounted on its shaft. The general performance of the primary heat transport pumps is satisfactory and as long as they run continuously they function normally without any problem. However, as and when the continuous operation was interrupted owing to power failure or gland coolant failure, we faced some problems with the pump mechanical seal and carbon bearings. These problems were:

(a) In the initial commissioning stages, phase unbalance in the supply voltage caused several trips on the motors.
(b) Owing to power failure and disturbance in gland cooling flows, damages were observed in the mechanical seal and in top and bottom carbon bearings.
(c) Vibrations were observed on the pump gland due to mechanical or electrical problems.
(d) Electrical noise was observed due to a rotor defect which also caused vibrations.
(e) Out of ten available motors, one was found defective in the 1974 annual inspection after two years of operation.
(f) One motor remained unserviceable owing to a peculiar electrical noise (sinusoidal) from the initial commissioning.

On one pump motor assembly, severe hunting vibration was noticed of the order of 1.2-2.5 mil. When the pump and motor were uncoupled, vibration was not noticed in the motor but there was some vibration in the pump.

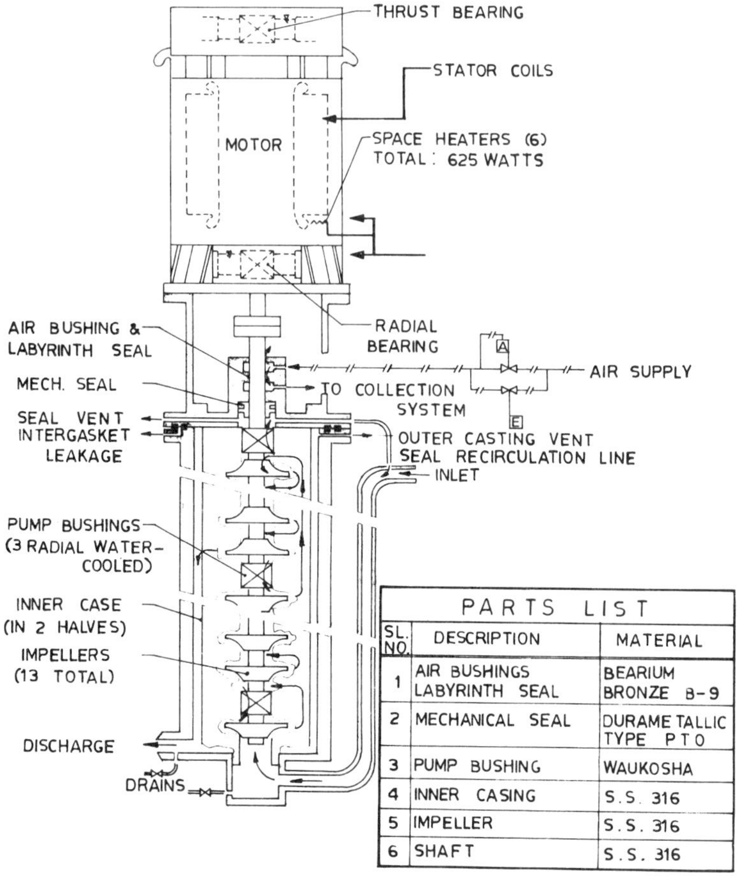

PARTS LIST

SL. NO.	DESCRIPTION	MATERIAL
1	AIR BUSHINGS LABYRINTH SEAL	BEARIUM BRONZE B-9
2	MECHANICAL SEAL	DURAMETALLIC TYPE PTO
3	PUMP BUSHING	WAUKOSHA
4	INNER CASING	S.S. 316
5	IMPELLER	S.S. 316
6	SHAFT	S.S. 316

FIG. 3. Main pressurizing pump.

A new pump assembly was interchanged with the same motor and, during operation of the pump, similar hunting vibration of the same order of magnitude was noticed. The motor was suspected to be defective and was dismantled for inspection. After dis-assembly, the stator was checked and found to be all right, but the rotor was found to have burning marks at two places on the rotor surface. Close inspection revealed two long crack marks running along the circumference perpendicular to the shaft on the copper bars in the rotor. One line crack ran along the entire circumference of the rotor. The second crack ran about three-quarters of the circumference with a hair-line thickness.

The reason for the cracking of the bars in the rotor could not be established. The manufacturer suggested that the long run-up time could cause this problem. In general, the run-up times have been higher than recommended (recommended <20 s; actual 29 s). The manufacturer suggests that there should not be a second restart within 24 h of the first restart. Evaluating the above data for all the motors, it could not be confirmed that the damage to the particular rotor was only due to the above reasons.

3.5. Problems with shutdown cooling pumps

For cooling the heat transport system below 300°F and for holding the system cold to perform maintenance work, a pump and a heat exchanger are connected between the inlet and outlet of the reactor headers. There are two pump motor units. The pumps are double-volute, double-suction, single-stage, horizontal-mounting types. The capacity is 2000 gal/min and the discharge head 70 ft. Motors are squirrel-cage induction type, 60 h.p., 1500 rev/min fitted with an anti-reverse rotation clutch on the motor shaft.

For the past two years, we have been facing the problem of repairing the mechanical seal. Either the inboard seal or the outboard seal leaked on several occasions. Since the shutdown cooling pumps and motors are situated in a congested location in the inaccessible area and leaks spread considerable quantity of tritium, this consumed several man·rem for replacement of seals. As most of the lines including the process water lines are of welded type, isolations involved cutting and rewelding of process water lines. Each seal replacement took about 16 h for completion of the job.

3.6. Problems with valves

The following are the types of valves that had problems very frequently: (1) double-packing high-pressure heavy-water valves, (2) motorized valves, (3) bellows seal valves, (4) relief valves for relieving heavy-water pressure. The problems are:

(a) Double-packing valves had excessive inter-seal leakage. Eventually, an external leak starts leaking heavy water. On some occasions, valves handling hot heavy water started leaking severely through inter-seal lines and flooded the leakage collection tank backing up hot water through other inter-seal lines and damaged the components of other equipment, spilling heavy water on the floor.

(b) Internal leaks through the valve plug, preventing isolation during maintenance. Ice plugs have to be formed on pipe lines having leaky valves for isolation purpose, thus involving more time and more man·rem.

(c) Cracks developed in the high-pressure valve body near weldings owing to vibration and pressure pulses and heavy water leaked.

(d) Frequent problems with limit switches on motorized valves, often limit switches failed to cut out the drive of the motor and damaged the valve stem. Sometimes, motors were overloaded because of the failure of the limit switch and were burnt.

(e) Heavy-water relief valves failed to reseat once they popped out and caused heavy-water leaks. Because of heavy-water leaks, we had to dismantle the valve, lap the plug and seat and reassemble them. Frequent damage of bellows in the relief valves and increase of back pressure was noticed. These bellows are imported and costly hardware which had to be replaced frequently.

(f) A problem with bellows seal helium gas-balance valves is frequent damage of the bellows, causing helium leakage from system to atmosphere. The consequence is loss of helium from the system and spread of tritium in the reactor building. These bellows were permanently replaced by double "O" rings on the valve stems. Because of the movement of the valve stem, these "O" rings wear out and leak helium.

Frequent maintenance was done to replace these "O" rings in the gas balance valves.

(g) Rubber-lined butterfly valves often have the rubber lining damaged owing to frequent operation. Some of the pieces of rubber lining escape into the rest of the system. This causes difficulty in using these valves for isolation as and when required, and also the escaped pieces get carried over and cause problems in maintaining the heavy-water level in the reactor vessel. On one occasion, the retrieval of these pieces caused an outage of 7 d.

3.7. Problems with heavy-water recovery driers

These driers are located in inaccessible areas and at congested locations. The following are the repetitive problems observed for the past two years of operation:

(a) Because of unbalanced air flow or low air flow in the ducts, the on-line heaters used to be burnt out frequently.
(b) Frequent failure of thermostats to cut out the electric supply to the heaters when the sheath temperature increases abnormally owing to low air flow.
(c) Seizure of bearings of fan motors on the ventilation ducts. On some occasions, the motors were burnt owing to the seizure of bearings.
(d) The cooling coils in the drier chiller unit fret and become damaged, leaking heavy water.
(e) The scrubber nozzles become choked up and have to be cleaned very frequently.

Considerable man·rem is being spent to maintain the heavy-water recovery driers and, in almost all poison outage periods, some of the maintainers have to work for the upkeep of the driers.

3.8. Channel temperature monitoring system

The temperature of the heavy water coolant at the exit of the coolant channels is monitored by platinum element resistance temperature detectors (RTD). For the purpose of this system, the 306 channels are divided into 8 zones – 4 inner and 4 outer. The temperature measurement of all RTDs in a zone is averaged and displayed as zone mean temperature and the deviation of each channel temperature from the corresponding zone mean is displayed as channel temperature deviation.

In view of the safety functions, the entire RTD system is duplicated. Controlled power reduction at 1% full power per second is initiated if zone mean temperature of a zone is high, and a trip (scram) is initiated if:

(a) Zone mean temperature of a zone is very high on both the installations or
(b) Channel temperature deviation of a channel is very high as seen on both the installations.

RTDs failed frequently and we had the maintenance problem of replacing them in feeder cabinets. The failure rate is about 20 per year. The RTDs

(2 per channel) are situated in the feeders connecting the channel exit to the reactor outlet header in insulation cabinets in the fuelling machine vaults. The temperature inside the cabinet when the heat transport system is hot does not permit access to the cabinets. To replace failed RTDs, access is needed to the insulation cabinets. Hence the heat transport system must be cooled down to around 50°C. The radiation in the insulation cabinets and the tritium concentration in the fuelling machine vault impose severe restrictions on the amount of time which is permissible to work on this job. Available experience to date indicates that an RTD replacement will entail an expenditure of 250 to 800 mrem because of gamma radiation only. The RTD tail penetrates through the specially provided opening in the insulation cabinets. The process of removing a failed RTD and installing a new one through the penetrations is cumbersome and time consuming. Access to the RTDs within the feeder cabinets is very difficult for an average individual in view of the limited spacing between feeders.

3.9. Problems with boiler feed pumps

There are two 100% duty boiler feed pumps (BFP) which are of horizontal, double-volute, 3-stage type, with a 4000 gal/min flow and discharge pressure of 660 lb/in^2 (g). The motor is 3000 rev/min and has a 2500 h.p. rating. The BFP recirculation system was originally designed on the basis of the recommended recirculation flow of 600 gal/min by the pump manufacturers. When the pumps were commissioned, there was heavy vibration of the order of 20 mil at the pump bearings. Hence, the recirculation flow was increased in stages up to 2500 gal/min in consultation with the manufacturer. At present, the recirculation flow is at 2500 gal/min, the vibration has come down to 2 mil and the pump is operating satisfactorily.

During the 1973 annual shutdown, the pump casings were opened and the impellers were inspected. The first-stage impellers of both BFPs were found in damaged condition due to erosion. In the 1974 annual shutdown, both the pumps were inspected again. The first-stage impeller of one pump had two successive vanes in damaged condition due to erosion. This problem is not yet resolved. The manufacturer has been approached for advice and modification.

3.10. Turbine performance and maintenance

The turbine is of Tandem compound type with one high-pressure stage and four low-pressure stages with speed 3000 rev/min. The turbine was rolled first on 30 September 1972, but we could not go above 1500 rev/min owing to vacuum problems in the condenser. On 4 October 1972, a speed of 3000 rev/min was reached. For about six months, i.e. from October 1972 to April 1973, we had major problems with the bearings. A high temperature of the bearings was observed during this period on several occasions and the turbine had to be shut down for inspection of the bearings. The bearing inspections revealed extensive scoring of the bearing metal on several occasions. The bearing surface area (twice length/diameter) was found to be only 0.87 which was considered inadequate. Hence the surface area was increased to 1.20 by machining the projection in the babbit metal. During this period, sand was observed in the oil system, which entered from the cast steel valves and the entire oil had to be flushed thoroughly for several days. On 6 April 1973, the turbine was loaded to 120 MW(e) after the above maintenance jobs. The turbine was operating satisfactorily for about six months. During this period, one bearing vibration was noticed to be high and of the order of 2.5 to 3.0 mil and had to be inspected. At the time of installation, the generator was kept 20 mil above the low-pressure two-stage turbine to accommodate for the upward expansion of the low-pressure exhaust on load. It was thought that this 20-mil correction was too much and the generator rotor stayed above the normal expected catenary of the rotor, which resulted in a lightly loaded bearing and caused vibration. In the 1973 annual shutdown, this was corrected and the generator bearings were brought down by 10 mil. The turbine was operating satisfactorily from December 1973 after the completion of annual shutdown jobs and it was loaded to 200 MW(e). In July 1974, the high-pressure rotor blade failed, bringing bearing vibration and the eccentricity of rotor to the alarm level, and the turbine was tripped manually. This failure caused an outage on the station for three and a half months. During this outage, the damaged rotor was replaced by Unit-II rotor and the Unit brought back on line.

3.10.1. Damage observed in high-pressure rotor in July 1974

One rotating blade was missing from its T-head (Fig. 4). The sheared head was left in the shaft slot. Two blades had broken along a length of $2\frac{1}{8}$ in. from the top. The root of another blade had cracked over a corner along a length of $\frac{3}{8}$ in. and the piece was left inside. The shrouding piece over the length of six blades was lost completely (Fig.5). Also, a small shrouding piece over the length of three blades had flown off. Stellite shields on six blades were missing. Lacing wire between two blades was missing and, in quite a number of blades, the brazing of lacing was disturbed or cracked. Deep dent marks were observed on 22 blades at the outlet angle.

A dye-penetrant test was carried out on the blades and on the rotor disc of the fourth-stage high-pressure rotor, and the following was observed:

(a) The slot of the entire rotor disc in stage 4 indicated no defect.
(b) All 130 blades were tested with dye-penetrant and 10 blades were found to have cracked near the base.

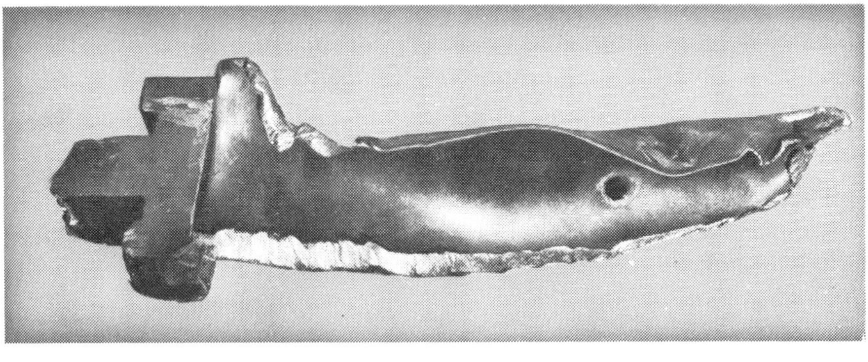

FIG. 4. Missing blade of high-pressure rotor when retrieved.

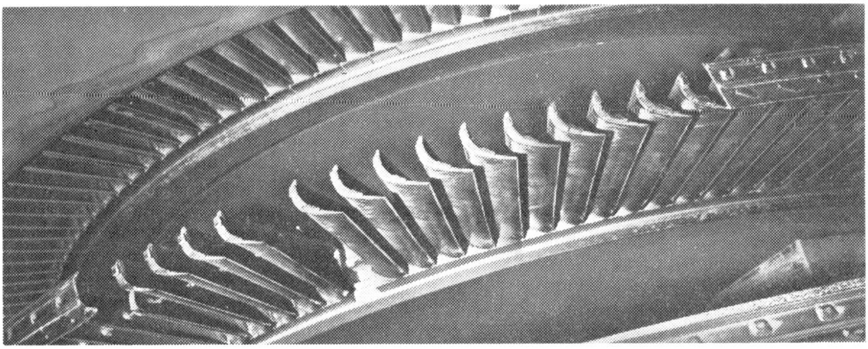

FIG. 5. Missing shrouding and blade on stage 4 of high-pressure rotor.

Out of 130 blades, 11 blades had cracked at the bottom and one more blade broke off at the same spot. The top third of two more blades were broken off. Evidently, there was a common problem in the blades that were fitted in the disc. All the cracks were identified more or less on one spot of the blades, indicating bad fitting inside the disc. The failure configuration of the broken blades indicated a fatigue-type of rupture. Detailed analysis is being done by the supplier with the available blades to establish the cause of damage.

4. STOCKING OF SPARE PARTS

Maintaining the requisite inventory of spare parts is a very important function of the maintenance unit. The problems are twofold:

(a) With the lapse of time, equipment becomes out-moded and procuring spare parts from the suppliers becomes very time consuming and expensive.
(b) Since, normally, the suppliers are reluctant to give fabrication drawings, making the components also becomes very difficult.

In view of the above, it is prudent to order spare parts to last for five years (in the judgement of the operator) and also to obtain the fabrication drawings at the time of ordering the original equipment.

METHODS OF SELECTION AND TRAINING OF PERSONNEL FOR THE RAJASTHAN ATOMIC POWER STATION

M. S. R. SARMA, V. K. WAGADARIKAR
Rajasthan Atomic Power Station,
Anushakti, Kota, Rajasthan,
India

Abstract

METHODS OF SELECTION AND TRAINING OF PERSONNEL FOR THE RAJASTHAN ATOMIC POWER STATION.

Personnel selected to work in a nuclear electric generating station rarely have the necessary knowledge and experience in all the related fields. A station can be operated and maintained and at the same time radiation doses absorbed by station personnel can be kept to a minimum only if the operating personnel are familiar with, and can be used for, all phases of station operation and the maintainers have more than one skill or trade. More technical knowledge and more diversified skills, in addition to those required in other industries, are needed because of the nature of the nuclear reactor and the associated radiation environment and high automation. A training programme has been developed at the Nuclear Training Centre (NTC) near the Rajasthan Atomic Power Station (RAPS), Kota, India, to cater to the needs of the operation and maintenance personnel for nuclear power stations including the Madras Atomic Power Station. This programme has been in operation for the last five years. The paper describes the method of recruitment/selection of various categories of personnel and the method of training them to meet the job requirements.

1. INTRODUCTION

The Rajasthan Atomic Power Station is a twin-unit station of the CANDU-PHW type. Each unit is rated for 220 MW(e) gross. The first reactor went critical on 11 August 1972. It was first synchronized to the grid on 30 November 1972 and has been operating since. The second reactor is expected to go critical in 1976. The number of personnel required to operate the station is estimated as 80 professionals and 450 non-professionals. The non-professionals will normally be designated as operators or maintainers, depending on the type of job they do.

The professionals are graduate engineers in mechanical, electrical or chemical engineering, and their work involves operation as well as maintenance. However, for instrument maintenance, we also take graduates in electronics or communications engineering.

Some of the non-professionals hold diplomas in mechanical, electrical or instruments engineering and would normally have undergone a 15-year course consisting of 12 years in school and three years in a polytechnic. They are employed for both operation and maintenance. In this category, we also take persons having a science degree (12 years in school plus 3 years in college). These personnel work in operations and also in technicians' jobs in the chemical control laboratory. Persons who have had 12 years of schooling are also employed, primarily for operations. Persons who have had 11 years of schooling and have obtained a trade certificate in a trade such as welder, machinist, electrician, instrument mechanic etc. from a technical training institute are preferred for maintenance.

It is seen that personnel requirements for operating and maintaining the station are large. The job of training such a large group was started in 1964 when the first group of trainees was sent to Canada. Ontario Hydro trained 30 engineers and 20 non-professionals in Canada between 1964 and 1969. They gave approximately 100 man-years of training to these people. The training of the remaining personnel was started in 1968 and this paper describes the methods of selecting these personnel and their training. For the purpose of training, we established a training centre near the RAPS Plant Site and have trained about 200 persons for RAPS and about 50 for the Madras Atomic Power Station.

2. SELECTION

Persons meant for operation and maintaining the station are normally selected under one of the five categories, viz, engineers, operators, mechanical maintainers, electrical maintainers and control maintainers.

2.1. Selection of engineers

Engineers with conventional experience are recruited from thermal power stations. Those with experience in research reactors are taken from the Bhabha Atomic Research Centre. The third category covers fresh graduates in engineering who have undergone a one year nuclear training course at the Bhabha Atomic Research Centre Training School in Bombay.

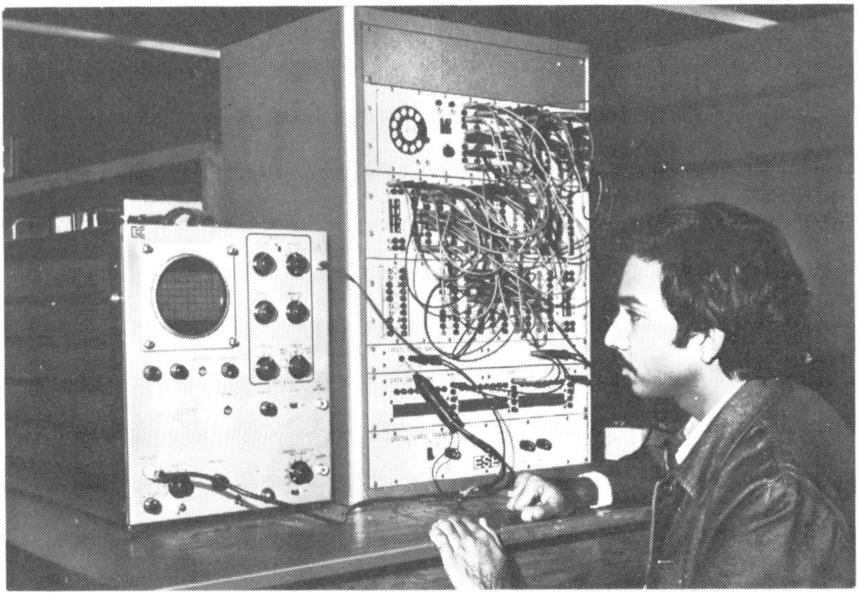

FIG. 1. Digital logic training at the NTC.

2.2. Selection of operators

The operators either hold diplomas in engineering or are graduates in science or have completed 12 years of schooling.

2.3. Selection of maintainers

Maintainers either have a diploma or a trade certificate given by a technical training institute.

At the time of selection, the operators and maintainers are given a written test and an interview to check their knowledge of physics, chemistry, mathematics and engineering. Both the written test and interview are carried out in English, as is the whole training programme.

It was difficult to obtain the right type of people by newspaper advertisements. Even those in the nearby areas did not respond to such advertisements. Therefore we recently started visiting institutes/schools of important district headquarters around the plant site to recruit persons suitable for our jobs directly from the institutions. This method has proved to be very satisfactory.

3. TRAINING PROGRAMME

The training programmes for different categories are described below:

3.1. Training of engineers

The training objectives are: to prepare engineers for the posts of shift charge engineers, mechanical, electrical and instruments maintenance engineers and to prepare physicists and chemists to carry out jobs of station physicist and station chemist.

A shift charge engineer is a person who has about eight to ten years' industrial experience and has obtained the qualifications to operate the station. Since we take both experienced and fresh graduates in engineering for training, the time spent by them in the different phases of training vary considerably. The following gives a general outline of the programme.

The engineers are given two-years' training consisting of six months in the Nuclear Training Centre followed by 18 months on-the-job training.

During the six months at the Nuclear Training Centre, the engineers take courses in nuclear theory and in description of the systems of the station (Fig.1). They also undergo radiation protection training and other safety courses. They take about 20 examinations to demonstrate their proficiency in the subjects.

The engineers, after going through the above-mentioned courses, are divided into two groups: operations engineers and maintenance engineers. The latter category, maintenance engineers, are then sent to the generating station to work in the maintenance sections. The field training programme for the operations engineers consists of working rotation shifts of the operating unit and in other units like planning and technical.

The operations engineers are given a field checklist of the jobs which they should know before they are declared competent to operate the station. They are expected to complete the checklist in two to four years' time, depending on their previous experience.

After completion of the field checklist, the operations engineers are given six written examinations to qualify to work as shift charge engineers in rotating shifts. On passing these examinations, they are given a field tour and a viva voce. They are then attached to rotating shifts as assistant shift charge engineers. If their progress is satisfactory, they are eligible to work as shift charge engineers and are given charge of one crew on shift duties.

Appendix I gives brief syllabi of courses given to all categories of trainees at the Nuclear Training Centre.

3.2. Training of operators

Each shift crew would have approximately 20 operators for operating the station. The operators' training programme aims at enabling these personnel to operate both the field equipment and the control panels in the control room (Figs 2 and 3). The training programme for operators is similar to that for the engineers, except that the courses given to them are not as intensive. An operator who passes the examination is assigned responsibility as first operator. There are normally two first operators in each shift, one in the control room and one in the field. The first operators carry out the operation of the station from the control panels and also organize the field operation with the help of the other operators. The first operator reports to the assistant shift charge engineer. Fresh trainees begin with service systems like compressors, chillers etc. As and when they gain proficiency in one area, they are moved to the next. A typical trainee takes about 10 to 12 years to come to the level of a first operator covering the whole programme.

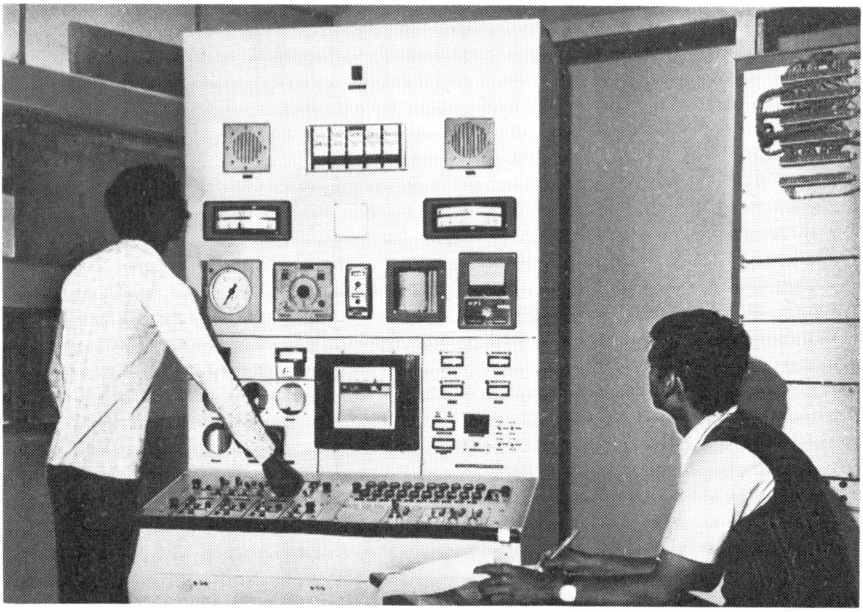

FIG. 2. Training in control panel operation at the NTC.

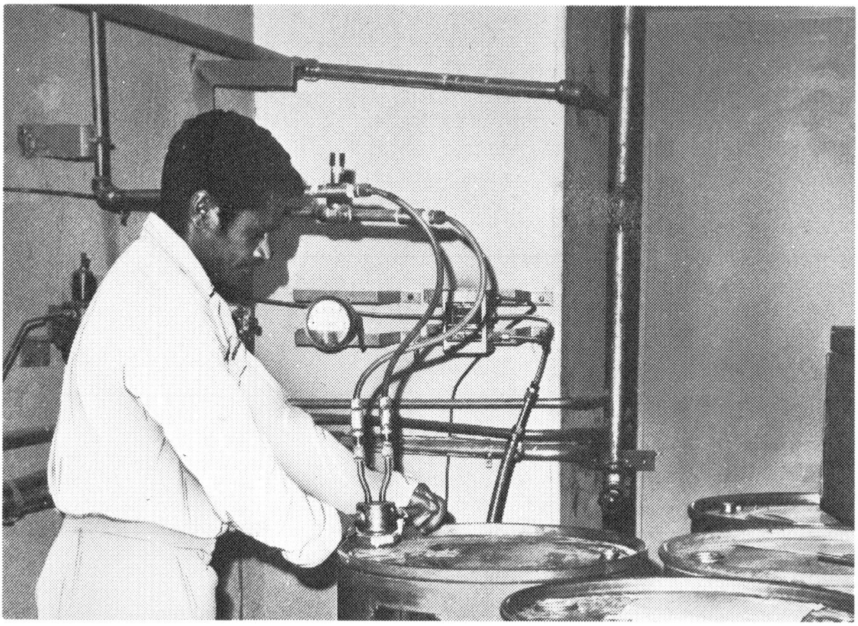

FIG. 3. Field training — addition of heavy water to system.

Like the engineers, the time taken by a person to achieve the position of first operator may vary considerably. We expect a first operator to have a minimum of 10 years' industrial experience before he is assigned duties in the control room.

The operational staff, namely the operations engineers and operators, cannot be trained in the shops and their training always remains incomplete unless they are trained in the actual control room. We are exploring feasibility of a nuclear power station simulator to train the operations staff.

3.3. Training of maintainers

The maintainers are divided into three groups: mechanical, electrical and control.

The maintainers' training is planned to ensure that a person acquires more than one skill or trade. For example, a machinist is also taught the jobs of a millwright so that the radiation dose can be evenly spread out. In fact, we call such personnel by a common name, "mechanical maintainer".

Training is given to all maintainers. The objectives are to acquaint the trainees with the power station and its basic equipment, to teach them the basic skills and to familiarize them with the work methods.

3.4. Training of mechanical maintainers

The trainee spends six months in the Nuclear Training Centre shops followed by an 18-month on-the-job training. During his training at the Training Centre, he is given class-room lectures and extensive skills training. The trainee is given courses in documentation, plant equipment, heavy-water handling and the protection training courses. He spends about 80% of his time learning the relevant skills such as fitting, millwright practice, machining, pipe and tube fitting, welding and rigging in the training shops. His progress in the shop training is monitored by means of equipment-oriented checklists. (Typical equipment-oriented checklists are appended). Field training at the station is monitored by checklists. The trainee is rotated through various areas to ensure that competency is acquired in each area.

The trainee is taken for advanced training after his 18 months' stay at the station and is given advanced courses in skills in mechanical maintenance. After completion of this training, he is sent to the maintenance shop. His responsibilities then depend on his educational qualifications and experience. He is initially asked to work under the supervision of an experienced maintainer and later carries out the jobs independently.

3.5. Training of electrical and control maintainers

The training of these personnel is similar to that of the mechanical maintainers, except that they are taught the skills pertaining to their field. A control maintainer is given skills training on basic electronic components, hardware and fabrication techniques, basic instrumentation, solid state logics, control instruments, nucleonic instruments, data processing etc. The electrical maintainers are given courses in the use of measuring instruments, fault finding and trouble shooting, wiring, batteries etc. (Fig.4).

FIG. 4. Motor control centre busbar checks at the NTC.

3.6. Field checklists

The field training is monitored by means of field checklists. Two typical checklists, one for operator and one for electrical maintainer are given as Appendices II and III, respectively, and show the method in which these checklists are prepared and enforced. Each person is given a set of these checklists. When the person demonstrates his competency in a particular job, his supervisor, authenticates this fact. It is the responsibility of the person to whom the checklist belongs to get this signed out. We make use of the checklist both to ensure his training progress and his suitability for promotion to higher grades.

4. RETRAINING

The Nuclear Training Centre carries out extensive retraining of persons already employed at RAPS. The retraining is normally an extension of the basic training and is in addition to the basic training mentioned previously. The main objectives of the retraining programme are to improve the person's skills and technical ability, to give him specialized training in certain jobs or to qualify him for certain positions such as first operator's position. The retraining continues throughout the man's stay with us and may even take several months in a year. Retraining is always done if new equipment is brought into the station. A typical example of retraining is qualification for work on special fittings, tungsten inert gas (TIG) welding etc.

The training section spends about half of its time on retraining people as the demand for such retraining increases with the indigenizing of equipment.

5. NUCLEAR TRAINING CENTRE

The Training Centre has a floor space of 3000 m^2 covering various training shops, lecture halls, office space and other ancilliary services. These shops have equipment worth about two million rupees. Wherever possible one piece of each equipment that is used in the station is available in these shops to familiarize the trainees with the jobs he will be doing at the station. The Training Centre has a staff of 3 training officers and 10 instructors and other ancilliary staff. The Training Centre also requisitions help from the station staff for conducting some of the training courses, particularly the systems training, thereby augmenting its own staff strength. The Centre also has its own hostels and residential buildings for its trainees.

ACKNOWLEDGEMENTS

The Hydro Electric Power Commission of Ontario, Canada, has established a Nuclear Training Centre at Rolphton, Ontario, Canada, to train persons to operate their Nuclear Power Stations. The Training Centre that we have described in this paper has been patterned on Ontario Hydro's Training Centre at Rolphton. In fact, we have reproduced some of their lessons and use them and have drawn heavily on their experience in preparation of the shop facilities. The authors had an opportunity to work at the Nuclear Training Centre, Rolphton, and are grateful for the help that Ontario Hydro gave in this respect.

APPENDIX I

BRIEF SYLLABI OF COURSES AT THE NUCLEAR TRAINING CENTRE

ENGINEERS

1. Nuclear theory
2. Sites and buildings for nuclear power stations
3. Reactor boiler and auxiliaries
4. Turbine generator and auxiliaries
5. Electrical systems (includes northern grid systems control)
6. Instrumentation and control systems
7. PERT (programme evaluation and review technique)
8. Radiation protection
9. RAPS systems training
10. RAPS work methods
11. RAPS operating principles and policies
12. Standard protection code

QUALIFICATION PROGRAMME

A. Six written examinations

1. Nuclear (general)
2. Conventional (general)
3. Radiation protection (general)
4. Radiation protection (specific)
5. Nuclear (specific)
6. Conventional (specific)

B. Oral examination in the field

C. Interview

OPERATORS

Syllabus similar to above

MAINTAINERS

Basic Training Courses

1. RAPS documentation
2. Station description
4. Plant equipment (mechanical, electrical and instruments)
4. Protection training courses

5. Heavy-water handling
6. Skills training

   (a) Mechanical skills training in fitting, millwright practice, machining, pipe and tube fittings, welding, rigging etc.
   (b) Control skills training in basic components, trade tools, measuring tools, hardware and fabrication techniques, installation, test equipment, basic instrumentation, basic electronics and solid state logics.
   (c) Electrical skills training in motor and associated controls, breakers maintenance, protective relaying, metering, interlocks, test instruments, batteries etc.

The trainee would spend about 80% of his time on skills training.

APPENDIX II

TYPICAL CHECKLIST FOR OPERATORS

RAPS system – field checkouts

SYSTEM: Data logger and recording annunciator LEVEL:

USI No. 6640 NAME:

Serial No.	Item	Checked out by	Date
1.	How many inputs can the data-logger scan?		
2.	How many process parameters are recorded on the station log? What is the method of their identification?		
3.	How many parameters are recorded on the mass log?		
4.	Reject a protective channel and read out the readings on recording annunciator print out.		
5.	....		
6..	....,		

APPENDIX III

TYPICAL CHECKLIST FOR AN ELECTRICAL MAINTAINER

RAPS system – field checkouts

SYSTEM: 3.3-kV Switchgear Name:

Serial No.	Item	Checked out by	Date
1.	Rack out a circuit breaker.		
2.	Identify (a) 52x, 52y, 52 etc. (b) Arc chutes (c) Blow-out coils (d) Charging spring and motor (e) Main contact and arcing contact (f) Switches 88-1, 88-2, 88-3. (g) Barrier stack		
3.	Dismantle the arc chute assembly.		
4.	Lubricate the appropriate parts of the breaker.		
5.	Remove and replace barrier stack.		
6.	Show how the auxiliary switch is adjusted.		
7.	Slow-close the breaker.		
8.	Close the breaker manually by charging the spring.		
9.	Close the breaker electrically.		
10.	Check the closing and tripping timings.		
11.	Identify the current transformers (CTs) in the switchgear and tell their functions.		
12.	Identify the incoming and outgoing sides of the breaker.		
13.	Rack in the breaker in test position.		
14.	....		
15.	....		

APPENDIX IV

TYPICAL CHECKLIST FOR SHOP FLOOR TRAINING

RAPS motor control centre

Checklist No. 104		Name of trainee: Date:
Serial No.	**Assignments**	**Report by the trainee**
1.	What is a motor control centre and what are the advantages in adopting motor control centre (MCC) type of motor control?	
2.	What are the MCCs controlled by the "manually" operated supply breakers?	
3.	How are the MCCs and their starter modules in the station marked and identified?	
4.	In case of undervoltage in a 415-V bus, do the MCC supply breakers trip and lock out in the case of (a) Class IV MCCs (b) Class III MCCs (c) Class II MCCs?	
5.	You are asked to check the control of the pump-house travelling screen motor. Which MCC will you go in for fault finding?	
6.	While attending to the fault for item (5) above, suppose you discover that the magnetic contactor needs replacement; how will you know the size and specifications of the contactor in case its name plate data are missing on the equipment?	
7.	....	
8. ⋮	....	
16.	....	

(cont.)

Serial No.	Assignments	Report by the trainee
17.	Learn and practice constructing a motor control logic circuit for the following logic of pump operation: Two cooling water pumps are provided for the two diesel generators in RAPS. The pumps can be selected on "RUN", "AUTO" or "Standby". If "AUTO" is selected, the lead pump starts automatically when either diesel engine starts. In the event of lead pump failure and the pressure in the header dropping to below 40 lb/n^2 (g), the standby pump starts automatically.	

Checked by:

APPENDIX V

SIMPLIFIED RAPS ORGANIZATION CHART

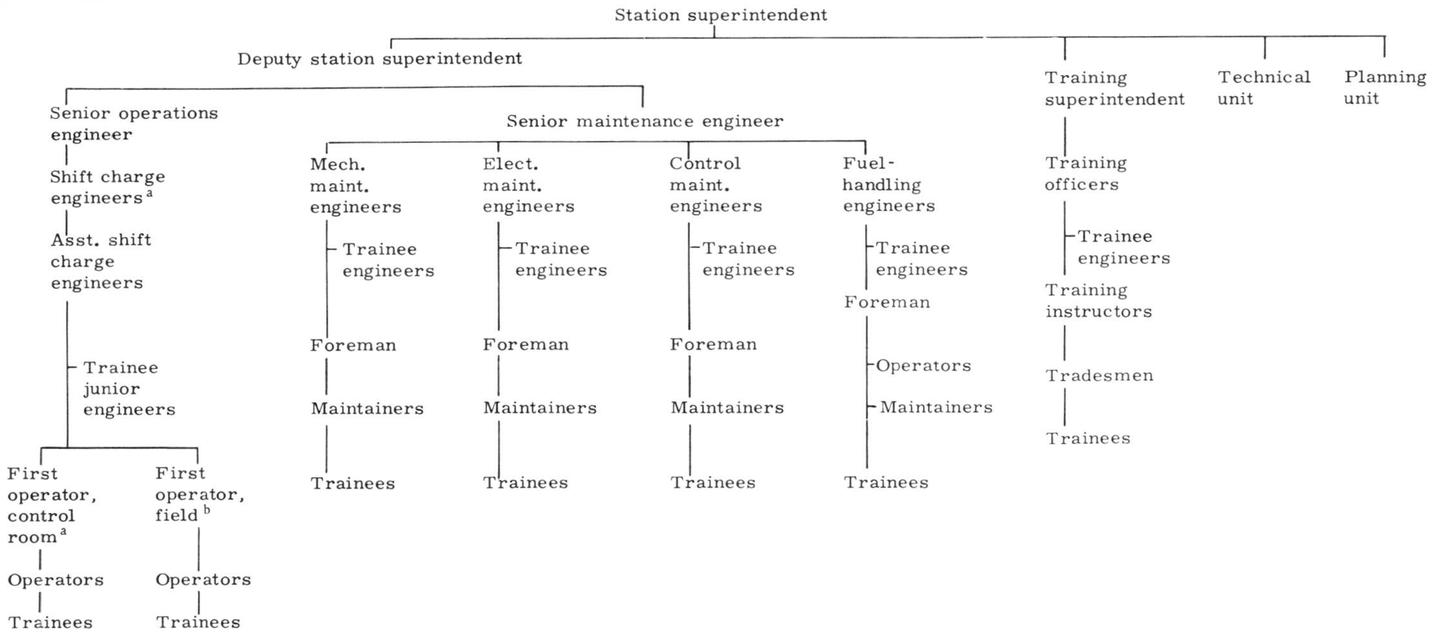

[a] These positions require shift charge engineer's qualifications.
[b] These positions require first operator's qualifications.

OPERATING EXPERIENCE OF KANUPP

S. D. HUSEINI
Karachi Nuclear Power Plant,
Paradise Point,
Karachi,
Pakistan

Abstract

OPERATING EXPERIENCE OF KANUPP.

The paper outlines the performance of Kanupp during its initial two years of operation, 1973 and 1974, with particular reference to its availability for power generation. Factors adversely affecting the plant availability are detailed individually. The paper also gives reference to the sources of heavy water leakages. The performance of the fuel and the fuelling machines is briefly described.

1. GENERAL

Karachi Nuclear Power Plant (KANUPP), owned and operated by the Pakistan Atomic Energy Commission, is situated near the city of Karachi on the Arabian Sea Coast. Based on the Canadian CANDU concept of natural uranium fuel and heavy water moderated reactors, it was designed and built by the Canadian General Electric Co. over a period of 6 years. It was formally inaugurated during the latter part of 1972 and since then has been feeding electric power into the Karachi grid owned by the local utility, the Karachi Electric Supply Corporation.

Kanupp is a heavy water moderated and cooled, natural uranium fuelled, horizontal pressure tube reactor with once-through on-power bi-directional fuelling. The plant output is 137 MW(e) gross and 125 MW(e) net, operating with almost dry steam at pressure of 550 lbf/in^2 (g), reactor primary cooling conditions of 1500 lbf/in^2 (g), and maximum temperature around 563°F, and no allowance for bulk boiling in the core. Reactor heat is transported by coolant from fuel channels via individual feeders into two outlet headers located on either side of the reactor. Each header provides coolant to three boilers; six pumps arranged in two sets provide the necessary head for circulation of coolant through the loop. A feed and bleed system supplemented by a vapour-filled surge tank is provided for maintaining the pressure of the primary loop within a narrow band under conditions of varying load. The reactor and all heavy water systems are contained in a prestressed concrete containment building and, except for part of the building, the containment is not normally accessible during power operation.

Plant control is based on the use of direct digital control (DDC), utilizing two digital computers with two independent control channels. The reactor is controlled in a load-following manner with the thermal power of the reactor adjusted to match the power being drawn by the turbine unit from the steam generators. The steam pressure error signal, coupled with the steam flow demand, positions the control valve in the helium system to the correct moderator pump-up rate in the reactor calandria. The correction to the calandria moderator level is provided by the control rods within a narrow band. The control valve of one of the regulating channels is normally kept in the shut position by demanding higher steam pressure, while the other channel provides control of the steam pressure to compensate for variations in the turbine load.

Kanupp is connected via two independent transmission lines for increased reliability to the Karachi utility electric grid with a firm installed capacity of its own of 350 MW(e) and normal system load varying between 350 and 160 MW(e) over a period of a week. Feeding a relatively smaller grid, Kanupp has been required to operate at lower than full power output, mainly because of limitations on the load demand and integrity of the Karachi grid. Load variations over a period of 24 hours are not uncommon and Kanupp cannot accept such load variations because of the limited excess reactivity available to meet large power variations.

2. PLANT AVAILABILITY

The plant operated during its first two years of operation at an average availability of 77% and a gross generation of 458 GW·h in 1973 and 584 GW·h in 1974, thus providing about 23 and 33% of the total utility load for these years (Fig.1). Within this period the plant was shut down for a period of 12 weeks to carry out modifications to the steam turbine to meet contractual obligations. Taking advantage of this particular period of long shut-down, however, general inspection and overhaul of the other major plant equipment were also undertaken. Except for this period of 12 weeks, the plant has been operating constantly to meet the power demands of the utility, barring interruptions caused by plant outages as outlined in the Table I.

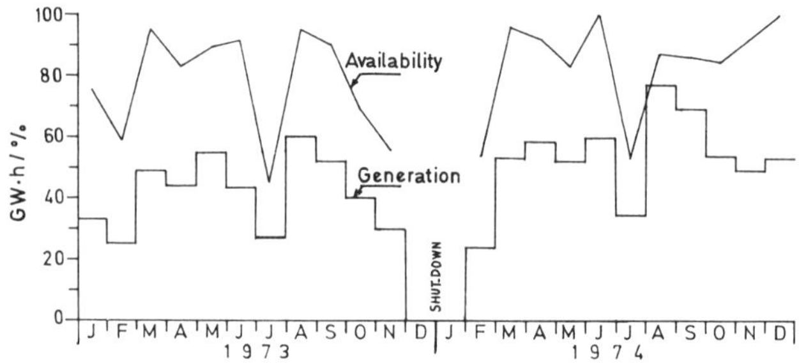

FIG. 1. Kanupp availability percentage and gross generation.

Although the plant has operated at a high availability factor, the frequency of power interruption caused due to trips or forced shut-downs over the first two years of operations was more than anticipated. The two longest uninterrupted runs achieved during these two years were of 51 and 53 days. The number of plant outages caused due to trips and forced shut-downs was 17 in 1974, as compared to 26 during the previous year, 35% of these were of short duration of 30 to 40 minutes. The outages of a minor nature were mainly caused by malfunctioning of the plant computers and

TABLE I. ANALYSIS OF OUTAGES DURING 1973 AND 1974

Cause of outage	1973 No. of outages		1973 %	1974 No. of outages		1974 %
1. Equipment failure		13 (10)	50 (38.5)		13 (3)	76.5 (17.7)
1. Computers	5 (3)			5 (3)		
2. Instrumentation	4 (4)			2		
3. Electrical	2 (2)			2		
4. Mechanical	1 (1)			1		
5. Condenser	-			2		
6. Seaweed	1			1		
2. Heavy water leaks		5	19.3		-	0
3. Grid problems		3	11.5		1	6
4. Errors		3 (1)	11.5		3 (1)	17.7
5. Tests etc.		2	7.7		-	0
		26 (11)			17 (4)	

Note: Numbers within brackets denote successful starts within poison-out periods.

plant control instrumentation. The reactor has been provided with reactivity booster elements (enriched uranium control rods) to provide excess reactivity up to 10 mK, sufficient to override xenon build-up resulting from a reactor trip within a period of 30 to 40 minutes. Failure to reload the reactor to at least 65% of its original power within this period will cause core 'poison out' and the reactor cannot be restarted for the next 25 to 36 hours to allow for the decay of xenon.

About 60% of the plant outages were the result of failure or malfunctioning of the plant equipment and more than half of these were due to the malfunctioning of the plant computers and plant control instrumentation. There were five occasions during 1973 when the plant was derated to cope with uncontrolled heavy water leaks from the reactor cooling system. The situation in 1974 showed a great improvement from the point of view of heavy water leakage. The plant tripped four times as a result of low frequency and undervoltage of the Karachi utility system. On all these occasions the electric power was not restored for more than thirty minutes, resulting in failure to start-up within the poison-out time. Human errors accounted for 14% of the trips, 3 in each year.

3. EQUIPMENT FAILURES

3.1. Plant computers

Although the DDC was initially expected to be more reliable, the experience during the first two years of operation was far from satisfactory;

as many as ten plant trips could be attributed to the malfunctioning of the controlling computers. Frequent failures of transistors in the analog input-hardware, malfunctioning of high gain multiplexer amplifiers and frequent breakdown of parallel communication links between the two computers are some of the faults that have been clearly identified. These are under constant review and improvements and modifications to the hardware have been made and are expected to be carried out in phases in future. The size of the existing core and bulk memory limit any change in the software programmes towards achieving increased availability. As and when they are expanded, software improvements are expected to be undertaken where necessary within a period of a year or so.

3.2. Instrumentation

Continuous drift in the control band/control set point was observed in some plant control loops and this resulted in several plant trips. The drift of the set point of the steam turbine exhaust hood temperature caused turbine trips on two occasions. The measurement of the calandria moderator level is one of the most important variables from the point of view of reactor control and at times gave faulty readings, resulting in undesirable control rod movements introducing fast negative reactivity or overshoot of the moderator level in the calandria. Prompt operator intervention might have prevented the eventual plant trip but this was not always possible because of the very short time available in occurrences of this nature. The drift in the positioners of the regulating valves resulted in a slow pumping rate of the calandria moderator level and reduced the positive reactivity rate to make control sluggish, resulting in large steam pressure variations for small power changes.

Boiler feed water level control is one of the main problems that needed greater attention in view of its instability, particularly during power transients. Due to non-mixing of reactor coolant in the headers, the boilers receive hot water on a preferential basis. Non-uniformity in the radial power distribution of the reactor core is thus reflected in differences in the streaming rates of individual boilers, making boiler level control during power changes most unstable. Frequent failure of the controlling operators of feed water control valves, primarily due to severe vibrations experienced in the boiler feed water lines, aggravated the level control problem, causing plant trip on two occasions.

3.3. Electrical faults

The station electrical load is provided via alternative circuits using the concept of 'starting' and 'running' transformers, the former being used during periods of shut-down and plant start-up and the latter usually supplying the station equipment after the plant has been synchronized with the grid and is generating net power. While transferring load the close transfer did not take place on three different occasions, resulting in loss of power to the main circulating pumps and subsequent reactor trip. The problem was traced to the breakage of the operating rod of one of the phases of a breaker, which prevented contact being made. Similar rod failures were found on two other breakers. The rods were replaced by modified ones and the spring tension was adjusted to give uniform tension on all the operating rods of the breaker.

Kanupp experienced low frequency and undervoltage due to the tripping of one or more generating units of the Karachi Grid on three different occasions, causing reactor trips on slowing down of the primary coolant circulating pumps. This problem, however, was solved by providing low frequency relays on the grid to limit decreases in the system frequency by automatically shedding individual system loads in the event of partial loss of generation. Two outages, however, occurred during 1974 through transmission line faults; one of these was caused by heavy grounding of one of the transmission lines of the grid and the other due to a defective choke of the coupling capacitor of the relay system of the Kanupp transmission line.

3.4. Mechanical equipment

The gas compressors, used for helium pumping in the calandria cover gas system, are of the centrifugal liquid ring type using moderator water to provide sealing rings. Three in number, they are arranged to run in parallel with two of them provided with larger H.P. motors to achieve the required pump-up rates. Because of the incompatability of the driving motor and the driven compressor, however, the latter experienced high starting torques, resulting in shaft breakage on two occasions and several failures of mechanical seals. On one occasion the trip of the compressor on loss of power was followed by the complete failure of the mechanical seal, resulting in a large spill of highly tritiated heavy water and subsequent increase in tritium concentration of up to 4000 times the maximum permissible concentration in the boiler room atmosphere. In view of their frequent failures the plant was forced to be operated on two compressors only at the cost of reduced pump-up rate and a consequent slow response of the plant regulating system.

During the annual inspection of the circulating pumps of the reactor cooling system their 'floating' throttling assemblies, which provide sealing of the reactor water at the pump shaft, were found to be damaged. It appears that the carbon bushes of the assemblies failed initially, which resulted in unbalanced forces on the bushing causing serious damage to its supporting pins, springs and nuts. The supports were later strengthened and the material of the carbon bushes improved in consultation with the manufacturer. The seal water of the circulating pumps is filtered in a 5 μm filter assembly and on one occasion the filter assembly collapsed and its nylon threads escaped into the pump seal cavity causing minor cracks in the throttling bush of one of the pumps.

There has been one serious case of valve failure, which caused a breach of integrity of the reactor cooling system. A casting defect in the form of a spiral-shaped pin hole was discovered in a 6 in.valve of the reactor cooling system, which was later repaired by welding.

3.5. Seawater intake

Apart from its use in the turbine condenser, seawater provides cooling to the process water through a series of heat exchangers. The cooling salt water is taken from the shallow sea coast, which quite often results in the intake of silt, seaweed and fish into the intake channel at an uncontrolled rate. This intake of seaweed was a source of two complete plant derations and several partial derations during the two years of operating experience. Stalling of the travelling screens, provided for the retention of seaweed etc., is a frequent occurrence during seaweed inrush. In spite of the screens provided, seaweed of finer size does escape into the heat exchangers causing total choking of their tubes.

The failure rate of tubes, which are of aluminium brass, is unusually high in the turbine condenser and the other heat exchangers and is primarily due to erosion caused by heavy concentrations of silt in the cooling seawater. All the process water heat exchangers are being retubed and the tubing of the turbine condenser may have to be taken in hand if the present rate of tube failure cannot be checked. Although the total number of failed tubes so far is too small to have any adverse effect on the condenser vacuum at present, the frequency of their failure during operation is a factor to be reckoned with from the point of view of plant availability. Although techniques have been devised to detect leaking condenser tubes on line, improvement is needed so as to detect even minor leaks.

4. HEAVY WATER LEAKS

Control of heavy water leakage from the reactor cooling system into the boiler room atmosphere is one of the main factors governing plant availability. The plant experienced as many as five plant outages of comparatively longer duration during the year 1973 as a result of abnormal leakage rates of heavy water from the primary cooling system. The leakage rates were reduced by 36% during the following year and no plant outage occurred on this account. The ambient air of the boiler room is dried by means of a closed circuit and air dryers and the condensate thus collected is usually downgraded heavy water, the light water contribution mainly coming from leaks in the boiler feed-water and steam systems. The heavy water isotopic of the condensate was in the range of 2 to 5% during 1973 and improved to about 15 to 20% in the following year.

Reactor end-fittings of the fuel channels, which are 208 in number on each (north/south) side of the reactor and are provided with removable screw-in type end-plugs to facilitate on-power fuelling, are the main source of heavy water leaks from the reactor coolant system. The end-fittings are fixed on the north end of the reactor and left floating at the other end to allow for expansion of the pressure tubes. On experiencing power transients and cycling, quite a few of the end-fittings start leaking at the end-plugs and the failure rate is more predominant at the floating end of the fuel channels. Since most of the leaks of this type are very minor in nature, their remote detection during on-power operation remained a problem for quite some time. However, during the early part of 1974 a leak detection device was developed and installed on each of the fuelling machines to provide scanning of reactor end-fittings in a routine manner. Since then there has been no case of plant outage due to any uncontrolled leakage from reactor faces, while as many as 86 (63 of them on the floating end) end-fittings were found to be leaking during the year 1974.

There were three cases of stainless steel tubing failures in the year 1973 at the welds of the activity monitoring of the reactor coolant. All 3 failures were due to poor welds.

Except for the valve pin-hole failure reported previously, the piping and pipe fittings of the primary coolant have remained trouble-free. However, leakages from the valve glands have been a significant source of atmospheric leakages from the system. During the period of commissioning when shredded asbestos fibre was used as packing material, pieces of packing leaked into the system and the material failed miserably to maintain the integrity of the gland packing system. This was later replaced by a braided type and then by shredded Teflon with precompression of 5 to 10 tons, which improved the performance of the valves considerably from the point of view of atmospheric leakages in both the heavy and light water systems, as evident from the comparatively controlled leakage rates during 1974. Two of the valves have been provided with live-loading packing as an initial step to improve the situation further.

5. INADVERTENT OPERATIONS

Nearly 14% of the outages resulted from inadvertent operations, which should be avoided in future by proper planning of maintainance jobs and training of the working crew. Some of the more significant operator errors are listed below:

(1) Wrong assembling of the fine control butterfly valve of the boiler level control. The stem has been modified to avoid recurrence.

(2) The common signal of generator power for plant regulation, tested under power. Two independent signals have now been provided for each of the controlling computers.

(3) While changing an indicator lamp the control bus grounded.

(4) The control rod control was inadvertently left on the computer during maintenance.

(5) The surge tank was vented inadvertently.

(6) Both protective channels were opened inadvertently during routine testing.

6. REACTOR FUEL AND FUELLING MACHINES

Kanupp uses standard CANDU-type fuel elements and its performance has been quite satisfactory, except for failure of a few bundles during load cycling in the later part of 1973. It was later realized that failures of this nature could have been avoided if proper reconditioning of fuel had been done at higher power levels.

The performance of the fuelling machines has been quite satisfactory. Only two cases of serious failure of the fuelling machines occurred during on-power fuelling and both of them were related to the failure of charge-tube drives. In one case the shaft key came off its slot and in the other the universal shaft coupling sheared off. In both cases the charge tube was made ineffective while the machines were locked-on to the reactor end-fitting and the end-plug removed. In the earlier case the machine contained two spent fuel bundles, resulting in radiation fields in the range of 20 to 30 R/h. A detailed survey of the fields around the reactor showed that the repair job could be accomplished with the plant on-line for a period of 7 to 10 minutes to bring the machines into service without exposing the personnel to excessive radiation doses. An external dose of 3.7 man • rem was received during the maintenance of the machine containing the two spent fuel bundles. This was shared between 7 persons.

7. RADIATION CONTROL

The tritium level in the primary reactor coolant increased by a factor of 4 at the end of 1974. In general, the ^{131}I level in the primary coolant remained at 1 to 2 μCi/l during 1973 but its concentration increased to a value between 10 and 200 μCi/l during 1974. Radiation fields in the boiler room, 24 hours after reactor shut-down, were found to be generally of the order of 25-50 mR/h, except near boilers and the end-fittings where these were of the order of 300 mrem/h.

The distribution of radiation doses among the workers of the plant during its two years of operation is illustrated in Fig.2. As can be seen, individual doses are fairly evenly distributed among the workers during 1974 as compared to 1973. Since the bulk of the radiation exposures occur during work in the boiler room, the comparatively uniform dose distribution was possible by allocation of work involving radiation exposure fairly uniformly among all workers at the plant rather than depending exclusively on a few specialized groups of workers.

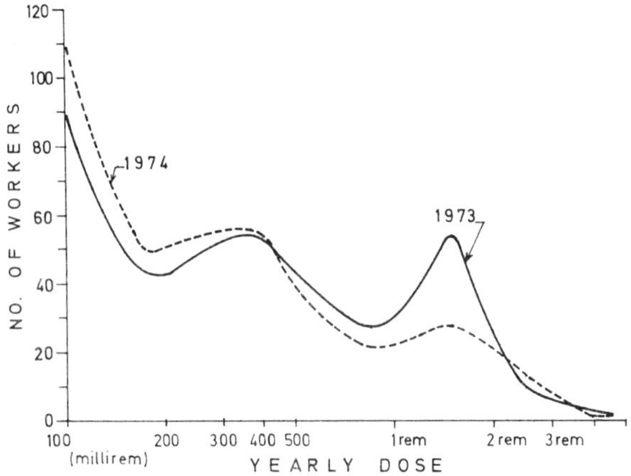

FIG. 2. Radiation dose distribution.

8. CONCLUSIONS

In summary it may be stated that the performance of Kanupp during its initial two years of operation with an availability of 77% compares well with the other nuclear power plants of the world, in spite of its somewhat higher frequency of failure. With the implementation of the first phase of improvements of plant equipment the failure rate was reduced by 35% within a period of one year and with further improvements, particularly of plant computers and the seawater intake system equipment, better control over heavy and light water leakage and more emphasis on training of operating personnel, Kanupp is likely to achieve still better performance in future.

DISCUSSION

(on papers IAEA-SM-195/38, IAEA-SM-195/21, IAEA-SM-195/34, IAEA-SM-195/35 and IAEA-SM-195/1)

G. BOGANI: I have a number of questions for Mr. Sarma in relation to the paper on maintenance experience at Rajasthan Atomic Power Station (IAEA-SM-195/34). Firstly, what is the maximum mechanical permanent deformation due to creep that you allow in pressure tubes (as a percentage of diameter)?

M.S.R. SARMA: About 2%.

G. BOGANI: Do you plan eventually to use electronic devices for inner diameter measurements?

M.S.R. SARMA: We propose to gauge the cool pressure tubes periodically, and the equipment which will be used for that purpose is on order.

G. BOGANI: Have you never had fretting corrosion problems?

M.S.R. SARMA: We have not experienced any so far, but the first pressure tube inspection may reveal something.

G. BOGANI: Could you say a little bit more about the fuel defects you have found — their size and probable cause, the history of the fuel and so on?

M.S.R. SARMA: The defects have been mainly of the pinhole type, as revealed by visual inspection. No detailed examination has been carried out so far, but this is to be done soon at Bhabha Atomic Research Centre in Bombay. No particular cause has been attributed to the failures; they seem to me to be of a random nature.

J.A. SUAREZ-ESTRADA: I have a question concerning the organizational arrangements at Rajasthan Atomic Power Station (IAEA-SM-195/35). What is the relationship of the Radiological Protection Section, the Chemical Section and the Nuclear Section to the Station Superintendent?

M.S.R. SARMA: The Radiological Protection Section does not report to Station Management; rather, it reports directly to the Health Physics Division of Bhabha Atomic Research Centre in Bombay, which deals with health and safety problems for all India's nuclear establishments. The other sections — Chemical, Nuclear, Conventional, Fuel Accounting and Heavy Water Accounting — all come under the Technical Unit as shown in the organizational chart.

J.A. SUAREZ-ESTRADA: And how many staff are employed in the Radiological Protection Section?

M.S.R. SARMA: About 30; they perform area surveys for contamination and radiation fields and give appropriate advice to the station staff.

J.A. SUAREZ-ESTRADA: In paper IAEA-SM-195/34 you say that the station dose in 1974 was 350 man•rem, and that that value was considered low. What value do you expect to find in 1975?

M.S.R. SARMA: It is estimated that the station dose for 1975 will be about 800 man•rem. According to our estimates and our experience, this is not an excessive figure.

G. BOGANI: I wonder if Mr. Huseini could tell us exactly what is meant by 'proper reconditioning of fuel during load cycling'.

S.D. HUSEINI: The full rated power of the plant is 137 MW(e). On one occasion, when the plant had been operating at loads between 70 and 90 MW(e), power was raised from this level to about 115 MW(e). This resulted in the failure of a few fuel bundles. On enquiring of the fuel manufacturer what might be the cause of the failure, we were told that the fuel needed 'reconditioning' for powers beyond 100 MW(e). There are no restrictions for load variations within 100 MW(e) but beyond this level the reactor has to be maintained at 100 MW(e) for at least 10 hours and the power then increased in steps of 1 MW(e) for every $1\frac{1}{2}$ hours until the desired power is reached. These limitations are necessary to limit the rate of increase of fission product pressures within the fuel.

A. PEREZ RODRIGUEZ: I have a question for Mrs. Kontoleon. Could you clarify the concept of partial failure? Do you not consider a transitional state between partial failure and correct operation?

Nadia KONTOLEON: Partial failures I have already defined as failures which are catastrophic to the individual component but not to the system. It is assumed here that partial failures are discovered only when a test procedure is carried out. Thus there is no transitional state between that in which a partial failure is present and the correct working state in which no failure is present.

B. CAFFREY: I have a question for Mr. Robinson. Is shift maintenance normally worked at N Reactor? If it is, how many maintenance staff do you have per shift?

R.K. ROBINSON: Each of the four rotating shifts which man the plant 24 hours a day and seven days a week includes maintenance staff who perform routine maintenance, engineered maintenance and corrective maintenance, in both outage and non-outage conditions. Their numbers and qualifications are such that corrective maintenance can be initiated for any system malfunction on any shift. A non-rotating day shift handles a large share of the maintenance work during planned and extended reactor outages. Each rotating shift normally has three craftsmen in each of the four disciplines: pipefitters, millwrights, electricians and instrument specialists.

B. CAFFREY: Presumably one of the main objectives of the engineered maintenance programme is to reduce occupational exposure to radiation. How successful have you been in this?

R.K. ROBINSON: The greatest risk of exposure comes when the ractor is shut down and personnel are able to enter the high radiation zones to perform maintenance and other duties. During reactor operation these zones are inaccessible. Consequently, any programme which reduces the number of unscheduled outages requiring access to and work in the high radiation zones makes a valuable contribution to safety. The fact that this phase of our programme has been successful is evident from the charts I showed you, indicating the reduction in the number of unscheduled or forced equipment outages.

But unscheduled outages are of course not the only cause of occupational exposure to radiation. In fact, most exposure is incurred during planned maintenance outages. United Nuclear Industries have an important programme whose purpose is to reduce the exposure of individuals to radiation through better training, improved job planning and manpower utilization, and improved techniques and protection devices. The anticipation of corrective maintenance and avoidance of breakdown maintenance permits more effective job planning and thereby reduces occupational exposure.

S.N. MOBERG: Could you give us a rough idea what resources, in man-years, were needed to develop and implement the engineered maintenance programme?

R.K. ROBINSON: The programme was started in 1970 and has been kept up and improved ever since. The man-years of effort needed to develop and implement it were approximately 10 in 1970, 10 in 1971, 9 in 1972 and 6 in 1973.

S.N. MOBERG: What are the hardware requirements of the programme? I have computer capacity particularly in mind.

R.K. ROBINSON: The minimum computer capacity required for utilizing the engineered maintenance programme is a mini-computer with a 32 000 word core memory, employing either cassette tape or disk storage. The maximum configuration cost is $75 000. Such a system is also capable of using software to provide analysis for test frequency corrections, and it can serve other purposes as well.

Because it is readily available, United Nuclear Industries use a Control Data Cyber 74-18 (75 000 word memory) with two magnetic tapes and FORTRAN language.

S. HARTWIG: Could you indicate what percentage of outage hours is due to inefficient performance of operators?

R.K. ROBINSON: Table A for the fiscal years 1972 to 1975 shows the approximate percentage of unscheduled outage hours due to faulty performance in operation or maintenance.

TABLE A. DETAILS OF UNSCHEDULED OUTAGE HOURS

Fiscal year	Total unscheduled outage hours as % of hours in campain	% of unscheduled outage hours due to avoidable errors	% of unscheduled outage hours due to equipment
1972	13	6	7
1973	6	2	4
1974	8	6	2
1975	7	a	a

[a] Data for fiscal year 1975 not yet analysed.

CHAIRMEN OF SESSIONS

Session I	A. NENTWICH	Austria
Session II	D. OKRENT	United States of America
Session III	S. E. BARDEN	United Kingdom
Session IV	J.-A. VAUJOUR	France
Session V	A. V. KARPOV	Union of Soviet Socialist Republics
Session VI	M. S. R. SARMA	India
Session VII	K. EFFAT	Egypt
Session VIII	L. ALVAREZ DE BUERGO	Spain
Session IX	L. ALVAREZ DE BUERGO	Spain
Session X	K. EFFAT	Egypt

SECRETARIAT OF THE SYMPOSIUM

Scientific Secretary	N. RAIŠIČ	Division of Nuclear Power and Reactors, IAEA, Vienna
Administrative Secretary	Caroline DE MOL VAN OTTERLOO	Division of External Relations, IAEA, Vienna
Editor	Brigitte KAUFMANN	Division of Publications, IAEA, Vienna
Records Officer	S. EDMINSTER	Division of Languages and Policy-making Organs, IAEA, Vienna

LIST OF PARTICIPANTS

AUSTRALIA

McDonald, N.R.	Australian Atomic Energy Commission, P.O. Box 41, Coogee, N.S.W. 2034

AUSTRIA

Bolhar-Nordenkampf, F.G.	Bundesministerium für Handel, Gewerbe und Industrie, Stubenring 1, A-1011 Vienna
Cap, F.	Universität Innsbruck, Innrain 52, A-6020 Innsbruck
Edler, A.	Institut für Dampfkesselbau, Reaktortechnik und Wärmewirtschaft, Technische Hochschule Graz, Kopernikusgasse 24/IV, A-8010 Graz
Gerzabek, H.H.	Bundesministerium für Inneres, Abteilung III/2, Vorgartenstrasse 223, A-1020 Vienna
Hillebrand, G.	Österreichische Studiengesellschaft für Atomenergie GmbH, Lenaugasse 10, A-1082 Vienna
Holzer, H.	Österreichische Studiengesellschaft für Atomenergie GmbH, Lenaugasse 10, A-1082 Vienna
Nentwich, A.	Gemeinschaftskernkraftwerk Tullnerfeld GmbH, Marc Aurel Strasse 4, A-1010 Vienna
Oszuszky, F.	Österreichische Elektrizitätswirtschafts AG, (Verbundgesellschaft), Am Hof 6A, A-1010 Vienna
Pinter, M.	Kernkraftwerk Planungs GmbH (KKWP), Jacquingasse 16-18, A-1030 Vienna

Powondra, F.	Österreichische Elektrizitätswirtschafts AG, (Verbundgesellschaft), Am Hof 6a, A-1010 Vienna
Pusch, W.	Bundesministerium für Gesundheit und Umweltschutz, Stubenring 1, A-1010 Vienna
Sperr, W.R.	Institut für Leichtbau, Technische Hochschule Wien, Getreidemarkt 9, A-1060 Vienna
Szeless, A.	Kernkraftwerk Planungs GmbH (KKWP), Jacquingasse 16-18, A-1030 Vienna
Viertl, R.	Institut für Statistik, Technische Hochschule Wien, Argentinierstrasse 8/7, A-1040 Vienna

BELGIUM

Cremades, J.A.	Westinghouse Nuclear, Europe, 73, Rue de Stalle, 1180 Brussels
Crustin, J.	Belgonucléaire S.A., 25, Rue du Champ de Mars, 1050 Brussels
Gillon, L.P.	Université de Louvain, 2, Chemin du Cyclotron, 1348 Louvain La Neuve
Mara, G.	Westinghouse Nuclear, Europe, 73, Rue de Stalle, 1180 Brussels
Remeysen, J.H.	Traction et Electricité, 31, Rue de la Science, 1040 Brussels
Schepens, A.J. (see also UNIPEDE)	S.A. EBES (Génération et Transport), 31, Rue de la Science, 1040 Brussels

BRAZIL

Magalhães, M.T.	Furnas — Centrais Elétricas S.A., Rua Real Grandeza 219, Rio de Janeiro

CANADA

Krasnodebski, J.	Ontario Hydro, 620 University Ave., Toronto, Ont. M1W 1L7
Wolfe, W.A.	Atomic Energy of Canada Ltd., Chalk River Nuclear Laboratories, Chalk River, Ont. K0J 1J0

CZECHOSLOVAKIA

Hatle, Z.	Czechoslovak Atomic Energy Commission, Slezská ul.9, Prague 2
Hrubec, J.	Sigma Trust/Pumping Equipment and Valves Manufacturing Works, Research Institute, Pasteurova 8a, 77211 Olomouc
Keher, J.	Atomic Power Plant A 1, Bohunice, Trnava
Sedláček, J.	Technical University Prague, Břehova 7, Prague
Ševčik, A.	Federal Ministry of Fuel and Power, Stepánská 18, Prague 1
Zbehlik, J.	Institute for Nuclear Research, 250 68 Řež

DENMARK

Kongsø, H.E.	Danish Atomic Energy Commission, Research Establishment Risø, DK-4000 Roskilde
Taylor, J.R.	Danish Atomic Energy Commission, Research Establishment Risø, DK-4000 Roskilde

EGYPT

Abaza, M.M.	Ministry of Electricity, Nasr City, Cairo
Effat, K.	Atomic Energy Establishment, 101, Kasr El Eini St., Cairo

FINLAND

Blomqvist, H.B.	Institute of Radiation Protection, Kalevankatu 44, SF-00280 Helsinki 18
Ervamaa, J.	Technical Research Centre of Finland, 02152 Espoo 15, Helsinki
Vaurio, J.K.	Imatran Power Company, Pl. 138, 00101 Helsinki 10

FRANCE

Allanic, J.	CIRNA, CEA, Centre d'études nucléaires de Saclay, B.P. n°27, 91190 Gif-sur-Yvette
Arlaud, J.-C.	CEA, Centre d'études nucléaires de Cadarache, B.P. n°1, 1315 Saint-Paul-lez-Durance
Blin, A.	CEA, Centre d'études nucléaires de Saclay, B.P. n°2, 91190 Gif-sur-Yvette
Carnino, Annick	CEA, Centre d'études nucléaires de Saclay, B.P. n°2, 91190 Gif-sur-Yvette
Charon, M.	CIRNA (GAAA), CEA, Centre d'études nucléaires de Saclay, B.P. n°18, 91190 Gif-sur-Yvette
Even, G.	Groupement pour les activités atomiques et avancées, 20 av. Edouard Herriot, 92350 Le Plessis-Robinson
Ferriole, G.	Electricité de France, Service de la production thermique, 3 rue de Messine, Paris 8e
Frydman, Alberte	Technicatome, CEA, Centre d'études nucléaires de Saclay, B.P. n°18, 91190 Gif-sur-Yvette
Gachot, B.	Electricité de France, SEPTEN, Tour EDF, Cédex n°8, 92400 Paris La Défense

Gonnot, R.	Framatome, 77/81 rue du Mans, 92400 Courbevoie
Greppo, J.-F.	Electricité de France, Sercice de la production thermique, 3 rue de Messine, Paris 8e
Grisollet, J.	Technicatome, CEA, Centre d'études nucléaires de Saclay, B.P. n°18, 91190 Gif-sur-Yvette
Israel, Sylvette	Electricité de France, 1 pl. du Général de Gaulle, 92140 Clamart
Jubault, G.F.	Groupement pour les activités atomiques et avancées, 20 av. Edouard Herriot, 92350 Le Plessis-Robinson
Labat, G.H.	Electricité de France, SEPTEN, Tour EDF, Cedex n°8, 92080 Paris La Défense
Larminaux, R.	Electricité de France, SEPTEN, Tour EDF, Cedex n°8, 92808 Paris La Défense
Lucenet, G.F.	Electricité de France, NERSA-EDF, 177 rue Garibaldi, 69006 Lyon
Mollard, Christine C.	Groupement pour les activités atomiques et avancées, 20 av. Edouard Herriot, 92350 Le Plessis-Robinson
Normand, J.	Technicatome, CEA, Centre d'études nucléaires de Saclay, B.P. n°18, 91190 Gif-sur-Yvette
Philippe, J.-M.	Electricité de France, 24 bd de la Libération, 93206 Saint-Denis
Pignatelli, R.	NIRA, Projet Super-Phenix, CEA, Centre d'études nucléaires de Saclay, B.P. n°2, 91190 Gif-sur-Yvette
Procaccia, H.	Electricité de France, 24 bd de la Libération, 93206 Saint-Denis
Prot, A.C.	CEA, Centre d'études nucléaires de Saclay, B.P. n°2, 91190 Gif-sur-Yvette

Quénée, R. — CEA, Centre d'études nucléaires de Saclay,
B.P. n°2,
91190 Gif-sur-Yvette

Refoubelet, P.M. — CGEE ALSTHOM,
13 rue Antonin Raynaud,
92130 Levallois Perret

Signoret, J.-P. — CEA, Centre d'études nucléaires de Saclay,
B.P. n°2,
91190 Gif-sur-Yvette

Vaujour, J.-A. — Electricité de France, SCF,
12 pl. des Etats-Unis,
B.P. n°801-16,
75783 Paris, Cedex 16

Verrière, P. — CEA, Centre d'études nucléaires de Cadarache,
B.P. n°1,
1315 Saint-Paul-lez-Durance

Vincent, F. — Electricité de France,
1 av. du Général de Gaulle,
92140 Clamart

Zaoui, R. — Framatome,
76 rue du Mans,
92400 Courbevoie

GERMAN DEMOCRATIC REPUBLIC

Ackermann, G. — Ingenieurhochschule Zittau,
Strasse der Jungen Pioniere 3,
88 Zittau

Rabold, H. — Staatliches Amt für Atomsicherheit und Strahlenschutz,
Waldowallee 117,
1157 Berlin

Sitzlack, G. — Staatliches Amt für Atomsicherheit und Strahlenschutz,
Waldowallee 117,
1157 Berlin

GERMANY, FEDERAL REPUBLIC OF

Antoni, R. — Institut für Reaktorsicherheit der Technischen Überwachungs-Vereine e.V.,
Glockengasse 2,
5 Köln 80

Badics, A.A. — Kraftwerk-Union AG,
Berliner Strasse 297-299,
605 Offenbach

Böhm, B. — Kraftwerk-Union AG,
Hammerbacherstrasse 12, 14,
852 Erlangen

Breiling, G.	Babcock – Brown Boveri Reaktor GmbH, Heppenheimer Strasse 27-29, 6800 Mannheim
Bremer, F.	Fried. Krupp GmbH, Krupp Universalbau, Postfach 932, 43 Essen 1
Cramer, H.V.	KKS, Postfach 1780, 216 Stade
Dobschütz, P. von	Institut für Reaktorsicherheit der Technischen Überwachungs-Vereine e.V., Glockengasse 2, 5 Köln 1
Drake, W.	Technischer Überwachungs-Verein e.V., Grosse Bahnstrasse 31, 2 Hamburg 54
Dressler, E.	Laboratorium für Reaktorregelung und Anlagensicherung Garching, Technische Universität München, 8046 Garching
Evers, H.O.	Technischer Überwachungs-Verein e.V., Grosse Bahnstrasse 31, 2 Hamburg 54
Fürste, W.	Rheinisch-Westfälischer Technischer Überwachungs-Verein e.V., Steubenstrasse 53, 43 Essen
Gödel, W.	Technischer Überwachungs-Verein e.V., Eichstätterstrasse 5, 8 München 21
Groha, D.F.O.	Kraftwerk Union AG, Berlinerstrasse 296-298, 605 Offenbach
Gutmann, G.G.	Fa. Krautkrämer GmbH, Luxemburgerstrasse 449, 5 Köln 41
Halmy, A.	Technischer Überwachungs-Verein Rheinland, Am Grauen Stein, 5 Köln 91
Hartwig, S.	Battelle-Institut, Am Römerhof 35, Frankfurt/Main
Hauck, D.	Babcock – Brown Boveri Reaktor GmbH, Heppenheimerstrasse 27-29, 6800 Mannheim

Hawickhorst, W.	Laboratorium für Reaktorregelung und Anlagensicherung Garching, Technische Universität München, 8046 Garching
Hennings, U.	Babcock – Brown Boveri Reaktor GmbH, Heppenheimerstrasse 27-29, 6800 Mannheim
Herrmann, H.J.	Institut für Reaktorsicherheit der Technischen Überwachungs-Vereine e.V., Glockengasse 2, 5 Köln 1
Hinrichs, J.	Hochtemperatur Reaktorbau GmbH, Gottlieb-Daimler Strasse 8, 6800 Mannheim
Hömke, P.	Institut für Reaktorsicherheit der Technischen Überwachungs-Vereine e.V., Glockengasse 2, Köln 5
Hoppe, H.C.	Technischer Überwachungs-Verein Stuttgart e.V., Bebelstrasse 48, 7 Stuttgart 1
Jacquemin, J.P.	Ministerium für Arbeit, Gesundheit und Soziales des Landes Nordrhein-Westfalen, Postfach 1134, 4 Düsseldorf
Jäger, E.	Ministerium für Soziales, Gesundheit und Sport, Rheinland-Pfalz, Bahnhofstr. 4, 65 Mainz
Kamarinopoulos, L.	Technische Universität Berlin, Institut für Kerntechnik, Berlin
Klein, K.W.	Badenwerk AG, Badenwerkstrasse 2, 75 Karlsruhe
Kügele, B.	Fa. Fichtner, Grazerstrasse 22, 7 Stuttgart 30
Kuhlow, B.	Bundesanstalt für Materialprüfung, Unter den Eichen 87, 1 Berlin 45
Lindner, J.P.	Maschinenfabrik Augsburg-Nürnberg Aktiengesellschaft, (M.A.N.) Postfach 85 Nürnberg 115
Marth, W.	Gesellschaft für Kernforschung mbH, Postfach, 7500 Karlsruhe

Meyer, F.A.	Technischer Überwachungs-Verein Baden e.V., Richard-Wagner-Strasse 2, 6800 Mannheim
Mühlhoff, W.H.	Interatom, Friedrich-Ebert-Strasse, 506 Bensberg/Köln
Oehme, H.	Interatom, Friedrich-Ebert-Strasse, 506 Bensberg/Köln
Pucher, M.	Ministerium für Arbeit, Gesundheit und Sozialordnung Baden-Württemberg, Rotebühlplatz 30, 7 Stuttgart
Sartori, R.	Rheinisch-Westfälischer Technischer Überwachungs-Verein e.V., Steubenstrasse 53, 43 Essen
Schallopp, B.	Büro Schallopp Berlin, Ringstrasse 71, 1 Berlin 45
Schempp, H.	Technischer Überwachungs-Verein Stuttgart e.V., Bebelstrasse 48, 7 Stuttgart
Spindler, H.	Laboratorium für Reaktorregelung und Anlagensicherung Garching, Technische Universität München, 8046 Garching
Thorade, K.A.	Bureau Veritas, Hopfenmarkt 33, 2000 Hamburg 11
Witulski, H.	Ministerium für Wirtschaft, Mittelstand und Verkehr des Landes Nordrhein-Westfalen, Haroldstrasse 4, 4 Düsseldorf
Wüstenberg, H.	Bundesanstalt für Materialprüfung, Unter den Eichen 87, 1 Berlin 45
Zeibig, H.	INTERATOM, Friedrich-Ebert-Strasse, 506 Bensberg/Köln

GREECE

Kontoleon, Nadia	Reactor Department, Nuclear Research Center 'Demokritos', Aghia Paraskevi Attiki, Athens

HUNGARY

Balogh, E.	Hungarian National Atomic Energy Commission, Kossuth ul.4, Budapest V
Baranyai, A.	Central Research Institute for Physics, P.O. Box 49, 1525 Budapest
Deme, S.	Central Research Institute for Physics, P.O. Box 49, 1525 Budapest
Raszl, J.	Hungarian National Atomic Energy Commission, Kossuth ul.4, Budapest V

INDIA

Sarma, M.S.R.	Rajasthan Atomic Power Station, Anushakti, (Via), Kota, Rajasthan

INDONESIA

Iljas, J.	Badan Tenaga Atom Nasional, Jalan Falatehan I/26, Kebayoran Baru, Jakarta, Selatan

IRAQ

Al-Ammar, M.A.	Nuclear Research Institute, Tuwaitha, Baghdad
Al-Chalabi, Z.M.	Nuclear Research Institute, Tuwaitha, Baghdad
Alkital, R.A.	College of Science, Baghdad University, Adamyah, Baghdad
Alnaib, A.B.	State Organization of Electricity, Jamhoria St., Baghdad
Kamil, G.M.	State Organization of Electricity, Jamhoria St., Baghdad

IRELAND

Caffrey, B.	Electricity Supply Board, 18/21 St.Stephen's Green, Dublin 2

ISRAEL

Borenstein, S.	Israel Electric Corp., P.O. Box 10, Haifa
Yiftah, S.	Israel AEC, Soreq Nuclear Research Centre, Yavne

ITALY

Bazzani, G.	Facoltà di Ingegneria, Istituto di Impianti Nucleari, Università di Pisa, Via Diotisalvi 2, 56100 Pisa
Bogani, G.	Centro Informazioni Studi Esperienze (C.I.S.E.), c.p. 3986, Segrate, Milan
Ceritali, G.	Istituto di Macchine, Università, Via Valerio 10, Trieste
Di Sapia, R.	NIRA, Piazza Carignano 2, Genoa
Martini, M.	Comitato Nazionale per l'Energia Nucleare (CNEN), Viale Regina Margherita, 00100 Rome
Tomassetti, G.	Comitato Nazionale per l'Energia Nucleare (CNEN), CSN Casaccia, c.p. 2400, Rome
Valentini, V.	NIRA, Piazza Carignano 2, Genoa

JAPAN

Miyashiro, S.	Toshiba Düsseldorf Office, Emmastrasse 24, 4 Düsseldorf, Federal Republic of Germany

LUXEMBOURG

Kayser, P. — Service de Radioprotection,
37, rue Glesener,
Luxembourg

MALAYSIA

Ong, Y.-F. — National Electricity Board,
P.O. Box 1003,
Kuala Lumpur

Ramanath, A. — National Electricity Board,
P.O. Box 1003,
Kuala Lumpur

MEXICO

Reinking, A.G. — Instituto Nacional de Energía Nuclear,
Av. Insurgentes Sur 1079,
Mexico 18, D.F.

MOROCCO

Mekki-Berrada, A. — Office National d'Electricité,
65, rue Aspirant Lafuente,
Casablanca

NETHERLANDS

Heshuysen, F.W. — Reactor Centrum Nederland,
Westerduinweg,
Petten

Van Otterloo, R.W. — KEMA,
Utrechtseweg 310,
Arnhem

Weber, J. — Ministry of Public Health and Environmental Hygiene,
Leidschendam

NORWAY

Jonstad, O. — Norwegian Water Resources and Electricity Board,
P.O. Box 5091,
Oslo 3

Løken, P.C. — Norwegian Water Resources and Electricity Board,
Middeltunsgt. 29,
Oslo 3

Michelsen, H.M.	Nuclear Energy Safety Authority, Pottemakervn. 4, Oslo 5
Moelsaeter, M.	Nuclear Energy Safety Authority, Pottemakervn. 4, Oslo 5
Örjasaeter, O.	Institutt for Atomenergi, P.O. Box 40, 2007 Kjeller

PAKISTAN

Huseini, S.D.	Karachi Nuclear Power Plant, P.O. Box 3183, Paradise Point, Karachi 29

POLAND

Czosnowski, St.	Ministry of Science, Higher Education and Technology, ul. Miodowa 6/8, 00-251 Warsaw
Zmyslowski, A.	Polish Atomic Energy Authority, Palac Kultury i Nauki 18 p., Warsaw

PORTUGAL

Cabral, J.R.	Companhia Portuguesa de Electricidade, CPE, R. Mouzinho Silveira 10, Lisbon

SAUDI ARABIA

Melibary, A.R.M.	Atomic Energy Section, Ministry of Petroleum and Mineral Resources, P.O. Box 345, Jeddah

SOUTH AFRICA

Tattersall, J.O.	Atomic Energy Board, Private Bag X256, Pretoria

SPAIN

Alvarez de Buergo, L.	Junta de Energía Nuclear, Avenida Complutense 22, Madrid 3
Perez Rodriguez, A.	Junta de Energía Nuclear, Avenida Complutense 22, Madrid 3
Suarez-Estrada, J.A.	Electra de Viesgo S.A., Apartado 35, Santander

SWEDEN

Andersson, B.	AB ASEA-Atom, P.O. Box 53, S-72104 Västerås
Hallstensson, L.	Swedish State Power Board, Fack, S-16287 Vällingby 1
Hilding, S.	Oskarshamn Power Group, S-57092 Figeholm, Stockholm
Moberg, S.N.	Statens Vattenfallsverk Värmekraftproduktion, S-16282 Vällingby 1
Nilsson, E.T.R.	Swedish Nuclear Power Inspectorate, P.O. Box 43058, S-10072 Stockholm 43

SWITZERLAND

Wenger, H.E.	Nordostschweizerische Kraftwerke AG, CH-5401 Baden

TURKEY

Kilicøglu, I.	Turkish Atomic Energy Commission, Bestekar Sok. 29, Kavaklidere, Ankara
Yesin, O.	Mechanical Engineering Department, Middle East Technical University, Ankara

UNION OF SOVIET SOCIALIST REPUBLICS

Karpov, A.V.	Institute of Power and Physics, Obninsk

UNITED KINGDOM

Adams, A.L.	United Kingdom Atomic Energy Authority, Atomic Energy Establishment Winfrith, Dorchester, Dorset
Atkinson, R.V.	Hinkley Point Nuclear Power Station, Bridgwater, Somerset
Barden, S.E.	Central Electricity Generating Board, Concorde House, 10/12 London Road, Maidstone, Kent ME16 8QB
Batty, A.C.	Nuclear Installations Inspectorate, Health and Safety, Thames House South, Millbank, London SW1P 4QJ
Butterfield, J.M.	Central Electricity Generating Board, Berkeley Nuclear Laboratories, Berkeley, Glos., GL13 9PB
Cave, L.	Pollution Prevention (Consultants) Ltd., Crown House, Copthorne Bank, Crawley, Sussex, RH10 3JG
Fothergill, C.D.H.	United Kingdom Atomic Energy Authority, Systems Reliability Service, Wigshaw Lane, Culcheth, Warrington, Cheshire WA3 3NE
Gow, S.	Central Electricity Generating Board, 24 Cathedral Place, London EC4P 4EP
Jones, P.H.	Rolls-Royce and Associates Ltd., P.O. Box 31, Derby DE2 8BJ
Lewis, G.	Nuclear Installation Inspectorate, Thames House South, Millbank, London SW1P 4JQ
Nichols, P.	Central Electricity Generating Board, Nuclear Health and Safety Department, Courtenay House, Warwick Lane, London EC4P 4EB
Norman, D.	Central Electricity Generating Board, GDCD, Barnwood, Gloucester GL4 7RS
Osgood, C.	Central Electricity Generating Board, Berkeley Nuclear Laboratories, Berkeley, Glous. GL13 OPB

Peacock, B. — Central Electricity Generating Board,
15-23 Oakfield Grove,
Clifton, Bristol

Pyne, H.W. — Ministry of Defense,
Block C, Room 85,
Foxhill, Bath, Avon BA1 5AB

Smith, P.G. — Central Electricity Generating Board,
Newgate Street,
London EC1

Snaith, E.R. — United Kingdom Atomic Energy Authority,
Safety and Reliability Directorate,
Wigshaw Lane, Culcheth,
Warrington, Cheshire

Snarey, V.L. — Royal Naval College,
Greenwich,
London SE10 9NN

Thomas, T. — Central Electricity Generating Board,
Sudbury House,
15 Newgate St.,
London EC1A 7AU

Tweedy, J.N. — South of Scotland Electricity Board,
Cathcart House, Inverlair Ave.,
Glasgow

UNITED STATES OF AMERICA

Ferriss, W.E. — Bechtel Power Corporation,
3621 S. State,
Ann Arbor, Mich.48106

Hopkins, L.A. — Boeing Engineering and Construction Company,
P.O. Box 3707,
Seattle, Wash.98124

Lehmann, P.H. — Carnegie-Mellon University,
Pittsburgh, Pa. 15213

Okrent, D. — University of California,
5532 Boelter Hall,
Los Angeles, Calif. 90024

Robinson, R.K. — United Nuclear Industries Inc.,
P.O. Box 490,
Richland, Wash. 99352

Waller, R.A. — Los Alamos Scientific Laboratory,
Group C-5 MS 254,
P.O. Box 1663,
Los Alamos, N.Mex. 87544

Wise, M.J. — Southwest Research Institute,
8500 Culebra Road,
San Antonio, Tex. 78213

Zavadoski, R.W.	U.S.E.R.D.A., Brussels's Office, 23, Avenue des Arts, Brussels, Belgium

YUGOSLAVIA

Đurđek, I.D.	NPP Krško, ul. 4 Julija 38, 68270 Krško
Močnik, R.A.	NPP Krško, ul. 4 Julija 38, 68270 Krško
Svilar, R.	NPP Krško, ul. 4 Julija 38, 68270 Krško
Vojnović, B.	Institute Ruder Bošković, Bijenička 54, 41000 Zagreb

ORGANIZATIONS

CEC (COMMISSION OF EUROPEAN COMMUNITIES)

Colombo, A.G.	Joint Research Centre Euratom-Cetis, 21020 Ispra (Varese), Italy
Francocci, G.F.	Joint Research Centre Euratom-Cetis, 21020 Ispra (Varese), Italy
Hampe, E.A.	Direction Protection Sanitaire, Rue Aldringen 29, Luxembourg
Kaut, W.K.	200, rue de la Loi, 1040 Brussels
Lamy, J.-L.G.	200, rue de la Loi, 1040 Brussels
Luisi, T.	Joint Research Centre Euratom-Cetis, 21020 Ispra (Varese), Italy

CMEA (COUNCIL FOR MUTUAL ECONOMICAL ASSISTANCE)

Barchenkov, A.P.	Av. Kalinina 56, Moscow, Union of Soviet Socialist Republics
Creutzburg, M.	Av. Kalinina 56, Moscow, Union of Soviet Socialist Republics

ECE (ECONOMIC COMMISSION FOR EUROPE)

Constantinescu, E.	Palais des Nations, CH-1211 Geneva 10, Switzerland

IAEA (INTERNATIONAL ATOMIC ENERGY AGENCY)

Fattah, A.	Division of Nuclear Power and Reactors, P.O. Box 590, A-1011 Vienna, Austria
Polliart, A. -J.	Division of Nuclear Power and Reactors, P.O. Box 590, A-1011 Vienna, Austria

ISO (INTERNATIONAL ORGANIZATION FOR STANDARDIZATION)

Ervamaa, J. (see also FINLAND)	Technical Research Centre of Finland, SF-02150 Espoo 15, Finland

OECD/NEA (ORGANIZATION FOR ECONOMIC CO-OPERATION AND DEVELOPMENT/ NUCLEAR ENERGY AGENCY)

Chapman, B.G.	Dragon Project, Atomic Energy Establishment Winfrith, Dorchester, Dorset, United Kingdom

UNIPEDE (INTERNATIONAL UNION OF PRODUCERS AND DISTRIBUTORS OF ELECTRICAL ENERGY)

Bensid, D.	Sonelgaz, 2, Boulevard Salah, Bouakouir, Algiers, Algeria
Ouahes, S.	Sonelgaz, 2, Boulevard Salah, Bouakouir, Algiers, Algeria
Schepens, A.J. (see also BELGIUM)	EBES (Génération et Transport), rue de la Science, 1040 Brussels, Belgium

WEC (WORLD ENERGY CONFERENCE)

Vibert, P.M.	5 Bury Street, St. James's, London SW1Y 6AB, United Kingdom

AUTHOR INDEX

(including participants in discussions)

Numerals underlined refer to the first page of the paper by the author concerned. Further numerals denote comments and questions in the discussions.

TRANSLITERATION INDEX

The following conversion table is provided for the convenience of readers and to encourage the use of SI units.

FACTORS FOR CONVERTING UNITS TO SI SYSTEM EQUIVALENTS *

SI base units are the metre (m), kilogram (kg), second (s), ampere (A), kelvin (K), candela (cd) and mole (mol).
[For further information, see International Standards ISO 1000 (1973), and ISO 31/0 (1974) and its several parts]

Multiply		*by*	*to obtain*
Mass			
pound mass (avoirdupois)	1 lbm	= 4.536×10^{-1}	kg
ounce mass (avoirdupois)	1 ozm	= 2.835×10^{1}	g
ton (long) (= 2240 lbm)	1 ton	= 1.016×10^{3}	kg
ton (short) (= 2000 lbm)	1 short ton	= 9.072×10^{2}	kg
tonne (= metric ton)	1 t	= 1.00×10^{3}	kg
Length			
statute mile	1 mile	= 1.609×10^{0}	km
yard	1 yd	= 9.144×10^{-1}	m
foot	1 ft	= 3.048×10^{-1}	m
inch	1 in	= 2.54×10^{-2}	m
mil (= 10^{-3} in)	1 mil	= 2.54×10^{-2}	mm
Area			
hectare	1 ha	= 1.00×10^{4}	m^2
(statute mile)^2	1 $mile^2$	= 2.590×10^{0}	km^2
acre	1 acre	= 4.047×10^{3}	m^2
yard^2	1 yd^2	= 8.361×10^{-1}	m^2
foot^2	1 ft^2	= 9.290×10^{-2}	m^2
inch^2	1 in^2	= 6.452×10^{2}	mm^2
Volume			
yard^3	1 yd^3	= 7.646×10^{-1}	m^3
foot^3	1 ft^3	= 2.832×10^{-2}	m^3
inch^3	1 in^3	= 1.639×10^{4}	mm^3
gallon (Brit. or Imp.)	1 gal (Brit)	= 4.546×10^{-3}	m^3
gallon (US liquid)	1 gal (US)	= 3.785×10^{-3}	m^3
litre	1 l	= 1.00×10^{-3}	m^3
Force			
dyne	1 dyn	= 1.00×10^{-5}	N
kilogram force	1 kgf	= 9.807×10^{0}	N
poundal	1 pdl	= 1.383×10^{-1}	N
pound force (avoirdupois)	1 lbf	= 4.448×10^{0}	N
ounce force (avoirdupois)	1 ozf	= 2.780×10^{-1}	N
Power			
British thermal unit/second	1 Btu/s	= 1.054×10^{3}	W
calorie/second	1 cal/s	= 4.184×10^{0}	W
foot-pound force/second	1 ft·lbf/s	= 1.356×10^{0}	W
horsepower (electric)	1 hp	= 7.46×10^{2}	W
horsepower (metric) (= ps)	1 ps	= 7.355×10^{2}	W
horsepower (550 ft·lbf/s)	1 hp	= 7.457×10^{2}	W

* Factors are given exactly or to a maximum of 4 significant figures

Multiply		*by*	*to obtain*
Density			
pound mass/inch^3	1 lbm/in^3	= 2.768×10^4	kg/m^3
pound mass/foot^3	1 lbm/ft^3	= 1.602×10^1	kg/m^3
Energy			
British thermal unit	1 Btu	= 1.054×10^3	J
calorie	1 cal	= 4.184×10^0	J
electron-volt	1 eV	$\simeq$ 1.602×10^{-19}	J
erg	1 erg	= 1.00×10^{-7}	J
foot-pound force	1 ft·lbf	= 1.356×10^0	J
kilowatt-hour	1 kW·h	= 3.60×10^6	J
Pressure			
newtons/metre^2	1 N/m^2	= 1.00	Pa
atmosphere[a]	1 atm	= 1.013×10^5	Pa
bar	1 bar	= 1.00×10^5	Pa
centimetres of mercury (0°C)	1 cmHg	= 1.333×10^3	Pa
dyne/centimetre^2	1 dyn/cm^2	= 1.00×10^{-1}	Pa
feet of water (4°C)	1 ftH$_2$O	= 2.989×10^3	Pa
inches of mercury (0°C)	1 inHg	= 3.386×10^3	Pa
inches of water (4°C)	1 inH$_2$O	= 2.491×10^2	Pa
kilogram force/centimetre^2	1 kgf/cm^2	= 9.807×10^4	Pa
pound force/foot^2	1 lbf/ft^2	= 4.788×10^1	Pa
pound force/inch^2 (= psi)[b]	1 lbf/in^2	= 6.895×10^3	Pa
torr (0°C) (= mmHg)	1 torr	= 1.333×10^2	Pa
Velocity, acceleration			
inch/second	1 in/s	= 2.54×10^1	mm/s
foot/second (= fps)	1 ft/s	= 3.048×10^{-1}	m/s
foot/minute	1 ft/min	= 5.08×10^{-3}	m/s
mile/hour (= mph)	1 mile/h	= 4.470×10^{-1}	m/s
		1.609×10^0	km/h
knot	1 knot	= 1.852×10^0	km/h
free fall, standard (= g)		= 9.807×10^0	m/s^2
foot/second^2	1 ft/s^2	= 3.048×10^{-1}	m/s^2
Temperature, thermal conductivity, energy/area·time			
Fahrenheit, degrees − 32	°F − 32	$\frac{5}{9}$	°C
Rankine	°R		K
1 Btu·in/ft^2·s·°F		= 5.189×10^2	W/m·K
1 Btu/ft·s·°F		= 6.226×10^1	W/m·K
1 cal/cm·s·°C		= 4.184×10^2	W/m·K
1 Btu/ft^2·s		= 1.135×10^4	W/m^2
1 cal/cm^2·min		= 6.973×10^2	W/m^2
Miscellaneous			
foot^3/second	1 ft^3/s	= 2.832×10^{-2}	m^3/s
foot^3/minute	1 ft^3/min	= 4.719×10^{-4}	m^3/s
rad	rad	= 1.00×10^{-2}	J/kg
roentgen	R	= 2.580×10^{-4}	C/kg
curie	Ci	= 3.70×10^{10}	disintegration/s

[a] atm abs: atmospheres absolute;
atm (g): atmospheres gauge.

[b] lbf/in^2 (g) (= psig): gauge pressure;
lbf/in^2 abs (= psia): absolute pressure.

HOW TO ORDER IAEA PUBLICATIONS

Exclusive sales agents for IAEA publications, to whom all orders and inquiries should be addressed, have been appointed in the following countries:

UNITED KINGDOM	Her Majesty's Stationery Office, P.O. Box 569, London SE 1 9NH
UNITED STATES OF AMERICA	UNIPUB, Inc., P.O. Box 433, Murray Hill Station, New York, N.Y. 10016

In the following countries IAEA publications may be purchased from the sales agents or booksellers listed or through your major local booksellers. Payment can be made in local currency or with UNESCO coupons.

ARGENTINA	Comisión Nacional de Energía Atómica, Avenida del Libertador 8250, Buenos Aires
AUSTRALIA	Hunter Publications, 58 A Gipps Street, Collingwood, Victoria 3066
BELGIUM	Service du Courrier de l'UNESCO, 112, Rue du Trône, B-1050 Brussels
CANADA	Information Canada, 171 Slater Street, Ottawa, Ont. K 1 A OS 9
C.S.S.R.	S.N.T.L., Spálená 51, CS-110 00 Prague Alfa, Publishers, Hurbanovo námestie 6, CS-800 00 Bratislava
FRANCE	Office International de Documentation et Librairie, 48, rue Gay-Lussac, F-75005 Paris
HUNGARY	Kultura, Hungarian Trading Company for Books and Newspapers, P.O. Box 149, H-1011 Budapest 62
INDIA	Oxford Book and Stationery Comp., 17, Park Street, Calcutta 16
ISRAEL	Heiliger and Co., 3, Nathan Strauss Str., Jerusalem
ITALY	Libreria Scientifica, Dott. de Biasio Lucio "aeiou", Via Meravigli 16, I-20123 Milan
JAPAN	Maruzen Company, Ltd., P.O.Box 5050, 100-31 Tokyo International
NETHERLANDS	Marinus Nijhoff N.V., Lange Voorhout 9-11, P.O. Box 269, The Hague
PAKISTAN	Mirza Book Agency, 65, The Mall, P.O.Box 729, Lahore-3
POLAND	Ars Polona, Centrala Handlu Zagranicznego, Krakowskie Przedmiescie 7, PL-00-068 Warsaw
ROMANIA	Cartimex, 3-5 13 Decembrie Street, P.O.Box 134-135, Bucarest
SOUTH AFRICA	Van Schaik's Bookstore, P.O.Box 724, Pretoria Universitas Books (Pty) Ltd., P.O.Box 1557, Pretoria
SPAIN	Nautrónica, S.A., Pérez Ayuso 16, Madrid-2
SWEDEN	C.E. Fritzes Kungl. Hovbokhandel, Fredsgatan 2, S-103 07 Stockholm
U.S.S.R.	Mezhdunarodnaya Kniga, Smolenskaya-Sennaya 32-34, Moscow G-200
YUGOSLAVIA	Jugoslovenska Knjiga, Terazije 27, YU-11000 Belgrade

Orders from countries where sales agents have not yet been appointed and requests for information should be addressed directly to:

Publishing Section,
International Atomic Energy Agency,
Kärntner Ring 11, P.O.Box 590, A-1011 Vienna, Austria

75-07595